DICTIONARY
OF
HERPETOLOGY

DICTIONARY OF HERPETOLOGY

Harvey B. Lillywhite

Illustrations by

Kenneth Paul Wray III

KRIEGER PUBLISHING COMPANY
Malabar, Florida
2008

Original Edition 2008

Printed and Published by
KRIEGER PUBLISHING COMPANY
KRIEGER DRIVE
MALABAR, FLORIDA 32950

Copyright © 2008 by Krieger Publishing Company

All rights reserved. No part of this book may be reproduced in any form or by any means, electronic or mechanical, including information storage and retrieval systems without permission in writing from the publisher.
No liability is assumed with respect to the use of the information contained herein.
Printed in the United States of America.

FROM A DECLARATION OF PRINCIPLES JOINTLY ADOPTED BY A COMMITTEE OF THE AMERICAN BAR ASSOCIATION AND A COMMITTEE OF PUBLISHERS:
This publication is designed to provide accurate and authoritative information in regard to the subject matter covered. It is sold with the understanding that the publisher is not engaged in rendering legal, accounting, or other professional service. If legal advice or other expert assistance is required, the services of a competent professional person should be sought.

Library of Congress Cataloging-In-Publication Data

Lillywhite, Harvey B., 1943-
 Dictionary of herpetology / Harvey B. Lillywhite. — Original ed.
   p. cm.
 ISBN-13: 978-1-57524-023-7 (alk. paper)
 ISBN-10: 1-57524-023-8 (alk. paper)
 1. Herpetology—Dictionaries. I. Title.
 QL640.7.L55 2008
  597.903—dc22

2007048337

10 9 8 7 6 5 4 3 2

Contents

Foreword vi

Preface and a Brief Guide to Format vii

Acknowledgments viii

About the Author viii

Dictionary of Herpetology
   A 1
   B 31
   C 48
   D 79
   E 94
   F 112
   G 124
   H 137
   I 152
   J 165
   K 167
   L 170
   M 182
   N 203
   O 214
   P 224
   Q 262
   R 264
   S 279
   T 315
   U 332
   V 336
   W 343
   X 347
   Y 349
   Z 350

Tables 353

Figures 356

Foreword

Besides meaning what we say, what could be more important that saying what we mean? James A. Peters published the first *Dictionary of Herpetology* almost 45 years ago and the need for an up-to-date dictionary is both powerful and obvious. Several thousand people now count themselves as professional herpetologists, tens of thousands more are interested in reptile and amphibian biology, and those numbers are growing rapidly even as the natural world shrinks in response to human impact. Those teaching herpetology courses have two nominal textbooks in English from which to choose, and each year many other volumes and thousands of scientific publications cover the subject. Not surprisingly, our vocabulary has grown correspondingly in almost half a century. Peters' dictionary boasted more than 3,000 words and abbreviations, whereas Lillywhite's encompasses 11,358 entries. Here the first ("Å. See *Angstrom unit*") and last ("zymogen, zymogen granules. *n*. See *proenzyme*") accounts are short, as are many of the others, whereas the longest (for "species") gets 687 words.

It is worth asking, what kind of person would devote himself or herself to a project like this? The late Jim Peters, author of the first dictionary and curator at the Smithsonian Institution, was a generalist whose professional contributions spanned a genetics textbook, systematic monographs, and innovative forays into computer-assisted collections management and biogeographical analysis. This new one is likewise a labor of love, as if years of spare-time obsession with words were a hobby, completed by a distinguished behavioral and physiological ecologist. Harvey Lillywhite grew up chasing lizards and snakes in California, obtained his Ph.D. from the University of California, Los Angeles, and has taught first at the University of Kansas and now the University of Florida, where he is Professor of Zoology and Director of the Seahorse Key Marine Laboratory. Harvey's research accomplishments—encompassing more than 200 publications—have benefited from his extensive travels and field perspectives, and he brings that breadth of experience and perspectives to this volume.

Research and education are at the core of conservation, and thus Harvey Lillywhite's new *Dictionary of Herpetology* will immeasurably aid our work as well as enhance the long-term survival of amphibians and reptiles.

April 3, 2008

Harry W. Greene, Cornell University, author of *Snakes: the Evolution of Mystery in Nature*

Preface and a Brief Guide to Format

Three circumstances contributed to the inspiration for producing this book. First was the fact that James Peters's *Dictionary of Herpetology* was out of print, and there has been no publication to replace this book since it became unavailable in 1964. Second was the invitation and challenge to produce a new dictionary by the publisher, Robert Krieger. Third was a perceived need—reinforced by the opinions of many others—for a new herpetological dictionary that included updated terminologies related to modern investigations, including the tools, methods and concepts from a variety of biological disciplines, as applied to the study of amphibians and reptiles. The resulting breadth of terminology is considerably greater than that of the earlier *Dictionary* by Peters, thus accounting for the increased size of the present volume.

Needless to say, an undertaking such as this *Dictionary* is almost limitless in its possible inclusions, so I have tried to be judiciously selective with respect to all additions and omissions of words. Generally, I have avoided inclusion of jargon, and I have omitted some antiquated terminology, unless it seemed useful in historical contexts or for interpretation of a reasonably large body of literature. With respect to the earlier *Dictionary*, I have omitted a number of Peters's terms that I judged to be either obsolete or nonuseful. A small fraction of such words are, in fact, included here, but with stated qualification in cases where such terms are to varying degrees prevalent in literature or in conversation. In a few cases, I have ventured my opinion regarding the usefulness of a term. I have also included some words that have arisen in various disciplines but are not yet common in herpetological literature. I have done so looking forward with anticipation that the relevance and utility of such terms will increase in the future. Where it seems useful, selected words that are used to define a term, and which themselves have entries elsewhere in the *Dictionary*, are italicized for cross-referencing purposes.

For various terms I have provided literature citations that are potentially useful. In some cases these references are related to the origin of the term, but in other cases they are intended to provide information relating either to terminological controversy or to more extensive explanation of a classic, complex, new, or important item of subject matter.

I have elected not to include the formal names of professionals who have been influential in herpetology. Biographical accounts are available elsewhere, and the possibility of omissions always risks offense. Further, I have not included full listings of species names, either scientific or common, owing to the rich diversity of amphibians and reptiles and the practical limitations of space. Species listings are available elsewhere, and their inclusion here is beyond the scope of this work. However, in my judgment it seemed useful to include the names of principal systematic groupings above the level of genus, and I have also included identifications of select, vernacular names of species that are commonly encountered in literature or that might be significant for scientific, historical, or other reasons. Realizing there are variable formats of usage with respect to capitalization of common names, I have chosen to capitalize all nonhyphenated words that comprise a common name, both for uniformity and distinction. Again, because of space limitations I have exercised due judgment in limiting the number of illustrations, literature references, and terms that are associated with medical, veterinary, or paleontology terminologies. There are numerous anatomical terms such as those of muscles, nerves, and blood vessels that are far too numerous to include in full. A substantial literature exists on the comparative usage of these terms, and such is greatly complicated by variations of terminologies that have been used by many different authors. Sorting all of these with comments on validity, precedence, current acceptance, and controversy is beyond the scope of this work.

Recent molecular studies have made it clear that birds should be considered as reptiles, so I have included a small amount of terminology related to avian biology. However, again because of space limitations I have not included a fuller coverage of avian terms owing to the fact that most persons interested to utilize this *Dictionary* will be those interested in amphibians and/or nonavian reptiles.

As a result of the expansionary and discretionary processes I have identified here, my hope for this new *Dictionary* is that it will be useful to a breadth of persons who might be interested in herpetology for one reason or another. I hope this volume will be useful to professionals who might be involved with reading or contributing to scientific literature. I also fervently desire that this book will assist students in the progression of their investigations or training related to herpetology. Additionally, numerous lay persons, clinicians, herpetoculturists, and others may find this book to be a useful guide to many aspects of terminology they might encounter in various places.

Acknowledgments

I would like to acknowledge the generosity and interest of many persons who have assisted with this project in numerous ways, including reviews or comments from James Albert, Tamatha Barbeau, Kenneth Dodd, Bill Duellman, Harry Greene, Lou Guillette, Darryl Heard, Victor H. Hutchison, Elliott Jacobson, Wayne King, Kenny Krysko, Steven Lillywhite, Paul Maderson, Ryan McCleary, Max Nickerson, Greg Pryor, Alan Savitzky, Coleman Sheehy III, Michael B. Thompson, Linda Trueb, Kent Vliet, Richard Wassersug, and Bruce Young. Kenny Wray did an excellent job with producing all of the illustrations that are used to complement the text. Chris Samuelson provided valuable assistance in checking the accuracy of references to older literature. I am especially grateful to my wife Jamie who gave much advice and encouragement, and patiently endured the numerous evenings at home when the moving keys of my laptop computer were a familiar sound contributing to the tedious and piecemeal growth of the manuscript. Finally, I owe a tremendous debt of gratitude to those colleagues, students, and friends who have, in sundry and meaningful ways, contributed to my enthusiasm for our discipline and to my limited but ever-increasing knowledge of it. Beyond the sterile appearances of scientific terminologies are doors to horizons of new knowledge and appreciation for Earth's treasures.

About the Author

Harvey B. Lillywhite was born in Nogales, Arizona, and he grew up in California where he acquired a passionate interest in the natural world, especially snakes, as a very young child. He earned his PhD at UCLA and held a postdoctoral fellowship at UC Berkeley. His early career in teaching and research was at the University of Kansas, and he is currently a Professor in the Department of Zoology at the University of Florida. He is also Director of the Seahorse Key Marine Laboratory, a small field station located on one of the larger islands in Florida's Cedar Keys. Dr. Lillywhite's research interests focus generally on the physiological and behavioral ecology of amphibians and reptiles, with emphasis on understanding the adaptation of form and function related to ecology, behavior and evolution. Dr. Lillywhite has published numerous research articles and book chapters on a broad range of subjects, and his study organisms have included insects, elephants, and dinosaurs in addition to herpetological taxa. He has traveled extensively and conducted field research mostly in Australia, India, Fiji, the Philippines, Thailand, Taiwan, and southern Mexico. He held a Fulbright fellowship for studies of tree frogs in India and has been a visiting lecturer and scientist at two Australian universities, Scripps Institution of Oceanography, and the NASA-Ames Research Laboratory in California. He is the recipient of a professional excellence award from the University of Florida. He presently serves on Editorial Boards for five international journals or book series, the Advisory Board of the Florida Institute of Oceanography, and the Executive Committee for the World Congress of Herpetology.

Å. See *Angstrom unit.*

a. Abbreviation of the Latin *annum*, meaning year.

a-, ab-. Prefix meaning "without," "from" or "away from."

AAA Treatment. Medical treatment of envenomation that includes the use of antivenin, antiobiotics, and antitoxin. First used by H.M. Parrish et al. (*J. Am. Vet. Med. Assoc.* 130:551, 1957). Cf. *TCS Technique.*

AAZPA. See *American Association of Zoological Parks and Aquariums.*

abaxial. *adj.* Away from the axis.

Abbildung. (Abb.) *n.* An illustration (meaning from German).

abdomen. *n.* The posteriormost region of the body cavity, caudad from the lung and thoracic region and containing the stomach, intestine, and reproductive system. Also refers to the "belly" or ventral body surface between the pectoral and pelvic appendages, or underlying the stomach and intestine. Sometimes used synonymously with *venter.*

abdominal. *adj., n.* 1. Pertaining to the ventral surface of the body. 2. Fourth pair of laminae on turtle plastron (Fig. 6); or a more general reference to *ventrals* or scales on the abdomen of squamates (Fig. 13.).

abdominal cavity. The cavity of the abdomen, containing the stomach, intestine, reproductive organs, and kidneys.

abdominal gland. 1. An enlarged, tubular gland in the posterolateral wall of the cloaca of male urodeles, sometimes extending into the abdominal cavity of some species. A small papilla empties contents either within or outside the cloaca and stimulates the female during courtship. 2. Glands located beneath abdominal scales of Casque-headed Skinks (*Tribolonotus*), unique among lizards.

abdominal length. A morphological measurement of tadpoles, defined as the medial, straight line distance from the base of the vent tube to the spiracular wall. There is a large measurement error because neither morphological feature is consistently definable (K.M. Carr & R. Altig, *J. Morphol.* 208:271–277, 1991).

abdominal pore. Paired openings in the ventral body wall near the cloaca, each connected to the coelom by a canal. Present in turtles and crocodilians, but not amphibians. May be homologous with the posterior pair of segmental urogenital pores in some fishes.

abdominal rib. A term used by some authors to denote *gastralia* or *parasternal.*

abdominal sternum. See *ypsiloid bone.*

abdominal sucker. See *belly sucker.*

abducens. *adj.* (L. away + leading) Denoting the sixth cranial nerve.

abduction. *n.* (*v.* **abduct**) Movement of a part, such as a limb, away from the medial axis or plane of the body, or movement of two parts away from each other. Cf. *adduction.*

abductor. *n.* A muscle that abducts, moving a part away from the sagittal plane or separating two parts. See *abduction.*

aberrant. *adj.* A feature or condition that departs from the usual or normal.

abience. *n.* Withdrawal from a stimulus, or an avoidance reaction. Cf. *adience.*

abiotic. *adj.* Reference to the nonliving components of the environment. Cf. *biotic.*

abiotic factors. Reference to nonliving characteristics of the environment such as physical and chemical components.

aboral. *adj.* Opposite to, or away from, the mouth.

abort. *v.* To arrest development.

abscess. *n.* A localized collection of pus in a cavity formed by disintegration of tissue.

abscissa. *n.* The horizontal or x- axis of a graph. Cf. *ordinate.*

abscission. *n.* (*v.* **abscise**) The natural separation (or process of separation) by two parts of an organism.

absenteeism. *n.* In herpetological literature, this term refers to the behavior expressed in crocodilians when a parent leaves its offspring. This term is misleading, however, because it implies that crocodilians typically remain with their young and that a missing parent is somehow abnormal. In fact, this is not the case; females remaining with young are highly variable, both in occurrence and in length of time the parent remains near the young.

absolute abundance. In ecology, the precise number of individuals of a taxon in a given area, population, or other measure of space. Cf. *absolute density, relative abundance.*

absolute age. The age of an object as established by some precise dating method, such as radiometric dating. Cf. *relative age.*

absolute density. The real number of individuals per unit of surface or volume of habitat. Cf. *absolute abundance, apparent density.*

absolute fitness. See *fitness.*

absolute humidity. See *humidity.*

absolute ranking. In systematics, the assignment of formal rank to a monophyletic group based on the criterion of age of origin rather than degree of divergence. Cf. *relative ranking*.

absolute temperature. Temperature measured from absolute zero, the state of no atomic or molecular thermal motion. The absolute scale is divided into kelvins (K), with 1K being equivalent to 1 Celsius degree. 0K is equal to –273.15 °C or –459.67 °F.

absorbance, absorptance. *n.* A measure of the loss of intensity of radiation passing through an absorbing medium. In spectrophotometry, it is defined by the relation log (I_o/I), where I_o = the intensity of radiation entering the medium, and I = the intensity after traversing the medium. See *Beer-Lambert Law*.

absorptiometry. *n.* A rapid, noninvasive technique that allows identification of major components of the body and properties of tissues by means of attenuation of photons as they pass through an absorber. Historically, such analyses involved single-photon absorptiometry (SPA) to measure density of bones (*densitometry*), or other tissue-like materials. Dual-energy X-ray absorptiometry (DXA) allows for measurement of total and regional bone mineral, bone mineral density fat mass, and bone-free lean tissue mass in various animals *in vivo*.

absorption threshold. The threshold level of substrate moisture from which a dehydrated amphibian can gain water, and below which there is a net loss of body water. Abbreviated "AT" by H. Heatwole (*Ecology* 43:460–472, 1962). See also *critical level*.

absorptivity. *n.* The fraction of incident radiation absorbed by a body or surface. The fraction of incident *radiant flux* at a given wavelength that is absorbed by a material or body. Cf. *emissivity*.

abundance. *n.* The number of individuals belonging to a population. See *absolute abundance* and *relative abundance*.

abundismus. *n.* A type of melanism in which an animal is not entirely black, but displays an increase in blackened markings in areas that are not normally black in the species.

abyssal. *adj.* Reference to great depths within Earth, or (more usually) to the deeper zone of ocean or lake waters below the permanent thermocline. Cf. *bathyl*.

acanthoid. *adj.* Spiny or spine-like.

acariasis. *n.* Infestation by mites or tick.

acaricide. *n.* A chemical that kills mites and ticks.

acarid. *n.* A mite or tick: small arachnid of the order Acarina.

acaridomatium. (*pl.* **-tia**) *n.* The pocket or infolding behind limb insertions of certain lizards, wherein mite infestations characteristically occur. Literally a "mite house"; used in earlier literature. (Syn. *dermal pocket, postfemoral dermal pocket*)

acaudal. *adj.* Without a tail.

acceleration. *n.* 1. From evolutionary biology, an evolutionary change in which the descendant species has a faster rate of growth or ontogenetic development than the ancestral species. 2. From physics, the rate of change of velocity of an object, for example in m/s^2.

accelerator muscle. The muscle used to propel a chameleon's tongue forward.

accepted name. Reference to a name that is adopted by an author as the correct name for a taxon.

accessory bladder. See *cloacal bursa*.

accessory chorioidal eye. A pigmented, hollow vesicle located dorsally on the choroid plexus, relatively distant from the pineal organ or parietal eye. First distinguished by A. Prenant (*Bibliog. Anat.* 2:223–229, 1894) in the lizard *Anguis fragilis*. Considered an infrequent accessory organ which disappears in the adult.

accessory fang. See *replacement fang*.

accessory gland. A gland attached to the anterior surface of the primary venom gland of venomous snakes. The venom duct drains the primary gland and runs through the accessory gland before fusing with the fang sheath. Mole Vipers lack a separate accessory gland. See Fig. 22.

accessory lateral plate. An obliquely directed process immediately below the neural spine on caudal vertebrae of the snake *Eryx,* extending forward to fit a groove on the preceding vertebra. Similar in function to *zygosphenes*, which coexist on anterior caudal vertebrae but are absent posteriorly. First investigated by M.S. Sood (*Proc. Indian Acad. Sci.* 14:390–394, 1941) and used alternately with the term *accessory lateral process*. See also *parapophysial articulation*.

accessory lateral process. See *accessory lateral plate*.

accessory organ. Any small structure appearing in connection with a pineal organ or parietal eye. Prenant, *Bibiog. Anat.* 1 (1895), distinguished four such organs in the lizard *Anguis*: epiphyseal eye, interparietal-epiphyseal eye, interparietal eye, and accessory chorioidal eye.

accessory palpebral. Used by S.J. Copland (*Proc. Linn. Soc. N.S.W.* 70:62–92, 1945) to describe scales lying between the preocular and true palpebral scales on the head of certain lizards.

accessory process. A spine-like process that projects laterally below the *prezygapophysis* and is more or less continuous with the buttress of the prezygapophysis. Restricted to snake vertebrae according to W. Auffenberg (*Tulane Stud. Zool.* 10:131–216, 1963).

accessory thyroid gland. See *ultimobranchial glands or body*.

accessory vesicle. A term for *cloacal bursa*, as used by Cuvier, according to H.M. Smith and L.F. James (*Trans. Kans Acad. Sci.* 61:86–96, 1958).

acclimation. *n.* A reversible change in morphology, behavior or physiology that occurs in response to a prolonged change of environmental condition(s), e.g. temperature, in the laboratory. Sometimes used interchangeably with *acclimatization*. Should not be confused with evolutionary *adaptation*. Cf. *acclimatization*.

acclimatization. *n.* A reversible change in morphology, behavior, or physiology that occurs in response to prolonged change of conditions in the natural environment. Cf. *acclimation*.

accommodation. *n.* 1. The eye's action or ability to bring an object into focus. 2. A change in the firing pattern of a sensory neuron during the prolonged application of a stimulus, typically due to an increase in threshold during the course of stimulation. This is a form of *sensory adaptation* and results in a decrease in sensation or response.

accretion. *n.* A process in which the growth or buildup of some structure or mass occurs by accumulation of fresh layers of material on top of existing layers. The term is used in contexts related to both anatomy and geology.

accuracy. (absolute accuracy) *n.* 1. The maximum difference between the measured value from a data acquisition device and the true signal (or voltage) applied to the input, typically specified as a ± voltage. 2. In statistics, the degree to which a measured quantity corresponds to the true value of what is being measured. Cf. *precision*.

acellular. *adj.* Noncellular; not composed of cells.

acelous, acoelous. *adj.* A vertebral condition in which the ends are flat, although some may appear to be biconcave to allow room for the intervertebral disc. (Syn. *amphicyrtean, platyan*)

acentric. *adj.* Having no body or center, in reference to vertebrae of amphibians.

acephalous. *adj.* Lacking, or having the appearance of being without, a distinct head.

acetabular cup or **fossa.** See *acetabulum*.

acetabulum. *n.* A socket of the pelvic girdle that articulates with the rounded head of the femur. (Syn. *acetabular cup, acetabular fossa*)

acetone. *n.* A volatile ketone used as a solvent and as an agent in the production of other organic compounds.

acetylcholine. (Ach) *n.* An acetic acid ester of choline, important as a synaptic transmitter in many different types of neurons in most animal species and the principal activator of vertebrate skeletal muscle.

acetylcholinesterase. *n.* 1. An enzyme that helps degrade the neurotransmitter *acetylcholine*, found naturally in the synaptic cleft. 2. This enzyme is a component of elapid snake venoms, although it is present in extremely small concentrations.

acetyl-coenzyme A. See *coenzyme A*.

achromatic. *adj.* Non-pigmented, or without color.

acicular. *adj.* Needle-shaped.

acid-base balance. Reference to processes and mechanisms that maintain the pH of body fluids within a narrow, regulated range.

acidic. *adj.* Of, or pertaining to, low pH (<7). Cf. *alkaline, basic*.

acidic dye. An organic anion that binds to, and stains, positively charged macromolecules.

acidophilic, acidophilous. *adj.* (*n.* **acidophily, acidophile**) Seeking or thriving in an acidic environment. Cf. *acidophobic*.

acidophobic. *adj.* (*n.* **acidophoby, acidophobe**) Intolerant or avoiding of acidic environments. Cf. *acidophilic*.

acidosis. *n.* Excessive body acidity, typically measured in body fluids such as blood.

acid precipitation, acid rain. Reference to deposition of airborne acids in precipitation, including snow, fog, and dry acidic particles, in addition to rainfall.

acid tide. In physiology, reference to acidification of the blood following pancreatic secretion.

acinar. *adj.* Having rounded compartments (acini), as in acinar testis.

acinose. *adj.* Made up of *acini*.

acinous. *adj.* Pertaining to a gland that is subdivided into small sacs or cavities, as in *nuptial pads* of anurans. Or, shaped like a grape.

acinus. (*pl.* **-ni**) *n.* 1. A small, saclike dilatation; particularly one found in a gland. 2. Any of the smaller lobules of a compound gland.

Acontiinae. *n.* A subfamily of *Scincidae*.

Acosmanura. *n.* 1. A taxon of frogs, sister with *Costata* within the newly designated *Sokolanura* (see D.R. Frost et al., *Bull. Amer. Mus. Nat. Hist.* 297:1–370, 2006). It is composed of the two clades *Anomocoela* and *Neobatrachia*. 2. A synonym for Orton's (1953) Type IV tadpole (from P.H. Starrett, pp. 251–271 in J.L. Vial, ed., *Evolutionary Biology of the Anurans. Contemporary Research on Major Problems*, 1973 equivalent to *ranoid* of O.M. Sokol, *Copeia* 1975:1–23, 1975).

acoustic. *adj.* Pertaining to sound, as in *acoustic* stimuli.

acoustico-lateralis system. See *lateral line system*.

acoustics. *n.* The study of sound.

acre. *n.* A fps unit of area equal to 4.046×10^3 meters.

acre-foot. n. A unit reference to the quantity of water sufficient to cover 1 acre of land surface to a depth of 1 foot (ca. 1234 cubic meters).

acridophagous. *adj.* (*n.* **acridophage, acridophagy**) The habit or condition of feeding on grasshoppers.

acrochordate. *adj.* Having large cone-like projections that appear to be hairy. Reference to a pattern of epidermal microsculpturing. See R.M. Price *J. Herpetol.* 16:294–306, 1982.

Acrochordidae. *n.* A clade (family) of highly unusual and totally aquatic snakes with coarsely keeled scales and tubercles. A single genus and three species are distributed from coastal India through southern Asia to the Solomon Islands and northern Australia, mostly in coastal rivers, estuaries, and mangroves. Known as *Wart Snakes* or *File Snakes*.

Acrochordoidea. *n.* A clade (superfamily) of snakes that includes the family *Acrochordidae*.

acrodendrophilous. *adj.* Thriving in tree tops or canopy habitats.

acrodont. *adj.* A condition of tooth attachment in which rootless teeth are fused to elevated apical surfaces of bone, characteristically in amphibians, tuatara, and certain squamates. (Syn. *euacrodont*) See also *subacrodont*.

Acrodonta. *n.* A basal clade of *Iguania* that includes lizards of the families *Agamidae* and *Chamaeleonidae*.

acrodont ankylosis. Attachment of a tooth at the horizontal surface or apex of the tooth-bearing bone. Cf. *thecodont, pleurodont, protothecodont ankyloses* (J.W. Osborn, *Symp. Zool. Soc. Lond.* 52:549–574, 1984).

acromial. *n.* A synonym for *clavicle* in anurans, according to S.D. Hsiao (*Peking Nat. Hist. Bull.* 8:169–204, 1934). See Fig. 3.

acromial process. A small, anteriorly projected extension of the cranial border of the scapula in reptiles including birds. In chelonians, this structure is large and long, and forms a prominent part of the pectoral girdle. (Syn. *acromion, acromion process*; sometimes = *clavicula; precoracoid; prescapular process; procoracoid; proscapula; proscapular process;* principally in turtles)

acromion, acromion process. See *acromial process*.

acrophilic. *adj.* (*n.* **acrophile, acrophily**) Reference to organisms that tend to settle or live on surfaces of other organisms.

Acta Herpetologica. A herpetological journal published by the Societas Herpetologica Italica.

ACTH. *n.* See *adrenocorticotropic hormone*.

actin. *n.* A ubiquitous protein that is the major constituent of 7-nm-wide microfilaments of cells and participates in muscle contraction and other forms of cellular motility.

actin-binding proteins. A large class of proteins that form complexes with actin.

actinobiology, actinology. *n.* Study of the effects of radiation on living organisms.

action force. Any *force* actively exerted upon an object or substrate by an animal. Cf. *reaction force*. See Fig. 25.

action group. A group of behaviors that are characterized by a description of their action.

action potential. A rapid, self-amplifying change (reversal) in the electric potential across the membrane of nerve, muscle or glandular cells, produced by regenerative inward current across the membrane. This is the "nerve impulse" that propagates along the cell membrane or nerve fiber without loss of amplitude, and is the unit of information that is transmitted and processed by the nervous system. (Syn. *nerve impulse, nerve signal*) Cf. *graded potential*.

action spectrum. The pattern of response to each of many different wavelengths of incident light.

action system. Reference to an animal's behavior pattern.

-activase II. See *arin*.

-activase X (etc.). Suffix meaning "activator of factor X, etc.," used in a nomenclatural scheme for describing exogenous hemostatic factors in snake venoms. By adding a portion or designated abbreviation of a snake's scientific name to the suffix, one obtains a designation for the fraction being identified. For example, using the name "*gabonica*" (species name for the Gaboon Viper) one obtains *gabonactivase X*.

activating enzyme. An enzyme that catalyzes a reaction involving ATP and a specific amino acid.

activation. *n.* In developmental biology, the initiation of a developmental program in a fertilized egg.

activation energy. The energy required for a chemical reaction to proceed.

active center. A flexible portion of an enzyme (protein) that binds to the substrate and converts it into the reaction product, or, in the case of carrier and receptor proteins, the portion of the molecule that interacts with the specific target compound.

active forager. A predator which moves through the environment and actively seeks prey. Cf. *ambush predator*.

active immunity. Immunity resulting from the body's own immunological defenses following exposure to an antigen.

active process. Used to denote any process that requires expenditure of metabolic energy. Cf. *passive process*.

active site. The portion of a protein that must be maintained in a specific shape and amino acid content to be functional. The binding and catalytic region of an enzyme molecule.

active state. With reference to muscle fibers, the condition in which myosin cross bridges are attached to actin and produce active force or tension in the muscle.

active transport. Movement of materials against a concentration or energy gradient; a process that requires energy and involves carrier molecules.

active uptake. Reference to absorption of ions by processes that require expenditure of metabolic energy.

activity. *n.* 1. A term commonly used in behavioral and ecological studies, having reference usually to locomotor movements. 2. The capacity of a substance to react with another substance. 3. The effective concentration of an ionic species in the free state.

activity area. Used by K.L. Hansen (*Copeia* 1957:274–277,1957), following Dice (1952), for *home range*.

activity pattern or rhythm. Periodic and repeated pattern of locomotory activity, usually associated with diel or seasonal variation.

activity temperature range. (ATR) The range of body temperatures at which free-ranging animals engage in ordinary or routine behavior. (Derived from the *normal activity range* of R.B. Cowles and C.M. Bogert, 1944; other synonyms: *TAR* or *thermo-activity range* Cf. *preferred optimum temperature zone*).

actomyosin. *n.* A complex of muscle proteins formed when myosin cross-bridges bind to actin.

acuity. *n.* Resolving power.

acuminate. *adj.* Tapering to a point.

acute. *adj.* Reference to exposure of relatively short duration, in contrast to *chronic*.

ad-. Prefix meaning "toward."

ad. Abbreviation for *adult*.

adaptability. *n.* The degree to which an organism or species can remain or become adapted to a wide range of environments by physiological or genetic means (after Th. Dobzhansky, *Evol. Biol.* 2:1–33, 1968; Pp. 109–121 in R.C. Lewontin, ed., *Population Biology and Evolution*, 1968).

adaptation or **adaption.** *n.* 1. The evolutionary process of adjustment to environmental conditions by means of natural selection acting to alter gene frequencies. The process of becoming *adapted*. 2. Outside evolutionary biology, the term includes any process of adjustment by an organism or its part to a particular environmental condition. 3. The term is also used to indicate any feature of an organism that enhances survival, thereby confounding the end product with the process. The term is restricted to a character that was shaped by natural selection for the function in which it is currently enhancing fitness. Cf. *exaptation, fitness*. See also *adaptive trait, aptation, sensory adaptation*.

adaptedness. *n.* The extent of adaptation to environment. The degree to which an organsism is able to live and reproduce in a given set of environments; the state of being adapted (after Th. Dobzhansky, *Evol. Biol.* 2:1–33, 1968; Pp. 109–121 in R.C. Lewontin, ed., *Population Biology and Evolution*, 1968).

adaption. *n.* See *adaptation*.

adaptive capacity. Reference to physiological tolerance of an organism determined by genetics.

adaptive enzyme. An enzyme that is formed by an organism in response to an outside stimulus.

adaptive landscape. A three-dimensional graph that shows the frequencies of two genes, each present in allelic forms and plotted against the average fitness for a given set of environmental conditions. A comparable conceptual plot may be constructed in multidimensional space to accommodate more than two loci. Fit genotypes (or species) able to occupy particular ecological niches are depicted as adaptive peaks separated by adaptive valleys representing lesser or unfit gene combinations. The concept of an adaptive landscape for phenotypic characters was first proposed by G.G. Simpson (*Tempo and Mode in Evolution*, 1944) and derived from S. Wright's concept of an adaptive landscape in which population fitness is a function of gene frequencies (*Proc. VI Internat. Congr. Genetics*, 1:356–366, 1932). Recently, the concept of adaptive landscape has been extended to incorporate considerations of *performance* (S.J. Arnold, *Integr. Comp. Biol.* 43:367–375, 2003). (Syn. *adaptive surface, adaptive topography*)

adaptive norm. The array of genotypes possessed by a given population of a species, which is compatible with the demands of the environment.

adaptive peak. One or one of several high points on an *adaptive landscape*, from which movement in any planar direction (representing changes in gene frequencies) results in lower average fitness. A combination of allele frequencies (or phenotype) that reflects locally maximal fitness of a population.

adaptive radiation. Evolutionary diversification of lineages from a generalized common ancestor, producing species that specialize for different ecological niches. The phenomenon occurs usually over a relatively short time and produces a variety of new forms. (Syn. *evolutionary radiation*)

adaptive significance. The value of a given character or character state to an organism, usually inferred.

adaptive strategy. Reference to a suite or sum total of adaptations of a taxon to the environment.

adaptive surface. See *adaptive landscape*.

adaptive topography. See *adaptive landscape*.

adaptive trait. A character or aspect of developmental pattern of an organism which enables or enhances the probability of that organism surviving and reproducing. (sensu Th. Dobzhansky *Amer. Nat.* 90:337–347, 1956). A heritable trait that has evolved to serve a particular function or purpose.

adaptive valley. A low point or valley on an *adaptive landscape*.

adaptive value. The property of a genotype that confers fitness to an organism in comparison with other genotypes in a given environment. The comparative fitness of different genotypes. (Syn. *Darwinian fitness*)

adaptive zone. In evolutionary biology, a concept of ecological pathways along which taxa evolve, essentially similar to the fundamental niche at the level of species.

Adder. *n.* Collective vernacular name for several genera and various species of Old World viperid snakes, especially *Vipera berus*.

additive factor. Any of a group of non-allelic genes that affect the same phenotypic character. Each gene enhances expression of the other in the phenotype.

additive genetic variance. Reference to the additive effects of alleles on the genotype, or the average effects of substituting one allele for another at a given locus or at multiple loci governing a polygenic trait. This component of variance determines the magnitude of response to selection of quantitative traits. Cf. *dominance genetic variance*.

additive hypothesis. In systematics, the concept that there is a value corresponding to each link on a given phylogenetic tree, and that the observed distance between a given pair of contemporary taxonomic categories is equal to the sum of the values for the links connecting the pair of taxonomic categories.

addled. *adj.* Term used to describe eggs that fail to hatch.

adduction. *n.* (*v.* **adduct**) Movement of a part, such as a limb, toward the medial axis or plane of the body, or movement of two parts together. Cf. *abduction.*

adductor. *n.* A muscle that adducts, moving a part toward the sagittal plane or drawing two parts together. See *adduction.*

adductor fossa. A fossa accomodating the major adductor muscles and opening into the lower jaw of reptiles. It is bordered by the articular, surangular, prearticular, and coronoid bones. Used by A.S. Romer (*Osteology of the Reptiles*, p. 199, 1956). (Syn. *mandibular fossa, Meckelian orifice, Meckelian vacuity*)

adductor mandibulae. The muscle or muscles closing the mouth in amphibians and reptiles.

adelphotaxa. *n.* Sister taxa.

adenohypophysis. *n.* The anterior, intermediate (nonneural), and secretory part of the *hypophysis* or *pituitary gland* (derived from the hypophyseal pouch). (Syn. *anterior lobe, anterior pituitary*)

adenosine diphosphate. (ADP) A nucleotide formed by hydrolysis of *ATP*, with the release of one high-energy bond.

adenosine triphosphate. (ATP) An energy-rich nucleotide used as a common energy currency in all living cells.

adenosine triphosphatase. An enzyme probably ubiquitous in all snake venoms, acting to destroy ATP and likely an important factor in the rapid production of hypotension and shock that are seen in many victims of snakebite.

adepidermal reticular network. A network of fibers forming nodules at crossing points, unique in the subepidermal basement membrane of *Triturus* according to A.J. Schmidt (*J. Morph.* 111:275–285, 1962).

adetoglossal. *adj.* A condition in which all edges of the tongue are free. Coined by T.M. Uzzell (*Copeia* 1961:78–86, 1961) with reference to the *boletoid* tongue of plethodontid salamanders. Cf. *detoglossal.*

ADH. *n.* See *antidiuretic hormone.*

adherent. *adj.* 1. Joined with, attached, or connected. 2. An ecomorphological guild of lotic tadpoles that live in flowing water and have small, complete marginal papillae and inhabit faster water than tadpoles in the *clasping* guild, maintaining position by means of an *oral disc*

adhesion. *n.* 1. Firm attachment, attraction or constant close proximity between two surfaces. The molecular attraction exerted between surfaces in contact. 2. The attraction of water molecules to other molecules in the walls of interstices of a sediment.

adhesion surface. Part of a subdigital lamella of lizards, including the free edge and ventral surface, which bears numerous fibrous keratinized structures (digital setae) having variously split, spatulate, or otherwise elaborated endings collectively forming a fibrous bunch. Originally described by B.C. Mahendra (*Proc. Indian Acad. Sci.* 13:288-306,1941).

adhesive apparatus. See *adhesive organ.*

adhesive gland. See *adhesive organ.*

adhesive molecules. Any complementary cell-surface molecules that bind specifically to one another, thereby causing cells to adhere to one another.

adhesive organ. Folds at the hindmost border of the mouth in embryonic tadpoles of anurans, variously shaped and covered internally with mucus-secreting cells. These structures enable tadpoles to adhere to submerged objects, stabilize hatchlings before swimming abilities develop, and are lost before metamorphosis. Studies have suggested these structures are not homologous with *balancers* of salamander larvae, and that connotations of "sucker" should not be used in most cases. (Syn. *adhesive apparatus; adhesive gland; cement organ; holder; oral sucker; sucker; sucking disc*)

adhesive pad. Tissue on the toes of certain lizards that enables them to adhere to smooth vertical surfaces.

adiabatic. *adj.* Reference to a thermodynamic process in which heat is neither gained nor lost nor transferred. Atmospheric air undergoes adiabatic cooling as it rises and expands with decreasing atmospheric pressure; adiabatic warming occurs as it descends and is compressed by increasing atmospheric pressure.

adiabatic lapse rate. See *lapse rate.*

adience. *n.* Movement toward a stimulus. Cf. *abience.*

adipose tissue. Body fat or lipids. See *fat body.* (Syn. *fat*)

adjuvant. *n.* 1. In immunology, any substance that enhances an immune response when given before or with an antigen. 2. Any substance added to a drug to enhance its action.

adnate. *adj.* Refers to two unlike parts that are closely connected, attached, or growing on, such as the tongue and floor of the mouth in various amphibians, chelonians, and crocodilians.

adnexus. (*pl.* -a) *n.* (*adj.* –al) An ancillary or supplementary part to a principal organ or structure. E.g., the trachea and primary bronchi are adnexae to the lung. (Not recommended as a useful term.)

adpress. *v.* To press close or hold against.

adrenal, adrenal gland. *n.* An endocrine gland located near the kidney or gonads, sometimes incorporated into the associated *mesentery*. The *adrenal cortex* produces glucocorticoids, mineralcorticoids, and weak androgens involved in sexual maturation, stress responses, and regulation of body fluids, whereas the *adrenal medulla* (deeper tissue) produces catecholamines.

adrenal corticosteroid. Reference to steroid hormones formed in the adrenal cortex. There are several such hormones in various species, the most common being *aldosterone, corticosterone,* or *cortisol,* and all are synthesized from cholesterol by cortical cells that have been stimulated by *adrenocorticotropic hormone*. (Note the weak androgens produced are usually not included in this term; aldosterone seldom is.)

adrenalin, adrenaline. *n.* See *epinephrine*.

adrenergic. *adj.* Pertaining to epinephrine and norepinephrine, or neurons that release them.

adrenergic receptors. (= **adrenoreceptors**) Receptors on cell surfaces that bind norepinephrine and epinephrine. The binding activates a G protein in the cell.

adrenocorticotropic hormone. (ACTH) A peptide hormone released by cells in the anterior pituitary that acts mainly on the adrenal cortex, stimulating synthesis and secretion of steroid hormones from that organ. (Syn. *adrenocorticotropin, corticotropin*)

adrenocorticotropin. *n.* See *adrenocorticotropic hormone*.

adrenoreceptors. *n.* See *adrenergic receptors*.

Adriosaurus microbrachis. Scientific name of the earliest fossil lizard exhibiting limb loss (vestigial limbs). The newly named species is a marine lizard that existed about 95 million years ago and offers clues to the evolution of terrestrial lizards as they transitioned to water. Some also think this fossil might provide a snapshot of the morphological changes that occurred during the transition from lizards to snakes. See A. Palci & M.W. Caldwell, *J. Vert. Paleontol.* 27:1–7, 2007.

adscapulum. *n.* A term for the *suprascapula* in Salientia, according to S.D. Hsiao (*Peking N.H. Bull.* 8:169–201, 1934).

adsorption. *n.* (*v.* **adsorb**) The adhesion of molecules on the surface of solids or fluids, or the exchange of substances between a solution and the surface of particles.

adspersed. *adj.* Widely scattered or distributed.

adult. (ad.) *n.*, *adj.* An individual animal that is fully developed, has reached sexual maturity, and has attained a life stage wherein its general form and appearance does not change markedly.

advanced. *adj.* Pertaining to a modified or derived evolutionary stage in a phylogenetic lineage. Cf. *primitive*.

advection. *n.* The transfer of heat or matter by convection or movement, such as horizontal movement of a water mass.

adventive. *adj.* Non-native, used in reference to an organism that is transported by natural or artificial means into a new habitat.

advertisement. *n., adj.* A conspicuous display, usually associated with males and serving to attract females and/or deter rival males.

advertisement call. Reference to vocalization of male anurans used in courtship and territorial behavior.

aeolation, eolation. *n.* Erosion of a land surface due to wind-blown sand or dust.

aeolian, eolian. *adj.* Related to, or caused by, the action of wind. Dispersed by the wind.

aeolian deposit. An alluvial or wind-borne soil deposit. Cf. *alluvial deposit, colluvial deposit*.

aer-. Prefix that indicates air or atmosphere.

aerial. *adj.* Airborne.

aerobe. *n.* An organism that lives in contact with air and is metabolically dependent on molecular oxygen.

aerobic. *adj.* A process or organism that only functions fully in the presence of free oxygen. Refers to aerobic metabolism in which nutrient macromolecules are ultimately oxidized by molecular oxygen in cellular respiration, yielding carbon dioxide and water as end products. Cf. *anaerobic*.

aerobic metabolic scope. The ratio of the maximum sustainable metabolic rate to the basal or standard metabolic rate. The maximum rate is usually measured as result of maximal intensive exercise or locomotion. (Syn. *metabolic scope, factorial scope for locomotion, scope for activity*)

aerohygrophilous. *adj.* Thriving in high atmospheric humidity.

Aeromonas. *n.* A genus of Gram-negative bacteria that is considered an opportunistic pathogen, i.e. a secondary invader. It has been isolated from abscesses, the oral cavity of animals with *stomatitis*, and from anurans with *red-leg* disease.

aestival. *adj.* Of or pertaining to the early summer season. Cf. *autumnal, hibernal, vernal*.

aestivation, estivation. *n.* (*v.,* **estivate**) A state of torpor characterized by reduced activity and metabolism associated with summer or periods of high temperatures or drought. Cf. *brumation*.

AFDW. See *ash-free dry weight*.

AFLDM. See *ash-free lean dry mass*.

AFLP. See *amplified fragment length polymorphism*.

afferent. *adj.* Leading, transporting, or conducting toward a central region. In sensory systems, this has reference to flow of nerve traffic to the central nervous system (*afferent fibers*). Cf. *efferent*.

affinity. *n.* 1. In immunology, the innate binding power of an antibody combining site with a single antigen binding site. 2. In respiratory physiology, the innate binding power of hemoglobin with oxygen, usually expressed as the P_{50} or partial pressure of oxygen at which a solution of blood or hemoglobin is half saturated with oxygen. 3. More generally, in physiology, the strength of binding of any receptor for its ligand. 4. In evolutionary biology, a close relationship or common ancestry of organisms, suggested by resemblances in their morphology, molecular properties, or other features.

affinity index. A measure of the similarity of species composition between two communities, calculated as $A = c/(aXb)^{1/2}$, where a and b are numbers of species occurring in communities A and B, and c is the number of species common to both.

afibrinogenemia. *n.* A blood disorder characterized by deficiency of fibrinogen and inability to clot.

African Bullfrog. Vernacular name for the African ranid frog

Pyxicephalus edulis (formerly treated as a subspecies of *P. adspersus*). The name bullfrog also is applied to any member of the genus. See *Bullfrog*.

African Bush Viper. See *Bush Vipers*.

African Centipede-eater. Collective vernacular name for species of African snakes belonging to the genus *Aparallactus* (Atractaspididae).

African Clawed Frog. Vernacular name for the pipid species *Xenopus laevis*. See also *Clawed Frog*.

African Dwarf Crocodile. Vernacular name for the crocodilian *Osteolaemus tetraspis*.

African Egg-eating Snake. See *Egg-eating Snake*.

African Forest Snake. Vernacular name for colubrid species of snakes belonging to the genus *Geodipsas*, distributed in Africa and Madagascar.

African Gartersnake. Vernacular name for African elapid snakes belonging to the genus *Elapsoidea*.

African Hairy Frog. Vernacular name for the African arthroleptid frog species *Trichobatrachus robustus*, named for the vascular hair-like appendages grown on the posterior flanks and thighs of males during the breeding season.

African House Snake. See *House Snake*.

African Journal of Herpetology. A journal published by the Herpetological Association of Africa, Durban, South Africa.

African Reed Frog. See *Reed Frog*.

African River Frog. See *River Frog*.

African Swamp Frog. Vernacular name for the ranid species *Dimorphognathus africanus*, endemic to west equatorial Africa.

Africanura. *n.* A new taxon of anurans, sister to the new taxon *Saukrobatrachia* within the new taxon *Ametrobatrachia*, and containing the African *Phrynobatrachidae* and *Pyxicephaloidea* (D.R. Frost et al., *Bull. Amer. Mus. Nat. Hist.* 297:1–370, 2006).

Afrobatrachia. *n.* A new monophyletic taxon of anurans, sister to *Microhylidae* within the new taxon *Allodapanura* and composed of the new taxa *Xenosyneunitanura* and *Laurentobatrachia* (D.R. Frost et al., *Bull. Amer. Mus. Nat. Hist.* 297:1–370, 2006).

Afrotropic. *n.* One of eight *biogeographical realms* including sub-Saharan Africa. (Syn. *Ethiopean region*)

Agama, Agamid. *n.* Common name for lizard species belonging to the Agamidae, especially the genus *Agama*.

Agamidae. *n.* A clade (family) of moderate- to large-sized, diurnal and primarily terrestrial lizards. Approximately 45 genera and more than 380 species are known from Africa, southern and central Asia, and the Indoaustralian Archipelago to Australia, New Guinea and the Solomon Islands.

agar. *n.* A polysaccharide extract of certain seaweeds used as a solidifying agent in culture media for growth of cells, especially bacteria. Agar mixtures also have been molded to form models of amphibians that evaporate as a free water surface.

Agastorophrynia. *n.* A new monophyletic taxon of anurans, sister to *Cycloramphidae* within the new taxon *Hesticobatrachia*, and composed of *Dendrobatoidea* and *Bufonidae*. See D.R. Frost et al., *Bull. Amer. Mus. Nat. Hist.* 297:1–370, 2006.

age. *n.* 1. The amount of time that has passed since an individual's birth. Newborns are classified as age 0, not age 1. 2. A period of geological history characterized by a dominant taxon or taxa. 3. A geological time unit shorter than an *epoch*. 4. The position of an organism or event in the geological time scale. 5. (*v.*) To become old or to attain maturity.

age class. A category in which individuals within a certain age interval are classified within a population. A *cohort*.

age-dependent selection. Selection in which the values for relative fitness of different genotypes or phenotypes vary with the age of the individual.

age distribution. See *age structure*.

Agent Orange. A mixture of organochlorine herbicides used as a defoliant by the U.S. military.

Age of Amphibians. A popular name for the *Carboniferous* and *Permian* periods (Table 1) when amphibians were dominant. The Age of Amphibians lasted about 100 million years (estimates vary).

age of differentiation. With respect to phylogeny, the age of the latest common ancestor of a given modern group. Cf. *age of origin*.

age of origin. With respect to phylogeny, the age of event that produced a given modern group and its sister group. Cf. *age of differentiation*.

Age of Reptiles. A popular name for the *Mesozoic* Era (Table 1), during which numerous groups of reptiles, including spectacular dinosaurs, diversified and radiated into virtually all ecological niches occupied by terrestrial vertebrates living today. The Age of Reptiles lasted about 200 million years (estimates vary).

age-specific death rate. The death rate of a given age class or cohort in a population, expressed as the number of individuals dying within an age class divided by the number of individuals that attain such age.

age-specific fecundity rate. The average number of eggs or young produced per time by an individual of specified age.

age-specific survival rate. The number or proportion of individuals from an initial cohort surviving to a specified age.

age structure. The number or proportion of individuals in each age class of a population. (Syn. *age distribution* or *composition*)

agglutinin. *n.* An antibody or hemotoxin that causes clumping of red blood cells (not equivalent to clotting).

aggregate, aggregation. (agg.) *n.* 1. A massing together of several or more individuals of a species in the same place, as at a hibernaculum, breeding chorus, etc. With respect to tadpoles, the term describes a group that is congregated in response to factors other than social interactions. Cf. *school*. 2. See *species aggregate*.

aggression. *n.* Reference to any hostile act or display to protect territory, offspring, or establish dominance.

aggressive mimicry. Resemblance of a predator to a harmless animal to facilitate approach to the prey.

Aglaioanura. *n.* A new monophyletic taxon of anurans, sister to *Dicroglossidae* within the new taxon *Saukrobatrachia*, and composed of *Rhacophoroidea* and *Ranoidea* (D.R. Frost et al., *Bull. Amer. Mus. Nat. Hist.* 297:1–370, 2006).

aglossal. *adj.* Having no tongue, as in pipid frogs. Cf. *phaneroglossal*.

aglyph. *n.* (*adj.* **–ous**) Reference to snakes in which teeth lack a groove or canal, used usually in reference to some colubroids in which the posterior fang is unspecialized or nonexistent. These teeth are solid and unable to transfer venom effectively, in contradistinction to *opisthoglyphous* snakes. Cf. *opisthoglyph, proteroglyph, solenoglyph*.

aglyphic, aglyphodont, aglyphous. *adj.* Tooth condition in which no groove or canal is present. Also used in reference to full teeth not linked to salivary or venom glands (fangless).

agonist. *n.* 1. A substance that can interact with receptor molecules and mimic the action of an endogenous signaling molecule. 2. Any substance that activates a receptor. 3. Reference to a muscle moving a part. Cf. *antagonist*.

agonistic behavior. Reference to social interactions between members of the same species, involving aggression, threat, or combative behavior in response to social conflict.

agregoserpentin. *n.* A class of proteins from snake venoms that act on blood platelets or thrombocytes.

air capillary. A small terminal air space in which respiratory gas exchange occurs in the unique *parabronchial lung* of birds. Each unit of air capillaries form intimate contact with blood capillaries along the pathway of the *parabronchus*. See also *crosscurrent exchange*.

air sac. 1. The posterior, non-respiratory segment of the snake lung. This structure is thin, relatively nonvascular, and varies in length in different species. It functions in various ways, including air storage, buoyancy, defensive displays through body inflation, and ventilation (as a bellow) of the lung. See Fig. 41. 2. The term may also be used in reference to *tracheal diverticula*, which are accessory outgrowths of the tracheal airway and function in defensive displays or sound production in various Asian species of snakes. 3. Any of numerous, mostly paired, compliant structures that ventilate the *parabronchial lung* of birds.

akinesis. *n.* (*adj.* **-ic**) Without mobility, as in reference to skull bones that are connected to one another to form a rigid unit.

akinetic skull. A rigid skull, characteristic of turtles and crocodilians, in which individual elements are without independent mobility. A skull lacking *cranial kinesis*.

ala. (*pl.* **alae**) *n.* A wing or wing-like process, often used with reference to bony outgrowths.

alactacid oxygen deficit. Reference to excess oxygen consumption above resting levels and following activity, utilized to replenish oxygen and phosphagen stores consumed during activity. Cf. *lactacid oxygen deficit*.

alar. *adj.* Winglike.

alarm call. 1. A sound produced by an animal when danger threatens or is perceived at some distance. 2. A sound given by either or both a male or female anuran while in amplexus if the pair is grasped by another male. The sound informs the interfering individual that pairing has already taken place and causes it to release its grasp. See B. Martof and E.F. Thompson, Jr., *Behaviour* 13:243–258, 1958. (Syn. *male release call, protective call*)

alarm reaction, alarm responses. Responses elicited by sudden exposure to threatening stimuli, such as rapidly moving objects or images in the immediate vicinity. Examples include color change, urination, heart rate changes, and expulsion of mucus.

alar plate. 1. Sensory portion of the *brain stem*, oriented vertically and located dorsal to the sulcus along the lateral ventricular wall. Cf. *basal plate*. 2. An accessory flattened extension from the middle of the zygapophysial ridge on posterior caudal vertebrae of the sand boa *Eryx*. First described by M.S. Sood (*Proc. Indian Acad. Sci.*, 14:390–394,1941).

alary cartilage. See *alinasal*.

alary, alary process, alar process. *n.* A cartilagenous process on the anterolateral edge of the hyoid plate or copus. May be absent or variably shaped in different species. (Syn. *processus alares, wing process*)

albedo. *n.* The fraction of incident radiation that is reflected by a surface or body, usually expressed as a percentage or decimal fraction. A measure of *reflectivity*.

albinism. *n.* (*adj.* **albinistic**) The absence of melanin in the eyes and skin, usually due to a deficiency of tyrosinase. Owing to a genetic mutation, an albinistic animal fails to develop dark coloration, so that its appearance is white or pinkish.

albino. *n., adj.* An animal lacking pigmentation.

albumen, albumin. *n.* 1. Simple proteins that are soluble in water or saline. Albumin is the most abundant of protein components in blood plasma and is synthesized and secreted by liver cells in vertebrates. It functions as a transport protein for numerous endogenous and exogenous compounds in the circulation, including metals, fatty acids, amino acids, and other metabolites. The most important physiological role of albumin is to deliver such solutes in the circulation to their target organs, as well as to maintain the pH and osmotic pressure of plasma. Recently it was discovered that albumin is expressed in skin of the frog *Bombina maxima* (Y.X. Zhang et al., *Comp. Biochem. Physiol.* B 143:153–159, 2006). 2. The watery protein solution between the shell and yolk of reptilian eggs, and equivalent to egg "white."

albuminuria. *n.* The presence of albumen in urine, characteristic of envenomation by Russell's viper (*Daboia russelli*).

alcian blue. A copper-containing dye for staining acid mucopolysaccharides that may be combined with periodic acide-Schiff reagent.

Aldabra Tortoise. A "giant" and long-lived tortoise species (*Geochelone gigantea*) endemic to the Aldabra Islands.

aldosterone. *n.* A mineralcorticoid secreted by the adrenal cortex; an important electrolyte-controlling steroid. Aldosterone is the primary mineralcorticoid synthesized by amphibians and reptiles, which also produce 18-hydroxycorticosterone—a precursor to aldosterone. This hormone is involved in water, ion, and energy balance.

Alethinophidia. *n.* A clade (infraorder) that includes all "true snakes" or "advanced snakes" including most modern lineages except the *Scolecophidia*. Roughly 2500 species are described in 16 families containing 450 genera. See *Caenophidia, Henophidia, Scolecophidia*.

algorithm. *n.* A set of mathematical procedures or series of logical steps that are followed in a specific order to solve a particular numerical or algebraic problem in a finite period of time.

alien. *adj.* Non-native, exotic.

aliform process. *n.* A wing-like posterior projection on the lateral margin of neural arch in vertebrae of *Siren*. Used by C.W. Gilmore (*Geol. Soc. Amer. Spec. Pap.* 9:1–96, 1938); C.J. Goin & W. Auffenberg (*Bull. Mus. Comp. Zool.* 113:497–514, 1955). See also *pterapophysis*.

alignment. *n.* In molecular biology, reference to matching of two nucleotide or amino acid sequences such that they share the same element at a specified position. This is used to detect homology between sequences.

alimentary canal. A hollow, tubular cavity extending through an animal and open at both ends. This functions in ingestion, digestion, and absorption of food materials. (Syn. *digestive tract, gastrointestinal tract, gut*)

alinasal; alinasal cartilage. *n.* A cup-shaped cartilage that supports the anterior margin of the external naris in anurans. The cartilage is associated with the posterior surface of the alary process of the premaxilla via the superior prenasal cartilage. Named by G.M. Higgins (*Illinois Biol. Monogr.* 6:7–91, 1920). (Syn. *alary cartilage, cartilago alaris*)

aliphatic. *adj.* Designating molecules made up of linear chains of carbon atoms.

aliquot. *n.* An integral part or representative sample that divides the whole without a remainder. Sometimes used loosely for any fraction or portion.

alisphenoid. *n.* A mammalian skull bone of unknown homology, although the term has been applied to various bones in the same general region of the reptilian skull (ventrolateral braincase near the pituitary). The reptilian epipterygoid is possibly the true homolog according to some authors. See *pleurosphenoid, laterosphenoid, orbitosphenoid*.

alkaline. *adj.* Of, or pertaining to a basic or high pH (>7). Cf. *acidic*.

alkaline tide. A period of increased body and urinary alkalinity associated with increased gastric secretion of acid (HCl) during digestion.

alkaloids. *n.* A class of nitrogen-containing organic bases found in plants or, secondarily, in animals. These substances are usually pharmacologically active, being bitter and often toxic. Alkaloids are important components of toxic skin secretions of many amphibians and include steroidal, monocyclic, bicyclic, tricyclic, pyridine, indole, guanidine (*tetrodotoxin*), and morphine alkaloids. Many of these occur only in trace amounts.

alkalosis. *n.* Excessive body alkalinity, usually measured in body fluids such as blood.

allanto-chorion. *n.* Combined allantoic and chorionic membranes which function as a surface for gas exchange with air through porous reptilian shells.

allantoic gill. *n.* A respiratory structure providing surfaces for gas exchange during embryonic stages of terrestrial plethodontid salamanders. The structure atrophies when the animal hatches from the egg capsule.

allantoic placenta. See *allantoplacenta*.

allantois. *n.* One of three embryonic membranes arising from endoderm near the posterior end of the gut in reptiles, birds, and mammals. This membrane functions in gas exchange, and the cavity it forms serves as an embryonic bladder to excrete nitrogenous wastes during development.

allantoplacenta. *n.* Apposition of the chorion and outer allantoic membrane to the uterine wall. (Syn. *allantoic placenta, chorioallantoic placenta*) For terminology discussion see J.R. Stewart & D.G. Blackburn, *Copeia* 1988: 839–852, 1988.

Allee effect. The phenomenon of manifest increase in the instantaneous birth rate or a decrease in the instantaneous death rate as the size of a population increases. In most populations, birth rates decrease and death rates increase as the population grows. Allee effects can cause simple models to exhibit complex dynamics, such as a minimum sustainable population size. The Allee effect is named after the ecologist W.C. Allee, who popularized the idea in an influential textbook, *Principles of Animal Ecology*, 1949.

allele. *n.* A shorthand form of *allelomorph*, representing one of a series of possible alternate (heritable) forms of a given gene. Alleles differ in DNA sequence and affect the functioning of RNA and/or protein. This term is sometimes used interchangeably with *gene*.

allelic frequency. The percentage of all alleles represented by a particular allele at a given locus in a population gene pool. See *gene frequency*.

allelomorph. *n.* See *allele*.

allelopathy. *n.* Reference to biogenic toxicity, an interaction in which chemicals introduced into the environment by one species suppress the growth or reproduction of another species. This term is commonly used in reference to plant interactions, but also applies to interspecific effects of growth inhibitory substances released by certain developing tadpoles. (Syn. *antibiosis*)

Allen's rule. A "bioclimatic rule" which postulates that heat loss is reduced for cold-adapted endotherms by a reduction in size of their appendages. With respect to ectotherms, this rule is problematic and generally inapplicable. See also *Bergmann's*, *Cope's rule*, and *Gloger's rule*.

allesthetic trait. Reference to any individual character that has adaptive function in relation to the nervous system of other organisms (e.g. courtship displays, mating calls, odors, etc.).

Alligator. *n.* Vernacular name for *American* and *Chinese Alligator*, characterized by a broad, blunt snout and teeth in the lower jaw that lie inside the closed mouth. See *Alligatoridae*.

Alligatoridae. *n.* A clade (family or subfamily, depending on classification scheme) that includes Alligators and Caimans. Three genera and eight species are currently recognized, mostly with distributions in Central and South America but also including southern United States and parts of China.

Alligator Lizards. Common name for various New World anguid species belonging to the genera *Barisia*, *Elgaria*, *Gerrhonotus*, *Mesaspis* and *Abronia* (*Arboreal Alligator Lizards*).

Alligator Snapping Turtle. Common name for the turtle species *Macrochelys temminckii*, a highly aquatic and well-known member of *Chelydridae*.

allo-. Prefix meaning "other."

allochroic. *adj.* Exhibiting variation of color or ability to change color.

allochronic, allochronous. *adj.* Existing at different times. Cf. *synchronic*.

allochronic speciation. 1. Speciation by means of separation in breeding seasons or patterns, without geographical isolation. 2. Speciation by means of sequential replacement of species through time.

allochronic species. In paleontology, species that do not occur in the same time horizon or geological period.

allochthonous. *adj.* 1. Exogenous; refers to an animal that has migrated or been transported into an area that is not its usual habitat. Cf. *autochthonous*. 2. Not originating at its present site; formed elsewhere outside a system, usually in reference to food materials.

allocryptic. *adj.* Reference to organisms that conceal themselves beneath a covering of other material, living or nonliving.

Allodapanura. *n.* A new monophyletic group of anurans, sister to the new taxon *Natatanura* within the new taxon *Ranoides*, and comprised of *Microhylidae* and a new taxon *Afrobatrachia* (D.R. Frost et al., *Bull. Amer. Mus. Nat. Hist.* 297:1–370, 2006).

alloenzyme. *n.* See *allozyme*.

allogamy. *n.* (*adj.* –**ous**) Cross-fertilization.

allograft. *n.* A graft or transplant of tissue from one organism to another individual of the same species.

allometric coefficient. The ratio of relative growth rates.

allometric growth. Reference to differential growth of body parts, expressed by the equation $y = bx^a$ where x and y are different body parts; a and b are fitted constants. See *allometry*. Cf. *isometric growth*.

allometric index. Ratio of tail length divided by hind-limb length in anuran tadpoles. First used by Roth (*Bull. Amer. Mus. Nat. Hist.* 11:99, 1929) according to J.L. Dolphin & E. Frieden (*J. Biol. Chem.* 217:735-744, 1955). (The reciprocal of this value is termed the *morphological index*.)

allometry. *n.* (*adj.* -**ic**) Systematic change in body proportions with increasing body size. The relative proportion of a body part changes in relation to the rate of growth of the whole or of another part as total body size increases. The term also refers to the study of such changes: the correlation between form and size. Cf. *isometry*.

allomone. *n.* A compound produced by one organism that affects the behavior of a member of another species, usually detrimentally. Interspecific *pheromones*.

alloparapatric speciation. A mode of speciation in which new species gradually originate through populations that are initially allopatric, but become parapatric before evolving completely effective reproductive isolation. Cf. *parapatric speciation*.

allopatric speciation. The evolution of distinct species through differentiation of geographically isolated populations. The evolution of a population into a new species while isolated geographically from the parental species. (Syn. *geographic speciation*)

allopatry. *n.* (*adj.* -**ic**) A condition of geographic distribution referring to species or populations of organisms that occur in non-overlapping, but usually adjacent, areas. Populations of such species live in different geographic locations separated by distance alone or by some barrier such as water or a mountain range. Describing geographically separated populations; disjunction. Cf. *parapatry*, *sympatry*.

Allophrynidae. *n.* A family of anurans that includes a single species, *Allophryne ruthveni*, inhabiting the Guayanan region of South America, northeastern and western Brazil. This may be a sister taxon to the Centrolenidae. *Ruthven's Frog* lacks teeth, and intercalary elements are present in the digits.

allopolyploid. *adj.* A *polyploid* organism arising from the combination of chromosome sets that are genetically distinct (from two different species).

all-or-none character. A *binary character*.

allosematic. *adj.* Reference to coloration or markings that imitate warning patterns of other characteristically noxious or dangerous organisms.

allospecies. *n.* A *semispecies*; component species of a *superspecies*.

allosteric effectors. Small molecules that bind reversibly to proteins at a site different from the active site and thereby cause an *allosteric* effect.

allosteric site. An area of an enzyme (other than its active site) that binds a substance other than the substrate, thereby changing the conformation of the protein such as to alter the catalytic effectiveness of the active site.

allostery. *n.* (*adj.* –**ic**) The reversible interaction of a protein molecule with another small molecule, which leads to alterations of shape of the protein and consequent change in the interaction of the protein with a third molecule.

allotherm. *n.* See *ectotherm*. Cf. *autotherm*.

allotopic. *adj.* Reference to populations or species that occur in different macrohabitats. Cf. *syntopic*.

allotrophic. *adj.* (*n.* **allotrophy**) 1. Pertaining to a state in which nourishment is obtained from other organisms. 2. Reference to influx of nutrients into an ecosystem from outside the system.

allotype. *n.* 1. In taxonomy, a *paratype* originally designated by the author and having sex opposite to that of the *holotype*. 2. A protein that is a product of a different allele of the same gene, when compared with other products of the gene.

allozymes. *n.* Allelic forms of an enzyme that differ slightly in amino acid sequence and can be distinguished by electrophoresis, as compared with the more general term *isozyme*. (Syn. *alloenzyme*)

alluvial deposit, alluvium. A deposit of sediment transported by water. Cf. *aeolian deposit, colluvial deposit*.

alluvial fan. A large, fan-shaped deposit of sediment that usually forms where a stream's velocity decreases as it emerges from a narrow canyon onto a flat plain at the foot of a mountain range.

Alopoglossinae. *n.* A subfamily of *Gymnophthalmidae*.

α-adrenergic receptor. A receptor on cell surfaces that binds norepinephrine and, less effectively, epinephrine. The binding activates a G protein.

α-bungarotoxin. *n.* See *bungarotoxin*.

α-cobrotoxin. *n.* A neurotoxic component of cobra venom, which binds reversibly to acetylcholine receptor sites.

alpha diversity, α-diversity. Reference to the diversity or richness of a species within a particular sample, habitat, community, or local area. Cf. *beta diversity, gamma diversity*.

α-keratin, α-keratin layer, α-cells. *n.* See *keratin*.

alphanumeric. *adj.* A code comprising both letters and numbers.

α-stimulating hormone. See *melanocyte stimulating hormone*.

alpha taxonomy. The aspect of taxonomy concerned with the description and naming of species, typically based on morphological characters. Cf. *beta taxonomy, gamma taxonomy*.

α-toxins. *n.* See *postsynaptic toxins*.

alpine. *adj.* Reference to habitats or organisms that occur between tree line and snow line in mountains.

alternate tooth replacement. Synonym for *intercalary replacement*, as used by C. Gans (*Breviora* 70:1-12, 1957).

alternating calls. Calls by male anurans spaced such that they do not overlap with calls of nearby males.

alternative hypothesis. In statistics, a hypothesis that is accepted when the *null hypothesis* is rejected.

alternative stable states. In ecology, the potential for two or more different types of communities (or ecosystem *equilibria*) following a disturbance.

altitude. *n.* Elevation or vertical distance above a datum point. (Syn. *elevation*)

altricial. *adj.* Reference to a newborn animal that is relatively undeveloped and totally reliant on parental care or nourishment for a period following birth or hatching. Cf. *precocial*.

altruism. *n.* (*adj.* **altruistic**) Behavior of an individual that benefits or increases fitness of others, usually reducing fitness of the individual simultaneously. Mutually beneficial behavior is known as *reciprocal altruism*. See *inclusive fitness*.

alveolar. *adj.* 1. Containing numerous small, blind sacs or depressions. 2. Pertaining to the jaw margin where teeth might be located. The functional, or biting, part of the jaw.

alveolar gland. Multicellular cutaneous glands of amphibians having numerous cavities or hollows which open into a central lumen or canal. Characteristic gland types are mucous glands and granular (poison) glands.

alveolar ridge. *n.* The ridge on the crushing surface of the upper jaw of turtles.

alveolar surface. Area of the jawbone usually occupied by teeth. The chewing surfaces of the jaws of turtles. See also *alveolar ridge*.

alveolus. (*pl.* **–i**) *n.* 1. A small blind cavity, as in the respiratory surfaces of the mammalian lung. Cf. *faveolus*. 2.

Refers to a cavity in a gland where secretion is produced. 3. Also refers to a tooth socket in reference to the jaw.

Alytidae. *n.* A family of frogs (formerly *Discoglossidae*), sister taxon with *Bombinatoridae* within the *Costata* and containing the genera *Alytes* and *Discoglossus*. See *Discoglossidae*.

Amazonian melanism. A tendency for several coral snake subspecies, species, and mimics of coral snakes to display melanism in the Amazon. Coined by J. Roze (*Coral Snakes of the Americas. Biology, Identification and Venom*, pp. 27, 211, 1966).

ambient. *adj.* Reference to surrounding environmental conditions.

ambient temperature. The temperature of the environment, used in reference to the medium that surrounds an animal The term usually references air temperature, but it may refer to the average temperature of fluid (air or water) outside the boundary layer in the immediate vicinity of an animal.

ambush predator. A predator that remains motionless and waits for prey to come within the range of a strike or lunge, in contrast to *active foragers* which move through the environment and actively seek prey. (Syn. *sit-and-wait predator*)

Ambystomatidae. *n.* A clade (family) of moderate-sized salamanders having many ancestral characters and commonly known as "Mole Salamanders" because many live in burrows much of their lives. Metamorphosis is facultative or obligate. About 30 species of a single genus (*Ambystoma*) occur in North America, including the Mexican Axolotl (*Ambystoma mexicanum*) that is used extensively in developmental and experimental biology. *Dicamptodon* has recently been included within the Ambystomatidae (D.R. Frost et al., *Bull. Amer. Mus. Nat. Hist.* 297:1–370, 2006). See *Dicamptodontidae*.

Ameivas. *n.* Collective vernacular name for Middle and South American species of teiid lizards belonging to the genus *Ameiva*.

amelanistic. *adj.* Lacking the melanin pigment which produces black, brown, and other dark colors.

amensalism. *n.* An interaction in which one species is adversely affected while another is unaffected (e.g., inhibition by a toxin).

American Alligator. Vernacular name for the alligator *Alligator mississippiensis*.

American Association of Zoological Parks and Aquariums. (**AAZPA**) See *Association of Zoos and Aquariums*.

American Bullfrog. See *Bullfrog*.

American Crocodile. Vernacular name for the crocodilian *Crocodylus acutus*.

American Gartersnake. See *Gartersnakes*.

American Gecko. Vernacular name for species of geckos belonging to the genus *Gonatodes*.

American Society of Ichthyologists and Herpetologists. (**ASIH**) A scientific society, established in 1913 by J.T. Nichols, dedicated to the study of fishes, crocodilians, turtles, reptiles, and tuataras. The Society publishes the journal *Copeia*. Current URL: http://www.asih.org

American Toad. Vernacular name for the bufonid species *Anaxyrus* (formerly *Bufo*) *americanus*. For recent taxonomic revision see D.R. Frost et al., *Bull. Amer. Mus. Nat. Hist.* 297:1–370, 2006.

American Watersnakes. See *Watersnakes*.

American Zoo and Aquarium Association. See *Association of Zoos and Aquariums*.

ameristic. *adj.* Not divided into unitary parts. Cf. *meristic*.

Ametrobatrachia. *n.* A new monophyletic taxon of anurans, sister to *Micrixalidae* within the new taxon *Telmatobatrachia*, and containing new taxa *Africanura* and *Saukrobatrachia* (D.R. Frost et al., *Bull. Amer. Mus. Nat. Hist.* 297:1–370, 2006).

amine. *n.* A derivative of ammonia in which at least one hydrogen atom is replaced by an organic group.

amino acids. A class of organic compounds containing at least one carboxyl group and one amino group. The alpha-amino acids, $RCH(NH_2)COOH$, are components of proteins and peptides.

amino acid sequence. The order of *amino acid* residues in a polypeptide chain.

aminobenzoic acid ethyl ester. See *tricaine methanesulfonate*.

amino group. ($-NH_2$) A chemical group which can form $-NH_3^+$ with addition of a proton.

ammodytoxin. *n.* A phospholipase from *Vipera ammodytes*.

ammonotelism. *n.* (*adj.* –**telic**) Physiological condition in which the principal form of nitrogenous excretion is ammonia. This tends to be characteristic of aquatic and amphibious species which have access to abundant free water. Cf. *ureotelism* and *uricotelism*; see also *deamination*.

amniochorion. *n.* A term used by A. Fisk and M. Tribe (*Proc. Zool. Soc. London*, 119:83–114, 1949) to refer to the two principal extraembryonic membranes, amnion and chorion.

amnion. *n.* 1. A membrane-bound, liquid-filled cavity in which a reptilian embryo develops. 2. The term also refers to the innermost, extraembronic membrane bounding this cavity in reptiles, birds, and mammals. (Syn. *amnion cavity; amniotic membrane*)

Amniota. *n.* A clade that includes the last common ancestor of mammals, birds, and all of its descendants including turtles, squamates, tuatara, and crocodilians.

amniote. *n.* A vertebrate whose embryo develops within a fluid-filled amnion. A member of the Amniota, which includes reptiles, birds, and mammals.

amniotic cavity or pouch. The sac or cavity that contains amniotic fluid.

amniotic membrane. See *amnion*.

amoebiasis. *n.* Infection by amoebic protozoans or protists, particularly in snakes, lizards, and chelonians. May cause slimy feces, refusal to eat, inadequate digestion, regurgitation, and frequent drinking. (Syn. *entamoebiasis, amoebic dysentery*)

amoebic dysentery. See *amoebiasis.*

amoeboid. *adj.* Resembling an amoeba in form or movement.

ampere. (A) *n.* The standard SI base unit of electric current.

amphi-. Prefix meaning "both," or "on both sides."

amphiarthrosis. *n.* 1. Articular connection or surface that permits only slight movement. 2. The point of union between two bones that are separated by an intervening substance, thereby allowing some movement.

Amphibia. *n.* A taxon (class) including the extant clades *Caudata* (salamanders), *Anura* (frogs), and *Gymnophiona* (caecilians), together forming *Lissamphibia* of Gadow (1901) and most recent authors. See *Lissamphibia.*

amphibian. *n.* Any member of the vertebrate class Amphibia, a monophyletic group of which living members include the newts and salamanders (*Caudata*), caecilians (*Gymnophiona*), and frogs and toads (*Anura*). See *Amphibia.*

Amphibian and Reptile Conservation. An international journal dedicated to the global preservation and management of amphibian and reptilian diversity. Published in Modesto, California.

amphibian decline. A global phenomenon of real or apparent decline of numerous amphibian populations, noted by many herpetologists and a subject of numerous investigations into multiple, often anthropogenic causes. Attention was first focused on this topic at the first World Congress of Herpetology held at Canterbury in 1989.

amphibian malaria. A mosquito-transmitted disease caused by a protozoan, *Plasmodium spp.*, that parasitizes anurans. (Syn. *frog malaria*)

amphibian papilla. See *papilla amphibiorum.*

Amphibian Research and Monitoring Initiative. (ARMI) A national (USA) research and monitoring program initiated by the U.S. Department of Interior to respond to indications of worldwide declines in amphibian populations. See R.J. Hall and C.A. Langtimm, *George Wright Forum* 18:14–25, 2001.

amphibian Ringer's. See *Ringer's solution.*

Amphibian Tree of Life. An important recent publication that proposes a new taxonomy of living amphibians (D.R. Frost et al., *Bull. Amer. Mus. Nat. Hist.* 297:1–370, 2006).

Amphibia-Reptilia. A herpetological journal published by Brill Academic Publishers.

AmphibiaWeb. An online system that enables search and retrieval of information relating to amphibian biology and conservation. Created in conjunction with the Digital Library Project at the University of California, Berkeley, one goal is to establish a "home page" for every species of amphibian in the world. There also is access to recent scientific publications searchable on a month-by-month basis. Current URL: http://amphibiaweb.org

amphibious. *adj.* Living, or capable of living, both on land and in water. (Syn. *semiaquatic*)

amphicyrtean. *adj.* See *acelous.*

amphicoelous, amphycoelous. *adj.* Condition of vertebra in which the centra are biconcave (hollowed at both ends), as in most fish, a few salamanders and caecilians, the Tuatara and gekkonid lizards. Developmentally, the condition might arise in several ways and does not imply common phylogeny. (Antiquated syn. *amphicoelian*). Cf. *opisthocoelous, procoelous.*

amphicondylous. *adj.* 1. With two condyles. 2. Refers to double knob on occipital area in articulation of the skull with vertebral column in amphibians (H. Gadow, *Amphibians and Reptiles*, p. 651, 1901).

Amphignathodontidae. *n.* A taxon (family) of anurans, sister to the new taxon *Athesphatanura* within the new taxon *Tinctanura*, and containing the genera *Flectonotus* and *Gastrotheca* distributed in Central and South America, Trinidad and Tobago. See D.R. Frost et al., *Bull. Amer. Mus. Nat. Hist.* 297:1–370, 2006.

amphigonia retarda. 1. Delayed fertilization. First used by F. Kopstein, *Bull Raffles Mus.* 14:81–167 (1938). 2. May also refer to female storage of viable sperm for use in subsequent ovulations (J.A. Oliver, *The Natural History of North American Amphibians and Reptiles*, p. 329, 1955).

amphigyrinid. *adj.* The condition of having two lateral *spiracles* in an anuran tadpole, as in pipids, rhinophrynids, and *Lepidobatrachus* (leptodactylid), although the latter case is not homologous with the former two. Cf. *laevogyrinid, mediogyrinid, paragyrinid.*

amphikinesis. *n.* (*adj.* **-ic**) A condition of *cranial kinesis* in which there are mobile joints (movement) between dermatocranium and neurocranium or between various elements of the dermatocranium.

amphipathic. *adj.* Reference to a molecule that has distinct polar and nonpolar segments (e.g., phospholipid).

amphiplatyan. *adj.* Condition of vertebrae which are fully ossified and flat at both ends.

Amphisbaena. *n.* 1. A poisonous serpent of classical mythology having a head at both ends and ability to move forward or backward. 2. A genus of *amphisbaenian.*

Amphisbaenia. *n.* A clade (suborder) of elongate, generally limbless squamates (except for *Bipes*) characterized by heavily ossified skulls modified for subterranean digging. Four families (*Amphisbaenidae, Rhineuridae, Trogonophidae, Bipedidae*), approximately 25 genera, and 160 species are found in the West Indies, South America, sub-Saharan Africa, and disjunct circum-Mediterranean areas.

Amphisbaenidae. *n.* A family of *amphisbaenians.*

amphithermic. *adj.* Reference to a broad or variable tolerance to a range of environmental temperature, reflected in existence of clines or subspecies.

amphitopic. *adj.* Reference to a broad or variable tolerance of environment or habitat, reflected in existence of clines or subspecies.

Amphiuma. *n.* Collective vernacular name for salamanders belonging to the family Amphiumidae and the genus *Amphiuma*.

Amphiumidae. *n.* Family of elongate, paedomorphic, aquatic salamanders of southeastern U.S. that lack eyelids. These are commonly called "Amphiumas" and were formerly known as "Congo Eels." This family is a sister taxon with *Plethodontidae* and recently assigned to the more inclusive new taxon *Xenosalamandroidei* (D.R. Frost et al., *Bull. Amer. Mus. Nat. Hist.* 297:1–370, 2006).

amphoteric compound. A molecule that can act as both an acid and a base (e.g., proteins).

amplecant, amplectant, amplectic. *adj.* Term used to describe amphibians engaged in amplexus.

amplex. *v.* To perform amplexus.

amplexation. *n.* See *amplexus*.

amplexus. *n.* Precopulation pairing; usually used in reference to sexual embrace of amphibians, with grip of male on female being usually either *axillary* (*pectoral*) or *inguinal* (*pelvic*) in position. Other amplectic behaviors include *cephalic* (on head or neck), *straddled* (male rests on shoulders of female while both bodies are vertical and sperm flows down the female's dorsum), *glued* (male is attached to a female's back by adhesive substance), or none. (Syn. *amplexation, clasping, embrace, sexual embrace*)

amplification. *n.* 1. The process whereby the strength of a recorded signal is increased or made more intense. 2. Increased structural or functional complexity during ontogeny or phylogeny. 3. See *gene amplification*.

amplified fragment length polymorphism. (AFLP) A DNA fingerprinting technique that examines variation at potentially thousands of loci distributed over a genome. The technique is simple and robust, and it is particularly useful for resolving relationships at the level of species and populations.

amplitude. *n.* 1. The extreme range of a fluctuating quantity, measured from trough to peak in a graphic display of an oscillating variable. The range of movement or displacement. 2. In a population cycle, the amplitude is the difference between the maximum population size and the population size at its midpoint, stated in numbers of individuals. 3. In ecosystem dynamics, amplitude refers to the range over which an ecosystem can fluctuate and still return to an earlier condition.

ampoule. *n.* See *ampule*.

ampule (ampoule, ampul). *n.* A hermetically sealed glass or plastic flask or bulb containing a solution for hypodermic injection.

ampulla. *n.* Generally, an inflated portion of a tubular structure, or dilation of a canal. 1. The term is applied to the expanded, thick-walled segment of the amphibian oviduct, which contains endocrine glands involved in the deposition of outer gelatinous egg envelopes. 2. The ampulla of the semicircular canals is the inflated osseous region at the base of the semicircular canals that contains the hair cells.

ampulla ductus deferens. Expanded and often glandular section of the *ductus deferens* that can serve as an area for sperm storage. Often referred to simply as "ampulla" and sometimes synonymous with *ampulla ureteris*. (Syn. *seminal vesicle*, but not homologous with the structure of this name in mammals; *vesicula seminalis*)

ampulla ureteris. A swollen and expanded area in the ureter of snakes that serves for storage of sperm during the mating season. Also has been called *ampulla ductus deferens* by various authors.

ampullary organ, ampullary receptor. *n.* Epidermal sense organ evidently derived from sunken *neuromasts*. These are similar to *neuromasts* in general structure except they are sunken in the epidermis, have a long neck, lack a *cupula*, and have one cluster of microvilli per sensory cell. They are restricted to the head where they are commonly associated in parallel with rows of neuromasts in larval caecilians and aquatic salamanders. Ampullary organs function as *electroreceptors* in at least some species. See Fig. 35.

amyotrophin. *n.* A toxin in cobra venom that destroys nerve cells and leads to muscle atrophy ("amyotrophy"). Used by F. Wall (*Poisonous Terrestrial Snakes of India*, pp 1–149, 1917).

an-. Prefix meaning "without" or "not."

anabatic. *adj.* Reference to winds being caused by upward movement of heated air.

anabolism. *n.* (*adj.* –ic) Synthesis by living cells of complex molecules from simpler precursors, with consequent utilization of energy. Cf. *catabolism*.

Anacondas. *n.* Vernacular name for large, highly aquatic, constricting Boa (*Boinae*) belonging to the genus *Eunectes*, inhabiting freshwater systems in South America.

anaerobic. *adj.* 1. Not requiring or involving the presence of oxygen. 2. Reference to environments that are significantly depleted or devoid of oxygen. (Syn. *anaerobiotic*)

anaerobiotic. *adj.* See *anaerobic*.

anaerobe. *n.* A cell or organism that can live without molecular oxygen. A strict anaerobe cannot live in the presence of oxygen. Some amphibians and reptiles may become *anaerobic* for short periods of time, but they cannot be considered true anaerobes.

anaesthesia, anesthesia. *n.* Loss of sensation.

anagenesis. *n. Phyletic evolution* within a single lineage, without subdivision or splitting. Modified forms replace one another in continuous succession without branching into new taxa. Cf. *cladogenesis*.

anagotoxic. *adj.* Used to describe an ability to counteract or neutralize the effects of venom. (Historically, reference is to the sulphurous waters of a fountain in São Pedro, Brazil: W.A. Prado and E.C. Arantes, *Mem. Inst. Butantan*, 14:157–165, 1940).

anal. *n., adj.* Reference to the anus. 1. Terminal plate in the ventral series in snakes, usually larger than the ventral scales and may be single or divided. (Syn.: *preanal, postabdominal*) See Fig. 13. 2. The posteriormost lamina, usually paired, on the plastron of turtles, or, in other usage, on the carapace. (Syn. *postcentral, subcaudal*) See Figs. 4, 6.

anal bladder or **bursa.** See *cloacal bursa*.

anal claw. *n.* See *pelvic spur*.

anal flap. See *anal fold, vent flap*.

anal fold. A fold of skin lying dorsolaterally to the anus, or laterally and paired, in some species of frogs.

analgesia, analgesic. *n., adj.* A pain-relieving drug.

anal gland. 1. Usually paired, elongate structures present in the base of the tail in snakes. Ducts open to the cloacal wall near the anus and release strong, usually foul-smelling secretions, often voided with cloacal contents upon handling or sometimes can be sprayed some distance. These glands are present in both sexes but are usually more prominent in females. They are used in defense, but may also function in sex recognition and trailing. (Syn. *anal sac, musk gland, postanal gland, postcloacal gland, scent gland, tail gland*) 2. This term is sometimes applied to the *hemiclitoris* of female snakes by snake systematists.

analog. *n.* 1. A compound related to, but slightly different structurally from, a biologically significant molecule such as a hormone. 2. A thing or part that is analogous.

analogy. *n.* (*adj.* **–ous**) Similarity of function applicable to structures that have evolved convergently and are of different evolutionary origin. Structural correspondence based on common function. See *homoplasy*. Cf. *homology, homologous*.

anal plate. The plate or scute which covers the vent of snakes. This is a misnomer because the digestive tract lacks an opening (anus) separate from the cloaca.

anal ridge. Refers to keeled scales above the cloaca of sexually mature male snakes of some species having smooth scales elsewhere on the body. These are not a reliable indicator of sex, for they may appear in some females and immature males, or be absent from some mature males.

anal sac. Synonym for *anal gland* in snakes; *cloacal bursa* in turtles.

anal spur. See *pelvic spur*.

analysis of covariance. (ANCOVA) A statistical method that determines whether the functional relationships described by two or more regression equations are the same (i.e., representing populations having the same slope). The method is a combined application of *linear regression* and *analysis of variance*.

analysis of variance. (ANOVA) A statistical method that partitions the total variation observed in an experiment among several statistically independent possible causes of variation. This is particularly useful in judging which sources of uncontrolled variation in an experiment need to be allowed for in testing treatment effects.

analytical centrifugation. High-speed centrifuge technique that is capable of characterizing soluble molecules according to their rate or extent of sedimentation.

anamniote. *n., adj.* Any vertebrate that lacks an *amnion* (characteristic of fishes and amphibians).

anaphylaxis. *n.* A severe hypersensitivity reaction which can cause circulatory, respiratory, and neurological distress that are potentially fatal. Involves reexposure to an antigen after an initial sensitizing exposure.

anapsid. *adj., n.* 1. A vertebrate skull lacking an opening in the temporal region of the skull (e.g., turtles among extant reptiles). 2. Any member of a group of reptiles that has no temporal fenestra. A member of *Anapsida*. See also *postorbital fenestrae*.

Anapsida. *n.* A group (subclass) of amniotes including the oldest known forms, distinguished by a skull having a complete covering of dermal bones and no openings (fenestrae) in the side of the skull behind the orbit. This group includes turtles and a number of fossil groups.

anastomosis. (*pl.* **–es**) *n.* A network of connections between blood vessels, or the joining of two or more cell processes or tubular vessels to form a network or branching system. Coming together.

anathermal. *adj.* Reference to a period of increasing temperature. Cf. *catathermal*.

anatomy. *n.* (*adj.* **–ical**) The study of structure of organisms, usually with reference to gross structure as revealed by dissection.

anatoxin. *n.* See *anavenom*.

anavenin. *n.* See *anavenom*.

anavenom. *n.* Snake venom that has been treated to destroy its toxic properties while retaining its ability to stimulate formation of antibodies. When injected into blood of a living animal it results in formation of antibodies and immunity to the anavenom. Blood serum from such an animal can then be used in treatment of snake bite. (Syn. *anatoxin; anavenin; toxoid*)

Anbesol®. *n.* A topical anesthetic that has been used to immobilize amphibians for photography (H. Kaiser & D.M. Green, *Herpetol. Rev.* 32:93–94, 2001). The product is an over-the-counter medication normally sold for alleviating pain in teething infants. See also *Orajel®*.

ancestor. *n.* (*adj.* **ancestral**) Any preceding member of a lineage, or progenitor of a more recent descendant taxon. The adjective may be used in reference to a primitive or plesiomorphic character state present or assumed to be present in an ancestor.

ancient DNA. (aDNA) Any DNA recovered from biological samples that have not been preserved specifically for later DNA analyses. Examples include the analysis of DNA recovered from ancient skeletal materials, ice, permafrost cores, archaeological historical materials, etc. The first aDNA study was that of R. Higuchi and colleagues at Berkeley, who extracted and sequenced traces of DNA from a museum specimen of an extinct horse (*Nature* 312:282–284, 1984).

ancipital. *adj.* Archaic term used by early authors to refer to double-edged (laterally compressed) tails of various lizards or salamanders.

ancocolous or **ancophilus**. *adj.* Living or thriving in canyons.

ANCOVA. See *analysis of covariance*.

andric. *adj.* Male. Cf. *gynic*.

androgen. *n.* (*adj.* –ic) A class of sex steroid hormones that stimulate male characteristics such as male secondary sex characteristics and promote protein synthesis (anabolism). In female reptiles, androgens can be converted in the central nervous system to estrogens (this also occurs in males) or other steroids, and may play a possible role in follicular maturation and oviductal maintenance. See *testosterone, dihydrotestosterone-5α*.

Anelytropsidae. *n.* Former taxonomic term used to designate a family of Mexican "snake lizards" now classified within the *Dibamidae*.

anemia. *n.* (*adj.* –ic) A pathologic decrease in hemoglobin per unit volume of blood.

anemo-. Prefix meaning "wind."

anemometer. *n.* An instrument used to measure wind velocity and sometimes direction.

anerythristic. *adj.* Refers to any specimen lacking all red pigment so that red markings are absent.

angiogenesis. *n.* The formation of blood vessels.

angiotensin II. (ang II) *n.* A blood protein that functions to stimulate aldosterone secretion and, at relatively high levels, contraction of blood vessels. It is converted from *ang I* by *angiotensin converting enzyme*, and ang I is converted from *angiotensinogen* by the action of *rennin*. There is some evidence that ang II increases rates of blood flow through the seat patch of anurans. See also *angiotensin converting enzyme*.

angiotensin converting enzyme. (ACE) A dipeptidyl carboxypeptidase found on the luminal surfaces of cell membranes in a variety of cells and in blood plasma and some other biological fluids. It is known primarily for its role in hydromineral metabolism and in regulation of blood pressure. ACE catalyzes the formation of angiotensin II from angiotensin I and can hydrolyze a wide range of other endogenous bioactive peptides.

angiotensinogen. *n.* See *angiotensin II*.

angle of the jaw. The angle formed by the articulation between the upper and lower jaws.

angle of the mouth. The angle formed by the connective tissues at the point of divergence of upper and lower jaws.

angstrom unit. (Å) A derived unit of length equal to 10^{-10} meter, or 0.1 nanometer. The unit was named in honor of the Swedish physicist Anders Jonas Ångstrom.

anguid. *n.* Pertaining to, or a member of, the *Anguidae*.

Anguidae. *n.* A clade (family) of lizards widely distributed in temperate and tropical regions, with many species evolving limb reduction and loss. There are 13 genera and approximately 110 species with disjunct distributions in North, Central, and South America, the West Indies, Eurasia, northwest Africa, southeast Asia, and islands of the Sunda Shelf. Four subfamilies are recognized: *Anguinae, Anniellinae, Diploglossinae,* and *Gerrhonotinae*.

Anguinae. *n.* A subfamily of *Anguidae*.

anguiform. *adj.* Snake-like, or shaped like a snake. (Syn. *anguiniform*)

Anguimorpha. *n.* A sister clade to *Scincomorpha* within the *Autarchoglossa*.

anguimorph replacement. Synonym for *intercalary replacement* used by C. Gans (*Breviora* 70:1–12, 1957).

anguiniform. *adj.* See *anguiform*.

angular. *n.* A dermal bone at the lower posterior edge of the mandible in most vertebrates. It is present in most or all reptiles, according to most authors, while absent in frogs and most salamanders.

angular acceleration. The rate of change of velocity around a point of rotation.

angular gland. A *musk gland* located on the inner sides of both halves of the lower jaw in crocodiles and some turtles.

angulate. *adj.* Having angles or an angular shape.

angulosplenial. *n.* Term sometimes used to describe a lower jaw, especially if its derivation appears to be from fusion of the *angular* and *splenial* elements. The term also has been used to refer to the *prearticular* of amphibians.

Aniliidae. *n.* A family containing a single snake species (*Anilius scytale*) found in the Amazon basin of northern South America.

Anilioidea. *n.* A superfamily of small, tropical fossorial or secretive snakes including about 60 species in 12 genera within the families *Uropeltidae, Cylindrophiidae, Anomochilidae,* and *Aniliidae*.

animal. *n.* Any member of the kingdom *Animalia*.

animal cap. The pigmented animal hemisphere of the amphibian *blastula*.

Animalia. See *kingdom*.

animal pole. The position or "pole" in animal oocytes that contains the most cytoplasm and the least yolk. This pole also is defined as where the nucleus is located and metabolic activity is highest. In reptiles, this is the clear region of the egg that will develop into the blastodisc. Cf. *vegetal pole*.

animal rights. The concept that animals have intrinsic and inalienable rights, similar to those ascribed to humans. Cf. *animal welfare*.

animal welfare. The principle that animals are entitled to humane treatment for comfort and well-being. Cf. *animal rights*.

anion. *n.* A negatively charged ion. Cf. *cation*.

aniso-. Prefix meaning "unequal."

anisodont. *adj.* Having teeth of irregular lengths. Coined by F. Wall (*Snakes of Ceylon*, p. xvii, 1921).

anisogamy. *n.* (*adj.* –**etic**) Reference to sexual reproduction in which the gametes are of different sizes, characteristic of the majority of animal species. Cf. *isogamy*.

anisomerism. *n.* 1. Serial repetition of unequal or different parts. 2. A reduction in the number and degree of differentiation of serially homologous structures.

anisomorphic. *adj.* (*n.* –**phy**) Reference to differences in shape or size.

anisotropy. *n.* (*adj.* –**ic**) Reference to a directional property of crystals and fibers having a characteristic high degree of molecular orientation. Muscle fibers, for example, have the property of transmitting light unequally in different directions (doubly polarizing). This property of anisotropic materials is called *birefringence*. Materials showing no birefringence are said to be *isotropic*.

Ankylosauria. *n.* An extinct ornithischian infraorder containing a genus of heavily armored, herbivorous quadripedal dinosaurs flourishing in the late Cretaceous period.

ankylose. *v.* To grow together into one, as in the fusion of two bones.

ankylosis. *n.* A rigid union between two parts, usually with reference to the union of tooth and jaw by hard mineralized tissue (*'bone of attachment'*) in most toothed reptiles (except crocodilians). This tissue is resorbed when a tooth is shed and redevelops during maturation of a replacement tooth. Cf. *gomphosis*.

anlage. (*pl.* **anlagen**) *n.* Embryological term adopted from German and used to designate an aggregation of primordial cells which precedes formation of a specific, distinguishable structure or organ. The mitotic progenitors of an organ during development. (Syn. *primordium*)

Anniellidae. *n.* Former family incorporating legless lizards, now included in the Anguidae.

Anniellinae. *n.* A subfamily of *Anguidae*.

annual. *adj.* 1. Having a yearly periodicity. 2. Living for one year. Cf. *biennial, perrenial*.

annual cycle. Repetition of biological events that recur each year (e.g., reproduction).

annual ring. See *growth ring*.

annual turnover. Reference to the total biomass produced in 1 year.

annular groove. A groove in the integument of caecilians that borders and defines an *annulus*. Dermal scales are present in the annular grooves in some species of caecilians.

annular pad. A thick ring of tissue surrounding the rim of the lens in eyes of certain reptiles, assisting in compression of the lens to achieve rapid accommodation in focusing.

annulment. *n.* Suppression by the *ICZN* of an available name as unavailable for purposes of priority and homonymy, and the ruling of a work as unavailable.

annulus. (*pl.* –**i**) *n.* 1. Any ring-like structure or marking. See *growth ring, lines of arrested growth*. 2. The body segment of an amphisbaenian. *Primary annuli* are defined by integumentary grooves that completely encircle the body, whereas those of *secondary annuli* do not.

anode. *n.* The positive electrode to which negative ions are attracted. Cf. *cathode*.

anododont. *adj.* Having a continuous, uninterrupted series of teeth. Coined by F. Wall (*Snakes of Ceylon*, p. xvii 1921).

Anoles. *n.* Collective vernacular name for numerous, diverse, and largely neotropical species of polychrotine lizards belonging to the genus *Anolis*. The name is also used in reference to other species of iguanid lizards in the genera *Enyalius, Leiosaurus, Phenacosaurus, Polychrus,* and *Pristidactylus*. See also *Polychrotinae*.

anolis unit. An antiquated measure of the mass of *melanocyte stimulating hormone* that will stimulate an average "stage 1 response" of color index when injected into 10 hypophysectomized *Anolis* lizards. Defined by Kleinholz and Rahn (*Anat. Rec.* 76:157–172, 1940). This term is seldom used in recent literature or research.

Anomalepididae. *n.* Family of *Blind Snakes* (*Dawn Blind Snakes*) inhabiting tropical Central and South America. These differ from typhlopids in lacking pelvic vestiges and having prefrontal bones that extend posteriorly over the orbits. Four genera and approximately 15 species are recognized.

anomalous. *adj.* Abnormal; departing from characteristic type.

anomaly. *n.* Abnormality or birth defect.

Anomochilidae. *n.* A clade (family) of snakes hypothesized to be a sister taxon to other *Alethinophidia* among extant snakes and therefore placed in its own family. A single genus and two species occur in peninsular Malaysia, Sumatra, and Borneo. Sometimes known as *Dwarf Pipe Snakes* or *Stump Heads*.

Anomocoela. *n.* A monophyletic clade of anurans, sister with *Neobatrachia* within the *Acosmanura* and containing *Pelobatoidea* (equivalent taxon of some authors) and *Pelodytoidea*. See D.R. Frost et al., *Bull. Amer. Mus. Nat. Hist.* 297:1–370, 2006.

anomocoelous. *adj.* Refers to a vertebral column that may have one of several different articular arrangements between vertebral centra, and is not *opisthocoelous* or *procoelous*.

anophthalmia. *n.* Lacking eyes as a result of a congenital deformity or anomaly in which eyes fail to develop.

anorexia. *n.* Loss of appetite.

anorthogenesis. *n.* Changes in the direction of evolutionary change based on preadaptation.

anosmic. *adj.* Having no sense of smell.

ANOVA. See *analysis of variance.*

anoxemia. *n.* A deficiency of oxygen in the blood.

anoxia. *n.* (*adj.* **–ic**) Lack of oxygen. Cf. *hypoxia.*

antagonist. (**-ism**) *n.* 1. An agent or molecule that inhibits, counteracts, competes for binding sites, or blocks an effect. 2. A muscle that produces an action that opposes forces by other muscles. 3. Inhibition of one species by the actions of another. Cf. *synergist, fixator.*

antagonistic pleiotropy. Reference to a negative correlation between two traits caused by the fixation of pleiotropic genes that affect both traits in adaptive directions; those pleiotropic genes that affect one trait in the adaptive direction and the other in the maladaptive direction contribute the genetic variation and covariation in the system.

antagonistic resources. See *complementary resources.*

Antarctic. *n.* One of eight *biogeographic realms* (also a *zoogeographical realm*) including the Antarctic, southern tip of South America, and adjacent islands.

Antarctogaea. *n.* The Australian *zoogeographical region*, excluding New Zealand and Polynesia.

ante-. Prefix meaning "before," "in front of."

ante-anal lamella. Antiquated synonym for the *preanal* in lizards.

antebrachial. *n., adj.* 1. Scales found on forearm or *antebrachium* and continuing upward along the frontal margin of the upper forelimb (*brachium*). See *prebrachial.* 2. Pertaining to, or located on, the forearm or epipodial region of the forelimb.

antebrachial bone. See *radioulna.*

antebrachium. *n.* The lower section of forelimb that contains the radius and ulna bones (forearm).

antefemoral. *adj.* Anterior to the femoral region.

antehumeral. *adj.* Anterior to the femoral region.

antenatal. *adj.* During pregnancy or before birth.

anteocular. *n.* See *preocular.*

anteorbital. *n., adj.* 1. Synonym for *preocular.* See Fig. 11. 2. Bordering the orbit of eye in front.

anterior. *adj.* At or near the front (head) end of the body. The front part of an organ. Opposite to *posterior.* (Syn. *cranial, rostral*)

anterior articular process. See *cardiac process.*

anterior coracoid. See *procoracoid.*

anterior cornu. See *ceratohyal.*

anterior epiphysis. See *parapineal organ.*

anterior frontal. Unpaired scale lying midline between the prefrontals in some snakes. Used by A.H. Wright and A.A. Wright (*Handbook of Snakes of the United States and Canada*, p. 1063, 1957). See *internasal.*

anterior lobe. 1. The section of turtle plastron lying anterior to the axillary notches. 2. The *pars distalis* of the pituitary; considered inappropriate by B.A. Houssay (*Quart. Rev. Biol.* 24:1–27, 1949) because it lies posteriorly in many amphibians.

anterior lung. See *tracheal lung.*

anterior orbital. Synonym for *preocular* in snakes. See Fig. 11.

anterior parietal. Denotes the first large scale behind the ocular in blind snakes of Anomalepididae. Used by A.H. Wright and A.A. Wright (*Handbook of Snakes of the United States and Canada*, p. 1063, 1957).

anterior pituitary. See *adenohypophysis.*

anterior-posterior axis. Reference to the body axis of a bilaterally symmetrical animal, extending from the anterior to posterior aspect (or poles of an embryo).

anterior process. A vague term used by various authors for projections arising at the anterolateral corners of the hyoid plate in anurans.

anterior *vena cava*. See *precaval vein.*

anterograde. *adv.* Reference to movement in a forward or anterior direction.

anterolateral. *adj.* Of or pertaining to the anterior region of the lateral surface of a body or structure. Oriented forward and to the side.

anteroventral. *adj.* Of or pertaining to the anterior region of the ventral surface of a body structure. Oriented forward and downward.

anthelmintic. *n., adj.* A drug or treatment that kills or expels worms usually of the intestines.

anthracosaur. *n.* A member of *Anthracosauroidea*, which includes non-amniote reptilomorph tetrapods, now extinct.

Anthracosauroidea. *n.* A diverse, long-lived group of non-amniote *reptilomorphs*, characterized by domed skulls with some kinetic ability and a five-fingered hand. Amniotes possibly originated within this group.

anthropomorphic. *adj.* Attributing to animals the expression or possession of human qualities or emotions.

anti-. (also **ant-**) Prefix meaning "against," "opposite."

antiandrogenic chemical. See *antiestrogenic chemical.*

antibiosis. *n.* See *allelopathy.*

antibiotic. *n.* A substance that inhibits the growth of, or kills, bacteria or other microorganisms.

antibody. *n.* An immunoglobulin or four-chain protein of specific amino acid sequence that will interact with a specific antigen or foreign substance that induced its synthesis (or one very similar to it).

antiboreal. *adj.* Reference to cool or cold temperate regions of the southern hemisphere. (Syn. *austral*) Cf. *boreal.*

antibothrophic. *adj.* Refers to antivenins effective against venoms that are characteristic of neotropical pit vipers other than rattlesnakes (e.g., *Bothrops*, etc.).

anticholinesterase. *n.* Component of some venoms that inactivates or inhibits the enzyme cholinesterase.

anticlinal. *adj.* Reference to a thoracic vertebra having its neural spine transitional between backward-leaning and forward-leaning.

anticoagulant. *n.* A substance that inhibits the normal coagulation or clotting cascade of blood.

anticoagulin. *n.* A generic term for any hemotoxin that functions to inhibit the blood clotting process. See also *hemocoagulin.*

antidiuretic hormone. (ADH) A nonapeptide hormone synthesized in the hypothalamus and released from storage in the posterior pituitary. This hormone in reptiles is *arginine vasotocin* (note there is a lysine vasotocin), which promotes water reabsorption by the animal and also plays a role in sexual behavior and the expulsion of eggs from the oviduct of turtles, lizards, and Tuatara. In amphibians, the posterior pituitary secretes the peptides *vasotocin* and *mesotocin* that both stimulate osmotic permeability of membranes. Of the two, vasotocin plays a major role in controlling the osmotic permeability of the skin, especially in semi-aquatic and terrestrial species, in addition to effects on the kidney. In response to dehydration, vasotocin decreases glomerular filtration and increases tubular reabsorption of water. In amphibians and reptiles, vasotocin also has a pressor action and increases blood pressure by means of vasoconstriction under certain conditions. Its role in cardiovascular regulation is not well understood. Vasotocin also influences sexual behavior in some amphibians.

antiestrogenic chemical, antiandrogenic chemical. Exogenous or steroid mimicking chemicals that interact with estrogen or androgen receptors by binding and activating or blocking the respective receptor mediated cellular pathways. The activated receptor complex then induces or represses the estrogen or androgen regulated genes via specific promotor sequences, and the resulting changed pattern of gene expression may influence several functions within the animal, including steroidogenesis, gonadal development, or sexual differentiation. See *endocrine disrupting compounds* or *contaminants.*

antifibrin. *n.* An anticoagulin in snake venoms that blocks conversion of fibrinogen to fibrin in the blood clotting process.

antigen. *n.* Any substance capable, under appropriate conditions, of inducing a specific immune response or reacting with the products of that response. Antigens may be soluble or particulate substances such as bacteria or cell particles, but are usually protein or carbohydrate toxins or enzymes. See also *antibody.*

antigen-antibody reaction. The formation of an insoluble complex between an antigen and its specific antibody.

antigenicity. *n.* The capacity to stimulate production of antibodies or the capacity to react with the antibody.

Antillean Subregion. A subdivision of the Neotropical *zoogeographical region* that includes the West Indian Islands.

antimetabolite. *n.* Any substance that inhibits a key enzyme in metabolism, used experimentally to suppress cellular activity.

antimicrobial peptides. Peptides with antimicrobial activity shown to be present in the skin of various frogs, studied largely in contexts of therapeutic potential and evolutionary relationships of frog species.

antioxidant enzymes. Enzymes that prevent the accumulation of reactive oxygen molecules in cells. See *reactive oxygen species.*

antipalmar. *adj.* Pertaining to dorsal surface of the forefoot (opposite the palm).

antiphonal alernation. Alternation of calling by two neighboring anuran males.

antiplantar. *adj.* Pertaining to dorsal surface of the hindfoot (opposite the sole).

antipredator behaviors or **responses.** Reference to various defensive behaviors that are elicited when an animal is confronted by a predator. Numerous categorical terms have been used in the literature to denote subclasses of behaviors to deter predators. Recently, a generalized terminology with consideration of snakes has been proposed by A. Mori and G.M. Burghardt (*Herpetol. J.* 14:79–87, 2004), characterizing responses in three contexts. First, responses of prey are considered as either *approach, neutral,* or *withdrawal* depending on whether prey animals move toward or away from a predator. Second, responses are considered to be either *locomotive, active-in-place,* or *static* depending on whether movements involve locomotion, body movement without locomotion, or immobilization. Third, responses are considered to be either *threatening, cryptic,* or *escape* depending on the apparent function.

antipredator blood-squirting. Reference to defensive squirting of blood from the orbital sinuses of Horned Lizards.

antiseptic. *n., adj.* 1. Preventing infection. 2. Any substance that prevents the growth of microorganisms.

antiserum. *n.* A serum that contains antibodies. See *antivenin.*

antispasmodic. *n., adj.* A substance administered to prevent or stop muscle spasms.

antitropic wind. A local wind such as fall winds or sea breezes.

antivenin. *n.* Serum prepared from blood of an animal that is immunized against the venom of a particular snake species by means of sequential doses of venom or anavenom. (Syn. *antivenene, antivenom, snake antitoxin, snakebite serum, antiserum, serum, serum antivenenosum, snake venom antitoxin*)

antivenom. *n.* See *antivenin.*

antocular. *n.* Synonym for *preocular,* as used by J. Van

Denburgh (*Reptiles of Western North America*, Vol. I, p. 44, 1922). See Fig. 11.

antorbital. *adj.* Relating to, or located in, the area in front of the orbit.

anucleate. *adj.* Without a nucleus.

Anura. *n.* A monophyletic clade (order) containing all living tailless amphibians that includes frogs and toads, the most numerous and diverse of amphibians. See also *Salientia*.

anuran. *adj., n.* Of or pertaining to the Anura or to members of that order. Sometimes used as a noun.

anus. *n.* The external orifice of the *alimentary canal* and the opening of the gut through which feces are expelled in mammals and some other vertebrates. Because amphibians and reptiles have a cloaca, this term should not be used. See *vent, vent tube*.

Aodaisho. *n.* Vernacular name for the Japanese Ratsnake (*Elaphe climacophora*).

aorta. *n.* The principal systemic outflow vessel of the heart, characteristically dorsal in position within the body cavity and called *dorsal aorta* in most amphibians and reptiles. The main arterial trunk. In crocodilians, the *left aorta* originates from the right ventricle and conveys blood to the posterior viscera, whereas the *right aorta* originates from the left ventricle and conveys blood to the anterior and posterior parts of the body.

aortic arch. An outflow vessel from the ventricle that contributes to, or forms, the *aorta*. In ancestral vertebrates and in fishes, aortic arches are multiple paired, lateral vessels that join the ventral and dorsal aorta. Typically, there are six such pairs of arches that form in all embryonic vertebrates. Three or four such arches persist in adult amphibians, and arch IV, called *systemic arch*, provides the main blood flow from the heart to the body. In modern reptiles, only three vessels open forward from the heart to form the right and left *systemic (aortic) arches* and the pulmonary artery. See Fig. 36.

aorticopulmonary septum. A wall of cardiac tissue projecting caudally into the ventricular lumen and separating the ostia of the pulmonary artery and the aortae. This is a prominent feature of the ophidian heart.

aortic valves. Opposing cusps of cardiac tissue at the base of the aortic outflow tracts that prevent backflow of blood from aortae into the ventricle during ventricular filling (diastole).

ap. Abbreviation for Latin *apud*, meaning "in the work of." This abbreviation is used when citing the work of an author contained within a publication of another author.

Aparallactinae. *n.* A subfamily of Colubridae consisting of relatively small, nocturnal, and usually secretive snakes distributed in sub-Saharan Africa and the Middle East. Some bear resemblance to *Atractaspis* (*Atractaspididae*), and the monophyly of this family is not well established.

apatetic coloration. Misleading coloration or camouflage, such as that resembling physical features of the environment.

apatite. *n.* A phosphate mineral that is a major constituent of the bones and teeth of vertebrates and also sedimentary phosphate rocks.

aperiodicity. *n.* (*adj.* –**ic**) Reference to the irregular occurrence of a phenomenon.

aperture. *n.* An opening or orifice.

aphagia. *n.* (*adj.* –**ic**) Absence of feeding; starvation.

aphotic. *adj.* Of or pertaining to an environment without light of biologically significant intensity. In reference to pelagic zones of water bodies, the *aphotic zone* is where light essentially is absent and photosynthesis does not occur. Cf. *euphotic*.

aphototaxis. *n.* (*adj.* –**ic**) Negative phototaxis, or absence of a directed response to a light stimulus by a motile organism.

aphototropism. *n.* (*adj.* –**ic**) Negative phototropism, or absence of orientation to light.

apical. *n., adj.* 1. An alternative term for the *rostral* in snakes of the genus *Vipera*. Used by E. Kramer (*Rev. Suisse Zool.* 68:627–725, 1961). See Fig. 10. 2. Pertaining to the apex or uppermost point on a conical or pointed object, opposite the base.

apical bristle. See *tactile bristle*. Used by H.M. Smith (*Evolution of Chordate Structure*, p. 117, 1960).

apical lobe. The distal, attenuated, or more slender segment of a hemipenis. Cf. *basal lobe*.

apical pit. A sense organ recessed in a small depression usually on the posterior tip of the dorsal scales of certain reptiles, particularly snakes. (Syn. *apical pore, apical scale pit, scale fossa, scale pit*)

apical pore. See *apical pit*.

apical scale pit. See *apical pit*.

aplacental viviparity. Viviparity in which the embryo receives nutrition from the mother without a placental connection between them.

aplasia. *n.* The failure of an organ to develop.

apnea. *n.* Cessation of breathing.

apneumonic. *adj.* Without lungs.

apo-. Prefix meaning "away from."

apocrine secretion. Secretion attributable to sloughing of the apical portion of a secretory cell.

Apoda. *n.* An older term for limbless amphibians. See *Gymnophiona*.

apodal. *adj.* Lacking feet or limbs.

apoenzyme. *n.* The protein portion of an enzyme which combines with a *coenzyme* to form a functioning enzyme.

apomorph, apomorphy. *n.* (*adj.* -**ic**) Reference to those characters of species that have evolved only within the taxonomic group in question. Reference to traits derived from, but no longer the same as, an ancestral character. Cf. *plesiomorphy, stasimorphic*.

aponeurosis. *n.* A broad, flat tendon or sheet of tough connective tissue serving to distribute the tension of a muscle.

apophysis. (*pl.* **-es**) *n.* Refers to various bony processes found on vertebrae. Specific processes are indicated by a prefix (e.g., *zygapophysis,* etc.). The term is sometimes applied to any bony outgrowth or protuberance.

apoptosis. *n.* Programmed death of cells, specific with respect both to tissue and timing during embryogenesis, metamorphosis, and during cell turnover in adult tissues. (Syn. *programmed cell death*)

aposematic. *adj.* Reference to characters, especially coloration, that advertise an organism as dangerous or unpleasant to potential predators ("warning" characters). Cf. *episematic.*

apostatic selection. Positive frequency-dependent selection for a common morph. Apostatic selection occurs when a predator concentrates disproportionately on the common varieties of nonmimetic polymorphic prey species. This is suggested to be a primary mechanism for maintaining stable prey polymorphism.

a posteriori. Use of a method of reasoning that seeks to establish a cause or general rule based on effect or experience. Descriptive of biometric or statistical tests in which the comparisons of interest are unplanned and become evident only after experimental results are obtained. Cf. *a priori.*

apotypic. *adj.* Divergent from the typical or basic form.

apotypic state. Reference to a derived character state in a transformation series.

apparent assimilation efficiency. (AAE) A term suggested in place of *assimilation efficiency* because feces include bacteria, sloughed gut lining, and other components not originating with a meal, and therefore the energy content of feces overestimates the energy of undigested food used to calculate the assimilation efficiency. Thus, calculated or "apparent" assimilation efficiencies are inherently an underestimation of the actual AE. See *assimilation efficiency.*

apparent competition. An indirect interaction that results in a decrease in the population growth or average abundances of two or more prey species that may not compete for the same resource but do share a common natural enemy (predator, parasite, or pathogen). This process may lead to the exclusion of one of the prey species (e.g., the one with lower intrinsic growth rate). Short-term apparent competition is the negative indirect effect that one prey species has on another prey species acting via its effects on predator foraging behavior. It is argued that alternative prey species in the diet of a food-limited generalist predator should reduce each other's equilibrial abundances, whether or not they compete directly. Term introduced by R.D. Holt (*Theor. Pop. Biol.* 12:197–229, 1977).

apparent density. The relative number of individuals per unit of surface or volume of habitat. This measure is a function of the data collection method. Cf. *absolute density.*

apparent digestive efficiency. (ADE) A term suggested in place of *digestive efficiency* because feces include bacteria, sloughed gut lining, and other components not originating with a meal, and therefore the energy content of feces overestimates the energy of undigested food used to calculate the digestive efficiency. Thus, calculated or "apparent" digestive efficiencies are inherently an underestimation of the actual DE. See *digestive efficiency.*

appeasement display. A behavioral display that functions to prevent attack or aggression between members of the same species, usually by releasing nonaggressive behaviors that replace the opponent's attack drive.

appendage. *n.* A subordinate portion or outgrowth of a structure, such as a tail or limb. See *integumentary appendage.*

appendicular. *adj.* Reference to a limb.

appendicular skeleton. Reference to skeletal elements of the limbs and associated pelvic and pectoral girdles.

appendix epididymis. Synonym for *stalked hydatid* according to H.M. Smith (*Evolution of Chordate Structure,* p. 333, 1960).

appendix fimbriae. Vestigial tip of the Wolffian duct in female metanephric animals, remaining after degeneration of the duct except the Gartner's duct. (Syn. *appendix vesiculosa*)

appendix testis. See *sessile hydatid.*

appendix vesiculosa. See *appendix fimbriae.*

appetitive behavior. Reference to behaviors that help to satisfy a need related to an internal condition. The term is also used to describe an initial, exploratory phase of an instinctive behavior pattern.

apposed. *v.* Being in immediate proximity or side-by-side.

apposition. *n.* Placement of things in immediate proximity, or application of adjoining layers. (Syn. *juxtaposition*)

applexation. *n.* See *amplexus.*

Applied Herpetology. A new international journal that focuses on issues and data related to biodiversity, conservation, environmental monitoring, farming, natural products development, and wildlife management. The journal was defined at the Fourth World Congress of Herpetology, Sri Lanka, 2002, and is published by Brill Academic Publishers.

a priori. Use of a method of reasoning that begins with cause or theory to establish an effect. Descriptive of biometric or statistical tests in which the comparisons of interest are determined on theoretical grounds in advance of experimentation. Also used in reference to a conclusion about a specific instance that derives from knowledge of the relevant general facts or conditions. Cf. *a posteriori.*

apron. *n.* A prominent fold of skin on the lower throat of certain bufonid species, which rests upon folds of the deflated vocal sac. Used by A.A. Wright and A.H. Wright (*Handbook of Frogs*, pp. 23 and 88, 1935).

aptation. *n.* Reference to any character currently subject to selection (contributes to fitness), regardless of whether natural selection shaped the character to function in its current role (*adaptation*) or whether a character has been coopted by selection for a new and still current role different from its origin (*exaptation*). See S.J. Gould & E.S. Vrba, *Paleobiology* 8:4-15, 1982.

AQP. See *aquaporin*.

aqua-. An element in compound words meaning "water."

aquaculture. The practice of raising commercial organisms in a controlled environment for commercial purposes.

aquaporin. (AQP) *n.* A protein that, in tetramers, forms water channels in membranes. AQP subgroups are selectively permeable to water and have been identified in the kidney of amphibians. See A.S. Verkman and A.K. Mitra, *Am. J. Physiol.* 278:F13–F28, 2000.

aqua-terrarium. *n.* An aquarium that is part terrestrial for housing amphibious amphibians or reptiles.

aquatic. *adj.* Living in water.

Aquatic Caecilians. Collective vernacular name for five genera of caecilians in the family Typhlonectidae, all being viviparous and facultatively or obligatorily aquatic.

aqueous humor. Aqueous fluid in the chambers of the eye between the lens and cornea.

aquifer. *n.* A body of saturated sediment or rock through which water can move readily.

Aquilonian region. A biogeographical region comprising Europe, Asia north of the Himalayas, Africa north of the Tropic of Cancer, and America north of latitude 45° North.

arachnid. *n.* A member of the class of invertebrates that includes mites and ticks, which are common parasites of reptiles.

ARAZPA. See *Australian Regional Association of Zoological Parks and Aquaria*.

Arber's law. The idea that any structure that disappears during the course of evolution of a lineage is never regained by descendants of that line. Cf. *Dollo's law*.

arboreal. *adj.* Living exclusively or mostly above ground in association with foliage of trees or shrubs.

Arboreal Alligator Lizards. Vernacular name for various Middle American species of anguid lizards belonging to the genus *Abronia*.

arboricolous. *adj.* Living predominantly in trees or woody shrubs.

arborize. *v.* To branch or extend in a pattern similar to the branching of a tree.

arbusticolous. *adj.* Living predominantly on scattered shrubs or shrublike perennial herbs.

arcade. *n.* A bony bridge across a skull opening.

arch. *n.* 1. A bridge of bone in the skull (e.g., *zygomatic arch*). 2. Also may be used with reference to "arching" behaviors of snakes that raise segments of the body to form arcs during aggression or social displays.

Archaeobatrachia. *n.* A clade of anurans that includes the more primitive families. Cf. *Neobatrachia*.

Archaeopteryx. *n.* The generic name for a late Jurassic vertebrate that possessed feathers and other avian features, but mostly had reptilian characteristics including claws on three of the five fingers, a long, flexible tail, and teeth.

archaic. *adj.* Reference to the oldest members of a lineage.

archallaxis. *n.* The addition of a new feature during the early period of embryonic morphogenesis, usually important to the direction of subsequent development.

arch centers. In vertebrate embryology, *chordal centers* that become cartilaginous *anlagen* giving rise to the dorsal and ventral arches of the vertebrae. See *arcuale, chordal centers*.

arche-, archi-. Prefix meaning "first," "primitive," "ancestral," or "beginning."

Archean. The earlier of the two subdivisions of *Precambrian* time. Also called *Archeozoic*. See Table 1.

archenteron. *n.* The embryonic digestive tube, formed by the endoderm during gastrulation.

Archeozoic. See *Archean*.

archepodium. *n.* Part of the tetrapod limb containing proximal elements that were derived from the ancestral fin elements. Term was proposed by T.S. Westoll, according to G.L. Orton (*Science* 120:1042–1043, 1954). Cf. *neopodium*.

archetype. *n.* The earliest common or hypothetical ancestor.

archinephric duct. The collecting duct of the *archinephros* (= *pronephros*) or primitive kidney, which appears during early development of all vertebrates and passes posteriorly to the cloaca. A general term for the urogenital duct. See Fig. 38 and *Wolffian duct*.

archinephros. *n.* The hypothetical primitive kidney of vertebrates, formed from the entire *nephrotome*. Cf. *holonephros, mesonephros, metanephros, opisthonephros, pronephros*.

archipelago. *n.* A group of islands.

Archosauria. *n.* A clade (subclass) inclusive of dinosaurs, crocodilians, and birds and their common ancestors, and comprising one of the major radiations of terrestrial vertebrates, a sister taxon to *Lepidosauria*. The name, meaning "ruling reptile," was created by E.D. Cope in 1869.

Archosauromorpha. *n.* One of two primary divisions of *Diapsida* that includes crocodilians and birds in addition to fossil relatives and other dinosaurs.

arcifery. *n.* (*adj.* **-al, -ous**) Condition of midventral overlap between free elements of the pectoral girdle (*epicoracoids* in Anura; *coracoids* in Urodela). See Fig. 3. Cf. *firmisterny*. An intermediate condition is found in several diverse genera of frogs in which the epicoracoid cartilages are fused anteriorly but free and overlap posteriorly (*arcifero-*

firmisterny, first used by H.W. Parker, *Annals Mag. Nat. Hist.* 18:201–203, 1926).

Arctogaea. (Arctogea) *n.* An inclusive *biogeographical region* including the *Ethiopian, Oriental, Palaearctic,* and *Nearctic* regions. (Syn. *Megagaea*) Cf. *Neogaea, Notogaea, Palaeogaea.*

arcuale, arcualium. (*pl.* **-ia**) *n.* Paired blocks of tissue which become cartilaginous and eventually ossify to form vertebrae, according to *Gadow's theory* of vertebral formation (H. Gadow & E.C. Abbott, *Philos. Trans.* Series B 186:163–221, 1895; H. Gadow, *Philos. Trans.* Series B 187:1–57, 1896). Such developmental steps occur in elasmobranches and many primitive bony fishes, but they are absent in tetrapods. Gadow proposed, however, that arcualia formed the underlying pattern of vertebral development in all later derived groups. See *arch centers.*

arcuate. *adj.* Bow-shaped, or in the form of arches.

are. (a) A metric unit of area equal to 100 m^2. Cf. *hectare.*

area effect. In island biogeography, the principle that species number increases with island area.

Area of Special Conservation Interest. (ASCI) Geographic areas of special concern in relation to conservation of wildlife or related natural resources.

area-species curve. See *species-area curve.*

arenicolous. *adj.* Living in association with sand or sandy areas. (Syn. *sabulicolous*)

arenobufagenin. *n.* See *arenobufagin.*

arenobufagin. *n.* A *bufogenin* isolated from *Bufo arenarum.*

arenobufotoxin. *n.* A *bufotoxin* first isolated from *Bufo arenarum* and formed by conjugation of *arenobufagin* with *suberylarginine.*

areola. *n.* 1. A region of small, elevated, and rounded lumps on the skin, frequently seen on the belly of various frogs (more usually referred to as *granular*). 2. A small non-pigmented area in a color pattern. Used by H.M. Smith (*Handbook of Amphibians and Reptiles of Kansas*, p. 313, 1950). 3. The central region of a lamina on the shell of a turtle, present at birth and surrounded by new growth that proceeds outward. See *laminar nucleus.*

areolate. *adj.* Covered with globular and closely set prominences. Descriptive of the granular condition of the underside of various frogs.

arginine ester hydrolase. A noncholinesterase enzyme found in many viper and pit viper venoms, believed to be involved in bradykinin releasing and clotting activities of these venoms.

arginine vasotocin. *n.* See *antidiuretic hormone.*

ARG UK. A network of wildlife volunteer groups which aim to protect and conserve the native amphibians and reptiles of the UK. Current URL: http://www.arg-uk.org.uk

arhythmic. *adj.* Nonrhythmic, or lack of diurnal periodicity.

arid. *adj.* Geographic or climatic term referring to the condition of having little or no rainfall (dry), sometimes defined as an area with less than 25 cm of rain per year. Evaporation exceeds precipitation, and vegetation is sparse.

aridity index. A measure of effective moisture supply used in classification of soils or climates.

-arin. Suffix meaning "prothrombin activating," used in a nomenclatural scheme for describing exogenous hemostatic factors in snake venoms. By adding a portion or designated abbreviation of a snake's scientific name to the suffix, one obtains a designation for the fraction being identified. For example, using the name "*gabonica*" (species name for the Gaboon Viper) one obtains *gabonarin*. (Syn. *activase II*)

arithmetic growth. Growth of an organism or population by linear increase in size or number of individuals. Morphologically, a constant is added to the length of a body part in each time interval. Cf. *geometric growth.*

arithmetic mean. The number that results from dividing the sum of a series by the number of items in the series, giving an "average" value for the series. See *mean.*

arousal. *n.* Reversal of torpor, or increased responsiveness to sensory stimuli.

arrenoidism. *n.* Condition in which typically male characteristics occur in females. Coined by A.R. Hoge et al. (*Mem. Inst. Butantan* 29:17–88, 1959) with reference to *Bothrops insularis* in which the hemipenis is present in a percentage of reproducing females. This term has not enjoyed wide usage.

arrested growth, lines of. See *lines of arrested growth (LAGs).*

Arrhenius plot. Generally, a plot of the logarithm of a reaction rate (e.g., growth rate, enzymatic reaction velocity) versus the reciprocal of the absolute temperature. The Arrhenius plot is a mathematical approach to exploring the impact of temperature on macromolecular processes, described in the late 1800s by the Swedish physical chemist Svante Arrhenius.

arribada. *n.* Refers to mass emergence of marine turtles onto a beach to deposit their eggs. Copulating pairs of turtles congregate in large numbers, followed by mass nesting of females generally over a period of several days. (Syn. *arribazons, morrinas, flotas*)

arribazons. See *arribada.*

Arroyo Toad. Vernacular name for the bufonid species *Anaxyrus* (formerly *Bufo*) *californicus*. For recent taxonomic revision see D.R. Frost et al., *Bull. Amer. Mus. Nat. Hist.* 297:1–370, 2006.

artenkreis. *n.* A German term that refers to a group of closely related species.

arteriole. *n.* A smaller artery immediately preceding capillaries.

artery. *n.* A blood vessel that transports blood away from the the heart to capillaries in tissues.

Arthroleptidae. *n.* A clade (family) of anurans that includes the Hairy Frog (*Trichobatrachus robustus*). Seven genera and about 76 species occur in sub-Saharan Africa. Of two subfamilies, *Arthroleptinae* and *Astylosterninae*, the latter is recognized as a separate family by some authors. D.R. Frost et al. (*Bull. Amer. Mus. Nat. Hist.* 297:1–370, 2006) recognize two subfamilies (*Arthroleptinae* and *Leptopelinae*) and also incorporate *Astylosterninae*.

Arthroleptinae. *n.* A subfamily of anurans belonging to the *Arthroleptidae*.

articular, articulare. *n.* A bone of the lower jaw which ossifies from the posterior *Meckel's cartilage* to form the articulating surface meeting the skull at the quadrate bone.

articular process. A protuberance situated between the pharyngeal processes of the amphibian cricoid cartilage. Used by M. Frazier (*J. Morphol. Physiol.*, 39:285–291, 1924).

articulate. *v.* (*n.* **articulation**) 1. In anatomy, to join or form a joint. 2. Reference to the soft tissue junction connecting two bones. When movable, the articulation forms a joint. 3. In fossilized bones, to remain joined as would have been the case in real life.

artifact, artefact. n. Any structure that is not typical of the actual specimen, usually resulting from cytological preparation, postmortem aging, etc. Not natural.

artificial character. A character that is arbitrarily selected, without consideration of phylogenetic relationships.

artificial insemination. Artificial introduction of semen into the reproductive tract of a female, or the mixing of male and female gametes by other than natural means.

artificial selection. The selection by humans of genotypes that contribute to the gene pool of successive generations of an organism. Selective breeding by humans.

artificial taxon. A group of organisms not corresponding to a natural unit of evolution. Cf. *natural taxon*.

arytenoid cartilage, arytenoids. Either of paired, semicircular, dorsal cartilaginous elements of the *hyoid apparatus*, connected to the larynx. These support the glottis and are an integral part of the sound-production system in anurans.

Ascaphidae. *n.* Family of "tailed frogs" (Anura), a single extant species inhabiting the Pacific northwest of North America. This family was recently synonymized with the family *Leiopelmatidae* (D. Frost et al., *Bull. Amer. Mus. Nat. Hist.* 297:1–370, 2006).

Ascaridoidea. *n.* Endoparasitic nematodes (roundworms) found in many lizards and boid snakes.

Ascaroidea. *n.* former name for *Ascaridoidea*.

ASCI. See *Area of Special Conservation Interest*.

ascorbic acid. Vitamin C, required in the diet of many captive reptiles.

asexual reproduction. Reproduction without sexual processes or formation of gametes. Occurs in some amphibians and reptiles. See *parthenogenesis*.

ash. *n.* A general term for the residues of combustion or incineration, representing inorganic mineral content of animal tissue.

ash-free dry weight. (AFDW) Qualification for units of tissue or animal mass measured by subtracting the water mass (determined by drying) and ash mass from the tissue wet mass. The ash component of tissue represents its inorganic mineral content and is determined by combustion of dry tissue.

ash-free lean dry mass. (AFLDM) Qualification for units of tissue or animal mass measured by subtracting the ash mass determined by combustion from the lean dry mass obtained through fat extraction. The ash component of tissue represents its inorganic mineral content.

Asian Black Spined Toad. Vernacular name for the Asian bufonid species *Duttaphrynus* (formerly *Bufo*) *melanostictus*. For taxonomic revision see D.R. Frost et al., *Bull. Amer. Mus. Nat. Hist.* 297:1–370, 2006.

Asian Box Turtle. Vernacular name for species of turtle belonging to the emydid genus *Cuora*.

Asian Coral Snakes. Vernacular name for species of Asian elapid snakes belonging to the genus *Calliophis*.

Asian Giant Toad. Vernacular name for the Asian *Bufo asper*.

Asian House Gecko. Vernacular name for the Asian gekkonid species *Cosymbotus platyurus*. See also *House Gecko*.

Asian Pipe Snake. Vernacular name for uropeltid (or cylindrophiid) snakes belonging to the genus *Cylindrophis*. See *Pipe Snakes*.

Asian Rat Snake. Vernacular name for colubrid snakes belonging to the Asian genus *Ptyas*.

Asian Salamanders. Collective vernacular name for various salamanders belonging to the family Hynobiidae and the genus *Hynobius* in particular.

Asian Toads. Vernacular name for species of anurans belonging to the family Megophryidae.

Asian Tree Pitviper, Asian Tree Viper. Vernacular name for various Asian pitvipers belonging to the speciose and ecologically diverse genus *Trimeresurus*.

ASIH. See *American Society of Ichthyologists and Herpetologists*.

Asp. *n.* A generic vernacular name that is sometimes used with reference to various venomous snakes of Africa, the Middle East, and Europe, especially African atractaspidid species of the genus *Atractaspis,* and the viper *Vipera aspis.* See *Burrowing Asp*.

aspect. *n.* 1. The degree of exposure of a site to environmental factors. 2. Reference to seasonal changes in the appearance of vegetation.

aspect ratio. Total length divided by body width.

asperity. (*pl.* **-ies**) *n.* A roughened outgrowth or surface, used sometimes for the nuptial pad that develops in male anurans.

asphyxia. *n.* Death from lack of oxygen.

aspidospondylous. *adj.* Refers to primitive amphibian vertebrae in which centra are ossified from cartilaginous arches and the elements are separate. Cf. *holospondylous*.

aspirate. *v.* The act of sucking or withdrawing fluids (including gases) from a compartment.

aspiration. *n.* 1. Mode of lung ventilation characteristic of most tetrapod vertebrates (including reptiles) in which expansion of the thoracic and pleural cavities lowers pulmonary pressures below atmospheric. 2. Various systems have been used in mechanical aspiration of venom as a treatment for snake bite. Research has shown the various methods (Aspivenin® , Extractor® , Venom-Ex®) to have low efficacy. See *black stone.*

aspiration pump. Term used for air-breathing mechanisms in reptiles (and mammals) where expansion of the rib cage or thoracic region creates low pressures around the lungs. Air is "sucked" (or aspirated) into the lungs. See *aspiration;* Cf. *buccal force pump.*

aspirator. *n.* A suction device or apparatus. See *aspiration.*

aspondyly. *n.* A condition in which the centra are absent from vertebrae.

ASRA. See *Association for the Study of Reptilia and Amphibia.*

assemblage. *n.* 1. A large group of fossils or other items found in the same location and regarded as being from the same time period. 2. A group of organisms found together in the same community or ecosystem.

assemblage zone. An aggregation of fossils in a body of sedimentary rock.

assembly rules. Theoretic principles by which a community of organisms evolves structure, as, for example, following the initial colonization of an island or disturbed habitat.

assertion display. A "low intensity" or casual display in lizards, not necessarily directed at another individual.

assimilation. *n.* With reference to ecological energetics, that part of consumption by an organism, population, or trophic level that is incorporated into metabolism for maintenance or production. Cf. *metabolizable energy.*

assimilation efficiency. In ecological energetics, the ratio of total food absorbed (or metabolized) to total food ingested, calculated as $[(C - (F + U)/C] \times 100$, where C = energy consumed, F = energy of fecal waste, and U = energy of urinary waste. See also *apparent assimilation efficiency.*

association. *n.* 1. In general usage, reference to a large assemblage of organisms in a particular area, usually with one or two dominant species. 2. A climax plant community comprising a ranked category in classification of vegetation.

Association for the Study of Reptilia and Amphibia. (ASRA) A national organization of the UK. There is no current website.

Association of Zoos and Aquariums. (AZA) A nonprofit organization dedicated to the advancement of zoos and aquariums in areas of conservation, education, science, and recreation. The AZA was founded in 1924 as the *American Association of Zoological Parks and Aquariums.* Also previously called the *American Zoo and Aquarium Association.* Current URL: http://www.aza.org

assortative mating. Reference to sexual reproduction in which pairing of males and females is not random, i.e. particular females tend to mate with particular males. Cf. *panmixis.*

astatic. *adj.* Characterized by sudden or great change.

asterisk. (*) In taxonomy, a symbol used to designate categories of infraspecific rank, usually placed between the second and third elements of a *trinomial name.*

Asterophryinae. *n.* A subfamily of *Microhylidae.*

asthenia. *n.* Weakness.

astragalocalcaneum. *n.* Used with reference to the tarsal region of frogs, where the *astragalus* and the *calcaneum* are partly fused to each other.

astragalus. *n.* Mammalian name for the inner of the proximal bones of the tarsus or ankle articulating with the tibia. In reptiles the astragalus appears to be a fusion of *tibiale* with *intermedium* and a central element. (Syn. *talus, tibiale*)

astrocyte. *n.* A *neuroglial* cell that passes nutrients between blood capillaries and neurons.

Astylosterninae. *n.* A subfamily of anurans belonging to the *Arthroleptidae.*

asymmetry. *n.* Skewness.

asymptomatic. *adj.* That which exhibits no symptoms.

asynchronous breeding. Reference to females arriving at the breeding site irregularly throughout the breeding period and sometimes returning to mate several times.

atavism. *n.* (*adj.* **atavistic**) The reappearance, in individual members of a species, of characters that once were possessed by all members of an ancestor. Reference is to old structures that appear anew, not newly evolved ones. Examples are elements of the hind limb skeleton in snakes.

ataxia. *n.* Failure of muscle coordination.

Athesphatanura. *n.* A new monophyletic taxon, sister to *Amphignathodontidae* within the new taxon *Tinctanura,* and containing *Hylidae* and the new taxon *Leptodactyliformes* (D.R. Frost et al., *Bull. Amer. Mus. Nat. Hist.* 297:1–370, 2006).

atlantal processes. Paired lateral processes on the atlas that articulate with the occipital condyles in modern amphibians.

atlas. *n.* The first cervical vertebra that articulates with the cranium. This provides a "cradle" in which the skull can "rock."

atmosphere. *n.* (*adj.* **–ic**) 1. The gaseous envelope surrounding Earth. 2. A derived unit of air pressure at sea level ($= 1.01325 \times 10^5$ N m^{-2} = 101.325 kPa = 1013.25 mb).

atoll. *n.* A circular reef surrounding a deep lagoon.

ATP. *n.* See *adenosine triphosphate.*

ATPase. *n.* Adenosine triphosphatase, a class of enzymes that catalyze the hydrolysis of ATP.

ATR. See *activity temperature range.*

Atractaspididae. *n.* A clade (family) of slender-bodied, venomous African snakes with unresolved relationships with other snakes. A single genus and 18 species occur in Africa and the Middle East. In the vernacular, known as *Burrowing Asps*, *Mole Vipers*, or *Stiletto Snakes*.

atresia. *n.* Congenital absence of a normal passageway or open lumen. See also *follicular atresia*.

atretic follicles. Egg follicles that have regressed before ovulation.

atrial natriuretic peptide. (ANP) A peptide hormone synthesized in the atria of the heart, functioning to increase urine output, sodium excretion, and receptor-mediated vasodilation, the net effect of which is to lower blood pressure. (Syn. *atrial natriuretic factor, ANF*)

atrial natriuretic factor. (ANF) See *atrial natriuretic peptide*.

atrioventricular bundle. Closely interwoven muscle fibers at the junction of atria and ventricle of the heart, functioning as part of the conducting pathway for pacemaker stimulation of cardiac contraction.

atrium. (*pl.* **-ia**) *n.* 1. A cavity or chamber. 2. Usual reference is to one of two receiving chambers of the heart, filling with blood from the sinus venosus and emptying to the ventricle. (Syn. *auricle*) See Figs. 36, 37.

atrophy. *v., n.* Degeneration or reduction in size of an organ or part, resulting from disease, disuse, or poor nutrition.

attenuation. *n.* A reduction in strength or intensity.

attenuate. *adj.* Slender, thin, or tapering.

attribute. *n.* A character.

atypical. *adj.* Unusual, or out of the norm.

auctorum. *n.* Of authors.

audiogenic. *adj.* A response to an auditory stimulus.

audiospectrogram. *n.* The graphic representation of vocalization produced by a sonograph consisting of a *sonogram* (frequency vs. time) and a *spectrogram* (amplitude vs. frequency).

audition. *n.* The sense of sound perception or *hearing*. (Syn. *hearing*)

auditory. *adj.* Pertaining to the sensory reception of sound.

auditory canal. See *external auditory meatus*.

auditory capsule. See *otic capsule*.

auditory cortex. Regions of the cerebral cortex associated with hearing.

auditory cup. A concavity on the external lateral surface of the quadrate which attaches at its rim to the tympanic membrane in lizards.

auditory mechanism. Reference to sound-conducting mechanisms described by A. Tumarkin (*Evol.* 9:221–243, 1955). Five of six such mechanisms are as follow. 1. *Vestibulo-hyoid*: A bone-conducting system in which the floor of the mouth may act to transmit vibrational stimuli to the columella. An essentially bone-conducting mechanism found in *Sphenodon* and some archosaurs. 2. *Vestibulo-quadrate*: A bone conducting system in which the columella extends from the quadrate to the oval window and vibrations are transmitted through the lower jaw and articulo-quadrate to quadrate. This mechanism is found in snakes, some lizards, and many extinct reptiles. 3. *Vestibulo-scapular*: A system in which vibrations pass from the forelimb to the scapula and thence to the operculum in oval window by means of the opercularis muscle, which inserts on the scapula. This mechanism is found in some amphibians. 4. *Vestibulo-squamosal*: This is a system found in some urodeles wherein sound is conducted from substrate to skull via the columella, which extends between the squamosal and oval window with no tympanic membrane or middle ear cavity. 5. *Vestibulo-tympanic*: A more familiar system including a tympanic membrane that receives sound stimuli from air and transmits vibrations via a columella that extends to the oval window. This mechanism is found in anurans, many reptiles and birds. (The mammalian tri-ossicle system was termed *vestibulo-ossicular*.)

auditory nerve. A nerve conducting information from the inner ear to the brain.

auditory ossicle. Term used to denote the bone of the middle ear (*columella*) that rests in the oval window or connects the tympanic membrane with the oval window. The term *auditory ossicles* is usually used with reference to the three middle ear ossicles of mammals.

aural. *adj.* Pertaining to the ear.

auricle. *n.* See *atrium*.

auricular. *adj.* 1. Pertaining to the *atria* (*auricles*) of the heart. 2. Pertaining to, or emanating from, the ear, especially features of the outer ear. In lizards, the term refers to large or modified scales projecting over the anterior ear opening. (Syn. *auricular lobule*)

auricular fold. The fold of skin that lies behind or over the ears of certain lizards and bears spines or other projections.

auricular lobule. See *auricular*.

auriculo-ventricular ring. Fibers forming a complete ring of nodal tissue around the opening between the right atrium and ventricle. Thought to be ancestral to the mammalian Bundle of His by A. Keith and I. MacKenzie (*Lancet* 1:101–103, 1910).

austral. *adj.* Southerly and pertaining to cooler regions of the Southern Hemisphere.

Australian Crowned Snakes. Vernacular name for Australian elapid snakes belonging to the genus *Drysdalia*, especially *D. coronata*.

Australian Freshwater Crocodile. Vernacular name for the crocodilian *Crocodylus johnstoni*.

Australian Froglets. Collective vernacular name for anurans belonging to the family *Myobatrachidae*.

Australasia or **Australasian region.** *n.* 1. One of the eight primary *biogeographic realms* of the world, including Australia, the Celebes, Papua New Guinea and islands eastward to the Solomons, New Zealand, and Fiji. The northern boundary of this region is known as *Wallace's Line*. 2. Also a formal zoogeographical realm, exclusive of New Zealand and Fiji.

Australian Regional Association of Zoological Parks and Aquaria. (ARAZPA) An organization established in 1990 that links over 70 zoos and aquariums across Australia, New Zealand, and the South Pacific, forming a network for wildlife conservation, environmental education, and wildlife research. Current URL: http://www.arazpa.org.au/

Australian Swamp Frog. Vernacular name for species of Australian frogs belonging to the myobatrachid genus *Limnodynastes*.

Australobatrachia. *n*. A newly designated monophyletic taxon of frogs, sister with *Nobleobatrachia* within *Notogaeanura* and composed of *Batrachophrynidae* and *Myobatrachoidea* (D.R. Frost et al., *Bull. Amer. Mus. Nat. Hist.* 297:1–370, 2006).

autapomorphic character. 1. A character derived from a plesiomorphic character state in the immediate ancestor of a single species. A derived character that is unique or occurs in a single species or taxon. 2. A uniquely derived character shared by several synapomorphous taxa.

autapomorphy. *n*. The possession of a unique derived character by a species or monophyletic taxon. Cf. *synapomorphy, symplesiomorphy*.

Autarchoglossa. *n*. A basal clade of *Scleroglossa* and sister group to *Gekkonidae*.

autecology. *n*. The ecology of individual organisms or species.

authority. *n*. The author citation for a scientific name.

auto-. Prefix meaning "self," "self-operating," "same one," or "automatic."

autocentral, autocentrous. *adj*. Term coined by G.J. Romanes, according to E.E. Williams (*Quart. Rev. Biol.* 34:1–32, 1959), who references the centra of tetrapodal vertebrae as independent formations of bone (not derived from the bases of the dorsal and ventral arches, as proposed by Gadow).

autochthonous. *adj*. 1. Reference to a species that has evolved within its native range. Of local origin. Cf. *allochthonous*. 2. Found at the place where it originated.

autocoid hormone or **autocoids.** A general term used to describe various active endogenous substances (e.g., histamine and serotonin) that do not yet fit into existing functional classifications.

autocrine secretion. Reference to secretions that affect the secreting cell itself. Cf. *endocrine secretion, paracrine secretion, exocrine secretion*.

autodiastyly. *n*. A condition of *autostyly* in which the palatoquadrate articulates with the skull and is movable upon it.

autogenous control. Regulation of gene expression by means of a product (protein or regulatory factor) either inhibiting or enhancing its own activity.

autograft. *n*. A transplantation of an organism's tissue from one part of its body to another part of the same organism.

autohaemorrhage, autohemorrhage. *n*. Condition of self-induced bleeding, characteristic of defense mechanisms in some snakes and lizards when disturbed or threatened.

autoimmunity. *n*. 1. A condition of immune response against the constituents of the body's own tissues (autoantigens). 2. In herpetology, this term has been used in reference to the supposed ability of a snake to survive the effects of its own venom or that of conspecifics.

autologous. *adj*. 1. Reference to a graft from one region to another in the same animal. 2. Reference to proteins or genes from the same individual.

autolysis. *n*. (*adj*. **autolytic**) Breakdown of a cell due to action of its own enzymes, usually following death of the cell.

automimicry. *n*. A form of mimicry in which mimics and their models are members of the same species. Members of the species are polymorphic with respect to their palatability or susceptibility to predators.

autonomic. *adj*. Self-controlled or spontaneous, usually with reference to involuntary nervous control of body functions. That part of the peripheral nervous system that supplies visceral motor nerves to various involuntary organs.

autonomous. *adj*. Self-regulating, with reference to intrinsic regulation of population size.

autonomic nervous system. The efferent pathways of the nervous system that control involuntary visceral functions, consisting of *sympathetic* and *parasympathetic* systems.

autoplastic. *adj*. Refers to autograft.

autopodium. *n*. The distal end of a limb, comprising the wrist or ankle and its digits. Cf. *stylopodium, zeugopodium*.

autoploid. *adj, n*. Reference to an organism having the characteristic chromosome set of its species.

autopsy. *n*. Postmortem examination of a body to determine the cause of death.

autoradiography. *n*. The process of making a photographic record of the internal structures of a tissue by means of radiation emitted from incorporated radioactive material. Radioactively labeled molecules are localized by applying a photosensitive emulsion to the surface of a radioactive specimen. (Syn. *radioautography*)

autosome. *n*. Any chromosome other than a sex chromosome.

autostyly. *n*. The condition of jaw suspension in which the mandible is attached directly to the braincase through the *quadrate*, without participation of the hyomandibula. (Syn. *metautostyly*) See also *autodiastyly, autosystyly*.

autosystyly. *n*. A condition of autostyly in which the palatoquadrate is fused with the cranium, and the junction permits no movements (as in urodeles and anurans).

autotherm. *n*. See *endotherm*. Cf. *allotherm*.

autotomy. *n*. (*adj*. **-ic, -ous**) Self division or fracture, as in the ability to release a body part such as the tail breakage that occurs in many lizard species and snake species when the animal is seized by a predator (*caudal autotomy*). The definition has been proposed as descriptive of taxa having intravertebral breakage followed by tail regeneration (J.M. Savage & J.B. Slowinski, *Biol. J. Linn. Soc.* 57:129–194, 1996). (Syn. *tail-dropping*) Cf. *pseudautotomy*, *urotomy*.

autotomy plane or **septum.** The breakage plane in which *autotomy* occurs in caudal vertebrae, usually consisting of soft tissue located in the centers of vertebrae. (Syn. *fracture plane, fracture septum, breakage plane*)

autotoxin. *n.* (*adj.* **-ic**) Any substance produced within an organism that is toxic to itself.

autotroph. *n.* See *primary producer*.

autotype. *n.* The original designated type of a specimen. Also used to describe a specimen that is designated by the author of a species as being identical to the holotype subsequent to the original publication.

autumnal. *adj.* Pertaining to the autumn. Cf. *aestival, hibernal, vernal, serotinal*.

auxiliary character. A character that is used to confirm the identification of a taxon. A confirming character. Cf. *diagnostic character*.

available name. Any name proposed for a taxon that meets the requirements of the *International Code of Zoological Nomenclature*. Cf. *unavailable name*.

avascular. *adj.* Without a blood supply.

average. *n.* The *mean, arithmetic mean*. (It is better to avoid using this sometimes vague term. It usually refers to the arithmetic mean, but is sometimes used to signify the median, mode, geometric mean, weighted mean, or other things.)

Aves. *n.* The birds, a clade (class) of amniote vertebrates considered by many to be reptiles. These are derived diapsid vertebrates, most closely related to crocodiles and considered by most to have originated within the *theropods* and therefore part of the dinosaur radiation. Alternate views consider birds arising earlier, perhaps from *saurischians* where birds might share a common ancestor with dinosaurs.

avian. *adj.* Of or pertaining to birds.

avidity. *n.* 1. Generally, the total combining power of an antibody with an antigen. The concept incorporates both the affinity of each binding site and the number of binding sites per antibody and antigen molecule. 2. In herpetology, the term has been used for the speed with which an antivenin neutralizes a venom.

AVIT. *n.* A family of proteins that are homologues of protein originally isolated from snake venom and frog secretions and present in many vertebrates. The venom component has many functions, including stimulation of muscle contraction, growth factor, and binding with G-coupled receptor proteins. The protein is not in itself toxic. The homologues contain 80-90 amino acids, of which 10 are cysteines with identical spacing. Various names have been given to these proteins, including *mamba intestinal protein 1, Bv8* (after *Bombina variegata*), *prokineticins*, and *endocrine-gland vascular endothelial growth factor*.

avitaminosis. *n.* See *hypovitaminosis*.

awn. *n.* See *terminal awn*.

axial. *adj.* Toward the axis.

axial bifurcation. Duplication of a body part along the body length. See also *dichotomy*.

axial mesoderm. The mesodermal tissue of an embryo that contributes to the notochord and somites.

axial skeleton. Reference to skeletal elements of the vertebral column and associated ribs.

axil, axilla, axille. *n.* The armpit or cavity beneath the insertion of an anterior appendage.

axilla-groin. *n.* Straight-line distance from the posterior margin of the forelimb insertion to the anterior margin of the hindlimb insertion. (Syn. *interlimb length*)

axillary. *adj., n.* 1. Of, relating to, or located near the axilla. 2. The cavity beneath the junction of a forelimb and the body. 3. The scute or scutes on the frontal border of the bridge in chelonians. (Syn. *axillary lamina, axillary plate*) See Fig. 6.

axillary amplexus. See *pectoral amplexus*.

axillary buttress. See *buttress*.

axillary embrace. See *pectoral amplexus*.

axillary gland. A glandular region on the chest of pelobatoid frogs of the genus *Megophrys*, located at the insertion of the forelimb.

axillary lamina. See *axillary*.

axillary notch. The notch in the anterior turtle shell that accommodates a front leg.

axillary plate. See *axillary*.

axillary pocket. Infolding of skin which forms a cavity behind the insertion of the forelimb in lizards. (Syn. *axillary pouch*)

axillary pouch. See *axillary pocket*.

axillary web. See *patagium*.

axillar wing. See *patagium*.

axiom. *n.* An established principle or self-evident truth.

axis. *n.* 1. A line through the center of a structure or body. 2. The second cervical vertebra of amniote vertebrates (sometimes the first neck vertebra of amphibians). 3. A fulcrum or point around which something turns.

Axolotl. *n.* A common name (from the Aztec) for the permanently aquatic, paedomorphic salamander *Ambystoma mexicanum* of the family Ambystomatidae found in mountain lakes of Mexico. See *Ambystomatidae*.

axon. *n.* The elongated, cylindrical process of a neuron that propagates the nerve impulse (action potential) away from the nerve cell body to the next level of information transfer (i.e., synapse with another neuron or effector). (Syn. *nerve fiber*)

axoneme. *n.* A shaft of microtubules that extends the length of a cilium, flagellum, or pseudopod of a cell.

axon terminal. The distal end of an *axon*. Typically, the site where signals are passed to another cell.

axoplasm. *n.* The cytoplasm within an axon.

AZA. See *Association of Zoos and Aquariums*.

Azemiopinae. *n.* A viperid subfamily that includes the southeast Asian viper *Azemiops feae*. This is a pitless viper seemingly more related to crotalines than to viperines. It is unusual in having colubridlike head scales, smooth body scales, no loreal pit organ, and no tracheal lung.

azotemia. *n.* An excess of urea or nitrogenous products in the blood.

azygous. *adj.* 1. Occurring singly and median in position. Frequently used in reference to plates on the head of lizards or snakes or bones of the skull table in some anurans. (Syn. *azygous plate, azygous scale*) 2. Unpaired. Used, for example, in nomenclature for blood vessels or in reference to unpaired scales on the midline of a structure and thus lacking a counterpart.

azygous plate. See *azygous*.

azygous scale. See *azygous*.

B

bachelor. *n.* An unmated male.

Bachias. *n.* Vernacular name for species of neotropical gymnophthalmid lizards belonging to the genus *Bachia*.

backbone. *n.* See *vertebral column*.

backcross. *n.* An individual or individuals resulting from *backcrossing*.

backcrossing. *n.* The process of crossing a hybrid with an individual having the same genetic complement as one of the hybrid's parental species.

backcross parent. The parent of a hybrid with which it is again crossed or with which it is repeatedly crossed. A backcross may also involve individuals having a genotype that is identical to the parent rather than the parent itself.

back-fanged. *adj.* See *rear-fanged*.

background radiation. Natural radiation comprised of cosmic radiation (from space) and terrestrial radiation (from decay of naturally occurring isotopes).

back mutation. A reverse mutation.

backswamp. *n.* An area of waterlogged land adjacent to a river.

bacteriocide. *n.* An agent that kills bacteria.

bacteriostasis. *n.* Inhibition of the growth of bacteria without killing them.

bacteriostat. *n.* An agent that retards the growth of bacteria, rather than killing them outright.

baculum. *n.* A bony rod reinforcing the penis of some varanid lizards.

badland. *n.* Reference to areas that are arid or semiarid with marked surface erosion and sparse vegetation.

bajada. *n.* A broad, gently sloping, depositional surface formed at the base of a mountain range by coalescence of individual alluvial fans in a dry region.

balance. *n.* In physiology, a concept related to equilibrium where two opposing forces are equal. The concept implies only that of an equilibrial state and does not specify a particular value, amount, or regulated level. Examples of usage are *water balance, thermal balance,* etc.

balanced polymorphism. See *polymorphism*.

balanced or **balancing selection.** Selection that favors a heterozygote and acts to maintain genetic variation and produce a *balanced polymorphism*.

balancer. *n.* Rodlike projections developing from the mandibular arch on each side of the head of larval, pond-dwelling salamanders. These structures produce a sticky, mucous secretion and assist the developing animal to keep from sinking and to maintain balance until the forelimbs develop (at which time the balancers degenerate). (Syn. *haltere, stabilizer*) See also *adhesive organ*.

Baldwin effect. The reinforcement or replacement of environmentally induced phenotypic characters by similar inherited characters under the influence of selection.

bale. *n.* A collective term for a group of turtles.

ball. *n.* 1. The rounded part of the centrum or body of a vertebra. 2. See *mating ball*.

ballast. *n.* 2. A structure that gives stability. 2. Reference to an object carried in water or air that gives stability or altitude.

ballistic tongue projection. The launch of a sticky tongue (salamanders, chameleons) like a projectile during prey capture, attributable to an elaborate tongue skeleton and specialized tongue muscles. When projected from the mouth, the tongue of plethodontid salamanders leaves the body completely and adheres to targeted insects, which are then reeled rapidly back into the mouth. Protraction and retraction of the tongue occurs within 20 ms. Plethodontid salamanders are the only known vertebrates with the ability to shoot part of the visceral skeleton entirely out of the body. Recent work demonstrates that the tongue projector muscles (*subarcualis rectus*, or SAR muscles) generate the largest force exerted by any vertebrate (S.M. Deban et al., *J. Exp. Biol.* 210:655–667, 2007).

ball position, balling, or **balling posture.** A defensive posture of some snakes, particularly boids, in which the body is coiled tightly into a ball and the head is protected. Typically the head is hidden within the ball, while the tip of the tail is protruded. The term should not be confused with mating aggregations that have traditionally been regarded as "balls." See *mating ball*.

Ball Python. Vernacular name for the stout West African boid *Python regius,* named for the defensive behavior of coiling into a tight spherical ball.

Bamboo Pit Viper. Vernacular name for the Asian species *Trimeresurus gramineus.* Cf. *Green Bamboo Viper*.

Bamboo Viper. See *Green Bamboo Viper*.

Banana Frogs. Collective vernacular name for African species of frogs belonging to the hyperoliid genus *Afrixalus,* and the microhylid genus *Hoplophryne*.

band. *n.* With reference to coloration, broad area of contrasting color that runs transverse to the vertebral axis and may or may not completely encircle the body. (Syn. *crossband, ring*)

Banded Geckos. Vernacular name for various American species of gekkonid lizards belonging to the genus *Coleonyx*.

Banded Sea Snakes. Vernacular name for several species of sea snakes belonging to the genera *Hydrophis* and *Laticauda*. See *Sea Kraits*.

Banded Skinks. Vernacular name for species of skinks belonging to the genus *Scincopus*.

Banded Water Snake. Vernacular name for the American natricine species *Nerodia fasciata fasciata*. See also *Southern Water Snake*.

banding. *n.* 1. Reference to presence of *bands* on an animal. 2. Reference to alternate light and dark transverse striations in red blood cells of salamanders, apparently as consequence of a corrugated surface.

Bandy-Bandy. *n.* Vernacular name for two species of elapid snakes belonging to the genus *Vermicella*, inhabiting many parts of Australia and characteristically colored with alternating black and white rings that usually encircle the entire body.

bar. *n.* 1. A ridge of alluvial deposit in shallow water, resulting from the actions of wind or water currents. 2. A derived *cgs* unit of pressure equal to 1×10^6 dynes cm^{-2}, or 10^5 newtons m^{-2}.

Barba Amarilla. Spanish name for a large and dangerous species of pit viper (*Bothrops asper*) responsible for a high incidence of snake bite throughout its range in Central and South America. (Syn. *Terciopelo, Fer de Lance, Yellow-jaw Tommygoff* in Belize)

barbel. *n.* A fleshy extension of skin (*filiform papilla*) usually on the head or neck and functioning as a tactile organ. Used in reference to sensory projections on the chin or throat of turtles, and cutaneous projections found in pipid and rhinophrynid larval amphibians. (Syn. *gular tentacle*)

Barking Treefrog. Vernacular name for the North American hylid species *Hyla gratiosa*.

baroreceptor. *n.* A sensory receptor that is stimulated by changes in pressure. Typically, this term refers to receptors in the walls of central blood vessels where they detect distension of the vessel wall due to increased blood pressure.

barrage lake. A lake formed by natural damming of a watercourse.

barred. *adj.* In reference to coloration, the presence of vertical markings of contrasting colors on the side of the body.

barren. *adj.* Devoid of vegetation, or of fossils.

barrier. *n.* In ecological meaning, any feature that restricts movement of individuals from one place to another, thereby preventing successful dispersal and establishment of a species.

barrier island. A ridge of sand extending above sea level parallel to a shoreline.

barrier reef. A reef separated from the shoreline by deeper water of a lagoon.

basal. *n.* 1. Basic or fundamental. 2. Generally, located at, or related to, the base or foundation. Arising from the base of a stem. 3. Reference to the earliest lineage of a group to have evolved. 4. A collective term for basidorsals and basiventrals (= *basalia*), two of the *arcualia*. Used by E.S. Goodrich (*Studies on the Structure and Development of Vertebrates,* p. 70, 1930).

basal area. The total area of ground covered by trees measured at breast height (1.4 m above ground), or the actual surface area of soil covered by plants measured close to the ground. (Syn. *basal cover, ground cover*)

basal cover. See *basal area*.

basal generative layer. See *stratum germinativum*.

basal group. A group that is outside a more derived clade.

basal hook. An enlarged spine (usually multiple) on the proximal portion of the hemipenis of snakes.

basal lamina. See *basement membrane*.

basal layer. See *stratum germinativum*.

basal lobe. Fleshy enlargement at the base of a hemipenis. Cf. *apical lobe*.

basal plate. 1. Portion of the chondrocranium lying between the foramen magnum and the hypophysis, usually in the form of a shallow bowl, which ossifies to form part of the base or floor of the skull. 2. The *copula* of the hyoid, as used by E.S. Goodrich (*Studies on the Structure and Development of Vertebrates,* p. 433, Fig. 471, 1930). 3. Motor portion of the brain stem, oriented horizontally and located ventral to the sulcus. Cf. *alar plate*.

basal tooth replacement. Tooth replacement in which the old tooth is pushed out by the new tooth directly from below.

basale. *n.* The bone in skull of caecilians formed by fusion of the occipital, otic, and sphenoid bones. (Used by T.S. Parsons and E.E. Williams (*Quar. Rev. Biol.* 38:26–53, 1963).

basement membrane. 1. Generally, a noncellular membrane that underlies most animal epithelia. At a level of electron microscope, the basement membrane is comprised of two structures with separate origins: a *basal lamina*, derived from epithelium, and a *reticular lamina*, derived from connective tissue. 2. A noncellular layer of collagenous fibers embedded in an amorphous ground substance, which secures the basal epidermal layer of skin to underlying connective tissue of the dermis. (Syn. *basal lamina, limiting membrane, limiting basal layer, limiting basal zone, lamina terminalis, membrana basilaris, membrana prima, membrana propria, membrana terminans*)

basibranchial. *n.* 1. The ventral or basal element of the visceral arch skeleton. In larval amphibians, this term has reference to any one of a series of median, unpaired elements of the visceral skeleton, some disappearing and other being incorporated into the copula or branchial plate at metamorphosis. 2. The *copula* of the *hyoid*, as both larval and adult structures. (Syn. *basihyal, copula*)

basicranial length. A measurement of the distance from the tip of snout to posterior edge of the occipital condyle in turtles. Used by L. Stejneger (*Bull. Mus. Comp. Zool.* 94:1–75, 1944) as a standard length measurement for comparing skulls of different sizes.

basicranium. *n.* The base of the skull.

basic. *adj.* Alkaline (high pH), or rich in alkaline minerals. Cf. *acidic*.

basic dye. Any of a number of organic cations that bind to, and stain, negatively charged macromolecules, such as nucleic acids.

basic type. The *primary type* of a taxon.

basidorsal. (*pl.* **-lia**) *n.* One of the pair of *arcualia* arising from posterior edge of a myotome, above the notochord. This becomes incorporated into developing vertebrae and forms the neural arch in amphibians.

basihyal. *n.* 1. The unpaired ventral component of the second visceral arch in developing embryos, eventually transforming to the *copula* of the adult hyoid. 2. The *copula* of the *hyoid*, as used by various authors. 3. The second component of the hyoid arch in *Bufo*, disappearing during early metamorphosis.

basihyal arch. The curve formed in the posterior margin of the *copula* of *hyoid* when the postero-lateral margins are elongated into arms.

basihyal shoe. The outwardly directed, flat-bottomed part of the *basihyal arch* in the lizard *Amphisbaena*. Used by W. Beebe (*Zoologica* 30:7–32, 1945).

basihyal valve. The lower or ventral of two fleshy folds in the posterior mouth of crocodilians. These act together to prevent water from entering the glottis when the mouth is opened beneath water. See *velum palatinum*.

basihyobranchial. *n.* See *copula*.

basihyoid. *n.* See *copula*.

basilar membrane. The delicate strand of elastic tissue bearing the auditory hair cells in the inner ear. In amphibians the hair cells are fixed to an apparently immobile cartilage shelf, and some authors do not recognize a basilar membrane in these taxa.

basilar papilla. See *papilla basilaris*.

Basilisk. *n.* 1. A legendary monster, part snake and part cock. 2. An American lizard of the genus *Basiliscus* (Corytophaninae), known for its ability to run bipedally across water.

basilingual plate. Synonym for *copula* in crocodilians.

basioccipital. *n.* An endochondral bone forming part of the floor of the cranium immediately behind the *basisphenoid* and bordering the foramen magnum below. This element forms part of the single *occipital condyle* in reptiles and is absent in modern amphibians.

basionym. (**basinym, basonym**) *n.* The original name of a taxon that is subsequently replaced by another, using the same stem, as a result of a change in rank or position.

basipterygoid joint. Moving joint of a kinetic skull which lies between the basipterygoid process and the pterygoid notch, permitting movement between the occipital and maxillary segments of the skull.

basipterygoid process. One of a pair of projections lying on the lateral surface of the basisphenoid bone. Together these make up the basal articulation with the palatal complex of the reptilian skull.

basisphenoid. *n.* A neurocranial bone forming part of the midventral floor of the brain case immediately anterior to the *basioccipital* and articulating laterally with the *pterygoids*.

basisphenoid rostrum. See *cultriform process*.

basiventral. *n.* One of the pair of *arcualia* arising from posterior part of a myotome below the notochord. It fuses with arcualia above it and in the anterior part of the next posterior body segment to form a vertebra. The *intercentrum* or *hypocentrum* is derived from the basiventrals.

bask. *v.,* **basking.** *n., adj.* A general term to describe an animal's behavior of positioning itself in relation to available microclimate so that it receives heat from an external source. The term for many users connotes absorption of solar radiation, but any heat source might legitimately be sought to elevate body temperature in context of this term.

basking range. Formerly used to denote the range of temperatures at which a reptile tends to remain inactive although alert while in direct sunlight. From R.B. Cowles and C.M. Bogert (*Bull. Amer. Mus. Nat. Hist.* 83:265–296, 1944).

basophilic. *adj.* Reference to any acidic compound that stains readily with basic dyes.

Bataguridae. *n.* A clade (family) of mostly freshwater aquatic or semi-aquatic turtles distributed in central and northern South America, Europe, northern Africa, the Middle East, and southern Asia. The greatest diversity is in southern Asia. This family is synonymized with *Geoemydidae*.

Batesian mimicry. A form of mimicry where a palatable mimic resembles an unpalatable model. Named after Henry W. Bates. Cf. *Mertensian mimicry*.

bathyal. *n.* See *bathyl*.

bathybenthic. *adj.* Reference to the depth zone of the ocean floor between 200 and 4000 m.

bathyl. *adj.* Reference to the upper zone of ocean water below the permanent thermocline, extending from about 100–300 meters to 1000–3000 meters, depending on latitude. This water overlies the *abyssal* zone of water. (Syn. *bathyal, bathybic*)

bathybic. *n.* See *bathyl*.

bathypelagic. *adj.* The depth zone of the oceanic water column below the level of light penetration, extending from 1000 to 2500 m.

Batrachia. *n.* A clade within Lissamphibia that includes an-

batrachian. cestral anurans and urodeles and their descendants, a sister taxon of Gymnophiona.

batrachian. *n.* An archaic term for any amphibian, especially a frog or toad (of the clade Batrachia).

Batrachomorpha. *n.* A major stem group of tetrapods that includes the nonamniote *temnospondyls*, which may have given rise to some of the living amphibians. Cf. *Reptilomorpha*.

Batrachophrynidae. *n.* A clade (family) of anurans, sister with *Myobatrachoidea* within the *Australobatrachia*, and composed of *Batrachophrynus, Caudiverbera,* and *Telmatobufo* genera (D.R. Frost et al., *Bull. Amer. Mus. Nat. Hist.* 297:1–370, 2006).

batrachotoxins. *n.* Toxic steroidal alkaloids known only from the skin of frogs of the genus *Phyllobates* and the feathers of one species of New Guinea bird. These substances depolarize membranes by increasing permeability to sodium ions, lead to irreversible depolarization of nerve and muscle cells, uncontrolled muscle spasms, depletion of neurotransmitter, and they act extremely fast. Batrachotoxins are also extremely potent cardiotoxins, causing cardiac arrhythmias, ventricular fibrillation and eventually cardiac arrest. These toxins are among the more powerful animal toxins known and are 250 times more potent than strychnine.

bauchstuck. n. Part of an amphibian egg that remains when all of the organizer area has been removed. During early cleavage, the dorsal half of the developing embryo is capable of neural axis formation and development to a full embryo, whereas the ventral half only gives rise to a 'belly-piece' or *bauchstuck*, not capable of neural axis formation.

Bauplan. (*pl. Baupläne*) *n.* German for building plan or blueprint. 1. An accepted German equivalent for *body plan*, widely used and coined in a biological sense in 1945 by the embryologist and philosopher J.H. Woodger. Reference is to a biological plan built on a common arrangement or similarities in biological architecture. The generalized or idealized archetypal body plan of a major taxon. 2. Reference to common and homologous properties of the members of a systematic taxon or group.

baymouth bar. A ridge of sediment that separates a bay from the ocean.

Bayesian analysis or **method.** A statistical method characterized by its incorporation of prior knowledge to determine the probability that a hypothesis is true. In systematics, the method is used to calculate the probability that a given unknown organism belongs to a specified taxon. It selects the ancestral trait value with the highest *posterior* probability, given the probabilities of "priors" (external evidence) and assumptions of trait evolution (defined by the user). The method was developed by Thomas Bayes, a Presbyterian minister, in the mid-18th century.

Bd. See *chytrid fungus*.

beach. *n.* A strip of sediment, usually sand, that extends from a low-water line inland to a cliff or zone of permanent vegetation.

beach face. The section of beach that is exposed to wave action.

bead. *n.* The narrow, outwardly directed curve on the lateral anterior edge of a segment of a rattlesnake rattle. Functions to prevent splitting or tearing and to produce noise when it strikes against the shoulder of the next anterior segment during rattling.

Beaded Lizard. Common name for either of two species of *Helodermatidae*, characterized by stout, cylindrical, and heavy bodies, nodular non-overlapping scales, short limbs, relatively thick tail, blunt head, and modified salivary glands that secrete venom from the lower jaws. See also *Gila Monster, Mexican Beaded Lizard*.

beak. *n.* 1. The horney covering of the jaws, also called *rhamphotheca* or *tomium* in turtles. 2. The ventral recurved extension of the premaxillary bones in the *Rhynchocephalia*. 3. The horny mouthparts of tadpoles. See *jaw sheath*.

Beaked Caecilians. Collective vernacular name for species of caecilians belonging to the family Rhinatrematidae, and especially the genus *Epicrionops*.

Beaked Sea Snakes. Vernacular name for elapid sea snake species belonging to the genus *Enhydrina*.

Beaked Snakes. Vernacular name for species of typhlopid snakes belonging to the genus *Rhinotyphlops*.

Beaked Toads. Collective vernacular name for species of South American bufonid anurans belonging to the genus *Rhamphophryne*.

beaker cell. See *flask cell*.

beakshield. *n.* See *rostral*.

Bearded Dragons. Vernacular name for several species of Australian agamid lizards belonging to the genus *Pogona*, especially *P. barbata*. The "beard" consists of a fringe of soft spines around the neck, which are displayed during defensive encounters.

Beauty Snakes. Vernacular name for species of snakes belonging to the genus *Orthriophis* (previously *Elaphe*).

bed load. In geology, reference to heavy or large particles of sediment that travel near or on the bed of a stream.

bed rock. In geology, reference to solid rock that underlies soil.

Beer-Lambert Law. The principle that absorption of light by a solution is a function of the concentration of the solute. It is a commonly used principle in photometry.

behavioral color change. Reference to exposure of colors by rapid postural adjustments or movement of scales to reveal surfaces differing in color from those currently in view.

behavioral fever. In *ectotherms*, the behavioral elevation of body temperature (achieved by increased basking) above normally selected levels due to disease. The response reflects a change in thermoregulatory set point induced by *exogenous pyrogens* (endotoxins produced by gram-negative bacteria) or, secondarily, *endogenous pyrogens* (heat-labile proteins released from tissues in response to circulating exogenous pyrogens). The resulting elevation of body temperature confers protection against bacterial infection in the afflicted animal.

behavioral isolation. A premating isolating mechanism wherein two species do not mate because of differences in courtship or mating behavior. (Syn. *ethological isolation*)

behavioral thermoregulation. Regulation of body temperature principally by means of behaviors that regulate the magnitude and direction of net heat flux between an animal and its environment. Characteristic of ectotherms, relevant behaviors involve shuttling movements to or away from a heat source as well as secondary postural adjustments.

bell. *n.* An individual segment of a Rattlesnake's rattle.

Bell Frog. Vernacular name for the Australian hylid frog, *Litoria aurea*, also called the Green and Golden Bell Frog.

bell gill. The enlarged gill of certain larval hylid frogs which carry their larvae in dorsal brood pouches. These structures, found in the genera *Cryptobatrachus, Hemiphractus,* and *Gastrotheca,* function to facilitate respiratory gas exchange.

belly flap. An outfolding of lateral, posterior, or both parts of the body wall to form a flap that is probably used for attachment by *semiterrestrial tadpoles*. See *belly sucker*.

belly gland. An aggregation of single-celled epidermal glands on midventral skin of males in most species of the frog genus *Kaloula*. This structure appears to function as an aid to amplexus, producing secretions which "stick" the male and female together (R.F. Inger, *Fieldiana Zool.* 33:185–531, 1954).

belly patch. A term sometimes used to describe a ventral area of color contrasting strongly with an animal's dorsal ground color. Cf. *pelvic patch*.

belly sucker. A modified abdominal region of a *gastromyzophorous tadpole*, including an elevated rim and musculature to raise the roof of the sucker to form negative pressure. (Syn. *abdominal sucker*; see also *belly gland, belly flap*) For discussion of terminologies, see R.W. McDiarmid & R. Altig, *Tadpoles. The Biology of Anuran Larvae*, 1999.

belt transect. A method for sampling organisms in a given habitat, based on encounters of individuals and species within a "belt" marked out across the habitat, usually 1 m wide. Cf. *line transect*.

bending. *n.* Deformation of an object due to opposing parallel forces. See Fig. 24.

bends. *n.* See *caisson disease*.

benign. *adj.* In veterinary medicine, nonaggressive or not of immediate threat. Cf. *malignant*.

benthic. *adj., n.* 1. Pertaining to the bottom under a body of water. Bottom-dwelling. 2. An *ecomorphological guild* of tadpoles, including lentic or lotic forms that rasp food from submerged surfaces using keratinized mouthparts, mostly at or near the bottom of pools and backwater sites.

Bergmann's rule. A "bioclimatic rule," originally formulated with reference to endotherms, currently defined as an intraspecific tendency for increasing body size with increasing latitude or decreasing environmental temperature (see also *Allen's rule, Cope's rule,* and *Gloger's rule*). Several authors have claimed that ectothermic vertebrates follow Bergmann's rule, whereas others have contested this concept and claim the converse is true. Recently, K.G. Ashton and C.R. Feldman concluded that chelonians follow Bergmann's rule, whereas squamates follow the converse of it (*Evolution* 57:1151–1163, 2005).

Bernoulli effect. The principle that pressure in a fluid decreases as velocity increases.

β-adrenergic receptor. A receptor on cell surfaces that binds *epinephrine* and *norepinephrine* equally well, normally leading to the activation of adenylate cyclase. Cf. *α-adrenergic receptor*.

β-Bungarotoxin. (β-Bgt) *n.* One of the more intensely investigated *presynaptic paralyzing toxins* isolated from the venom of Taiwan Banded Krait, *Bungarus multicinctus*.

Betadine. *n.* A topical antiseptic used to swab minor cuts or surgical sites on humans and animals, popular for a variety of veterinary and research applications in herpetology. It is a strong, broad-spectrum antibiotic available as a solution or as swabsticks. Most over-the-counter solutions are 10% providone-iodine. Betadine® brand first aid antibiotics combine Polymyxin B sufate and Bacitracin zinc with moisturizers to control infection and promote healthy healing of damaged human skin.

beta diversity. (β-diversity) A measure of the rate and extent of change in species along a gradient from one habitat to another. Cf. *alpha diversity, gamma diversity*.

β-gland. *n.* A type of *generation gland* in which glandular material derives from the outermost (first-formed) elements of an *epidermal generation* (the *Oberhautchen* and/or the *β-layer*) on posterior abdomen of several lizard taxa. The term was first applied by P.F.A. Maderson in relation to specialized structures in *Lygodactylus* (*Breviora* 228:1–35, 1968). Cf. *escutcheon scale*.

β-keratin, β-keratin layer, β-cells. *n.* See *keratin*.

β-neurotoxin. Presynaptic blocking toxin components of snake venoms.

beta taxonomy. The arrangement of species into hierarchical systems of higher categories or taxa. Identification of natural groups and phylogenetic reconstruction. (Syn. *macrotaxonomy*) Cf. *alpha taxonomy, gamma taxonomy*.

β-toxins. See *presynaptic paralyzing toxins*.

bezoar. *n.* A hard mass formed in the stomach by compaction that does not pass into the intestine. Formerly, these were used in the treatment of snake bite.

BHS. See *British Herpetological Society*.

bi-. Prefix meaning "two" or "double."

bias. *n.* A systematic error in sampling that is inherent in the sampling technique.

bibliographic reference. See *reference*.

bibliography. *n.* A listing of references to a given subject, commonly used in publications as a synonym for *references* or *literature cited*.

Bibliotheca Herpetologica. A journal of the history and bibliography of herpetology, published by the International Society for the History and Bibliography of Herpetology (ISHBH), containing articles, essays, bibliographies, and news of people and events in the herpetological community.

bicarinate. *adj.* Having two keels. See *carinate*.

bicentric. *adj.* Having two centers of distribution or evolution.

bicephalous. *adj.* Having two heads.

bicipital. *adj.* Having two heads.

bicipital ribs. Ribs having two heads that articulated with the vertebrae, characteristic of primitive tetrapods.

bicolor. *n.* (*adj*., **-ed**) A term meaning "two color" and usually relating to alternating ring or band colors of snakes, or to a lateral view of tadpole tails in which the muscle is dark dorsally and pale ventrally.

biconodont. *adj.* Reference to teeth having two cusps. (Syn. *bicuspid*)

bicuspid. *adj.* 1. Of or pertaining to having two toothlike projections. 2. Reference to the paired terminal toothlike projections on the jaw surfaces of some turtles.

Bidder's canal or **duct**. A longitudinal duct running parallel with the inner margin of the kidney and associated with the *ductuli efferentia* in amphibians, fusing together in some species.

Bidder's organ. The upper cortical lobe of the progonad in the Bufonidae, which retains its ovarial nature in males even though the rest of the gonad differentiates into a testis. This rudimentary ovarian tissue develops on the anterior end of the larval testes, and is considered to be a *paedomorphic* trait in many adult bufonids. Absence of this structure is considered to be primitive, and its presence in bufonids is derived according to W.E. Duellman and L. Trueb (*Biology of Amphibians*, p. 472, 1986). (Syn. *cortical lobe*)

biennial. *adj.* Lasting for two years, or occurring on a cycle every two years. Cf. *annual, perrenial*.

biennial breeding or **reproductive cycle**. Reproduction every two years, generally in reference to females of various snake species in temperate climates.

bifid. *adj.* Divided into two lobes or parts by a median cleft.

biflagellate spermatozoa. Spermatozoa having two free flagella

bifurcate, bifurcated. *adj.* Divided into two branches or projections, as the tongue of snakes and some lizards.

big bang reproduction. See *semelparity*.

Bighead Turtle. Vernacular name for the Asian chelydrid *Platysternon megacephalum*.

bilateral. *adj.* Pertaining to two sides.

bilateral symmetry. A form of symmetry in which the body can be divided into two equal parts along a longitudinal plane. The two parts are mirror images of each other. This is characteristic of most vertebrates.

bile. *n.* A secretion of the liver that is stored in the *gall bladder* prior to release into the upper small intestine where it acts to emulsify fats preparatory to their digestion and absorption.

bile salts. Acids in bile that promote emulsification and solubilization of intestinal fats.

biliary. *adj.* Pertaining to the bile or bile duct system.

bilobed. *adj.* Divided into two lobes.

bimodal. *adj.* A state of having two modes or peaks in a frequency distribution. See *bimodal distribution*. Cf. *unimodal, polymodal*.

bimodal distribution or **population**. A population or set of values in which the measurements of a given character are clustered around two values (modes or peaks).

bimodal foraging. Having two modes of foraging.

binary character. A two-state, all-or-none character, recorded as present or absent.

binary name. Binomial name. See *binomial nomenclature*.

binding site. The specific region of an enzyme that binds to the substrate.

binocellate, biocellate. *adj.* Having two eyespots, as the rear of certain leptodactylid frogs, used as a defensive display.

binocular parallax. See *parallax*.

binocular vision. Overlapping visual fields involving two eyes, such that images are focused on both of two retinas simultaneously, thereby aiding perception of depth and distance. Such vision is well developed in certain species of arboreal snakes capable of judging distances and slight movements during prey capture.

binomen. *n.* In systematics, the complete binomial name of a species. See *binomial nomenclature*.

binomial distribution. A probability function expressing the probabilities that an event will or will not occur, n, $n-1$, $n-2$ ..., 0 times are given by the successive coefficients in the binomial expansion $(a + b)^n$. Because a and b are the probabilities of occurrence and nonoccurrence, respectively, their sum equals 1.

binomial nomenclature. The current system (developed by C. Linnaeus or Karl von Linné) used for assigning organisms two Latin names, one designating the genus and the second indicating species (Syn. *binomen, binomial name, binominal name, binary name*).

binominal. *adj.* Reference to binomial nomenclature.

-bio-. Combining form meaning "life" or "living."

bioassay. *n.* Determination of the effect of a compound or substance using living organisms or tissues, often with comparison to some standard.

biocellate. *adj.* See *binocellate.*

biochemical (or **biological**) **oxygen demand.** (**BOD**) The amount of oxygen required to degrade organic material and oxidize reduced substances in a water sample. Used as a measure of oxygen requirement of bacterial populations and as an index of water pollution.

biochronology. *n.* The relative dating of rocks and geological events by the use of fossil evidence (biostratigraphic data).

bioclast. *n.* A single fossil fragment.

bioclimate. *n.* See *microclimate.*

bioclimatology. *n.* The study of the manner in which climate relates to and affects the activities and characteristics of organisms. Cf. *ecoclimatology.*

biocoenosis. *n.* A community or natural assemblage of organisms, excluding the physical aspects of the environment.

bioconcentration. n. Reference to the phenomenon whereby a compound present in an aquatic environment is accumulated in the biomass of organisms living in that environment, calculated usually as the ratio of the concentration of a compound in an organism to its concentration in the water. The concept also extends to accumulation of substances in food chains.

bioconversion. *n.* Conversion of metabolic substrate to cellular biomass.

biodemography. *n.* The integrated study of ecology and genetics of populations.

biodiversity. *n.* A term coined to substitute for *biological diversity*, thus denoting the variety of living organisms (species diversity) and, connotatively, their genetic diversity, ecological diversity, etc. The total variety of life and its processes. See also *species diversity.*

biodiversity hotspot. An area noted for its high degree of diversity and endemism (N. Myers et al., *Nature* 403:853–858, 2000).

bioenergetics. *n.* Ecological energetics, or the study of energy flow through ecosystems.

biogenetic law. The now discredited, earlier idea of Ernst Haeckel that ontogeny recapitulates phylogeny.

biogenic. *adj.* 1. Resulting from the actions of living organisms. 2. Necessary for life and living processes.

biogenic amines. Basic compounds related by decarboxylation to parent amino acids, having important functions as neurotransmitters (e.g. epinephrine, norepinephrine, serotonin, dopamine) and present in secretions of amphibian skin.

biogeochemical cycle. A cyclic system in which a given chemical element (e.g., carbon or nitrogen) is transferred between biotic and abiotic components of the biosphere.

biogeochemistry. *n.* The study of the interrelationships between organisms and the chemical features of the Earth.

biogeographic (-al) realms. Reference to divisions of the world land masses according to their distinctive biotas. See Australian, Ethiopian, Nearctic, Neotropical, Oriental, Palearctic.

biogeographical realm or **region.** Reference to any geographic region characterized by a distinctive biota, the terminology originating from a biogeographical and ecological land classification system first formally proposed by M. Udvardy (*A classification of the biogeographical provinces of the world. IUCN Occ. Paper* No. 18, 1975). There are eight such biogeographical regions (or *ecozones*): *Nearctic, Palearctic, Afrotropic, Indomalaya, Australasia, Neotropic, Oceania,* and *Antactic.* (Syn. *ecozone*)

biogeography. *n.* Science dealing with the description and interpretation of broad-scale geographical distributions of organisms. See *dispersal biogeography* and *vicariance biogeography.*

bioindicator. *n.* 1. Any measurement of a contaminant accumulating in body tissues of organisms exposed to contaminants, indicating potential environmental perturbations due to contamination. An integrator of contaminant loads on a system. Bioindicators (alternatively *biomarkers, biomonitors, sentinel organisms*) have been used commonly for detection of toxicants such as heavy metals or PCBs. 2. An organism, species, population, or biological process whose change in numbers, function, or status is used as an indicator of changes in the integrity or quality of the environment or ecosystem. A bioindicator is an anthropogenically-induced response in organisms at various levels of organization (molecular, physiological, etc.) that is causally linked to effects of the environment. Cf. *biomarker.*

bioinformatics. *n.* A broad interdisciplinary science that includes both conceptual and practical tools for understanding the generation and processing of biological information. (Syn. *biological computing*)

biological. *adj.* Pertaining to living organisms or living processes.

biological amplification. A process in which retained substances become more concentrated in biological tissues with each successive level in a food chain. (Syn. *biological magnification, biomagnification*)

biological clock. Refers to internal (*endogenous*) mechanisms generating rhythmic behaviors or functions synchronized with, or entrained by, environmental conditions. The term refers to any mechanism that allows expression of specific genes at periodic intervals, or to any aspect of physiology that regulates body rhythms.

biological computing. See *bioinformatics.*

biological control. The use and planned manipulation of living organisms for the purpose of controlling pest populations.

biological diversity. See *biodiversity*.

biological efficiency. The ratios of productivity of an organism or population to the gross energy consumed.

biological evolution. See *evolution*.

biological fitness. See *fitness*.

biological magnification. See *biological amplification*.

biological oxygen demand. See *biochemical oxygen demand*.

biological races. Sympatric populations of a species which are similar morphologically but which do not interbreed due to other biological differences such as behavior cycles.

biological role. The actions or use of a structure or part of an organism in contexts of adaptation, behavior, and environment. Cf. *function*.

biological rhythm. Reference to regular periodicity exhibited by any biological process.

biological species. One or more populations of similar organisms that have the potential to interbreed but are genetically isolated due to reproductive isolating mechanisms. See *species*.

biological species concept. See *species*.

biological stress response. See *stress response*.

biology. *n.* The study of living organisms, their parts, and systems.

bioluminescence. *n.* Light produced by living organisms, and the emission of light of biological origin. There has been a long herpetological debate concerning a so-called luminous lizard, *Proctoporus shrevei* of Trinidad, based on natural history accounts from the 1930s suggesting that males of this species are capable of producing light. Recent investigations indicate that scales of this lizard are highly reflective, but there is no evidence for bioluminescence (C.M. Knight, W.H.N. Gutzke, & V.C. Quesnel, *Caribbean J. Sci.* 40:422–426, 2004).

biomagnification. *n.* See *biological amplification*.

biomarker. *n.* 1. Any of numerous biochemical and/or physiological changes in organisms exposed to contaminants, representing initial responses to environmental perturbations and contamination. Biomarkers may be general or specific, but are generally regarded as more sensitive than *bioindicators* at higher levels of biological hierarchy. They offer more complete and biologically more relevant information on the potential impact of contaminants on the health of organisms. Cf. *bioindicator*. 2. In biochemistry and medicine, a characteristic or specific physical or chemical trait that is objectively measured and evaluated as an indicator of normal biologic or pathogenic processes, condition, or pharmacological responses to a therapeutic intervention.

biomass. *n.* The total mass of organic matter in a particular sample, population, region, trophic level, etc., measured as mass, volume, or energy.

biome. *n.* A category of ecological community or ecosystem usually classified by dominant vegetation and characterized by adaptations of organisms to that particular environment extending over a large area. (E.g., savanna, tundra, boreal forest, etc.). Generally, biomes are subdivisions of *ecozones* or *biogeographical regions*.

biomechanical couple. Reference to two parallel and equal forces acting on an object in opposite directions, thereby tending to cause rotation.

biomechanics. *n.* Study of the movements of plants or animals in relation to physical principles and mechanical laws. Application of the principles of mechanics to the structure and function of organisms.

biometeorology. *n.* Study of the effects of weather or atmospheric conditions on living organisms.

biometrics. *n.* See *biometry*.

biometry. *n.* The application of statistics to biological problems; biological statistics. (Syn. *biometrics*)

biomineralization. *n.* A process in which animal or plant material beomes converted to mineral material.

biomonitoring. *n.* The use of living organisms to monitor conditions of the environment. See *bioindicator*.

biophilia. *n.* A natural affinity or bond of humans for life and living things, espoused by E.O. Wilson (*Biophilia*, 1984). Wilson finds biophilia in our archetypal fascination with snakes or serpents.

biophysical ecology. A subdiscipline of ecology in which investigators seek to define the physical boundaries within which organisms operate. The origins of the field date to early studies of reflectivity and energy exchange of the leaves of plants during the 1950s; interest in applications to animal ecology were stimulated largely by W.P. Porter and D.M. Gates (*Ecol. Monogr.* 39:245–270, 1969).

biophysics. *n.* The application of physics to studies of living organisms or systems.

biopsy. *n.* The removal for examination of tissue from a living body.

bioremediation. *n.* A process that uses microorganisms to degrade and detoxify chemically contaminated soil or water.

biorhythm. *n.* A recurring event or process in the functioning of organisms, usually at regular periodic intervals. E.g., daily cycle of sleep or activity, reproductive behaviors, etc. See also *circadian rhythm*.

biosocial aggregation. A general term that describes a group of organisms that are congregated due to some social interaction.

biospace. *n.* A term for a *realized niche*. See *ecological niche*.

biosphere. *n.* The sum of all the planet's living communities and ecosystems, equaling the entire part of Earth inhabited by life; the global ecosystem.

biosphere reserve. Any one of numerous sites having conservation, ecological, or biodiversity importance and designated by UNESCO in effort to establish a system of internationally protected areas.

biostratigraphy. *n.* Study of the distribution of fossils in distinct layers of rock.

biosynthesis. *n.* The production of chemical compounds by a living organism.

biosystematics. *n.* See *systematics.*

biosystematic species. See *species.*

biota. *n.* A general term for all the organisms of all species living in a given area or region.

biotelemetry. *n.* Study of the location, movements, behavior, and physiology of organisms using remote detection and transmission equipment. (Syn. *radiotelemetry, radio tracking*)

biotic. *adj.* Reference to living organisms in the environment. Cf. *abiotic.*

biotic community. A biological community or association.

biotic factors. Attributes of the environment that result from activities of living organisms.

biotic potential. See *reproductive potential.*

biotic province. A region inhabited by a characteristic set of taxa (species, families, orders), bounded by barriers that prevent the spread of the distinctive kinds of life to other regions and the immigration of foreign species.

biotope. *n.* A specific ecological area where an animal or population lives. (E.g., sand dune, coral reef, etc.) The smallest geographical unit of the biosphere that can be delimited and characterized by a distinct biota.

bioturbation. *n.* The disturbance of the soil surface or subsurface by living organisms, e.g. as in burrowing by animals or extension of plant roots.

biotype. *n.* 1. A subset of a species or population that resemble, but differ physiologically from, other members of the species or population. 2. A group of genetically identical individuals.

biparental. *adj.* Having two parents.

biparous. *adj.* (*n.* –**ity**) Reference to production of two individuals at one birth. See *parity.* Cf. *uniparous, multiparous.*

bipedal. *adj.* Reference to condition or locomotion involving two feet (E.g., said of lizards that run with use of the two hind legs while the forelimbs are elevated and held off the ground).

bipedalism, bipedality. *n.* The habit or ability of walking on two legs.

Bipedidae. *n.* Family of amphisbaenians consisting of the Mexican genus *Bipes*, which have retained only forelimbs.

bird. *n.* Any member of the class *Aves.*

Bird-voiced Treefrog. Vernacular name for the North American hylid species *Hyla avivoca.*

Bird Snakes. Vernacular name for neotropical species of snakes belonging to the genus *Pseustes.* (Also known as *Puffing Snakes.*)

birefringence. *n.* See *anisotropy.*

birth. *n.* 1. The release of some form of an immature individual from the reproductive tract of its mother, including viviparous and ovoviviparous species. 2. Also used in reference to the analogous release of a post-embryonic individual from nonoviductal sites in a parent's body (e.g., various amphibians such as *Gastrotheca, Rheobatrachus,* and *Rhinoderma*). Cf. *hatch.*

birth plate. The lamina or scale of turtles that is present on the shell during the period of hatching. This structure is either lost shortly after hatching or persists as a smooth or granulated area surrounded by growth rings.

birth rate. (B) The change in the number of births in a population measured over a short time interval (units are births/time). See also *instantaneous birth rate.*

biserial. *adj.* Describes structures that occur in two series at a site, e.g. two rows of labial teeth per tooth ridge in tadpoles. Cf. *multiserial, uniserial.*

bisexual. *adj.* 1. Reference to a species having individuals of both sexes. 2. Reference to an individual animal that has both ovaries and testes (or both stamens and pistils in a plant).

bisexual reproduction. Reproduction in which each offspring receives half its genome from its mother and half from its father, both being members of the same species.

bisexual species. A species that consists of both male and female individuals.

bit. *n.* The fundamental quantitative unit of information used in digital computers, giving the solution to a binary choice.

bite force. The force generated during occlusion or biting of the jaws. Measurements are made using force transducers or strain gauges and have provided measures of biomechanical performance of the jaws. Bite force measured in *Alligator mississippiensis* is the highest measured for any living taxon and exceeds that estimated for some theropod dinosaurs (>9000 N). See G.M. Erickson et al., *J. Zool., Lond.* 260:317–327, 2003.

bitistatin. *n.* A *disintegrin* from the venom of *Bitis arietans.*

bitmap. *n.* A means of storing or processing monochrome images or graphics by converting image densities to densities of dots.

bitypic. *adj.* Reference to a taxon comprised of only two immediately subordinate principal taxa. Cf. *monotypic, polytypic.*

bivalent. *adj., n.* A pair of homologous, synapsed chromosomes during the first meiotic division.

bivariate. *adj.* Descriptive of a condition of a variable that occurs simultaneously with another variable, usually depicted in a graphic plot of the two variables.

bivariate analysis. The simultaneous analysis of two variables. See *bivariate.*

Black and Yellow Sea Snake. See *Yellow-bellied Sea Snake.*

blackbody. *n., adj.* Any body or object with a surface *emissiv-*

ity of 1, absorbing 100% of *thermal radiation* striking it and emitting all of the thermal radiation it can. Most physical objects and animals have surface emissivities less than 1 and do not have strictly blackbody surface properties, although the latter can provide a useful reference for comparisons. Cf. *graybody.*

black box. A conceptual entity whose function can be evaluated without specifying the content.

Black Caiman. Vernacular name for the crocodilian *Melanosuchus niger.*

Black Cobra. See *Desert Blacksnake.*

blackening hormone. Older term for *intermedin, melanocyte stimulating hormone* (MSH), or *melanotropin* produced in the pars intermedia of the pituitary.

Black-headed Python. Vernacular name for the north Australian boid species *Aspidites melanocephalus.*

Black-headed Snakes. Collective vernacular name for various colubrid snakes belonging to the Asian genus *Sibynophis* and American genus *Tantilla.* Various species of the latter are also called *Centipede Snakes* or *Crowned Snakes.*

Black-necked Garter Snake. Vernacular name for the American colubrid species *Thamnophis cyrtopsis.*

Blacksnake. *n.* Collective vernacular name for various species of snake belonging to several taxa including the colubrid genus *Coluber,* elapid genera *Pseudechis* and *Walterinnesia,* and the atractaspidid genus *Macrelaps.*

black stone. A fragment of bone toasted or burnt by craftsmen and used to extract venom from snake bite victims. The black stone originated in India and became popularized as a Far Eastern custom in Europe and elsewhere. The remedy has strong symbolism and is still traded and used in developing countries. The method is a traditional therapy but not recommended for serious treatment of snake bite. Originally both the black stone and Chinese musk would have come from the head of a venomous snake and placed on the wound or site of snake bite.

Black-striped Snakes. Vernacular name for the American dipsadine species *Coniophanes imperialis.*

Blacktail Rattlesnake. Vernacular name for the American pitviper *Crotalus molossus.*

Black Toad. Vernacular name for the American bufonid species *Anaxyrus* (formerly *Bufo*) *exsul.* For recent taxonomic revision see D.R. Frost et al., *Bull. Amer. Mus. Nat. Hist.* 297:1–370, 2006.

blackwater. *n.* Reference to water that is rich in humic acids and low in nutrient content.

bladder. *n.* See *urinary bladder.*

Blanding's Turtle. Vernacular name for a well-known species of emydid turtle, *Emydoidea blandingii.*

BLAST. (Basic Local Alignment Search Tool) An algorithm that is widely employed for determining similarity between nucleic acid or protein sequences, using sequences that are present in data bases.

blastema. *n.* A small protuberance consisting of competent cells from which an animal organ or appendage begins to regenerate.

blastocoel. *n.* The fluid-filled cavity of the blastula that forms in the embryo after the morula stage.

blastocyst. *n.* A newly fertilized oocyte.

blastodisc. *n.* The cap or disk of embryonic cells resting on the uncleaved mass of yolk at the animal pole of an egg. Typically produced by meroblastic cleavage.

blastomere. *n.* One of the embryonic cells produced during the first few cleavages of the fertilized ovum.

blastoporal lip. The dorsal rim of the amphibian blastopore, which functions as the organizer to induce formation of the neural tube.

blastoporal pigment line. The first visible indication on the surface of the amphibian embryo that *gastrulation* is underway, formed as a result of the apical constriction of *bottle cells,* thereby concentrating pigment granules near the apex of each of the bottle cells and creating a crescent of pigment.

blastopore. *n.* The opening into the gastrocoel or primitive gut formed at gastrulation.

blastula. *n.* An early embryonic stage in animals near the end of *cleavage* but before *gastrulation,* consisting of a hollow sphere and comprised of one tissue layer of several hundred cells.

bleb. *n.* A small blister containing blood or serum, sometimes found on the ventral scales of snakes that are housed in overly moist conditions.

blepharitis. *n.* Inflammation of the eyelids.

Blind Lizards. Common name for various species of Asian and Mexican lizards belonging to the *Dibamidae.*

blindsack. *n.* Term applied by T.S. Parsons (*Bull. Mus. Comp. Zool.* 120:103–277, 1959) to the *prechoanal sac* or *sinus lateralis nasi,* applied to amphibians to avoid calling the structure Jacobson's organ.

Blind Salamanders. Vernacular name for species of salamanders belonging to the plethodontid genera *Haideotriton* and *Typhlomolge.*

Blind Skinks. Vernacular name for southeast Asian lizards belonging to the dibamid genus *Dibamus* (not true skinks).

Blind Snakes. *n.* Collective common name for various species of snakes belonging to the Scolecophidia, and especially Typhlopidae. All are burrowing snakes having very rigid skulls adapted for pushing through soil, slender body shape, cycloid scales, and eyes that are either reduced or absent. Also called *Thread Snakes* or *Worm Snakes.*

block fiber of Gaskell. Specialized muscle fibers at the sinoatrial and atrioventricular junctions of reptilian heart. Characterized by a rich sarcoplasmic content, these fibers evidently function to delay conduction of contractile stimuli from one chamber to the next and thereby enhancing filling prior to contraction.

blood. *n.* The liquid that moves through vessels of the circulatory (or transport) system of animals. It consists of both a liquid (plasma) and cellular fraction, the latter containing red blood cells that transport oxygen, smaller amounts of CO_2, and play an important role in buffering respiratory acid.

blood clotting or **coagulation.** A cascade of enzyme-mediated reactions in blood plasma that produces strands of fibrin to stop bleeding from injured vessels.

blood plasma. The liquid portion of the blood, containing water and solutes but not the cell fraction of whole blood. (Syn. *plasma*)

Blood Python. Vernacular name for the south Asian boid species *Python curtus*.

Bloodsuckers. *n.* Vernacular name for various Asian species of agamid lizards belonging to the genera *Bronchocela* (Common Bloodsuckers), *Calotes,* and *Pseudocalotes* (False Bloodsuckers).

blood vascular system. The cardiovascular system, including the heart, vessels, and blood contained therein. See *cardiovascular system.*

blood volume. The quantity of blood contained in the circulatory systsem of an animal at any point in time.

blood volume deficit. The volume of blood that is lost from the circulation, often expressed as a percentage of the original blood volume.

bloom. *n.* A term applied by C. McCann (*Dom. Mus. Bull.,* 17:1–127, 1955) to describe *iridescence* of the scales of Australian geckos. Usage not recommended.

blotch. *n.* An area of color differing from the ground color, usually dark, rounded, square, or diamond (saddle) and often having a contrasting border.

blotting. *n.* Transfer of DNA, RNA, or proteins from a two-dimensional separation medium to another surface of higher affinity. See *northern blotting, Southern blotting, western blotting.*

Bluetongue Skink. Vernacular name for species of Australian scincid lizards belonging to the genus *Tiliqua*.

Blunthead Slug Snakes. See *Slugsnakes.*

Blunthead or **Blunt-headed Snake** or **Tree Snake.** Vernacular name for some species of the neotropical colubrid genus *Imantodes*.

Boa Constrictor. Vernacular name for the neotropical boa *Boa constrictor.*

Boas. *n.* Collective common name for numerous snake species belonging to the family *Boidae*. See also *Dwarf Boa, Ground Boa, Pacific Boas, Sand Boa,* and *Tree Boa.*

bob. *v., n.* See *headbob display.*

bobbing. *n.* Reference to the act of amphibian larvae swimming to the surface of a pond or other body of water for air. See R.J. Wassersug & E.A. Seibert, *Copeia* 1975:86–103, 1975. See also *headbob display.*

Bocourt's Water Snake. Vernacular name for the Asian homalopsine species *Enhydris bocourti.*

BOD. See *biochemical oxygen demand.*

body. *n.* Generally refers to the trunk of an animal, exclusive of head and tail, but is used variously to refer to different parts of the animal. May also refer to an organ or organelle not possessing a specific name (e.g., Golgi body, polar body, etc.).

body axis. Imaginary line of reference, which in lateral view extends from the tip of the snout to the posterior *body terminus.*

body cavity. Colloquial for *coelom* or *coelomic cavity.*

body condition, body condition index. (CI) Reference to the condition of an animal relative to others in a population, usually considered as equivalent to body composition. There are various methods that have been used to calculate some form of "condition index." Two types of condition indices are in common use: ratios and residuals. Ratio indices are one measure of size divided by a second measure of size and include (i) mass divided by length, (ii) mass divided by length cubed, (iii) mass divided by length raised to an emirically determined power, (iv) the cube root of mass divided by length, (v) mass divided by mass predicted from length, (vi) log mass divided by log length, and (vii) log mass divided by log mass predicted from log length. Residual indices involve similar measurements but are based on residuals from regressions of mass on length. These diverse condition indices may be flawed in their use but are widely used in ecological literature. For a review of various condition indices, including those based on body composition as well as morphology, see R.D. Stevenson & W.A. Woods Jr., *Integr. Comp. Biol.* 46:1169–1190, 2006. (Syn. *body mass index, build index, condition index, slenderness index*)

body fields. The regions between the body stripes on the dorsum of an organism.

body groove. A lateral fold of skin on the body surface.

body length. Straight-line measurement from the tip of the snout to the posterior body terminus. In literature and practice, this is a potentially ambiguous measurement, so reported values should specify whether the measurement is of *total body length* (*head-tail* or *snout-tail*) or of *snout-vent length.*

body mass index. See *body condition index.*

body pit. A depression dug by female sea turtles before digging the actual nest below and posterior to the pit. Sometimes additional pits are dug at other sites after the eggs are laid, presumably to deceive predators.

body ratio. The total length (body plus tail) divided by body length.

body temperature. Reference to the "average" temperature of an animal, theoretically the value defined as "the sum of the products of the heat capacity and temperature of all the tissues of the body divided by the total heat capacity of the organism" (J. Bligh & K.G. Johnson, *J. Appl. Physiol.* 35:941–961, 1973). No single temperature mea-

body temperature during activity. See *Activity Temperature Range*.

body terminus. The intersection of the posterior body wall and the axis of the caudal myotomes, often equated with the vent in adult amphibians and reptiles.

bog. *n.* A wet area having a peat substrate rich in organic debris but low in mineral content, characterized by vegetation of ericaceous shrubs, sedges, and mosses.

Bog Turtle. Common name for *Glyptemys* (formerly *Clemmys*) *muhlenbergii,* a well-known member of *Emydidae*.

Bohr effect. (Bohr shift) A change in the oxygen affinity of hemoglobin due to a shift in pH. Normally in terrestrial vertebrates there is observed a decrease in affinity with increased acidity (lower pH or increased P_{CO2}) of the blood, reflected in a right shift of the *oxygen dissociation curve.* In some species (usually aquatic environments) the opposite may occur and is referred to as a *negative Bohr shift* (left-shift with increased blood acidity).

boid. *n.* A member of, or pertaining to, *Boidae*.

Boidae. *n.* A clade (family) that includes snakes commonly known as boas, sand boas, and pythons. There are approximately 20 genera and about 75 species distributed in North, Central, and South America, West Indies, southwest Pacific islands, Africa, Madagascar, the Middle East, southern Europe, and southern Asia from India throughout the Indoaustralian Archipelago and Australia. Three subfamilies are recognized: *Boinae, Pythoninae,* and *Erycinae*. The taxonomy of boids differs among various workers, some recognizing boas and pythons as two separate families. See *Booidea*.

boiling school. Behavior attributed to a group of tadpoles (e.g. *Scaphiopus*) which, in aggregation, swim up rapidly near the center of the mass, break through the surface film, swing to one side and swim to the bottom, after which the action is repeated. The activity stirs up large quantities of mud and possibly increases the food available (A. Bragg & O.M. King, *Wasmann J. Biol.* 18:273–289, 1960).

Boinae. *n.* A subfamily of *Boidae,* comprising about 29 species distributed from western North America through subtropical South America and the West Indies.

Bolitoglossini. *n.* A clade of *plethodontid* salamanders that includes all extant forms except for the basal desmognathines, *Hemidactyliini* and *Plethodontini*.

boletoid. *adj.* Mushroom-shaped, as the tongue of certain salamanders.

bolus. *n.* A small, round lump or mass, usually with reference to a discrete plug or collection of food material moving through the gut.

Bolyeriidae. *n.* A clade (family) of snakes that includes two unusual species, unique among tetrapods in having divided maxillary bones with movable anterior and posterior parts. These snakes are known as *Round Island Boas* (or *Split-jawed Boas*) and are confined to that island in the Indian Ocean. One of the species (*Bolyeria multocarinata*) is thought to have likely gone extinct sometime since 1975.

Bolyerioidea. *n.* A clade (superfamily) of snakes including the sole family *Bolyeriidae*.

bombesin. *n.* A family of related peptides (*bombesin-like peptides*) having C-terminal homology, studied in detail using the anuran *Bombina orientalis* as a source. These peptides demonstrate antimicrobial actions and have no appreciable hemolytic activity, thus giving them considerable potential as therapeutic agents. They are present in skin extracts of anurans and have numerous effects on gastrointestinal activities, systemic blood pressure, lung function, thermoregulation, kidney function, immune system, and numerous other effects in mammals. The peptides *ranatensin, phyllolitorin,* and *litorin,* found in the brain and nervous system of frogs, are also in this family.

Bombinanura. *n.* A clade of anurans that includes all extant forms except *Ascaphidae* and *Leiopelmatidae*.

Bombinatoridae. *n.* A clade (family) of anurans, sister with *Alytidae* within the *Costata,* and characterized by derived features of the skull and hyoid. Two genera, including *Bombina* (fire-bellied toads) and *Barbourula,* and eight species are distributed in Europe and southeastern Asia.

bombinins. *n.* A large peptide family with antimicrobial and hemolytic properties, first isolated from skin of the European anuran *Bombina variegata* (G. Kiss & H. Michl, *Toxicon* 1:33–39, 1962; A. Csordas & H. Michl, *Monatshefte fur Chemie* 101:182–189, 1970).

bonding. *n.* Formation of a close relationship between two or more individuals.

bone. *n.* A hard and tough material produced by the deposition of a specific calcium phosphate salt, hydroxyapatite, into a fibrous or cartilaginous matrix.

bone of attachment. A poorly defined term that is traditionally used to describe the tissue attaching the teeth of all nonmammalian, nonthecodont amniotes to the jawbone. It is a bone-like tissue attaching the tooth to the tooth-bearing bone, probably homologous to alveolar bone.

bony scale. See *dermal scale*.

bony style. The presence of ossification in the central, hindmost section of the pectoral girdle.

Boodontini. *n.* A clade of colubrid snakes encompassing an uncertain number of moderate-sized, nocturnal constricting snakes including African Housesnakes (*Lamprophis*), Watersnakes (*Lycodonomorphus*), and Filesnakes (*Mehelya*). There are perhaps 15 genera and about 45 species of boodontine snakes.

Booidea. *n.* Formerly a clade (superfamily) of snakes that included all Boas and Pythons. See *Boidae*.

booid. *n.* A member of, or pertaining to, the former *Booidae*. Not to be confused with *boid*.

Boomslang. *n.* Common name for an arboreal species of African colubrid snake (*Dispholidus typus*), having swift movements and an extremely toxic venom. The common name means "Tree Snake" in Afrikaans.

Boophinae. *n.* A subfamily of *Mantellidae.*

bootstrap. *n.* 1. A technique for estimating an unknown distribution by repeated sampling from a sample distribution, used to arrive at "confidence" limits for evolutionary trees generated by cladistic methods. 2. A method using computer simulations to estimate species richness from quadrat samples.

bora or **bora wind.** *n.* A cold wind descending from a mountain slope and blowing onto lowland or coastal areas.

boreal. *adj.* Northward, pertaining to cold, temperate regions of the northern hemisphere comprising a northern zone of plant and animal life characterized by coniferous forest and taiga.

boss. *n.* 1. A rounded protuberance or swelling, forming a knob-like process. 2. A swollen bump on the head of certain toads.

botany. *n.* The study of plants.

bothrojaracin. *n.* A protein from the venom of *Bothrops jararaca* that causes aggregation of blood platelets.

bottle cell. 1. Synonym for *beaker cell* or *flask cell*. 2. Bottle-shaped cells (owing to contraction of their apical margins) on the lip of the amphibian blastopore, apparently functional during infolding of the gastrula.

bottleneck effect. See *genetic bottleneck*.

boulder. *n.* A large particle of sediment greater than 256 mm in diameter and too large to be easily handled.

boundary layer. Reference to a layer of fluid (air or water) adjacent to a physical boundary, in which motion of the fluid is affected by the boundary and exhibits a mean velocity that is less than the free stream velocity (away from the boundary). The region within which shearing forces occur, and the region responsible for frictional drag of moving objects. (Syn. *film, unstirred layer*)

boundary layer resistance (R_b). See *resistance*.

Bow-fingered Geckos. Vernacular name for various Eurasian species of gekkonid lizards belonging to the genus *Cyrtodactylus*.

Bowman's capsule. See *renal corpuscle*.

Bowman's gland. Multicellular glands that are responsible for secreting the layer of mucus present on the olfactory epithelium in terrestrial vertebrates, in conjunction with secretory supporting cells. In most frogs and caecilians, the appearance of Bowman's glands is associated with metamorphosis. Mucus secretion protects the epithelium from desiccation and also secretes odorant-binding proteins to aid in transporting volatile molecules across the layer of mucus.

Box Turtles. Collective vernacular name for American species of the emydid genus *Terrapene*. See also *Asian Box Turtles*.

Boyle's Law. The principle that, at a given temperature, the product of the pressure and volume of a given mass of gas is constant.

bp or **BP.** Abbreviation for "before present", usually used in context of geological or evolutionary time. Reference to before the present time.

brachial. *n., adj.* 1. Pertaining to the upper or humeral part of the forelimb, or *brachium*. 2. Used as name for any scale on the brachium of lizards, and for the *humeral* of the turtle plastron. See Fig. 6.

brachium. *n.* 1. The propodial region of the pectoral limb, or upper forelimb containing the humerus. 2. A limb corresponding to the human arm.

brachy-. Prefix meaning "short."

Brachycephalidae. *n.* A clade (family) of tiny, leaf-litter frogs (<16 mm) lacking a sternum and possessing reduced digits. Recently placed within the new taxon *Meridianura* and sister with *Cladophrynia* by D.R. Frost et al. (*Bull. Amer. Mus. Nat. Hist.* 297:1–370, 2006). Contains 17 genera and numerous species distributed in tropical North and South America and the Antilles. Formerly contained *Eleutherodactylinae* + *Brachycephalus*.

bracket *or* **bracketed key.** A dichotomous key in which contrasting parts of a couplet are numbered and presented together. (Syn. *parallel key*; cf. *indented key*)

brackish. *adj.* Refers to water that is partly saline, but less salty than pure seawater.

brady-. Prefix meaning "slow."

bradycardia. *n.* A reduction of heart rate from the normal level, often of occurrence in diving animals. Cf. *tachycardia*.

bradykinin. *n.* A nonapeptide kinin formed from a precursor normally circulating in blood. It is a potent cutaneous vasodilator and increases capillary permeability, in addition to constricting smooth muscle and stimulating pain receptors.

bradymetabolism. *n.* Reference to comparatively low levels of resting metabolism characteristic of amphibians and reptiles, in contrast with higher levels of basal metabolism characteristic of birds and mammals (*tachymetabolism*).

bradytelic evolution, bradytely. Reference to a lower than average rate of evolution. Cf. *horotelic, tachytelic evolution*.

brain. *n.* The main neuronal center in the body, located in the head as the anterior expanded mass of the central nervous system.

braincase. *n.* See *neurocranium*.

brain stem. The posterior part of the brain comprising the *midbrain, pons*, and *medulla*.

branch. *n.* In systematics and evolutionary biology, this term refers to a lineage or line on a phylogenetic tree; *or* a line connecting a branch point (node) to a terminal taxon. *Branch length* refers to the distance between a taxon and a given node in a cladogram, used to estimate evolutionary distances between taxa.

branchia. *n.* External gills, present at larval or adult life stages in various amphibians.

branchial. *adj.* Pertaining to gills or gill-like structures that function in respiratory gas exchange.

branchial aperture. See *spiracle*.

branchial arch. Any of several visceral arches lying posterior to the *hyoid arch*, supporting gills in tadpoles, and giving rise to parts of the hyoid apparatus.

branchial artery. Gill artery.

branchial basket. The cartilaginous structures supporting the gills of amphibian tadpoles.

branchial chamber. See *gill chamber*.

branchial cleft. Synonym for *gill slit* (= gill or branchial opening).

branchial fissure. A gill slit on the neck of certain aquatic salamanders (e.g., *Amphiuma*) that lack external gills.

branchial food trap. A glandular region located between the ventral velum and filter plates of microphagous anuran larvae. Typically, these are small, striated crescentic structures covered with rows of mucous glands, located on the ventral surface of the ventral velum or on the buccal floor in those suspension-feeding tadpoles that do not have a ventral velum. The structures function as surfaces for collecting food (R. Wassersug, *Occas. Pap. Mus. Nat. Hist. Univ. Kansas* 48:1–23 and 49:1–24, 1976). See also *pharyngobranchial tract*. (Syn. *collecting organ*)

branchial process. *n.* See *rachis*.

branchiate. *adj.* Having gills.

branchihyal. *n.* Either of two lateral arms on the posterior end of the hyoid apparatus in snakes and certain lizards (*Anniella*). Also called thyrohyal, ceratohyal, or hypohyal by various authors.

branch length. See *branch*.

breakage plane. The septum or partition of soft tissue in the center of tail vertebrae where breakage or separation occurs in autotomy. (Syn. *autotomy plane, breaking joint, breaking point, fracture plane*)

breaking joint or **point.** See *Breakage plane*.

breaking strength. The maximum force a structure endures just before it breaks or fails.

Breathing pore. A *spiracle*.

breed. *n., v.* 1. To reproduce. 2. To propagate animals under controlled, artificial conditions. 3. A group of animals related by descent from a common ancestor.

breeding barrier. See *reproductive isolating mechanism*.

breeding pattern. The timing and ecological conditions related to reproductive cycles, usually with reference to amphibians in earlier literature. See also *reproductive mode*.

breeding size. Reference to breeding population size, or the number of individuals in a population that are reproductive during a given generation.

brev-. Prefix meaning "short."

Brevicipitidae. *n.* A taxon (family) of anurans characterized by a short head and direct development, distributed in sub-Saharan East and southern Africa. The group is sister to *Hemisotidae* within the new taxon *Xenosyneunitanura* and contains five genera. Formerly a subfamily in Microhylidae. For taxonomic revision see D.R. Frost et al., *Bull. Amer. Mus. Nat. Hist.* 297:1–370, 2006.

Brevicipitinae. *n.* A former anuran subfamily of *Microhylidae*, now recognized with family status. See *Brevicipitidae*.

brevinins. *n.* A name assigned to two unique antimicrobial peptides isolated from the skin of Japanese ranid frogs.

bridge. *n.* 1. That part of the shell connecting the carapace and plastron in turtles. 2. The bony arch across the temporal region of the skull in some reptiles, as used by H. Gadow (*Amphibia & Reptiles*, p. 295, 1901).

brille. *n.* The transparent scale, or *spectacle*, that covers the cornea of snakes and some lizards, derived from fusion of the eyelids. See *spectacle*.

Brillouin index. An index of specific diversity of a system derived from information theory: $D = 1/N \times \log_2 N!/(n_1!n_2!\ldots n_s!)$, where N is the total number of individuals observed in a sample and $n_1, n_2 \ldots n_s$ are the number of individuals of the species 1, 2 …s.

bristle. *n.* See *tactile bristle*.

British Herpetological Society. A society which undertakes conservation activities to benefit amphibians and reptiles, particularly British indigenous species. Current URL: http://www.thebhs.org/

brittleness. *n.* In biomechanics, a measure of how sensitive a material is to cracking.

Broad-headed Skink. Vernacular name for the American scincid species *Eumeces laticeps*.

Broad-headed or **Pale-headed Snakes.** Vernacular names for Australian elapid snakes belonging to the genus *Hoplocephalus*.

broad heritability. See *heritability*.

broad spectrum. 1. Refers to lighting for vivaria that approximates the spectrum and quality of natural sunlight (especially the inclusion of UV wavelengths). 2. Refers to antibiotics or vermifuge drugs that are effective in destroying a range of taxa of microorganisms or intestinal worms.

bronchi. (*pl.*) *n.* Conducting airways in the lung, arising as secondary branches from the *trachea*. See also *bronchioles*.

bronchial columella. Fibrous strands connecting bronchi with the round window of the otic capsule. Described by

E. Witschi (*Zeitschrift f. Naturforschung* 4B:230–242, 1949) who suggested these function to conduct vibrations from the lungs to the inner ear in larvae of several species of ranid frogs. Similarly, a bronchial extension known as "*bronchial diverticulum*" has been suggested to function as auxiliary acoustic organs in *Xenopus* tadpoles. These appear to affect responses to pressure changes.

bronchial diverticulum. See *bronchial columella*.

bronchial process. A lateral process of the cricoid cartilage of the amphibian larynx, extending laterally and ventrally, sometimes to, or partly around, the lungs.

bronchial sac. *n.* Synonym for *bronchial diverticulum* used by W.V. Bergeijk (*Anat. Rec.*, 120:754, 1954).

bronchiole. *n.* Smaller conducting airways in the lung, arising as tertiary and greater levels of branches from the bronchi.

bronchus. (*pl.* –i) *n.* The airway within the lung that is supported by cartilage.

Bronzebacks, Bronzeback Tree Snakes. *n.* Vernacular name for about ten species of slender, diurnal Asian snakes belonging to the colubrid genus *Dendrelaphis*.

brood, brooding *n., v.* 1. The offspring from a single birth or from a single clutch of eggs. 2. To incubate, cover, remain with, and perhaps care for a clutch of eggs. The term describes behaviors of a parent while attending its nest and progeny closely, and does not include territorial behavior or nest site defense (sensu L.A. Somma, *Parental Behavior in Lepidosaurian and Testudinian Reptiles*, 2001). Usually used in reference to parental behaviors of Pythons, certain lizards, many amphibians.

brood care. The attention given by parental animals to eggs and young.

brood pouch. Any body sac or cavity in which eggs can be placed for part or all of development. Examples include egg cavities in the back of female pipid frogs, dorsal pockets of some female hylids (e.g., *Gastrotheca*), and gular sacs of male *Rhinoderma*.

Brook Salamanders. Vernacular name for plethdontid salamanders belonging to the American genus *Eurycea*.

Brown Anole. Vernacular name for the Caribbean species *Anolis sagrei*, which is an invasive species in Florida, Hawaii, Taiwan, and other areas.

Brownian motion or **movement.** Reference to a continuous, random motion of microscopic particles dispersed in a fluid, resulting from their bombardment by molecules of the fluid. The phenomenon was first observed by the botanist Robert Brown in 1827 when studying pollen particles. The term also refers to models of movement or mixing based on the physics of these small particles, and has been applied to some models of evolutionary change.

Brown Snakes. Vernacular name for American colubrid species of snakes belonging to the genus *Storeria*, especially *S. dekayi*, and for Australian elapid snakes belonging to the genus *Pseudonaja*. See also *Brown Tree Snake*, *Littersnakes*, and *Madagascan Hog-nosed Snakes*.

Brown Tree Snake. Vernacular name for the colubrid species *Boiga irregularis*. Common to Australia, New Guinea and nearby islands, this species was accidentally introduced to the Pacific island of Guam, where it has exterminated native birds and remains a nuisance invasive species.

Brown Water Snake. Vernacular name for the American natricine species *Nerodia taxispilota*.

browsing. *n.* Feeding on parts of plants such as leaves, shoots, or twigs. Reptilian examples include various herbivorous turtles and lizards. Recently, R. Shine et al. (*Func. Ecol.* 18:16–24, 2004) have compared the foraging mode of the sea snake *Emydocephalus annulatus* to herbivorous browsing of mammals, due to slow, continous movement with frequent ingestion of small, immobile, defenseless food items. Cf. *grazing*.

brow spot. A small, lightly pigmented and comparatively inconspicuous mark on the head of some frogs, located between or slightly anterior to the eyes. This is an external remnant of the embryonic connection of structures of the *parapineal organ* with the surface. (Syn. *parietal spot* or *fleck*)

brumation. *n.* A condition of torpor during extended periods of low temperature (winter dormancy), intended to distinguish such states of inactivity of amphibians and reptiles from the term "hibernation" that is used commonly in reference to birds and mammals. Term coined by W. Mayhew (*Comp. Biochem. Physiol.* 16:103–119, 1965).

Brunn effect. Reference to the increased osmotic permeability of amphibian skin in response to antidiuretic hormone (vasopressin), first described by F. Brunn (*Zeitschrift für die Gesampten Experimentellen Medizin* 25:170–175, 1921).

brush border. A free epithelial cell surface that bears numerous microvilli.

Brush Lizard. Vernacular name for some American iguanid species of *Urosaurus*.

bryocolous. *adj.* Living on or in moss.

Bubble-nest Frogs. Collective vernacular name for numerous species of Asian frogs belonging to the rhacophorid genus *Philautus*. Also called *Bush Frogs*.

buccal. *adj.* Pertaining to the cheek or mouth, usually the mouth cavity.

buccal apparatus. Reference to all structures in the buccopharyngeal cavity of larval anurans, considered as a unit. This term does not refer to any structures of the mouthparts according to R.W. McDiarmid & R. Altig (*Tadpoles. The Biology of Anuran Larvae*, 1999).

buccal cavity. The oral cavity.

buccal force pump. Reference to mechanics of breathing by means of raising the floor of the mouth to force air into the lungs by means positive pressure, characteristic of amphibans. See also *buccal pump;* Cf. *aspiration pump.*

buccal incubation. See *Rhinodermatidae.*

buccal papillae. A collective term for the papillae in the buccal cavity of anuran larvae.

buccal pump. A muscle-powered mechanism that draws water into the buccal cavity of a tadpole. See also *buccal force pump.*

buccopharyngeal respiration. Reference to respiratory gas exchange across the vascular lining of mouth and pharynx, employed by some turtles and amphibians.

buccopharynx. *n.* A collective term for the buccal cavity and pharynx, considered together as a unit.

buccopulmonary respiration. Reference to ventilation of the lungs by raising and lowering the floor of the mouth, as occurs in all living amphibians. See *buccal force pump* and *respiration* for further clarification.

buckler. *n.* Literally, a small shield, used in older literature to refer to turtle shell, carapace, or fused dermal scutes.

bufadienolide. *n.* See *bufodienolides.*

bufagin. *n.* A toxic steroid first found in *Bufo agua* (= *Chaunus marinus*) by J.J. Abel & D.I. Macht (*J. Pharmacol. Exp. Therap.* 3:319–378, 1912) and related to several compounds having names based on various species from which these were isolated. Same as *bufogenin* in modern usage.

buffer. *n., v.* 1. A chemical system that stabilizes the concentration of a substance. Acid-base systems serve as buffers for pH and prevent large changes in the concentration of hydrogen ions. 2. In context of landscape ecology, a buffer is a multi-use transition area designed and managed to protect a core reserve or critical corridor from destructive impacts of human activities. 3. To protect a system from change due to external factors.

bufo-chrome 1, 2. *n.* A series of fluorescent pterin substances found in skin of *Bufo vulgaris.*

bufodienolides. *n.* Polyhydroxyl C24 steroids, which constitute the fundamental carbon ring structure of *bufogenins.* These compounds are present in bufonid skin secretions and exert cardiotonic activity, and some exhibit antitumor, insecticidal, and antimicrobial properties. (Syn. *bufadienolides*)

bufogenin. *n.* Any of a series of toxic steroids from secretions of bufonid skin glands, resembling digitalis in physiological activity while differing in placement or number of hydroxyl, acetyl, acetoxyl, or aldehyde groups. Synonymous with *bufodienolides.*

bufonin. *n.* Name from an early description of a "poison" conditioning the milky appearance of gland secretions from toads (G.K. Noble, *The Biology of the Amphibia,* p. 134, 1931).

Bufonidae. *n.* A recently revised clade (family) of anurans having a Bidder's organ, prominent parotoid glands, and terrestrial habits in most species. In colloquial terminology, bufonids are often called "toads" or "true toads." There are 33 or more genera and approximately 455 species that are cosmopolitan in temperate and tropical regions except east of Wallace's Line, Madagascar, and oceanic islands. Numerous generic changes have been proposed recently for bufonid taxa, and the revised family is composed of 48 genera (D.R. Frost et al., *Bull. Amer. Mus. Nat. Hist.* 297:1–370, 2006). *Chaunus* (formerly *Bufo*) *marinus* has been introduced in many places including Florida, Australia, New Guinea, and various islands where it has become a serious pest.

Bufonoidea. *n.* A clade of anurans that includes relatively derived families likely comprising the sister group to Ranoidea within Neobatrachia. This clade is termed *Hyloidea* in some literature. See C.R. Darst & D.C. Cannatella, *Mol. Phylogenet. Evol.* 31:462–475, 2004. For more recent revision see *Hyloides.*

bufotenines. *n.* Various indoalkylamine present in the toxic alkaloid skin secretions of bufonid toads. These are a derivative of serotonin and produce hallucinogenic sensations in humans.

bufotoxin. *n.* Various toad venom derivatives generally associated with toxic steroid bufogenins. The name also applies more generally to various toxins or products secreted from skin glands of toads, especially the genus *Bufo* (*and Chaunus*), and named by combining the word with specific names of a species (e.g., alvarobufotoxin). Symptoms of poisoning by bufotoxins include salivation, shortness of breath, cardiac arrhythmias, cyanosis, acid-base abnormalities, and seizures.

bufo-yellow 1,2. A series of yellow fluorescent substances found in small quantities in the skin of *Bufo vulgaris.*

build index. An early method of estimating *body condition,* first used by I.W. Wilder (*J.Exp. Zool.*, 40:1–112, 1924), who defined the following formula:

$$\text{Build index} = \frac{(\text{Mass in mg})(10{,}000)}{(\text{Length in mm})^3}$$

bulbus. *n.* A heart chamber.

bulbus arteriosus. A term that, in older literature, has been applied to the *conus arteriosus* or muscular base of the ventral aorta of amphibians.

bulbis cordis. A term from older literature that was equivalent to the *conus arteriosus* of amphibians.

bull. *n.* In older literature this term is occasionally used for an adult male crocodilian or turtle.

bulla. *n.* A hollow knob or disc, or a large blister typically containing blood.

bullen. *n.* A net attached to a metal ring that is used to capture sea turtles in deep water.

Bulletin of the Association of Reptilian and Amphibian Veterinarians. A publication of the said Association.

Bulletin of the Chicago Herpetological Society. A monthly journal with original articles, photographs, columns, book reviews, and advertisements in addition to news of the Society.

Bulletin of the Herpetological Society of Japan. A Japanese journal of the said Society.

Bulletin of Zoological Nomenclature. The official organ of the International Commission on Zoological Nomenclature.

Bullfrog. *n.* Vernacular name for relatively large species of frogs belonging to the ranid genera *Pyxicephalus* (African), *Lithobates* (*Rana*) (North America: *Lithobates catesbeianus*; formerly *Rana catesbeiana*), and *Tomopterna* (southern Asia and Africa); the name also is applied to neotropical leptodactylid species of the genus *Adenomera* and to microhylid frogs of the Asian genus *Kaloula*. For recent taxonomic revision see D.R. Frost et al., *Bull. Amer. Mus. Nat. Hist.* 297:1–370, 2006.

Bullsnake. *n.* Vernacular name for the North American colubrid snakes *Pituophis catenifer sayi* and *Pituophis deppei*.

Bunch Grass Lizard. Vernacular name for the American iguanid lizard species *Sceloporus scalaris*.

bungarotoxin. (BuTX) *n.* A blocking agent composed of a group of neurotoxins isolated from the venom of members of the snake genus *Bungarus* (Kraits). One component of bungarotoxin, α-BuTX, binds selectively and irreversibly to nicotinic acetylcholine receptors.

Bunsen solubility coefficient. The quantity of gas at STPD that will dissolve in a given volume of liquid per unit partial pressure of the gas in the gas phase. The coefficient is used only for gases that do not react chemically with the solvent.

buoyancy. *n.* The tendency to float or sink in a fluid environment, water or air. The buoyant force equals the force on a mass of fluid having equivalent volume to the object. Objects float if they have positive buoyancy and tend to sink if they have negative buoyancy.

Burmese Python. Vernacular name for the Asian species *Molurus bivittatus*, popular in the pet trade.

bursa. *n.* Any pouch or sac.

burrow. *n., v.* 1. A hole or tunnel dug below ground and used by an animal for shelter or refuge, sometimes associated with reproductive activities such as oviposition. 2. To move underground or through substrate by digging or specialized locomotion.

Burrowing Adder or Asp. A vernacular name for snakes belonging to the African Atractaspididae, also known as *Mole Vipers* or *Stiletto Snakes*.

Burrowing Frog or **Treefrog.** Vernacular name for species of hylid frogs belonging to the Middle American genus *Pternyhyla*. See also *Giant Burrowing Frogs*. See also *Burrowing Toad*.

Burrowing Python. See *Sunbeam Snake*.

Burrowing Toad. Vernacular name for the American genus *Rhinophrynus*. See also *Burrowing Frogs*.

Bush Frogs. See *Bubble-nest Frogs*.

Bushmaster. *n.* Vernacular name for the large neotropical pitviper, *Lachesis muta*.

Bush Vipers. Collective vernacular name for various species of African vipers belonging to the genus *Atheris*.

Butler's Garter Snake. Vernacular name for the American colubrid species *Thamnophis butleri*.

Butterfly Lizards. Vernacular name for agamid lizards belonging to the genus *Leiolepis*.

button. *n.* 1. In common usage, the first permanent and distal segment on the rattle of a Rattlesnake, the prebutton being lost at first ecdysis. In older literature the term was applied to various different segments of the rattle. 2. The *osteoderm* present in the scales of crocodilians. There may be one or two interlocking osteoderms under each scute. 3. See *iButton*.

Button Frogs. Vernacular name for species of South American frogs belonging to the leptodactylid genus *Cycloramphus*.

buttress. *n.* A supportive structure that arises from a bone in the turtle plastron to ankylose with a pleural bone in the carapace.

Bv8. See *AVIT*.

C

c. Latin abbreviation for *cum*, meaning "with."

C, °C. Degrees *Celsius* (centigrade).

ca. Latin abbreviation for *circa*, meaning "about."

cable. *n.* An elastic, gelatinous string by which eggs of certain plethodontid salamanders attach eggs to submerged objects (C.H. Pope, *Amer. Mus. Novit.* 153:1–15, 1924).

CABS. See *Center for Applied Biodiversity Science.*

cachexia. *n.* A condition of wasting or chronic debility.

Cacos. *n.* Vernacular name for African species of ranid frogs belonging to the genus *Cacosternum.*

caduceus. *n.* The wand of Hermes or Mercury, used as a symbol of the medical profession. The staff or rod has an entwining snake, symbolic of the god of healing. (Syn. *staff of Aesculapius*)

caducibranchiate. *adj.* Having temporary gills, as in many larval amphibians. (ant. *perennibranchiate*)

caecilians. *n.* See *Gymnophiona.*

Caeciliidae. *n.* A clade (family) of burrowing caecilians including 21 genera and 109 species distributed in Central and South America, Africa, and parts of southern Asia.

caeciliid. *n, adj.* A member of, or pertaining to, *Caeciliidae.* Not to be confused with *caecilians*, which is a more inclusive term.

caecum. *n.* See *cecum.*

Caenogaea. *n.* A zoogeographical region that includes the *Nearctic, Oriental,* and *Palaearctic.* Cf. *Eogaea.*

caenogenetic. *adj.* (*n.* **caenogenesis**) 1. Of recent origin. 2. Reference to transitory adaptations appearing early in the ontogenetic development of an organism.

caenomorphism. *n.* Changing of organisms from a complex to simpler form.

Caenophidia. *n.* A speciose clade of "advanced snakes" that includes over 2500 species in combined *Acrochordidae* and *Colubroidea.* However, based on molecular studies, P. Gravlund (*Biol. J. Linn. Soc.* 72:99–114, 2001) concludes that *Acrochordus* is not a caenophidian.

Caenozoic. *n.* See *Cenozoic.*

caeridins, caerins. *n.* Peptide molecules isolated from secretions of parotoid glands of the Australian frog *Litoria splendida*, some of which possess potent antimicrobial and antiviral activity.

caeruleins. *n.* A family of decapeptides first isolated from skin extracts of the Australian frog *Pelodryas caerulea* (formerly *Litoria caerulea*) and later from the skin of pipid and leptodactylid species of frogs. Biological effects include a variety of actions related to gastrointestinal motility, smooth muscle, gastrointestinal secretions, stimulation of endocrine glands, trophic effects, systemic blood pressure, and central nervous system.

Caiman. *n.* Collective vernacular name for three genera of alligatorids (*Caiman, Melanosuchus, Paleosuchus*) distinguished by broad, blunt snouts and distributed in Central and South America. See *Jacare.*

caisson disease. The formation of small gas bubbles during decompression. Rapid ascent from diving can cause such bubbles, especially of nitrogen, which can cause blockage of blood vessels, joint pain, paralysis, and even death. (Syn. *bends, decompression sickness*)

Calamarinae. *n.* A clade (subfamily) of Asian colubrid snakes represented by nine genera and about 50 species, commonly called *Reedsnakes.* Characterized by slender bodies, short tail, secretive, semi-burrowing habits, and resembling some venomous elapid snakes in colors and defensive behaviors.

calcaneum, calcaneus. *n.* The larger of the tarsal bones, which, in anurans, is elongated and fused with the astragalus to form the extra segment of the hindlimb. (Syn. *fibulare*)

calcar. *n.* A fleshy projection on the heels of some frogs, often referred to as a *spur.*

calcareous. *adj.* Containing calcium carbonate, or rich in calcium salts. (Syn. *calcified*)

calcification. *n.* The process of calcium deposition in living tissue, usually a matrix of connective tissue.

calcified. *adj.* See *calcareous.*

calcineurin. *n.* A Ca^{2+}-calmodulin-dependent phosphatase important for myotoxicity driven by neurotoxic phospholipase components of snake venoms.

calcitonin. *n.* A protein hormone secreted by the *ultimobranchial glands* of amphibians and reptiles, in response to elevated levels of blood calcium. Calcitonin plays various roles in calcium metabolism, and in most cases the varied roles are not well understood.

calcule. *n.* Any of small, convex, and smooth scales located on the dorsum of the Chinese Crocodile Lizard (*Shinisaurus*). First used by S.B. McDowell & C.M. Bogert (*Bull. Amer. Mus. Nat. Hist.* 105:7–142, 1954).

calculi. *n.* Any abnormal stone-like deposit formed in the body.

caldera. *n.* A volcanic depression much larger than the original crater.

California King Snake. Vernacular name for the American colubrid snake *Lampropeltis getula californiae*.

California Legless Lizard. Vernacular name for a limbless species (*Anniella pulchra*) of anguid lizard.

California Mountain King Snake. Vernacular name for the American colubrid species *Lampropeltis zonata*.

California Newt. Vernacular name for the salamandrid species *Taricha torosa*. See also *Pacific Newt*.

California Slender Salamander. Vernacular name for the Pacific Coast plethodontid species *Batrachoseps attenuatus*.

caliology. *n.* The study of burrows, nests, tubes, and other domiciles constructed by animals.

calipash. *n.* The dorsal layer of gelatinous fat in the body and flippers of sea turtles. This substance is dull green in color and is the source of the common name of the Green Turtle. This substance is used for making soup and is esteemed as a delicacy in various species of turtles.

calipee. *n.* The gelatinous cartilage from the ventral body and attached to the plastron of Green Sea Turtles (*Chelonia*) and some other species of turtles.

call. *n.* The vocalized sound, song, or trill of an anuran, usually associated with males and breeding choruses. (Syn. *vocalization*)

call group. One or more call notes.

call note. A single note, whether in a single- or multiple-call group.

callosity or **callus.** *n.* An area of thickened, horny skin, as the *tubercle* on the foot of anurans. Historical usage includes reference to hardened areas on the plastron of soft-shelled turtles. Generally, a roughened area of skin, sometimes with superficial, sculptured bone either exposed or just beneath the surface.

calmodulin. *n.* An intracellular calcium-binding regulatory protein found in virtually all tissues and mediates most calcium-regulated processes. It is important in the contraction of smooth muscles, as well as a wide spectrum of enzymatic and cellular functions. It appears to be the commonest translator of intracellular calcium messages.

calorie. *n.* A derived unit of heat defined as the amount of heat required to raise the temperature of 1 g of water 1 °C at one atmosphere of pressure. Equivalent to 4.186 joules at 15 °C. (Syn. *gram calorie*)

calorimetry. *n.* The measurement of heat production in an animal. (Cf. *indirect calorimetry*)

calyx. *n.* (*pl.* **calyces**; *adj.* **calyculate**) 1. Generally, a cup-shaped structure. 2. Small pockets or sunken interstices between the retiform ridges on the hemipenis of some snakes.

calyculate. *adj.* Covered by cup-shaped structures.

Cambrian. The earliest Period of the *Paleozoic Era* of geologic time. See Table 1.

camouflage. *n.* Characters that render an animal visually inconspicuous, thereby avoiding detection by predators or prey. See *crypsis*.

cAMP. See *cyclic AMP*.

campo. *n.* Vernacular for South American habitats characterized by grassy plains, scattered bushes, and small trees.

Canadian Toad. Vernacular name for the North American bufonid species *Anaxyrus* (formerly *Bufo*) *hemiophrys*. For recent taxonomic revision see D.R. Frost et al., *Bull. Amer. Mus. Nat. Hist.* 297:1–370, 2006.

canaliculate. *adj.* Channeled with longitudinal ridges. Reference to a pattern of *epidermal microsculpturing*. See R.M. Price *J. Herpetol.* 16:294–306, 1982.

canaliculus. (*pl.* **–i**) *n.* A groove or tubular channel.

canalization. *n.* The phenomenon of having developmental pathways leading to a standard phenotype in spite of genetic or environmental disturbances. A canalized character is a trait having variability restricted within narrow boundaries even when the organism is subjected to mutations or a disturbing environment. The term was proposed by C.H. Waddington (*J. Genetics* 39:75–139, 1940; *Nature* 150:563–565, 1942; *The Strategy of the Genes*, 1957). (Syn. *developmental homeostasis*)

canalizing selection. Reference to selection that eliminates genotypes rendering developing individuals sensitive to environmental fluctuations.

cancellous bone. A type of bone structure that is spongy, not having tissues that are tightly packed. It is a highly vascularized bone tissue that forms as a primary or secondary tissue in the innermost region of a bone. Cancellous bone may be well organized and may show growth lines. Cf. *compact bone*.

candela. *n.* The standard SI unit of luminous intensity.

candling. *v.* Transillumination. Passing light through a structure and toward an observer (e.g., through an egg).

Canebrake Rattlesnake. Vernacular name sometimes used for southern populations of the Timber Rattlesnake (*Crotalus horridus*).

Cane Toad. See *Giant Toad*.

canker. *n.* Ulceration of the labial region and oral cavity, commonly known as "mouth rot." See *stomatitis*.

cannibalism. *n.* The act of feeding upon individuals of the same species.

cannula. *n.* (*pl.* **–ae**) A flexible tube that can be inserted into a cavity, duct, or blood vessel to introduce a drug or withdraw fluid. During insertion the lumen is usually occupied by a *trocar*. Cf. *catheter*.

canopy. *n.* The uppermost stratum of foliage formed by the crowns of trees in forest vegetation.

canthal. *adj., n.* 1. Of or pertaining to the canthal ridge, or the region of the head between the tip of the snout and the eye. See Fig. 2. 2. Scales located along the upper surface of the *canthal ridge*. These lie behind the level of the *prenasal* and *postnasal* suture and in front of the *supraocular*. When these scales are large and in contact along the midline, they are referred to as *prefrontals*. Fig. 10, 14.

canthal ridge. The sharply angled or gently rounded juncture of the side of the snout and top of the head. The angle of the head from the tip of the snout to the anterior corner of the eye of amphibians, or the anterior end of the eybrow in reptiles. (Syn. *canthus rostralis*)

canthus. *n.* Either corner of the eye where the eyelids meet. The term has been used as synonymous with *canthus rostralis*, but the terms are not equivalent.

canthus rostralis. See *canthal ridge*.

Cantil. *n.* Vernacular name for the neotropical pit viper species *Agkistrodon bilineatus*, distributed from Mexico through tropical Central America to Belize.

cantilever. *n.* A projecting beam or body that is supported at only one end.

Canyon Lizard. Vernacular name for the American iguanid lizard species *Sceloporus merriami*.

Canyon Treefrog. Vernacular name for the North American hylid species *Hyla arenicolor*.

Cape region. Reference to the South African phytogeographical region at or near the Cape of Good Hope.

Cape Tortoises. Vernacular name for South African tortoises belonging to the genus *Homopus*.

capillary. *n.* The smallest (microscopic) blood vessels that are lined by highly permeable endothelium and exchange gases and chemical substances with tissues through which they pass. Capillaries arborize extensively to provide a large exchange surface and connect smaller arteries (*arterioles*) with smaller veins (*venules*).

capillary action. The drawing of a liquid into small openings or spaces as a result of cohesive and surface tension forces acting where the surface of the liquid is in contact with a solid. This is responsible for the tendency of liquids to move along narrow spaces between objects such as soil particles or in narrow tubes, without external pressure being applied.

capillary water. Water that is held by capillary forces as a film around soil particles, available to plants and soil organisms. Cf. *osmotic water*.

capitate. *adj.* Possessing a head or headlike structure. This term has been applied to certain structural variants of the snake hemipenis.

capitellum. *n.* The rounded protrusion on the undersurface of the reptilian humerus, adjacent to the trochlea and on which the radius moves.

capitulum. *n.* 1. The head proper of a rib, projecting ventrally with respect to the tuberculum and articulating with a vertebral *centrum* (the hypocentrum or intercentrum of the vertebral column in primitive tetrapods). Many extant groups of amphibians and reptiles have departed from this primitive pattern. Cf. *tuberculum*. 2. The apical area of the *hemipenis* in squamate reptiles that is separated from the rest of the organ by a distinct groove.

capsule. *n.* An enclosing structure.

capsular cavity or **chamber.** *n.* The liquid-filled space between the amphibian egg and its surrounding jelly envelopes. (Syn. *perivitelline space*)

capture efficiency. (α) The effect of a predator on the per capita growth rate of a prey population.

carapace. *n.* The domelike dorsal component of the shell of turtles, usually comprised of both epidermal and bony plates. Cf. *plastron*.

carapace depth or **height.** Equivalent measurements of a turtle's shell, taken as the maximum straight line height of the shell, perpendicular to the plastron or to a flat surface on which it rests. See Fig. 8.

carapace length. *n.* Length measurement of carapace, sometimes called *straight length*, extending in a straight line from the anterior margin of the nuchal plate to the posterior margin of the supracaudal plate. See Fig. 9.

carapace width. *n.* The maximum straightline width of carapace, sometimes measured at centers of fourth marginals. Also called *straight width*. See Fig. 9.

carbohydrase. *n.* An enzyme that catalyzes the breakdown of carbohydrates.

carbohydrate. *n.* A category of food macromolecules, consisting of aldehyde or ketone derivatives of alcohol, utilized by animals primarily for the storage and supply of chemical energy. The sugars and starches are the more important categories of carbohydrates.

carbon. *n.* The third most abundant of the biologically important elements.

carbon cycle. Reference to the storage and movement of carbon on and near the surface of Earth. See *mineral cycles*.

carbon dating. Measurement of the radioactive isotope C-14 as a method for estimating age of fossils.

carbonic anhydrase. An important enzyme that catalyzes the reversible interconversion of carbonic acid to carbon dioxide and water. This process is important for the blood transport of CO_2 and, indirectly, O_2.

Carboniferous. A geologic Period of the *Paleozoic Era* following the *Devonian* and preceding the *Permian*, characterized by swamp formation and deposition of plant remains that later hardened into coal. See Table 1.

carbonyl group. A doubly bonded carbon-oxygen group (C=O), present in a variety of organic molecules.

carboxyl group. A chemical group (COOH), present in a variety of organic molecules, and characteristically acidic because it becomes negatively charged when a proton dissociates from its hydroxyl group.

carcinogenic. *adj.* (*n.* **carcinogen**) Capable of causing or producing cancer.

cardiac. *adj.* Relating to the heart.

cardiac output. The volume rate of blood outflow from the heart, expressed as volume per time. In reference to mammals, the term conventionally denotes the outflow of systemic blood from the left ventricle, which is in series and

equal over time to the right pulmonary outflow. In amphibians and non-crocodilian reptiles having an undivided ventricle, it is best to qualify the cardiac outflow as systemic, pulmonary, or total.

cardiac process. The rounded protuberance from the anterior margin of the amphibian cricoid cartilage, located where it crosses the lateral edge of the arytenoid cartilage. (Syn. *anterior articular process*)

cardiac shunt. See *intracardiac shunt*.

cardinal, cardinal vein. *adj., n.* Relating to the primitive system of veins that drain the head, dorsal body wall, and kidney.

cardiomyopathy. *n.* A pathological alteration of cardiac muscle.

cardiotoxin. *n.* (*adj.* –**ic**) Any toxin or blood-borne pathogen having its primary effect on the heart.

cardiovascular. *adj.* Of or relating to the circulatory system, including heart and vessels.

cardiovascular facilative movement. A term proposed to describe caudocephalic waves in snakes, specifically in context of blood pressure control and circulation. These movements are lateral undulations or body "waves" that move characteristically in an anterior direction in nonlocomoting snakes and involve either the neck or most of the body (beginning at posterior end). These movements have been shown to occur in response to lowered blood pressure and evidently function to increase venous return to the heart, which increases cardiac output and elevates arterial pressure. See H.B. Lillywhite, *Physiol. Zool.* 58:159–165, 1985. Cf. *caudocephalic wave*.

cardiovascular system. The circulatory system, heart and associated blood vessels.

Carettochelyidae. *n.* A clade (family) of turtles, with a single species, *Carettochelys insculpta*, distributed in southern New Guinea and extreme northern Australia. This turtle inhabits fresh water rivers, lagoons, and brackish estuaries.

carey. *n.* A Spanish word, probably derived from an Indian term, use to denote tortoise shell.

Caribbean Toads. Collective vernacular name for species of American bufonid anurans belonging to the genus *Peltophryne*.

carina. *n.* A *keel*.

carinate. *adj.* Possessing keels (carinas), as occurs on scales of various squamates. A keel is characterized by an elevated, longitudinal, and straight ridge. Scales may be *unicarinate* (single keel), *bicarinate* (two keels), *tricarinate* (three keels), etc.

carnivore. *n.* (*adj.* **-ous**). Any animal that feeds exclusively or largely on animal flesh.

carotid artery. One of two arteries that originates from an aortic arch, courses to the head via the neck, and supplies the brain (*internal carotid*) and facial tissues (*external carotid*) with oxygenated blood. In some species (e.g., certain snakes) there may be only a single carotid artery.

carotid body or **gland.** A spongy swelling at the base of the external and internal carotid artery in amphibians (= *carotid labyrinth*) or reptiles (= *carotid "sinus"*). This structure is neither homologous nor identical with the carotid body of mammals.

carpal. *adj., n.* 1. Pertaining to the *carpus*. 2. A bone of the wrist.

Carpet Python. Collective vernacular name for boid species belonging to the genus *Morelia*, especially the species *Morelia spilota*.

carpus. *n.* 1. Any bone in the distal region of the forelimb of tetrapods, situated between the radius or ulna and the metacarpals (the wrist). Also refers to the segment of the limb corresponding to this region. 2. A scale situated between the digits and articulation with the forearm, often used with a prefix to indicate position (supra, infra, etc.).

carr. *n.* A fen woodland or mire supporting scrub vegetation.

Carrier's Constraint. Reference to the circumstance that living amphibians and reptiles cannot run and breathe normally at the same time. The sprawling gait of reptiles especially flexes the trunk and alternately compresses the lung with stepping movements, first right side followed by left side, etc., alternating with expansion when the opposite side is compressed. This distortion interferes with normal lung ventilation, which becomes impossible when the cycling is rapid during running. This was called "Carrier's Constraint" by R. Cowen (*History of Life*, 1990) to honor the work of D.R. Carrier related to this subject (*Paleobiology* 13:326–341, 1987; *Amer. Zoologist* 31:644–654, 1991).

carrion. *n.* Dead and decaying flesh.

carrying capacity. (K) The maximum number of individuals that can be supported in a population that is growing according to the logistic growth equation and is in stable equilibrium with the other biota of a community. The units are numbers of individuals and reflect the availability of food, space, and other resources in a given area of habitat.

cartilage. *n.* (*adj.* **cartilaginous**) A translucent elastic tissue that forms the embryonic skeleton and is variously converted to bone in adult vertebrates. See *elastic cartilage, fibrocartilage, hyaline cartilage*.

cartilago alaris. See *alinasal*.

caruncle. *n.* A temporary keratinized protuberance on the apical snout of turtles, crocodilians, and tuataras, used to cut the eggshell at time of hatching. This structure is not calcified and thus differs from the egg tooth of hatchling squamates. (Syn. *egg-caruncle, shell-breaker*)

Cascabel, Cascavel. *n.* Spanish word meaning bell or rattle and used generally to describe *Rattlesnakes* in Central and South American countries.

cascque. *n.* An "armored" head of certain lizards and amphibians which variously involves proliferation of dermal roofing bones, thickened skin, coossification of skin and skull, and skull edges forming sculptured ridges, crests, or tubercles. Sometimes said to resemble a "helmet."

Casque-headed Frogs. Vernacular name for species of South American frogs belonging to the hylid genus *Aparasphenodon*.

Casque-headed Treefrogs. Vernacular name for species of South American frogs belonging to the hylid genus *Trachycephalus*.

cast or **casting.** *n.* 1. An outdated term for slough, shedding, or ecdysis. 2. See *cast-corrosion*.

cast-corrosion. *adj., n.* Reference to process or techniques for visualizing vasculature or other anatomical spaces, involving injection of a fluid material (e.g., see *Microfil®*) that flows into the spaces of interest and subsequently solidifies. The surrounding tissues are then removed by corrosion methods to produce a solid cast or model of the tissue space(s).

Castellana. *n.* Vernacular name for the pitviper *Agkistrodon bilineatus howardgloydi*.

casting cycle. A shedding cycle, usually with reference to the interval between successive ecdysis events.

cata-. Prefix meaning "down" or "against."

catabatic. *adj.* Reference to winds that are blowing downslope.

catabolism. *n.* Disassembly of complex molecules into simpler ones, accompanied by the release of energy. Cf. *anabolism*.

catadont. *adj.* Having teeth in the lower jaw only.

catalogue. *n.* An index or compilation of taxonomic literature within an alphabetical listing of species.

catalysis. *n.* An increase in the rate of a chemical reaction, promoted by a substance—the *catalyst*—that is not consumed in the reaction.

catalyst. *n.* A substance that increases the rate of a reaction without being used up in the reaction.

cataract. *n.* Opacity of the crystalline lens or capsule of the eye.

catastrophism. *n.* A current view that certain mass extinctions resulted from cataclysmic events such as the impact of very large meteorites.

catathermal. *adj.* Reference to a period of decreasing temperature. Cf. *anathermal*.

catch-mark-recatch method. See *Lincoln index* and *mark-recapture*.

catchment. *n.* See *watershed*.

catecholamines. *n.* A group of sympathomimetic amines that act as both neurotransmitters and hormones, including epinephrine, norepinephrine, and dopamine. Catecholamines are present as constituents of glandular skin secretions in some species of bufonid amphibians.

category. *n.* In taxonomy, any of the levels in a taxonomic hierarchy.

Cat-eyed Snake. Vernacular name for neotropical species of colubrid snakes belonging to the dipsadine genus *Leptodeira*.

cathartic. *adj.* Causing gastrointestinal tract evacuation. (Syn. *purgative*)

catheter. *n.* A rigid or usually flexible tube that can be inserted into a cavity, duct, or blood vessel and used to pass fluids in or out. Typically, in animal research these are small, consist of polyethylene, Teflon, or similar flexible material, and used commonly to measure blood pressure or withdraw fluids. Cf. *cannula*.

cathode. *n.* The negative electrode to which positive ions are attracted. Cf. *anode*.

cation. *n.* A positively charged ion. Cf. *anion*.

CAT scan. A *computerized axial tomography* scan, resulting from the technique of using a computer to process data from a tomograph in order to display a reconstruction of an organism's body in cross section. See also *tomography*.

Catsnakes, Cat Snakes. *n.* Collective vernacular name for various Old World snakes belonging to the genera *Boiga* and *Telescopus* (also called *Tiger Snakes*).

caudad, caudally. *adv.* In the direction of the tail.

caudal. *n., adj.* 1. Of, pertaining to, or located toward the tail. Used in herpetology for naming any scale, plate, or structure on the tail of a reptile or amphibian. See Fig. 13. 2. Synonym for the *postcentral* of turtles. See Fig. 4.

caudal autotomy. See *autotomy*.

caudal disk. The blunted region at the end of the tail in certain species of snakes, especially those in the Uropeltidae. (Syn. *caudal shield*)

caudal fin. The flattened, membranous dorsal or ventral extensions of the tail of amphibians, used in swimming.

caudal luring. Use of the tail as a lure for prey by juveniles of many species of snakes. The tail tip is typically of bright contrasting color and is wiggled, usually while elevated and in a wave-like motion. (Syn. *tail waving*)

caudal marginal. See *postcentral*, and Fig. 4.

caudal rib. A true rib articulating with a caudal vertebra.

caudal shield. See *caudal disk*.

caudal vertebra. A vertebra of the tail.

caudal whorls. A series of differentiated caudal scales arranged in whorls encircling the tail (e.g., those in the Spiny-tailed Iguanas, *Ctenosaura*).

Caudata. *n.* The monophyletic clade that includes all living salamanders (having four limbs and a long tail). An alternative name is *Urodela*.

caudocephalic wave. A horizontal undulating wave of the body that results in gentle body contacts between a female snake and a courting male that produces the movement while lying beside or upon her. The undulation moves from posterior to anterior and is thought to stimulate mating. Used by J.A. Oliver (*The Natural History of*

North American Amphibians and Reptiles, p. 225, 1955). Cf. *cardiovascular facilitative movement*.

cave or **cavern**. *n*. A naturally formed underground chamber.

cavernarius. *adj*. See *cavernicolous*.

cavernicolous. *adj*. Living in subterranean caves or passages. (Syn. *cavernarius*)

cavernosus. *adj*. Having internal cavities.

Cave Salamanders. Vernacular name for several species of plethodontid salamanders belonging to the genus *Gyrinophilus*.

cavum arteriosum. A dorsal subcompartment of the undivided ventricle of non-crocodilian reptiles, located dorsal to the *cavum pulmonale* and *cavum venosum*. This compartment receives oxygenated blood from the left atrium and communicates with the *cavum venosum* through the *interventricular canal*. See Fig. 36.

cavum dorsale. The larger of the two cavities in the single ventricle of noncrocodilian reptiles, lying to one side of the *interventricular septum* (= *muscular ridge*) and named by P.N. Mathur (*Proc. Indian Acad. Sci.* B, 20:1–29, 1944). This is the dorsal cavitation into which the atrioventricular orifices empty and from which the right and left aortic arches arise. It is further subdivided into the *cavum arteriosum* and the *cavum venosum*.

cavum pulmonale. The smaller of the two principal cavities in the single ventricle of noncrocodilian reptiles, lying on the left side of the *interventricular septum*. The *pulmonary artery* opens from the cavum pulmonale. May be synonymized with *cavum ventrale*. See Fig. 36.

cavum ventrale. The ventral cavity of the heart of noncrocodilian reptiles, bounded by a *muscular ridge* on the right, and from which the pulmonary artery is derived. The term originated with descriptions by P.N. Mathur (*Proc. Indian Acad. Sci.* B, 20:1–29, 1944) and may be synonymized with *cavum pulmonale* (F.N. White, *Amer. Zoologist* 8:211–219, 1968).

cavum venosum. A subcompartment of the single ventricle of noncrocodilian reptiles, which lies to the right side of the *interventricular septum* (= *muscular ridge*), receives oxygenated blood from the cavum arteriosum via the *interventricular canal*, and ejects blood into the two aortic arches. See Fig. 36.

CBSG. See *Captive Breeding Specialist Group*.

cc. Cubic centimeter. See *milliliter*.

CCK, CCK-PZ. See *cholecystokinin, cholecystokinin-pancreozymin*.

cDNA. (copy DNA) Single-stranded, complementary DNA produced from an RNA template by the action of RNA-dependent, DNA polymerase (reverse transcriptase) *in vitro*.

cDNA library. A collection of *cDNA* molecules, representative of all the various mRNA molecules produced by a specific type of cell of a given species, spliced into a corresponding collection of cloning vectors such as plasmids.

CD-ROM. Compact Disc Read Only Memory. A disc that contains digitally encoded information readable only by a computer.

cecal fermentation. Reference to digestion of food by microorganisms in the ceca of intestines.

cecum, caecum. (*pl.* –a) *n*. (*adj*. -al) Any blindly ending pouch, but typically has reference to outgrowths of the gut or digestive tract.

celiotomy. *n*. Surgical opening and entry into the body (coelomic) cavity.

cell. *n*. The smallest, membrane-bound protoplasmic body capable of independent reproduction.

cell cycle. The sequence of events between one mitotic division of a cell and another.

cell division. The process by which two daughter cells are produced from a single parent cell.

cell line. A group of cells derived from a primary culture at the time of the first transfer.

cell mass. See *cell nest*.

cell nest. 1. Solid groups of cells arranged as a syncytium in the salamander intestine just below the epithelium. First described by Hoffman in 1878. S.F. Patten Jr. (*Exp. Cell. Res.* 20:638–641, 1960) demonstrated these cells furnish replacement cells to the epithelium, thus being analogous to the intestinal crypts of other tetrapods. (Syn. *cell mass, epithelial mass*) 2. The group of cells in the testis of frogs that gives rise to a bundle of sperm.

cellular. *adj*. Of, or pertaining to, cells.

cellular automata (CA) models. In ecological modeling, the use of modern computers to iterate the consequences of local interactions between entities (S. Wolfram, *Theory and Applications of Cellular Automata*, 1986). CA models have been used to model plant growth, population dynamics, community interactions, and community spatial dynamics.

cellular immunity. Immune responses carried out by active cells rather than by antibodies.

cellular signal transduction. The process and pathways by which a cell receives external signals and transmits, amplifies, and directs them internally.

cellulitis. *n*. Inflammation of connective tissue.

Celsius scale. (°C) A temperature scale with divisions of °C (equal to 5/9 [°F - 32]). The scale measures the freezing point of water at 0 °C and the boiling point of water at 100 °C at normal atmospheric pressure. Equivalent to *centigrade scale*, which was officially renamed Celsius in 1948. The Celsius scale is named after the Swedish astronomer Anders Celsius (1701-1744), who devised it in 1742 but in reverse (i.e. freezing point was 100° and boiling point 0°). Cf. *Fahrenheit scale*. See also *absolute zero*.

cement, cementum. *n*. A thin layer of calcified tissue that covers the enamel layer at the root of a tooth, aiding in holding the tooth in place in its socket. Cement resembles compact bone and attaches directly to the dentine of the tooth root. It has been observed rarely in small nonmammalian amniotes but is often present in large nonmammalian amniotes (e.g., sauropterygians, mosasaurs). J.W. Osborn (*Symp. Zool. Soc. Lond.* 52:549–574, 1984) identified a 'protocement' tissue covering the teeth of all nonmammalian, nonarchosaurian amniotes, but the tissue is probably synonymous with cement because there are no histological features distinguishing them.

cement gland cells. A packed group of apical, pigmented, ectodermal cells, situated ventral to the cranial neural folds of anuran larvae. They become about ten times the height of adjacent epithelial cells and are filled with secretory mucin or glycoprotein. They eventually lose their pigment and disappear.

cement organ. See *adhesive organ*.

cementosome. *n*. See *lamellar bodies*.

census. *n*. In ecology, a procedure for counting all members in a population.

Cenozoic. *n*. The most recent Era of the *Phanerozoic Eon* of geologic time, incompassing the *Tertiary* and *Quaternary* Periods. See Table 1. (Syn. *Caenozoic*)

Center for Applied Biodiversity Science. (CABS) An arm of *Conservation International* located in Washington, DC.

Center for North American Herpetology. (CNAH) A nonprofit organization founded by J.T. Collins in 1994 for the purpose of promoting research on the preservation and conservation of North American amphibians, crocodilians, reptiles, and turtles through education and dissemination of information. Headquartered in Lawrence, Kansas, the CNAH maintains an online list of current scientific and standard common names for North American herpetofauna. Current URL: http://www.naherpetology.org/

center of buoyancy. The point in an immersed body that represents the center of gravity of the displaced water, and the point through which the resultant force of buoyancy acts.

center of gravity. The point in a body through which the resultant force of gravity acts. Also the point from which a body can be suspended in equilibrium in any position.

center of origin. Reference to an area from which a given taxonomic group of organisms has originated and spread. The *hypothesis* of this name is the generalization that genetic variability is greatest in the area where a species arose. Conversely, marginal populations are likely to exhibit a more limited number of adaptations.

centi-. (c) Prefix used to denote x 10^{-2}.

centigrade scale. See *Celsius scale*.

centimeter. (cm) A metric (*cgs*) measure equal to one hundredth of a meter (10^{-2} meters).

Centipede Snakes. Vernacular name for American species of snakes belonging to the colubrid genus *Tantilla* (also called *Black-headed Snakes*).

central. *n*. 1. Refers to any of the unpaired, middorsal laminae on the carapace of turtles. (Syn. *vertebral, dorsal, neural*) See Fig. 4. 2. Used by J.E. De Kay (*Zoology of New York*, 1842, p. 37) in reference to the *frontal* plate in snakes.

Central American Coral Snake. Vernacular name for the coral snake species *Micrurus nigrocinctus*.

central chemoreceptors. Sensory structures in the brain that monitor pH and initiate appropriate changes in breathing.

central circulation. That part of the blood circulation that includes the heart and the major vessels associated with it.

central nervous system. (CNS) All neurons that comprise the brain and spinal cord. Cf. *peripheral nervous system*.

central pattern generator. A group of neurons that produces and maintains a rhythmic pattern of action, such as breathing, walking, chewing, etc.

centrifugal. *adj*. Acting in a direction outward from a center of rotation. Cf. *centripetal*.

centrifugal selection. See *disruptive selection*.

centripetal. *adj*. Acting in a direction inward toward a center of rotation. Cf. *centrifugal*.

centripetal selection. See *stabilizing selection*.

centrolateral. See *costal*, also Fig. 4.

Centrolenidae. *n*. A clade (family) of anurans sharing the presence of a medial process on the third metacarpal as a diagnostic character. Most centrolenids are small (<30 mm) and frequently possess a transparent venter, giving rise to the common name of *Glass Frogs*. Four genera, including *Allophryne*, and more than 135 species occur in Central and South America. See D.R. Frost et al., *Bull. Amer. Mus. Nat. Hist.* 297:1–370, 2006

centrolenid. *n*. A member of, or pertaining to, Centrolenidae.

centrotype. *n*. In numerical taxonomy, this term refers to the *operational taxonomic unit* that lies closest to the geometric center of the cluster of such units.

centrum. (*pl.* **centra**) *n*. The body or basal portion of a vertebra formed by replacement of the original notochord and distinct from the neural arch above it.

cephalic. *adj*. Of or pertaining to the head.

cephalic amplexus. See *amplexus*.

cephalic plate. The enlarged scales on the top of the head in reptiles.

cephalization. *n*. The anatomic and evolutionary tendency to concentrate neurosensory and feeding mechanisms at the head or head end of a body.

ceramides. *n*. A class of lipids consisting of a long-chain or sphingoid base linked to a fatty acid via an amide bond. Ceramides are formed as key intermediates in the biosynthesis of all complex sphingolipids, in which the terminal primary hydroxyl group is linked to carbohydrate, phosphate, etc. Ceramides are an important component of stratum corneum lipids, and together with free fatty acids and cholesterol contribute to the efficacy of the water permeability barrier.

Ceratobatrachidae. *n.* A taxon (family) of anurans, sister to the new taxon *Telmatobatrachia* within the new taxon *Victoranura*, and containing 6 genera with species distributed in Asia and various islands of the Pacific.

ceratobranchial. *n.* The principal element in the hyoid apparatus, derived from the branchial arches I-IV and occupying posterior lateral positions. In adult amphibians this element fuses with the *hypobranchial* to form a single cartilage. Also has been used to refer to posterior horns on the hyoid of some reptiles. (Syn. *epibranchial, ceratohyal, posterior cornu, thyrohyal*)

ceratohyal. *n.* Paired anteriolateral cartilaginous or ossified processes of the horn of the hyoid apparatus, derived from the hyoid arch. Sometimes there is attachment of this element with the hypohyal. When there is loss or fusion of parts of the hyoid horn, the single cartilage is usually called a ceratohyal. (Syn. *anterior cornu, hypohyal* in snakes, *principal horn* in anurans)

Ceratophryidae. *n.* A recently revised taxon (family) of anurans, sister to the new taxon *Hesticobatrachia* within the new taxon *Chthonobatrachia*, and containing seven South American genera. See D.R. Frost et al., *Bull. Amer. Mus. Nat. Hist.* 297:1–370, 2006.

Cercosaurinae. *n.* A subfamily of *Gymnophthalmidae*.

cerebellum. *n.* A part of the *hindbrain*, dorsal to the *medulla oblongata*, that is important in motor coordination. A derivative of the dorsal part of the metencephalon.

Cerebond™. *n.* A new custom, space-filling adhesive formulated to bond stainless fixtures to bone, but could also be applied to other uses. The new formulation is intended to replace methyl methacrylate dental cement, which has been reclassified by the U.S. Postal Service as a hazardous material. Cerebond contains no hazardous materials, is noninflammatory, odor-free, and will not irritate cut skin in either the cured or uncured form. This new product is offered by MyNeurolab.com.

cerebral artery. A branch of the internal carotid artery that supplies blood to the brain.

cerebral cortex. The outer layer of cells (grey matter) covering the two hemispheres of the cerebrum.

cerebral hemispheres. The large paired structures of the *cerebrum*.

cerebrospinal fluid. Fluid within the cavities of the brain and spinal cord, and within the subarachnoid space.

cerebrum. *n.* The largest part and highest center of the brain, evolved from the olfactory centers of lower vertebrates and most prominent in mammals. The hemispheres are derived from the telencephalon. (Syn. *telencephalon*)

cervical. *adj., n.* 1. Pertaining to the neck. 2. The *nuchal* lamina of turtles (Fig. 4), or *nuchal* plate in crocodiles.

cervical vertebra. A neck or presternal vertebra without ribs. This is a single element in amphibians and allows little head movement. Increased numbers (usually seven) of cervical vertebrae in reptiles increases flexibility of movement of the head.

CEWL. See *evaporative water loss*.

cf. Abbreviation of the Latin *confer*, meaning "compare (with)."

cgs system. A system of scientific units based on the centimeter-gram-second.

Chacoan Monkey Frog. Vernacular name for the phyllomedusine *Phyllomedusa sauvagii*. See *Monkey Frog*.

challenge display. A "high intensity" display associated with aggressive performance toward another male of the same species. C.C. Carpenter (*Copeia*, 1961:396–405, 1961) analyzed this display in reference to eight categories of *display behavior*. See also *assertion display*.

Chamaeleonidae. *n.* A clade (family) of lizards that includes the *True Chameleons*, easily recognized by possession of casques, horns, and crests on the head in most species. Most chameleons have laterally compressed bodies and the ability to change color. Six genera and approximately 130 species are recognized in Africa and Madagascar, the Middle East, India and Sri Lanka, and southern Spain.

Chameleon. *n.* Common name for various species of lizards belonging to the *Chamaeleonidae* ("True" Chameleons). See also *Dwarf Chameleons, Leaf Chameleons, Stumptail Chameleons*.

chaos theory, chaos. Behavioral description or analysis of certain nonlinear dynamical systems, which characteristically exhibit sensitivity to initial conditions (popularly referred to as the "butterfly effect") and behavior that appears to be random, even though the system is deterministic in the sense that it is well defined and contains no random parameters. The word chaos (derived from Greek) typically refers to unpredictability, in the metaphysical sense the opposite of law and order. Mathematically, chaos describes an aperiodic deterministic behavior that is very sensitive to its initial conditions.

chaparral. *n.* A dense shrubland community associated with temperate Mediterranean-climate regions and characterized by a long, warm dry season, low rainfall during the cool season, varied sclerophyllous evergreen shrubs, and a diverse fauna. The term is usually applied to regions of the west coast of North America, especially California, but may also be used in reference to similar communities in coastal Chile, South Africa, and the Mediterranean region where other local names also are in use.

chaperones. *n.* Proteins that assist various nascent polypeptide chains fold correctly into their tertiary shapes while stabilizing and protecting them in the process, or preventing them from making premature or inappropriate intermolecular associations. (Syn. *chaperonins, molecular chaperones*)

chaperonins. *n.* See *chaperones*.

character. *n.* Any observable phenotypic property of an organism transmitted from parent to offspring and used as a basis for comparison. In systematics, a character is a feature considered to vary independently of other features and to be homologous among the taxa of interest. (Syn. *phenotype, trait*)

character congruence. Agreement among characters for particular phylogenetic relationships. Cf. *character incongruence.*

character displacement. Reference to exaggeration in phenotypic differences between species in areas of *sympatry* relative to areas of *allopatry*. The phenomenon has an adaptive basis and functions to mitigate competition in regions of contact between the two species. See also *niche diversification.*

character gradient. Reference to a *cline.*

character hierarchy. A system of sequentially dependent characters.

character incongruence. Disagreement among characters for particular phylogenetic relationships. Cf. *character congruence.*

character independence. Reference to two or more characters that are uncorrelated across evolutionary time or, in *phylogenetics*, from node to node along the branches of a tree.

character index. A numerical value that quantifies the extent of difference between two taxa.

character polarity. The temporal direction of change between alternative (*primitive* and *derived*) states of a character.

character release. Reference to increase in variation of certain phenotypic characters associated with *ecological release* of a species.

character state. Any of a suite of different expressions of a character in different organisms. The particular condition or expression of a feature in a taxon. It may be any one of the alternative conditions of a character.

character state reconstruction. The process of estimating the ancestral or primitive condition of a character at a given node (branching point) in a *phylogenetic tree.*

character state tree. Reference to a branching or linear sequence of character states in a transformation series.

character weighting. The assignment of different values to taxonomic characters.

characteristic. *n.* A distinguishing or diagnostic feature, sometimes used loosely as a synonym for *character.*

characteristic fossil. See *index fossil.*

characteristic species. See *indicator species.*

Chekered Garter Snake. Vernacular name for the American colubrid species *Thamnophis marcianus.*

checklist. *n.* A list of species from a particular region or taxon, often published with identification keys.

cheironym. *n.* An unpublished (manuscript) scientific name.

chelation. *n.* Combination or binding of a substance with a metal ion.

Chelidae. *n.* A clade (family) of Sideneck Turtles characterized by extensive emargination of the cheekbones such that only a parietal-squamosal bar remains. These mostly inhabit swamps or slowly moving waters in South America, Australia, and New Guinea. Eleven genera and about 50 species are recognized.

Chelonia. *n.* See *Testudines.*

Chelonian Conservation and Biology. An international journal of turtle and tortoise research, first published in 1993.

Chelonian Research Foundation. A nonprofit organization founded in 1992 for the support of research on turtles and tortoises worldwide with emphasis on diversity and conservation. Current URL: http://www.chelonian.org/

Cheloniidae. *n.* A clade (family) of Sea Turtles, often treated together with *Dermochelyidae* as a clade relative to other turtles. Five genera and perhaps six species are recognized, with worldwide distribution in temperate and tropical oceans. Common names of species are *Kemp's Ridley, Olive Ridley, Flatback Seaturtle, Green Seaturtle, Hawksbill,* and *Loggerhead Seaturtle.*

Chelonioidea. *n.* A clade (superfamily) of extant turtles that includes *Cheloniidae* and *Dermochelyidae.*

Chelydridae. *n.* A clade (family) of turtles, largely aquatic and possessing heavy bodies, large heads, and powerful jaws, including the Snapping Turtles of North America. Three species are recognized in two subfamilies: *Chelydrinae* (North America to Central America and Ecuador) and *Platysternidae* (southeastern Asia).

Chelydrinae. *n.* A subfamily of *Chelydridae.*

chemical energy. Energy contained in the chemical bonds of molecules.

chemoreception. *n.* The transduction and perception of chemical stimuli, resulting in capacity to detect and differentiate certain chemicals in the environment.

chemoreceptor. *n.* A sensory receptor that is specifically sensitive to certain molecules, sampling chemicals in the environment that humans perceive as odor or taste.

chemosensory. *adj.* Of or pertaining to taste or olfaction, and the related function of the vomeronasal organs.

chemotactic peptides. See *mast cell disrupting peptides.*

chemotaxis, chemotropism. *n. (adj.* –**ic**) The directional attraction or repulsion of an organism toward or away from a diffusing chemical.

chemotherapy. *n.* Prevention or treatment of disease by means of drugs.

chevron. *n.* 1. A V-shaped marking, usually with the point of the V on the mid-dorsal line and contrasting with the ground color. 2. Small V-shaped bones attached to the ventral surface of most caudal vertebrae in reptiles and mammals. These protect caudal vessels and are probably homologous with the hemal arches of some amphibians. (Syn. *hemal arch*)

chiasma. *n.* A crossing of fibers (e.g., as of optic nerves).

Chicken Turtle. Vernacular name for emydid turtles belonging to the genus *Deirochelys.*

chief cells. The presumed source of secreted digestive enzymes in the *fundus* of the stomach.

Children's Python. Vernacular name for the boid species *Antaresia childreni*.

chimera. *n.* Individual formed by grafting together major parts of different species, or one that is formed from abnormal chromosome distribution after normal fertilization. In either case, the essense is that an individual is composed of a mixture of genetically different cells.

Chinese Alligator. Vernacular name for the Asian alligatorid species *Alligator sinensis*. See *Alligator*.

Chinese Crocodile Lizard. Vernacular name for the Asian lizard *Shinisaurus crocodilurus*, placed by some in a separate family, Shinisauridae.

Chinese Green Tree Viper. See *Green Bamboo Viper*.

chin gland. See *mental gland, musk gland*.

chin pad. See *mental gland*.

chin shield. 1. Any of paired, elongated scales on the lower jaw of snakes and situated behind the first pair of infralabials on either side of the midline. (Syn. *geneial, genial, inframaxillary, sublingual, submental, thin shield, mental*) See Fig. 12. 2. The term also applies to a series of paired scales on the lower jaw of lizards, posterior to the mental scale. (Syn. *submaxillary, submental*) See Fig. 15.

chiridium. (*pl.* –**ia**) *n.* A limb, or muscular appendage with well-defined joints bearing digits, and not a fin, at its end.

chiropterygium. (*pl.* **-ia**) *n.* The jointed, fingered, tetrapod limb of a vertebrate.

Chirping Frog. Vernacular name for African ranid frog species belonging to the genus *Arthroleptella*.

Chisel-teeth Lizards. An alternate common name for *agamids*, or lizard species belonging to the Agamidae.

chi-squared stastitic. (χ^2) A statistic calculated as the sum of a set of terms, each being the quotient of the squared difference between an observed frequency and the corresponding expected frequency divided by the expected frequency.

chi-squared test. A statistical procedure that allows an investigator to determine how closely an observed set of values fits a given theoretical expectation, using the chi-squared statistic. See also *goodness of fit*.

Chitty hypothesis. See *polymorphic behavior hypothesis*.

choana. (*pl.* **-ae**) *n.* A funnel-like opening, used in reference to the internal openings of nasal passages in the roof of the mouth or buccopharynx. See also *naris*.

choanal glands. Multicellular secretory glands in the palatal epithelium of probably most amphibians. These empty into the choanae and are very large in anurans, sometimes nearly occluding the lumen of the choanal canal.

choanal papilla. Papillate or fimbriate flaps of skin projecting from the lateral borders of the choanae of Green Sea Turtles, *Chelonia mydas* (T.S. Parsons, *Breviora*, 85:1–7, 1958) and Soft-shelled Turtles, *Amyda spp.* (R.G. Webb, *U. Kans. Publ. Mus. Nat. Hist.* 13:429–611, 1962). (Syn. *narial flap*)

cholecystokinin, cholecystokinin-pancreozymin. (**CCK, CCK-PZ**) *n.* A hormone and putative neurotransmitter, once thought to be two hormones (hence the hyphen), secreted from the intestinal mucosa, which stimulates ejection of bile and secretion of pancreatic digestive enzymes. Peptides homologous to mammalian CCK have been characterized in brain, stomach, and small intestine of amphibians and reptiles, and the CCK peptide appears to have originated early in vertebrate evolution.

cholesterol. *n.* A natural sterol and precursor to steroid hormones; also a ubiquitous component in skin and other tissues of tetrapods, important to membrane function.

cholinergic. *adj.* Reference to a nerve fiber that liberates acetylcholine as a neurotransmitter.

cholinesterase. *n.* An enzyme that catalyzes the hydrolysis of choline esters into choline and the appropriate acid (e.g., acetylcholinesterase). Such enzymes are components of some snake venoms.

chondral. *adj.* Pertaining to cartilage.

chondral bone. Bone that arises in or around cartilaginous precursors. See also *endochondral bone, perichondral bone*.

chondrification. *n.* The process of conversion into cartilage.

chondro-. Prefix pertaining to cartilage.

chondroblast. *n.* A cell which secretes the matrix of cartilage.

chondroclast. *n.* A cell which lyses the matrix of cartilage.

chondrocranium. *n.* The cartilaginous embryonic brain case, or the part of the adult brain case derived from it, including the fused or associated nasal capsules. The part of the head skeleton other than the *splanchnocranium* and consists of cartilage or replacement bone.

chondrocyte. *n.* Any cell in a cartilaginous matrix.

chorda dorsalis. Used by H. Gadow (*Amphibia and Reptiles*, 1901, p. 654) in reference to *notochord*.

chordal cartilage. Historically used in reference to *perichordal tube* and, more commonly, *autotomy plane*.

chordal centers, chordacentra. In vertebrate embryology, a chain of cartilaginous elements differentiated from the notochord and giving rise to *arch centers* that eventually form the vertebrae. Cf. *arcuale*.

chordal sheath. Historical synonym for *perichordal tube*.

chordamesoderm. *n.* The roof of the archenteron, which forms notochord and mesoderm and induces the neural plate.

Chordata. *n.* The phylum of animals characterized by possession of a notochord, a dorsal hollow nerve cord, and gill slits at some developmental stage. Amphibians and reptiles are members of the Chordata (*chordates*).

chordate. *n.* A member of the chordata.

chorioallantoic membrane, chorioallantois. *n.* An extraembryonic membrane formed by union of the chorion and allantois.

chorioallantoic placenta. The membrane resulting from fusion of chorion and outer allantoic membrane at the embryonic pole in viviparous vertebrates. The membrane associates with the uterine wall and functions as a placenta in allowing exchange between maternal and fetal circulations. Close association of maternal and embryonic vascular beds suggests this is the likely site for gas exchange. (Syn. *allantoplacenta, seroallantoic placenta*) Cf. *omphaloplacenta*. For terminology discussion see J.R. Stewart & D.G. Blackburn, *Copeia* 1988:839–852, 1988.

chorion. *n.* The outermost extraembryonic membrane in amniotes, which surrounds the embryo and yolk sac. It becomes vascularized by the allantoic vessels.

chorioplacenta. *n.* A term presented as a theoretical concept that invoked initial metabolic activities of the extraembryonic ectoderm in contact with the uterine epithelium. Emphasis is entirely on the ectodermal layer, and a true chorionic placenta has yet to be described in squamates. For terminology discussion see J.R. Stewart & D.G. Blackburn, *Copeia* 1988:839–852, 1988. Cf. *chorioallantoic placenta, choriovitelline placenta*.

choriovitelline placenta. A transitory "yolk sac placenta" forming a vascularized area around the embryo during early development. Apposition of the vascular trilaminar yolk sac and uterine epithelium. Present in all viviparous reptiles. For terminology discussion see J.R. Stewart & D.G. Blackburn, *Copeia* 1988:839–852, 1988.

choroid. *n.* The middle, vascular layer of the brain or eye, located between the *sclera* and *retina* and providing nutritional support for the ocular tissues.

choroid plexus. Infolding of pial blood vessels and ependyma that lie within the ventricles of the brain and is responsible for the production of cerebrospinal fluid.

chorus. *n.* Calls produced by an aggregation of male frogs, usually in association with breeding periods.

Chorus Frogs. Collective vernacular name for species of North American frogs belonging to the hylid genus *Pseudacris*.

chrom(o)-. Word element meaning "color."

chroma. *n.* The perceived difference (as a relative value) between a color and an achromatic percept of the same brightness. Cf. *saturation*.

chromaffin. *n.* Reference to endocrine tissue that is functionally related to the adrenal medulla, but may be diffuse. This tissue stains strongly with chromium salts and is the source of *catecholamines*.

chromatin. *n.* A complex of nucleic acids and proteins that comprises eukaryotic chromosomes.

chromatograph. *n.* An instrument used in chemical analysis of gases and liquids.

chromatography. *n.* A widely used technique for separating and identifying the components present in a mixture of molecules that have similar chemical and physical properties. The technique generally exploits the fact that different components in a sample will move at different rates through a substrate such as chromatography paper or another porous solid matrix.

chromatophore. *n.* A generalized term for a cell that contains pigment and imparts skin color. Principal types are *erythrophores* (red), *melanophores* (brown to black), *iridophores* (silvery white), and *xanthophores* (yellow). See Figs. 28, 30, 33, 34.

chromatophore unit. See *dermal chromatophore unit* and Fig. 33.

chrome. *n.* Refers to a pigment in biological usage.

chromomycosis. *n.* An infection caused by various pigmented fungi, affecting anurans.

chromosomal inheritance. Inheritance of characters by means of chromosomes of the nucleus. (Syn. *Mendelian inheritance*)

chromosome. *n.* one of a number of deeply staining nuclear threadlike structures comprised of *chromatin* that carries genetic information organized in a linear sequence. For a given organism, a *chromosome set* represents a genome and consists of one representative from each of the chromosome pairs characteristic of the somatic cells in a diploid species. Chromosomes are visible as morphological entities only during cell division and usually occur as pairs (2N) in somatic cells and are unpaired (1N) in the sex cells.

chromosome complement. The number of chromosomes in a nucleus.

chromosome mapping. Determination of the gene sequences on a chromosome.

chromosome set. See *chromosome*.

chromotaxis. *n.* Reference to animals seeking a matching background on which they are camouflaged.

chronic. *adj.* Reference to exposure of relatively long duration, in contrast to *acute*.

chrono-. Combining form meaning "time."

chronobiology. *n.* The study of biological rhythms.

chronocline. *n.* In paleontology, reference to a cline or character gradient in the time dimension.

chronospecies. *n.* 1. A species that can be studied from its fossil remains extending through a defined period of time. (Syn. *paleospecies*) 2. Successive species that replace each other in a phyletic lineage and are given ancestor-descendant status according to geological time sequence.

chronostratigraphy. *n.* The study of geologic history based on an analysis of the age of distinctive rock layers and their time sequence.

chronotropic. *adj.* Pertaining to rate or frequency, especially in reference to the heartbeat.

Chthonobatrachia. *n.* A new monophyletic taxon of anurans, sister to the new taxon *Diphyabatrachia* within the new taxon *Leptodactyliformes*, and composed of *Ceratophryidae* and the new taxon *Hesticobatrachia* (D.R. Frost et al., *Bull. Amer. Mus. Nat. Hist.* 297:1–370, 2006).

Chuckwalla. *n.* Vernacular name for American species of iguanid lizards belonging to the genus *Sauromalus*, especially the relatively large saxicolous, herbivorous species *Sauromalus ater* (formerly *Sauromalus obesus)* and *Sauromalus obesus*, common to arid regions of the southwestern United States and Mexico. These two species are synonymized by B.D. Hollingsworth (*Herpetol. Monogr.* 12:38–191,1998).

chyme. *n.* A liquefied mass of partially digested food after it enters the intestine. (Syn. *digesta*)

chymotrypsin. *n.* A class of protease enzymes produced in, and secreted from, the pancreas, and subsequently acting to hydrolyze peptide and ester bonds in the small intestine.

chytrid, chytrid fungus. *n.* In herpetological literature, reference to a species of fungus in the phylum Chytridiomycota (chytrids), specifically *Batrachochytrium dendrobatidis* (sometimes referred to as "Bd"), that has been shown to cause fatal skin disease in frogs and toads. Infections by this pathogen has devastated captive collections of amphibians and is believed to be a significant factor in the declines of many wild anuran populations. For a history of the identification of this pathogen in amphibian declines, see D.K. Nichols, *Herpetol. Rev.* 34:101–104, 2003. Recently, it has been suggested that the amphibian chytrid originated in Africa and was disseminated by the international trade in *Xenopus laevis* (C. Weldon et al., *Emerging Infectious Diseases* 10:2100–2105, 2004). See *chytridiomycosis*.

chytridiomycosis. *n.* The skin disease of amphibians caused by *Batrachochytrium dendrobatidis*, a chytrid fungus (L. Berger et al., *PNAS* 95:9031–9036, 1998). Chytridiomycosis may be treated with antifungal medication such as trimethoprimsulfadiazine (TMS), miconazole, or itraconazole. See *chytrid*.

CI. See *body condition index*.

cilia. (*s.* **–ium**) *n.* Populations of microscopic, hairlike processes that extend from the surfaces of various cells comprising a ciliated epithelium. These are motile organelles with a microtubular substructure, generally found in large numbers.

ciliary. *n.* 1. Small scales bordering the edge of the eyelid in lizards and crocodilians. 2. *adj.* Pertaining to, or associated with, the eye, eyelid or orbit, and frequently used with prefixes super-, supra-, and infra-, to indicate position of structure relative to the eye. 3. Relating to a hair-like structure.

ciliary body or **muscle.** A tiny ring of muscle that focuses the lens of the eye.

ciliary cushion. Structures hanging between the filter plates in the dorsolateral pharynx of anuran larvae, consisting of ciliary cells and goblet cells originating from the esophagus. Structurally, these are similar to *ciliary grooves* and in close contact with them, producing and transporting mucus with entrapped food particles to the esophagus (R. Wassersug, *Occas. Pap. Mus. Nat. Hist. Univ. Kansas* 48:1–23 and 49:1–24, 1976).

ciliary groove. Reference to a horizontal groove located at the margin of the pharyngeal roof, extending from the anterolateral corner of the ventral velum into the esophagus. The histology and function are similar to the ciliary cushion (R. Wassersug, *Occas. Pap. Mus. Nat. Hist. Univ. Kansas* 48:1–23 and 49:1–24, 1976).

ciliary muscle. See *ciliary body*.

ciliated. *adj.* Possessing cilia.

circa. (ca.) Latin, meaning "approximately" or "about."

circadian rhythm. A cycle of behavior or physiological process that is repeated daily on a 23–25 h interval, attributable to an "internal clock" but synchronized to time-related factors in the environment. *Circadian* refers to more or less daily repetition (e.g., 23–25 h). (Syn. *diurnal rhythm*)

circannual cycle or **rhythm.** A biological rhythm with a yearly cycle, usually in reference to annually cyclic behaviors such as breeding, hibernation, migration, etc., based in part on physiological changes linked to exogenous factors such as day length.

circular backward burrowers. Reference to anurans that construct burrows by digging with the feet while rotating in a corkscrew pattern.

circular folds. See *folds of Kerckring*.

circular smooth muscle. The inner layer of smooth muscle that encircles the small intestine.

circulatory system. See *cardiovascular system*.

circum-. Prefix meaning "about" or "around."

circumaustral. *adj.* Reference to distribution around the high latitudes of the southern hemisphere. Cf. *circumboreal*.

circumboreal. *adj.* Reference to distribution around the high latitudes of the northern hemisphere. Cf. *circumaustral*.

circummarginal groove of disk. The deep indentation bordering the outer margin of the disk on expanded finger tips of certain species of *Rana*. Used by R.F. Inger (*Fieldiana, Zool.* 33:15–754, 1954).

circumnarial. *n.* See *circumnasal*.

circumnasal. *n.* 1. One of the row of scales that surrounds the nasal protuberance in crocodilians. 2. Also used to describe large scales surrounding the nostril of geckos, except for the *rostral* and *labial* scales. (Syn. *circumnarial*)

circumocular series. Used to describe scales surrounding the eye of certain lizards.

circumorbital. *n.* See *supraorbital semicircle*.

circumscription. *n.* In systematics, the defined limits of a taxon, or the sum of individuals within those limits, as defined by an author.

circumtropical. *adj.* Encircling Earth in a zone within the tropics.

cirque. *n.* A steep-sided, amphitheater-like hollow carved into a mountain at the head of a glacial valley.

cirrate. *adj.* Coiled, or forming a slender spiraling coil. Used with reference to the hemipenis of snakes.

cirrus, cirrhus. (*pl.* -ri) *n.* A slender, flexible appendage. For example, the projection that extends from the upper lip below the nostril, traversed by the nasolabial groove, in some plethodontid salamanders. Also synonymous with *tentacle* of caecilians or *Xenopus*.

cis-. Prefix meaning "on the same side as."

cismontane. *adj.* Reference to this (the speaker's) side of the mountains. In California, the reference is to geographic regions west of the Peninsular Ranges. Used extensively, for example, by G. Pickwell (*Amphibians and Reptiles of the Pacific States*, 1947). Cf. *transmontane*.

cisterna. *n.* A flattened, fluid-filled reservoir enclosed by a membrane.

cistron. *n.* A gene or transcriptional unit of DNA. The original term (S. Benzer, 1955) referred to the segment of DNA that specified the formation of a specific polypeptide chain, but was subsequently expanded to include the transcriptional start and stop signals.

CITES. Abbreviation for *Convention on International Trade in Endangered Species of Wild Fauna and Flora*. CITES regulates the international trade in specimens of wild fauna and flora.

citrate cycle. See *citric acid cycle*.

citric acid cycle. A cycle of enzyme-controlled reactions in which energy released in the oxidation of fats, proteins, and carbohydrates is used to form ATP from ADP; molecules of CO_2 and H_2O are formed as by-products. The enzymes are localized in mitochondria. (Syn. *citrate cycle, Krebs cycle, tricarboxylic acid cycle*)

clade. *n.* 1. Any group of organisms defined by characters exclusive to all its members and that distinguish the group from all others. 2. In evolutionary biology, a taxon or other group consisting of a single species and its descendents, representing a distinct branch on a *cladogram* or *phylogenetic tree*. A *monophyletic group*. See *cladogram, phylogenetic tree*.

cladist. *n.* Informally, a systematist who is a follower of the Hennigian tradition of systematics methodology.

cladistic distance. Reference to the number of branching points between any two points on a phylogenetic tree.

cladistic evolution. See *cladogenesis*.

cladistics. (cladism, cladistic method) *n.* A method of classification that reconstructs phylogenetic sequences by deductive processes that analyze primitive and derived character states of related organisms to generate dichotomously branched sister groups. Evolutionary relationships are the basis for classification, and the criterion for establishing groups of organisms is the recency of common ancestry, based on the identification of shared, derived characters (*synapomorphies*). The graphic representation of such an analysis is the *cladogram*. Cf. *evolutionary method, omnispective method, phenetic method*.

cladistic species concept. See *species*.

clado-. Prefix meaning "branch" or "offshoot."

cladogenesis. *n.* Branching evolution, with lineages splitting into two or more lineages. Cf. *anagenesis, phyletic evolution*. (Syn. *cladistic evolution, dendritic evolution*)

cladogram. *n.* A branching diagram constructed by cladistic methods. A cladogram represents a diagrammatic analysis of phylogenetic relationships among organisms that relates them according to suites of common characters. The construction attempts to represent the true evolutionary branchings of a lineage through time from its ancestral taxon, based on shared characters derived from a common ancestor. Cladograms show common ancestry, but do not indicate the amount of evolutionary time separating taxa. Cf. *phylogram, phylogenetic tree*.

Cladophrynia. *n.* A newly designated monophyletic taxon considered to be sister with *Brachycephalidae* within the new taxon *Meridianura* and containing the new taxa *Cryptobatrachidae* and *Tinctanura* (D.R. Frost et al., *Bull. Amer. Mus. Nat. Hist.* 297:1–370, 2006).

Clark's Spiny Lizard. Vernacular name for the American iguanid species *Sceloporus clarkii*.

clasping. *n., adj.* 1. See *amplexus*. 2. An ecomorphological guild of anuran larvae that includes lotic tadpoles living in water with medium to slow currents and possessing marginal papillae with an anterior gap, in addition to other features of dentition and use of oral disc.

clasping reflex. The spasmodic grasping of any object by a male anuran when touched on the underside of the forelimb or in the region of the chest, during periods of mating activity. The reflex is a means of sex recognition and probably serves as a stimulus and aid to egg laying due to the pressure applied to a clasped female. (Syn. *Goltz's clasping reflex*)

clasping spell. The *total response time* between the first contact with a female frog and the unclasping of a male frog.

class. *n.* 1. A grouping used in classification of organisms, representing a subdivision of a *phylum*. In turn, a class is subdivided into *orders*. 2. In statistics, a category or group of like observations.

classical. *adj.* In taxonomy, reference to a name that is derived from Latin or ancient Greek.

classification. *n.* 1. An arrangement of living organisms into hierarchical groups according to their similarities and evolved ancestry. 2. A node-based name based on such an arrangement. 3. The process of grouping organisms according to features they have in common, or on the basis of their ancestry, or both.

clavate. *adj.* Club-shaped or thickened at one end.

clavicle. *n.* A dermal bone connecting the *scapula* to *sternum* in the pectoral girdle. This element is usually prominent in anurans, lizards, and tuatara, but can be reduced or absent. See Fig. 3.

claw. *n.* A horny point or tip on the end of a digit. In herpetological literature, this includes true claws of reptiles as well as the bony tip that penetrates the skin on several of the digits in some African ranid frogs and thickened, cornified caps on tips of toes in some other amphibians.

Clawed Frogs. Collective vernacular name for species of frogs belonging to the family *Pipidae.*

clay. *n.* Sediment composed of fine particles (diameter 2–4 μm) having colloidal properties. Sometimes this term is used for all sediment particles smaller than 4 μm in diameter.

clearance time. Reference to the period of time between ingestion of food and the elimination of feces derived from that food. (Syn. *passage time*)

clear cutting. A method of harvesting timber in which all trees within a given area are felled at one time, regardless of their size or age.

clear layer. Term given to innermost living cells of an *outer epidermal generation* in shedding squamates. The layer is so-named because the cytoplasm does not stain, and the nuclei appear to lie in a vacuole. The clear layer is part of the *stratum intermedium* where the *outer epidermal generation* and newly forming *inner epidermal generation* split at sloughing. For further discussion of terminology see P.F.A. Maderson, *J. Zool.* 146:98–113 (1965). See Fig. 31.

clear spot. See *light spot.*

cleavage. *n.* The process by which a dividing egg cell gives rise to all cells in an organism. Reference to the cell divisions that convert the zygote to a blastula.

cleavage initiating substance. Any factor capable of inducing cleavage in an amphibian egg, whether it continues to develop or not.

cleavage plane. See *autotomy plane.*

cleidoic egg. An egg of oviparous reptiles or birds that is enclosed within a protective shell or membrane permitting gaseous exchange but restricting water loss in a terrestrial situation. In reality, many of these eggs can absorb or lose water, and they are not truly waterproof. The "closed-box" definition of this term is attributable to presence of the shell and embryonic membranes (*amnion, chorion,* and *allantois*), but is not literal because exchange occurs between the egg and its environment.

cleithrum. *n.* A dermal element or "extra bone" lying alongside and fused with the chondral bones of the pectoral girdle in some amphibians, especially prominent in fossil forms.

climagraph, climograph. *n.* Common reference to a bivariate climatic chart or graph.

climate. *n.* The long-term average conditions of weather in a specified region. Cf. *weather.*

climate (or **climatic**) **change.** Literally a change in climate. Climate change can refer to differences between climatic conditions of two locations in the same season, or to changes over time at the same place. Current usage of this term often is in reference to changes in climatic conditions associated with *global warming.*

climate space. A term from *biophysical ecology*, denoting all limits of physical climatic and microclimatic variables determining conditions in which an animal can live. The climate space is a function of ambient temperature, absorbed radiation, convection, and metabolic heat production and demarcates a combination of limits related to maximum and minimum allowable body temperature, metabolic rate, conductance, and latent heat loss. See W.P. Porter & D.M. Gates, *Ecol. Monogr.* 39:245–270, 1969; W.P. Porter et al., *Aust. J. Zool.* 42:125–162, 1994.

climatic province or **region.** A region or division of Earth's surface characterized by a particular set of climatic conditions.

climatic rule. Reference to any generalization that describes a trend in geographical variation of animals that can be correlated with a gradient of climate.

climatology. *n.* The study of climate, including its effects on living organisms.

climax community. The endpoint of a classic facilitation model in which the community is diverse, self-replacing, and relatively stable. Disturbances can remove the climax community and restart the successional sequence. See *facilitation.*

climax phenomena. Reference to metamorphosis of larval anurans beginning at the time of forelimb emergence and including the period of tail resorption, tympanum development, and mouth modification. First used by W. Etkin (*Physiol. Zool.* 7:129–148, 1934).

climax species. A plant characteristic of a climax community.

Climbing Salamanders. Collective vernacular name for North American salamanders belonging to the genus *Aneides.*

cline. *n.* (*adj.* **–al**) A gradient or change in a variable over distance, as in thermocline, topocline, etc. Common usage in ecology is with reference to a continuous or gradual change in a character among members of a population or species over a given geographical or ecological range. (Syn. *character gradient*)

clinal variation. Reference to a geographic character gradient.

clinodeme. *n.* A deme that is part of a graded series of demes distributed over a given geographical region.

clinoid process. Pronounced wings extending up the basisphenoid bone on either side of the braincase wall in *Sphenodon.*

clitoris. *n.* An erectile rod of tissue that is the female equivalent of the male penis, as found in certain reptiles.

cloaca (*pl.* **-cae**). *n.* The common chamber and passageway into which the reproductive, urinary, and digestive ducts or canals release their contents, opening to the exterior through the vent.

cloacal bone. See *postanal bone*.

cloacal bursa. (*pl.* **-ae**) Literally a purse, but used to denote a pouch or sac. In herpetological literature, this term refers to respiratory expansions of the walls of the cloaca, arising from the *urodeum* in many aquatic turtles. These structures are highly vascularized and provide accessory sites for gas exchange when irrigated with water that is circulated in and out of the vent. (Syn. *accessory bladder, accessory vesicle, anal sac, anal bursa, cloacal sac, cloacal gill*)

cloacal capsule. The thickened wall of the cloaca in females of snake species in which the hemipenis of males is densely covered with spines. This structure is sometimes felt as a distinct and firm lump that can be used to identify the sex of certain snake species.

cloacal cartilage. See *hypoischium*.

cloacal egg. An amphibian egg that has passed down the oviduct in preparation for fertilization. Such an egg has peripheral yolk plates and is covered with jelly envelopes, distinct from *coelomic eggs*.

cloacal gaping. Sexual behavior in some female snakes consisting of everting the cloaca while the tail is raised as indication of receptivity to males.

cloacal gill. See *cloacal bursa*.

cloacal gland. 1. The largest part of a large, tubular gland surrounding the opening to the anus in male salamanders. The gland empties into the cloaca through rows of papillae along the cloacal wall and plays a role in pheromone production and probably spermatophore production. 2. The *musk gland* of crocodilians and some turtles. 3. Each of paired glands opening on each side of the cloacal margin in the *Tuatara*.

cloacal papilla. Any of numerous small villi on the wall of the cloaca in male salamanders, containing ducts from cloacal glands and secreting the jelly spermatophore base.

cloacal popping. Rapid extrusion and retraction of the cloaca in certain North American snakes (e.g., *Ficimia, Micrurus*). The action produces a clearly audible snapping or popping sound while the tail is held aloft as a defensive action. First described by E.H. Taylor (*Copeia* 1931:4–7, 1931).

cloacal probe. A blunt instrument with a tapered or rounded end used to determine the sex of a snake by insertion at the posterior end of the cloaca, usually after lubrication. In males, the probe penetrates the base of the tail with minimum force due to the presence of hemipenes.

cloacal quick-reading thermometer. A small, narrow-bulb mercury thermometer that registers temperature (0 – 50 °C) quickly and can be carried in the field. Formerly known as *Schultheis*, these precision thermometers were used to record body temperatures in many early studies of reptilian thermoregulation. Available from, and manufactured by, Miller and Weber, Inc., Queens, New York. (Syn. *Schultheis thermometer, Miller-Weber thermometer*)

cloacal sac. See *cloacal bursa*.

cloacal sacculus. See *cloacal bursa*.

cloacal spur. One of a pair of upward-curving projections at the base of the tail in certain male geckos. See also *pelvic spur, spur*.

cloacal swelling. The noticeable swelling around the vent of uredeles, especially in breeding adults.

cloacal temperature. A common measurement of "core" body temperature where the cloaca is the conventional site for placing the thermometer or sensor. This does not necessarily represent body temperture elsewhere in the animal, especially if the body is elongate.

cloacal vertebra. (*pl.* **-ae**) This term has been used to denote vertebrae in the vicinity of the pelvis, with short transverse processes but without ribs or ventral processes and without direct connection to the pelvis. In snakes these are vertebrae intervening between trunk and caudal vertebrae.

clone. *n.* 1. A population of genetically identical cells or organisms, all derived from a single original cell or organism. 2. Genetically engineered replicas of DNA sequences. 3. In colloquial use, an organism that has been artificially brought to life by the extraction of a cell or cells from another organism.

cloned library. A collection of cloned DNA sequences representative of the genome of the organism under study.

closed population. A population of organisms in which there is no immigration or emigration, and thus there is no input of genetic variation except that introduced by mutation. Cf. *open population*.

closed system respirometry. An experimental technique for measuring oxygen consumption and/or carbon dioxide release of an animal that is confined to a closed water- or air-filled chamber. The measurements are achieved by quantifying pressure or volume changes (in air) or by sensing changes in concentration of the respective gases in samples of the medium withdrawn from the chamber. Cf. *open system respirometry*.

clotting factor or **principle.** Used synonymously for *hemocoagulin*.

clotting time. The time required for whole blood to clot, normally 4–8 minutes.

cloud forest. A montane, high-altitude forest characterized by high humidity, frequent heavy mists, cool temperatures, and dense plant life. Normally these occur on windward slopes of tropical mountains at heights above about 1000 m.

Clown Treefrog. Vernacular name for the several hylid species of *Dendropsophus*, especially *Dendropsophus leucophyllatus*, commonly found across the Amazon and the Guyanas. Formerly *Hyla leucophyllata*.

club cell. See *Leydig cells*.

clumped distribution. A pattern of distribution in which observations or individuals are more aggregated or clustered than would be the case in a random distribution. The presence of one individual or value increases the probability that another will occur nearby. (Syn. *contagious distribution*) Cf. *continuous distribution, random distribution, uniform distribution*. See *lumpiness*.

cluster analysis. Grouping of variables within a set of variables that are highly correlated, excluding those that are negatively correlated or uncorrelated. Such analysis is used in numerical taxonomy to arrange *operational taxonomic units* into homogeneous clusters based on mutual similarities.

clutch, clutch size. *n.* The collective number of eggs ovulated and laid by a single female at one time.

cm. See *centimeter*.

CNAH. See *Center for North American Herpetology*.

cnemial. *adj.* Pertaining to the tibia. *n.* The enlarged, projecting scales on the posterior border of both fore and hind limbs of crocodilians (collectively equivalent to *cnemial fringe*).

CNS. See *central nervous system*.

co-. Prefix meaning "together," "with," or "sharing."

Coachwhip. *n.* Vernacular name for the large North American Whipsnake, *Masticophis flagellum*.

coadaptation. *n.* 1. A process of selection which tends to accumulate harmoniously interacting genes in the gene pool of a population. 2. Reference to evolution of mutually advantageous adaptations in two or more interactive species.

coagulant. *n.* An agent that promotes blood *coagulation*.

coagulation. *n.* The congealing of blood clot from liquid blood.

coagulopathy. *n.* A state of impaired blood coagulation.

coarse-grained environment. In ecology, reference to an environment experienced as sets of alternative, disconnected conditions. Cf. *fine-grained environment*.

coarse-grained resource. Reference to resources distributed in such a manner that a consumer organism encounters them disproportionately to their relative abundance in the environment.

coast. *n.* The land near the sea, including the beach and a strip of land inland from the beach.

coat. *n.* The outer layer of material which envelopes an amphibian egg; often used in reference to the *jelly envelope*.

cobble. *n.* A sediment particle with a diameter of 64 to 256 mm.

Cobra. *n.* Collective vernacular name for various species of Asian and African snakes belonging to the family Elapidae, and especially the genus *Naja*. These are front-fanged venomous snakes that flatten the neck to spread a "hood" in defensive displays. See also *Spitting Cobras, Water Cobras, King Cobra*.

cobra venom factor. (**CVF**) A component of Cobra venom consisting of several oligosaccharide chains and a polypeptide chain linked together by disulfide bridges. Activates the serum complement.

coccidiosis. *n.* A disease caused by protozoans that affect the digestive tract and blood.

coccygeal stripe. A longitudinal stripe at the rear of some frogs, crossing the vertebral column posterior to the sacrum.

coccygeum. *n.* The body region posterior to the sacrum. (Not the same as *coccyx*.) This term was coined by B.C. Mahendra (*Okajimas Folia Anat. Jpn.* 28:243–255, 1956) who preferred use of *coccygeal vertebrae* in place of *caudal vertebrae*.

coccyx. *n.* Fused vertebrae posterior to the sacrum, as in frogs, and replacing "tail" in an osteological sense. The "tail" is restricted to the postanal region.

cochlear duct. A sac-like extension or evagination from the posterio-inferior portion of the sacculus. The reptilian cochlear duct contains two separate sensory areas: the *papilla basilaris* and a *macula lagena*. The structure never elongates into a coiled cochlea as in mammals.

cochleophagus. *adj.* Of or pertaining to a predator that eats snails.

cocoon. *n.* 1. Generally, a tough, protective covering. 2. Amphibian "cocoons" are formed facultatively in certain species (mostly anurans) that estivate in response to drought and consist of multiple layers of shed epidermis that cover the body while animals are dug into drying soil. The numerous built-up layers of shed epidermis cover the entire body except for openings at the nares.

Code, Code of the International Commission on Zoological Nomenclature. *n.* A system of rules and recommendations intended to regulate nomenclature, published in the International Code of Zoological Nomenclature. Often referred to as simply the "Code."

code. *n.* In phylogenetic systematics, the numerical value for a particular character state.

codes for institutional resource collections. See *standard symbolic codes for institutional resource collections*.

codominant. *adj.* Designating genes for which both alleles of a pair are fully expressed in the heterozygote.

codon. *n.* The nucleotide triplet in messenger RNA that specifies an amino acid that is to be inserted in a specific position during translation and formation of a polypeptide. (Syn. *coding triplet*)

coefficient of selection. (s) See *selection coefficient*.

coefficient of variation. In probability theory and statistics, a measure of the amount of variation within a population, or the dispersion of a probability distribution. The coefficient of variation is defined as the ratio of the standard deviation to the mean, often multiplied by 100 to express the measure as a percentage.

coelodont. *adj.* Condition of teeth being hollow, with a permanent pulp-cavity and attachment to the outer wall of the jaw, leaving the base free.

coelom or **coelomic cavity.** *n.* The fluid-filled body cavity, limited at all surfaces by mesoderm and housing the digestive tract and visceral organs. See *peritoneal cavity*.

coelomic egg. An amphibian egg that has been ovulated from the ovary into the coelomic cavity.

coelomoduct. *n.* See *nephrostome*.

coeno-. Prefix meaning "in common," or "sharing."

coenzyme. *n.* An organic molecule that must be associated with an *apoenzyme* to form a functioning enzyme.

coenzyme A. A coenzyme, of which the acetylated form plays a central role in the *citric acid cycle*.

coevolution. *n.* Evolutionary change in one or more species that occurs in relatively close synchrony with another species as a consequence of their interactions or interdependence. It is a process of reciprocal evolution between interacting species driven by natural selection. Reciprocal adaptations between predators and their prey are examples.

coexistence. *n.* The persistence together of two or more species that share the same habitat and are usually potential competitors.

cohesion species concept. See *species*.

cofactor. *n.* An atom, ion, or molecule that combines with an enzyme to activate it.

cog-tooth valve. Reference to a specialized valve in crocodilians, derived from connective tissue and situated at the base of the pulmonary artery. The valve can close and restrict blood flow to the lungs, thus contributing to *intracardiac shunts* in these reptiles. See Fig. 37.

cohort. *n.* 1. A group of individuals that are all born and recruited into a population at the same time; thus, an age class. 2. A group of animals of the same species, possessing a common characteristic, which are studied over a period of time as part of a scientific investigation.

cohort life table. See *horizontal life table*.

coil. *n.* One or a series of spirals, used in reference to the usual position of an inactive snake when it is not outstretched.

coincidental evolution. See *concerted evolution*.

coition. *n.* See *coitus*.

coitus. *n.* Copulation or sexual union in vertebrates. (Syn. *coition, copulation, sexual intercourse*)

cold-blooded. *adj.* An obsolete term sometimes used inappropriately to describe the nonhomeostatic condition of body temperatures in amphibians and reptiles. This term should be replaced by *ectothermic* or *poikilothermic*. See *ectothermy*.

cold hardiness or **hardening.** Tolerance or resistance of an organism to low, usually freezing, temperatures, enhanced by previous or repeated exposure to the cold conditions. (Syn. *cold resistant*)

cold resistant. See *cold hardiness*.

cold temperate zone. A zone of latitude extending from 45° to 58° in both northern and southern hemispheres.

colic. *adj.* Of or pertaining to the large intestine (colon).

collagen. *n.* Structural fibers of protein found in skin, tendon, bone, and cartilage.

collagenase. *n.* A proteinase component of snake venoms that digests collagen matrices.

collagenoblast. *n.* A precursor cell to production of collagen, arising from a fibroblast.

collagenolysis. *n.* (*adj.* **–lytic**) The dissolution or lysis of collagen.

collar. *n.* Synonym for *gular fold*, or reference to a band of color across the dorsum of the neck.

Collared Lizard. Vernacular name for species of iguanid lizards belonging to the genus *Crotaphytus*, common to southwestern United States and Mexico.

collateral. *n, adj.* 1. Reference to a side of a nerve or blood vessel. In the blood circulation, *collaterals* are vessels that branch and service tissue to the side of a main vessel from which they are derived. 2. Ancillary or subordinate. (Syn. *side branch*)

collateral type. In systematics, a specimen used in the description of a species other than the primary type.

collecting organ. See *branchial food trap*.

colligative properties. Characteristics of a solution that depend on the number of molecules in a given volume. These properties include changes of freezing point, vapor pressure, osmotic pressure, and boiling point related to addition or reduction of solute particles in a solution.

collective species. See *superspecies*.

colloid. *n.* (*adj.* **–al**) A system in which fine particles are suspended in a liquid. Also may refer to a gelatinous substance.

collum scapulae. See *scapula*.

colluvial deposit, colluvium. A deposit of natural materials by gravity, as rocks at the base of a slope. Cf. *aeolian deposit, alluvial deposit*.

coloniality. *n.* A behavior pattern in which a large number of individuals of a single species congregate in the same limited area, as for breeding.

colonize. *v.* To invade a new area and establish a breeding population.

colony. *n.* A group of organisms that have become established in a new area or that live together in close proximity at a particular site.

Colorado River toad. Vernacular name for the American bufonid anuran species *Cranopsis alvaria* (formerly *Bufo alvarius*). For taxonomic revision see D.R. Frost et al., *Bull. Amer. Mus. Nat. Hist.* 297:1–370, 2006.

coloration. *n.* The *color pattern* of an animal, usually with reference to colors attributable to skin pigments of chromatophores.

color change. See *behavioral color change, morphological color change,* and *physiological color change*.

color index. Assignment of numerical values to stages in progressive color change. See *melanophore index*.

color pattern. See *coloration*.

-colous. Suffix meaning "to inhabit," or "inhabiting."

colubercholinesterase. *n.* A cholinesterase enzyme component of elapid snake venoms, causing the hydrolysis of acetylcholine. (Syn. *acetylase, ophiocholinesterase*)

Colubridae. *n.* A speciose clade (family) of snakes that includes many commonly known "harmless" and rear-fanged species with poorly resolved systematics within the clade. Approximately 320 genera and more than 1800 species are distributed worldwide. At least 7 subfamilies are recognized: *Aparallactinae, Colubrinae, Dipsadinae, Homalopsinae, Natricinae, Pareatinae,* and *Xenodontinae.* Some of these subfamilies are elevated to family level in alternative classifications, which may also include the additional subfamilies *Xenodermatinae, Nothopsinae, Lamprophiinae,* and *Dasypeltinae.*

colubrid. *n.* Collective vernacular name for any species of snake belonging to the family Colubridae, sometimes also referred to as *advanced snakes*.

Colubrinae. *n.* A subfamily and the largest clade of Colubridae, having many species that are prevalent in North American fauna and familiar to North American students. Contains approximately 700 species in approximately 300 genera.

Colubroidea. *n.* A clade (superfamily) of "advanced" or "higher" snakes that includes *Atractaspididae, Colubridae, Elapidae,* and *Viperidae.*

columella. *n.* The auditory ossicle (a bone or cartilagenous element) that crosses the middle ear cavity and transmits sound or vibrational stimuli from the tympanic membrane to the oval window. In some species the columella may form in two or more pieces and develop processes connecting to the skull or hyoid arch when a tympanic membrane is absent. This element evolved from the hyomandibular bone of fishes. (Syn. *columellar apparatus, stapes*) See also *extracolumella*.

columella auris. Term used to denote the proximal stapes or *columella* together with the *extracolumella* distally.

columella-Meckelian cartilage. A band of cartilage connecting the *extracolumella* with Meckel's cartilage in crocodilians. (Syn. *stylohyoid*)

columella process. A prominent protrusion of the *extracolumella* through the tympanic membrane of certain frogs.

columellar apparatus. See *columella*.

columellar joint. The diarthrosis between the ventral end of the epipterygoid and columellar fossa of the pterygoid, permitting slight anteroposterior and mediolateral motions of a kinetic skull.

columnar. *adj.* Resembling a column.

column chromatography. A method for separating organic compounds by percolating a liquid containing the compounds through a porous material, which may be an ion exchange resin, in a cylinder.

comb. *n.* 1. A term sometimes applied to the fringe of elongated scale appendages on the lateral border of the toes of desert lizards. 2. The femoral pores of male lizards. The latter exude wax substances, especially during the mating season, and collectively resemble the teeth of a comb.

combat dance. A challenge behavior between two male snakes, usually of the same species, involving vertical postures, intertwining of bodies, and pushing of the opponent in attempt to knock the individual to the ground. See *ritual combat*. See Fig. 39.

combined description. Description of a new species that is assigned to a monotypic genus and uses the same character states to diagnose both the species and genus.

combined parsimony analysis. In phylogenetic studies, a cladistic analysis in which all characters and taxa are merged into one data matrix and are analyzed simultaneously. See A. Kluge, *Syst. Zool.* 38:7–25, 1989; K. Nixon & J. Carpenter, *Cladistics* 12:221–241, 1999. (Syn. *simultaneous parsimony analysis*)

comm. The Latin abbreviation of *communicavit*, meaning "communicated."

command neuron or **system.** A neuron or set of neurons that, when stimulated, elicit a set pattern of coordinated movements. A command neuron has the ability to organize and produce an entire behavior by controlling the activities of many other neurons.

commensal. See *commensalism*.

commensalism. *n.* Reference to the association of different species wherein one species (the *commensal*) derives a benefit, but the other is unaffected.

Commission. *n.* See *International Commission on Zoological Nomenclature*.

commissure. *n.* 1. A joint, seam, or line of closure, as between the two jaws when the mouth is closed. 2. A tract joining equivalent structures on the two sides of the central nervous sytem.

commonality, principle of. In systematics, the accepted consideration that a character state having the widest distribution among taxa comprising a higher taxon is the more primitive among others.

common ancestor. Reference to an ancestral species that is in common among more recent species that are being discussed. A species that during some past time split into two or more species, each of which gave rise to one of the clades under consideration.

Common Blind Snake. Collective vernacular name for species of snakes belonging to the typhlopid genus *Ramphotyphlops*.

Common Caecilian. Collective vernacular name for numerous species of caecilians belonging to the family Caeciliidae, and especially the genus *Caecilia*.

Common Coral Snake. Vernacular name for New World elapid species of snakes belonging to the genus *Micrurus*.

common carotid artery. See *carotid artery*.

Common European Toad. Vernacular name for the bufonid anuran species *Bufo bufo*.

Common Garter Snake. Vernacular name for the American colubrid species *Thamnophis sirtalis*. (Syn. *Red-sided Garter Snake*)

Common Grass Snake. Vernacular name for the European colubrid species *Natrix natrix*. See *Grass Snake*.

Common House Gecko. See *House Gecko*.

Common Iguana. Vernacular name for the species of lizards in the iguanid genus *Iguana*. See also *Green iguana*.

Common Kingsnake. Vernacular name for various subspecies of American colubrid snakes belonging to the species *Lampropeltis getula*.

Common Lancehead. Vernacular name for the pitviper *Bothrops atrox*, native to tropical South America east of the Andes and to Trinidad. See *Lancehead*.

common name. Colloquial or vernacular name (in contrast to *scientific name*) of a species.

Common Sea Krait. See *Sea Krait*.

Common Sea Snakes. Vernacular name for various species of elapid sea snakes belonging to the genera *Disteira* and *Hydrophis*. The latter genus is the most speciose taxon of true sea snakes.

Common Toad. Vernacular name for the bufonic species *Bufo bufo*.

Common Treefrogs. Collective vernacular name for numerous species of tree frogs belonging to the genus *Hyla*. See *Tree Frogs*.

Common Viper or **Common European Viper.** Vernacular name for the well-known European viper *Vipera berus*.

Common Worm Snakes. Collective vernacular name for species of typhlopid snakes belonging to the genus *Typhlops*. (Also called *Blind Snakes*.)

communal nesting or **oviposition.** Oviposition by more than one conspecific or heterospecific female in the same nest cavity. Communal oviposition was defined by R.E. Espinoza & F. Lobo (*Herpetol. Nat. Hist.* 41:65–68, 1996) as "the nonincidental deposition of eggs at a shared nest cavity by two or more conspecifics. This distinguishes communal oviposition from colonial nesting behaviors in which nests are constructed adjacent to one another, but the eggs are generally not deposited in the same nest cavity."

communication. *n*. Any act of an organism that passes information to another organism and modifies its behavior.

community. *n*. See *ecological community*.

community matrix. See *guild matrix*.

compact bone. A type of bone having dense, closely packed tissue, with little interstitial space. Cf. *cancellous bone*.

compacted coarse-cancellous bone. A bone type related to remodeling in which spongy medullary bone undergoes a process of compaction by deposition of bone within cancellous spaces. This yields an irregular structure that is easily recognized in all reptiles.

comparative methods. Statistical and procedural techniques used to analyze data from multiple taxa of organisms for ecological or evolutionary patterns.

comparative physiology. A subdiscipline of physiology in which diverse species are compared to discern physiological principles and evolutionary patterns related to functional attributes of organisms.

compass orientation. The ability to orient in a specific direction without following landmarks.

compensation. *n*. In physiology, the tendency for a character or rate process to return toward its previous value during acclimation to a change of environment that acutely changes the variable. E.g., assume the temperature of a frog is changed suddenly from 25 °C to 10 °C, and by Q_{10} effect the metabolic rate decreases. In time, however, *compensation* or compensatory acclimation will elevate the metabolic rate such that it returns to, or near, the previous value at 25 °C.

compensatory motion or **movement.** The movement observed when a stationary animal is moved and its head or anterior body moves the opposite direction to maintain orientation toward a fixed point.

competence. *n*. A characteristic of a part of an embryo that enables it to react to a given morphogenetic stimulus by differentiation in a given direction.

competition. *n*. A major force that structures ecological communities, acting directly or indirectly. 1. The interactions of individuals or species resulting from utilization of essential resources that are in limited or potentially short supply (*exploitation competition*). 2. The detrimental interaction between two or more individuals or species that seek a common resource that is not limiting (*interference competition*). See also *apparent competition, contest competition, scramble competition*.

competition coefficient. (α) A dimensionless number in the Lotka-Volterra competition model that specifies the per capita effect of species 2 on the population growth rate of species 1, measured relative to the effect of species 1.

competition coefficient. (β) A dimensionless number in the Lotka-Volterra competition model that specifies the per capita effect of species 1 on the population growth rate of species 2, measured relative to the effect of species 2.

competitive exclusion. The exclusion of one species by another when both compete for a common resource that is in limited supply. See *competitive exclusion principle*.

competitive exclusion princple. A principle in ecology that complete competitors, equivalent to two species of organisms characterized by the same ecological requirements or theoretical niche, cannot coexist indefinitely. In other words, there must be some difference between two coexisting species in terms of resource utilization.

The principle implies that resources are limiting, which is not always the case. (Syn. *Gause's Law*)

competitive interactions. Reference to species negatively affecting each other's population growth rate and depressing each other's population size.

competitive release. Expansion of range, habitat use or prey utilization by a species due to a reduction in intensity of interspecific competition (sometimes by removal of another species).

complementary resources. In ecology, reference to a pair of resources for which increased consumption of one by an organism leads to a decreased demand for the other resource. (Syn. *antagonistic resources*)

complete dominance. Condition in which a trait produced by one allele is fully expressed (dominant) over the trait of another (recessive). Cf. *incomplete dominance*.

complex. *n.* In taxonomy, reference to a group or assemblage of related taxa (e.g., species complex), often used when the status of individual units is unresolved.

complex character. A phenotypic character that is determined by more than one gene and is not transmitted to offspring as a simple unit.

complexity. *n.* In ecological sciences, this term has recently become a metaphor "deployed to advance knowledge of fundamental questions in ecology, including the relationship between parts and wholes, and between order and disorder" (J.D. Proctor & B.M.H. Larson, *BioScience* 55:1065–1068, 2005).

complex life cycle. A life history that includes an abrupt ontogenetic change in an individual's morphology, physiology, and behavior, usually associated with a change in habitat. The life cycle involves two or more distinct phases and is characteristic of many amphibians (H.M. Wilbur, *Ann. Rev. Ecol. Syst.* 11:67–93, 1980; N.A. Moran, *Ann. Rev. Ecol. Syst.* 25:573–600, 1994).

compliance. *n.* The change in volume of a compartment or structure per unit change in applied pressure.

complimentale. *n.* One of nine investing bones of the lower jaw, formed in the embryo of certain caecilians.

component analysis. A type of multivariate analysis that represents a *p*-dimensional variation as due to a number of components such that as few components account for as much of the variation as possible. See also *principle components analysis*.

composite material. In biomechanics, reference to a material having two or more components of different *stiffness*, which resists cracking. Bone and skin are tough because they contain composite fibers or crystals of small diameter.

composite species. 1. A polytypic species with two or more subspecies. 2. In paleontology, a species represented by specimens obtained from two or more localities of different geological age.

compound. *adj.* In taxonomy, a name formed by the union of two or more words.

compound bone. The lower jaw element resulting from fusion of the articular, prearticular, and surangular bones in certain lizards and snakes.

compressed. *adj.* Flattened or pressed together. Usually and preferentially used in reference to vertical flattening of the body, as in some displays. Cf. *depressed*.

compression. *n.* Stress in an elastic solid resulting from a load that is directed toward the object and perpendicular to its surface. See *stress* and Fig. 24.

compressive force. A force that is directed toward an object such that it tends to press or squeeze the object together. See *compression*.

computerized or **computed axial tomography**. See *CAT scan*.

concave. *adj.* Curved inward. Often used in reference to the plastron of some male turtles. Cf. *convex*.

concealed surface. Any surface of the body that is normally concealed from view, especially while an animal is in a resting position. Many such surfaces in amphibians and reptiles are brilliantly colored and serve to startle potential predators when the animal moves.

concentration refuge. With reference to tadpole ecology, the concentration or density of food below which a larval anuran ceases to feed and thus the remaining prey survive and reproduce.

concentric contraction. Isotonic contraction of muscle in which the muscle shortens. Cf. *eccentric contraction*.

conceptual model. See *model*.

concerted evolution. The production and maintenance of homogeneity within families of repetitive DNAs, e.g. the homogeneity of nucleotide sequences in each of the hundreds of tandemly arranged rRNA genes of *Xenopus laevis* (D.D. Brown et al., *J. Mol. Biol.* 63:57–73, 1972).

concertina movement. A basic type of locomotion in snakes, characterized by sequential extension and contraction of the body from one anchored or stationary site to the next as the animal moves with accordion-like appearance in one direction.

conch, concha. (*pl.* **-ae**) *n.* A protuberance on the outer lateral wall of the nasal chamber, often restricted to simple lamellar processes. See T.S. Parsons, *Bull. Mus. Comp. Zool.* 120:104–277, 1959. Conchae are present in crocodilians, Tuataras, lizards, and snakes. Cf. *pseudoconcha, nasal turbinates*.

concordance. *n.* Reference to characters that are the same in the operational taxonomic units under consideration.

concordant. *adj.* 1. Occurring at the same time. 2. In systematics, reference to members of a group that share a given trait. Cf. *discordant*.

concurrent. *adj.* Flow of adjacent fluids in the same direction. Cf. *countercurrent*.

condensation. *n.* 1. Formation of water on an object when its temperature falls below the dew point of surrounding air.

2. In evolutionary biology, reference to an increase in the rate of development during phylogeny, either due to accelerated development or deletion of ontogenetic stages.

condition index. (CI) See *body condition index*.

conditioning. *n.* In ethology, a process whereby behavior of an organism is altered in response to a stimulus such that an action normally evoked by a different stimulus occurs when the two stimuli are presented together with repetition.

conductance, electrical. (G) A measure of the ease with which a conductor carries an electric current, expressed as the flow of charge per unit of voltage gradient; also the inverse of electrical resistance or impedance, and measured in siemens.

conductance, gaseous. (G) The concept of conductance applied to gases, in physiology expressing the ease with which a gas passes across a membrane or barrier. Calculated as the molar or, usually, volume rate of air flow per unit of partial pressure driving force for the specific gas being considered (i.e., flux divided by partial pressure difference across the membrane). Also the inverse of *resistance*.

conductance, thermal. (C) The concept of conductance applied to movement of heat down a temperature gradient across an object or substance. In metabolic experiments, this term is often calculated as a "dry" conductance that ignores evaporative heat transfer, expressed as rate of heat flow per unit of temperature difference between object and air (or between two compartments of substance). Also called *conductive heat transfer coefficient* and given as $J \cdot s^{-1} \cdot °C^{-1}$, $W \cdot °C^{-1}$, or various other units. This measurement is most commonly encountered in literature related to *endotherms* rather than amphibians and reptiles. Cf. *thermal conductivity*.

conduction. *n.* Transfer of heat directly from one elementary particle to another, thus between objects that are in direct physical contact.

conductive heat transfer coefficient. See *conductance, thermal*.

conductivity, electrical. The intrinsic ability of a substance to conduct electrical current; the reciprocal of *resistivity*.

conductivity, thermal. (k) The intrinsic ability of a substance to conduct heat, expressed as $J \cdot s^{-1} \cdot m^{-1} \cdot °C^{-1}$, or various other units.

condyle. *n.* In anatomy, a rounded projection at the end of a bone; usually a rounded protrusion that forms an articulating surface.

condylobasal length. Skull measurement from anteriormost point of premaxillary bone to tip of occipital condyle. Used by S.B. McDowell and C.M. Bogert (*Bull. Amer. Mus. Nat. Hist.* 105:1–142, 1954).

cone. *n.* A photoreceptor cell in the vertebrate retina, functioning in bright light and differentially sensitive to light of different wavelengths. Cones are thus responsible for color vision. See also *rod*.

confidence interval. A range of values that are set to include the parameter being estimated a specified percentage of the time (usually 95%).

confidence limits. The upper and lower extremes of a confidence interval encompassing the true value of a population parameter (such as the mean) within which sample values (e.g., means) may fall, and from which inferences may be made, with any desired degree of confidence, such that it is unlikely that a disproportionate number of rejections of a true hypothesis should occur. E.g., 95% confidence limits indicate there is a 95% probability that the parameter being estimated lies within these limits.

confocal scanning microscope. A microscope using a focused laser beam to rapidly scan different areas of the specimen in a single plane. Light reflected or emitted from this plane is assembled by a computer into a composite image of the specimen.

conformer. *n.* An animal whose internal conditions tend to vary with those of the environment (thermoconformer, osmoconformer, etc.). Cf. *regulator*.

confusing coloration. A form of protective coloration in which the color of an animal tends to confuse a predator by virtue of a different appearance depending on whether the possessor is at rest or in motion.

congener, congeneric. *n., adj.* Belonging to the same genus.

congenital. *adj.* 1. Existing before or at birth. 2. A character state such as an abnormality or disease that is present as such at birth.

Congo Eel. See *Amphiumidae*.

congregate. *v.* To collect together or form a group.

congruence. *n.* 1. In cladistics, congruent characters refer to shared features whose distribution among organisms corresponds to that in the cladistic grouping of said organisms. The most likely cladogram provides maximum congruence of all characters involved. 2. The degree of correspondence between different classifications.

congruent. *adj.* Matching or coincident.

conical. *adj.* Resembling or shaped like a cone.

coni circumfossati. A spikelike projection of amphibian skin surrounded by a ditch or depression (*fossa* = ditch) (H. Elias and J. Shapiro, *Amer. Mus. Novitates*, No. 1820:1–27, 1957).

conjugate acid-base pair. Two molecules related by gain or loss of a proton (H^+ ion).

coniferous forest. Forest dominated by needle-leaf conifers such as spruce, fir, hemlock, and pine, forming great belts in subarctic regions of North America and Eurasia between 50° and 65° N latitude. See *taiga*.

conjunctiva. *n.* The membrane covering the front of the eyeball.

conjunctivitis. *n.* An infection of the inner eyelids, often used

connective tissue. Generally, tissues that include bone, cartilage, fibrous tissue, adipose tissue, and blood. These tissues have a variety of functions in diverse contexts.

in herpetology to identify a more generalized eye infection.

connectivity. *n.* 1. In ecology, the interrelationships between different components or compartments of an ecosystem. 2. In information technology, reference to connections of computers to others in a system or to the Internet.

consensus tree. In cladistics, a single tree that combines information about groupings contained elsewhere in two or more different trees.

conservation. *n.* Generally, reference to all activities associated with preservation of an extant species (or other resource) in its natural habitat, involving planned management of natural resources. In more specific usage, the use of all procedures and methods necessary to bring an endangered species or threatened species to the point at which the measures taken pursuant to the *Endangered Species Act* are no longer necessary.

conservative. *adj.* 1. Tending to remain unchanged. 2. In evolutionary biology, showing a relatively slow rate or small extent of evolutionary change. Taxa retaining many ancestral characters and similar to an ancestral group.

conserved, conserved character. *adj.* Inherited from a common ancestor and retained in contemporary, related species. Conserved features may or may not be subjected to selection.

conspecific. *adj., n.* Belonging to the same species. Sometimes used as a noun. Cf. *heterospecific.*

constriction. *n.* A method used by certain snakes to immobilize and kill prey by squeezing a captured animal within progressively tightening body coils. The constriction by coils of the snake's body disable lung ventilation, thereby causing suffocation, and may also interfere with blood circulation.

constrictor. *n.* 1. Term applied to any snake that subdues prey by constriction. 2. A muscle that constricts or closes. Cf. *dilator.*

consumer. *n.* Any organism that feeds on another organism or on existing organic matter.

consumption. (C) *n.* Used in ecological energetics to identify the total intake of food or energy by a heterotrophic individual, population, or trophic unit per unit time. The term (C) equals the sum of *respiration* (R) + *production* (P) + *rejecta* (F + U, feces and urinary wastes).

consumption efficiency. The percentage of energy transferred from one trophic level to the next.

contagious distribution. See *clumped distribution.*

Contemporary Herpetology. An electronic journal designed to facilitate the dissemination of peer-reviewed articles covering all aspects of herpetology, including reviews, monographs, points-of-view, and faunistic surveys of poorly known areas.

contest competition. Competition in which a resource is unequally partitioned between competitors such that a successful organism obtains all it requires while an unsuccessful one obtains insufficient resource for survival and reproduction. Cf. *scramble competition.*

contiguous. *adj.* Having boundaries that make contact, while areas do not overlap.

continental. *adj.* Of or pertaining to a continent or to an island that was separated from a continental margin because of geological faulting.

continental climate. A climate having a difference between summer and winter temperatures that is greater than the average range for that latitude owing to distance from an ocean or sea.

continental drift. The well-accepted theory that the world's continents once formed a single land mass and have drifted apart to arrive at their present positions (A. Wegener, 1912). This theory was refined by the concept of *plate tectonics* (A.L. du Toit, 1927) that places the continents on larger sections of Earth's crust that are in motion.

continental island. An island that was once part of a mainland land mass and became isolated by rising seas or subsidence of low-lying intervening areas. Cf. *oceanic island*

continentality. *n.* An overall ratio of land area to sea area.

contingency table. A table having two or more rows and columns of data in which observations or individuals are classified according to either of two variables. Tests such as the χ^2 are used to measure the relationship between the variables.

continuity. *n.* In taxonomy, a principle that continuity of usage of a particular name should take precedence over the priority of publication in determining which of two or more competing scientific names should be adopted.

continuous breeder. An animal that may breed at any time of the year.

continuous distribution. A data set that yields a continuous spectrum of values. E.g., measurement of the mass of individual animals, carried out to one or more decimal places. Cf. *discontinuous distribution.*

continuous population growth. Reference to a population that has births and deaths occurring steadily such that the trajectory of population size reflects continuous growth and resembles a perfectly smooth curve.

continuous trait or **character.** A quantitatively defined feature with no easily distinguished boundaries between phenotypes (e.g., size, cell counts, gene expression).

continuous variable. A variable that can assume theoretically any value between two given limits. See also *random variable.* Cf. *discrete variable.*

continuous variation. Phenotypic variation exhibited by quantitative characters, such as body mass or length, that vary by imperceptible degrees in progression from one extreme to another. Cf. *discontinuous variation.*

continuum. *n.* A gradual or imperceptible transition between two or more extreme values.

contour line. A line on a *topographic map* connecting points of equal elevation.

contra-. Prefix meaning "against" or "opposite."

contralateral. *adj.* Pertaining to the opposite side. Cf. *ipsilateral.*

control. *n., adj.* 1. In physiology, the capability to modulate or change a variable. Cf. *regulation.* 2. A standard of comparison. In experimental work, a control refers to a condition that is not changed and is used as a comparison against a condition that is changed by the experimenter to examine cause and effect in context of the outcome.

controlled variable. The variable that is regulated in a feedback control system.

conus. *n.* The most anterior of primitive heart chambers.

conus apicalis. Refers to the thickened, cornified, and usually blackened sharp apex of individual tubercles on the skin of anurans, characterized by the absence of a surrounding fossa at the base. Used by H. Elias and J. Shapiro (*Amer. Mus. Nov.* 1819:1–27, 1957). Cf. *conus circumfossatus.*

conus arteriosus. A thickened part of the amphibian heart between the ventricle and distributing vessels of the systemic and pulmocutaneous arches. A spiral fold within this structure facilitates separation of oxygenated and deoxygenated blood. Sometimes called *truncus arteriosus* by some authors. The conus is not present in reptiles.

conus circumfossatus. Refers to the thickened, cornified, and usually blackened sharp apex of individual tubercles on the skin of anurans, characterized by presence of a surrounding fossa at the base. Used by H. Elias and J. Shapiro (*Amer. Mus. Nov.* 1819:1–27, 1957). Cf. *conus apicalis.*

conus papillaris. A comblike, vascular projection from the optic nerve head into the cavity of the vitreous humor of the eye. This structure presumably provides supplemental nutritional support for the deeper ocular tissues. (Syn. *conus papillaris, marsupium, papillary cone*; the term *pecten* is used, but applies correctly to a similar structure in birds)

convection. *n.* 1. The mass transfer of heat due to mass movement of a gas or liquid. A very slow circulation of a substance may be driven by differences in temperature and density within that substance. *Free convection* refers to such conditions, with zero wind speed or water current. 2. Generally, movement of a fluid by mass flow.

convective precipitation or **rain.** Rainfall that results from convection caused by solar radiation heating the ground. Cf. *orographic precipitation* or *rain.*

conventional behavior. Any display or behavior by which members of a population reveal their presence and thus allow others a means to assess the density of the population.

convergence. *n.* 1. The independent evolution of similar character states in distantly related or unrelated lineages, usually living in similar environmental conditions or experiencing a similar mode of life. *Convergent evolution* occurs when these organisms experience similar selection pressures related to similarity of one or more factors in the environment. Common, but separate, evolution. See *homoplasy.* (Syn. *convergent adaptation*) 2. In physiology, a pattern in which inputs from many different neurons impinge upon a single neuron. 3. In development, reference to convergence of an epithelial sheet toward a central site.

convergent adaptation. See *convergence.*

convergent evolution. See *convergence.*

conversion efficiency. 1. The ability of predators to convert each prey item captured into additional per capita growth rate. 2. The ability of the digestive system to convert food to assimilated energy. The ratio of mass of food ingested to body mass increase. Cf. *assimilation efficiency.*

convex. *adj.* Curving outward. Cf. *concave.*

cooperative breeding. A breeding system in which parents are assisted in the care of their young by other adult or subadult animals.

coordinates. *n.* Numbers which locate the position of a variable point in space.

coossification. *n.* (adj. **coossified**) A condition in which the integument is fused to the underlying bones. Examples are the "helmets" of Casque-headed Tree Frogs.

Cooters, River Cooters. *n.* Common name for various species of turtles belonging to the genus *Pseudemys* (family *Emydidae*).

Copeia. *n.* A journal of herpetology (and ichthyology) published by the American Society of Ichthyologists and Herpetologists.

Cope's Gray Treefrog. Vernacular name for the North American hylid species *Hyla chrysoscelis.*

Cope's "law of the unspecialized". A theory formulated by the 19th-century paleontologist E.D. Cope that evolutionary novelties associated with new major taxa are more likely to originate from a generalized, rather than specialized, member of an ancestral taxon.

Cope's rule. A "bioclimatic rule" which suggests there is a general tendency for animals to increase in body size during the course of phyletic evolution. There are numerous exceptions to this "rule," both for endotherms and ectotherms. See also *Allen's rule, Bergmann's rule,* and *Gloger's rule.*

Cophylinae. *n.* A subfamily of *Microhylidae.*

Copperhead. *n.* A vernacular name for the North American pit viper *Agkistrodon contortrix,* and the Australian elapids in the genus *Austrelaps,* especially *Austrelaps superbus.* See also *Moccasin.*

coprodeum, coprodaeum. *n.* The anterior or innermost portion of the cloaca that receives feces from the intestine.

coprolite. *n.* fossilized fecal material.

coprophagy. *n.* (*adj.* **–ous**) Reference to ingestion of feces. This phenomenon has been reported in larval amphibians and a few turtles and snakes. It is probably more widespread than currently known and is advantageous in the transfer or acquisition of friendly symbiotic gut microorganisms.

copula. *n.* Literally a connection, applied to the unpaired cartilage at the anterior, ventromedial end of the hyoid. In amphibians this structure is derived from the larval *basibranchials* and may become incorporated into the hyoid plate (anurans). In reptiles the structure develops directly. This element may form the basal cartilage of the tongue (salamanders) or may have an elongate extension, the *entoglossal*, into the tongue (lizards). See also *radial*. (Syn. *basihyobranchial, basihyoid, basilingual plate, copular plate* or *series, corpus hyale, corpus hyoideum, hyoid copula*)

copular plate or **series.** See *copula*.

copulation. *n.* (*v.* **copulate**) Coitus involving an intromittent organ.

copulation path. The route followed by sperm from the point of egg penetration to the nucleus. (Syn. *penetration path*)

copulatory plug. See *seminal plug*.

copulexus. *n.* A new term introduced to describe the unique form of mating that results in internal fertilization when the turgid or engorged "tail" of the male frog *Ascaphus truei* is inserted into the female and introduces sperm by passage in a sulcus that is formed by the tail. Such copulation has been assumed to be an adaptation that ensures fertilization in fast-moving water. Term introduced by D.M. Sever et al. (*J. Morphol.* 248:1–21, 2001).

Coqui. *n.* Vernacular name for the Puerto Rican leptodactylid frog *Eleutherodactylus coqui*.

coracoid. *n.* A ventral and usually well-ossified element of the shoulder girdle, which, together with the scapula, forms the site of articulation with each forelimb. A true coracoid evolved in therapsid dinosaurs leading to mammals. The element in living amphibians and reptiles is frequently called a coracoid but is better termed the *anterior coracoid* or *procoracoid*. This term also has been used for the *scapula* of turtles in older literature. See Fig. 3.

coracoid cartilage. Synonym for *epicoracoid*. See Fig. 3.

coracoid fenestra. An emarginated region of the coracoid.

coracoid foramen. An opening in the coracoid bone of the pectoral girdle.

coracoid plate. A broad area of ossification, curving across the chest below the *glenoid fossa*, where the coracoid and scapula are fused as a single unit. See *scapulocoracoid*.

coral knoll. A column of coral within the lagoon of an atoll.

coral reef. Dynamic, wave-resistant structures built by living communities of corals containing symbiotic algae. Coral reefs are among the richest of marine ecosystems in species, productivity, biomass, and structural complexity. Over a quarter of all marine fish are found in reefs, which are also home to numerous species of sea snakes.

Coral Snakes. Collective vernacular name for various elapid species of snakes belonging to the genera *Calliophis, Micruroides, Micrurus,* and *Sinomicrurus*.

cordate. *adj.* Heart-shaped.

cordiform foramen. A large opening in the pelvic girdle, which separates the pubis and ischium and turns the pelvic girdle into a tripartite structure in turtles, *Sphenodon*, and lizards. Used by N.G. Stephenson (*J. Linn. Soc. Lond., Zool.* 44:278–299, 1960). (Syn. *pubo-ischiadic fenestra, thyroid fenestra, obdurator foramen*)

cordillera. *n.* An entire mountain system, including subordinate ranges, interior basins, and plateaus.

cordon. *n.* An individual string of eggs oviposited by a bufonid anuran.

cordylid. *n, adj.* A member of, or pertaining to, Cordylidae.

Cordylidae. *n.* A clade (family) of lizards characterized by scales that are arranged in transverse circles around the body, often with keeled or spiny tails and osteoderms. Four genera and approximately 42 species occur in eastern and southern Africa.

core area. The area within a *home range* that is most commonly utilized by an animal.

core reserve. An extensive area of wilderness or wild lands that are managed to permit ecological processes to function as naturally as possible.

core temperature. A body temperature measurement that should represent the average temperature of the internal (vs. peripheral) body mass, theoretically at a depth below which it is affected by a change in the gradient of temperature through the peripheral tissues. This situation usually does not exist in small ectotherms such as most amphibians and reptiles. See also *deep body temperature*.

coriaceous. *adj.* Having a leathery texture.

Coriolis force. The deflecting force of the Earth's rotation causing the water in the air masses and oceans to move clockwise in the Northern Hemisphere and counter-clockwise in the Southern Hemisphere.

corkscrew vein or **valve.** A corkscrew-shaped segment of the portal vein just posterior of the liver in terrestrial species of snakes. It is speculated this structure functions as a bi-directional valve to minimize gravitational surges of retrograde blood movement. See H.B. Lillywhite, *Amer. Zool.*, 27:81–95, 1987.

cornea. *n.* The transparent surface covering the iris and pupil, through which light passes as it enters the eye.

corneal layer. Descriptive term sometimes applied to the *stratum corneum* of the epidermis.

corneous or **cornified.** *adj.* Reference to keratinized epithelium, as the *stratum corneum* of the epidermis.

corniculum. *n.* See *principal horn.*
cornified. *adj.* See *corneous.*
corn-mark. *n.* Reference to lumps or projections on the inner surfaces of alligator skin. (Syn. *button*)
Corn Snake. Vernacular name for the North American colubrid species *Pantherophis guttata* (formerly *Elaphe guttata*). Also called *Red Rat Snake.*
cornua (*pl. –ae*) *n.* General term for various projecting parts of the hyoid apparatus.
cornuate. *adj.* Having one or more horns or projections. The noun *cornu* (pl. *cornua*) has been widely used in the past to refer to a horn or horn-like projection.
coronary artery. A small artery that provides blood flow to the cardiac muscle. Reptiles are the first vertebrates to have evolved a well-developed coronary artery system. Coronary artery blood flow supplements the direct exchange of oxygen and other metabolites from blood contacting the spongy internal chambers of the heart.
coronoid. *n.* A dermal bone of the lower jaw, located dorsally between the *dentary* and *surangular* elements, nearer to the back of the jaw and lacking teeth. This element does not occur in anurans and some other amphibians.
coronoid process. A dorsal projection of the lower jaw, associated with the *coronoid* or *dentary.*
corpora adiposa. Synonym for *fat bodies.*
corpora cavernosa. See *penis.*
corpus. *n.* Latin term meaning body and used in conjunction with various anatomical designations. Reference to any mass or solid part of an organ.
corpus albicans. A term for the regressed form of the *corpus luteum,* which becomes a white scar after some time if fertilization and pregnancy does not occur following ovulation. This is a degenerate, nonfunctional structure.
corpus hyale. See *copula.*
corpus hyoideum. See *copula.*
corpus luteum. (*pl. corpora lutea*) The remnants of a ruptured egg follicle following ovulation, which secrete relatively large volumes of progesterone until late pregnancy in viviparous reptiles. The follicle is transformed into an endocrine gland in reptiles, some mammals, and some sharks. Following oviposition or parturition, the *corpora lutea* regress and become *corpora albicans*, which may persist for months or years. Cf. *corpus albicans.*
corpus sterni. See *epicoracoid.*
corpus Y. See *ultimobranchial body.*
correlation. *n.* The degree to which statistical variables vary together. The extent of correlation is measured by the *correlation coefficient*, *r*, which has a value from zero (no correlation) to –1 or +1 (perfect negative or perfect positive correlation). The relationship between variables is not one of dependence, and both variables are theoretically attributable to random factors. Cf. *regression* and *linear regression.*
correlation coefficient. (*r*) See *correlation.*

corridor. *n.* A route of migration whereby members of a species can disperse from one area to another. In conservation biology, corridors are considered to be important features of habitat that mitigate extinction of populations occurring in patchy habitats. Ecological corridors are linkages of landscape designed to permit large-scale and long-term ecological processes to continue operating within fragmented ecosystems. (Syn. *wildlife corridor*)
corrigendum. (*pl. –a*) *n.* A term used in reference to journals to indicate an error has been discovered in a printed work (after printing) and a substantive correction has been made to a previous issue (sometime shown with a correction on a separate sheet). Cf. *erratum.*
corrosion cast. A cast or replica of internal anatomic spaces such as blood vessels, produced by injecting a material (usually plastic) that subsequently hardens and is exposed by chemical erosion of surrounding tissues. *Corrosion casting* is the process by which a corrosion cast is made.
cortex. *n.* The outer portion of an organ.
cortexin. *n.* See *corticin.*
cortical. *adj.* Pertaining to the cortex or outermost portion of an organ. Cf. *medullary.*
cortical bone. Reference to the outer boundary or *cortex* of a bone. Cf. *medullary bone.*
cortical cord. See *sex cord.*
cortical lobe. See *Bidder's organ.*
corticin. *n.* A paracrine, diffusible secretion released by the cortex of the undifferentiated early embryonic gonad. It acts to repress the development of the medulla while stimulating the cortex of the ovary, leading to differentiation of the female gonad and secondary characters. E. Witschi (*Biol. Rev.* 9:460–488, 1934) first theorized from amphibian experiments that corticomedullary inductors, *cortexin* and *medullarin,* induced sex differentiation in amphibians. Cf. *medullarin.*
corticosteroid. *n.* A steroid produced by the adrenal cortex.
corticosterone. *n.* A corticosteroid hormone secreted by the adrenal gland and influential with respect to glucose metabolism, water, ion and energy balance, and sexual behavior. Corticosterone generally is the dominant glucocorticoid in amphibians and reptiles, and circulating levels are increased by stress. However, the dominant glucocorticoid in amphibians varies with terrestrial or aquatic habits. E.g., corticosterone dominates in terrestrial amphibians while cortisol dominates in metamorphosing ranid tadpoles, in aquatic anurans like *Xenopus laevis*, and in some aquatic urodeles. See D.O. Norris, *Vertebrate Endocrinology* 4th ed., 2006. Recently, it was found that embryonic corticosterone accelerates hatching events in some oviparous lizards.
corticotropin. *n.* See *adrenocorticotropic hormone.*
corticotropin-releasing factor. (CRF) A brain peptide that is

released from neuronal endings and stimulates release of *adrenocorticotropic hormones* from the anterior pituitary.

cortisol. *n.* A corticosteroid hormone secreted by the adrenal gland. This metabolic steroid regulates protein and carbohydrate metabolism and is used as an indicator of general stress.

coryphodont. *adj.* A condition of dentition in which teeth increase in length progressively from the anterior to posterior parts of the mouth.

Corytophaninae. *n.* A clade (subfamily) of iguanid lizards that possess well-developed head crests and casques and includes the spectacular *Basiliscus*. Three genera and nine species occur from central Mexico to northwestern South America.

cosmopolitan. *adj.* Generally, reference to widespread distribution of a species (i.e., throughout all biogeographic regions). Having a wordwide or very broad distribution.

costal. *n., adj.* 1. A plate on the carapace of turtles, located between the *central* (*vertebral*) and *marginal* plates and overlying the *pleural* bones. Also a synonym for *pleural* bones. (Syn. *centrolateral, lateral*) See Figs. 4, 5. 2. Of or pertaining to a rib or structure in thoracic region of the body.

costal bones. The bones of the turtle carapace lying between the neural and peripheral bones. See Fig. 5.

costal fold or **groove.** Deep grooves or the tissue between them (folds) that occur on the lateral aspects of the bodies of salamanders, usually indicating rib positions. (Syn. *costal furrow, costal interspace, costal plica, costal space, intercostal fold* of various workers)

costal furrow. See *costal fold*.

costal interspace. See *costal fold*.

costal plica. See *costal fold*.

costal space. See *costal fold*.

Costata. *n.* A monophyletic taxon of anurans, sister with *Acosmanura* within the newly designated *Sokolanura*, and containing *Bombinatoridae* and *Alytidae* (formerly *Discoglossidae*) See D.R. Frost et al., *Bull. Amer. Mus. Nat. Hist.* 297:1–370, 2006.

costiform. *adj.* Riblike.

costiform process. 1. Elongated, lateral projections from the nuchal bone of the carapace of trionychid and chelydrid turtles. 2. Any projection that bears resemblance to a rib.

coterminous. *adj.* Reference to organisms having similar distributions.

cotransmitter. *n.* A second neurotransmitter molecule synthesized in and released from an axon terminal along with a small transmitter molecule such as Ach or GABA. The cotransmitter modulates the action of the other transmitter molecule.

cotransport. *n.* Carrier-mediated transport across a membrane in which two dissimilar molecules bind to specific sites on a carrier protein for transport in the same direction.

Cottonmouth. *n.* Common name for *Agkistrodon piscivorus*, a pitviper common to the southeastern United States. (Syn. *Moccasin, Water Moccasin*)

cotyle, cotyla, cotylus. *n.* A cuplike cavity; used in reference to acetabulum, or the concave surface on the centrum of vertebra, into which the condyle of the neighboring vertebra fits.

cotyloid. *adj.* Shaped like a cup, or cavity.

Cotylosauria, cotylosaurs. *n.* A loosely used term coined originally to designate the basal group of *amniotes*. The cotylosaurs, meaning literally "stem reptiles," were envisioned to be the basal group of amniotes from which all later groups evolved. Now extinct, these were massive primitive reptiles of the Carboniferous and Permian periods and are often called *stem reptiles.*

cotype. *n.* One of a pair of specimens that constitutes the type series of a taxon.

counter-. Prefix meaning "the opposite direction" or "moving against."

counteracting selection. The operation of selection pressures at two or more levels of organization such as individual, family, or population, such that certain genes are favored at one level but not favored at another level. Cf. *reinforcing selection*.

countercurrent. *adj.* Flow of adjacent fluids in opposite directions. Cf. *concurrent*.

countercurrent exchange. A specialized parallel arrangement of incoming and outgoing blood vessels that essentially uncouples the direction of flow of some exchangeable entity (e.g., heat, solutes) from that of the blood. The arrangement maximizes concentration or temperature differences at the opposite ends of the exchanger. Cf. *crosscurrent exchange*.

countercurrent heat exchange. A specific reference to countercurrent exchange of heat (see *countercurrent exchanger*). Usually this involves movement of heat from an artery or arterial system containing warm blood to an adjacent vein or venous system containing cooler blood being returned from the periphery of an organisms to the body core. Such a mechanism reduces heat loss and maintains a relatively high core temperature, as in the Leatherback Sea Turtle (*Dermochelys*).

countershading. *n.* A color pattern in which the more brightly illuminated aspect of the body (normally the dorsal surface) is darker than is the less illuminated (ventral) surface. The effect is to disrupt the silhouette in a gradient of downstreaming light and to obscure the appearance of a shadow on the body and thereby render the animal less distinguishable from its background.

countersunk. *adj.* Positioned below the margins, as the lower jaw of many burrowing snakes that fits within the margins of the upper jaw.

countertransport. *n.* The uphill transport of a substance across a membrane driven by the downhill diffusion of a different substance.

counterweight. *n.* An equivalent weight.

coupled reactions. *n.* Chemical reactions having a common intermediate and thereby a means of transferring energy from one to another.

coupled transport. Simultaneous transport of two substances in the same or opposite directions, via a membrane protein.

courtship. *n.* Generally has reference to behaviors that attract and pair opposite sexes during periods when a species is receptive to copulation and reproduction. See *courtship ritual*.

courtship bob. A type of bob behavior described in *Anolis* lizards and used in contexts of courtship.

courtship call. A characteristic vocalization produced by a male anuran when a female approaches his calling site.

courtship ritual. Reference to a characteristic genetically determined behavior pattern involving the production and reception of a complex sequence of visual, auditory, or chemical stimuli by the male and female prior to mating. See *courtship*.

covariance. *n.* A measure of the strength of association between two varables. The *correlation coefficient* between two varables equals the sum of $(x - \bar{x})(y - \bar{y})$ over all pairs of values for the variables x and y, where $\bar{x}$ is the mean of the x values and $\bar{y}$ is the mean of all y values.

cover type. The existing or dominant vegetation of an area.

cow. *n.* A female terrapin that is sexually mature.

cowl. *n.* Refers to a hood and has been used in reference to the "hood" display of a cobra.

coxal. *adj.* Toward or pertaining to the hip.

Crab-eating Frog. Vernacular name for the Southeast Asian species of ranid frog *Rana cancrivora*. This species inhabits estuarine mangroves and tolerates saline habitats. It is also called *Mangrove Frog*.

Crab-eating Water Snake. Vernacular name for the Asian homalopsine species of *Fordonia*. Also called *Mangrove Snakes*.

cranial. *adj.* Relating to the skull or cranium and sometimes used synonymously with *anterior* to specify direction.

cranial club. A modified cranial crest with enlargement near the parietal spur of *Anaxyrus* (formerly *Bufo*) *terrestris*. First used by O. Sanders (*Herpetologica* 17:145–156, 1961).

cranial crest. A raised bony ridge on the head of certain species of bufonid toads, generally between and behind the eyes. (Syn. *otoparietal ridge, cranial ridge,* with reference to certain salamandrids) See Fig. 2.

cranial kinesis. Independent movement of bones, or groups of bones, in the skull. Especially characteristic of squamate reptiles and used in reference to movement between the upper jaw and braincase about joints that are between them. See *akinetic skull, amphikinesis, mesokinesis, metakinesis, prokinesis, streptostyly,* and Figs. 18, 19.

cranial nerve. Any one of about 12 (reptiles) to 17 (most amphibians) nerves entering or exiting the brain. Cranial nerves originate in the brain stem and supply sensory and motor function to various parts of the upper body. Cf. *spinal nerve*.

cranial ridge. See *cranial crest*.

cranial skeleton. Reference to skeletal elements of the head, or skull.

craniofacial angle. The angle formed by the braincase and muzzle of the skull, used to differentiate certain amphisbaenid genera in which the angle is strong.

craniology. *n.* Study of the physical structure of the cranium.

cranioquadrate passage. A channel lying between the quadrate ramus of the pterygoid and lateral wall of the reptilian braincase. Used by A.S. Romer (*Osteology of the Reptiles,* 1956, p. 63).

cranium. *n.* The bones forming the enclosure of the brain, excluding lower jaw. (Syn. *skull*)

cranterian. *adj.* Refers to teeth located behind the diastema on the posterior part of the maxillary bone.

crash. *n.* A precipitous decline in the size of a population.

Crawfish Frog. Vernacular name for the American ranid species *Lithobates areolatus* (formerly *Rana areolata*). For recent taxonomic revision see D.R. Frost et al., *Bull. Amer. Mus. Nat. Hist.* 297:1-370, 2006.

crawl. *n.* Used in reference to sea turtle movements on nesting beaches during onshore periods of oviposition. The tracks of a turtle on the beach. See *false crawl*.

Crayfish Snake. Vernacular name for American species of snakes belonging to the colubrid genus *Regina*. See also *Queen Snake*.

crease. *n.* Used by L.M. Klauber (1956) in reference to scale separations (*suture, hinge,* or *interscale tissue*) in the head scalation of snakes.

creatine phosphate. A phosphorylated nitrogenous compound that acts as a storage form of high-energy phosphate for rapid phosphorylation of ADP to ATP, primarily in muscle. (Syn. *phosphocreatine*)

creche. *n.* A nursery for very young animals.

crenate. *adj.* Having a notched or scalloped edge, as in the edge of a scallop shell.

crenated. *adj.* Shrunken.

crenulate. *adj.* Minutely notched or crenate, thus having a wavy outline. An edge with a wavy margin.

crepitaculum. *n.* An antiquated term for the rattle of rattlesnakes, not uncommon in older literature.

crepuscular. *adj.* Behaviorally active at dusk or dawn (twilight).

crest. *n.* Any elevated, prominent ridge or fold of skin, usually dorsal and median and associated with the tail or back of many lizards, or with the tails or backs of male salamanders during the breeding season.

crestal. *n.* Any of the large, convex, and sharply keeled scales comprising the dorsal scaly crests of the Chinese Crocodile Lizard (*Shinisaurus*).

Cretaceous. *n. adj.* The third and most recent Period of the Mesozoic Era, encompassing the appearance of various herbivorous dinosaurs, evolution of mammals, and extinctions of dinosaurs at the end of the Period. See Table I.

crevicular. *adj.* Inhabiting crevices.

Cribo. *n.* Vernacular name of the Central and South American subspecies of *Drymarchon corais,* a species that also includes the Indigo Snakes of North America. See also *Mussurana.*

Cricket Frogs. Vernacular name for species of hylid frogs belonging to the American genus *Acris.*

cricoid cartilage. The cartilaginous ring that supports the base of the larynx. (Syn. *cricothyroid*)

cricothyroid. See *cricoid cartilage.*

crista. (*pl.* –ae) *n.* 1. Sensory hair cells (or patches of these cells) within the *ampullae* of semicircular canals of the inner ear, specialized for detecting angular acceleration. Cf. *macula.* 2. The folds of the inner mitochondrial membrane.

crista glandularis. Any of various glandular ridges found in various species of the frog genus *Rana* and containing large, granular glands. (Syn. *dermal fold*)

crista introversa. Projections of the epidermis inward into the dermis, as described on the dorsum of several species of *Bufo.*

crista parotica. A horizontal ridge on the lateral surface of the otic capsule.

cristate. *adj.* Having comb-like ridges or crests. Reference to a pattern of epidermal microsculpturing. See R.M. Price *J. Herpetol.* 16:294–306, 1982.

criterion. *n.* A predetermined standard on which a decision might be based.

critical level. The level of substrate moisture defining the *absorption threshold,* below which amphibians confined to a substrate experience sharp increases in rates of water loss.

critical oxygen tension. (P_c) The level of PO_2 below which the rate of oxygen consumption of a metabolic oxygen regulator becomes dependent on the ambient PO_2.

critical tadpole. A larval anuran that is starved when entering, or during the period of, metamorphosis. Cf. *stasis tadpole.* Term from D'Angelo et al. (*J. Exp. Zool.* 87:262, 1941).

critical thermal maximum. (CTM, CT_{max}) The mean of collective temperatures at which a species, when heated gradually, becomes incapacitated such that long-term survival is not possible at this or at higher temperatures. Typically, locomotor activity becomes disorganized, and the animal loses its ability to escape from conditions that will promptly lead to its death. Originally the *critical maximum,* used by R.B. Cowles and C.M. Bogert (*Bull. Amer. Mus. Nat. Hist.* 83:263–296, 1944) and modified by C.H. Lowe and V.J. Vance (*Science,* 122:73–74, 1955) and V.L. Hutchison (*Physiol. Zool.* 34:92–125, 1961; *Amer. J. Physiol.* 237:R367–R368, 1980). Cf. *critical thermal minimum.*

critical thermal minimum. (Ct_{min}) The temperature at which a species, when cooled gradually, becomes incapacitated such that long-term survival is not possible at this or lower temperatures. The low temperature produces cold narcosis and prevents locomotion. (Origin of term is similar to *critical thermal maximum.*)

critically endangered. A IUCN criterion for threatened species when a taxon is known to be at immediate and extremely high risk of extinction in the wild. Cf. *endangered, vulnerable.*

croak. *n., v.* The guttural sound made by vocalizing male anurans.

croak reflex. A guttural trill that vibrates the sides of the body when a male anuran is clasped by another male, providing sex recognition and release by the clasping male.

croc. *n.* Slang term for a crocodilian. Usage should be discouraged in formal writing.

Crocodile. *n.* Collective vernacular name for various species of crocodilians belonging to the family *Crocodylidae.* See *Crocodylidae.*

Crocodile Specialist Group. (CSG) An international network of herpetologists devoted to conservation of crocodilians worldwide. Current URL: http://www.flmnh.ufl.edu/natsci/herpetology/crocs.htm

Crocodile Specialist Group Newsletter. The official publication of CSG, providing current information about crocodilians, their conservation, status, and management worldwide.

crocodilian. *n.* A member of the *Crocodylia.* Living forms include 23 species of Crocodiles and Alligators; all are large, semiaquatic predators.

Crocodylia. *n.* The clade (order) that includes all extant crocodilians.

Crocodylidae. *n.* A clade (family or subfamily, depending on classification scheme) that includes two genera and 13 species of Crocodiles distributed largely in worldwide tropical locations across the Western Hemisphere, Africa, and Asia, and the Indoaustralian Archipelago to New Guinea and northern Australia.

crossband. *n.* An area of color transverse to the long axis of the body and extending down the sides to the ventral aspect of the animal, but not extending across the belly. Thus, a crossband is different from a *ring* in that its two ends do not join ventrally.

crossbreeding. *n.* See *outbreeding.*

cross-bridges. *n.* Molecular projections from myosin thick filaments that bind to sites on actin thin filaments during muscle contraction. Binding of the two filament parts forms the cross-bridge.

crosscurrent exchange. A gas exchange system characteristic of birds in which each blood capillary branch stepwise passes across an *air capillary* at roughly right angles and collects oxygen, the levels of which rise serially in the departing blood. Cf. *countercurrent exchange.*

cross feeding. In husbandry, reference to transferring an uneaten food item from one captive animal's cage to another.

crosshatched. *adj.* Having parallel ridges crossing longitudinal grooves. Reference to a pattern of *epidermal microsculpturing.* See R.M. Price *J. Herpetol.* 16:294–306, 1982.

cross infection. Transfer of a disease organism or agent from one individual to another of the same species.

crossing. *n.* Mating of individuals of different races or strains in order to promote genetic recombination.

crossing over. The exchange of genetic material between homologous chromosomes during meiosis or mitosis.

cross-reactivity. *n.* The effect of an antivenin prepared from venom of a species of snake on the venom of a different species of snake.

crotactin. *n.* A lethal component of *crotoxin* isolated by W.P. Neumann & E. Habermann (*Biochem. Zeits.* 327:170–185, 1955).

crotalid. *n.* Any pitviper belonging to the family Crotalidae.

crotalin. *n.* A term used for venoms from pitvipers, especially Rattlesnakes, or the active principle of such venoms.

Crotalinae. *n.* A subfamily of *Viperidae,* popularly known as *pit vipers* or *pitvipers*.

crotamine. *n.* An isolated component of Rattlesnake venom, belonging to the protein family of myotoxins, less active than crotactin but similarly associated with crotoxin.

Crotaphytinae. *n.* A clade (subfamily) of iguanid lizards with spectacularly colored species ("collared lizards") inhabiting mesic and desert areas of North America. Two genera and 12 species occur in the United States and Mexico.

crotch. *n.* The point of bifurcation of the hemipenis in certain snakes.

crotoxin. **(CTX)** *n.* A venom component isolated from the venom of the South American Rattlesnake (*Crotalus durissus terrificus*) in 1937 and subsequently characterized as a phospholipase having a variety of neurotoxic and myotoxic actions. See *calcineurin.*

crown. *n.* The top part of the head.

crown canopy or **cover.** The canopy formed by forest trees.

Crowned Snake. Vernacular name for various American colubrid snakes belonging to the genus *Tantilla,* especially *Tantilla relicta,* and for Australian elapid snakes belonging to the genus *Drysdalia.*

crown group. A group of organisms consisting of the last common ancestor of the group and all of the living descendents. See *stem group.*

crown-rump length. The distance between the crown of the head and the rump area of embryonic alligators, effectively measured as the maximum shell diameter since the embryos lie in a coiled position within the egg.

crown shield. Antiquated term for the enlarged scales on the dorsum of the head in snakes.

cruciform. *adj.* Reference to a cross shape but sometimes used less correctly for X-shapes as well.

crural. *adj.* Of or pertaining to the *crus*, but often extended in usage to include other parts of the leg.

crural valve. See *femoral valve.*

crus. *n.* The part of the leg extending between the knee and ankle, containing the tibia and fibula. The lower leg; used in reference to a limb that is shaped like the human leg. (Syn. *shank, tibial segment*)

cryo-. Prefix meaning "cold," or "pertaining to cold."

cryophase. *n.* Reference to the part of a thermocycle during which colder temperatures prevail. Cf. *thermophase.*

cryosection. *n., v.* A technique used in histology when tissues are frozen and cut using a *cryostat.*

cryostat. *n.* A cold chamber containing a microtome to section frozen tissues.

crypsis. *n.* (*adj.* **–ic**) Matching between the color and pattern of an organism and a random sample of the background it is viewed against, as perceived by another organism. Concealment. See J.A. Endler, *Evol. Biol.* 11:319–364, 1978. (Syn. *camouflage, cryptic coloration*) Cf. *semasis.*

cryptic coloration. Concealing coloration that renders an animal less susceptible to visual detection when viewed against its background, which it resembles ideally in both color(s) and pattern. See *crypsis*; cf. *phaneric coloration.* (Syn. *protective coloration*)

cryptic species. See *sibling species.*

cryptobiosis. *n.* A condition in which all external signs of metabolic activity are absent from a dormant organism.

Cryptobranchidae. *n.* A clade (family) of completely aquatic, giant, paedomorphic salamanders having flattened bodies and heads, incomplete metamorphosis, lidless eyes, and absence of a tongue pad. These are the largest extant salamanders (1.8 m in the Japanese *Andrias*) and inhabit cold streams. Three species are currently recognized.

cryptobranchid. *n, adj.* A member of, or pertaining to, Cryptobranchidae.

Cryptobranchoidei. *n.* A monophyletic clade (superfamily) that includes cryptobranchid and hynobiid salamanders. Based on molecular data, some authors propose that this clade was the sister group to other salamanders.

Cryptodeira, Cryptodira. *n.* A clade (suborder) of extant turtles distinguished by retraction of the neck in a vertical plane. This suborder includes most families of extant turtles. Cf. *Pleurodira.*

cryptodirous. *adj.* Reference to a turtle's capability for withdrawing the head beneath the carapace by means of bending the neck in the vertical plane.

cryptosporidiosis. *n.* A disease of lizards and snakes caused by the internal protozoan *Cryptosporidium,* which is a coccidian having a direct life cycle. Symptoms include weight loss, regurgitation of food, and eventual death without treatment.

cryptozoic. *adj.* Reference to habitation of secluded and dark places, such as beneath stones or in holes, crevices or leaf litter.

Cryptozoic. *n.* See *Precambrian*.

CSG. See *Crocodile Specialist Group*.

ctenii. (*pl.*) *n.* Structures resembling the teeth of a comb.

Ctenosaur. *n.* Alternative common name for Spinytail Iguanas, genus *Ctenosaura*.

CTM. See *critical thermal maximum*.

CT$_{max}$. See *critical thermal maximum*.

CT$_{min}$. See *critical thermal minimum*.

CT scan. See *CAT scan*.

CTX. See *Crotoxin*.

Cuban Crocodile. Vernacular name for the crocodilian *Crocodylus rhombifer*.

Cuban Tree Frog. Vernacular name for the West Indian hylid species *Osteopilus septentrionalis*. This species is endemic to Cuba and has been introduced elsewhere in the Lesser Antilles, Puerto Rican Bank, and Florida.

cuboidal. *adj.* Shaped like a cube.

cuesta. *n.* A ridge with a steep slope on one side and a gentle slope on the other side.

cuirass. *n.* A leathery armor, used in reference to the bony shields and plates of crocodilians and turtles. (Syn. *lorica*)

cultriform process. A slender extension of the *parasphenoid bone*, shaped like a V and running anteriorly beneath the braincase and between the pterygoids. (Syn. *basisphenoid rostrum, parasphenoid rostrum*)

culture. *n.* The growth of microorganisms on artificial media.

cumulative distribution. A distribution comprised of a number of observations above or below a given value.

cuneate. *adj., n.* 1. Wedge-shaped. 2. This name is applied to aberrant or sporadic wedge-shaped scales that lie between, but do not entirely separate, adjacent labial scales.

cuneiform. *adj., n.* 1. Wedge-shaped. 2. A small bone situated next to the distal end of the ulna where it articulates.

cupula. *n.* A gelatinous structure in which the sensors of a *neuromast* are imbedded.

curare. *n.* A South American arrow poison that blocks synaptic transmission at the neuromuscular junction by competitive inhibition of nicotinic acetylcholine receptors. Generally, the term *curare* includes a variety of highly toxic extracts from various botanical sources. (Syn. *D-tubocurarine*, which is the active ingredient of curare)

curarimimetic. *adj.* Descriptive of various toxins that produce effects similar to that of curare.

curarization. *n.* Administration of curare (usually *tubocurarine*) to induce muscular relaxation by means of blockade at the neuromuscular junction.

Curl Snake. Vernacular name for the Australian elapid species *Suta suta*. Also called *Myall Snake*.

Curlytail Lizards. Vernacular name for species of West Indies iguanid lizards belonging to the genus *Leiocephalus*.

current. *n.* A nontidal horizontal or continuous flow of water.

Current Herpetology. A journal of herpetology, formerly the *Japanese Journal of Herpetology*. It is the official journal of the Herpetological Society of Japan, Kyoto.

curretage. *n.* The process of scraping or cleaning tissues from body surfaces.

cursorial. *adj.* Reference to ability to move rapidly over ground, or to limbs and digits adapted for running. Cf. *graviportal*.

cusp. *n.* A pointed tip or toothlike projection, as the tip of a tooth or the edge of a jaw. This term does not refer to serrations on *jaw sheaths* of tadpoles.

cuspation. *n.* The distribution and developmental state of cusps, as applied to patterns on teeth.

cuspid, cuspate, cuspidate. *adj.* Possessing a cusp or cusps.

cutaneous. *adj.* Pertaining to integument or skin.

cutaneous evaporative water loss. (CEWL) See *evaporative water loss*.

cutaneous drinking. Reference to water uptake by anurans across ventral integument exposed to a wet or moist substrate (S.D. Hillyard, *J. Exp. Zool.* 283:662–674, 1999).

cutaneous gas exchange. See *cutaneous respiration*.

cutaneous respiration. An imprecise terminology to describe exchange of respiratory gases (O_2, CO_2) between the blood and the environment via the skin. Characteristic of many amphibians and some aquatic reptiles, especially marine snakes. (Contrary to the implication of this term, it is seldom used to refer explicitly to cellular respiration within the skin tissue.) (Syn. *cutaneous gas exchange*)

cutting edge. 1. A distinct ridge on the distal end of the fang in many venomous snakes. Two such edges often oppose each other laterally, but may vary in position. (Syn. *wearing edge, reinforcement ridge*) 2. In *Heloderma*, a sharp flange beside the open groove present on both anterior and posterior aspects of teeth in the lower jaw. 3. See *tomium*.

CVF. See *cobra venom factor*.

CWL. See *evaporative water loss*.

cyanomorph. *n.* A color morph in which the dominant color is blue.

cyanosis. *n.* (*adj.* **cyanotic**) A bluish discoloration of the skin due to an abnormal amount of reduced or deoxygenated hemoglobin in the blood.

cybernetics. *n.* Study of control and communication systems.

Cyclanorbinae. *n.* A subfamily of *Trionychidae*.

cyclical selection. Selection in one direction followed by reversed selection (opposite direction) attributable to cyclical fluctuations in environment, e.g. seasonal temperature changes.

cyclical stability. In ecosystem dynamics, the property of a system to oscillate across a focal point or range.

cyclic AMP. (cAMP) A ubiquitous cyclic nucleotide produced from ATP, acting as an important regulatory agent that functions in cells as a second messenger for many hormones and transmitters.

cyclic GMP. (cGMP) A cyclic nucleotide analogous to cAMP but present in cells at a far lower concentration and producing target cell responses that are often opposite to those of cAMP.

cycloid. *adj.* Reference to reptilian scalation where scales have free, rounded, or evenly curved borders.

Cycloramphidae. *n.* A recently reformulated family of anurans, sister to the new taxon *Agastorophrynia* within the new taxon *Hesticobatrachia*, and containing 14 genera distributed in southern South America. See D.R. Frost et al., *Bull. Amer. Mus. Nat. Hist.* 297:1-370, 2006.

cyclotreme. *adj.* Possessing a rounded anal opening, as in turtles and crocodilians.

cylindrical. *adj.* Reference to something that is nearly equidimensional in cross section, or nearly circular (neither *depressed* nor *compressed*). This term is usually used in reference to body shape. Cf. *fusiform.*

Cylindrophiidae. *n.* A family of snakes known as *Pipe Snakes,* sometimes classified as a subfamily of *Uropeltidae.*

Cylindrophiinae. *n.* A subfamily of *Uropeltidae,* considered by some to warrant family status.

cyst. *n.* 1. A saclike structure that is part of the life cycle of certain parasites which form a protective wall. 2. A saclike structure within the body that becomes abnormally filled with fluid or diseased matter.

cystic. *adj.* Relating to a bladder or pouch.

cystic spermatogenesis. A pattern of spermatogenesis characteristic of fish and amphibians in which cysts representing different stages in spermatogenesis coexist within a testis lobule.

-cytin. Suffix meaning "platelet aggregating," used in a nomenclatural scheme for describing exogenous hemostatic factors in snake venoms. By adding a portion or designated abbreviation of a snake's scientific name to the suffix, one obtains a designation for the fraction being identified. For example, using the name "*gabonica*" (species name for the Gaboon Viper) one obtains *gabonicytin*.

cytochromes. *n.* A group of proteins containing iron and functioning in the electron transport chain in aerobic cells by accepting and transferring electrons.

cytogenetics. *n.* The study of chromosomal behaviors and mechanisms and their influence on inheritance and evolution.

cytology. *n.* A subdiscipline of biology concerned with the structure, function, and life histories of cells.

cytolysin. *n.* A substance capable of *cytolysis,* such as enzymes or components of snake venoms.

cytolysis. *n.* (*adj.* **cytolytic**) The destruction or dissolution of cells.

cytoplasm. *n.* Intracellular aqueous fluid or gel that contributes to cell mass and suspends the intracellular organelles, exclusive of the nucleus.

cytoplasmic folds. *n.* See *microridges.*

cyplasmic inheritance. See *extranuclear inheritance.*

cytosol. *n.* The unstructured aqueous phase of the cytoplasm between the structured organelles.

cytotoxin. *n.* (*adj.* **–ic**) A toxin or component of venom that affects the walls of blood vessels and soft tissues, causing leakage of fluid that results in swelling and potential localized necrosis.

cytotypes. *n.* Different karyotypes within a nominal species or population.

D

dactyl. *n.* A finger or toe; a digit.

dagger. *n.* An enlarged, spikelike structure on the inside of the forefoot of certain anurans.

Dalton. *n.* A unit named after J. Dalton (who developed the atomic theory of matter) and equal to the mass of the hydrogen atom, or 1.0 on the atomic mass scale.

Dalton's law. The principle that the partial pressure of a gas is independent of other gases present in a mixture. The total pressure of a mixture is the sum of the partial pressures of all the gases present.

dam. *n.* The female parent in animal breeding terminology. Cf. *sire.*

damped oscillations. Oscillations in which the period of the fluctuations becomes smaller with time, converging on a stable equilibrium point.

DAPTF. See *Declining Amphibian Populations Task Force.*

DAQ. See *data acquisition.*

dark field microscope. A microscope designed so that viewed objects appear bright upon a dark background, achieved by blackening out entering light rays and directing peripheral rays against the object from the side.

Darling effect. See *Fraser-Darling effect.*

Darlington's rule. A rule of thumb that on oceanic islands a tenfold increase in island area results in a doubling of species number (P.J. Darlington, *Zoogeography: The Geographical Distribution of Animals*, 1957).

darwin. *n.* A unit measure of evolutionary change, defined as an increase or decrease in any given character by a factor of 2.7 per million years. See *Haldane's evolutionary unit.*

Darwinian. *adj.* Of or relating to the research and theories of Charles Darwin, English naturalist (1809–1882) who became influential for his theory of evolution by natural selection. See *Darwinian evolution.*

Darwinian evolution or **Darwinism.** The theory that biological evolution involves natural selection of adaptive variations.

Darwinian fitness. See *adaptive value.*

Darwinian selection. See *positive selection.*

Darwin's Frogs. Vernacular name for species of South American anurans belonging to the family Rhinodermatidae, especially *Rhinoderma darwinii.*

Darwin's Sea Snakes. Vernacular name for elapid sea snakes belonging to the genus *Hydrelaps.*

Dasypeltinae. *n.* A subfamily of African and Asian *Egg-eating Snakes*, alternatively classified in *Colubrinae.*

data. (*s. datum*) *n.* Observations or units of information resulting from experiment or study. (Note that data are plural. Many writers use it incorrectly as singular.)

data acquisition. (DAQ) Collecting and measuring electrical signals from sensors, transducers, and test probes or fixtures and inputting them to a computer for processing.

data matrix. A tabulation of data illustrating differences between categories, often in machine readable form.

data transformation. Commonly used applications of mathematical modification to the values of a variable in a data set. One reason for transformation of data is to improve the normality of variables, which is important for statistical analysis. There are multiple options for transforming non-normal data. See *inverse transformation, log transformation, square root transformation.*

daughter. *n.* Reference to offspring of a given generation, applicable to both sexes.

daughter cells. Two cells that result from division of a single cell.

daughter species. A recognizably new species that evolves from a parent species.

Dawn Blind Snakes. Blind snakes belonging to the family *Anomalepididae.*

day. *n.* 1. *Sidereal day* refers to the mean time taken for one revolution of the Earth. 2. *Solar day* refers to the mean time taken for the sun to reach any given position on consecutive days. The mean time interval between consecutive sunrises, or any specified position of the sun (24 h). 3. *Lunar day* is the time interval between consecutive moonrises (24.8 h).

Day Geckos. Vernacular name for species of Geckos belonging to the genus *Phelsuma.*

dazzle effect. A phenomenon where a light-colored object (e.g. a desert lizard) seems to disappear against light and reflective environments (e.g. open desert sands) due to the high reflectance of light from the immediate environment of the object. The effect has significance because it may occur even when the object is not entirely color-matched to its background.

dB. See *decibel*

dbh. Abbreviation of "diameter at breast height" (1.4 m), used by foresters in evaluating tree size.

de-. Prefix meaning "away," down," or "from."

dead space. In physiology, reference to the volume of air that is ventilated (or breathed) by an animal but does not participate in gas exchange. That is, oxygen is not extracted from this air, which, for example, might move in and out of conducting airways without contacting the exchange parenchyma of the lung.

deamination. *n.* The oxidative removal of NH_2 groups from amino acids to form ammonia. In the metabolism of proteins, most amphibians and reptiles convert the ammonia to either urea or uric acid, or both. These conversions are not necessary provided an animal has abundant water with which to remove (excrete) the ammonia, which is cytotoxic. Aquatic amphibians are generally *ammonotelic*, and a number of amphibians and reptiles are more ammonotelic than formerly supposed.

Death Adder. Collective vernacular name for several species of Australian elapid species of snakes (genus *Acanthophis*) characterized by a number of features that are convergent with those of viperid snakes.

death feint. A defensive reflex involving assumption of a usually motionless position similar to that of a dead individual. (Syn. *tonic immobility, hypnotic reflex, thanatosis*)

death rate. In the exponential growth model, the death rate (D) is the change in the number of deaths in a population measured over a short time interval (units are deaths/time). In the Lotka-Volterra model, the death rate (q) is the instantaneous death rate for the predator population in the absence of the prey population. The death rate q is identical to d, the instantaneous death rate in the exponential growth model. Units are predators/(predator • time).

debridement. *n.* (*v.* **debride**) Surgical cleansing of a wound or removal of necrotic tissue.

debris. *n.* Any unconsolidated plant or animal material at Earth's surface.

deca-. Prefix meaning "ten" or "tenfold." Used to denote unit x 10. (Syn. *deka-*)

decahydroquinolines. *n.* A class of alkaloids present in the skin of certain species of dendrobatid frogs. The first of this class was originally designated as *pumiliotoxin-C*.

decalcification. *n.* Loss of calcium salts from bone or tooth, or the process of removing calcareous matter.

decapitation. *n.* Removal of the head.

decarboxylation. *n.* The loss or removal of a carboxyl group from an organic compound and resultant formation of CO_2.

decay. *n.* Organic decomposition. (Syn. *putrefaction*)

deci-. Prefix used to denote unit x 10^{-1}.

decibar. *n.* A measure of pressure equal to one tenth of normal atmospheric pressure (1 bar). This unit also equals a change of depth in sea water equal to 1 m.

decibel. (dB) The unit for expressing a logarithmic measure of the ratio of 2 signal levels: $dB = 20 \log_{10} V_1/V_2$, for signals in volts. The unit is commonly used in relation to measures of sound intensity.

deciduous. *adj.* Of or pertaining to the act of shedding, usually at at regular intervals (e.g., loss of leaves during the dry season).

deciduous forests. Forests in which most of the trees lose their leaves for several months each year in response to seasonal climatic stress.

Declining Amphibian Populations Task Force. (DAPTF) A worldwide network of over 100 working groups of scientists, conservationists, and amateur naturalists, established in 1991 by the IUCN Species Survival Commission to collect and monitor data on amphibian populations and to assess their geographic distribution and possible causes of decline. The DAPTF publishes a Newsletter. Current URL: http://www.open.ac.uk/daptf

decompression sickness. See *caisson disease.*

decussation. *n.* Crossing over or passing from one side to the other, usually used in relation to afferent nerve tracts carrying sensory information to the brain.

dedifferentiation. *n.* The loss of differentiation, or regression to a less specialized condition.

deduction or **deductive method.** A method of reasoning that involves testing of predictions that are deduced from general principles or theory. Cf. *inductive method.* (Syn. *Popperian method*)

deep body temperature. A measurement of core body temperature at a point farther within the body than the cloaca. The site of measurement should be specified. (Syn. *core temperature*)

deep ecology. A nonscientific term expressing a philosophical movement which looks for the fundamental facets of human culture that lead to degradation of the environment and demands changes in basic ideas underlying civilization, such that nature will be respected as valuable in itself and a part of the human identity. The term was coined by the Norwegian philosopher Arne Naess in 1972.

defecation. *n.* (*v.* **defecate**) The process of expelling feces from the digestive tract.

defense-fight reaction. A defensive reaction of some toads in which the body is elevated off the ground by extended limbs, while the head is thrust forward in aggressive movements toward a predator.

defense reaction. Reference to defensive posture of anurans which inflate the lungs and bow the head low when disturbed.

defensive behavior. Any activity engaged in by an animal to avoid falling prey to a predator or being disadvantaged by a competitor.

defensive male reaction. A behavior in *Pseudacris* where one or both of two approaching males become motionless at short distances, evidently reflecting a state of watchfulness.

deferent. *adj.* A duct that carries away, as the sperm duct.

definition. *n.* In taxonomy, a diagnostic statement of characters that distinguish a given taxon.

definitive host. A host in which sexual maturation of a parasite occurs.

definitive kidney. The posterior, thickened, functional kidney of salamanders, distinct in development from the *sexual kidney*.

deforestation. The permanent removal of trees from forests, usually for commercial purposes, or conversion of forests for nonforest uses such as pasture or croplands.

degeneration. *n.* Loss or reduction of structure or function during the course of ontogeny or evolution.

deglutition. *n.* The process of swallowing.

degradation. *n.* Reduction of complexity by breakdown into smaller or simpler parts.

degree day. A measurement of thermal exposure calculated as the product of time and temperature averaged over a specified interval.

degrees of freedom. (df) In statistics, this term refers to a measure of the number of independent pieces of information on which the precision of a parameter estimate is based, e.g., the number of independent deviations from the mean on which an estimate of population variance can be based. The *df* for an estimate equals the number of observations (values) minus the number of additional parameters estimated for the calculation.

dehydrate. *v.* (*n.* **dehydration**) The act of drying, or replacement of water in a specimen by a nonaqueous medium. To remove or lose water. Cf. *desiccate*.

dehydrobufotenine. *n.* An inactive compound that yields bufotenine upon hydrogenation or bufothionine upon esterification of the phenolic group.

dehydrogenase inhibitor. A substance, including various components of snake venoms, that inhibits dehydrogenase enzymes associated with glycolysis.

dehydrogenation. *n.* Oxidation of a compound by removal of equal numbers of protons and electrons, usually two of each.

deimatic behavior. Intimidating postures or actions that occur in an animal which cannot flee very fast, or which has been caught by a pursuing predator.

deka-. See *deca-*.

delamination. *n.* The formation of a tissue layer by means of separation and subsequent aggregation of cells from a preexisting embryonic tissue layer. Splitting of embryonic tissue sheets.

delayed fertilization. The union of egg and sperm following a period of prolonged sperm storage in the genital tract of the female following copulation. This phenomenon is not uncommon and has been reported for periods as long as several years in various reptiles. See *amphigonia retarda* (synonym).

deleterious. *adj.* Having an adverse effect; said of a mutation that reduces survival and fitness.

deletion. *n.* The loss of a segment of genetic material from a chromosome.

deletion mutation. A mutation that removes one or more nucleotides from a genome.

deliquescent. *adj.* Becoming fluid by uptake of water from air.

delta. *n.* A body of sediment deposited at the mouth of a river where the river velocity decreases as it flows into a standing body of water.

deltoid crest. See *deltopectoral crest*.

deltopectoral crest. A process extending ventrally from the proximal head of the reptilian humerus, with its two faces attaching to the pectoralis and deltoid musculature, respectively. This structure may be comprised of separately projecting expansions termed the *deltoid crest* and *pectoral crest*, respectively.

deltorphins. *n.* See *opioid peptides*.

deme. *n.* An isolated or otherwise distinct population within a species, having defined characteristics and localized structure. (Syn. *local population*)

demersal. *adj.* 1. Inhabiting the bottom or lowest layer of a lake or sea, while having capacity for active swimming. 2. Eggs of amphibians that are denser than water and therefore sink.

demographic stochasticity. Uncertainty due to variation in the sequence of births and deaths in a population, causing population numbers to vary unpredictably. Demographic stochasticity can generate a substantial risk of extinction in small populations, whereas in large populations the sources of random variation tend to average out over the long run.

demography. *n.* The study of populations with particular reference to growth rates and age structure. Reference to the statistical characteristics of a population such as size, birth and death rates, density, movement or migration, etc.

den. *n.* A secluded, usually subterranean site where numerous animals are gathered together for retreat or overwintering. The term is commonly applied to aggregations of snakes.

denaturation. *n.* 1. Destruction or alteration of the normal nature of a substance by physical or chemical means. 2. Alteration of the structural properties of a protein, resulting in loss of enzymatic function or other activity.

dendrites. *n.* Fine processes of a neuron that characteristically provide the principal receptive area for synaptic inputs from other neurons and carry impulses toward the cell body.

dendritic evolution. See *cladogenesis*.

dendroaspin. *n.* See *mambin*.

Dendrobatidae. *n.* A clade (family) of anurans characterized by a pair of dermal scutes on the dorsal surfaces of the fingers (also observed in a few *leptodactylids* and *myobatrachids*), mostly diurnal with terrestrial habits, complex parental care, and toxic skin secretions. These are known in popular literature as Poison Frogs or Dart

Poison Frogs; however, not all species are highly toxic. Eight genera and about 205 species occur in Central and South America, with highest diversity in Colombia and Ecuador. Recently, the number of genera has been increased to 14, and this group is classified as a sister taxon to a new family *Thoropidae*, together comprising the *Dendrobatoidea* (D.R. Frost et al., *Bull. Amer. Mus. Nat. Hist.* 297:1–370, 2006).

dendrobatid. *n, adj.* A member of, or pertaining to, *Dendrobatidae*.

Dendrobatoidea. *n.* A taxon (superfamily) of anurans, sister to *Bufonidae* within the new taxon *Agastorophrynia*, and containing *Dendrobatidae* and the new family *Thoropidae*. See D.R. Frost et al., *Bull. Amer. Mus. Nat. Hist.* 297:1–370, 2006.

dendrogram. *n.* A branching diagram resembling a tree and used to depict relationships or history of a group of organisms, or the resemblance of characters. See *cladogram*.

dendropeptin. *n.* A peptide isolated from the venom of *Dendroaspis angusticeps*. It facilitates liberation of second messengers in endothelium and has actions related to cardiac diuretic factors.

dendrotoxin. *n. Presynaptic facilitating toxins* first discovered in venom of the Mamba, *Dendroaspis*. They facilitate the release of acetylcholine in the neuromuscular junction and cause a strong depolarization of the postsynaptic membrane.

denervate. *v.* (*n.* **denervation**) To eliminate the nerve supply to a tissue or organ, either chemically or physically.

de novo. From Latin, meaning arising anew from an unknown source, or denoting synthesis from very simple precursors.

densitometry. *n.* See *absorptiometry*.

density. *n.* 1. The number of individuals or other item of observation per unit of area or volume. 2. The ratio of mass to volume of a fluid or object.

density dependence. Population processes in which the instantaneous birth and death rates (b and d) are influenced by the density or size of the population. Population growth tends to be enhanced by density dependent factors when population size decreases, or is retarded when population size increases. Cf. *density independence, inverse density dependence*.

density dependent selection. Selection in which the values for relative fitness depend on the density of the population.

density independence. Population processes that are not affected by the current density or size of the population. Cf. *density dependence*.

dental formula. Shorthand expression of the characteristic number and type of teeth present in an animal (most often used in reference to mammals).

dental gland. Secretory oral glands having openings into pockets at the base of teeth, present in many squamates and thought to have given rise to *venom* and *Duvernoy's glands* based on strong developmental evidence. See also *gland of Gabe*. See Fig. 20.

dental plate. See *tooth ridge*.

dentary. *n.* A principal dermal bone of the lower jaw, present in all reptiles and amphibians and bearing teeth in squamates and crocodilians.

dentary pseudo-teeth. Tooth-like structures situated at the tip of the lower jaw in some anurans, e.g. the tusked frogs of the genus *Adelotus*. (Syn. *dentiform process*)

dentate. *adj.* Toothed or serrated.

denticle. *n.* 1. A small tooth or tooth-like projection. (Syn. *denticule, dentiform process*) 2. Formerly, one of several keratinized, toothlike structures surrounding the mouth in most tadpoles. However, R.W. McDiarmid and R. Altig (*Tadpoles. The Biology of Anuran Larvae*, 1999) suggest this term should be discarded in reference to the *labial teeth* of tadpoles. See Fig. 1.

denticulate. *adj.* Having fine tooth-like spines or projections. Generally used with reference to the edge of a scale. (Syn. *serrate, serrated*)

denticulus corneus. Tiny subcellular jags of the *stratum corneum* observed in the skin of some amphibians.

denticulus pluricellularis. Irregularly shaped, multicellular projections of epidermis observed in some frogs.

dentiform process. See *dentary pseudo-teeth*.

dentigerous process. That portion of a bone bearing teeth.

dentin or **dentine.** A hard, bonelike tissue that is the main substance of teeth. The material is softer than enamel, but harder than bone.

dentition. *n.* Quantitative and qualitative features of the teeth of a particular species. Usually the reference is to the type of teeth and their physical properties and arrangement.

deoxygenated. *adj.* Depleted of oxygen. Cf. *oxygenated*.

deoxygenation. *n.* Removal or depletion of oxygen.

deoxyhemoglobin. *n.* Hemoglobin that has no oxygen bound to the iron binding sites of the heme groups.

deoxyribonuclease. (DNase) *n.* An enzyme component of snake venoms that degrades DNA.

deoxyribonucleic acid. (DNA) The class of nucleic acids responsible for hereditary transmission and for the coding of amino acid sequences of proteins. The sequence of nucleotides adenine, guanine, cytosine, and thymine—contained in two polynucleotide chains forming a double helix—is the basis of the genetic code.

deoxyribose. *n.* A five-carbon sugar that forms a major structural component of DNA.

depauperate. *adj.* Of or pertaining to a region with a low number of species. Impoverished.

dependence. *n.* A condition in which a variable is controlled or influenced by another variable. Cf. *independence*.

dependent character. A character whose state or definition depends on the simultaneous occurrence of another character.

depolarization. *n.* A reduction or reversal of the potential difference that exists across a membrane when compared with the resting membrane potential.

depredate. *v.* To capture prey.

depredation. *n.* The act of capturing or subduing prey.

depressed. *adj.* Reference to a structure that is flattened in the dorsoventral dimension or horizontal plane. Cf. *compressed.*

depressor. *n.* A muscle that depresses or lowers a structure by pulling it down.

depressor mandibulae. The muscles that open the mouth in amphibians and reptiles.

derived. *adj.* Reference to the more recent characters or conditions in an evolutionary lineage. Not original; coming later; specialized. The opposite of *primitive.* See *apomorph.*

derived character. A character or feature that departs from the ancestral condition for the taxon in question. See *derived.*

derived similarity. See *synapomorphy.*

dermal. *adj.* Reference to skin generally, but correct and more specific usage should pertain to the *dermis.*

dermal armor. Reference to protective *dermal bone* on the surface structures of some vertebrates, such as the composite shell of turtles.

dermal bone. Bone that ossifies directly from mesenchyme without replacing cartilage. See *osteoderm.*

dermal chromatophore unit. Reference to dermal chromatophore unit of amphibians or reptiles. These are discrete structures in which a *xanthophore* overlies an *iridophore,* and a *melanophore* is situated beneath both of these. Extensions of the melanophore radiate upward to wrap and overlie the top of the xanthophore, and migrating melanin can thereby mask (darken) or modify the other colors when stimulated to disperse. See Fig. 33.

dermal fold. 1. A fold of skin. 2. See *crista glandularis.*

dermal layer. *Dermis.*

dermal ossicle. See *osteoderm.*

dermal papilla. A part of the tooth-forming primordium that is derived from the neural crest cells and differentiates into odontoblasts that secrete dentin.

dermal plate. Any of the *lamina* or flattened external plates on the shell of a turtle.

dermal platelet. Any of small dermal ossifications in the carapace of hatchling and juvenile Leatherback Sea Turtles (*Dermochelys coriacea*).

dermal pocket. See *acaridomatium.*

dermal pressure receptor. A more recent term describing *integumentary sense organs* on the upper and lower jaws of crocodilians, renamed to emphasize the supposed mechanosensory function of pressure detection. See *integumentary sense organ.*

dermal scale. One of numerous and patterned small, thin plates of bone found in many bony fishes and caecilians. It has been suggested by R. Ruibal and V. Shoemaker (*J. Herpetol.* 18:313–328, 1984) that this term be restricted to structures in fishes and caecilians, and not used to describe *osteoderms* of anurans and reptiles. Cf. *osteoderm.* (Syn. *bony scale*)

dermal supraoccipital. See *postparietal.*

dermal tumor. A tumor formed in the skin. See *pox.*

dermarticular. *n.* See *prearticular.*

dermaseptin. *n.* A linear peptide isolated from skin extracts of *Phyllomedusa sauvagei.* This peptide displays potent inhibitory activity against pathogenic fungi.

Dermatemydidae. *n.* A clade (family) of turtles with a single species distributed in large rivers, lakes and pools from southern Mexico to northern Hondurus.

dermatitis. *n.* A condition of inflamed skin, which may blister and become necrotic in severe cases.

dermatocranium. *n.* Parts of the bony skull that are derived from dermal bone. This is the major part of the reptilian skull and includes the nasals, prefrontals, frontals, and parietals. The premaxillae, maxillae, and mandibles also derived from the dermatocranium.

dermatoglyphics. *n.* Study of the ridged or surface features of the skin. See *epidermal microsculpturing.*

dermatology. *n.* The study of skin and its diseases.

dermatome. *n.* Populations of embryonic *somite* cells that contribute to skin musculature.

dermatophytosis. *n.* Fungal infection of the skin.

dermatozoon. *n.* An animal parasite of the skin.

dermis. *n.* The inner layer of skin beneath the epidermis, derived from embryonic mesoderm and containing fibrous connective tissue, blood vessels, nerves, chromatophores, glands (especially amphibians), and ossicles in some reptiles. The dermis of turtles and tortoises is unique in that much of it is ossified. (Syn. *dermal layer*) See Figs. 28–31.

Dermochelyidae. *n.* A clade (family) of Sea Turtles, often treated together with Cheloniidae as a clade relative to other turtles. A single species (Leatherback) is recognized, with worldwide distribution in temperate and tropical oceans.

dermorphins. *n.* Hepta-peptide amides isolated from skin of certain phyllomedusine frogs, are interactive preferentially with opiate receptors. See *opioid peptides.*

dermo-supraoccipital. See *postparietal.*

derotrematous. *adj.* Reference to salamanders that lose gills but retain gill slits as adults.

descendants. *n.* All individuals that are derived from reproduction by two individuals.

description. *n.* In systematics, a full statement of the diagnostic character states for a taxon.

desert. *n.* A region with low precipitation, usually defined as < 25 cm per year, and characterized by very sparse vegetation Also referred to accurately as *arid lands,* deserts are defined climatically as regions where a combination of high temperatures and low rainfall cause evaporation to exceed precipitation.

Desert Blacksnake. Vernacular name for the Middle Eastern elapid species *Walterinnesia aegyptia*. Also known as *Black Cobra*.

deserticolous. *adj.* Living in arid or desert regions, mostly on open ground.

desertification. *n.* As defined by the United Nations Environment Programme, this term refers to degradation of land in arid, semiarid, or dry subhumid areas resulting mainly from adverse human impact. Three processes are generally involved: degradation of vegetation, degradation of soils, and salinization. The word was first used by the French forester Aubréville in 1949 to describe deforestation in tropical and subtropical Africa.

Desert Iguanas. Vernacular name for species of lizards belonging to the iguanid genera *Dipsosaurus* and *Ctenoblepharis*. The American species *Dipsosaurus dorsalis* has been the subject of much ecological and physiological research.

Desert Night Lizard. Vernacular name for the American xantusiid species *Xantusia vigilis*.

Desert Night Snakes. See *Night Snakes*.

desert pavement. A thin layer of closely packed gravel that protects the underlying sediment from further removal by wind.

Desert Rat Snake. Vernacular name for colubrid species of snakes belonging to the American genus *Bogertophis*. See also *Trans-Pecos Rat Snake*.

Desert Spiny Lizard. Vernacular name for the phrynosomatine iguanid species of American lizard *Sceloporus magister*.

Desert Tortoise. A well-known American species of tortoise (*Gopherus agassizii*) common to southwestern deserts.

desiccant. *n.* An agent that promotes drying, usually used in reference to chemical products that absorb moisture. Cf. *dehydration*.

desiccate. *v.* (*n.* **desiccation**) Reference to total drying. Cf. *dehydrate*. (Note this word is frequently misspelled as "dessicate".)

design. *n.* Reference to the structure and functional organization of a character related to its *biological role*.

desmosome. *n.* A type of cell junction that serves primarily to secure the structural bonding between neighboring cells. The structure for intercellular attachment consists of two dense plaques on opposing cell surfaces. (Syn. *macula adherens*)

desquamate. *v.* (*n.* **desquamation**) Reference to ecdysis. Literally to flake or peel off.

determinate growth. Growth in which an animal reaches a characteristic size, and then growth essentially ceases during the lifetime of an individual. Cf. *indeterminate growth*.

deterministic model. A model in which the parameters are constant and do not vary unpredictably with time. Therefore, a given input produces one exact prediction as an output. Cf. *stochastic model*.

detoglossal. *adj.* Having the margins of the tongue partly attached to part of the lower jaw, as in the majority of amphibians. Cf. *adetoglossal*.

detritivore. *n.* An organism that feeds on fragmented particulate organic matter (*detritus*).

detritus. *n.* 1. Dead or decaying organic matter. 2. Disintegrated material.

deuterium. *n.* See *heavy hydrogen*.

deuterostome. *n.* An animal in which the anus forms from, or near, the embryonic blastopore, while the mouth forms at the opposite end of the embryo. All chordates are deuterostomes.

deuterotype. *n.* A replacement for a type specimen.

development. *n.* An orderly sequence of progressive changes —growth and differentiation—resulting in an increased complexity of a biological system. The term is usually used in reference to embryonic development.

developmental genetics. The study of mutations that produce developmental abnormalities in order to gain understanding of how normal genes control development.

developmental habitat. The place where immature turtles feed and grow prior to reaching adult size. The term could also be applied to other taxa.

developmental homeostasis. See *canalization*.

developmental homology. Similarity of anatomical features derived from a common embryological source.

developmental pathway. The fate of a cell or cellular region during embryonic development.

developmental plasticity. The ability of an organism to respond to changes in the environment by evoking different developmental programs that result in different adult phenotypes. Cf. *phenotypic plasticity*. See S.J. Smith-Gill, *Am. Zool.* 23:47–55, 1983.

developmental rate. Reference to a change in form, stage, or differentiation through time.

developmental reaction norm. (**DRN**) The set of developmental or ontogenetic trajectories produced by a genotype in response to naturally occurring, or experimentally imposed, environmental variation. The concept recognizes the ability of a genotype to produce different phenotypes in different environments (*developmental plasticity*).

devolution. *n.* Regressive evolution.

Devonian. *n.* The fourth Period of the Paleozoic Era, following the Silurian and preceding the Carboniferous Periods. See Table 1.

dewlap. *n.* A vertical and often distensible fold or flap of skin on the throat, capable of being extended or retracted by means of action of the hyoid apparatus. The term is usually applied to structures that are well developed in certain lizards, especially *Anolis spp.*, and used in territorial and other behavioral displays. (Syn. *gular expansion, gular fan, gular pouch, throat fan*) See Fig. 40.

dew point. The temperature at which air having a given water vapor content becomes saturated with water vapor and condensation occurs.

dextral. *adj.* Of or pertaining to the right side. Cf. *sinistral.*

df. See *degrees of freedom.*

DHT. See *dihydrotestosterone-5α.*

di-. Prefix meaning "two" or "twice."

diacranterian. *adj.* Reference to posterior maxillary teeth that are well separated from others by means of an abrupt enlargement of size or an edentulous space. See also *syncranterian.*

diad. *n.* See *dyad.*

Diadectosalamandroidei. *n.* A newly described monophyletic taxon of salamanders representing the majority of living salamander diversity. A sister taxon with *Cryptobranchoidei.* Inclusive taxa are newly designated *Hydatinosalamandroidei* and *Plethosalamandroidei* (D.R. Frost et al., *Bull. Amer. Mus. Nat. Hist.* 297:1–370, 2006).

Diadem Snake, Diadem Ratsnake. Vernacular name for species of snakes belonging to the colubrid genus *Spalerosophis,* distributed throughout the Middle East, North Africa, and India.

diagenesis. *n.* From geology and paleontology, the various processes affecting deposits and organic remains after deposition and before recovery.

diagnosis. *n.* A formal description of the character states that distinguish one taxon from another.

diagnostic, diagnostic character. *adj., n* Distinguishing; used in reference to specific character states that distinguish unambiguously a particular species or clade of animals from others.

diagnostic fossil. See *index fossil.*

diagonal sequence gait. Locomotion where the forefoot strikes the ground after the hind foot on the opposite side.

diagonal type. A type of scale row arrangement in which longitudinal rows run downward and backward in a long curve.

Diamondback Rattlesnake. Collective vernacular name for either Eastern Diamondback Rattlesnake (*Crotalus adamanteus*) or Western Diamondback Rattlesnake (*Crotalus atrox*). The name also is applied sometimes loosely to other Rattlesnakes such as the Red Diamond Rattlesnake (*Crotalus ruber*) and, erroneously, to any Rattlesnake that has a dorsal pattern of blotches resembling a diamond pattern.

Diamondback Terrapins. Vernacular name for a well-known species of emydid turtle, *Malaclemys terrapin.*

Diamondback Water Snake. Vernacular name for the American natricine species *Nerodia rhombifera.*

Diamond Python. Vernacular name for the common Australian boid species *Morelia spilota.*

diaphragm. *n.* The wall or septum separating the viscera from the buccopharynx of a tadpole. (Syn. *peribranchial wall*) (Not to be confused with the structure of the same name in mammals.)

diaphyseal center. A region of osteogenesis in the mid-shaft of a long bone.

diaphysis. (*pl. –es*) *n.* The shaft or central length of a long bone. The portion of bone between the extremities (*epiphyses*).

diapophysis. *n.* A *transverse process* arising from the neural arch of a vertebra and acting as a surface for articulation with the upper head of a bicipital rib. In ribs with a single head and single articulation, the diapophysis and *parapophysis* may be fused (see *synapophysis*). R.C. Stebbins (*Amphibians of Western North America,* 1954, p. 517) used this term for the nonrib-bearing lateral projections from the sacrum of frogs, articulating with the ilium.

diapromegadont. *adj.* See *promegadont.*

diapsid. *adj., n.* 1. Reference to a reptilian skull in which there are two temporal openings (*postorbital fenestrae, temporal fenestrae*). Characteristic of extant reptiles except turtles. See also *postorbital fenestrae.* 2. A member of *Diapsida.*

Diapsida. *n.* A clade that includes most, and possibly all, extant reptiles distinguished by the presence of two openings in the side of the skull behind the orbit. The relative placement of turtles with respect to Diapsida is controversial.

diarthrosis. *n.* A freely movable joint having a joint capsule or cavity with articular cartilages on joined elements. See *synovial joint.*

diastem. *n.* In paleontology, a gap in the fossil record.

diastema. *n.* 1. A space or cleft. 2. An edentulous gap or space between teeth on the jaw.

diastole. *n.* The phase of the heart cycle during which the myocardium is relaxed and the chambers are filling with blood. Cf. *systole.*

diastolic pressure. The lowest arterial blood pressure measured during the diastolic phase of the cardiac cycle.

Dibamidae. *n.* A clade (family) of small, attenuated, burrowing lizards having vestigial eyes covered by a scale. Two genera and 10 species are distributed in southeast Asia east through the southern Philippines and islands of the Sunda Shelf to western New Guinea.

Dicamptodontidae. *n.* A clade (family) of moderate to large salamanders including four species representing a single genus (*Dicamptodon*) distributed in northwestern North America. This family was recently synonymized with the family *Ambystomatidae* (D. Frost et al., *Bull. Amer. Mus. Nat. Hist.* 297:1–370, 2006).

dicephalous. *adj.* Possessing two heads. (Syn. *bicephalous, bicipital*)

dicephalus tetrabrachinus. Used by R. Rugh (*Experimental*

Embryology, 1948, p. 460) to define a condition in which the first furrow of an amphibian egg coincides with the sagittal plane, thereby exaggerating constriction and resulting in duplication of chorda, auditory vesicles, and fore limbs.

dichocephalous. *adj.* Two-headed ribs. (Syn. *bicipital, bicephalous, dicephalous*)

dichopatric. *adj.* A condition where populations or species have geographical ranges separated such that individuals from the two different populations never meet and thus gene flow is impossible.

dichopatric speciation. A form of *allopatric speciation* in which a parental species is separated into two separate parts by development of a geographic or other extensive barrier. Cf. *peripatric speciation*.

dichotomous character. A character that exists in only two states.

dichotomous data. Binary data representing a present or absent state.

dichotomous key. An identification key constructed as a sequence of paired alternatives. (Syn. *sequential key*)

dichotomy. *n.* A bifurcation or division into two equal parts.

dichromatic. *adj.* Having two varieties of coloration, independent of age and sex. See also *sexual dichromatism*.

dichromatism. *n.* A condition in which members of a species exhibit one of two distinct color forms.

Dicroglossidae. *n.* A taxon (family) of anurans, sister to the new taxon *Aglaioanura* within the *Saukrobatrachia*, and composed of two subfamilies (*Dicroglossinae, Occidozyginae*) with species distributed in Africa, Asia, and some Pacific islands. See D.R. Frost et al., *Bull. Amer. Mus. Nat. Hist.* 297:1–370, 2006 for taxonomy.

Dicroglossinae. *n.* A subfamily of the anuran family *Dicroglossidae*.

didactyl. *adj.* Having two digits, either fingers or toes, on a foot.

diel. *adj.* Reference to a chronological day (24 h), distinct from a daylight period of varying seasonal length. A diel cycle has a period of approximately 24 h. Cf. *circadian, diurnal*.

diencephalon. *n.* The posterior unpaired segment of the embryonic forebrain; contains the *thalamus, epithalamus,* and *hypothalamus*.

diet-induced thermogenesis. See *postprandial calorigenesis*.

differentiation. *n.* The process whereby descendants of a cell or group of cells (tissue) achieve and maintain specializations of structure and function. A process by which unspecialized structures become modified and specialized for the performance of specific functions. See *morphogenesis*.

differentiation rate. Generally, the rate of development related to changes in structures, or between two different stages (in tadpoles).

diffraction. *n.* The process by which light forms a pattern of colors when passing through an aperture or past the edge of an opaque object.

diffuse competition. In ecology, reference to simultaneous interspecific competition between numerous species that have small degrees of niche overlap with other species. Cf. *intense competition*.

diffuse coevolution. Coevolution process that involves interactions of many organisms as contrasted with strictly pairwise interactions.

diffusion. *n.* Dispersion of atoms, molecules, or ions as a result of random thermal motion. The net diffusive movement of molecules is from a region of higher to lower concentration.

diffusion coefficient. A coefficient relating the rate of diffusional flux to concentration gradient, path length, and the area across which diffusion occurs.

digametic. *adj.* Characterized by having two kinds of gametes, producing male or female offspring. See *heterogamety*.

digesta. *n.* See *chyme*.

digested energy. In ecological energetics, reference to the total energy ingested by an animal less the energy lost in feces (= consumption – egesta).

digestion. *n.* The mechanical and chemical breakdown of food, converting it to chemical substances that can be absorbed into blood across the gut wall.

digestive efficiency. In ecological energetics, the efficiency of energy absorbed through the gut, calculated as [(C – F)/C] x 100, where C = energy consumed, and F = energy of fecal waste. See also *apparent digestive efficiency*.

digestive energy. In ecological energetics, the energy absorbed through the gut.

digestive enzymes. Enzymes secreted by the alimentary canal to aid in chemical digestion. Some of the enzymes in snake venoms may also function to assist digestion.

digestive epithelium. Epithelium lining the small intestine.

digestive system or **tract.** See *alimentary canal*.

digger. *n.* The "spade" or tubercle on the hind foot of *Spadefoot Toads* (*Pelobatidae*).

digit. *n.* A finger or toe, or other such homologous structure.

digital. *n., adj.* 1. Any scale on a digit, situated between the palm or sole and the claw. 2. Reference to a digit. 3. Reference to data in the form of numbers.

digital fringe. Reference to enlarged, free-ending scales protruding from the lateral margins of the longer toes in certain desert species of lizards. These give the edge of the toe a comb-like appearance and probably assist locomotion on substrates of loose sand.

digital groove. In taxonomic usage, reference to a distinct deep groove bordering the margins of the digital pads in some species of frogs.

digital pad. 1. Fleshy structures located on the tips of the digits in certain amphibians. 2. Also used to describe the elaborate collections of digital setae that form lamellae and pad-like structures on the ventral surfaces of the digits of geckoes and anoles.

digital web. An interdigital membrane connecting the toes of aquatic amphibians and reptiles and used as an aid in swimming.

dihydrotestosterone-5α. (**DHT**) An androgen synthesized from testosterone by the action of 5α-reductase (NADPH required as cofactor) within various tissues. DHT is important for development of reproductive glands (prostate and bulbo-urethral) in mammals, phallus development in crocodilians, and possibly the differentiation of wolffian ducts into male gonads in amphibians.

dikinesis. *n.* (*adj.* **dikinetic**) A condition of *cranial kinesis* in which there are two joints in the skull. See *amphikinesis*.

dilator. *n.* A muscle that opens or dilates. Cf. *constrictor*.

dimer. *n.* A molecule consisting of two joined, simpler molecules (monomers) of the same kind.

dimorphism. *n.* (*adj.* **-ic**) The occurrence of two different kinds of individuals within a species, as in *sexual dimorphism* where the sexes differ in body size, color, etc. The term usually applies to genetically determined, discontinuous morphological characters.

DIN color system. A system developed as a German standard for color specification, based on principles similar to the *Munsell color ordering system*. The three perceptual variables of the DIN system are hue, saturation, and darkness. See M. Richter & K. Witt, *Color Res. Application* 11:138–145, 1986.

dinosaur. *n.* Informal and popular reference to extinct, fossil reptiles, or members of the *Dinosauria*. The term was first used in 1842 by the British comparative anatomist and paleontologist Richard Owen who applied it to three impressively large fossil reptiles from the English countryside (*Rep. Br. Assoc. Adv. Sci.* 1841:60–294, 1842). See *Dinosauria*.

Dinosauria. *n.* A clade that includes the higher taxa *Ornithischia* and *Saurischia*, commonly known as dinosaurs. Dinosauria has been defined as all descendants of the most recent common ancestor of birds and *Triceratops* (K. Padian and C.L. May, *New Mexico Mus. Nat. Hist. Sci. Bull.* 3:379–381, 1993).

dioecious. *adj.* Having male gonads in one individual and the female gonads in another individual.

dioxins. *n.* A family of aromatic chemical compounds formed as impurities in the process of chlorinating phenols used in producing herbicides or during heating of polychlorinated biphenyls. The term commonly refers to the compound 2,3,7,8-tetrachlorodibenzo-*para*-dioxin, also known as TCDD or tetrachlorodioxin. These compounds are exceptionally toxic, environmentally persistent, and comprise an important class of pollutants.

diphasic. *adj.* Having two phases.

Diphyabatrachia. *n.* A new monophyletic taxon of anurans, sister to the new taxon *Chthonobatrachia* within the new taxon *Leptodactyliformes*, and containing *Centrolenidae* and *Leptodactylidae*. See D.R. Frost et al., *Bull. Amer. Mus. Nat. Hist.* 297:1–370, 2006.

diphycercal tail. Reference to a tail in which caudal fins extend symmetrically above and below the vertebral column for its full length, as in various larval amphibians.

diplasiocoely. *n.* Reference to a vertebral column in which the first seven vertebrae are procoelous, the eighth is biconcave, and the sacral vertebra is convex anteriorly with a double condyle posteriorly. Used by G.K. Noble (*The Biology of Amphibia*, 1931, p. 514) as a classification character for anurans.

diplo-. Prefix meaning "double" or "twofold."

Diplodactylinae. *n.* A subfamily of *Gekkonidae*.

Diploglossinae. *n.* A subfamily of *Anguidae*.

diploid. (**2N**) *adj.* (**diploidy, diploid number,** *n.*) The condition of having two sets of homologous chromosomes, one derived from each parent, in somatic cells of most organisms. Having two copies of a gene per locus per genome (may or may not be the same *allele*).

dipole. *n.* A molecule having separate regions of net negative and net positive charge, such that one end acts as a positive pole and the other as a negative pole.

Dipsadidae. *n.* See *Dipsadinae*.

Dipsadinae. *n.* A clade of Colubridae including semi-arboreal snakes that are widespread and diverse in Central and South American, specialized for feeding on snails and slugs. Alternate classification schemes consider this group to be a family (e.g., J.T. Collins, *J. Kans. Herpetol.* 19:18–20, 2006; N. Vidal et al., *Comptes Rendus Biologies* 330:182–187, 2007).

direct competition. Competition in which one individual or species excludes another from a resource by direct aggressive behavior or use of toxins. Cf. *indirect competition*.

direct development. Complete embryonic development within the egg of some amphibians, and thus a free-living larval stage is absent.

directional selection. A mode of natural selection in which individuals at or near one end of the range of phenotypic variation are favored relative to those individuals toward the other end of the range. This selection process has the effect of moving the mean value of a character in the direction of the most fit phenotypes from generation to generation. Cf. *disruptive selection, stabilizing selection*. (Syn. *dynamic selection, progressive selection*)

directive coloration. Color patterns or surface markings that divert attention of a predator to a nonvital part of a prey animal's body.

dirolent. *adj.* An olfactory movement observed in tortoises, wherein the head is moved in a straight, forward motion. Term coined by A. Eglis (*Herpetologica*, 18:1-8, 1962). Cf. *latolent*.

disaccharide sugar. A double sugar formed when two *mono-*

saccharides (single sugars) are joined together by dehydration synthesis.

disarticulation. *n.* Reference to the separation of a joint.

disc. *n.* Descriptive term used for a variety of flat, circular or plate-like anatomical structures in herpetology. Most common usage is with reference to digital pads of amphibians or lizards, used to adhere to smooth surfaces and assist climbing, and the *adhesive organ* or *oral disc* of amphibians.

discharge orifice. The distal end of the fang of a venomous snake through which the venom is expelled from the venom duct and gland.

Discoglossanura. *n.* A clade of anurans that includes all extant forms except the basal groups of *Ascaphidae, Leiopelmatidae,* and *Bombinatoridae.*

Discoglossidae. *n.* A former clade (family) of small to moderate size anurans (Midwife Toads and Painted Frogs) with aquatic eggs and tadpoles, except in *Alytes*—unusual frogs in which males carry fertilized eggs on their backs and hind limbs until they are deposited in water at hatching. Two genera and 10 species are distributed in western Europe, northwestern Africa, and the Middle East. This group is now designated to be *Alytidae.* See D.R. Frost et al., *Bull. Amer. Mus. Nat. Hist.* 297:1–370, 2006. See *Alytidae.*

discoidal cleavage. Reference to early mitotic divisions of an embryo that are restricted to the animal pole; an extreme case of meroblastic cleavage and characteristic of reptiles.

discontinuous distribution. 1. A data set recorded as whole numbers, and thus not yielding a continuous spectrum of values. E.g., the number of head scales per individual in a population of lizards. Cf. *continuous distribution.* 2. Occurrence of a species in two or more areas while absent from intervening regions.

discontinuous variation. Phenotypic variation of characters that fall into two or more non-overlapping classes. Cf. *continuous variation.*

Discophinae. *n.* A subfamily of *Microhylidae.*

discordant. *adj.* 1. Not parallel to any layering or parallel planes. 2. In systematics, reference to members of a group that do not share a given trait. Cf. *concordant.*

discrete character. A qualitative or categorical character. See also *discrete variable.*

discrete diagnosability. The ability to differentiate between populations by nonoverlapping morphological characteristics within their range of variation.

discrete signal. A signal used in communication that is either on or off and does not show intermediate values. Cf. *graded signal.*

discrete trait or **character.** A qualitatively defined feature with only a few distinct phenotypes (e.g., polymorphism; presence vs. absence).

discrete variable. A variable that can assume only certain values. Cf. *continuous variable.* See also *random variable.*

discretionary resource, discretionary energy. The total amount of assimilated resource (energy) available for allocation to growth, storage, or reproduction over some biologically meaningful time period in a life history (B.W. Grant & W.P. Porter, *Amer. Zool.* 32:154–178, 1992).

discriminant analysis. Multivariate methods that are employed to assess the adequacy of classification into groups of various objects under study, or to assign objects to one of a number of defined groups of objects. Thus, discriminate analysis may be used in connection with a descriptive or a predictive objective. The methods used are Multiple Discriminant Analysis, Fisher's Linear Discriminant Analysis, or K-Nearest Neighbors Discriminant Analysis.

disinfectant. *n.* Any substance used to eradicate pathogens.

disintegrins. *n.* Toxins (peptides) found in the venoms of viperid snakes that inhibit *integrins* (trans-membrane proteins that allow the transfer of extracellular messages to the cytoplasm).

disjunct. *adj.* Disjoined or spatially separated, as in a noncontinuous geographic distribution.

disk. *n.* See *disc.*

disorientation. *n.* The result of using inappropriate cues for moving in a particular direction. E.g., hatchling sea turtles will move inland toward street lights instead of (correctly) toward the sea.

dispersal. (dispersion) *n.* Movements of individuals away from a specific location (*emigration*) to another (*immigration*), related to various factors such as territoriality, departure from natal or denning sites, resources, weather, or seasonal phenomena, etc.

dispersal biogeography. Explanations of biogeographic patterns based on movements across barriers with subsequent divergence of new species and ancestral species into descendant species. Replicated patterns of divergence require individual explanations for each clade. Cf. *vacariance biogeography.*

dispersion. *n.* 1. The distribution of individual entities in space. 2. Reference to purposeful movement of individuals into or out of an area or population, usually involving short distances. 3. In statistics, the scatter of observations or values around the mean or central value.

displacement. *n.* Reference to shifts in the location of ribs relative to association with dermal elements of the carapace during development or evolutionary history of turtles.

displacement activity. Reference to one or more behaviors that occur outside the functional context to which it is normally related, typically triggered by stimulus from another behavioral act that is denied consummation.

display. *n.* A stereotyped behavior that is used in contexts of courtship, territorial advertisement, defense, etc., and communicates to other animals, often intraspecifically. Quantitative analysis of display behaviors addresses components such as posture, movement, sequence, cadence, etc.

display action pattern. A graphic representation of display movements, usually illustrating a species-specific pattern. Such action patterns for *Anolis* lizards, for example, include a tracing that represents movements of the body or head along with parallel tracing depicting expansion and contraction of the gular flap.

dispondylous vertebra. A condition of vertebrae in which a vertebral segment is composed of two centra. Cf. *monospondylous vertebra.*

disruptive coloration. Animal colors that disrupt the shape or outline of an animal when viewed against its background. The coloration disguises the body by breaking it up into visually distinct parts.

disruptive selection. A mode of *natural selection* in which individuals at the extremes of the range of phenotypic variation are favored relative to individuals near the mean, which are at a disadvantage. Disruptive selection increases the variation but tends to not change the mean value of a trait. Cf. *directional selection, stabilizing selection.* (Syn. *centrifugal selection; diversifying selection*)

dissection. *n.* The careful exposure of anatomical parts of an organism for instructional or research purposes.

disseminated intravascular coagulopathy. (DIC) An abnormal state of coagulation that depletes the stores of blood clotting factors and platelets, thereby rendering blood non-coagulable and creating a risk of severe bleeding.

dissociation. *n.* Separation, as in thermal agitation or solvation of a substance into simpler constituents.

dissociation constant. (K′) The empirical measure of the degree of dissociation of a conjugate acid-base pair in solution.

dissolved oxygen. Reference to oxygen dissolved in water, primarily from atmospheric sources or plant photosynthesis. The amount of oxygen in water is small, being limited by a low solubility. The maximum amount that will dissolve is between 8 and 11 mg/L, and the saturation level is decreased by higher temperatures and dissolved salts.

distal. *adj.* Located away from the base, origin, or point of attachment of a structure, or away from the central axis of the body. Farther from a body's center, or toward the tip of an extremity. Cf. *proximal.*

distance effect. In island biogeography, species number decreases with distance or isolation from a source pool of colonists.

distance methods. In phylogenetic inference, distance methods find phylogenetic trees whose branch lengths most closely reflect the actual "distances" that are observed among all possible pairs of species. The lengths of the branches measure the amount of evolutionary change reasoned to have occurred between an ancestor and a descendant. The methods utilize overall measures of similarity or difference without regard to how the patterns arise. Thus, distinctions between shared versus derived traits may be lost. Cf. *parsimony methods, maximum likelihood methods.*

distance-Wagner tree. A *cladogram* that displays differences of character states between taxa as variable branch lengths.

distress call. A loud, piercing cry given by some anurans to advertise alarm when they are startled or suddenly grasped. (Syn. *fright cry, protective call*)

distributary. *n.* Small shifting river channel that carries water away from the main river channel and distributes it over a delta's surface.

distribution. *n.* 1. The total geographic range in which a given species or group of organisms occurs. 2. The spatial pattern related to members of a population or group. 3. See *frequency distribution.*

disulfide linkage. A bond between sulfide groups that determines protein tertiary structure by linking together portions of polypeptide chains.

ditypic. *adj.* Dimorphic; having two distinct morphs.

diuretic. *n.* Any drug or substance that increases the flow of urine.

diurnal. *adj.* Active during daylight hours. Cf. *nocturnal.*

diurnal rhythm. An activity cycle that approximates a 24-h cycle of repetition. (Syn. *circadian rhythm*)

divergence, divergent evolution. *n.* Acquisition of dissimilar characters by related organisms.

diversifying selection. See *disruptive selection.*

diversity. *n.* See *biodiversity* and *species diversity.*

diversity index. A measure of the number of species or relative abundance of species in a community. See *Shannon-Weiner index, richness index, Brillouin index, Simpson index.*

diverticulum. (*pl. –a*) *n.* A circumscribed pouch or sac.

divided. *adj.* 1. As used in anatomy, literally divided into two parts with reference to an otherwise single structure (e.g. a scale) recognized in other species. 2. Used by H.G. Dowling and J.M. Savage (*Zoologica* 45:17–28, 1960) to denote a snake hemipenis in which the basal undivided part is shorter or equal to the length of the apical segment.

divisive classification. A statistical method that divides a body of data into groups according to predetermined criteria.

DNA. See *deoxyribonucleic acid*.

DNA arrays. See *DNA chip*.

DNA chip. Small sheets of glass or silicon that contain numerous (hundreds to millions) of sites, called *features*, at each of which are firmly attached millions of copies of a single-stranded DNA segment or probe. Fluorescent markers are used to detect single-nucleotide mispairings, and this technology can be used to screen entire genomes for mutations in numerous genes simultaneously. (Syn. *microarray*)

DNA clock. See *molecular clock*.

DNA-DNA hybridization. A molecular technique to determine stability of two complementary strands of DNA "hybridized" from two different organisms, assumed to correlate with sequence homology and therefore relatedness.

DNA fingerprinting technique. A technique for determining relationships or molecular genetic variation by generating a pattern of DNA restriction fragments that are unique to an individual or species, depending on the presence of complementary sequences scattered throughout the genome. (Syn. *DNA typing, genetic fingerprinting*)

DNA homology. Reference to the degree of similarity between base sequences in different DNA molecules.

DNA library. See *genomic library*.

DNA microarrays. See *DNA chip*.

DNA polymerase. An enzyme that catalyzes the formation of DNA from deoxyribonucleoside triphosphates, using single-stranded DNA as a template.

DNA probe. See *probe*.

DNA replication. See *replication*.

DNA sequencing techniques. Any of a number of sequential protocols that selectively cleave DNA, separate the fragments by electrophoresis, and identify the components by autoradiography.

DNA typing. See *DNA fingerprinting technique*.

Dog-faced Water Snakes. Vernacular name for the Asian homalopsine snakes belonging to the genus *Cerberus*.

doldrums. *n.* Reference to equatorial region having low atmospheric pressure and between 5° N and 5° S. The region is characterized by light, variable winds and changing weather.

Dollo's Law. The proposition that the evolution of a particular lineage is essentially irreversible, such that structures and functions lost are not regained. Cf. *Arber's law*.

DOM. Abbreviation for *dissolved organic matter*.

domain. *n.* 1. Highest formal category in the classification of living organisms. 2. In molecular and cellular biology, reference to a structurally or functionally discrete region of a molecule, membrane, or cellular structure.

dome. *n.* Reference to "hunching" the back into an arc or "dome" when female *Sceloporus* lizards are stimulated by presence of male lizards and elevate the body by standing high on outstretched legs.

dominance. *n.* 1. The phenomenon whereby one allele suppresses the expression of another allele. Cf. *recessive*. 2. See *species diversity*. 3. See *social dominance*.

dominance genetic variance. The component of genetic variance attributable to nonadditive effects of genes, such as dominance. Cf. *additive genetic variance*. (Syn. *nonadditive genetic variance*)

dominance hierarchy. In ethology, reference to a rank-order of dominance-submission relationships within a group of animals.

dominance threshold. The point, in relation to changing resources, at which an individual accepts subordination rather than competes for an individual territory. Presumably, after experience in a social environment, an individual may act to maximize the benefits and minimize the costs of a particular life history option.

dominant frequency. In bioacoustics, the frequency within which the greatest amount of acoustic energy is concentrated.

dominant species. A species that predominates in a community because of its abundance, size, or coverage as each relates to fitness of associated species.

Döppler principle. Increasing or decreasing frequency of a signal received from a moving object as the object moves closer or away, respectively.

dormancy. *n.* A general term for states of reduced body activity and metabolic quiescence such as sleep, torpor, hibernation, and estivation. See also *brumation*.

dorsad. *adv.* Toward the dorsum.

dorsal. *adj., n.* 1. Reference to the upper surfaces of the body; the back or spinal regions of vertebrates. Cf. *ventral*. 2. Used to identify the enlarged scales on the backs of crocodilians, any of the small scales on the backs of lizards or snakes, and the *central* lamina of the turtle carapace. See Fig. 4. 3. Name given to vertebrae between cervical and sacral vertebrae.

dorsal aorta. Major artery formed by union of two aortic arches and functions to transport blood posteriorad along the middorsal line.

dorsolateral. *adj.* A position that is intermediate between dorsal and lateral.

dorsal gland. The *hatching gland* of tadpoles.

dorsal hollow nerve cord. A diagnostic feature of *chordates*, the nerve cord lies above the gut and is tubular with a fluid-filled central canal.

dorsal integumentary gland. See *integumentary gland*.

dorsal plica. Earlier reference to *dorsolateral fold* (also *dorsolateral dermal plica*).

dorsal vertebra. Any of the presacral vertebrae in reptiles, except for cervicals (A.S. Romer, *Osteology of the Reptiles*, 1956, p. 227). (Syn. *thoracicolumbar vertebra, thoracolumbar vertebra*)

dorsolateral. *adj.* Reference to characters situated between the center of the back and the side of an animal.

dorsolateral dermal plica. See *dorsolateral fold*.

dorsolateral fields. The area on the body between the paravertebral and lateral body stripes. See *body fields*.

dorsolateral fold. A longitudinal, glandular ridge running between the dorsal and lateral aspects of the body in certain anurans. (Earlier syn. *dorsal plica, dorsolateral dermal plica, dorsolateral ridge*)

dorsolateral ridge. See *dorsolateral fold*.

dorsoposterior. *adj.* Of or pertaining to the upper portion of the rear of a body structure.

dorsoventral (dorsal ventral) *adj.* Extending in a direction from the dorsal to ventral surface of a body.

dorsoventrally compressed. Of or pertaining to a body structure that is flattened from top to bottom.

dorsum. *n.* Reference to the upper surface of an animal.

dose-response curve. A graphic plot showing the relationship between a biological response and the administered dose of radiation, drug, or other chemical.

double clutching. Reference to amphibians or reptiles that produce two clutches of eggs or offspring during the same season. (The term is often used in reference to *boas* and *pythons*.)

doubling time. The amount of time required for a population (of cells or animals) to double in size. If a population is growing exponentially, it will double in size after a constant amount of time, no matter how large it has become. The doubling time is generally longer than the generation time and will equal the generation time only if every individual in the population gives rise to two others.

doubly labeled water technique. Reference to a method using isotopically labeled water for determination of water and energy budgets of free-living animals. Gross water fluxes can be measured with tritium- or deuterium-labeled water, and metabolic rates can be measured using deuterium- and ^{18}O-labeled water (doubly labeled water). An individual animal is injected with an appropriate amount of labeled water, which equilibrates with the animal's body water, and the specific activities of the isotopic labels are measured in blood samples collected from the animal at varying intervals of recapture following release into the wild. The hydrogen isotope is lost from the animal primarily in the form of water. Because the oxygen of body water is lost as water and as CO_2, the difference between the washout rates of hydrogen isotope and oxygen isotope is a measure of the rate of CO_2 production, hence metabolic rate (N. Lifson & R. McClintock, *J. Theor. Biol.* 12:46–74, 1966; K.A. Nagy, pp. 227–245 in N.F. Hadley, ed., *Environmental Physiology of Desert Organisms*, 1975). Thus, measurement of the decrease in specific activity of these isotopes with time provides an estimate of the rates of water flux and energy metabolism. In some circumstances, it is possible to partition water input and threby obtain an estimate of dietary water input and, hence, feeding rate. Equations to describe isotope fluxes and improved applications of isotopic methods can be found in physiological literature. See also *field metabolic rate*.

DOW. See *dry organic weight*.

down-regulation. Control of physiological activity by a decrease in receptor density in a target cell membrane. Cf. *up-regulation*.

drag. *n.* The resistance to movement of an object through a fluid, which increases with the viscosity and density of the medium and varies with the shape and surface area and properties of the object. The drag force acts parallel to the relative movement of fluid and opposite to the direction of movement. The total drag is the sum of *induced drag* and *parasite drag*, the latter being the sum of *pressure drag* and *skin friction*.

Dragon Lizards, Dragons. *n.* Vernacular name for various species of agamid lizards, especially those belonging to the genera *Amphibolurus, Chelosania, Chlamydosaurus, Ctenophorus, Diporiphora, Draco, Gonocephalus, Hypsilurus, Lophognathus, Mictopholis, Physignathus, Pogona,* and *Tympanocryptis*.

drainage basin. The total area drained by a stream and its tributaries.

draw. *v.* To remove a fang forcibly from a venomous snake. Fang removal is a practice of some snake charmers in various parts of the world.

drift. *n.* See *genetic drift*.

drift fence. An upright structure of various designs that intercepts movements of animals and guides them into traps, generally increasing capture rates. These structures are commonly used in conjunction with *pitfall traps* as an effective means to sample smaller species of amphibians and reptiles, usually in terrestrial (but sometimes aquatic) habitats.

DRN. See *developmental reaction norm*.

drosopterin. *n.* A highly fluorescent pteridine pigment present in the dewlap colors of *Anolis* lizards (originally discovered in *Drosophila* fruitflies). (Syn. *neodrosopterin*)

"dry" bite. Colloquial reference to a bite of a venomous snake in which no venom is actually injected into the victim.

dry forest. Forested habitat characterized by low deciduous trees, thick and thorny shrubs, and numerous types of grasses. Tropical dry forests are characterized by a dry season and high temperatures.

dry organ weight. (DOW) Dry weight of a (laboratory desiccated) organ, usually expressed as *ash-free dry weight*.

Dtellas. *n.* Vernacular name for largely Asian Geckos belonging to the genus *Gehyra*.

D-tubocurarine. *n.* See *curare*.

duct. *n.* A tube or channel through which excretory fluids are moved.

ductuli efferentes or ***ductulus efferens*.** Embryonic tubules leading from testis to kidney. In some amphibians these fuse to form Bidder's duct. Sperm pass through these tubules to reach the archinephric duct.

ductuli epididymidis or ***ductulus epididymidis*.** Conducting tubules of the kidney that connect the efferent ductules of the testis with the former archinephric duct, or ducts of the epididymis.

***ductus Botalli*.** See *ductus arteriosus*.

***ductus arteriosus*.** An embryonic connection between the pulmonary artery and the dorsal aorta, and is the dorsal segment of the sixth aortic arch. This vessel serves as a shunt during fetal development, leading pulmonary blood into the systemic circulation while the lung is not yet functional. This vessel disappears after metamorphosis or birth in frogs and most amniotes, but persists in reduced form in adult salamanders, apodans, tuataras and some turtles. (Syn. *ductus Botalli*)

ductus caroticus. Part of the dorsal aorta lying between the third and fourth aortic arches, tending to disappear in adult amniotes. Cranial and body segments of the original aorta are completely separated by disappearance of this vessel. However, it tends to persist in apodans, some salamanders, tuataras, and some lizards.

ductus cuvieri. The common entry of venous blood returning to the sinus venosus of the heart via the venae cavae. The vessel conveys blood returning from both anterior and posterior parts of the body. (Syn. *common cardinal vein*)

ductus deferens. The duct used for sperm transport and derived from the embryonic archinephric duct in amniotes, connecting with the testis at its proximal or anterior end. In amphibians, a part of this duct functions to transport both sperm and urine to the cloaca. (Syn. *mesonephric duct, nephric duct, segmental duct, urogenital duct, vas deferens, Wolffian duct*) See Fig. 38.

ductus epididymidis. A duct that transports sperm from the *ductuli epididymidis* to the *ductus deferens*, running along the *epididymis* as a single tube.

ductus spermaticus. See *sulcus spermaticus*.

dud. *n.* A vernacular term applied to an intrauterine egg that has stopped development and subsequently shrinks and hardens within the oviduct. The condition is distinguished from eggs undergoing resorption. The term is jargon and should not be used in formal writing.

duff layer. Organic matter in various stages of decomposition on a forest floor.

duodenum. *n.* The anteriorad segment of the small intestine, continuing posteriorly from its junction with the stomach.

dura or **dura mater.** *n.* A tough, fibrous membrane that forms the outermost of two protective meninges covering the brain and spinal cord.

durability. *n.* In ecology, the probability that a population or species will be represented by living descendants after an extended period. Cf. *resilience*.

durophagous. *adj.* Reference to eating hard-bodied prey. (This term is often used in herpetology as descriptive of snakes and lizards that prey on skinks or other lizards that are armored with osteoderms.)

Dusky Salamander. Collective vernacular name for plethodontid salamanders belonging to the genus *Desmognathus*.

Duvernoy's gland. Paired exocrine or secretory glands associated with the upper jaws of some members of Colubridae and related families of snakes (Fig. 20.). Named after D.M. Duvernoy, a French anatomist, the gland arises embryonically from the rear of the maxillary dental lamina, together with the posterior pair of maxillary teeth, and is homologous with the venom gland of front-fanged snakes. The term was used originally by A.M. Taub (*J. Morphol.* 118:529–542, 1966) to replace the term "parotid gland" in reptiles (not homologous with that of mammals) and has traditionally been used to indicate the gland of oral secretion for colubrid snakes. In various species, the glands produce proteins or a combination of mucus and proteins. Nearly half of colubrid species, including the rear-fanged species, produce venom in Duvernoy's gland that immobilizes prey or may be simply digestive in function. The function of many ophidian oral secretions is unknown, and there is a current trend to eliminate the term Duvernoy's gland and refer to all as venom glands. See S.P. Mackessy, *J. Toxicol.-Toxin Rev.* 21:33–63, 2002.

Dwarf Boas. Vernacular name for several species of Boas belonging to the family *Tropidophiidae*. Also called *Wood Snakes*.

Dwarf Chameleons. Vernacular name for African Chameleons belonging to the genus *Bradypodion*. The name also is used by some for the Madagascan genus *Brookesia*.

Dwarf Crocodile. See *African Dwarf Crocodile*.

Dwarf Frogs. Vernacular name for species of leptodactylid frogs belonging to the genus *Physalaemus*, distributed in Central and South America.

Dwarf Geckos. Vernacular name for the neotropical gekkonid species belonging to the genus *Sphaerodactylus*. (Syn. *Reef Geckos*)

Dwarf Pipe Snakes. Vernacular name for snakes belonging to the family *Anomochilidae*, also called *Stump Heads*.

Dwarf Iguanas. Vernacular name for hoplocercid species of lizards belonging to the genus *Enyalioides*.

Dwarf Sirens. Collective vernacular name for sirenid salamanders belonging to the genus *Pseudobranchus*. These are paedomorphic, eel-like, freshwater animals with prominent external gills and no hind legs. Cf. *Sirens*.

Dwarf Snakes, Dwarf Racers. Vernacular name for Eurasian species of colubrid snakes belonging to the genus *Eirenis*.

Dwarf Wolf Snakes. Vernacular name for African colubrid snakes belonging to the genus *Cryptolycus*.

dyad, diad. *n.* A double or paired character.

dymantic. *adj.* Startling coloration, being readily noticed but without immediately evident purpose.

dynamic ecotrophic coefficient. The ratio of the consumption of a prey species by a predator to the production of that prey species.

dynamic equilibrium. See *steady state*.

dynamic range. The range of energy over which a sensory system is responsive and can encode information related to stimulus intensity.

dynamic regime. A concept in ecology based in systems theory and the nonlinear behaviors of communities and ecosystems, defined as a stable basin of attraction in a state space, in which the attraction is formed by internal relationships between species and their environments. Ecosystems can have many possible regimes, or states, and the size, shape and nature of those regimes are primarily dictated by internal relationships, which if changed can trigger shifts in the regime. The concept is used in ecosystem management and restoration, usually in context of the response of ecosystem patterns and processes to disturbance or environmental change. See M. Scheffer & S.R. Carpenter, *Trends Ecol. Evol.* 18:648-656, 2003.

dynamic selection. See *directional selection*.

dynamic steady state. See *steady state*.

dyne. (dyn) The *cgs* unit of force, equivalent to the force required to accelerate a mass of 1 g by 1 cm per s per s ($= 10^{-5}$N).

dys-. Prefix meaning "bad," "insufficient," "abnormal," or "difficult."

dysecdysis. *n.* A condition of impaired or abnormal skin shedding.

dysfunction. *n.* Impairment of normal function.

dysplasia. *n.* 1. Abnormality of development, as revealed by pathology, size, shape, cellular organization, etc. 2. Disorganized growth that arises from embryonic tissues being implanted into adult hosts. N.S. Rafferty (*J. Exp. Zool.* 147:33–41, 1961) implores this term be used in place of *teratoma*.

dyspnea. *n.* Labored or difficult breathing, usually in response to low oxygen levels in blood.

E

e. 1. A symbol for the electric charge on one electron. 2. A mathematical symbol used as the base of natural (Napierian) logarithms and exponentials. The real number *e* is irrational, which means that its decimal expansion is infinite and nonrepeating (to five significant digits, $e = 2.7183$). The symbol *e* was introduced by the Swiss mathematician Leonard Euler (1707–1783), probably because it is the first letter of the word "exponential." It is sometimes called the Euler number and continues to be used in his honor.

ear flap or **ear lid.** Either of an upper and lower skin fold situated at the external opening of the auditory canal of crocodilians. The flaps prevent water entry when an animal is in water, whereas lifting of the upper lid opens a narrow slit and permits passage of sound when the animal is out of water.

Earless Lizards. Collective vernacular name for various species of lizards belonging to the agamid genera *Aphaniotis, Cophotis, Tympanocryptis*, and the phrynosomatid (or iguanid, depending on classification scheme) genus *Holbrookia*.

Earless Monitors. Vernacular name for species of Asian Monitor Lizard belonging to the genus *Lanthanotus*.

Earthsnakes. *n.* Vernacular name for various species of secretive or burrowing colubrid snakes belonging to the New World genera *Virginia, Conopsis,* and *Geophis,* and also for various snakes belonging to various clades within the Uropeltidae.

Eastern Alligator Lizard. Vernacular name for American anguid species belonging to the genus *Gerrhonotus*. Cf. *Western Alligator Lizard*.

Eastern Brown Snake. Vernacular name for the Australian elapid species *Pseudonaja textilis*.

Eastern Coral Snake. Vernacular name for the American Coral Snake species *Micrurus fulvius*.

Eastern Diamondback Rattlesnake. Vernacular name for the American pitviper *Crotalus adamanteus*.

Eastern Newts. Collective vernacular name for North American salamandrid species belonging to the genus *Notophthalmus*. See also *Newt*.

Eastern Racer. Vernacular name for the American colubrid species *Coluber constrictor*.

Eastern Ribbon Snake. Vernacular name for the American colubrid species *Thamnophis sauritus*.

ebb current. A tidal current produced by a receding tide.

ebb tide. A receding tide.

Eberth's body. Fine threads or system of extracellular fibrils developing into amorphous masses lying in close contact with the basal membrane of the epidermis in early stages of tadpoles.

Eberth-Kastschenko layer. A calcified layer located in the dermis of certain species of anuran amphibians, situated between the *stratum spongiosum* and *stratum compactum* and consisting of proteoglycans, glycosaminoglycans, and calcium phosphate deposits with a crystalline structure comparable to that of hydroxyapatite. The layer is thought to be protective or important for calcium storage and mobilization, but little is known definitively about its function. Term is based on N. Kastschenko (*Archmikranatentwgesh* 21:357–386, 1882).

eburnification. *n.* A condition of bone when it is dense and hardened to resemble ivory. This term has been applied to parts of the vertebrae used in crushing egg shells in some egg-eating snakes.

ec-. Prefix meaning "out" or "outside."

ecarinate. *adj.* Reference to scales that are smooth and without a keel.

eccentric contraction. Isotonic contraction of muscle in which the muscle is stretched. Cf. *concentric contraction*.

ecchymosis. *n.* (*adj.* **ecchymotic**) The oozing of blood or small hemorrhages from a vessel into surrounding tissue, or the red, black, or blue discoloration of tissue that results from this condition (small dispersed blood clots). Such discoloration is a typical result of snake bite involving venom with hemorrhagic components.

eccritic temperature. The original definition of this term referred to ambient temperature (D.L. Gunn & C.A. Cosway, *J. Exp. Biol.* 15:555–563, 1938), but common usage in herpetology is synonymous with *Selected Body Temperature*.

ecdemic. *adj.* Non-native. Cf. *endemic*.

ecderonic. *adj.* Reference to a tooth having epidermal origin. Cf. *enderonic*.

ecdysis. *n.* The sloughing or shedding of the *stratum corneum* of the epidermis, related to periodic renewal of this *outer generation* layer by a replacement *inner generation*. In snakes and in some other squamates and amphibians, ecdysis is a synchronous, pan-body event, whereas in other reptiles and amphibians the skin may be sloughed in patches. (Syn. *molting, shedding, skin shedding, sloughing, exfoliation*) See also *zip fastener model*.

EC$_{50}$. The concentration of an agonist (drug or compound) that produces 50% of the maximum possible response for that agonist, *in vivo*. It is commonly used as a measure of drug potency. Cf. *IC$_{50}$*.

ECG. See *electrocardiogram*.

echinate. *adj.* Having prominent sharp spines. Reference to a pattern of *epidermal microsculpturing*. See R.M. Price *J. Herpetol.* 16:294–306, 1982.

echinoreticulate. *adj.* Having irregular spines that anastomose into a reticulum. Reference to a pattern of *epidermal microsculpturing*. See R.M. Price *J. Herpetol.* 16:294–306, 1982.

echinulate. *adj.* Bearing small spines. Reference to a pattern of *epidermal microsculpturing*. See R.M. Price *J. Herpetol.* 16:294–306, 1982.

echistatin. *n.* A *disintegrin* component of venom from the Saw-scaled Viper (*Echis carinatus*), which is found to prevent bone loss that occurs as a result of osteoclast activity. This may serve as an important new drug in treatment of osteoporosis and metastatic bone disease.

eclectic. *adj.* Being comprised of theories, beliefs, or principles gathered from a number of different sources.

ecochronology. *n.* Dating of biological events with use of paleoecological evidences.

ecoclimatology. *n.* The study of organisms in relation to climate. (Syn. *bioclimatology*)

ecocline. *n.* 1. A *cline* or gradient of *ecotypes*; i.e., adaptive character variation in a sequence of populations distributed along an ecological gradient. 2. Changes in community structure with respect to slope along a mountain or ridge.

ecodeme. *n.* A *deme* associated with a specific habitat.

ecogeographical divergence. The evolution of two or more different species from a single ancestral species, each exhibiting divergence associated with adaptation to a local habitat in different geographic areas.

ecogeographical rule. Any generalization that describes a trend of character variation that is correlated with geographic or environmental conditions.

ecography. *n.* Descriptive *ecology*.

ecological amplitude. The range of an environmental factor within which an organism or process can function; i.e. range of tolerance.

ecological assessment. A comprehensive assessment of an entire region proposed for development or mitigation, including evaluation of both biotic and abiotic components of the area in question.

ecological barrier. Factors of the environment or landscape that interrupt dispersal.

ecological community. An assemblage of interacting species (plants and animals) that occur together in space and time.

ecological efficiency. A measure of the efficiency of energy transfer from one trophic level to the next. A coefficient of *ecological efficiency* is the ratio of energy assimilated at one trophic level to that assimilated at the previous trophic level. (Syn. *gross ecological efficiency, food chain efficiency*)

ecological equivalents. Distantly or unrelated species that have similar ecological roles in different geographical areas.

ecological genetics. A qualifying term that concerns observations of *microevolution* in the wild.

ecological indicators. See *indicator species*.

ecological isolation. A premating isolating mechanism in which members of different species seldom have contact because of differences in spatial or temporal utilization of habitat, or adaptation to different habitats or features of habitat.

ecological morphology. A relatively new field of investigation that examines the relationship between functional morphology and how animals have evolved in, and can inhabit, particular environments. The field incorporates interdisciplinary methods and integrative concepts (P.C. Wainwright & S.M. Reilly, eds., *Ecological Morphology: Integrative Organismal Biology*, 1994).

ecological niche. The ecological place and role of an organism in its environment or community. The niche is conceptualized as a multidimensional space, of which the coordinates are the various parameters representing the condition of existence of the species, to which it is restricted by the presence of competitor and predatory species. The conditions for existence include all resources and physical parameters that determine where an organism can live and how abundant it can be within its range. An extension of the niche concept is the distinction between the fundamental and realized niche. The *fundamental niche* of a species includes the total range of environmental conditions that are suitable for existence without the influence of interspecific competition or predation from other species. (Syn. *ecospace, prospective ecospace*) The *realized niche* describes that part of the fundamental niche actually occupied by the species. The term "niche" was coined by the naturalist Joseph Grinnell in 1917, given a working conceptual definition by Charles Sutherland Elton in 1927, and popularized by the zoologist G. Evelyn Hutchinson who in 1957 also distinguished the fundamental and realized niche concepts. See also *biospace*.

ecological pressure. The totality of all environmental factors that act as agents of natural selection.

ecological pyramid. A concept of the trophic structure of a community which takes the form of a pyramid in terms of numbers, biomass, or energy. Producers form the base of the pyramid, and successive levels represent consumers of higher trophic levels. See *food pyramid*.

ecological race. See *ecotype*.

ecological release. Expansion of the niche of a species, usually in terms of food resources or habitat, due to removal of a competitor or other restricting species. See also *competitive release*.

ecological segregation. A principle that populations evolve to reduce competition by using the environment in different ways. See *ecological isolation*.

ecological selection. (environmental selection, survival selection, individual selection, or **asexual selection)** Reference strictly to ecological processes that operate on the inherited traits of a species, equivalent to *natural selection* minus *sexual selection*.

ecological species. See *species*.

ecological succession. A gradual process of progressive replacement and change in a community that evolves to a stable climax, usually described in terms of its plant composition.

ecology. *n.* The study of the relationships between organisms and their environment, especially those factors and relationships that determine the abundance and distribution of organisms.

ecomorph. *n.* A morphology shared by relatively distantly related species adapted for similar ecological conditions.

ecomorphological guild. A categorization of species or populations from several taxa that share morphological features in common, suggesting some commonality in ecology. Animals with similar ecologies have convergent morphologies regardless of their taxonomic relationships.

ecomorphology. *n.* Study of the relationship between the form and morphology of an organism and its environment or ecology. See *ecomorphological guild*.

ecophenotype. *n.* A nongenetic modification of phenotype that occurs in response to a change in environmental conditions. See *phenotypic plasticity*.

ecophysiology. *n.* The study of physiological adaptations of organisms to their environment.

ecoregion. *n.* See *ecozone*.

ecospace. *n.* See *ecological niche*.

ecospecies. See *species*.

ecosystem. *n.* The physical environment and sum total of all interactions in a living community of organisms.

ecotone. *n.* A transition zone between two ecological conditions or community type.

ecotype. *n.* 1. Reference to a particular habitat within a larger geographic area. 2. A race within a species that is genetically adapted to a particular environment. (Syn. *ecological race*)

ecozone. *n.* The highest division or category in a hierarchical system of natural regions, representing global divisions that have characteristic interplay of climate, soil, and other living conditions for plants and animals. Ecozones are sometimes subdivided into independent subregions or *ecoregions* (which may be further subdivided into *ecoprovinces*, *ecodistricts*, etc.). Nine ecozones can be defined according to J. Schultz (*The Ecozones of the World*, 2005). (Syn. *biogeographical realm*)

ectethmoid. *n.* See *prefrontal*.

ecto-. Prefix meaning "outside" or "outer."

ectochordal. *adj.* A type of anuran centrum in which the entire perichordal sheath is succesively converted into cartilage and bone, and the definitive centrum becomes an ossified cylinder enclosing a persistent notochord.

ectoderm. *n.* The outermost primary germ layer of an embryo, which forms the epidermis, epidermal derivatives, and nervous system.

ectoglyph. *n.* 1. A venomous snake with an open groove in the fang. 2. A venomous snake having a fang in the posterior part of the mouth. (Syn. *opisthoglyph, opisthoglyphodont*. See also *pleuroectoglyph, proectoglyph*.

ectomesenchyme. *n.* A loose association of cells derived from the neural crest.

ectoparasite. *n.* Any parasite, such as a mite or tick, that attaches to the outer body of an animal.

ectopic. *adj.* Generally, reference to being out of place; in an abnormal position or location.

ectopterygoid. *n.* A dermal bone situated between the palatine and quadrate bone or subtemporal fossa in the palate of most reptiles and many extinct amphibians; absent in modern amphibians except for caecilians. (Syn. *external pterygoid*) See Fig. 21.

ectotherm. *n.* (*adj.* **-ic**) An organism whose body temperature is dependent on heat exchange with the environment. Thus, regulation of comparatively high body temperature depends on external heat sources such as sunlight or warm substrates. (Syn. *allotherm;* older, but not strictly equivalent terms, are *cold-blooded* and *poikilotherm*) Cf. *endotherm*.

ectothermy. *n.* The mode of temperature regulation in which body temperature depends primarily on absorption of heat energy from the environment.

ED$_{50}$. See *effective dose*.

edaphic. *adj.* Reference to the nature or influence of the soil.

edaphic factors. Reference to the physicochemical characteristics of soils or substrates.

EDC. See *endocrine disrupting compounds*.

edema. *n.* (*adj.* **-tous**) A condition of fluid accumulation and swelling in tissue spaces, usually attributable to excessive filtration of plasma from capillaries. Retention of interstitial fluid in organs or tissues.

edentate. *adj.* Of or pertaining to an absence of teeth.

edentulous. *adj.* Lacking teeth.

edge effect. The influence of adjoining communities on a marginal area or ecotone, which often contains more species of higher densities of individuals than either adjoining community.

edge species. A species that is found commonly or predominantly in the marginal area of a community.

ediculae. (*pl.*) *n.* A term that is used to describe the terminal gas exchange chambers in the reptilian lung if the chambers are wider than they are deep. Cf. *faveoli*.

Edible Frog. Vernacular name of the European frog *Rana esculenta*, which is a hybrid of *R. lessonae* and *R. ridibunda*.

EDMA. See *Euclidean Distance Matrix Analysis*.

effective breeding population. The actual number of reproducing individuals in a population.

effective dose. (ED_{50}) The dose required to produce a particular response in half the population of test organisms.

"effectiveness" of thermoregulation. A quantitative index used together with measures of precision and accuracy to evaluate or describe thermoregulation by ectothermic animals. The "effectiveness" is estimated as the extent to which body temperatures are closer on average to the *setpoint range* than are *operative temperatures*. Such procedures for evaluating temperature regulation by field-active ecotherms were introduced by P.E. Hertz, R.B.Huey, and R.D. Stevenson (*Amer. Naturalist* 142:796–818, 1993) and are sometimes referred to as the "*Hertz index*."

effective population size. (N_e) In population genetics, the concept of the average number of individual animals in an idealized population that contribute genes to succeeding generations. Introduced by S. Wright (*Genetics* 16:97–159, 1931; *Science* 87:430–431, 1938), it is a basic parameter in many models in population genetics.

effector. *n.* A muscle or gland that contracts or secretes, respectively, in direct response to nerve stimulation.

effector site. The precise site on a molecule where an activity is localized.

efferent. *adj.* Reference to transport or conduction away from a central region toward the periphery, e.g. from central nervous system to a peripheral organ or structure. Cf. *afferent*.

effete. *adj.* No longer functional.

efficacy index. A measure of the efficacy of any given antivenin, measured as the ratio of the LD_{50} of venom plus antivenin injection to the LD_{50} of venom injection alone. (Syn. *therapeutic index*)

effluent. *adj., n.* The outflow from a river or lake. Generally, discharge of liquid from any source.

efflux. *n.* The movement of solute or solvent out of a cell across the plasma membrane.

effodient. *adj.* Having a habit of digging.

eft. *n.* Antiquated, local name for various small aquatic salamanders or their immature larvae. The juvenile, terrestrial phase in the life cycle of a newt.

egesta. (F or F + U) *n.* In ecological energetics, the part of *consumption* (C) that is expelled as fecal waste, or is regurgitated, and therefore not absorbed by an animal. Some authors include both fecal and urinary wastes for this term (F + U).

egestion. *n.* The process of eliminating waste material from the digestive tract as feces.

egg. *n.* A female hapoloid reproductive cell at any stage prior to fertilization, and its derivatives following fertilization. (Syn. *ovum*)

egg binding. See *egg retention*.

egg brooding. See *brood, brooding*.

egg capsule. A term denoting the outer gelatinous envelope(s) that surround amphibian eggs. (Syn. *jelly envelope*)

egg caruncle or **callus**. See *caruncle*.

Egg-eating Snakes, Egg-eaters. Vernacular name for colubrid snakes of the African genus *Dasypeltis* and the Indian genus *Elachistodon*, which are specialized for eating avian eggs. The name also has been applied to the South American colubrid *Drepanoides anomalus*, which preys on lizard eggs.

egg envelope. Refers to the *jelly envelope* surrounding amphibian eggs, and is also used in developmental literature to describe the various membranes or substances that might associate with eggs.

Egg Frogs. Collective vernacular name for species of African arthroleptid frogs belonging to the genus *Leptodactylodon*.

egg guarding. Defense by one or both parents of its developing eggs against disturbance or predation.

egg mass. The sum total and volume of eggs deposited together by a single female amphibian.

egg membranes. The membranes produced around the egg as it develops in the ovary, or laid down around the egg as it passes through the oviduct. The former are *primary egg membranes*, while the latter are *tertiary egg membranes*. The tertiary egg membrane(s) form the egg capsule in amphibians, and a series of tertiary membranes including albumin layers and the shell membranes surround the ovum of reptiles. (The secondary egg membrane, or *zona pellucida* of mammals, does not occur in amphibians and reptiles.)

egg sac or **sack**. The common envelope or protective membrane and the adherent jelly envelope surrounding the eggs of certain salamanders.

egg retention. Any abnormal condition in which reptilian eggs are retained within the oviduct and fail to be laid. (Syn. *egg binding*)

egg or **tadpole attendance.** The presence of a parent or parents in the vicinity of developing eggs or larvae.

egg tooth. Generally, any structure on the tip of the snout which is used to break or cut through egg membranes at hatching. A more resticted usage refers to the deciduous, toothlike structure on the end of the premaxillary bone of squamates and used to cut through the leathery shell.

egress. *n.* The act of going out or leaving. Cf. *ingress*.

Egyptian Toad. Vernacular name for the North African bufonid species *Amietophrynus* (formerly *Bufo*) *regularis*. For taxonomic revision see D.R. Frost et al., *Bull. Amer. Mus. Nat. Hist.* 297:1–370, 2006.

EIA. See *environmental impact assessment*.

eicosanoids. *n.* Paracrine hormones derived from fatty acids, known to affect a wide range of physiological processes. See also *prostaglandins*.

eigenvalue. *n.* In *principle components analysis*, a name for the proportion of total variance that is accounted for by the corresponding principle component.

einstein, Einstein unit. A unit of light energy concentration sometimes used in physical chemistry, equal to the energy per mole of photons carried by a beam of monochromatic light.

EIS. See *environmental impact statement*.

ejaculatory pump. Term coined by D.L. Jameson (*Syst. Zool.* 4:105–119, 1955) to identify a stepwise behavioral sequence resulting in fertilization of eggs by amplexing frogs.

elapid. *n., adj.* A member of, or pertaining to, *Elapidae*.

Elapidae. *n.* A speciose clade of snakes that includes a large number of familiar terrestrial venomous species including Cobras, Mambas, Kraits, Coral Snakes, and the marine Sea Snakes. The distribution is cosmopolitan throughout much of the midlatitudes of the world. Two subfamilies, approximately 62 genera, and some 600 species are recognized. *Elapinae* are terrestsrial elapids, and *Hydrophiinae* includes Australopapuan elapids and Sea Snakes. Marine elapids were previously treated separately from the terrestrial elapids and were given family distinction. However, molecular and recent phylogenetic studies indicate that Australopapuan terrestrial elapids are the closest relatives of Sea Snakes, and these are a more recent radiation than Afro-Asian terrestrial elapids. Furthermore, some of the highly aquatic Sea Snakes (*Hydrophis* spp.) originated as a separate radiation from the sea kraits (*Laticauda* spp.).

Elapinae. *n.* A subfamily of *Elapidae,* widely distributed and including Cobras, Mambas, Kraits, Australian Tiger Snakes, and Coral Snakes.

elastic cartilage. Cartilage that is flexible and springy due to the presence of elastic fibers in the matrix. Cf. *fibrocartilage, hyaline cartilage*.

elasticity. *n.* (*adj.* **elastic**) 1. The property of being capable of distortion, stretching, or compression by a force, with subsequent spontaneous return to (or toward) original shape when the force is removed. 2. In ecosystem dynamics, this term refers to how quickly a system returns to an earlier state following a perturbation.

elastic modulus. A measure of the deformation caused per unit of load. *Stress* divided by *strain*.

elastic strain. Strain in which a deformed body recovers its original shape after the stress is released. A reversible change produced by stress.

elastin. *n.* A glycoprotein and principal component of elastic fibers found in a variety of tissues including tendons and the walls of blood vessels and pulmonary bronchi.

electivity. *n.* A measure of the preference, or lack of preference, of a consumer for a particular species of prey.

Electivity index. A measure of *electivity*: $E_i = r_i - n_i / r_i + n_i$, where E_i is Electivity index for prey organism i; r_i is percentage of i in the diet; and n_i is the percentage of i in the environment. Values from 0 to 1.0 indicate preference, whereas values from –1.0 to 0 indicate avoidance.

electrocardiogram. (**ECG**; sometimes **EKG**) *n.* A graphic plot of heart electrical activity against time.

electrode. *n.* Either terminal of an electrical apparatus.

electrolyte. *n.* A compound that dissociates into ions when dissolved in water.

electromagnetic fields. (**EMF**) The power-frequency (60 Hz) electric and/or magnetic fields produced by the transmission, distribution, and utilization of electric energy or associated with natural phenomena.

electromagnetic radiation or **spectrum.** A term that refers to all energy waves caused by the acceleration of electric charge. This spectrum includes radio waves and infrared radiation through visible radiation, to ultraviolet radiation and high-frequency cosmic rays.

electromyogram. *n.* An electrical recording of muscle contraction or activity.

electromyography. *n.* The recording and study of the electrical activity of muscle.

electron carrier. An enzyme or coenzyme that can acquire and lose electrons reversibly, such as the cytochromes.

electron-dense. A term used in reference to a dense area of an electron micrograph, which might indicate a region of highly concentrated macromolecules or the binding of electrons to heavy metals used as fixatives or stains for various structures.

electron microscope. An instrument that focuses a beam of electrons passed through a vacuum and focused by a series of magnetic lenses used as a magnifying system with a resolving power hundreds of time greater than the best optical microscopes. A *transmission electron microscope* (*TEM*) forms images by means of electrons that pass through the specimen, whereas a *scanning electron microscope* (*SEM*) forms images by means of electrons that are reflected back from the specimen. The latter is used for viewing the minute structures on the surfaces of objects.

electron transport chain. An orchestrated group of electron carriers that are localized in mitochondria and function in a cascade of reactions that transfer electrons from substrates to oxygen. The released energy is used to phosphorylate ADP to ATP.

electron volt. (**eV**) A unit of energy equivalent to the energy gained by an electron passing through a potential difference of one volt.

electrophoresis. *n.* (*adj.* **electrophoretic**) A technique for separating charged molecules such as proteins by means of differential movement in a solution subjected to an electric field. The medium consists of either a gel or a solution that is held by a porous support such as filter paper. The molecules are separated based on differences in electric charge and also by the size or geometry of the molecule.

electroreception. *n.* The ability to detect electric fields. *Ampullary organs* have been identified as *electroreceptors* in some aquatic amphibians.

electroreceptor. *n.* A sensory structure that responds to electric fields.

electrotonic potential. A graded electric potential generated locally by currents flowing across a membrane. Unlike a nerve action potential, an electrotonic potential is not all-or-none and is not actively propagated. Sensory stimuli activating a sensory structure typically produce an electrotonic potential that is conducted with decay to a region of membrane where action potentials are actively propagated.

element X. Either of a pair of bony flakes situated lateral to the basisphenoid-basioccipital suture and medial to the foramina ovalia in amphisbaenids.

element Y. Either of a pair of bony flakes lying anterior to the elements X, anterior to the foramina ovalia, and beneath the pterygoid, where it joins the basipterygoid process in certain amphisbaenids. This and *element X* have been comparaed with the *operculum* of amphibians.

Elephant Trunk Snake. *n.* Vernacular name for the acrochordid species *Acrochordus javanicus*. (Syn. *Filesnake, Wart Snake*)

Eleutherodactylinae. *n.* Former taxon (subfamily) of small frogs containing 12 genera and about 600 species of *Eleutherodactylus*. See *Brachycephalidae*.

eleuterognathine. *adj.* Used in reference to the lack of overlap of maxillae in front of the premaxillae of microhylid frogs. Cf. *symphygnathine*.

elevation. *n.* The vertical distance between a given point and a specified datum surface. (Syn. *altitude*)

elfin forest. A forest of higher elevations in warm, humid regions having stunted trees with numerous epiphytes. Cf. *krummholz*.

elimination. *n.* In ecological energetics, reference to losses of energy or biomass from a population or trophic unit per unit of time, area or volume. Such losses include mortality, predation, emigration, ecdysis.

ELISA. See *enzyme-linked immunosorbent assay*.

elliptical. *adj.* Having the form of an ellipse, or closed curve, usually with reference to shape of vertical pupils of some amphibians and reptiles.

El Niño Southern Oscillation or **El Niño.** A large oceanic current of warm water offshore of coastal Peru and Eduador, flowing southwards against the coastal and offshore currents. The easterly tradewinds cause the warm surface water to move away from the continent, resulting in an upwelling of nutrient-rich cold water. Episodes of intensification of El Niño due to weakened easterly winds occur at irregular intervals, although usually twice every decade. Changes in Pacific Ocean currents and global atmospheric conditions, due to their connections to intensifications of the El Niño current, are now referred to as El Niño. The warm ocean surface temperatures associated with El Niño have global climatic effects and are often associated with weather anomalies around the world. (From Spanish for "the child," meaning the Christ child, as the condition typically begins at Christmas time.) See K.E. Trenberth, *Bull. Amer. Meteor. Soc.* 78:2271–2277, 1997. Cf. *La Niña*.

eluvium. *n.* A sand-dune community.

elygium. *n.* A zone of pigment at the basal margin of the iris (= *ocular elygium*) or at the skin-cornea margin (=*epidermal elygium*), presumably shielding the eye from excessive light. See E.D. Van Dijk, *Ann. Natal Mus.* 18:231–286, 1966. Cf. *umbraculum*.

em. See *emend*.

emaciation. *n.* (*adj.* **emaciated**) Condition of being abnormally lean, usually from starvation or disease.

emarginate. *adj.* A scale border that is notched or scalloped in outline.

emargination. *n.* Reference to large notches in the bony braincase. Cf. *fenestra*.

embedding. *n., v., adj.* Immersing fixed tissue in a medium appropriate for histological sectioning and subsequent staining and microscopic examination. The most common embedding technique is to use wax, whereas epoxy resin or plastic is used for electron microscopy. Embedding can also be accomplished using frozen, non-fixed tissue in a freezing medium.

EMBL Reptile Database. A web-based database intended to provide information on the classification of all living reptiles, exhibited as listings of species and their pertinent higher taxa. The database covers all living snakes, lizards, turtles, amphisbaenians, tuataras, and crocodilians. Currently it is supported by the Systematics working group of the German Herpetological Society. Current URL: http://www.reptile-database.org

embolism. *n.* A sudden blockage or occlusion of a blood vessel due to a clot, bubble, or other object that has migrated from another part of the body. Cf. *thrombus*.

embolomerous. *adj.* A condition of *dispondylous* vertebra, found in extinct reptilian ancestors, in which the *pleurocentrum* and *intercentrum* are of roughly the same size, forming a double centrum.

embolus. (*pl.* **–i**) *n.* A clot, bubble, or object that migrates from one location to lodge in, and thereby block, a smaller blood vessel at another location. Cf. *thrombus*.

embrace. *n.* See *amplexus*.

embryo. *n.* An organism during early rapid development of the body, applicable generally to stages from fertilization to birth or hatching.

embryogenesis. *n.* The process of embryo formation and development.

embryology. *n.* The scientific study of embryos, especially the development and differentiation that occurs from fertilization to birth or hatching.

embryonic development. Reference to successive stages of growth and differentiation of an embryo preceding birth or hatching.

embryonic disc. The disc or tissue from which the embryo and its extraembryonic membranes develop.

embryonic induction. See *induction*.

embryonic shield. A term applied to the *scutes* of newly hatched chelonians. As a turtle grows, new keratin is formed in *seams* between the scutes and the formerly adjacent embryonic shields become separated.

embryotrophe. *n.* Nutritional materials supplied to a developing embryo, including yolk and additional materials contributed from maternal tissues.

emend. (em.) Abbreviation of the Latin *emendatus*, which means amended or altered by. In systematics, the abbreviation precedes the name of an author when he/she has changed the original spelling of a taxon without excluding the type of the name.

emerald network. Collective areas designated to be of special conservation interest and located throughout the territory of parties to the 1979 Bern Convention (on the Conservation of European Wildlife and Natural Habitats). The network was established in 1998 and operates in cooperation with the European Union's Natura 2000 programme.

Emerald Tree Boa. Vernacular name for the neotropical boid species *Corallus caninus*, popular among herpetoculturists.

emergent evolution. The appearance of novel and unpredictable characters due to rearrangement of existing potentialities.

emergent properties. Properties that are manifest at higher levels of a hierarchical system due to processes involving aggregations or interactions of lower-level units. E.g., populations have properties that are not expressed or identifiable in individuals that make up the population.

emesis. *n.* The act of vomiting.

emigration. *n.* Reference to individuals (*emigrant*) leaving a population and traveling to another location. Cf. *immigration*.

emissive power. The heat per unit time per unit area emitted by an object. For a *blackbody*, this is given by the *Stefan-Boltzmann relation*. Cf. *emissivity*.

emissivity, surface emissivity. (ε) *n.* The ratio of radiant energy emitted by a surface or body to that emitted by an ideal blackbody. The fraction of blackbody emittance at a given wavelength emitted by a material or body. By definition, the emissivity of a blackbody surface = 1. Cf. *absorptivity, emissive power*.

Emo Skinks. Vernacular name for species of scincid lizards belonging to the genus *Emoia*.

empathic learning. A response to the observed experience of another individual. E.g., avoidance of coral snakes by a predator that has observed a fatal or debilitating snakebite.

empirical. *adj.* (*n.* **empiricism**) Reference to that based on direct observation rather than theory or preconception.

empirical modeling. Correlative modeling based on mathematical description of relationships among variables from experiment and observation.

empirical taxonomy. Classification of organisms based on observed phenotypic similarity.

emulsify. *v.* To break up fats into smaller droplets.

Emydidae. *n.* A clade (family) of turtles, mostly freshwater and aquatic, with the exceptional species *Malaclemys terrapin* inhabiting coastal marine habitats. There are 12 genera and 40 species inhabiting North America, South America, the West Indies, Europe, northern Africa, and western Asia.

emydid. *n, adj.* A member of, or pertaining to, *Emydidae*.

enamel. *n.* The outermost hard, acellular, and crystalline substance deposited over dentine of the tooth crown of tetrapods. It consists almost entirely of calcium salts and is derived from ectoderm.

enamel prisms. Bundles of crystallites that extend from the enamel-dentine junction close to the outer enamel surface without interruption, and characteristic of occlusal dentition. True prisms are distinct microstructural domains and are bordered by a well defined interprism boundary plane comprising the prism "sheath." Prisms are of equal size and do not split or merge. Prism cross sections are rounded or variable in shape, but never polygonal. Prisms may decussate in layers, in groups, or individually. These microstructures are considered to be adaptations to resist abrasion or brittle failure in complex dentitions because enamel crystallites are more resistant to wear if they are arranged perpendicular to a surface rather than more approximately parallel. Prismatic enamel is characteristic of mammals and is present in the agamid lizard *Uromastyx*, which is the only extant reptile with mammal-like occlusal dentition used in mastication. See J.S. Cooper & D.F.G. Poole, *J. Zool.* 169:85–100, 1973. Cf. *prismless enamel*.

encapsulated sensory receptor. The terminus of a sensory neuron or nerve fiber that is enclosed within accessory tissue. Cf. *free sensory receptor*.

encephalitis. *n.* Inflammation of the brain.

encephalomalcia. *n.* Degeneration or softening of brain tissue.

encephalomyelitis. *n.* Inflammation of the brain and spinal cord.

encounter call. A vocalization produced by a male frog or toad when a rival male calls or approaches closely.

encyst. *v.* To form a *cyst*.

endangered. *adj.* 1. A IUCN criterion for threatened species where a taxon is known to be at high risk of extinction in the near future if threatening conditions continue to operate. Cf. *critically endangered, vulnerable*. 2. Used generally to denote a species or ecosystem that is threatened with extinction.

Endangered Species Act. United States legislation enacted in 1973 to protect animal and plant species threatened by extinction. The Act has been subject to periodic renewal and amendment by the U.S. Congress.

end-body. *n.* The fleshy terminus of the tail of rattlesnakes, giving rise to each new rattle. (Syn. *matrix*)

endemic. *adj., n.* Reference to a species or population that is native, or restricted naturally in distribution, to a particular region or habitat.

endemism. *n.* The state of being endemic; native to a particular region.

endergonic. *adj.* Characterized by a concomitant absorption of energy. Cf. *exergonic*.

enderonic. *adj.* Reference to a tooth that is derived from the dermis. Cf. *ecderonic*.

endo-. Prefix meaning "within," "inward," or "inside."

endobiotic. *adj.* Developing or situated within a living organism.

endocardium. *n.* The internal endothelial lining of the heart.

endochondral bone. Bone preformed in cartilage that ossifies during development.

endochondral ossification. The process in which bone is deposited in preexisting cartilage. Such bone is called *replacement bone*. Cf. *intramembranous ossification*.

endocrine. *adj., n.* Secreting (chemical messengers) into the blood circulation. Cf. *exocrine*.

endocrine disrupting compounds or **contaminants. (EDCs)** Xenobiotics that mimic naturally occurring hormones or are antagonistic to their modes of action. A variety of compounds occur in the environment and have been shown to be deleterious to wildlife. Some effects attributed to these contaminants include reduced fertility, abnormal reproductive development, reduced viability of offspring, impaired hormone activity, altered growth and metamorphosis, and altered sexual behavior. Thus, EDCs alter endocrine functioning in a variety of ways, and much research has focused on the interactions with steroid and thyroid hormones and their receptors. (Syn. *endocrine disrupters, antiandrogenic chemicals, antiestrogenic chemicals*)

endocrine gland. Ductless organs or tissues that secrete hormones into the blood circulation. Cf. *exocrine gland*.

endocrine secretion. Reference to a biologically active substance that is produced by a ductless gland and characteristically travels to target cells or tissues by blood circulation. Cf. *autocrine secretion, exocrine secretion, paracrine secretion*.

endocytosis. *n.* A process in which materials such as foreign bacteria or particles are engulfed by a cell.

endoderm. *n.* The innermost primary germ layer that lines the primitive gut of the early vertebrate embryo beginning in the gastrula stage. The endoderm forms the epithelial lining of the alimentary canal and all of its outgrowths.

endogamy. *n.* The selection of a mate from within a small kinship group (*inbreeding*). Cf. *exogamy*.

endogenous. *adj.* Arising within the body or within an organ. Cf. *exogenous*.

endogenous clock. Reference to any function or physiological system that shows an inherent sustained periodicity.

endoglyph. *n.* (*adj.* **-phous**) A venomous snake with a closed groove in the fang, or with a grooved tooth in the anterior part of the mouth (i.e., front-fanged snakes). (Syn. *proglyph; proglyphodont*)

endolymph. *n.* Extracellular fluid of the membranous labyrinth of the inner ear, contacting the sensory hair cell receptors. Cf. *perilymph*.

endolymphatic duct. A slender tube extending upward and inward from the *sacculus* to the *endolymphatic sac* of the inner ear.

endolymphatic sac. Terminus of the *endolymphatic duct* extending upward from the *sacculus* and ending in the braincase. (Syn. *saccus endolymphaticus*)

endoparasite. *n.* A parasite that lives inside an animal, e.g. lung flukes or tapeworms.

endoplasmic reticulum. A labyrinth of internal membrane organized into branching tubules and flattened sacs in the cytoplasm of most eukaryotic cells.

endoplastron. *n.* See *entoplastron*.

endorphins. *n.* Endogenous *neuropeptides* that exhibit morphine-like actions, found in the central nervous system of vertebrates.

endoskeleton. *n.* The internal skeleton of an animal, including skull, vertebral column, ribs, limb girdles and elements, excluding any part of the skeleton that is of dermal origin. Cf. *exoskeleton*.

endosteal bone. Bone that is preferentially deposited and reabsorbed from endosteal surfaces depending on the direction of growth in a given region of bone. Cf. *periosteal bone*.

endosteum. *n.* A fibrous connective tissue that coats the internal surface of bones and has potential to form new bone material. Cf. *periosteum*.

endothelium. *n.* The single-celled, inner lining of blood vessels and lymphatics.

endotherm. *n.* (*adj.* **-ic**) An organism in which the body temperature is determined largely by internal heat production of metabolism. The term usually applies to birds and mammals, but some reptiles are facultatively endothermic. Examples are brooding female pythons, large sea turtles while swimming, and certain newborn or hatchling snakes while assimilating yolk for a brief period following birth or hatching. There is controversy concerning whether extinct dinosaurs were endothermic, but larger ones have been termed *inertial endotherms* by virtue of high thermal inertia attributable to large body size. (Syn. *autotherm*) Cf. *ectotherm*.

endothermy. *n.* The mode of temperature regulation in which body temperature is determined largely by a balance of rates of metabolic heat production and its dissipation to the environment.

endotrophic. *adj.* Depending on maternal investment as the sole source of nutrition during development, as in lecithotrophic and viviparous forms. R. Altig & G.F. Johnston (*Smithsonian Herpetol. Info. Serv.* 67:1–75, 1989) defined six developmental guilds of endotrophic anurans: *viviparous, ovoviviparous, direct development, paraviviparous, exoviviparous,* and *nidicolous.*

endplate potential. (epp) A postsynaptic potential in a muscle fiber at the neuromuscular junction (or motor endplate).

end product. A molecule that is the final product in a chain or cascade of metabolic reactions.

end product inhibition. A phenomenon wherein the end product of a reaction chain inhibits the enzyme involved in the initial reaction(s) that lead to the product. See *feedback.*

end product repression. A phenomenon wherein the end product of a reaction chain functions as a corepressor and shuts down the operon turning out the enzymes that control the reaction chain.

endysis. *n.* Archaic term occasionally used to describe the genesis of a new epidermal layer that becomes functional following ecdysis of the previous generation. Defined by R. Rugh (*Experimental Embryology,* p. 462, 1948). Cf. *ecdysis.*

enemy reaction. A specific defensive or flight behavior stimulated by presence of a predator or conspecific rival.

enhancer. *n.* In molecular biology, a nucleotide sequence that potentiate the transcriptional activity of physically linked genes.

energetic equivalence rule. The proposition that population energy use is independent of species body mass (J. Damuth, *Nature,* 290:699-700, 1981; *Biol. J. Linn. Soc.* 31:193–246, 1987).

energetics. *n.* Reference to energy transformations within an organism, community, or system.

energy. *n.* The capacity to do work.

energy budget. 1. The balance of energy input and utilization within an organism, community or system. 2. The sum total of energy expended or required by an organism or population, often expressed on a per annum basis and subdivided by category of use (maintenance, reproduction, etc.).

energy flow. Reference to passage of energy in and out of an organism, population, community, or system. The passage of energy through different trophic levels of a food chain.

energy metabolism. The complex totality of biochemical reactions within cells that utilize or generate ATP and other high-energy compounds, which serve as the immediate source of energy for all biological events.

energy subsidy. Energy input to a system from an external source that reduces the *in situ* energy requirement for maintenance of that system.

enkephalins. *n.* Endogenous neuropeptides exhibiting morphinelike actions, found in the central nervous system of vertebrates.

Ensatina. *n.* Vernacular and generic name for plethodontid salamanders belonging to the genus *Ensatina.*

ental. *adj.* Inner, central, or internal.

entamoebiasis. *n.* See *amoebiasis.*

enteric. *adj.* Intestinal; of the *enteron* or alimentary canal.

enteric nervous system. Reference to the network of nerves associated with the digestive system.

enterocrinin. *n.* A hormone released by the intestinal mucosa and increases the production of intestinal secretions.

enterogastrone. *n.* A hormone released from the intestinal mucosa and acts to inhibit further gastric secretion and motility.

enterolith. *n.* A stone or concretion found in the intestine.

enthalpy. *n.* The heat produced or taken up by a chemical reaction.

entoglossal or **entoglossal process.** *n.* The long, slender, medial extension of the *copula* in reptilian hyoid, running anteriorly into the tongue and serving as the principal rod supporting its extension. (Syn. *lingual process, glossohyal, urohyal*)

entoglossum. *n.* See *hypoglossum.*

entoplastron. *n.* The unpaired, medial bone in the plastron of most turtles, derived from the clavicle and situated anteriorly between the paired *epiplastra* and *hypoplastra.* (Syn. *endoplastron, episternum*) See Fig. 7.

entosternal. *n.* Alternate term for the *epicoracoid* of frogs.

entrainment. *n.* Coupling of a biological rhythm to an external time source (*zeitgeber*).

entropy. *n.* A measure of the portion of energy not available for work in a closed system, and a reflection of molecular randomness.

envelope. *n.* See *jelly envelope, egg sac, egg envelope.*

envenomation. *n.* The act or state of having venom injected into tissues, as in venomous snake bite. (Syn. *snake bite, ophidiasis, ophiotoxemia, ophiotoxismus, ophitoxaemia venenation*)

environment. *n.* The complex of physical and biotic factors that surround an organism. The term can also be applied to the immediate surroundings of any object of discussion, e.g. an internal organ such as the heart.

environmental impact assessment. (EIA) A detailed survey and analysis of a proposed human action to determine the likely effect on the environment.

environmental impact statement. (EIS) A document filed in the United States whereby policy makers and the general public evaluate the potential impact of federally initiated or permitted actions and weigh all reasonable alternatives. The document is designed to ensure that all federal agencies are abiding by the goals and principles of the National Environmental Policy Act (NEPA), enacted by the United States Congress in 1970 as a body of law and guidelines for environmental protection.

environmental sex determination. A condition in which the sex of an individual is determined by external environmental factors (e.g., temperature) and not directly by syngamy alone. Cf. *genetic sex determination.* (Syn. *phenotypic sex determination*)

environmental stochasticity. A condition of uncertainty due to variation in environmental conditions.

environmental variance. A portion of the phenotypic variance attributable to variation of factors in the environments to which the individuals in a population have been exposed. Cf. *genetic variance, phenotypic variance.*

enzootic. *adj.* Reference to diseases that affect a geographically limited number of animals. Cf. *epizootic.*

enzyme. *n.* A protein with catalytic properties that initiates or facilitates a chemical reaction between other molecules. A protein catalyst.

enzyme or **enzymatic activity.** A measure of the catalytic efficacy of an enzyme, reflected in the number of substrate molecules that react per minute per enzyme molecule.

enzyme-linked immunosorbent assay. (ELISA) An immunochemical technique that uses an antibody-linked enzyme that can react with a substrate to produce a measurable signal. This assay avoids the hazards of using radiochemicals or the expense of fluorescence systems.

Eocene. *n.* The second Epoch of the Paleogene Period in the Cenozoic Era, noted for the appearance of modern mammalian fauna. See Table 1.

Eogaea. *n.* A zoogeographical region incorporating Africa, South America, and Australasia. Cf. *Caenogaea.*

eolation. *n.* An alternative spelling of *aeolation.*

eolian. *adj.* An alternative spelling of *aeolian.*

Eon, eon. *n.* 1. A unit of time equal to 10^3 years. 2. See *geological Eon.*

eosin. *n.* A commonly used histological stain that colors cytoplasm pink. See *H & E.*

eosinophilic. *adj.* Staining readily with eosin, a histologic stain.

Eosuchia. *n.* A clade of late Permian/early Triassic group of *lepidosaurs*, and likely the ancestors of modern lepidosaurs.

epanodont. *adj.* Having teeth in the lower jaw only.

epapophysis. *n.* A dorsomedial process of the centrum. See also *subneural process.*

epaxial. *adj.* 1. Above or upon an axis. 2. Reference to trunk muscle that lies, or originates, dorsal to the vertebral column or body axis. Cf. *hypaxial.*

ependyma. *n.* (*adj.* –**al**) The layer of cells that line the cavities of the chordate central nervous system, including the central canal of the spinal cord.

ephemeral. *adj.* Short-lived or transient.

epi-. Prefix meaning "upon," or "on the surface of."

epiboly. *n.* The spreading of surface cells during embryonic gastrulation.

epibranchial. *n.* 1. Cartilage situated at the distal end of a *ceratobranchial,* usually forming only a point or tip in adult forms. 2. The second prominent segment of the first branchial arch in *Eurycea,* distal but attached to the *ceratobranchial.* Used by L. Smith (*J. Morphol.* 33:527–583, 1920).

epicardium. *n.* The outer layer of heart, or inner layer of *pericardium.*

epichordal. *adj.* Patterns of vertebral development that have reference to the *centrum.*

epicondyle. *n.* A protrusion on the surface of a bone, above its condyle.

epicoracoid, epicoracoid cartilage. *n.* Cartilaginous elements of the pectoral girdle lying at the ventral end of the scapulocoracoidal plate or the coracoids, with variation in size and position in different species. The epicoracoid forms midventral cartilages that overlap or fuse in the midline of frogs, whereas in reptiles it may form large sheets of cartilage originating as unossified ventral and anterior margins of the *coracoid* (e.g., Geckos) or may be reduced to a small cartilage bordering the *sternum* (crocodilians). A synonym for *procoracoid* in fossil reptiles. See Fig. 3.

epidemic. *n.* Reference to a disease that afflicts a high proportion of a population over a wide area.

epidemiology. *n.* Study of the various factors related to the incidence, distribution, transmission, and control of diseases among individuals or populations.

epidermal appendage. Structures borne by reptilian scales such as glands, sense organs, and climbing footpads. The latter are expressions of the *oberhautchen's* ability to form *epidermal microsculpturing.* See also *feather.*

epidermal elygium. See *elygium.*

epidermal generation. A unit of lepidosaurian *epidermis* that is periodically formed and shed (L. Alibardi & P.F.A. Maderson, *J. Morphol.* 256:111–133, 2003). A complete, mature unit comprises six distinct cell types: when superficial in position, just preshed it is called an *outer epidermal generation.* During the renewal phase of the shedding cycle, the forming unit, lying beneath the outer generation that will soon be shed to the environment, is called an *inner epidermal generation.* At shedding (*sloughing, ecdysis*) the outer epidermal generation separates from the incomplete, inner generation by a split within the shedding complex. Earlier terminologies distinguished "old" (outer) from "new" or "young" (inner) generations of epidermis. For history of terminologies see P.F.A. Maderson, *J. Zool.* 146:98–113, 1965. Fig. 31.

epidermal microsculpturing. Reference to microscopic features of surface morphology present on scales of squamate reptiles. (Syn. *dermatoglyphics, epidermatoglyphics, microdermatoglyphics, microornamentaion, microsculpturing*)

epidermal scale. See *scale*.

epidermatoglyphics. *n.* See *epidermal microsculpturing*.

epidermis. *n.* (*adj.* –**al**) The outer layers of skin of vertebrates, derived from ectoderm and overlying the dermis. The outermost layers of epidermis (*stratum corneum*) are keratinized and protective, and they are renewed periodically by sloughing or ecdysis. See Figs. 28–31.

epididymis. *n.* The largely convoluted tubule that drains the efferent ducts of the testis into the urethra, derived from the embryonic *mesonephric* or *wolffian duct* and *efferent ducts*. This structure lies at the surface of the testis and functions to store spermatozoa until they are released to the exterior via the *vas deferens*. See Fig. 38.

epiglottis. *n.* The valvelike closure of the glottis.

epigamic. *n.* That which functions to stimulate members of the opposite sex during courtship. Sexually attractive characters.

epigamic selection. See *sexual selection*.

epigeal, epigean. *adj.* Living on the surface of the substrate, at or above the soil surface.

epigeic. *adj.* Terrestrial, or at the soil surface.

epigenesis. *n.* 1. The control of gene expression by mechanisms that are not under the immediate control of genes. 2. The concept that morphological complexity develops gradually during development from simple beginnings in an essentially formless egg, attributable to some vital force during the evolutionary debates of the 18th and early 19th centuries (epigenesis vs. preformationism).

epigenetics. *n.* 1. The mechanisms by which genes produce their *phenotypic* effects. 2. Study of the causal mechanisms of development with reference to events above the level of the genes.

epihyal. *n.* A short, cartilaginous process on the distal end of a *ceratohyal*, usually quite small or absent in adults, but forming a prominent part of the second hyoid arch in plethodontid salamanders.

epileon. *n.* The cartilaginous area on the dorsal extremity of the *ilium* in certain salamanders.

epilimnion. *n.* A vertical zone of a lake consisting of an upper mixed layer of relatively warm water above the thermocline. Cf. *metalimnion, hypolimnion*.

epimere. *n.* The dorsal, segmented division of the embryonic lateral mesoderm. (Syn. *somite*)

epimeric muscle. Muscles derived from the *epimere* (Syn. *epaxial* muscle).

epinephrine. *n.* A neurotransmitter and hormone of the adrenal medulla that rapidly increases metabolic and other activities, including heart rate and constriction of blood vessels. (Syn. *adrenalin, adrenaline*)

epiotic. *n.* Antiquated term that has been variously applied to skull bones of amphibians and the otic area formed from the *supraoccipital* in crocodilians.

epipelagic. *adj.* Of or pertaining to the upper layer of the marine water column where there is some light, extending from the surface to a depth of 100 to 200 m depending on the turbidity of the water.

epiphyllous. *adj.* Reference to organisms or eggs found on leaf surfaces.

epiphysis. (*pl.* -**es**) *n.* 1. Areas of accessory ossification located at either end of a long bone, characteristic of mammals but found also in tuataras, some lizards, and crocodilians. These may ossify the articular regions of bone before the growth of the shaft, or *diaphysis*, is completed. 2. The *pineal organ*, a dorsal outgrowth of the diencephalon.

epiphysis cerebri. See *pineal, pineal body*.

epiphyseal apparatus or **complex.** Antiquated term for two processes projecting dorsally from the diencephalon of the brain in reptiles and amphibians, forming the pineal and parapineal organs. The structures contain hydroxyindole-O-methyltransferase (HIOMT) activity and melatonin.

epiphyte. *n.* A plant which grows on another plant but is not parasitic (e.g., mosses, orchids, etc.).

epiplastron. (*pl.* -**stra**) *n.* Either of the paired anterior elements of the turtle plastron. In tryionychid turtles the anterior plastron is unpaired and usually called the *entoplastron*. See Fig. 7.

epipodial. *n.* Either of the bones forming the lower limbs of vertebrates (the *radius* and *ulna* in forelimb; the *fibula* and *tibia* in hindlimb).

epipterygoid. *n.* Either of a pair of membrane bones extending upwards as a rodlike structure from the dorsal side of the pterygoid to the wall of the cranium in nonmammalian vertebrates. It is reduced or absent in adult amphibians and is lost in snakes and turtles.

epipubic cartilage. See *ypsiloid bone* or *cartilage*.

epipubis. *n.* Any bony or cartilaginous structure lying immediately anterior to the pubis, either on the midline or closely lateral to it. (Syn. *prepubis* of crocodilians, *ypsiloid cartilage* in salamanders, *epipubic process* of some lizards and turtles)

epipygal. *n.* See *suprapygal*.

episcapulum. *n.* See *suprascapulum*.

episematic. *adj.* Reference to characters, especially coloration, that aid in recognition and serve as welcoming signals. Cf. *aposematic*.

epistasis. *n.* (*adj.* **epistatic**) The condition of having one gene that masks the effect of another to which it is nonallelic.

epistaxis. *n.* Bleeding from the nose.

episternum. *n.* 1. The cranial segment of the sternum. The cartilaginous part of the sternal apparatus of frogs, lying at the free end of the ossified *omosternum*. See Fig. 3. 2. Also used in reference to the omosternum by various authors. 3. Used synonymously with *interclavicle* of lizards and *entoplastron* of turtles. See Fig. 7.

epistropheus. *n.* Used in older herpetological literature to indicate the *axis*, or second cervical vertebra.

epithalamus. *n.* Part of the *diencephalon* posterior and superior to the *thalamus*.

epithecal. *n.* A term used in reference to *osteoderm* situated in turtle shell, primitively as laminae with bony cores that overlay dermal ossifications, being lost in modern turtles except for the bony mosaic in *Dermochelys*. Cf. *thecal*.

epithelial. *adj.* Relating to, or composing, the *epithelium*.

epithelium. (*pl.* **-ia**) *n.* Reference to tightly-packed, membranous tissue that covers, or lines, an organ, internal body cavity, and organism.

epithet. *n.* In taxonomy, a second word of a binomial name of a species, or the second and third words of a trinomial name of a subspecies.

epitope. *n.* A known structural component of an antigen molecule responsible for its specific interaction with antibody molecules elicited by the same or related antigen.

epizoic. *adj.* Living or growing on the external surfaces of an animal.

epizootic. *adj.* Reference to a disease that suddenly and temporarily affects a large number of animals. An epidemic disease in animals. Cf. *enzootic*.

epizygapophysial spine. A small spur on the posterior edge of the neural arch, immediately above the postzygapophysis. Coined by W. Auffenberg (*Tulane Stud. Zool.* 10:131–216, 1963) in reference to this "spur-like process."

Epoch. *n.* See *geological Epoch*.

eponym. *n.* A term or phrase derived from the name of a person or place. E.g., *Cope's Rule, Darwinian fitness*, etc.

Equatorial current. A warm surface ocean current that flows westward in the tropical Pacific Ocean.

Equatorial zone. A latitudinal zone that extends about 15° on either side of the equator.

equilibrial or **equilibrium community.** See *equilibrium*.

equilibrial or **equilibrium population.** See *equilibrium*.

equilibrium. *n.* 1. In physiology, the lowest energy state of a system in which opposing forces are balanced and no work can be gotten from the system. Cf. *steady state*. 2. In population biology, an *equilibrium population* is one in which the allelic frequencies of its gene pool do not change through successive generations. See *Hardy-Weinberg Law*. Or, reference can also be to a population in which death and emigration rates are balanced by birth and immigration rates. 3. In ecology, an *equilibrium community* is one in which the rate of extinction of species is equal to the rate of immigration of new species. Note, an equilibrium is *stable* if the system returns to its former state following a perturbation. An equilibrium is *unstable* if, when the system is perturbed, it goes to some other state. Note the use of "equilibria" to describe the states of ecosystems is viewed by some as misleading because the word implies a static character that ecosystems do not have. Cf. *dynamic regime*. See *alternative stable states*.

equilibrium model (of island biogeography). A condition that species richness on islands is an equilibrium between ongoing colonization of new species and extinction of resident species. This theory was developed by R.H. MacArthur and E.O. Wilson (*Evolution* 17:373–387, 1963; also *The Theory of Island Biogeography*, 1967).

equinox. *n.* Either of the two occasions each year when the sun crosses the equator, producing day and night of equal duration.

equipotent. *adj.* Having equal potential or effect.

equitability. *n.* A condition of a community or biological assemblage in which there is *evenness* in the distribution of species or their relative abundances. (Syn. *evenness*)

Equitability index. (J′) A measure of evenness, a component of species diversity that reflects the relative abundance of species, calculated as $J' = H'/\log S$, where H' is the *Shannon-Weiner index* of diversity, and S is the number of species. (Syn. *Pielou's evenness*)

Era. *n.* See *geological Era*.

erectile. *adj.* Capable of being raised or erected, as of a crest or penis.

Eremiainae. *n.* A lizard subfamily of the *Lacertidae*.

eremial, eremean. *adj.* Reference to areas or habitat where there is little rainfall and weather and soil conditions create desert-like conditions with absence of scrub or tree cover.

eremic. *adj.* Pertaining to a desert or sandy region.

eremology. *n.* The study of deserts.

eremophilous. *adj.* Thriving in desert regions.

erg. 1. The cgs unit of work, equal to a force of 1 dyne acting over a distance of 1 cm (= 10^{-7} J). 2. A sandy desert.

ericaceous. *adj.* Pertaining to a heath.

erosion. *n.* Weathering or removal of land surface by water, wind, ice, or other agents.

erratum. (*pl.* **–a**) *n.* 1. An error in printing, especially when such an error is noted in a list of corrections and bound into a book. Errata are most commonly issued relatively soon after the original text was printed. Cf. *corrigendum*. 2. Patches to security issues in a computer program are sometimes called errata.

error. *n.* In statistics, the deviation of an obtained value from an expected value, the latter being based on the entire population from which the statistical unit was chosen randomly. The word error, in this context, is a misnomer. See *Type I* and *Type II error*. Cf. *residual*.

eructation. *n.* Release of gas from the stomach via the esophagus.

Erycinae. *n.* A subfamily of *Boidae*, comprising about 14 species of medium-size semifossorial snakes commonly known as Sand Boas.

erythematous. *adj.* Reference to bruising or seepage of blood into surrounding tissues.

erythrism. *n.* Reference to dominance of reddish pigment in an individual or population, compared to normal coloration that is less red.

erythrocyte. *n.* A mature red blood cell. These cells are formed elements that contain hemoglobin and circulate in the blood of all vertebrates (except for certain Antarctic fishes). (Syn. *red blood cell*)

erythropoiesis. *n.* The formation of red blood cells, predominantly in kidney (embryos and adult amphibians), spleen (amphibians, reptiles), and bone marrow (predominantly in reptiles, some amphibians).

erythrophore. *n.* A *lipophore* in which the oil droplets have a red or reddish hue. Cf. *xanthophore*.

escarpment. *n.* A cliff or steep slope produced by erosion or faulting, typically separating two level or less sloped areas.

esculentin. *n.* An antimicrobial peptide isolated from skin secretions of the frog *Rana esculenta*.

escutcheon, escutcheon scales. *n.* Specialized gland-bearing scales in the pre- or peri-cloacal region of Sphaerodactyline and Eublepharine Geckos. Generally differentially pigmented than surrounding scales, they have "pits" on the outer surfaces that house holocrine, secretory material derived from the innermost (last-formed) elements of the modified epidermal generation (lacunar tissue and clear layer). Various taxa have epidermal modifications that are anlagen of functional units, or may be vestigial. The term was first used by C. Grant who described a patch of posterior abdominal scales in some New World gekkonids as forming an "escutcheon" (J. Dep. Agric. Porto Rico 15:199–213, 1931). Subsequent studies identified two distinct types of *generation glands*: *escutcheon scales* and so-called *β-glands*. In the latter, cells forming the glandular material derive from the outermost (first-formed) elements of the modified epidermal generation. Cf. *β-glands*.

esophagus. *n.* The region of the gut or alimentary canal that lies between the mouth and stomach, functioning to conduct food from the headgut to the digestive areas.

esophageal teeth. Term used in reference to the highly specialized hypapophyses of Egg-eating Snakes. These penetrate the esophagus and are capped with enamel for use in breaking egg shells (J.L. Wortman, p. 389 in W.F. Litch, ed., *American System of Dentistry, Vol. 1,* 1886).

esoteric. *adj.* Arising from within an organism or system. Restricted to a small group.

essential amino acids. Amino acids that cannot be synthesized from other components of diet and therefore must be included as an essential part of the diet.

essential element. A chemical element that is essential to the life of an organism.

estimate. *n.* The value of some property of a sample calculated with the intention of inferring the parametric value of that property of a population. (Syn. *estimator, sample statistic*)

estimator. *n.* See *estimate*.

estivation. *n.* See *aestivation*.

estradiol-17β. (E_2) *n.* The most common natural estrogen with the highest biological activity. In amphibians, E_2 is essential for oocyte development and maturation within ovarian follicles and regulates many aspects of reproductive function such as oviduct growth and secretions. In various reptiles, estrogens affect behavioral processes related to sexual receptivity, vitellogenesis and yolk deposition, oviductal growth, synthesis of oviductal secretions, and termination of pregnancy.

estrogens. *n.* A family of sex steroids synthesized primarily in the ovary, but some are produced in the adrenal cortex, brain, and testis. These hormones play a central role in stimulating the female system during reproductive cycles and development of secondary sexual characteristics, function to prepare the reproductive system for fertilization, and are essential for vitellogenesis and oocyte development in all species studied to date. In females, estrogens exhibit a pronounced seasonal cycle coincident with reproductive activity and appear to play a role in the differentiation of the ovary.

estrogen receptors. Receptors sensitive to estrogens that are present prior to sex determination in the genital ridge, higher in density in tissue destined to become an ovary than in tissues destined to become testis. These receptors bind to endogenous steroids and also environmental contaminants including various pesticides and pesticide metabolites having estrogenic activity.

estuarine. *adj.* Reference to *estuary*.

estuary. *n.* A region of brackish water where the sea meets a river and there is admixture of fresh and salt water. A drowned river mouth.

et. Latin, meaning "and;" often used in nomenclature to connect the names of coauthors.

et al. Abbreviation for Latin *et alii*, meaning "and others." This abbreviation is commonly used in author citations involving more than two authors.

Ethiopian region. See *Afrotropic*.

ethmoid. *adj., n.* A term used in reference to the skull region anterior to the *sphenethmoid* of amphibians.

ethmoid plate. See *planum basale*.

ethocline. *n.* A cline or graded series in the expression of a behavior within a group of related species.

ethogram. *n.* A list or abbreviated description of a behavioral pattern in a species.

ethological isolation. See *behavioral isolation*.

ethology. *n.* (*adj.* –**ical**) The study of animal behavior in contexts of adaptation and natural conditions.

ethospecies. *n.* Reference to species that are distinguished primarily by behavioral traits.

etiology. *n.* The cause or study of causes, especially in reference to diseases.

etymology. *n.* The study of the history of words; how their form and meaning have changed over time.

euacrodont. *adj.* See *acrodont*.

euautosystyly. *n.* A condition of *autosystyly* in which only part of the quadrate is fused to the cranium by means of special processes in urodeles and anurans.

eublepharid, eublepharine. *n., adj.* Reference to, or a member of, the gekkonid subfamily *Eublepharinae*.

Eublepharinae. *n.* A lizard subfamily of *Gekkonidae*.

Euclidean Distance Matrix Analysis. (EDMA) A method used in morphological studies to analyze shape differences using three-dimensional coordinate data, comparing landmarks in a coordinate-free environment. See S. Lele, *Mathematical Geology* 25:573–602, 1993.

Euclidean space. Three-dimensional space.

eucolumella. *n.* Reference to the *columella* of salamanders and caecilians, which is derived from the hyomandibular arch (H.M. Smith, *Evolution of Chordate Structure*, p. 214, 1960). Cf. *paracolumella, pseudocolumella*.

eucryptic. *adj.* Reference to a species that mimics the coloration of the substratum or surroundings, and not other species or structures.

eugenics. *n.* The science of breeding, especially with application to improvement of certain hereditary qualities of a race or species.

eukaryotes. *n.* One of three primary kingdoms of living organisms. These include all organisms consisting of one or more cells having compartmentalized internal structure and true nuclei bounded by nuclear envelopes.

eulittoral. *adj.* The zone between the highest and lowest seasonal water levels, or highest and lowest tidal reaches. Equivalent to the *intertidal* zone.

eumelanins. *n.* A group of black and brown pigments found in *melanophores* and *pterorhodophores*.

Eumetazoa. *n.* A subdivision of the animal kingdom including organisms that possess organ systems, a mouth, and digestive cavity.

euphotic. *adj.* That part of the pelagic zone of a body of water wherein light is sufficient to allow photosynthesis. Cf. *aphotic*.

eupleurodont. *adj.* A condition in *pleurodont* species where new teeth lie in a vertical row beneath functional teeth and move into empty sockets when the outer teeth are lost (a condition known as "*vertical replacement*").

eupnea. *n.* A term for normal breathing.

Eurasian Adders. Collective vernacular name for Eurasian species of vipers belonging to the genus *Vipera*. Also called *vipers*.

Eureptilia. *n.* One of two major sauropsid lineages, which includes the three major diapsid lineages: *Lepidosauromorpha, Archosauromorpha,* and *Euryapsida*. Cf. *Parareptilia*.

European Common Frog. Vernacular name for the ranid species *Rana temporaria*.

European Toad. See *Common European Toad*.

European Treefrog. Vernacular name for the hylid frog *Hyla arborea*.

eury-. Prefix meaning "wide." Cf. *steno-*.

euryapsid. *n., adj.* A type of skull, derived within *diapsids* and having a single pair of temporal openings, but the squamosal-postorbital arch forms the lower border of the paired openings. Two extinct groups of Mesozoid marine reptiles, the *plesiosaurs* and *ichthyosaurs*, possessed such a skull. Cf. *parapsid*.

Euryapsida. *n.* A group of extinct marine reptiles—*plesiosaurs* and *ichthyosaurs*—thought to be diapsid derivatives having a single temporal opening high on the skull.

eurybathic. *adj.* Tolerant of a wide range of depth. Cf. *stenobathic*.

eurycoenose. *adj.* Common or having a wide distribution. Cf. *stenocoenose*.

euryhaline. *adj.* Capable of living in a comparatively wide range of salinities. Cf. *stenohaline*.

euryhydric. *adj.* Tolerant of a wide range of moisture conditions. Cf. *stenohydric*.

euryphagic, europhagous. *adj.* Reference to animals that accept a relatively wide range of food items as prey. Cf. *stenophagic*.

euryplastic. *adj.* Characterized by breadth of phenotypic plasticity or developmental response to variation in environment. Cf. *stenoplastic*.

eurythermal, eurythermic. *adj.* Capable of tolerating a comparatively wide range of temperatures. Cf. *stenothermal*.

eurytopic. *adj.* Capable of living in a wide range of habitats and/or having a wide geographic distribution. Cf. *stenotopic*.

eustachian tube. *n.* A tube connecting the nasopharynx with the middle ear chamber, allowing transmission of vibrational energy from liquid of the inner ear to air. Found in anurans and most reptiles, while absent in all caecilians, salamanders, turtles, and snakes. (Syn. *internal auditory meatus*)

eustatic. *adj.* 1. Pertaining to global changes in sea level, such as might be attributed to formation or melting of glaciers, but excluding those due to subsidence or uplift of coastlines. 2. Changing very gradually and slowly.

euthanasia. *n.* The act or means of killing an animal in a humane manner. This is a legalized requirement for terminal experiments on vertebrates in most countries having animal welfare legislation. Painless death.

eutrophication. *n.* The process of nutrient enrichment in water, often related to runoff of nitrogen and phosphorus that stimulates algal blooms and growth of bacteria, which, in turn, contribute to depletion of oxygen in water.

eV. See *electron volt*.

evagination. *n.* An outpocketing.

Evans blue. A green or blue-green dye injected intravenously to determine blood volume and movement. Also called *T-1824*.

evaporation. *n.* Conversion of liquid water to vapor, and loss of moisture therefrom.

evaporative water loss. (EWL) Loss of water from an organism or other surface due to evaporation. In physiological and herpetological literature, cutaneous evaporative water loss from an animal is often represented as *CWL, CEWL* (= cutaneous evaporative water loss), or *TEWL* (= transepidermal evaporative water loss). See *TEWL* for alternate meaning.

evapotranspiration. *n.* The total water loss from the environment, including transpiration from the surface of plants and direct evaporation from plants, soil, and other objects.

evenness. *n.* See *species diversity, Equitability index*.

eversible. *adj.* Capable of being everted (turned inside out).

evo-devo. *n.* A new label now given to *evolutionary developmental genetics*, involving investigations of the evolution of major control genes in development.

evolution. *n.* 1. Generally, change through time. 2. In biological meaning, evolution is the change through time in the frequencies of different forms of a gene in the gene pool of a population. It results in a cumulative, heritable change in a population. See *anagenesis, cladogenesis, genetic drift, gradualism, natural selection, orthogenesis, punctuated equilibrium*.

evolutionary biology. An integrated science that incorporates data from several disciplines including systematics, ecology, and genetics to understand evolutionary phenomena.

evolutionary clock. See *molecular clock*.

evolutionary conservatism. Reference to preservation of common ancestral similarity in diverging species due to retention of a relatively high proportion of common ancestral alleles.

evolutionary developmental genetics. See *evo-devo*.

evolutionary game theory. Theory employed in mathematical models of species competition, relating to optimum strategies that depend on the more likely behaviors of two or more competitors. The theory marries the application of game theory with models of changes in gene frequencies inspired from *population genetics*. However, it differs from classical game theory by focusing on the dynamics of strategy changes moreso than properties of strategy equilibria. See *strategy*.

evolutionary method. A method of classification that employs hypothetical reconstructions of evolutionary history incorporating cladistic data for sequential branching events with morphological divergence. Cf. *cladistics, omnispective method, phenetic method*.

evolutionary morphology. A subdiscipline that investigates the causes and patterns of morphological changes through time, using morphological characters and the application of comparative and phylogenetic methods.

evolutionary novelty. See *novelty*.

evolutionary radiation. See *adaptive radiation*.

evolutionary rate. The rate of evolutionary change. Slow evolution is called *bradytelic*, rapid evolution is called *tachytelic*, and "average" rate of evolution is called *horotelic*.

evolutionary physiology. A subdiscipline that uses comparative phylogenetic methods to understand the evolution of animal function from a physiological viewpoint using physiological markers or characters.

evolutionary significant unit. (ESU) This concept is a response to how to define a species as distinct for purposes of conservation under the U.S. *Endangered Species Act* (R.S. Waples, *Act. Mar. Fish. Rev.,* 1991). The population must be reproductively isolated; occupies a distinct or unusual habitat; displays distinctive adaptation to the environment, and must represent an important component in the evolutionary legacy of a species. (Syn. *wildlife species*, a term used by COSEWIC, Committee on the Status of Endangered Wildlife in Canada)

evolutionary species concept. See *species*.

evolutionary stable strategy. (ESS) See *strategy*.

evolutionary synthesis. See *neo-Darwinism*.

evolutionary tree. See *phylogenetic tree*.

evolutionary trend. A gradual, adaptive change in the evolution of a character within a phyletic line.

evolve. *v.* To change over time. See *evolution*.

ex. Latin, meaning "from" or "according to."

ex-. Prefix meaning "out" or "beyond."

Ex. See *exploitation index*.

exanal. *n.* Any of the scales bordering the anus of the snake *Xenopeltis* (B.C. Mahendra, *Current Science* 6:559–560, 1938).

exaptation. *n.* A character that confers a selective advantage under current conditions, but originally had a different function and was not shaped by natural selection for the current role. E.g., parts of the reptilian jaw evolved to become the middle ear bones of mammals. The term *preadaptation* is also used for structures that were present earlier and then switched function in response to altered selective pressures. However, exaptation is the preferred term, as it does not suggest foresight or preplanning for the ultimate use of the structure. See *adaptation*. Cf. *nonaptation* (S.J. Gould & E.S. Vrba, *Paleobiology* 8:4–15, 1982).

excess post-exercise oxygen consumption. (EPOC) Reference to elevated metabolic rate above resting levels for extended periods during recovery from exercise. Cf. *oxygen debt*.

excision. *n.* (*v.* **excise**) Surgical removal of tissues.

excitability, excitable membrane. *n.* 1. In neurophysiology, the capacity for altered membrane conductance (and often membrane potential) in response to stimulation. The ability of a membrane to conduct an *action potential*. 2. The capacity of a living organism to respond to a stimulus.

excitation-contraction coupling. The process by which neural or electrical excitation of a muscle membrane leads to activation of the contractile process in muscle fibers.

excitement color reaction. Rapid color change in response to an external stimulus. The change may be described as *excitement darkening* or *excitement pallor*.

excitement secretion reaction. The swift and liberal secretion of skin mucus by certain amphibians in response to external stimuli.

excl. Abbreviation for the Latin *exclusus*, meaning "excluded." In taxonomy, this abbreviation is used to indicate components that are included in a given taxon by one author, but considered not to belong to that taxon by the present author (and thus excluded).

exclusion principle. Reference to theory that two species cannot coexist indefinitely in the same locality if they have identical ecological requirements. See *competitive exclusion principle*.

exclusive species. A species virtually totally confined to a particular community.

excrescence. *n.* Any natural outgrowth of the epidermis such as nuptial pads, tentacle-like extensions of the head, etc.

excreta. (U) *n.* 1. Waste matter eliminated from the body, especially urine and feces. 2. In ecological energetics, that part of assimilated energy that is eliminated from the body as excretion, secretion, or exudation (a component of *rejecta*).

excretion. *n.* The removal of metabolic waste products or contaminants, usually in the process of urine formation by the kidney.

excretory organ. Any organ that excretes wastes, excess metabolites, or ions from the body (e.g., *kidney, salt gland*).

exemplar. *n.* A random sample of a taxon.

exendin. *n.* A novel peptide family found exclusively in the venom of helodermatid lizards, produced in the salivary glands and related to *glucagon* hormones.

exergonic. *adj.* Characterized by a concomitant release of energy, often accompanied by a release of heat. Cf. *endergonic*.

exfoliation. *n.* 1. The act of molting or skin shedding. See *ecdysis*. 2. In geology, the stripping of concentric rock slabs from the outer surface of a rock mass due to a weathering process of chipping or flaking.

exfoliation dome. A large, rounded landform developed in a massive rock, such as granite, by the process of exfoliation.

exhaustive sampling. In statistics, reference to examination of the entire population.

exoccipital. *n.* Either of the pair of bones that lie on either side of the *foramen magnum* and derived from cartilage of the chondrocranium. (Syn. *lateral occipital, oto-occipital, parotic, pleuroccipital*)

exocrine secretion. Pertaining to organs or structures that secrete substances onto the surfaces of organs or the body via a duct. Cf. *endocrine secretion, autocrine secretion, paracrine secretion*.

exocrine gland. A gland that secretes a fluid via a duct. Cf. *endocrine gland*.

exocytosis. *n.* A process by which cellular products are expelled to the exterior of a cell by fusing vesicles with the plasma membrane, which subsequently expels the contents.

exogamy. *n.* The selection of mates by mating pairs that tend to be nonrelatives. Cf. *endogamy*.

exogenous. *adj.* 1. Reference to influential factors external to an animal, such as temperature, humidity, etc. 2. Originating from outside a system. (Syn. *allochthonous*)

exogenous DNA. DNA that originates outside an organism (e.g. from another cell or virus).

exopthalmos. *n.* (*adj.* **–ic**) Abnormal and excess protrusion of the eye.

exoskeleton. *n.* Hard, outer supportive or protective materials other than bone or cartilage, e.g. osteoderms, laminae, scales, etc., located close to, or on, the outside of the body. Cf. *endoskeleton*.

exostosis. *n.* (*adj.* **exostosed**) 1. A spur or broader bony outgrowth from a bone or root of a tooth, including surface ossifications that increase with age. 2. The rough, complexly patterned surfaces on the roof of the skull in some anurans, which may or may not fuse with skin.

exothermic. *adj.* Characterized by a release of heat.

exotic. *adj.* Not native or indigenous to a particular region. The term applies to many introduced or invasive species that become established in new environments, often as result of human introduction, and is sometimes used as a noun.

exotic species. See *exotic*.

exotrophic. *adj.* Having a free-living, feeding larval stage.

exoviviparous. *adj.* A reproductive mode in which the embryo develops using oogenic energy sources in a terrestrial egg before the hatchling moves to a site usually in or on a parent's body. See R. Altig & G.F. Johnston, *Smithsonian Herpetol. Info. Serv.* 67:1–75, 1989.

expected value. In statistics, the expected value of a *random variable* indicates the average or central value, estimated by the population *mean*.

experimental error. 1. Uncontrolled variation in an experiment. 2. Chance deviation of observed results from those

experimental group or **unit.** The grouping or subjects to which an experimental treatment is applied in an experiment. Cf. *control.*

expiration. *n.* The act of exhaling air from the lung. Cf. *inspiration.*

explant. *n.* An excised part from a tissue or organ used to initiate a culture *in vitro.*

exploitation competition. Competition that occurs because species use a shared resource that is in limited supply, and one species exploits the resource more efficiently than another. Cf. *interference competition.*

exploitation index. (Ex) *n.* An index of thermoregulation in which the amount of time an ectotherm spends in the setpoint range is expressed as a percentage of the time available for it to achieve body temperatures within the setpoint range for a given season. This is an index of the extent to which the animal exploits the available thermal environment. Term introduced by K.A. Christian & B.W. Weavers (*Ecol. Monogr.* 66:139–157, 1996). See also R.B. Huey et al., *Ecology* 70:931–944, 1989.

explosive breeders. Species having relatively short reproductive periods of a few days, with the result that a large number of animals in a local population mate at the same time.

explosive evolution or **radiation.** The splitting of a taxonomic unit into multiple lines of descent within a relatively short period of geological time.

exponential growth phase. See *logarithmic growth phase.*

exponential population growth. A simple model of population growth in which the population growth rate (dN/dt) is the product of the current population size (N) and the instantaneous rate of increase (r). Exponential growth implies no limit to population size and an accelerating rate of population growth. No population in nature exhibits exponential growth for very long; all populations have the potential to increase exponentially because each individual can leave more than one offspring in the next generation. The model of exponential growth is the foundation for most modeling efforts in population and community ecology. See *logistic equation.*

expression. *n.* See *gene expression.*

expressivity. *n.* Reference to the range of phenotypes that are expressed by a given genotype in relation to a specified range of environmental conditions.

exsanguination. *n.* Removal of blood, or total bleeding by severing the blood vessels.

extant. *adj.* Still in existence; currently living. Cf. *extinct.*

extant phylogenetic bracket. A method devised to assess nonskeletal features of extinct forms by consideration of occurrence of a trait in their closest living relatives (L.M. Witmer, pp. 19–33 in J.J. Thompson, ed., *Functional Morphology in Vertebrate Paleontology*, 1995).

extensile. *adj.* Capable of being extended. (Syn. *extensible*)

extensible. *adj.* See *extensile.*

extension. *n.* The opening of a joint, or movement that straightens the limb across a joint.

extensor. *n.* A muscle that opens a joint when it contracts, thereby extending, stretching, or straightening a limb or other extremity. Cf. *flexor.*

external. *adj.* Outside or away from the center of an object.

external auditory meatus. The external canal of the ear, usually extending to the eardrum. (Syn. *auditory canal, external ear*)

external carotid artery. *n.* An artery that branches from the common carotid artery and supplies blood to the anterior neck, floor of mouth, and facial tissues. Cf. *internal carotid artery*. See also *carotid artery.*

external development. *n.* Development of the embryo outside the body of the mother, as in many amphibians.

external ear. The external ear structures, usually consisting of a short canal (*external auditory meatus*) which opens to the surface through an external orifice. The external ear is present in some reptiles such as lizards and crocodilians, but is absent in amphibians. See Fig. 16.

external fertilization. *n.* Fertilization of the egg outside the body of the mother (as in many anurans).

external gill. Protruding respiratory structures on either side of the neck of larval amphibians. More generally, any specialized extension of the skin that functions in the aquatic exchange of respiratory gases (e.g., the "hairs" of the *African Hairy Frog*).

external naris. (*pl.* **-es**) Nasal openings on the outside of the body. Cf. *internal nares.*

external pterygoid. See *ectopterygoid.*

external respiration. Reference to exchange of oxygen and carbon dioxide at an animal's *gas exchanger*, in distinction to cellular *respiration*. See *respiration.*

exteroceptor. *n.* A sense organ that detects stimuli arriving at the body surface from a distance. Such a receptor responds to environmental stimuli. Cf. *proprioceptor, interoceptor.*

extinct. *adj.* Not in existence in a particular area; no longer living. Cf. *extant.*

extinction. *n.* Death or disappearance of a species or taxonomic group, regionally over a particular area or worldwide.

extinction coefficient. A measure of the attenuation of light within a water column, or within a plant canopy.

extinction curve. A graph of the total number of species in a given habitat or area against either time or the extinction rate.

extinction rate. The number of species in a given habitat or area that become extinct per unit time. In metapopulation models, the extinction rate is the proportion of sites occupied by populations that go extinct per unit time.

extirpation. *n.* 1. Surgical removal of a part. 2. Extermination or removal of a population of a species from a given area. Local extinction in contrast with global extinction of a species.

extracapsular perilymphatic sinus. A sinus of perilymphatic fluid located outside the auditory capsule in snakes and turtles.

extrabrillar fringe. *n.* A projecting border around the eye in some lizards, especially prominent in certain Geckos. Used by E.H. Taylor (*U. Kans. Sci. Bull.* 38:3–322, 1956).

extrachromosomal inheritance. See *extranuclear inheritance*.

extracolumella. *n.* A small bone located between the *columella* and *tympanic membrane* in certain amphibians and reptiles. This element forms a cartilaginous or bony structure lying across the tympanum, or, when the tympanum is absent, an enlarged connection to parts of the skull, which act to conduct sound. (Syn. *extrastapedial, extra-stapes* and inappropriately *malleus*)

extractive reserve. A wild habitat from which natural products are harvested on a sustainable basis and without extinction of native species or excessive harm to the environment.

extraembryonic. *adj.* Reference to structure formed by or around the embryo, but not retained by the adult body form.

extraembryonic membranes. Structures formed around the embryo of amniotes, which eliminate wastes (*allantois*), provide food (*yolk sac*) and oxygen, and protect the embryo (*amnion* and *chorion*).

extrafusal fiber. A contractile muscle fiber that makes up the bulk of skeletal muscle. Cf. *intrafusal fiber*.

extranuclear inheritance. Non-Mendelian heredity attributable to DNA in mitochondria. (Syn. *extrachromosomal inheritance, cytoplasmic inheritance, maternal inheritance*)

extraperitoneal. *adj.* Positioned in the body wall beneath the lining of the coelom, or peritoneum, in contrast to being suspended in the coelom within mesenteries.

extrapolation. *n.* The process of estimating a value beyond the range of actual values, usually based on an equation obtained from the actual values.

extraskeletal bones. See *heterotopic*.

extrastapedial. *n.* See *extracolumella*.

extra-stapes. *n.* See *extracolumella*.

extratropical. *adj.* Outside the tropics.

extravasation. *n.* The escape of blood from a vessel into surrounding tissues.

extrinsic. *adj.* Originating outside the body or part on which it acts. From the environment. Cf. *intrinsic*.

extrinsic isolating mechanism. Any environmental barrier that isolates potentially interbreeding populations. Cf. *intrinsic isolating mechanism*.

exude. *v.* To ooze or diffuse out.

exudate. *n.* Any exuded or discharged substance, e.g. fluids or cellular debris from wound sites.

exuvia. (*pl.* **-ae**) *n.* Any part of an animal that is shed or cast off. A shed skin.

exuviation. *n.* (*v.* **exuviate**) The act of shedding or *ecdysis*.

eye. *n.* An organ of visual reception that includes parts specialized for optical processing of light as well as photoreceptive neurons.

eye cap. See *spectacle*.

Eyelash Pitviper, Eyelash Palm Pitviper. Vernacular name for the neotropical species *Bothriechis schlegelii*.

Eyelash Sea Snake. See *Peron's sea snake*.

eyelid fringe scales. The series of outermost scales on the margins of the eyelids.

Eyelid Geckos. Vernacular name for Asian and Middle Eastern lizards belonging to the gekkonid genus *Eublepharis*.

Eyelid Skinks. Vernacular name for various species of Skinks belonging to the genus *Eumeces*.

eye shield. A scale covering of a rudimentary eye, as the *ocular* of Blind Snakes.

eye spot. A rounded area of contrasting color, but especially used in reference to *ocelli* on the rump or groin areas of anurans and displayed in defensive behaviors. In some species a raised knob or glandular area may comprise the center of the eye spot.

eye stripe. A line or band of color that passes "through" the eye of animal.

F

F, °F. See *Fahrenheit scale (°F)*.

F_1. The offspring that result from a first experimental crossing of an animal. The parental generation that begins a genetic experiment is called P_1.

F_2. The progeny produced by intercrossing F_1 individuals.

face-off posture. An aspect of lizard display behavior in which usually two males present their flanks to one another while facing in opposite directions. The posture may be accompanied by compression or inflation of the body.

facet. *n.* 1. A smooth flat or rounded surface that forms part of the articular surface of a joint. 2. A surface that forms on a tooth as a result of wear from abrasion with another tooth.

facial. *adj.* Pertaining to the face.

facial length. A measurement defined by S.B. McDowell and C.M. Bogert (*Bull. Amer. Mus. Nat. Hist.* 105:5–142, 1954) as the distance from the anteriormost point of the premaxilla to the frontoparietal suture.

facial nerve. The seventh cranial nerve, which innervates much of the mouth, including taste buds, and conducts sensory as well as motor impulses.

facial pit. See *pit organ*.

facilitated diffusion. Reference to diffusional transport of a solute across a membrane and down its concentration gradient, promoted by a passive carrier system that increases membrane permeability.

facilitation. *n.* 1. In cellular physiology, a phenomenon whereby activation of a synapse causes an increase in the efficacy of the synapse during subsequent stimulation. 2. A model of ecological succession in which each group of species that enters a patch alters the environment in a way that facilitates the entry of successive sets of species. The endpoint of the classic facilitation model is a self-replacing *climax community*. 3. *Social facilitation* refers to enhancement of behavior of an animal due to the presence or behaviors of other individuals.

factor. *n.* 1. Any causal agent. 2. In statistics, any variable considered to influence the variable under investigation.

factorial. (!) *n.* For a given positive integer N, N! is defined to be the product of the consecutive integers from 1 to N. e.g., $3! = 3 \times 2 \times 1 = 6$.

factorial analysis. In statistics, a linear combination of variables that are ranked according to each one's influence on variance observed in the data. The analysis is used for resolving complex relationships by isolating and ranking causal factors.

factorial scope for activity or **locomotion.** See *aerobic metabolic scope*.

facultative. *adj.* 1. Not obligatory. Possessing flexibility to function or live under more than one set of environmental conditions, or to express a characteristic capable of change according to environmental demands. 2. A behavior that may or may not occur, depending on circumstances. Cf. *obligate*.

Fahrenheit scale. (°F) A scale of temperature that specifies the freezing and boiling points of water at 32° and 212°, respectively (at sea level). Cf. *Celsius scale*.

failure. n. In biomechanics, the loss of functional integrity or ability to perform. A material, for example, may fail but not break. Cf. *fracture*.

falcate. *adj.* Hooked or curved, as a claw or limbs of sea turtles. (Syn. *falciform*)

falciform. *adj.* See *falcate*.

False Boas. Vernacular name for several species of neotropical xenodontine snakes belonging to the genus *Pseudoboa*.

False Brook Salamanders. Collective vernacular name for species of plethodontid salamanders belonging to the genus *Pseudoeurycea*.

false crawl. The tracks left by a sea turtle that has ascended a beach but returned to the sea without laying eggs. Cf. *crawl*.

False Cobra. Vernacular name for the Asian snake genus *Pseudoxenodon*.

False Coral Snakes. Vernacular name for South American species of xenodontine species of snakes belonging to the genus *Erythrolamprus*.

False Gharial. An unusual species of crocodilian, *Tomistoma schlegelii*, previously and morphologically classified in the family Crocodylidae; however, recent immunological studies have placed it in the family Gavialidae. (Syn. *Malayan Gharial*)

False or **Mimic Chameleons.** Vernacular name for species of iguanid lizards belonging to the genera *Chamaeleolis* and *Chamaelinorops*.

False Monitors. Vernacular name for species of neotropical teiid lizard belonging to the genus *Callopistes*.

False Pitvipers. Vernacular name for species of colubrid snakes belonging to the American colubrid genera *Waglerophis* and *Xenodon*.

false rib. A term for ribs that articulate with each other but not with the *sternum*. Cf. *floating rib*.

false roof. With reference to turtles, the secondary protective covering of bone wherein the temporal region consists of a chamber formed of the parietals, postfrontals, squamosals, quadrato-jugals, and jugals.

False Water Cobras. Vernacular name for large, aquatic South American xenodontine snakes of the genus *Hydrodynastes*. These snakes spread a hood in a defensive manner similar to that of Old World Cobras, but do not generally elevate the body when doing so. These snakes have a mildly toxic saliva that has variable effects on humans. (Also called *False Water Snakes*.)

False Water Snakes. See *False Water Cobras*.

family. *n.* In systematics and taxonomy, a group of related genera comprising a taxonomic level above *genus* and below *order*.

family name. The scientific name of a taxon at family rank, usually ending in *–idae* if the taxa are animals.

family tree. A diagrammatic representation of the lineage of a family or taxonomic group.

fan. *n.* 1. The elevated and compressed crest on the posterior part of the head in *Basiliscus* lizards. 2. Sometimes used in reference to a *dewlap*. See Fig. 40.

fanning. *vt.* Reference in herpetological literature to periodic movements of external gills or tails in aquatic amphibians, generally in contexts of gas exchange or sexual communication respectively.

fang. *n.* A specialized tooth used to conduct venom. Typically, a fang may be any of hollow, grooved, or elongated teeth located on the maxilla of the upper jaw of snakes.

farad. (F) The unit of electrical capacitance.

Farallon plate. A tectonic plate of the eastern Pacific Ocean from 50 to 25 Ma that was entirely subducted under the North American plate.

far infrared, or **far red light.** Electromagnetic radiation in the spectral range, usually regarded as between 3–10 and 1000 μm. See also *thermal radiation*.

farming. *n.* The practice of culturing sea turtles, crocodilians, or other amphibians or reptiles in a closed-cycle system for commercial purposes. Cf. *ranching*.

fascia. (*pl. –iae*) *n.* A band of connective tissue that forms a sheath, as around a muscle or blood vessel.

fascicle. *n.* 1. A small bundle of *axons* within a nerve, or a bundle of muscle fibers defined by a connective tissue sheath within a muscle. 2. In herpetological literature this term has been used in reference to the bundle of vomerine teeth in anurans.

fasciculation. *n.* 1. The formation of fascicles. 2. Localized, small involuntary muscle contractions observable beneath the skin.

fasciculins. *n.* Cholinesterase inhibitors that prevent the destruction of acetylcholine in the synaptic gap, first isolated from venom of the Mamba, *Dendroaspis*.

fasciotomy. *n.* A surgical procedure to incise *fascia* with the aim of reducing or preventing a deleterious buildup of pressure within enveloping fascia.

fast twitch fiber. See *twitch fibers*.

fat. *n.* 1. A glycerol ester with fatty acids. 2. *Adipose tissue*. 3. See also *lipids*.

fat body. Reference to masses of *adipose tissue* in the body cavity of amphibians and reptiles. These are generally assumed to function with respect to energy storage or production of sexual products. (Syn. *corpora adiposa*)

fathom. *n.* A fps unit of length used especially to measure depth of water. One fathom equals 6 ft or 1.8288 m.

fatigue fracture. Reference to a reduced breaking strength of an object following prolonged use.

fatty acid. An acid present in lipids, consisting of a carboxylic acid group and a long hydrocarbon chain, typically linear and having variable carbon content from C_2 to C_{34}.

fault-block mountain range. A range created by uplift along normal or vertical faults.

fauna. *n.* 1. Reference to the sum total of animal life in a given region. 2. Animal life in general, as distinct from plants. See *herpetofauna*.

faunal element. A species of animal that is a distinguishing element of a region.

faunal extinction. See *mass extinction*.

faunal province. A zoogeographical subregion containing a distinct fauna that is more or less isolated from others by migration barriers.

faunal region. A region that supports a characteristic fauna. See *zoogeographic realm*.

faunistics. *n.* Reference to study of a *fauna* in a particular region or locality.

faveolus. (*pl.* **-i**) *n.* A unit of reptilian lung parenchyma, representing a compartment with trabecular walls or an individual respiratory compartment or unit of the "honeycomb" within the lung. The term is from H-R Duncker (pp. 2–15 in J. Piiper, ed., *Respiratory Function in Birds, Adult and Embryonic*, Springer-Verlag, Berlin, 1978) who pointed out that units of reptilian parenchyma are not homologous with mammalian alveoli. The term faveoli is used if the air exchange chambers are deeper than they are wide. Cf. *ediculae*. See also *faviform*. See Fig. 41.

faviform. *adj.* Reference to terminal respiratory compartments of nonmammalian vertebrates in which the air spaces are subdivided by secondary and tertiary septa. This term should be used to distinguish such vertebrate lungs from the alveolar lungs of mammals. See also *faveolus*.

Fea's Viper. Vernacular name for *Azemiops feae*, an unusual species of Viper once thought to be an elapid and currently thought to be closely related to Pitvipers. This species is known only from mountainous regions of Myamar (Burma), southern China, and Vietnam.

feather. *n.* A specialized *epidermal appendage* and unique diagnostic character of *birds*.

Feathered Bush Viper. Vernacular name for the equatorial

African Viper *Atheris hispidus*, named for the elaborately keeled scales on the anterior body.

feces. *n.* Undigested material and bacteria eliminated from the hindgut.

fecund. *adj.* Capability for reproduction or producing large numbers of eggs.

fecundity. *n.* 1. The quality of producing offspring, often used in reference to the numbers of offspring. 2. Potential fertility or reproductive capacity, specifically the quantity of gametes, generally eggs, produced per individual over a specified period of time.

fecundity ratio. The ratio of pregnant females in a population to the total number of females present.

fecundity schedule. A table that gives the average number of offspring born per unit time to individual females of age x.

federally endangered. Reference to any species within the United States that is in danger of extinction throughout all or a significant part of its range.

federally threatened. Reference to any species within the United States that is likely to become an endangered species within the foreseeable future throughout all or a significant part of its range.

feedback, feedback control. *n.* In control systems, the return of output or information about the status of a controlled variable to the input part of a control system. The influence of the result of a process on the functioning of the process. In *negative feedback*, the sign of the output is inverted before it is fed back to the input so as to stabilize the output. In *positive feedback*, the output is unstable because it is returned to the input without a sign inversion. Thus, it becomes self-reinforcing or regenerative.

feedback inhibition. See *end product inhibition, feedback*.

feeding rate. The rate at which individual predators capture prey (units are victims/[time · predator]).

fell. *n.* A bare rocky hillside or mountain slope.

female. *n.* The egg producing form of a bisexual organism. Symbolized ♀. See *female symbol*.

female symbol. (♀) The zodiac sign for Venus, the goddess of love and beauty in Roman mythology.

femoral. *adj., n.* 1. Of or pertaining to the thigh or femur. 2. Scales on the femur of lizards. 3. Either of a pair of posteriorad plates or laminae on the plastron of turtles, positioned between the *abdominal* and *anal* plates. See Fig. 6.

femoral gland. A gland situated on an animal's thigh. Usually reference is to a hypertrophied, glandular region on the ventral side of the thigh in certain male anurans or to *preanal organs* of lizards. Cf. *precloacal gland*.

femoral organ. See *preanal gland, femoral gland, femoral pore*.

femoral pore. Small canal openings found on enlarged scales on the undersides of the femur in various lizards, usually better developed in males than in females. The pores are plugged with a wax-like substance containing cellular debris and, in breeding males, project from the openings to form a *comb*. See Fig. 17.

femur. (*pl.* **femora**) *n.* The proximal bone or element of the hind leg (= thigh).

femoral valve. A soft flap of skin on the margin of the plastron of the Softshell Turtle *Lissemys*. When compressed, this tissue presses against the carapace and hides the hind leg. (Syn. *crural valve*)

Fence lizard. Vernacular name for the American iguanid species of *Sceloporus undulatus*. See also *Western Fence Lizard*.

fenestra. (*pl.* **-ae**) *n.* A large opening or window, usually within or between bones. Cf. *emargination, foramen, fossa*.

fenestra ovale or *ovalis*. The opening through the otic capsule between the middle ear cavity and the inner ear and point through which mechanical vibrations are transmitted to the sensory hair cells involved in hearing. (Syn. *oval window, fenestra vestibuli*)

fenestra rotunda. See *round window*.

fenestration. *n.* An opening in the surface of a bone.

fenestra vestibuli. See *fenestra ovalis*.

feral. *adj.* Wild and not domesticated.

Fer-de-Lance. *n.* Vernacular name for several neotropical crotaline species: *Bothrops asper, B. atrox, B. caribbaeus,* and *B. lanceolatus*. Some authors prefer the name be restricted to the species found on Martinique, *B. lanceolatus*. The name is French (arguably Créole) for "spearhead" or "iron of the lance." See *Lancehead, Terciopelo, Barba Amarilla*.

fermentation. *n.* The process by which symbiotic microorganisms extract energy anaerobically by releasing cellulase enzymes that break down plant materials in the guts of vertebrates. Among amphibians and reptiles, fermentation takes place either in the stomach (*gastric* or *foregut fermentation*, e.g. tadpoles) or in specialized parts of the intestine (*intestinal* or *hindgut fermentation*, e.g. iguanine lizards, some turtles).

ferruginous. *adj.* Having the color of iron rust; reddish-brown.

fertility. *n.* 1. The ability to produce viable eggs or sufficient viable sperm. The capacity to conceive or induce conception. 2. In population ecology, the number of female offspring produced by females of age class i.

fertility ratio. The number of offspring in a population expressed as a ratio of the number of adult females.

fertilization. *n.* The union of two *gametes*, male and female, to produce a diploid (2N) *zygote*. *External fertilization* takes place outside the female's body, whereas in *internal fertilization* the sperm are introduced into the female's body.

fertilization membrane. *n.* See *vitelline membrane*.

festoon. *n.* Reference to the series of pointed scales with free margins encircling the border of the eyelid in the lizard genus *Uma*.

fetus. *n.* (*adj.* **fetal**) The unborn young of a *viviparous* animal in the later stages of development.

fever. *n.* See *behavioral fever*.

Feyliniinae. *n.* A lizard subfamily of the *Scincidae*.

fiber optics. Transmission of light through a bundle of very fine flexible glass or plastic fibers. The technology was developed in the late 1960s and early 1970s and has numerous applications in biomedical, physiological, and anatomical research and in communications.

-fibrase. Suffix meaning "fibrinogen digesting," used in a nomenclatural scheme for describing exogenous hemostatic factors in snake venoms. By adding a portion or designated abbreviation of a snake's scientific name to the suffix, one obtains a designation for the fraction being identified. For example, using the name "*gabonica*" (species name for the Gaboon Viper) one obtains *gabonifibrase*.

fibril. *n.* A small fiber or threadlike part of a fiber.

fibrilla. *n.* The thin, hairlike projections forming the functional exhange surfaces on the gills of salamanders. See *fimbria*.

fibrillar. *adj.* Of, or like, small fibers or fibrils.

fibrillation. *n.* Repeated short muscular contraction of low amplitude.

fibrin. *n.* A fibrous protein derived from *fibrinogen* by action of thrombin, and a principal component of blood clot formation.

fibrinogen. *n.* A protein element of blood plasma, converted to *fibrin* by the enzymatic action of thrombin.

firbrinogenolysis. *n.* (*adj.* **–lytic**) The proteolytic destruction of *fibrinogen* in circulating blood.

fibrinolysins. *n.* (*adj.* **-lytic**) Substances that break down blood clots and thereby promote bleeding. Enzymatic dissolution of *fibrin*.

fibrinolytic index. (FI) A comparative measure of the ability of a venom to lyse protein, with fibrin being used as the protein that is lysed. It is calculated as the ratio of the mg of fibrin lysed by 1 mg of test venom compared with the mg of fibrin lysed by 1 mg of "standard" venom from *Bothrops jararaca*. Higher values indicate higher comparative proteolytic activity.

fibroblast. *n.* A connective tissue cell that can differentiate into *chondroblasts, collagenoblasts*, or *osteoblasts* in connective tissue.

fibrocartilage. *n.* Cartilage that is reinforced with collagen fibers which enhance resistance to tensile or warping loads. Cf. *elastic cartilage, hyaline cartilage*.

fibrolamellar bone. Bone material that results from a pattern of development that involves continous growth at a rapid rate. Cf. *zonal bone*.

fibropapilloma. *n.* 1. Generally, a benign tumor that contains both fibrous and glandular elements. 2. A potentially fatal tumor-forming disease seen in sea turtles. See *fibropapillomatosis*.

fibropapillomatosis. *n.* The formation of multiple tumors containing much fibrous tissue, potentially caused by naturally produced tumor promoters (e.g., from toxic cyanobacteria).

fibrosis. *n.* An abnormal increase in the amount of fibrous connective tissue.

fibula. *n.* The outer and usually smaller of the two long bones that articulate with the femur and extend from the knee to ankle of the lower hind limb in most tetrapods.

fibulare. *n.* The *calcaneum*, one of the proximal elements of the ankle.

Fick diffusion equation. An equation defining the rate of solute diffusion through a solvent.

fictive swimming. Involuntary swimming induced in the laboratory for the purpose of investigating kinematics of an animal.

fiducial limits. 95% *confidence limits* around the mean.

field capacity. The amount of water that is retained in a previously saturated soil after free drainage has ceased, usually expressed as percentage of dry weight of the soil.

field metabolic rate. (FMR) The "averaged" rate of energy utilization during normal activities of an animal in the field. Energy expenditure related to a range of activities is integrated over a period of time and determined by the *doubly labeled water technique*. See K.A. Nagy, *Ecol. Monogr.* 57:111–128, 1987.

fig. Abbreviation of the Latin *figura*, meaning figure or illustration. In printed works this refers to a diagram, graph, or illustration.

fight bob. One of the bob displays by *Anolis* lizards, involving a series of stereotyped push-ups performed by a lizard maintaining a combative posture toward another lizard. First used by B. Greenberg and G.K. Noble. (*Physiol. Zool.* 17:392–439, 1944).

Figures of Eberth. Characteristic and extensive filamentous components of basal epidermal cells of larval amphibians in all extant Orders. Named after their discoverer in frog larvae (C.J. Eberth, *Arch. Mikrobiol. Anat.* 2:490–506, 1866), Figures are thought to function in relation to flexibility and the biomechanics of body movements in aquatic environments.

Fiji iguanas. Vernacular name for the iguanid species in the genus *Brachylophus,* endemic to the Fiji islands.

filament. *n.* 1. *Fibrilla* of gill. 2. See *tail filament*.

filaria. *n.* Parasitic, threadlike nematodes of the family Filariidae (or superfamily Filarioidea). Some of these parasitize reptiles, especially lizards and crocodilians, and are transmitted through bites of insects.

File Snake. *n.* Vernacular name for each of three species belonging to the clade (family) *Acrochordidae*. See also *Boodontini*.

filial cannibalism. Cannibalism of kin. Cf. *heterocannibalism*.

filial generations. Reference to any generation following a parental generation, symbolized F_1, F_2, etc.

filiform. *adj.* Threadlike, resembling a filament.

filiform papilla. 1. A type of papilla covering the tongue of most vertebrates. 2. Slender, pointed projections in the skin of bufonid toads, containing a slender core with a capillary loop.

filter bridge. In biogeography, a narrow corridor or temporary land bridge that acts as a selective filter for organisms attempting to migrate along or across it.

filter feeding. A mode of feeding that involves entrapment and ingestion of smaller and usually suspended particles in the environmental medium that is passed through a filtering process while feeding. Some larval amphibians strain water-borne particles from water that is irrigated through the buccal cavity. (Syn. *suspension feeding*)

fimbria. (*pl.* –ae) *n.* Soft, arborescent extensions of the skin, as seen on the head and neck of certain aquatic turtles or the gills of aquatic amphibians.

fimbriate. *adj.* 1. Having delicate hair-like projections. 2. Reference to a pattern of *epidermal microsculpturing.* See R.M. Price *J. Herpetol.* 16:294–306, 1982.

fimbriated peristome. See *umbrella.*

fin. *n.* Any flap or bladelike extension of the body used for balance or propulsion during aquatic locomotion.

final host. The *definitive host* in which the sexual phase of a parasite occurs.

final thermal preferendum. The temperature ultimately selected by an ectothermic animal after being placed in a laboratory thermal gradient, regardless of prior thermal experience.

fine-grained environment. In ecology, reference to an environment experienced as a succession of interconnected conditions. Cf. *coarse-grained environment.*

fine-grained resource. Any resource that is distributed in such a manner that a consumer organism encounters it in the proportion in which it actually occurs. Cf. *course-grained resource.*

finger pad. Padlike structures on the digits of some anurans and lizards.

fingerprinting technique. In molecular biology, this refers to methods employed to determine differences in amino acid sequences among related proteins. See also *DNA fingerprinting.*

finite rate of increase. A ratio measuring the proportional change in population size from one time step to the next in a discrete model of exponential population growth.

fiord. (fjord) *n.* A coastal inlet that is a glacially carved valley, the base of which is submerged.

Fire-bellied Toads. Collective vernacular name for Eurasian species of anurans belonging to the family Bombinatoridae and especially the genus *Bombina.*

Firebelly Newts. Vernacular name for species of salamandrid salamanders belonging to the Asian genus *Cynops.*

fire climax. Reference to a plant community having structure and composition that are dependent on more or less regular burning.

Fire Salamanders. Collective vernacular name for species of salamanders belonging to the genus *Salamandra,* especially the European *Salamandra salamandra* and, sometimes more generally, the family Salamandridae. Ancient people believed these salamanders were born of fire because they were hidden in firewood and emerged when the logs became too hot.

Fire Skink. Vernacular name for the lygosomine Skink species *Riopa fernandi,* included by some in the genera *Lygosoma* or *Mochlus.* This is a forest species native to tropical West Africa.

firmisternal. *adj.* Reference to a pectoral girdle that exhibits *firmisterny.* See Fig. 3.

firmisterny. *n.* A condition in which the *epicoracoids* of the anuran pectoral girdle are fused midventrally. The margins of the two coracoid plates abut one another, thereby bracing the animal against the jar of landing during leaping locomotion. See Fig. 3. Cf. *arcifery.*

first-order kinetics. A pattern of enzymatic reaction in which the rate of product formation is directly proportional to the prevailing substrate concentration. The rate slows gradually with the result that the reaction never goes to completion. Cf. *zero-order kinetics.*

Fischberg's method. A technique for producing tetraploidy in amphibians. Eggs are exposed to heat shock of 35.6–36.0 °C for ten minutes during time of closure of the first cleavage furrow at the vegetative pole of the egg.

Fish Caecilians. Collective vernacular name for species of caecilians belonging to the family Ichthyophiidae.

fissure. *n.* A groove or cleft, as on the ventral surface of the spinal cord.

fitness. *n.* The ability of an organism to contribute its genes to the next generation, roughly the number of offspring contributed to the next generation weighted by the probability of surviving to reproduce. *Absolute fitness* is the absolute contribution to the next generation, or the per capita growth rate of a given genotype. *Relative fitness* usually is calculated as fitness in relation to a reference genotype, usually the most fit or average fit genotype. Relative fitness also is measured by the average contribution to the breeding population by a phenotype, or by a class of phenotypes, relative to the contributions of other phenotypes. Cf. *adaptation.* Note that fitness predicts a trait's future, whereas the term *adaptation* implies something about the history of a trait (retrospective).

Five-lined Skink. Vernacular name for the scincid species *Eumeces fasciatus* and *E. inexpectatus.* Note that H.M. Smith (*J. Kansas Herpetol.* 14:15, 2005) recently proposed that *Plestiodon* be a replacement name for the genus *Eumeces* in North America.

5' nucleotidase. An active phosphatase and common component of many snake venoms.

fixation. *n.* The use of a specialized chemical such as formalin to kill and immobilize cells such that tissues are preserved for later examination.

fixative. *n.* A solution that is used for *fixation* of tissues, making permanent preparation for microscopic study.

fixator. *n.* A muscle that functions to stabilize a joint. Cf. *synergist, antagonist.*

fixed action pattern. Elemental and innate motor pattern forming a stereotypical unit in the behavioral repertoire of an animal. A behavior that is a stereotyped response to specific stimuli. (Syn. *modal action pattern*) See also *sign stimulus.*

fixed serum dose. A method to test the effectiveness of a specific amount of antivenin in neutralizing a venom. Various ratios of venom added to a fixed amount of antivenin are injected into white mice, and the amount of venom required to cause death of the mouse is divided by the minimum lethal dose of the venom. The higher the ratio, the greater the effectiveness of the antivenin. Used by T.S. Githens & N.O. Wolff (*J. Immunol.* 37:47–51, 1939).

fjord. *n.* See *fiord.*

flaccid. *adj.* Limp or lacking turgor or muscle tone.

flaccid paralysis. Paralysis with no muscle tone, characteristic of peripheral nerve damage.

flagellum. (*pl.* –a) *n.* 1. A propulsive organelle shaped as a whip-like outgrowth on spermatozoa and various microorganisms. 2. The term has been applied to slender, filamentous terminations of the tail in certain anuran larvae.

flap. *n.* 1. Generally, a flexible fold of skin, sometimes with muscle and having a free margin. 2. This term has also been applied to various structures in older herpetological literature.

flash color or **coloration.** Usually bright, contrasting color on a concealed surface or body region that is revealed when the animal moves, opens or extends the colored area or structure. The sudden appearance of the color serves as defensive, warning, or social communication. Normally hidden, flash colors can be exposed suddenly to deter a predator.

flask cell. A mitochondria-rich cell type considered to be a single-celled gland in the epidermis of amphibians. The orientation of flask cells is at right angles to the epidermal surface with the apex beneath the *stratum corneum.* They join adjacent cells by desmosomes, and a nerve fiber is commonly located near the neck. Flask cells have been considered as sensory, mechanical, or secretory structures, or involved in keratinization and sloughing. Flask cells appear in the epidermis during metamorphic climax and remain throughout life in anurans, urodeles, and apodans. Used by G.K. Noble (*The Biology of the Amphibia,* p. 139, 1931). (Syn. *beaker cell, bottle cell*)

Flatback Turtle, Flatback Seaturtle. Vernacular name for the marine turtle *Natator depressa* (Cheloniidae).

Flat Lizards. Vernacular name for African species of cordylid lizards belonging to the genus *Platysaurus.*

Flathead Toads. Collective vernacular name for species of bufonid anurans belonging to the South Asian genus *Pelophryne.*

flattened. *adj.* Strongly depressed with the long axis in the horizontal plane. Contrast with *compressed.*

flavin adenine dinucleotide. (FAD) A coenzyme that performs an important function in electron transport and is a prosthetic group for some enzymes.

flexion. *n.* Bending or closing of a joint.

flexor. *n.* A muscle that bends a limb, folds, or closes a joint when it contracts. Cf. *extensor.*

flicking. *n.* Rapid, often repetitive movements of the forelimbs. E.g., this term has been used to describe symmetrical and repeated movements in *Xenopus,* which throw both arms forward and then suddenly retract them.

flipper. *n.* 1. A paddle-like limb of an aquatic turtle. 2. Any of small pieces of skin cut from the legs of crocodilians in the commercial skin trade.

float. *n.* 1. A collective term for a group of crocodilians. 2. See *umbrella.*

floaters. *n.* Subordinate individuals that are forced to disperse to less favorable habitat because of inability to establish a territory.

floating reserve. Reference to individuals in a population of a territorial species that remain unmated and do not hold a territory, but which can fill spaces that might be vacated by death or removal of other individuals.

floating rib. A *false rib* that articulates with nothing ventrally. Cf. *false rib.*

flocculation. *n.* Contact and adhesion of fine particles to form larger and heavier particles, forming aggregations of particles in the dispersed phase of a colloid.

flood plane. A broad strip of land built up by sedimentation on either side of a stream channel, subject to periodic inundation at times of abnormally high flow.

flora. *n.* (*adj.* **floral, floristic**) The vegetation of a given geographic region or period of time. Plant life in general, as distinct from animals.

Florida Green Water Snake. Vernacular name for the American natricine species *Nerodia floridana.*

Florida King Snake. Vernacular name for the American colubrid species *Lampropeltis getula floridana.*

Florida Scrub Lizard. Vernacular name for the American iguanid lizard species *Sceloporus woodi.*

floristic composition. The plant species of a given region or habitat.

flotas. See *arribada.*

flounces. *n.* Fleshy transverse ridges on squamate *hemipenes.*

flow. *n.* Movement of fluid, or the quantity of fluid moving, usually expressed in volume per time (e.g. ml/min).

flow cytometry. A technology that sorts particles having specified properties by means of staining suspended particles with a fluorescent dye and passing them single file through a narrow laser beam. The laser excites the dye and emits fluorescent signals, which are subsequently amplified and transmitted to a computer.

Flower Pot Snake. Vernacular name for the Asian typhlopid species, *Ramphotyphlops braminus*, also referred to as *Common Blind Snake*.

fluctuate. *v.* To move back and forth, or exhibit an irregularly oscillating state.

fluctuating selection. Reference to variation in the adaptive value of a trait as a function of varying conditions of environment.

fluorescein staining. Use of flurane dyes to detect shallow ulceration of the cornea of the eye.

fluorescence. *n.* Emission of light from substances that absorb radiation from an outside source and continue to emit radiation after the source of exciting energy is inactivated. See *luminescence*.

fluorescence microscopy. Microscopic examination of a specimen that is self-luminous, usually attributable to the use of dyes which emit light when exposed to blue or ultraviolet light.

fluorescent antibody technique. A method for localizing a specific protein or other antigen by staining a tissue with a fluorescent-labeled antibody specific for the antigen.

fluorescent pigments. Fluorescent dyes used in field studies to track movement of reptiles and amphibians. The dyes are applied to the underside of animals and rub off onto the ground when the animal moves. These have been used for indirect tracking of small lizards, hatchling turtles, tortoises, and amphibians.

fluvial. *adj.* 1. Relating to, found in, or produced by, rivers. See also *lotic*.

flux. *n.* 1. The rate of flow of matter or energy across a unit area: quantity per unit area per time. 2. The quantity of organic matter transferred per unit time from one compartment of a food web to another.

Flying Dragons. Vernacular name for gliding species of agamid lizards belonging to the genus *Draco*.

Flying Frogs. Collective vernacular term for species of tree frogs belonging to the family Rhacophoridae, especially the genus *Rhacophorus*.

Flying Geckos. Vernacular name for species of southeast Asian Geckos belonging to the genus *Ptychozoon*. Aerial gliding is enhanced by possession of *patagia* in these lizards.

Flying Snakes, Flying Tree Snakes. Vernacular name for several species of Asian colubrid snakes belonging to the genus *Chrysopelea*. Although some other snakes dive out of trees, these snakes are specialized morphologically and behaviorally for controlled gliding.

Fly River Turtles. See *Pig-nosed Turtle*.

FMR. See *field metabolic rate*.

foam nest. A frothy, jelly-like structure that surrounds eggs and developing larvae of various tropical species of anurans. The foam nest is usually constructed by "whipping" skin secretions with kicking or paddling movements of the hind or front legs, and it either floats on water or is suspended on a leaf. Eggs are deposited within the foam, which has a protective function.

Foam Nest Frogs, Foam Nest Treefrogs. Collective vernacular name for African species of ranid frogs belonging to the genus *Chiromantis*. Eggs of these species are laid in foam nests attached to branches or other objects overhanging water at varying heights.

focal. *adj.* Confined to a single location.

foci. *n.* 1. Specific locations. 2. Reference to pathological locations.

foehn (föhn) wind. A warm dry wind on the leeward side of a mountain range.

fold. *n.* A projecting margin or edge caused usually by membrane or tissue doubling upon itself. Most often used in reference to structures on the body and tail of caecilians, glandular ridges behind the eyes of anurans, and lateral folds or grooves of salamanders. See *dorsolateral fold*, *costal fold*.

folds of Kerckring. Extensive folds of the intestinal mucosa. (Syn. *circular folds*)

foliform. *adj.* Leaflike, or having an extensive free edge (as in some scales).

folivore. *n.* (*adj.* –ous) An animal that feeds on leaves or leafy plant materials (foliage).

follicle. *n.* 1. A cavity or pouchlike depression within a tissue or organ, usually having a secretory or protective function. 2. The structure of the ovary that contains the developing ovum. *Follicular maturation* refers to the formation and maturation of egg follicles, eventually culminating in ovulation of the eggs.

follicle gland. See *integumentary sense organ*.

follicle pits. See *integumentary sense organ*.

follicle pores. See *integumentary sense organ*.

follicle stimulating hormone. (FSH) A *gonadotropin* secreted from the *anterior pituitary*, functioning to stimulate the development of ovarian follicles in females and spermatogenesis in males. Present in all vertebrates except the *Squamata*, which produce an FSH-like gonadotropin not specifically identical to FSH in mammals.

follicular atresia. Degeneration (*apoptosis*) and resorption of an ovarian follicle before it reaches maturity and ruptures.

follicular maturation. See *follicle*.

foliose. *adj.* Reference to foliage- or moss-like pattern of groups of chromatophores or shape of individual chromatophores.

food chain. A hierarchical series of organisms that feed one on another within a natural community. Energy is thereby transferred from one level to the next, and a food chain expresses these relations in linear form. Each stage or level is called a *trophic level*.

food chain efficiency. See *ecological efficiency*.

food pyramid. Visualization of related feeding patterns as a pyramid, where larger numbers of organisms are on the lower level and successively smaller layers occur in hierarchical order toward the top. The pyramid reflects the fact that relatively large numbers of plants and herbivores are required to support a smaller number of top predators at the upper part of the pyramid. (Syn. *ecological pyramid*)

food web. *n.* An interconnected series of *food chains*. The concept is an abstract connection between a group of species in a community describing predator-prey interactions (which one feeds on which).

foot. *n.* The fps unit of length, equal to 0.3048 m.

footfall. *n.* Reference to foot contact with the ground during locomotion.

Foothill Yellow-legged Frog. See *Yellow-legged Frog*.

foot length. A measurement of distance between the tip of the longest (fourth) toe to the base of the outer metatarsal tubercle, or, if this is absent, to the tarso-metatarsal articulation.

footplate. *n.* The expanded base of the *columella* providing ovate contact with the oval window of the otic capsule. (Syn. *stapedial footplate, stapedial plate, otostapes* in part)

foot stage. A premetamorphic developmental stage of anuran larvae in which characteristic features of the foot become evident. Term coined by A.C. Taylor & J.J. Kollros (*Anat. Rec.* 94:7–13, 1946).

foot stamp. 1. A submissive behavior in certain lizards, characterized by a loser of a conspecific confrontation or fight flattening its body to the ground and stamping its feet. This usually ends the conflict and allows subsequent escape of the submissive animal. 2. Thumping of the ground by certain turtles, evidently to stimulate emergence of earthworms that are subsequently eaten. See J.H. Kaufmann, *Copeia* 1986:1001–1004, 1986; *Nat. Hist.* 8:8–10, 1989.

foraging. *n.* Behavior associated with seeking, capturing, and consuming food.

foramen. (*pl.* **foramina**) *n.* A small natural opening or aperture that allows passage through a tissue wall. Smaller than a *fenestra*.

foramen magnum. The large aperture at the base of the skull allowing connection between the brain and spinal cord.

Foramen of Panizza, or *Foramen Panizzae*. An opening between the right and left systemic arches of crocodilians, allowing shunting of blood between these two vessels. See Fig. 37

forb. *n.* A broad-leaved herbaceous plant.

force. *n.* A push or pull that resists or causes motion (mass × acceleration). See *action force, reaction force,* and Figs. 24, 25.

force feeding. The act of forcibly feeding an animal a food item, often practiced in the husbandry of various reptiles (particularly captive snakes that do not eat voluntarily).

force plate or **platform.** A device of variable configurations that measures *reaction forces* of animals that contact it, usually in contexts of locomotion. In studies of locomotion, a force plate records the ground reaction forces exerted by an animal when its limb or other body part contacts the plate during production of active forces used in locomotion.

forebrain. *n.* The anterior part of the brain, including the *cerebral hemispheres* (*telencephalon*), *thalamus, hypothalamus,* and *epithalamus* (*diencephalon*). See *prosencephalon*.

foredune. *n.* A dune ridge, somewhat stabilized by vegetation.

foregut. *n.* The anterior embryonic gut, which develops into the adult pharynx, esophagus, stomach, and anterior intestine. Functions are involved with food conduction, storage, and digestion. Cf. *hindgut*.

foregut fermentation. See *fermentation*.

forelimb. *n.* Either of the anterior pair of limbs, each comprised of the upper arm, forearm, and *manus*.

fore reef. The outer part of a barrier reef or atoll.

forest. *n.* A relatively large area of trees that are closely canopied. In forest assessment, forest lands are considered to be those having a tree crown cover (standing density) in excess of 20%.

Forest Dragons. Collective vernacular name for various species of agamid lizards belonging to the genera *Gonocephalus* and *Hypsilurus*.

Forest Skinks. Vernacular name for species of Skinks belonging to the genus *Sphenomorphus*.

Forest Pitvipers. Vernacular name for the neotropical pitvipers belonging to the genus *Bothriopsis*.

form. *n.* 1. Reference to shape or size of an organism, as distinct from color, texture, etc. 2. Any minor variant or recognizable subset of a population. (Syn. *morph*) 3. A depression or cavity made by a turtle in soil or vegetation.

formalin or **formaldehyde.** *n.* A gaseous substance which in its liquid phase is used as a disinfectant and to fix or preserve tissues.

form drag. See *pressure drag*.

formed elements. Reference to the cellular products of blood, excluding the plasma.

fossa. (*pl.* –ae) *n.* A pit, cavity, or depressed area, as used in anatomy. Literally, a ditch or trough.

fossette. *n.* A small pit or depression.

fossil. *n.* Any remains or trace of past life that is preserved in rocks.

fossiliferous. *adj.* Fossil-bearing or yielding fossils.

fossil record. The cumulative evidence about organisms and conditions of the past that is provided by fossils.

fossorial. *adj.* Reference to habits or structural adaptations related to burrowing, digging, or subterranean life beneath ground vegetation or substratum.

founder effect. A phenomenon of likely evolutionary divergence when a small number of individuals from a larger population establish themselves in a new geographic area. Divergence is subsequently likely due to new and different selection pressures acting on a select fraction of the genetic diversity represented in the parental population. The term was introduced by E. Mayr in 1963 (*Animal Species and Evolution*). (Syn. *founder principle*)

founder effect speciation. See *peripatric* speciation.

founder principle. See *founder effect*.

Fourier's law. The rate of heat flow in a conducting body is proportional to its thermal conductance and to the temperature gradient.

Fourier transformation. Analysis based on *Fourier's theorem* that any periodic function can be reduced to a series of sine and cosine terms, each represented by an amplitude and a phase.

Four-toed Salamander. Vernacular name for the plethodontid species *Hemidactylium scutatum*.

fovea. *n.* 1. A small depression. 2. Reference to the *fovea centralis*, a small depression over a cluster of closely packed cones on the optical axis of the retina, where nerve fibers are bent aside to yield especially acute diurnal vision.

foveal. *n.* Any of small scales surrounding the *pit organ* of pit vipers. Usage and terminology of adjacent scales varies among authors. See *lacunal, pit scale*.

foveate. *adj.* 1. Bearing round pits separated by distinct walls. 2. Reference to a pattern of *epidermal microsculpturing*. See R.M. Price *J. Herpetol.* 16:294–306, 1982.

foveoreticulate. *adj.* 1. Having round pits whose walls anastomose irregularly into a network. 2. Reference to a pattern of *epidermal microsculpturing*. See R.M. Price *J. Herpetol.* 16:294–306, 1982.

fowlerobufagin. *n.* A *bufogenin* isolated from *Anaxyrus fowleri* (formerly *Bufo woodhousei fowleri*).

fowlerobufotoxin. *n.* A *bufotoxin* formed by conjugation of *fowlerobufagin* with suberylarginine.

Fowler's Toad. Vernacular name for the American bufonid species *Anaxyrus* (formerly *Bufo*) *fowleri*. For recent taxonomic revision see D.R. Frost et al., *Bull. Amer. Mus. Nat. Hist.* 297:1-370, 2006.

Fox Snake. Vernacular name for the North American colubrid snakes *Pantherophis gloydi* and *Pantherophis vulpine* (formerly *Elaphe gloydi* and *E. vulpine*).

fps system. A system of scientific units based on the *foot-pound-second* and also known as the British system.

fracture. *n.* In biomechanics, a break or loss of structural integrity, involving the separation of a material under a load. Cf. *failure*.

fracture plane or **septum.** *n.* See *autotomy plane*.

fragile tail. A tail that can be shed by *autotomy* when an animal is attacked.

Fraser-Darling effect. Stimulation of reproductive activity in a mating pair caused by the presence and behavior of other members of the species. (Syn. *Darling effect*)

frass. *n.* Fine animal debris including fecal matter that is found on the soil surface.

frazil. *n.* Minute ice crystals that form as water begins to freeze.

free body diagram. A drawing that depicts an isolated mechanical system and illustrates as vectors all external translational and rotational forces acting on the system.

free convection. See *convection*.

free energy. In thermodynamics, the energy available to do work at a given temperature and pressure.

free-living. Living independently of a host organism.

free radical. Any of a number of unstable and highly reactive molecules having an unshared electron. These are potent oxidizing agents capable of attacking a variety of organic structures indiscriminantly, including DNA. Cumulative damage to cellular organelles by reactive free radicals of oxygen is theorized to contribute to aging processes (*free radical theory*).

free radical theory. See *free radical*.

free-running cycle. Reference to the length of a circadian rhythm when an external time cue is absent.

free sensory receptor. The terminus of a "naked" sensory neuron lacking supportive associated structures. Cf. *encapsulated sensory receptor*.

freeze. *v.* 1. Transition of water from liquid to solid phase. 2. In ethology, to become rigid or motionless, thereby holding a fixed position for a significant length of time. Such behavior is assumed often related to fear.

freeze dry or **drying.** *v.* (*adj.* **freeze dried**) A process of desiccating a frozen organism or substance in a vacuum, used commonly for preserving snake venoms. See *lyophylize*.

freeze etching. A technique for preparing biological materials for electron microscopy using an instrument that allows frozen tissues to be sectioned in a vacuum, with the result that surface irregularities accentuate cellular structures.

freeze fracture. A technique in which frozen samples are fractured with a knife, and the complementary surfaces are cast in metal for examination with the electron microscope.

freezing resistance. Resistance of an animal to freezing, conferred by *supercooling* or by depression of freezing point of body fluids related to accumulation of "antifreeze" molecules such as glucose or glycerol. Freezing resistance is especially characteristic of some frogs and turtles.

freezing tolerance. *n.* Reference to the ability to withstand freezing without cellular damage. A number of species of amphibians have been shown to tolerate repeated freez-

frenal. *n.* An archaic term for *loreal*, used in older literature. See Fig. 11.

frenocular. *n.* In general, a scale lying between the loreal and the eye, or between the loreal and the preoculars. Usage of this term varies among authors.

frenonasal. *n.* See *postnasal*.

frenoorbital. *n.* See *preocular*.

frenulum. *n.* Band of connective tissue between the arytenoid cartilage and vocal chord of the anuran larynx.

frequency. *n.* 1. The number of times an action or occurrence is repeated in a given time period. 2. The number of periodic oscillations or waves per unit time, usually expressed in cycles per second. 3. The number of items belonging to a class or category in a population or sample, measured as a fraction of the entire collection. The frequency at which something occurs will necessarily equal its probability of occurrence IF the sample is the entire population, or the sampling process is repeated an infinite number of times.

frequency-dependent fitness. A condition wherein the adaptive value of a genotype varies with changes in allelic frequencies.

frequency-dependent selection. Selection that involves frequency-dependent fitness.

frequency distribution. A grouping of the possible values of a variable into broader classifications to indicate the frequency with which observed values in a particular interval occur. The collection of individual values in a population or sample, summed over the entire range of possible values. A distribution can be expressed as either a *frequency distribution* or a *probability distribution* depending on whether reference is to a sample or a population.

frequency modulation. Variation in pitch, or frequency, of sound vibration over time. This term is used in connection with call patterns in vocalization of anurans.

fresh water. Generally, water from lakes and rivers having a salinity less than 2 parts per thousand.

Freund's incomplete adjuvant. A carrier for snake venom that is injected into an animal in the preparation of antivenin.

friable. *adj.* Easily crumbled or broken apart.

friction. *n.* Mechanical resistance to relative motion. Dry *static friction* is the product of the normal force and the coefficient of friction, equal to the force that has to be overcome in order to induce motion along the contacting surfaces. The resistance to motion generated at contact between two surfaces that are in motion is termed *sliding friction*, which is the force that has to be overcome in order to keep two objects moving at a constant velocity.

friction disc, friction pad. Reference to *adhesive pads* and similar structures present on digits of anurans or lizards.

friction surface. Reference to an area that may be keratinized or finely papillate, or both, on the roof of the *belly sucker* of *gastromyzophorous* tadpoles.

Friedman's test. A nonparametric method comprising a two-way analysis of variance and used to test for differences in means between several samples, each containing the same number of counts.

fright cry. *n.* Sudden, loud *distress calls* of anurans when suddenly grasped or startled. (Syn. *pain cry, protective call, mercy cry, scream*)

frill. *n.* 1. The large extension of skin forming a large loose "ruff" or "frill" around the neck in the agamid frilled lizard, *Chlamydosaurus kingii*, of Australia and New Guinea. The frill normally lies in folds around the neck and shoulders, but is erected by means of cartilaginous extensions of the hyoid apparatus when the lizard is alarmed and the mouth is opened in a defensive display. The frill on each side overlaps on the crown of the head. 2. A distinctive, often large plate of bone extending upward and outward at the back of the skull in certain dinosaurs, formed by the parietal and squamosal bones. It possibly functioned as a structure used in sexual display.

Frilled Lizard, Frilled Dragon. Vernacular name for the Australian agamid lizard *Chlamydosaurus kingii*, known for its characteristic threat display of a prominent frill of scaled skin.

fringe. *n.* Any row of scales or epidermal derivatives projecting from a body surface. Commonly used in reference to elongated edges of scales on the digits of certain lizards that function as an aid to locomotion in loose sand. Also used in reference to the posteriormost scales that cover the proximal edge of the rattle in a rattlesnake.

Fringe-toed Lizards. Vernacular name for phrynosomatid (or iguanid) species belonging to the genus *Uma*, common to sandy regions of the southwestern United States and Mexico.

fringing reef. A coral reef attached directly to shore.

frog. *n.* A categorical name for the majority of anurans, once used to differentiate amphibious or aquatic anurans (frogs) from terrestrial ones (toads). However, the lines are blurred and there is no strict differentiation of frogs from toads. Therefore, this term refers properly to any member of the order Anura. The term *true frogs* is used in reference to the family Ranidae.

froglet. *n.* A small and sexually immature anuran produced either by metamorphosis of a tadpole or by direct development.

FrogLog. *n.* An official publication of the *Declining Amphibians Populations Task Force*, disseminating a range of articles on research, discoveries, or conservation news related to the phenomenon of *amphibian declines*.

frog malaria. See *amphibian malaria*.

frog toxin. Reference to any one of a number of poisonous substances secreted from the skin of anurans. (Syn. *batrachotoxins*)

FrogWatch USA. An educational anuran monitoring program started by the U.S. Geological Survey in 1999 and currently a partnership with the National Wildlife Federation. It relies on citizen volunteers to gather information on frog populations throughout the United States. Current URL: http://www.nwf.org/frogwatchUSA/

FrogWeb. A web-based resource developed by the U.S. Geological Survey National Biological Information Infrastructure and providing access to information and educational materials on amphibian declines and malformations. Current URL: http://www.frogweb.gov/

frontal. *n* or *adj.* 1. Of, or pertaining to, the front of the head of an animal. 2. See *frontal plane.* 3. Reference to the scale(s), space, or plates on mid-top of the head between the supraocular scales in squamates and turtles. (Syn. *central, vertebral, vertical*) See Figs. 10, 14. 4. The large and prominent dermal bone on the dorsal roof of the skull, lying between the eye orbits—usually paired and distinct, but sometimes fused with other bones. 5. Pertaining to a boundary layer (or front) that separates two adjacent air or water masses.

frontal cutaneous gland or **frontal gland.** Used in older literature to designate the *stirnorgan* or *hatching gland* of amphibians.

frontal organ. *n.* 1. Synonym for *parietal eye* of reptiles (F. Tilney & L.F. Warren, *Amer. Anat. Mem.*, Vol. 9, p. 33, 1919). 2. Synonym for *stirnorgan* of amphibians (R.M. Eakin, *Proc. Nat. Acad. Sci.* 47:1084–1088, 1961).

frontal plane. A plane passing through the body at right angles to the *sagittal plane.* The plane divides the body into dorsal and ventral parts. Cf. *transverse plane.*

frontal precipitation or **rain.** Precipitation attributable to the action of a front when warm, moist air is forced to rise over a colder, denser air mass. Cf. *convective precipitation, orographic precipitation.*

frontal ridge. The elevated ridge of many species of *Anolis* lizards, beginning on the snout and running posteriad to the anterior margin of the eyes.

front-fanged. *adj.* Synonym for *endoglyphous.*

front height. A measurement that quantifies a vertical dimension of the turtle shell, equivalent to the distance from a flat surface, on which the shell rests, to the notch at the *nuchal.* See Fig. 8.

frontispiece. *n.* The illustration facing the title page of a book.

frontonasal. *n.* 1. A scale or scales located between the *internasals, prefrontals,* and *loreals* in many lizards and turtles, more rarely in snakes. (Syn. *prefrontal*) See Figs. 10, 14. 2. In typhlopid snakes, these are a pair of middorsal scales lying above the nasal and posterior to the rostral.

frontoparietal. *n.* 1. Enlarged single, or paired, dorsal cephalic scale(s) situated between the *frontal* and *parietal* scales of certain turtles. 2. An enlarged single or paired scale(s) situated between the *frontal* and *parietal* scales of lizards. (Syn. *interparietal*) See Fig. 14. 3. A bone formed by fusion of the *frontal* and *parietal* bones in anurans.

frontoparietal foramen. A prominent opening between the paired frontoparietal bones of certain anurans.

frontosquamosal arch. The bony arch uniting the *frontal* and *squamosal* bones in certain salamanders.

frugivore. *n.* (*adj.* **–ous**) An animal that feeds on fruit. Fruit-eating.

FSH. See *follicle stimulating hormone.*

***F*-test.** In statistics, a test used to evaluate the probability that two sample variances are the same (i.e. drawn from the same population). (Syn. *variance ratio test*) See *analysis of variance.*

FU. A symbol for *rejecta* that is used in ecological energetics.

fugitive species. A species characteristic of temporary habitats due to its ability to disperse to newly disturbed habitats from other areas where it is an inferior competitor.

fulcrum. *n.* The point of pivot or the axis of rotation.

function. *n.* The dynamic properties of a structure, excluding the use or action of the structure which is more critically termed its *biological role.* See *biological role.*

functional constraint. Generally a limitation, but usually used in reference to limits for evolutionary change by natural selection due to physical or mechanical laws.

functional morphology. The study of form and function, with emphasis on the adaptive use of structure. Classically, this discipline has emphasized studies of locomotion and feeding processes, usually in evolutionary contexts, but also has a legitimate broader usage applicable to numerous areas that integrate anatomical design with the function it performs.

functional response. In ecology, a change in the rate of victim capture by a predator as a function of change in density of prey. Units are victims/(predator • time). Cf. *numerical response.*

fundamental frequency. In bioacoustics, the lowest frequency harmonic.

fundamental niche. See *ecological niche.* (Syn. *ecospace, prospective ecospace*) Cf. *realized niche.*

fundamental theorem of natural selection. A theorem, developed by R.A. Fisher, which holds that the rate of increase in fitness of a population at any given time is directly proportional to the genetic (or additive) variance in fitness of its members. See R.A. Fisher, *The Genetical Theory of Natural Selection*, 1930, 1958 (2nd ed.).

fundus. *n.* The anterior, usually larger bulk of the stomach that receives ingested food from the *esophagus* and contains the secretory glands of the glandular epithelium. (Syn. *corpus*) Cf. *pars pylorica.*

funnel. *n.* 1. A term used by some authors for the oral *umbrella* of tadpoles of *Megophrys.* 2. Reference to the nasal valve in tadpoles of *Ascaphus* (G.K. Noble, *Ann. N.Y. Acad. Sci.* 30:31–128, 1927). 3. The *infundibulum* of vertebrates.

funnel mouth. A term used in reference to upturned or enlarged *oral discs* present in some species of tadpoles. Usage is discouraged by R.W. McDiarmid & R. Altig (*Tadpoles. The Biology of Anuran Larvae*, 1999) to enhance clarification among the terms *mouthparts, oral disc,* and *mouth.*

funnel trap. Enclosures of various design that include inward facing funnels that allow small animals to enter the device, while having difficulty locating the small entry hole for exit. These traps have proven effective in collecting live specimens in field studies, including aquatic, terrestrial, and arboreal amphibians and reptiles.

furcula. *n.* An older alternate term for clavicle. See Fig. 3.

furcate. *adj.* Forked or branched. See also *bifurcate.*

furlong. *n.* A unit of length equal to 220 yd or 201.168 m.

fusiform. *adj.* Spindle-shaped; tapering towards each end.

fusion. *n.* In anatomy, the act or fact of being joined together.

fuzzies. *n., pl.* Vernacular jargon for newborn rodents that have just begun to grow hair.

fynbos. *n.* A term given to Mediterranean-climate shrublands in the Cape region of South Africa, dominated by shrub species peculiar to South Africa's southwestern and southern Cape. The vegetation is dominated by tall protea shrubs (proteoids), heath-like shrubs (ericoids), wiry, reed-like plants, and bulbous herbs. Fynbos is the smallest floral kingdom in the world, but has the largest number of plant species relative to its size.

G

GAA. See *Global Amphibian Assessment*.

GABA. (γ -aminobutryric acid) A widespread inhibitory neurotransmitter.

Gaboon Viper (or **Adder**). Vernacular name for the African viperid species *Bitis gabonica*. This snake is very stout with exceptionally long fangs.

Gadow's hypothesis or **theory** (of vertebral formation). The now discredited idea proposed by Hans Gadow that composition of vertebrae from four pairs of *arcualia* as a basic pattern in fishes was applicable to tetrapods, and further evolution in terrestrial vertebrates was by reduction or loss of elements of this basic structure (H. Gadow & E.C. Abbott, *Philos. Trans.* Series B 186:163–221, 1895; H. Gadow, *Philos. Trans.* Series B 187:1–57, 1896). Reassessment of this theory has shown Gadow's assumptions to be inapplicable to tetrapods, and "his terminology even in modified form should not be employed in tetrapods" (E.E. Williams, *Quart. Rev. Biol.* 34:1–32, 1959).

Gaia hypothesis. The notion that the surface of the Earth behaves as a highly integrated organism capable of controlling its own composition and its environment. The idea (named after the Greek goddess of the Earth) was put forth by J.E. Lovelock in *Gaia: A New Look at Life on Earth* (1987).

gain. *n.* The factor by which a signal is amplified, sometimes expressed in dB.

gait. *n.* The pattern or sequence of limb movement and foot placement during terrestrial walking or running.

Galapagos Land Iguanas. Vernacular name for species of iguanid lizards belonging to the genus *Conolophus*, endemic to the Galapagos Islands.

Galapagos Marine Iguana. A species of Iguana (*Amblyrhynchus cristatus*) endemic to the Galapagos Islands, unusual because it enters the sea to graze on marine algae.

Galapagos Tortoise. A well-known species of "giant" tortoise (*Geochelone nigra*) endemic to the Galapagos Islands.

gall bladder. A muscular sac that receives, stores, and concentrates *bile* that is transported from the liver. The concentrated bile is discharged into the anterior small intestine in reaction to the presence of food passing there from the stomach.

gallery forest. A strip of forest situated along the margins of a river in an otherwise unforested landscape.

Galliwasp. *n.* Collective vernacular name for anguid lizard species belonging to the genus *Diploglossus*, native to Middle America and the West Indies.

gallon. *n.* A unit of volume. An *Imperial gallon* is a fps unit equal to 4.54609 dm^3. A *US gallon* is equal to 3.785412 dm^3 (or 0.8327 Imperial gallon).

gallop. *n, v.* A gait characterized by high speed and an uneven pattern of footfall.

Gallotinae. *n.* A subfamily of *Lacertidae*.

GALT. See *gut-associated lymphoid tissue*.

gamabufagenin, gamabufogenin. *n.* A synonym for *gamabufotalin*.

gamabufagin. *n.* See *gamabufotalin*.

gamabufotalin. *n.* A *bufogenin* with hydroxyl groups at C_3, C_{11} and C_{14}, isolated from *Bufo formosus*. (Syn. *gamabufagenin, gamabufagin*)

gamabufotoxin. *n.* The bufotoxin formed by conjugation of *gamabufotalin* with suberylarginine, isolated from *Bufo formosus*.

game. *n.* In ecological contexts, a competitive interaction between organisms relating to the promotion of fitness.

gamete. *n.* A haploid germ or sex cell (egg or sperm) that fuses with another of the opposite sex to form an *embryo*. An egg or a sperm.

gametic pool. The total potential production of gametes in any given generation of a sexually reproducing population.

game theory. See *evolutionary game theory*.

gametocyte. *n.* A cell that will form gametes by division. A *spermatocyte* or *oocyte*.

gametogenesis. *n.* The process of gamete formation.

gametokinetic response. Production and discharge of spermatozoa in response to stimulation by exogenous gonadotropin.

gamma diversity. The richness of species across a range of habitats within a geographical area, or in widely separated areas. This measure depends on both *alpha diversity* of the local habitats and the extent of *beta diversity* between them.

gamma link. A dichotomously branching link in a food chain involving two predators feeding on one prey species. Cf. *lambda link, iota link*.

gamma taxonomy. Reference to aspects of taxonomy concerned with intraspecific variation, trait evolution, biogeography, evolutionary patterns and processes, and with phylogenetic trends. Cf. *alpha taxonomy, beta taxonomy*.

gamosematic. *adj.* Reference to behavior, coloration, or markings that assist members of a pair to locate each other.

-gamy. Suffix meaning "fertilization" or "sexual union."

ganglion. (*pl.* **–ia**) *n.* An anatomically distinct collection of nerve cell bodies within the peripheral nervous system.

gangrene. *n.* Reference to dead, blackened tissue caused by interruption of blood supply.

gap. *n.* 1. A break or discontinuity in a linear structure, used in reference to breaks in a tooth row. 2. A break or open space within a forest. 3. Discontinuities in a variable distribution. See *lumpiness*.

gape. *n., v.* 1. The mouth opening or aperture. 2. The act of opening the mouth.

gape limited. Reference to a predator that cannot ingest prey larger than a maximum size set by the functional size of its mouth.

gaping. *n.* A condition in which the mouth is continually open, *e.g.,* as a result of respiratory disease, thermoregulatory response, etc. This behavior is especially notable in crocodilians in thermoregulatory contexts. However, mouth gaping can occur at night and appears to be a social signal to conspecifics.

gap junctions. Specialized membrane junctions that connect cytosols and enable electrical coupling between cells. Plasma membranes at these junctions are about 2 nm apart and are linked by tubular assemblies of proteins called *connexons* in vertebrates.

Gartersnakes, Garter Snakes. *n.* Collective vernacular name for American colubrid snakes of the genus *Thamnophis*. The name also applies to about seven species of elapid snakes belonging to the genus *Elapsoidea* in sub-Saharan Africa. These are small- to medium-size burrowing, nocturnal snakes.

Gartner's duct. A short, blind, and functionless tubule, closed at both ends and representing the *ductus deferens* in female metanephric animals.

gas chromatography. A chromatographic technique based on the use of an inert gas to capture by sweep through a column the vapors of the materials to be separated.

gas exchange. Reference to uptake of oxygen and elimination of carbon dioxide between an animal and its environmental medium, air or water. (Syn. *external* respiration) Cf. *respiration*. See *cutaneous gas exchange* or *respiration*.

gas exchange partitioning. Reference to differential use of functional gas exchange surfaces available in the same animal (gill, lung, skin, etc.).

gas exchanger. Any body surface (gill, lung, skin, etc.) that is used for *gas exchange*.

gastralium. (*pl.* **-ia**) *n.* 1. Rib-like dermal bones found in the ventral abdominal wall between the last true rib and the pelvic girdle in crocodilians, the Tuatara, and some fossil reptiles. (Syn. *secondary rib*) Cf. *parasternal*. 2. Dermal ossifications that contribute to the plastron of turtles.

gastric. *adj.* Pertaining to the stomach.

Gastric Brooding Frog. Vernacular name for the Australian myobatrachid species *Rheobatrachus silus*, now considered to be extinct.

gastric fermentation. See *fermentation*.

gastric inhibitory peptide. (GIP) A gastrointestinal hormone released into the blood circulation from the duodenal mucosa, acting to inhibit gastric secretion and motility. GIP is also called *glucose-dependent insulinotropic polypeptide* because it is a potent releaser of insulin. GIP is important in postprandial digestion and is secreted in response to the presence of fat or glucose in the duodenum.

gastrin. *n.* A polypeptide hormone and putative neurotransmitter that is released by endocrine cells in the pyloric region of the gastric mucosa and acts to induce gastric secretion and motility. Peptides homologous to mammalian gastrin have been characterized in brain, stomach, and small intestine of amphibians and reptiles, and the gastrin peptide appears to have originated early in vertebrate evolution.

gastrocentrous. *adj.* A vertebra in which the *centrum* is composed principally of *pleurocentrum*, according to H. Gadow (*Amphibia and Reptiles*, pp. 282–284, 1901). The gastrocentrous vertebrae are divided into *temnospondylous* or *stereospondylous* types, depending on the presence or absence of a *hypocentrum*.

gastrocnemius muscle. The most prominent ventral muscle of the shank, so named in reptiles and in mammals.

gastrocoel. *n.* The cavity within the early embryonic gut of the *gastrula*.

gastroenteritis. *n.* 1. Inflammation of the stomach and intestine. 2. A reptilian disease caused by parasitic protozoans (genus *Entamoeba*) affecting the blood vessels and mucous membranes of the intestine. The condition is especially prevalent in snakes.

gastrointestinal peptide hormones. Hormones that regulate activity of smooth muscle and digestive secretions in the alimentary canal. Several of these hormones also occur as neuropeptides in the brain.

gastrointestinal tract. See *alimentary canal*.

gastrolith. *n.* A smooth stone or pebble found in the stomach of crocodilians and some fossil reptiles, thought to be swallowed deliberately as an aid to digestion or buoyancy control.

gastromyzophorous. *adj.* Bearing a ventral sucker, as in the tadpoles of frogs of the genera *Amolops* and *Atelopus*, used for maintenance of position in fast and sometimes turbulent water.

gastrostege. *n.* Synonym for *ventral* (or belly *scute*) of snakes, used commonly in earlier literature. (alternate spellings:

gasterostege, gastrostegite, gastrostiga) See *transverse ventral plates* and Fig. 13.

gastrula. *n.* A phase of the developing embryo produced by invagination of the *blastula* to form inner and outer germinal layers of the *archenteron*.

gastrulation. *n.* The embryonic process that forms the *gastrula* following the *blastula* stage.

gate. *n.* In cellular physiology, reference to a regulated channel by which ions pass selectively across the plasma membrane.

gating. *n.* Reference to on-off behaviors of gated membrane channels and therefore the switching on or off of a cellular function.

Gause's Law. See *competitive exclusion principle*.

Gaussian curve. A normal or symmetrical bell-shaped curve that reflects a frequency distribution for a normally distributed population.

Gavialidae. *n.* A clade (family or subfamily, depending on classification scheme) that includes two species of crocodilians distributed in northern India and neighboring countries east to Burma (Myanmar), southern Thailand and Malaysia, Sumatra, Borneo, and Java. See also *False Gharial*.

Gavial or **Gharial.** *n.* A member of the *Gavialidae*, and possibly the most aquatic of the crocodilians.

GBIF. See *Global Biodiversity Information Facility*.

Gecko. *n.* Collective common name for numerous species of lizards belonging to the *Gekkonidae*.

Gekko. *n.* The journal of the Global Gecko Association, publishing articles, book reviews, and other items related to the Association's interests.

gekkonid. *n, adj.* A member of, or pertaining to, *Gekkonidae*.

Gekkonidae. *n.* A clade (family) of lizards that includes Geckos, in addition to pygopods. Approximately 97 genera and 1050 species are classified among four families. *Eublepharinae* are distributed in North and Central America, east and west Africa, southern Asia, southeast Asia, and islands of the Sunda Shelf. *Gekkoninae* are cosmopolitan except for northern North America and Eurasia. *Diplodactylinae* are represented in the Australian region and New Zealand. *Pygopodinae* occur in Australia and New Guinea.

Gekkoninae. *n.* A subfamily of *Gekkonidae*.

gelatinous layer, **coat**, or **envelope.** See *jelly envelope*.

gen. Abbreviation of the Latin *genus*, meaning "genus."

gen. et sp. nov. Abbreviation of the Latin *genus et species nova*, meaning "new genus" and "species."

gen. rev. Abbreviation of the Latin *genus revivisco*, meaning "genus revised."

gene. *n.* The fundamental physical and functional unit of heredity consisting of a specific region of encoded information on the DNA molecule. The smallest, indivisible unit of heredity, which carries information from one generation to the next. Note that historically the definition of a gene has changed as more of its properties have become better understood.

genealogy. *n.* A lineage or pedigree.

gene activation. Induction of gene function by an inducer molecule, resulting in the transcription of one or more structural genes. (Syn. *genetic induction*)

gene amplification. Any process by which DNA sequences are replicated to a greater degree than their relative representation in the parent molecules. E.g., during development some ribosomal genes are amplified and become active in some amphibian oocytes during oogenesis.

gene bank. See *genomic library*.

gene cluster. A group of functionally related genes.

gene complex. The entire system of interacting genetic factors of an organism.

gene expression. The result of gene activation, or the synthesis of gene products that determine the phenotype.

gene flow. The exchange of genes between different populations attributable to emigration and immigration of individuals.

gene frequency. The percentages of all alleles at a specified locus, determined for a given population.

gene library. See *genomic library*.

gene mapping. Determination of the sequence of genes and their relative distances from one another on a specific chromosome.

gene mutation. A point mutation, or any heritable change in a single gene.

gene pool. The sum of all genetic information that is possessed by reproductive members of a population of sexually reproducing organisms.

gene product. Either a nontranslated RNA molecule or a polypeptide chain transcribed from a DNA molecule.

gene substitution. The replacement of one allele by a mutant allele.

generalist. *adj., n.* In ecology, reference to a species that has comparative breadth of resource utilization, such as a broad pattern of habitat tolerance or food preference. Cf. *specialist*.

generalized. *adj.* In evolutionary biology, reference to unspecialized characters or those having greater potential for evolving into alternative states relative to specialized ones.

generation. *n.* 1. Formation or production. 2. All individuals produced within a single life cycle of a population or species.

generation gland. A generic term proposed by P.F.A. Maderson & K.W. Chiu (*Herpetologica* 26:233–238, 1970) for holocrine epidermal glands borne on posterior abdominal scales in Geckos. Subsequently identified in other lizard taxa, secretory activity is associated with periodic skin-shedding. Cf. *preanal organ*.

generation time. The interval of time required for neonates or hatchlings to become sexually mature and produce young. The generation time is an estimate of the amount of time it takes one cohort to grow up and replace another, measured as the average duration of a life cycle between birth and reproduction.

generator potential. A change in the transmembrane electric potential within the receptive portion of a sensory neuron, the amplitude of which is graded in proportion to the stimulus intensity and is sufficiently large to produce action potentials at the spike generating zone of the membrane. Cf. *receptor potential*.

generic. *adj.* Of or pertaining to a *genus* or genera.

generic name. A scientific name of a taxon at *genus* rank, and the first word of a *binomen* or *trinomen*.

-genesis. Suffix meaning origin, descent, formation, or development.

genetic. *adj.* Of or pertaining to genetics or heredity.

genetic assimilation. Alteration of genetic material that reinforces phenotypic modifications which otherwise have no genetic basis.

genetic bottleneck. Reference to fluctuations in gene frequencies that occur when a large population contracts and then expands again with an altered gene pool, usually with reduced variability and a result of *genetic drift*. (Syn. *bottleneck effect, population bottleneck*) See *founder effect*.

genetic code. The consecutive nucleotide triplets (codons) of DNA and RNA that specify the sequence of amino acids for protein synthesis.

genetic correlation. The degree to which a gene(s) influences two or more traits simultaneously.

genetic distance. 1. Reference to the number of allelic substitutions per locus that have occurred during the independent evolution of two populations or species being compared. 2. The distance between linked genes in terms of recombination units or map units.

genetic differentiation. See *genetic divergence*.

genetic divergence. Reference to the accumulation of differences in allelic frequencies between populations that are evolving in geographical or ecological isolation. This phenomenon may also be called *genetic differentiation*.

genetic drift. Reference to random fluctuations of gene frequencies due to sampling errors, most evident in small populations. This phenomenon can give rise to evolutionary change that is independent of selection. (Syn. *Sewall Wright effect*. In 1931 S. Wright developed a detailed theory of genetic drift: *Genetics* 16:97–159.)

genetic engineering. Experimental alteration of the genetic makeup of an individual.

genetic equilibrium. See *Hardy-Weinberg law*.

genetic fingerprinting. See *DNA fingerprinting*.

genetic hitchhiking. Reference to the spread of a neutral allele (or in some cases deleterious alleles or mutations) in a population by means of it being closely linked to a beneficial allele that increases in frequency due to selection.

genetic induction. See *gene activation*.

genetic linkage. Association of genes that reside on the same chromosome, resulting in the nonindependent inheritance of two or more nonallelic genes (or particular alleles that are on the same chromosome).

genetic load. 1. The average number of deleterious recessive genes carried in heterozygous combination per individual in a population, multiplied by the mean probability that each recessive gene will cause premature death when in a homozygous condition. 2. The relative difference between the measured mean fitness of a population and the theoretic mean fitness that would exist if the most fit genotype presently in the population were to become ubiquitous.

genetic map. Reference to the specific linear array of mutable sites on a chromosome, as deduced from genetic recombination experiments.

genetic marker. Reference to a gene that can be employed to mark a nucleus, chromosome, or locus, or to otherwise identify a cell or individual that carries it, or is of known location and can be used as a reference point for mapping other genes. Such genes typically express a phenotype that is easily recognized.

genetic polymorphism. A prevalent or long-term condition in a population where two or more genotypes occur in frequencies that cannot be attributable to recurrent mutation.

genetic recombination. The process or condition in which combinations of genes different from parental genotypes are produced by *Independent assortment* or *crossing over*.

genetics. *n.* The study of heredity, concerned with the physical and chemical nature of genes, their products, control, replication, and consequences for development of organisms and evolutionary changes in populations.

genetic sex determination. (GSD) A process of sex determination in which the sex of an individual is established at the moment of conception by the particular complement of genes received from its parents. Cf. *environmental sex determination, phenotypic sex determination, temperature-dependent sex determination*. (Syn. *genotypic sex determination*)

genetic variance. The phenotypic variance of a trait in a population that is attributable to genetic heterogeneity.

genial (also **geneial**). *n.* An antiquated term commonly used in older literature for reference to the *chin shield* of squamates. See Fig. 12.

genioglossus muscle. A muscle involved in tongue rotation originating on the mandible and inserting on the tongue pad.

geniolabial. *n.* A scale situated between the *infralabials* and *chin shields* of *Gerrhonotus* (H.M. Smith & M. Alvarez del Toro, *Herpetologica* 19:100–105, 1963).

genitalia. *n.* Organs of the reproductive system involved in copulation. Such are secondary sex characters that result from the effects of hormones secreted by the male or female gonads.

genital ducts. See *mesonephric duct, Müllerian duct.*

genital pore. An external opening to the male or female reproductive structures.

genital ridge. That part of the embryonic *mesomere* that develops into gonads. (Syn. *gonadal ridge*)

genocline. *n.* A cline of genotype frequencies, sometimes resulting from hybridization between genetically distinct adjacent populations.

genome. *n.* 1. The total set of genes carried by a single gamete, which includes both nuclear and mitochondrial DNA. 2. The complete set of loci, or collection of genes, in an organism. 3. The complete DNA sequence of an organism.

genome size. The amount of DNA in the haploid genome, often measured in units of mass for eukaryotic organisms.

genomic library. A random collection of fragments of the DNA of a given species, which is then inserted into vectors and cloned in a suitable host. The collection must be large enough to include all the unique nucleotide sequences of the genome. (Syn. *DNA library*; Cf. *cDNA library*)

genomics. *n.* Studies of the structure and function of genomes, in context of species for which extensive nucleotide sequences are available. Current studies in genomics attempt to identify genes and to describe the functions they perform, the products they create, and their interactions with one another.

genomorph. *n.* A polyphyletic taxon of generic or subgeneric rank that includes superficially similar but not closely related species.

genotype. *n.* 1. The genetic makeup of an individual, which, along with other factors such as environment, determines the organism's *phenotype*. 2. The set of genes an organism possesses at a locus or loci of interest.

genotypic sex determination. See *genetic sex determination.*

genotypic species cluster concept. See *species.*

genotypic variance. Reference to the phenotypic variance of a given character in a population that is attributable to differences in genotype among individuals.

gentamicin. *n.* A bacteriocidal aminoglycoside antibiotic that is primarily effective against Gram-negative bacteria such as *Pseudomonas.*

genus. (*pl.* **genera**) *n.* A category of taxonomy consisting of a grouping of closely related species and comprising a level of classification above *species* and below *family.* The genus comprises the first name in a scientific name (*e.g.* The species *Vipera berus* belongs to the genus *Vipera*).

Genyophryninae. *n.* An anuran subfamily of *Microhylidae.*

geo-. Prefix meaning "earth," "terrestrial."

geochronology. *n.* The study of time measurement in relation to Earth's evolutionary history.

geocline. *n.* A graded sequence of character variation through a series of populations, correlated with spatial or topographical separation.

Geoemydidae. *n.* See *Bataguridae.*

geographical barrier. Any geographical feature that interrupts gene flow between populations.

geographical equivalents. Different taxa that have generally the same distributional pattern. Coined by A. Grobman (*N.Y. Acad. Sci.* 45:261–316, 1944).

geographic information system. (GIS) Computer-based technologies for storing, organizing, and analyzing any spatial array of data or geographically referenced information.

geographical isolation. Separation of potentially interbreeding populations by geographic barriers.

geographical race. A *race* that is geographically separated from other populations of the same species and frequently regarded as subspecies.

geographic speciation. See *allopatric speciation.*

geographic variation. Reference to any differences between spatially separated populations of a species.

geological Eon. The largest and most inclusive unit or division of geologic time. Eons are divided into Eras, which are in turn divided into Periods, Epochs, and Stages. See Table 1.

geological Epoch. A unit of geologic time and subdivision of a geologic *Period.* The use of this term is usually restricted to divisions of the *Tertiary* and *Quaternary* periods. See Table 1.

geological Era. A major period or interval of a *geological Eon.* See Table 1.

geological Period. A major period of interval of a *geological Era.* See Table 1.

geologic time. 1. Very long spans of time extending over millions of years. 2. The period of time from the formation of Earth to the beginning of recorded history. Prehistoric time.

geologic time divisions. See Table 1. (Syn. *stratigraphic time divisions*)

geology. *n.* The scientific study of Earth.

geomagnetotaxis. *n.* Magnetic compass taxis, an orientation and directed response of a motile organism to Earth's magnetic field. This has been demonstrated in a variety of animals including some amphibians.

geometric growth. Growth in which a population or the length of a body part is multiplied by a constant in each time interval. Cf. *arithmetic growth.*

geometric mean. The *n*th root of the product of a set of *n* positive numbers, yielding a measure of location lying between the *arithmetic mean* and the *harmonic mean.*

geometric morphometric sex estimation. A method using landmark-based *geometric morphometric* methods to detect sexual dimorphism, first demonstrated in relation to hatchling turtles. See N. Valenzuela et al., *Copeia* 2004:735–742, 2004.

geometric morphometrics, geometric morphometric analysis. Reference to morphometric methods that quanitify the shape of an object after the effects of nonshape variation in position, orientation, and scale have been held constant mathematically. These methods retain the geometry of shape throughout the analysis, have high statistical power, low bias and error of estimating mean shapes, and do not introduce patterns of covariation to the data. See F.J. Rohlf and L.F. Marcus, *Trends Ecol. Evol.* 8:129–132, 1993.

geometric progression or **series.** A series of values in which the ratio between successive values is the same (e.g., 10, 100, 1000, etc.).

geometric similarity. *Isometric growth*, or change in size without a change in relative shape or form. (Syn. *isometry*.)

geomorphology. *n.* Pertaining to the development, configuration, and distribution of the surface features of Earth.

geophagy. *n.* Reference to eating of earth or soil.

geophysics. *n.* The application of physical laws and principles to study of Earth.

Georgia Blind Salamander. Vernacular name for the species *Haideotriton wallacei*.

geotaxis. *n.* (*adj.* **geotactic**) A directed response to gravity by a motile organism, either away from or toward the ground.

geothermal. *adj.* Reference to heat energy derived from Earth's interior.

geotropism. *n.* (*adj.* **geotropic**) An orientation response to gravitational field. (Syn. *gravitropism*)

germ cell. A *gamete*.

germinal. *adj.* Reference to the nature of a germ cell or early stages of development.

germinal cells. Cells that produce gametes by meiosis. Cf. *germ cell*.

germinative layer. See *stratum germinativum*.

germ layers, germinal layers. Three primordial cell layers present in the early vertebrate embryo, from which all tissues and organs differentiate. See *ectoderm, endoderm, mesoderm*.

germ plasm. The hereditary material transmitted to offspring through the germ cells.

gerontology. *n.* The study of aging.

gerontomorphosis. *n.* Evolutionary change due to modifications of adult structures.

Gerrhonotinae. *n.* A lizard subfamily of the *Anguidae*.

Gerrhosauridae. *n.* A clade (family) of lizards having scales arranged in transverse rows and underlain by osteoderms. There is a prominent lateral fold along the body, and some species exhibit reduced limbs. Six genera and approximately 35 species occur in sub-Saharan Africa and Madagascar. These lizards are often included in the Cordylidae, their sister group, in which case they comprise the subfamily Gerrhosaurinae.

Gerrhosaurinae. *n.* In some classifications, a lizard subfamily of the *Cordylidae*. See *Gerrhosauridae*.

gestalt. *n.* A condition of biological phenomena in which the properties of the functional whole differ from those predicted from the sum of the component parts. (Syn. *holism*)

gestation, gestation period. *n.* The period of embryonic development from fertilization to birth of young in viviparous animals. Pregnancy. This term usually is used in reference to mammals.

ghara. *n.* The prominent protuberance at the dorsal tip of the snout of mature male Gharials. This structure develops to form a cover over the nostrils and serves as a chamber to produce a loud buzzing sound when the animal exhales during courtship and territorial advertisement. The nasal knob was named from the Hindi word for "mud pot" and is the feature for which the term gharial is derived. (Syn. *nasal knob*)

Gharial. *n.* See *Gavial*.

GH-inhibiting hormone, GIH. See *somatostatin*.

Ghost Frogs. Vernacular name for African species of heleophrynid frogs belonging to the genus *Heleophryne*.

ghost lineage. An assumed fossil record of a lineage not yet found, but inferred to exist based on reasonable evidence.

Giant Burrowing Frogs. Vernacular name for species of Australian frogs belonging to the myobatrachid genus *Heleioporus*.

Giant Garter Snake. Vernacular name for the American colubrid species *Thamnophis gigas*.

Giant Geckos. Vernacular name for largely Australian species of Geckos belonging to the genus *Rhacodactylus*.

Giant Salamanders. 1. Collective vernacular name for salamanders belonging to the family Cryptobranchidae and to the genus *Andrias* in particular (Chinese and Japanese Giant Salamanders). 2. The name also applies to several species in the American family Dicamptodontidae and the genus *Dicamptodon*.

Giant Toad. Vernacular name for the bufonid species *Chaunus* (formerly *Bufo*) *marinus*. A New World species, this anuran has been introduced in many tropical and subtropical parts of the world outside its native range, including Australia, Pacific islands and other parts of Asia, and the Americas. (Syn. *Cane Toad, Marine Toad*)

gibbose, gibbous. *adj.* Humped or protuberant.

giga-. (**G**) Prefix used to denote unit x 10^9.

gigabase. (**gb**, or **gbp** for gigabase pairs) *n.* A unit of length for DNA molecules, equivalent to one billion nucleotides.

gigantism. *n.* A condition of being significantly larger than the norm.

Gila Monster. Common name for the helodermatid lizard species *Heloderma suspectum*, common to southwestern

United States. The name also is applied to the family Helodermidae. See also *Beaded Lizards.*

Gilbert's Skink. Vernacular name for the American scincid species *Eumeces gilberti.* Note that H.M. Smith (*J. Kansas Herpetol.* 14:15, 2005) recently proposed that *Plestiodon* be a replacement name for the genus *Eumeces* in North America.

gill. *n.* Thin, membranous, and highly vascular structures used primarily for exchange of respiratory gases (oxygen, carbon dioxide), release of nitrogenous waste, and osmotic exchange between an aquatic organism and its water environment. Gill structures are characteristically located in the pharyngeal region of aquatic or larval amphibians and are either *external* or contained within a gill chamber, which is irrigated or ventilated.

gill arch. A bony or cartilaginous dorsoventrally curved bar situated between gill slits and supporting the internal gills of fishes and many larval amphibians.

gill chamber. The cavity which encloses the internal gills and through which water passes to exit at the *spiracle.* (Syn. *branchial chamber, opercular cavity, opercular chamber, opercular sac, peribranchial chamber*)

gill cleft. See *gill slit.*

gill filament. The fine processes of the gills of fishes and larval amphibians, collectively providing a large surface area and functioning to exchange respiratory gases.

gill filter rows. Dense rows of folded epithelium arising internally from the gill bars of tadpoles and used to collect particulate matter that is drawn into the mouth. These structures may be supported by cartilage at the base, but they do not have an osseous skeleton. (Syn. *gill ruffles*)

gill raker. Any of the numerous bony or cartilaginous projections from the internal surfaces of gill arches in fishes and larval amphibians, functioning to prevent solid objects from passing through the gill slits.

gill ruffles. See *gill filter rows.*

gill slit. A cleft or narrow opening at the base of the external gills and lying between the *gill arches,* providing a passage for excurrent water passing over the gills. (Syn. *branchial cleft* or *opening; gill cleft*)

gingiva. *n.* (*adj.* –**al**) The mucous membranes and underlying soft tissues that surround teeth and gums.

GIP. See *gastric inhibitory peptide.*

girdle. *n.* 1. In anatomy, a curved or circular structure, especially one that encircles another structure. 2. A group of connected bones that provide support for a pair of limbs. See *pectoral girdle* and *pelvic girdle.*

Girdled Lizards. Collective common name for various species of African lizards belonging to the *Cordylidae,* especially the genus *Pseudocordylus,* and also Madagascan species of the gerrhosaurid genus *Zonosaurus.*

girth. *n.* The measured circumference of an animal, usually at the body center or abdomen.

GIS. See *geographical information system.*

gizzard. *n.* A term sometimes applied to the crocodilian stomach, which has an enlarged and muscular fundic chamber containing gastroliths that are possibly involved in mechanical (grinding) reduction of food. Correct usage of this term should be reserved for the muscular posterior portion of the multichambered stomach of birds.

glacial epoch. See *Ice Age.*

gland. *n.* An aggregation of specialized cells that synthesize and secrete, or excrete, specific chemical compounds usually having various physiological actions in the body.

Gland of Gabe. A large, distinct seromucous dental gland on the lower jaw of some anguimorph lizards. See M. Gabe & H. Saint Girons, *Mem. Mus. Nat. Hist. Nat., Ser. A, Zool.,* 58:1–118, 1969; E. Kochva, pp. 43–161 in B.C. Gans & K.A. Gans, eds., *Biology of the Reptilia,* Vol. 8, Physiol. B. In *Heloderma* it is a functional venom gland associated with grooved teeth. (Syn. *mandibular gland, dental gland*)

glandula parapancreatica. See *parapancreas.*

glandular. *adj.* Of or pertaining to glands.

glandular scale. Sometimes used in reference to the *escutcheon* scales of lizards.

glans penis. The apical portion of the turtle or mammalian penis.

Glass Frogs. Collective vernacular name for numerous species of anurans belonging to the family *Centrolenidae,* especially the genera *Hyalinobatrachium* and *Centrolene.*

Glass Lizards. Collective common name for limbless species (*Ophisaurus* spp.) of anguid lizards, characterized by easily autotomized elongate tails.

glenoid. *n.* A socket or smooth, shallow depression, especially as part of a joint.

glenoid cavity, glenoid fossa. The concave joint surface that articulates with the humerus in the pectoral girdle, the femur in the pelvic girdle, or between the skull and mandible.

glia, glial cells. *n.* Nonexcitable supportive cells associated with neurons in nervous tissue, providing nutrition or guidance for migration during development. (Syn. *neuroglia*)

glide, gliding. *v., n.* Reference to largely passive but controlled movement through air, employed by certain arboreal amphibians and squamate reptiles that use a flattened body or appendages as airfoils to significantly reduce the angle of descent after jumping or falling from an elevated site. J.A. Oliver (*Amer. Nat.* 85:171–176, 1951) suggested this term be applied to movement through air involving a pathway more than 45° from the vertical (or less than 45° to the horizontal). See also *parachuting.*

Global Amphibian Assessment. (GAA) The first comprehensive assessment of the conservation status of the world's known species of frogs, toads, salamanders, and caecilians. The current website (http://www.globalamphibians.org) presents results of the assessments, including IUCN

Red List threat category, range maps, ecological information, and other data for every amphibian species.

Global Biodiversity Information Facility. (GBIF) A website project that facilitates digitization and global dissemination of primary biodiversity data. Current URL: http://www.gbif.org

global positioning system. (GPS) Reference to a hand-held or vehicle-mounted system that uses satellite communications to determine the geographical position and other navigational information.

global warming. Current reference is to past, present, and long-term projected warming of the planet due largely to increased emissions of *"greenhouse gases."* Increased atmospheric concentrations of greenhouse gases since the beginning of the industrial revolution have committed Earth to some future rise in temperature.

Global Zoo Directory. A worldwide comprehensive register of zoological collections. As of this writing, the Directory is no longer available electronically from the Conservation Breeding Specialist Group, which has discontinued collecting information from specific zoos and posting the Global Zoo Directory at their Web site.

globins. *n.* 1. Generally, a widespread group of respiratory proteins including the hemoglobins. 2. The protein or polypeptide constituent of hemoglobin.

globular. *adj.* Rounded or elliptical in shape.

globulins. *n.* A class of water-insoluble plasma proteins important to immune response, clotting, fluid balance, and other functions.

Gloger's rule. A "bioclimatic rule" which suggests that endotherms have lighter coats of insulation (fur or feathers) color in cold, wet climates and darker coats in warm, dry climates. This rule is not applicable to skin color of ectothermic vertebrates, and colors of amphibians and reptiles often change with temperature in converse to Gloger's rule. See also *Allen's rule*, *Bergmann's rule*, and *Cope's rule*.

glomerular filtration rate. (GFR) The rate of filtration of plasma from all nephron capillaries in the kidneys, equal to the total amount of fluid per minute entering the kidney tubules.

glomerulus. (*pl.* –i) *n.* A coiled mass of capillaries, usually used in reference to the filtering capillaries of the kidneys. Glomeruli of reptiles are not as vascular as those of birds and mammals. See *renal corpuscle*.

glomus. *n.* A collection of *glomeruli* grouped together. More generally, a histologically distinct body having numerous fine arterioles connecting directly to veins and often associated with a rich nerve supply.

glossal. *adj.* Reference to the tongue.

glossal skeleton. Reference used by H.M. Smith (*Evolution of Chordate Structure,* p. 170, 1960) to the *hyoid apparatus*.

glossohyal. *n.* See *entoglossal*.

glossopharyngeal, glossopharyngeal nerve. *n., adj.* The ninth cranial nerve, associated with normal swallowing and palatal movements.

Glossy Snake. Vernacular name for two American colubrid species belonging to the genus *Arizona*.

glottis. *n.* The opening from the *pharynx* to the *trachea* or *larynx*.

glove, gloving. *n.* Reference to the condition, or process, of developing pigmented, spinelike outgrowths on the ventral surface of the forelimb in mated male *Xenopus* frogs. Collectively, these structures assist the male in maintaining a firm grip on female frogs during *amplexus*. (Syn. *nuptial brush*)

glucagon. *n.* A peptide hormone released by pancreatic islet cells, induced by low blood sugar or growth hormone and acting to influence both steroidogenic and chromaffin tissues and to stimulate glycogenolysis in the liver. Mammalian glucagon produces effects in amphibians similar to those observed in fish and mammals, including modulation of adrenal gland, thyroid, and endocrine pancreas.

glucocorticoid. *n.* Any of a group of steroid hormones produced in the adrenal gland and serving a variety of functions, predominantly related to carbohydrate regulation and metabolism, inflammation and immune responses, and stress responses. Increased levels of endogenous glucocorticoids are found in embryonic crocodiles near the time of hatching and in larval amphibians that are near metamorphosis. In combination with thyroid hormone, glucocorticoids also stimulate surfactant production in developing lungs of some reptiles. The most important glucocorticoids in amphibians and reptiles are *corticosterone* and *cortisol*.

gluconeogenesis. *n.* The synthesis of carbohydrates from noncarbohydrate sources, e.g. amino acids.

glucose. *n.* A six-carbon sugar that provides the primary metabolic fuel in cells.

glucose-dependent insulinotropic polypeptide. See *gastric inhibitory peptide*.

glued amplexus. See *amplexus*.

glutamate. *n.* A common excitatory synaptic transmitter in the vertebrate central nervous system.

glycerol. *n.* A trihydric alcohol that combines with fatty acids to form fats.

glycocalyx. *n.* A meshwork of acid mucopolysaccharides and glycoprotein filaments associated with the microvilli of the intestinal *brush border*.

glycogen. *n.* A branched polymer of glucose found in animals and typically stored in liver and muscles.

glycogenesis. *n.* Synthesis of glycogen from carbohydrates.

glycogenolysis. *n.* The breakdown of glycogen to glucose.

glycolipid. *n.* A lipid that contains carbohydrate.

glycolysis. *n.* The anaerobic, enzymatic conversion of glucose to the simpler compounds of lactate or pyruvate, resulting in energy production as ATP. This constitutes the principal route of carbohydrate breakdown and oxidation and is an important aspect of activity metabolism in amphibians and reptiles that utilize glycolysis for ATP production in muscle during short but intense bursts of activity.

glycolytic. *adj.* Of, or pertaining to, *glycolysis*.

glyconeogenesis. *n.* See *gluconeogenesis*.

glycoprotein. *n.* Any of a class of conjugated proteins consisting of a compound of protein with a carbohydrate group. (Syn. *mucoprotein*)

glycoside. *n.* A compound that yields a sugar upon enzymatic hydrolysis.

glyphodont. *adj.* Possessing teeth which are grooved or channeled for passage of venom. *I.e.,* possessing *fangs*. (Syn. *glyphous*)

glyphous. *adj.* See *glyphodont*.

gnathostome. *n.* Any vertebrate having jaws.

Goannas. *n.* A common name for Monitor Lizards (*Varanidae*) in Australia.

goblet cell. *n.* A mucus-secreting cell found in most epithelia.

goitre, goiter. *n.* Abnormal enlargement of the thyroid gland, usually related to disturbance in iodine metabolism and hormone secretion. (In older herpetological literature, this term was sometimes used incorrectly to designate the *dewlap*.)

golden cell, golden pigment cell. See *xanthophore*.

Golden-eyed Treefrogs. Vernacular name for South American frogs belonging to the hylid genus *Phrynohyas*.

Golden Frogs, Golden Toad. Collective vernacular name for species of anurans belonging to the dendrobatid genus *Phyllobates* and the ranid genus *Mantella*.

Goltz's clasping reflex. See *clasping reflex*.

gomphodonty. *n.* See *gomphosis*.

gomphosis. *n.* Condition whereby roots of teeth are embedded in a socket to whose walls they are attached by unmineralized fibers. Among reptiles, crocodilians are the only group featuring a gomphosis-like connection between alveolar bone and the roots of teeth. In general usage, *gomphodonty* is synonymous with *thecodonty*, but the latter term is preferable.

gonad. *n.* A reproductive organ in which *gametes*, eggs or sperm, are produced: the *ovary* or *testis*, which are primary sexual characters.

gonadal recrudescence. The seasonal reestablishment of ovarian or testicular development after a period of gonadal quiescence.

gonadal ridge. See *genital ridge*.

gonadal steroids. Steroid hormones secreted from gonads. Principal gonadal steroids are *progesterone, estradiol-17β*, and *testosterone*. Gonadal steroids exert an autocrine or paracrine function by influencing localized tissues, and function as endocrine hormones when released into the bloodstream to affect distant target tissues.

gonadogenesis. *n.* The embryonic development of the gonads.

gonadotropins, gonadotropic hormones. *n.* Hormones that influence the activity of the gonads, particularly those secreted by the anterior pituitary, follicle stimulating hormone, and luteinizing hormone. See *pituitary gonadotropins*.

gonadotropin-releasing hormone. A hormone secreted from the *hypothalamus* and which stimulates the release of *gonadotropic hormone* from the *pituitary*.

gonaduct. *n.* A duct from gonad to exterior.

Gondwana or **Gondwanaland.** *n.* The southern supercontinent formed by the initial breakup of *Pangaea* in the Mesozoic and later fragmented to form South America, Africa, Arabia, India, Australia, New Zealand, and Antarctica. Cf. *Laurasia*.

gonial, goniale. *n.* See *prearticular*.

gonosomatic index. (GSI) The ratio of total gonad mass to total body mass of an individual, usually expressed as a percentage.

goodness of fit. Statistical agreement between a set of observed frequencies and a set of expected frequencies, usually measured by a *chi-squared test*.

Gopher Frog. Vernacular name for the North American ranid species *Lithobates* (formerly *Rana*) *capito* and *L.* (formerly *Rana*) *sevosa* (Dusky Gopher Frog). For recent taxonomic revision see D.R. Frost et al., *Bull. Amer. Mus. Nat. Hist.* 297:1–370, 2006.

Gopher Snake. Vernacular name for American colubrid snake species belonging to the genus *Pituophis*, especially *Pituophis catenifer* and subspecies. See also *Bullsnake* and *Pine Snakes*.

Gopher Tortoise. Vernacular name for tortoises belonging to the genus *Gopherus*, especially the well-known species *Gopherus polyphemus* of the southeastern United States.

Gosner stages. A staging table for developmental sequence of anuran embryos and larvae, described by K.L. Gosner (*Herpetologica* 16:183–190, 1960) and used extensively by embryologists. Cf. *Harrison stages;* see *stage*.

Gould's Goanna. See *Sand Monitor*.

gout. *n.* A disease in which there is excess *uric acid* in the blood and deposits of uric salts in various tissues, especially the viscera and joints.

G proteins. Guanine nucleotide-binding regulatory proteins that react with transmembrane receptor proteins that are activated by binding of a signaling ligand such as a hormone. Such activated G proteins dissociate from their receptors and, in turn, activate effector proteins that control levels of second messengers.

GPS. See *global positioning system*.

gracile. *adj.* Gracefully slender or slim.

gracilis muscle. A superficial sheet of muscle extending along the undersurface of the thigh and functions to flex the knee. The homologous muscle in amphibians and reptiles is generally called the *puboischiotibialis*.

grade. *n.* Reference to a group of organisms that possess a similar level of adaptive organization. A level of evolutionary attainment.

graded potential. A nerve impulse, or change in membrane potential, that is proportional to the intensity of a stimulus and which decays with the distance it is conducted over the membrane. Cf. *action potential*.

graded response. A response that increases in amplitude or intensity as a function of the energy of a stimulus. This is an incremental membrane response that is not all-or-none.

graded signal. A signal used in communication that varies in frequency and/or intensity and thus conveys quantitative information. Cf. *discrete signal*.

gradient. *n.* A change in a quantitative property of a system per unit of distance over which the property changes. Thus, the magnitude of a gradient depends on both the difference in value or level of the property and the distance over which such difference is measured. Also used more generally to describe a regularly increasing or decreasing change in a factor.

gradient analysis. 1. See *ordination*. 2. Examination of species distributions along environmental gradients to determine whether ranges of different species coincide to produce discrete communities.

gradualism. *n.* See *phyletic gradualism*.

graft. *n., v.* Reference to the transfer of a small piece of tissue from its normal position to another position in either the same or a different organism. See *allograft, autograft, heterograft, homograft, xenograft*.

Graham's Crayfish Snake. Vernacular name for the American colubrid *Regina grahamii*.

gram, gramme. (g) The *cgs* unit of mass, defined as 1×10^{-3} of SI base unit kilogram.

gram calorie. (gcal) Same as *calorie*.

graminivore. *n.* (*adj.* -ous) An animal that feeds on grasses.

Gram negative. Reference to microorganisms that appear pink microscopically following application of the Gram's staining process.

Granite Night Lizard. Vernacular name for the American xantusiid species *Xantusia henshawi*.

Granite Spiny Lizard. Vernacular name for the American iguanid lizard species *Sceloporus orcutti*.

granivore. *n.* (*adj.* –ous) An animal that feeds on seeds.

granular. *adj.* 1. Possessing granules. 2. Having a "pebbled" surface texture; small, convex, and nonoverlapping.

granular gland. A major type of alveolar gland found in amphibian skin, often concentrated into macroscopic clusters and typically producing defensive, toxic, multiple, holocrine secretions including amines, alkaloids, and peptides. (Syn. *poison gland*) See Fig. 28.

Granular Toad. Vernacular name for the Middle American bufonid species *Chaunus* (formerly *Bufo*) *granulosus*. For taxonomic revision see D.R. Frost et al., *Bull. Amer. Mus. Nat. Hist.* 297:1–370, 2006.

granulation. *n.* 1. The formation or division of a hard substance into smaller particles. 2. The formation in wounds of small rounded masses of tissues in healing, e.g. new capillaries on the surface of a wound.

granuliberin. *n.* A peptide isolated from *Rana rugosa* having potent chemotactic and disruptive actions on mammalian mast cells, which liberate histamine during inflammatory reactions. See also *mast cell disrupting peptides*.

granulocyte. *n.* A white blood cell characterized by cytoplasmic granules with affinity for acidophilic or basophilic stains.

granuloma. *n.* A mass or nodule of chronically inflamed tissue with *granulations*, usually associated with infection.

Grass Frogs. Collective vernacular name for various species of frogs belonging to the genera *Ptychadena, Rana, Limnodynastes, Pseudacris, Litoria, Fejervarya* (formerly *Limnodynastes* and *Rana*)*, Eleutherodactylus,* and *Ptychadena*.

grassland. *n.* A terrestrial biome where herbaceous grasses are the dominant vegetation.

Grass Lizards. Vernacular name for species of Asian lacertid lizards belonging to the genus *Takydromus*.

Grass Snake. Vernacular name for the European natricine *Natrix natrix* and several species of Afro-Asian *Psammophis*.

grass swimming. A mode of locomotion in limbless or nearly limbless reptiles that propel themselves through thick grass using sinuous curves of the body. In lizards, reduced limbs are pressed against the sides of the body during locomotion.

gravel. *n.* Sediment particles between 2 and 256 mm in diameter.

gravid. *adj.* The condition of a female being swollen from carrying eggs or accumulated embryos. (Syn. *pregnant*)

graviperception. *n.* The perception of *gravity*.

graviportal. *adj.* 1. Adapted for supporting great body weight. 2. Describing a terrestrial animal that is large and heavy-bodied, therefore slow-moving. Cf. *cursorial*.

gravitational pressure. The component of pressure in a fluid due to gravity. This pressure increases with depth or decreases with height below or above some arbitrary datum level. Often referred to as *hydrostatic pressure*.

gravitropism. *n.* See *geotropism*.

Gray. (Gy) A *SI* unit that is replacing the *rad*, equal to the amount of radiation causing 1 kg of tissue to absorb 1 joule of energy (= 100 rad).

graybody. *n., adj.* An object or body that emits only a fraction of the thermal energy emitted by an equivalent blackbody. By definition, a graybody has a surface emissivity < 1, and a surface reflectivity greater > 0. Cf. *blackbody*.

gray crescent. Marginal cytoplasm in the fertilized egg of amphibians, opposite the point of sperm penetration and exposed by movement of more intensely pigmented cytoplasm.

gray matter. Tissue of the brain and spinal cord that includes nerve cell bodies. Cf. *white matter.*

Gray Treefrog. Vernacular name for the North American hylid species *Hyla versicolor.*

grazing. *n.* Feeding on herbage, algae, or phytoplankton by cropping the entire surface growth or consuming the entire plant. Cf. *browsing.*

Greater Earless Lizard. Vernacular name for the American phrynosomatid (or iguanid, depending on classification scheme) species *Cophosaurus texanus.*

Great Ice Age. Reference to the *Pleistocene* Epoch. See Table 1.

Great Plains Skink. Vernacular name for the American scincid species *Eumeces obsoletus.* Note that H.M. Smith (*J. Kansas Herpetol.* 14:15, 2005) recently proposed that *Plestiodon* be a replacement name for the genus *Eumeces* in North America.

Great Plains Toad. Vernacular name for the American bufonid species *Anaxyrus* (formerly *Bufo*) *cognatus.* For taxonomic revision see D.R. Frost et al., *Bull. Amer. Mus. Nat. Hist.* 297:1–370, 2006.

Green Bamboo Viper. Vernacular name for the Chinese pitviper *Trimeresurus stejnegeri.* (Syn. *Bamboo Viper, Chinese Green Tree Viper*)

Green and Golden Bell Frog. Vernacular name for the Australian hylid species *Litoria aurea.*

green belt. Generally, any swath of open space or rural land that separates or interrupts urban development.

Green Frog. Vernacular name for the American ranid frog *Lithobates* (formerly *Rana*) *clamitans.* For recent taxonomic revision see D.R. Frost et al., *Bull. Amer. Mus. Nat. Hist.* 297:1–370, 2006.

greenhouse effect. Planetary warming that results from certain atmospheric gases allowing sunlight through to the Earth's surface while trapping energy radiated outward from the Earth's surface. Heat is absorbed by the lower layers of the atmosphere, and this heat is largely responsible for both long-term climate changes and short-term weather conditions. The principal *greenhouse gases* that contribute to this effect are water vapor, carbon dioxide, methane, chlorofluorocarbons (CFCs), hydrogenated chlorofluorocarbons (HCFCs), tropospheric ozone, and nitrous oxide.

Greenhouse Frog. Vernacular name for the frog *Eleutherodactylus planirostris*, endemic to Cuba and introduced elsewhere widely in the Americas including Mexico and the Atlantic Coastal Plain, including islands.

greenhouse gases. See *greenhouse effect.*

Green Iguana. A relatively large species of iguanid lizard, *Iguana iguana,* distributed throughout much of tropical Mexico and Central America. This species is very popular in the pet trade. See *Iguana Iguana.*

Green Parrotsnake. See *Parrotsnake, Lora.*

Green Rat Snake. Vernacular name for the American colubrid species *Senticolis triaspis,* also known as Mountain Rat Snake.

green rods. Photoreceptors in the retina of anurans and most salamanders that absorb wavelengths maximally at 432 nm.

Greensnakes, Green Snakes. *n.* Vernacular name for American colubrid snakes of the genera *Opheodrys* and *Liochlorophis.* Also vernacular for the Brazilian snake *Liophis viridis.* See *Rough Green Snake, Smooth Green Snake.*

Green Toad. Vernacular name for the American bufonid species *Anaxyrus* (formerly *Bufo*) *debilis, Anaxyrus* (formerly *Bufo*) *retiformis* (Sonoran Green Toad), and the European species *Pseudepidalea* (formerly *Bufo*) *viridis.* For taxonomic revision see D.R. Frost et al., *Bull. Amer. Mus. Nat. Hist.* 297:1–370, 2006.

Green Treefrog. Vernacular name for the North American hylid species *Hyla cinerea.*

Green Tree Python. Vernacular name for the Asian arboreal boid *Morelia viridis,* common to New Guinea.

Green Turtle, Green Seaturtle. Vernacular name for the marine turtle *Chelonia mydas* (Cheloniidae).

gregarious. *adj.* Associating with others of the same species; living with groups rather than in isolation.

Grey's Sea Snake. Vernacular name for the elapid Sea Snake species *Ephalophis greyi.*

grid-canaliculus. *n.* A visible line seen on the surfaces of certain bones and delimiting the margins of *growth zones.*

groin. *n.* The depression, angle, or cavity formed where the hind limbs join the abdomen.

grooming. *n.* The behavior of cleaning the body surface to remove foreign matter or parasitic organisms. Animals may groom themselves (*self grooming*) or others (*allogrooming*).

groove. *n.* A narrow, elongate depression or channel between two folds, used in reference to soft tissues, skin, or teeth.

gross ecological efficiency. See *ecological efficiency.*

gross primary production or **productivity. (GPP)** Reference to the total assimilation of organic matter by an autotrophic individual, population, or trophic unit, expressed per unit time per unit area or volume. Sometimes referred to simply as *gross production.* Cf. *net primary production.*

gross secondary production or **productivity.** Reference to total assimilation of organic matter by a primary consumer individual, population, or trophic unit, expressed per unit time per unit area or volume. Cf. *net secondary production.*

Grotto Salamander. Vernacular name for the North American plethodontid species *Typhlotriton spelaeus.*

Ground Boas. Vernacular name for several species of boid snakes belonging to genera *Boa, Candoia,* and *Casarea.*

ground color. The background or plain color on which more prominent aspects of color pattern or markings appear. The ground color is that part of a coloration not recognized as a specific type of pattern.

ground cover. See *basal area.*

Ground Skinks. Vernacular name for species of scincid lizards belonging to the genera *Leiolopisma* and *Scincella.*

ground plan. In systematics, this term has reference to the set of characters that are typical of relatively unmodified members of a clade. For example, the ground plan of lizards includes four limbs and a tail. Some lizards, including skinks and other groups, have evolved away from the ground plan, having modified tails and limbs that are reduced or absent.

Ground Snakes. Vernacular name for various species of neotropical burrowing snakes belonging to the colubrid genera *Atractus, Liophis,* and the American genus *Sonora.*

ground substance. The matrix of protein fibers surrounding *connective tissue,* varying from liquid to solid depending on the type of tissue.

ground truth. Correlation between aerial surveys and actual ground surveys of trackways or other evidence of animal presence to obtain estimates of numbers of nests, animals, etc. The ground estimates are compared with the aerial estimates to evaluate the accuracy of aerial surveys, which can be used to cover larger areas more easily or places where ground access is not feasible.

group. *n.* 1. An assemblage of related taxa, as in a species-group. 2. A general term for any interacting assemblage of organisms.

group behavior. Reference to the assemblage and interaction of a significant number of members of the same species for a period of time, usually for purposes related to breeding, predation, or protection.

group selection. Natural selection that acts on a group of individuals wherein characters are selected that benefit the group rather than the individual. See *kin selection.*

growth. *n.* Increase in size, number, or progressive complexity. When assessing size increases in animals, it is important to distinguish or specify *lean growth*—measured by changes in length or size—as distinct from increases in mass solely due to *storage.*

growth factor system. A family of proteins that function in regulating many cellular processes in virtually all tissues including cell proliferation, differentiation, and apoptosis.

growth hormone. (GH) A protein hormone that is secreted by the anterior pituitary and stimulates growth, directly influencing protein, fat, and carbohydrate metabolism and regulating growth rates. (Syn. *somatotropin*)

growth rate. The change in some measure of size over time.

growth ring or **zone.** Any one of a series of concentric regions seen in cross sections of bone from various vertebrates, representing a period of growth interrupted by lines representing periods of arrested or slower growth. Growth rings are also present in some reptilian laminae or scales. (Syn. *annual ring, annulus*; see also *lines of arrested growth*)

GSD. See *genetic sex determination.*

guanine. *n.* A purine and one of the four nitrogenous bases in DNA and RNA. Guanine forms the silver-white pigment in the platelets of *iridophores.*

guano. *n.* The white, pasty excretory product of birds, many reptiles, and a few amphibians that is rich in uric acid.

guanophore. *n.* See *iridophore.*

gubernaculum cordis. A ligamentous strand of tissue connecting the pericardium to the ventricular area of the heart of certain reptiles.

guild. *n.* A group of species having similar resource requirements and foraging strategies, thereby having similar roles in an ecological community.

guild matrix. A rectangular array of competitive coefficients of the member species of a guild, often referred to as the *community matrix.*

gular. *adj., n.* 1. Generally of, or pertaining to, the throat region. 2. Designation of scales on the lower jaw of snakes, situated between the lower *labials, chin shields,* and *ventrals,* and the foremost plate(s) on the plastron of most turtles (the latter being called *scapular* by some authors). See Figs. 6, 12, 15.

gular disc. A distinctly thickened, circular area of skin on the throat of many male anurans, overlying the vocal sac and expanding during throat inflation and calling.

gular expansion or **fan.** See *dewlap* and Fig. 40.

gular fold. A fold of skin running transversely across the throat immediately anterior to the insertion of the forelegs. See Fig. 17.

gular gland. See *musk gland.*

gular pouch. A pouch of bare skin between the lower mandibles. Antiquated term for *dewlap.*

gular sac. See *vocal sac.*

gular tentacle. An older, alternative term for *barbel.*

gular valve. See *palatal valve.*

gular vibration. The rapid movement of the buccal floor of various amphibians.

Gulf Coast Toad. Vernacular name for the bufonid species *Bufo valliceps.* Molecular work has suggested that northern and southern populations of this taxon are likely not conspecific, and the name associated with this taxon has been controversial since 1952.

gustation. *n.* The chemical sense of taste, involving chemosensory detection of ions and molecules in solution by means of specialized epithelial sensory receptors. Cf. *olfaction.*

gustatory receptors. Sensory structures that are involved in taste.

gut. *n.* The *alimentary canal,* especially the intestine.

gut-associated lymphoid tissue. (GALT) Glands associated with the mucosa of the posterior esophagus of snakes, sometimes organized into esophageal tonsils (boid snakes: E.R. Jacobson & B.R. Collins, *Develop. Comp. Immunol.* 4:703–711, 1980).

gutter. *n.* A deep groove running down either side of the vertebral keel, or the concave surface of the flared marginals in certain turtles.

guttural. *adj.* Produced in the throat and said of harsh or rasping sounds.

Gymnophiona. *n.* A monophyletic clade (order) of amphibians, collectively called *caecilians*, forming a sister taxon of Batrachia. These are elongate, legless (except in earliest fossil), burrowing, or aquatic animals representing 33 genera and approximately 170 species found in tropical habitats worldwide. Six families are recognized: see *Caeciliidae, Ichthyophiidae, Rhinatrematidae, Scolecomorphidae, Typhlonectidae, Uraeotyphlidae.* In a recently proposed taxonomic revision of amphibians (D.R. Frost et al., *Bull. Amer. Mus. Nat. Hist.* No. 297:1–370, 2006), Rhinatrematidae and *Stegokrotaphia* form two principal groups, with Ichthyophiidae and Caeciliidae comprising two families within the latter taxon. Uraeotyphlidae is included within the Ichthyophiidae, while Scolecomorphidae and Typhlonectidae are regarded as subfamilies within the Caeciliidae. (Syn. *Apoda*)

Gymnophthalmidae. *n.* A clade (family) that includes lizards collectively known as microteiids. Limb reductions and body elongation have evolved multiple times, and many of these lizards are specialized for burrowing. There are 35 genera and more than 150 species distributed from Mexico to southern Argentina, and in the Lesser Antilles. Four subfamilies are recognized: *Alopoglossinae, Gymnophthalminae, Rhacosaurinae,* and *Cercosaurinae.*

Gymnophthalminae. *n.* A subfamily of *Gymnophthalmidae.*

-gyn-. Combining affix meaning "female" or "ovary."

gynic. *adj.* Female. Cf. *andric.*

gynogenesis. *n.* Reproduction and developement in which the ovum is activated by sperm that degenerate without fusing with the nucleus; thus, the embryo contains only maternal chromosomes.

Gyr. A billion (10^9) years.

gyrus. *n.* A swollen ridge on the surface of the brain. Cf. *sulcus.*

H

H^2, h^2. See *heritability*.

habenula, habenular body. *n.* A group of small nuclei associated with the epithalamus of the *diencephalon*.

habit. *n.* 1. A pattern of behavior, usually inherited. 2. The characteristic form and physical appearance of an organism. (Syn. *habitus*)

habitat. *n.* The sum total physical and chemical properties of the environment of a particular species or population, usually typical of a region. See also *microhabitat*. (Syn. *oikos*)

habituation. *n.* A progressive loss of behavioral response probability following repetitive exposure to a stimulus.

habitus. *n.* See *habit*.

Habu. *n.* Vernacular name for any of several species of southeastern Asian pitvipers, especially the Japanese species *Trimeresurus flavoviridis* and *T. gracilis*.

haem-, haema-, haemat-, haemo-. See *hem-, hema*, etc.

haematemesis. *n.* Vomiting of blood.

haematuria. *n.* Presence of blood in the urine.

haemorrhagin. *n.* (*adj.*, **-ic**) A toxic protein that damages the walls of thin blood vessels and causes bleeding.

haemostasis. *n.* See *hemostasis*.

hair cell. A mechanosensory epithelial cell bearing minute hairlike structures known as *stereocilia* and, in some cases, a single *kinocilium*.

Hairy Frog. See *African Hairy Frog*.

haldane effect. A reduction of the total CO_2 content of the blood (at constant partial pressure of CO_2) that results from increased oxygenation of hemoglobin.

Haldane's evolutionary unit. A unit measure of increasing body size on an evolutionary time scale.

half-life. 1. In biology, the time required for an animal or organ system to eliminate one-half of the dose of a given chemical with which it has been inoculated. 2. The time required for half of the mass of a radioactive substance to decay into another substance. 3. The survival time for half the individual components of an unstable system.

half-moon track. A semicircular or similarly shaped track left by a sea turtle that emerged from the sea but turned around and returned almost immediately without nesting.

half-sibs. Individual organisms having only one parent in common.

halic. *adj.* Of or related to saline conditions.

hallux. *n.* The first, or innermost, digit on the hind foot.

haltere. *n.* See *balancer*.

hamada. *n.* A rocky desert.

Hamadryad. *n.* 1. A herpetological journal originated and published by the Centre for Herpetology, Madras Crocodile Bank Trust, located near Chennai (Madras), India. This journal publishes papers largely related to Asian herpetology. 2. A vernacular name for the *King Cobra*.

hamate. *adj.* Hooked.

handbook. *n.* A field guide or identification key.

handling time. The amount of time (per prey item) required for a predator to capture and eat a victim.

H & E. Abbreviation for the commonly used dual histological stains *hematoxylin* and *eosin*.

hanging valley. A smaller valley that terminates abruptly high above a main valley.

haplo-. Prefix meaning "simple" or "single."

haploid. (N) *adj.* Having half the usual complement of chromosomes charasteristically found in *diploid* somatic cells of organisms; one copy of a gene per locus per genome. Typical of the gametes of organisms whose union restores the diploid number.

haplotype. *n.* A single species that is included in a genus at the time of its designation, thereby becoming the type species of the genus.

Harderian gland. A gland associated with the front or medial margin of the orbit, secreting an oily fluid that functions to keep the cornea moist. In some snakes, however, the duct of this gland empties directly into the mouth or Jacobson's organ. In caecilians the gland is modified to provide fluid for transporting chemical particles from the tentacle to the vomeronasal organ. Generally, Harderian glands are more prominent than *lacrimal glands* in amphibians and reptiles. Cf. *lacrimal gland*. See Fig. 20.

hardiness. *n.* See *cold hardiness*.

hard pan. A compacted layer of soil, typically rich in deposited salts and restricting drainage and root penetration by plants.

Hardun, Hardun Agama. *n.* One of several subspecies of the agamid species *Laudakia* (*Agama*) *stellio*, inhabiting rocky habitats in southwestern Asia, Greece, and northeastern Africa.

hardware. *n.* The electronic and physical components of a computer. Cf. *software*.

Hardy-Weinberg principle, law or **equilibrium.** A qualified circumstance for an infinitely large, randomly interbreeding population wherein gene frequencies and genotype frequencies remain constant from generation to generation, provided there is no selection, migration or mutation. For a single pair of alleles, A and a, the frequencies of gametes carrying alleles A and a are, at equilibrium, p^2 (AA), $2pq$ (Aa), and q^2 (aa). Named after G.H. Hardy and W. Weinberg.

Harlequin Frog. Vernacular name for species of frogs belonging to the pseudid genus *Lysapsus*. The name also is used in reference to the colorful anurans of the genus *Atelopus* (family *Bufonidae*).

harmonic. *n.* In bioacoustics, an overtone, as present in some frog calls, whose frequency is an integral multiple of a fundamental frequency.

harmonic direction finder. A small tag that is useful for tracking small animals (e.g., small snakes). The tag consists of a diode and antenna that converts incoming radio signals to a harmonic frequency of the original, broadcast wavelength, and reradiates at the harmonic frequency (C. Engelstoft et al., *Herpetol. Rev.* 30:84–87, 1999). The tag does not require a power source, so it can be very small, weighing as little as 0.4 mg.

harmonic mean. The reciprocal of the arithmetic mean of the reciprocals of a set of observations.

Harrison stages. Developmental sequence of embryonic *Ambystoma*, described by R.G. Harrison and used extensively by embryologists. See pp. 44–66 in R.G. Harrison (ed.), *Organization and Develoment of the Embryo*, 1969. Cf. *Gosner stages.* See *stage*.

hastate. *adj.* Having triangular shape like an arrow.

hatch. *v.* The emergence of an embryo from an egg or extraembryonic membranes. In viviparous and ovoviviparous forms, hatching occurs in the oviduct.

hatchery. *n.* A human-made structure or enclosed area constructed for the incubation of eggs.

hatching. *n.* The process of departure from an enclosing eggshell after incubation and early development is complete.

hatching gland, hatching gland cell. Reference to any number of unicellular glands on the head or snout (sometimes nape) of embryonic anurans and urodeles (also fish embryos), functional with respect to production of protease enzymes that assist degeneration of the *vitelline membrane* and *jelly envelopes*. (Syn. *frontal gland, frontal cutaneous gland*)

hatchling. *n.* A newly hatched animal (just emerged from an egg). Given the potential size variation of conspecific hatchlings, the term is restricted by some to a size class approximately equivalent to neonatal age in which an age has been inferred only from indirect data based on size, mass, shape, or other benchmarks of development (D.J. Morafka, *Herpetol. Monogr.* 14:353–370, 2000). Cf. *neonate*.

Haversian bone. A type of secondary bone, either *compact* or *cancellous* (spongy), that is formed by internal reconstruction of preexisting bone (a process termed *Haversian remodeling*) and exhibits a series of distinctive vascular canals (Haversian canals). This bone type is lacking in squamates and some early fossil forms, while in others such as turtles and crocodilians, they are restricted to certain localized areas of the cortex. The compact form occurs only in endotherms (birds, mammals) among extant vertebrates, but is also common in cortical bone of the fossils of some dinosaurs and therapsids, suggesting they grew in similar ways. See *osteon*.

Haversian remodeling. See *Haversian bone*.

Hawksbill Turtle, Hawksbill Seaturtle. Vernacular name for marine turtles belonging to the genus *Eretmochelys* (Cheloniidae).

head. *n.* 1. The anterior or upper extremity of a structure, especially the anterior, cranial structure of an animal containing the brain and proximal sense organs. 2. The working surface of a *labial tooth* in tadpoles, often bearing cusps.

headbob display. Refers to rapid, jerky movements of the head, characteristically by repeated up-and-down movements. The term is usually applied to lizards or snakes in multiple contexts (territorial defense, courtship, same-sex aggressive encounters, signaling to predators), but also occurs in some turtles. (Syn. *bob, pushup*)

head-body length. See *snout-vent length*.

head cap. Area of black coloration that covers the snout and part of the dorsal head of Coral Snakes (*Micrurus* and *Micruroides*).

headgut. *n.* The cranial or anterior region of the alimentary canal that provides an external opening for the entry of food.

headland. *n.* Point of land along a coast.

head nod. One of the characteristic *bob* displays of *Anolis* lizards, consisting of a short series of up-and-down movements of the head by a smaller lizard when it is in the presence of a larger, dominant individual. Thought to be a gesture of submission. (Syn. *subordinate gesture*)

headslap. *n.* A display performed by mature crocodilians in which the head is held just above the water, the jaws are opened and clapped together, and the head slapped against the water. The sequence produces an audible "pop" followed by a tremendous splash, thought to be functional in attracting females during courtship and in deterring rival males.

head-starting. The experimental practice of hatching eggs in captivity and rearing hatchlings to a size such that, when they are released into the wild, they are less vulnerable to predation and other dangers. *Head-starting* is being used for turtles, crocodilians, lizards, amphibians, and Tuatara.

headwaters. *n.* See *stream headwaters*.

hearing. *n.* The perception of sound by the auditory system, involving vibrational stimulation of hair cells in the inner ear. The sensitivity to various sound frequencies varies among different taxa of vertebrates. (Syn. *audition*) See also *auditory mechanism*.

heart. *n.* The central muscular pump of the cardiovascular system, comprised of *atria* and *ventricle(s)*—the latter being anatomically divided only in birds and crocodilians among the reptiles.

heat. *n.* The energy of random molecular motion or vibration that is transferred down a thermal gradient by conduction, convection, or radiation. Cf. *temperature*.

heat capacity. The *specific heat* of a body multiplied by its mass, i.e. the amount of heat required to elevate temperature of the entire object by 1 °C (cal or joules per °C).

heat flux. The rate of heat flowing from or to an object, or past a reference datum (e.g., W/m^2).

heath or **heathland.** *n.* Vegetation community dominated by small leaved shrubs of Ericaceae and Myrtaceae, characteristic of low fertility, acidic, and poorly drained soils.

heat increment of feeding. (HIF) See *postprandial calorigenesis*.

heat of nutrient metabolism. See *postprandial calorigenesis*.

heat of vaporization. The heat per unit mass of a given liquid necessary to convert liquid to gas at its boiling point.

heat sensors. Sometimes used as a general terminology in reference to *loreal* or *labial pits*.

heat-shock proteins. Molecules involved in protein folding and cellular responses to nonlethal heat exposure or other stresses. Comparative studies demonstrate these proteins are highly conserved during evolution and function to prevent aggregation and denaturation of target proteins during heat exposure. See also *molecular chaperones*.

heavy hydrogen. Deuterium, an isotope of hydrogen having an atomic weight of 2.

heavy metals. Metallic elements of high specific gravity, such as mercury, lead, copper, arsenic, and others, that are toxic pollutants in various environments.

hectare. (ha) *n.* A metric unit of area, equal to 10^3 m^2 or 100 *are*.

hedge cells. Cells in the gas exchange parenchyma of some reptiles that contain microvilli and may be involved in reabsorption of fluid on the surface of the lung. These are located between ciliated epithelial cells and squamous (Type I) cells and cuboidal (Type II) cells.

hedonic. *adj.* In herpetology, used in reference to stimuli for sexual activity.

hedonic gland. Any gland which produces a secretion acting as a sexual stimulant.

heel appendage or **heel flap.** A triangular projection of skin located at the articulation of tibiofibula and tarsal bones in various species of frogs.

Heleophrynidae. *n.* A clade (family) of anurans with a single genus and six species that inhabit high mountain streams in extreme southern Africa.

heli-. Prefix meaning "sun."

heliophilic. *adj.* Sun-loving.

heliotaxis. *n.* Movement that is directed toward or away from sunlight. Cf. *phototaxis*.

heliotherm. *n.* (*adj.* **-ic**) An ectotherm whose principal heat source is radiant energy from the sun.

heliothermy. *n.* Thermoregulation achieved by basking in the sun.

Hellbender. *n.* Vernacular name for North American salamanders belonging to the genus *Cryptobranchus*, especially *C. alleganiensis*.

helmet. *n.* A term used to designate the bony extension of the skull in the lizard *Basiliscus* or the *casque* of various hylid frogs.

helmithiasis. *n.* Infestation with helminth parasitic worms.

Helodermatidae. *n.* A clade (family) of stout-bodied lizards with short, blunt tails, blunt heads, and modified salivary glands that secrete a toxic venom from the lower jaws. A single genus (*Heloderma*) is distributed from southwestern United States to Guatemala and represents two species that are the only venomous lizards known.

helophilous. *adj.* Thriving in marsh habitat.

helothermine. *n.* A toxic polypeptide isolated from venom of the Mexican Beaded Lizard (*Heloderma horridum*), unlike other polypeptides previously reported in literature and having, among other effects, capability to lower body temperature. Thus, helothermine is possibly a "hypothermic" toxin. Reported by J. Mocha-Morales et al. (*Toxicon* 28:299–309, 1990).

hemal arch. A ventrally directed arch of bone that surrounds and protects a blood vessel beneath caudal vertebrae of fish, salamanders, most reptiles, some birds, and many long-tailed mammals. (Syn. *ventral arch*) See also *chevron*. Cf. *neural arch*.

hemapophysis, hemopophysis. *n.* A term applied by various authors to the reptilian *chevron*, the *intercentrum* of cervical vertebrae, and various ventral arch elements or processes in the vertebral column.

hematocrit. *n.* The percentage of total blood volume occupied by red blood cells, frequently used as a measure of relative amounts of red cells and hemoglobin.

hematologic, hematological. *adj.* Of or relating to the blood.

hematopoiesis. *n.* (*adj.* **–tic**) The formation and development of blood cells. (Syn. *hemopoiesis, hemopoietic*)

hematoxylin. *n.* A commonly used histological stain that colors nuclei blue. See *H & E*.

heme. *n.* The iron porphyrin component of *hemoglobin*, responsible for binding of oxygen within a structural surrounding of carrier protein.

hemi-. Prefix meaning "half."

hemicelous, hemicoelous. *adj.* One of two basic conditions of *opisthocoely* in which the posterior half of the notochordal canal is hollow and the anterior half, with associated condyle, is solid bone. Cf. *holocelous* and *pseudocelous*.

hemiclitoris. *n.* A fully differentiated erectile structure in female Monitor Lizards (*Varanus* spp.) and some other squamates, distinguished from the male hemipenis by lack of spines. This structure is sometimes called the *anal gland* by snake systematists. See W. Bohme, *J. Zool. Syst. Evol. Res.* 33:129–132, 1995.

Hemidactyliini. *n.* A clade of *plethodontid* salamanders (not to be confused with *hemidactylus* Geckos) that includes members of *Plethodontinae* except for *Plethodontini* and *Bolitoglossini*.

hemipenial disc. A collar-like structure formed from modified *calyces* in the snake hemipenis. Observed in certain xenodontine snakes by H. Clark (*J. Iowa Acad. Sci.* 51:411–445, 1944).

hemipenis. (*pl.* –es) *n.* Either of paired copulatory organs lying in a cavity at the base of the tail in squamate reptiles.

Hemiphractidae. *n.* A clade (family) of anurans sister with *Meridianura* within the *Nobleobatrachia* and containing the Middle and South American genus *Hemiphractus*.

Hemiphractinae. *n.* A subfamily of the *Hylidae* with members in Panama and South America. Alternative classification to *Hemiphractidae*.

Hemisotidae. *n.* A clade (family) of anurans characterized by the absence of a sternum and a skull that is highly modified for burrowing (known as *Shovel-nosed Frogs*). The head is small, pointed, and delimited posteriorly by a transverse fold of skin. Unlike all other frogs, the snout is used to dig headfirst into soil. A single genus (*Hemisus*) with nine species occur in savanna regions of sub-Saharan Africa.

hemo-. *Word element* meaning "blood."

hemocoagulase, hemocoagulin. *n.* A substance or hemotoxin, and common component of snake venoms, that affects the coagulation of blood.

hemocoel. *n.* A blood-filled channel within connective tissue that lacks a continuous endothelial lining.

hemodynamic(s). *n, v.* Reference to the forces and flow patterns of blood circulating through vessels. Pertaining to actions of blood.

hemoglobin. *n.* The pigment of red blood cells responsible for reversible binding and transport of oxygen. In most vertebrates it is composed of four heme groups associated with four carrier polypeptide chains of globin.

hemolysin. *n.* A common component of snake venoms capable of breaking down red blood cells.

hemolysis. *n.* (*adj.* -lytic) The rupturing or dissolution of red blood cells, which liberates hemoglobin.

hemopathic. *adj.* 1. Reference to any disease of the blood. 2. Used, inappropriately, in herpetology for the effects of snake venoms that impair circulatory function. See *hemotoxic*.

hemopoiesis, hemopoietic. *n, adj.* See *hematopoiesis*.

hemorrhage. *n.* (*adj.* -gic) Bleeding, or escape of blood from vessels due to rupture or other injury.

hemorrhagin. *n.* A generic term that has been used to describe components of snake venom that have primary effects on blood vessels and lymphatic vessels, usually by means of decomposing the endothelial walls.

hemostasis. *n.* 1. The arrest of bleeding by vasoconstriction, coagulation, or other means. 2. Any phenomenon or action aimed at stopping a hemorrhage. 3. Interruption of blood flow.

hemostat. *n.* A small surgical clamp used for constricting blood vessels.

hemotoxic, hematoxic. *adj.* (*n.* **hemotoxin**) Reference to the effects of toxic chemicals or substances that impair circulatory function, or, specifically, those effects that act primarily on blood vessels and lymphatic vessels. The term also applies to toxins that cause *coagulopathy*. Used largely in reference to components of snake venoms in herpetological literature.

hemotumescence. *n.* Erection achieved by blood infiltration, as in erection of the *penis* and *hemipenes*.

Henderson-Hasselbalch equation. The formula widely used in acid-base physiology for calculation of the pH of a buffer solution: pH = pK' + log([H$^+$ acceptor]/[H$^+$ donor]). In common usage in respiratory physiology, pH = pK' + log ([HCO$_3^-$]/αP$_{CO_2}$), where α is the solubility coefficient for C$_{O_2}$ and P is the partial pressure of the gas.

Hennigian systematics. An approach to systematics that ignores morphological similarities in establishing phylogenetic relationships among taxa, except for synapomorphy. Recency of common ancestry is the fundamental criterion and is assessed primarily by recognition of shared derived character states. (Syn. *phylogenetic systematics*)

Hennig's deviation rule. A generalized principle that one of two daughter species diverges more strongly than another from the ancestral condition of a species that splits during its evolutionary history.

Henophidia. *n.* Boidlike snakes (= Booidea). Not a formal taxon.

Henry's Law. The principle that the quantity of gas that will dissolve in a liquid is proportional to the partial pressure of the gas in the gas phase.

heparin. *n.* An acidic mucopolysaccharide present in many tissues, especially the liver and lungs, having potent anticoagulant properties. Commercially available heparin solutions prepared from domestic animals are used with indwelling catheters or added to blood samples to prevent coagulation during studies of blood or hemodynamics.

heparinize. *v.* To render the blood unable to clot by use of heparin solutions.

hepatic. *adj.* Reference to the liver.

hepatic portal vein. A large vein that delivers blood directly from the intestine to the liver.

hepatocyte. *n.* A liver cell.

hepatotoxic. *adj.* Reference to a substance that is injurious or lethal to liver cells.

hepta-. Prefix meaning "seven" or "sevenfold."

herb. *n.* (*adj.* **herbaceous**) A nonwoody plant having stems that are not secondarily thickened and lignified, and which die down annually.

herbaceous stratum. The nonwoody growth of a forest floor.

herbivore. *n.* (*adj.* **–ous**) An animal that feeds on plants.

heredity. *n.* The transmission of genes, hence traits, from parents to offspring.

heritability. *n.* The proportion of total phenotypic variation that is attributable to genetic variation and thus expresses capacity for being inherited. *Broad heritability* (symbolized H^2) denotes the degree to which a trait is genetically determined and is expressed as the ratio of the total genetic variance to the phenotypic variance. Reference is to a polygenic trait in a given population. *Narrow heritability* (symbolized h^2) denotes the degree to which a trait is transmitted from parents to offspring and is expressed as the ratio of the additive genetic variance to the total phenotypic variance. See *additive genetic variance.*

Hermann's Tortoise. Vernacular name for the tortoise *Testudo hermanni.*

hermaphrodite. *n.* An animal that possesses both male and female sex characters. A *true hermaphrodite* possesses gonadal tissues of both sexes, whereas an individual of one genetic sex that possesses secondary sex characters of both sexes is termed a *pseudohermaphrodite.*

herp. *n.* A slang term for any individual amphibian or reptile. The usage is largely confined to amateur literature, and its use in formal writing should be discouraged. See *herper, herptile.*

herper. *n.* A slang term for one who collects, studies, or breeds amphibians and/or reptiles. Usage in formal writing should be discouraged. See *herp, herpetoculture.*

herpes. *n.* An inflammatory and highly contagious viral skin disease that affects turtles and various squamate reptiles.

herpesian. *adj.* Pertaining to amphibians and reptiles.

herpetoculture. *n.* A term used colloquially for the practice of breeding amphibians or reptiles, usually for commercial sale. A person who engages such activity is called a *herpetoculturist.*

herpetofauna. *n.* A biogeographic term to designate all populations of amphibians and reptiles that inhabit a particular region.

herpetogeny. *n.* The history of colonization and evolution in the establishment of modern amphibian and reptilian faunas.

Herpetologica. *n.* A quarterly journal published by the *Herpetologists' League*, established in 1936 by Chapman Grant.

Herpetological Conservation and Biology. A peer-reviewed journal publishing original research, reviews, perspectives, and correspondence on the conservation, ecology, and management of amphibians and reptiles.

Herpetological Conservation Trust. (HCT) A UK-based charity and charitable company established to further the conservation of amphibians and reptiles. Founded in 1989 by Vincent Weir and Ian Swingland and launched at the First World Congress of Herpetology in Canterbury. Current URL: http://www.herpconstrust.org.uk/

Herpetological Journal. A journal published quarterly by the British Herpetological Society.

Herpetological Monographs. A journal of the *Herpetologists' League* publishing lengthy research articles, syntheses, and special symposia.

Herpetological Review. A journal of the *SSAR* publishing shorter articles, range extensions, and technical notes of interest to the community of herpetologists.

herpetologist. *n.* A person who studies amphibians and/or reptiles as a profession.

Herpetologists' League. (HL) An international organization of people devoted to the scientific study of amphibians and reptiles (traditional usage; see *reptile*). The society was established by Major C. Grant in 1936. Current URL: http://www.inhs.uiuc.edu/cbd/HL/HL.html

herpetology. *n.* The scientific study of amphibians and reptiles, spanning all biological disciplines as applied to these specific taxa. One who studies amphibians or reptiles is called a *herpetologist.*

Herpetozoa. *n.* A quarterly journal devoted to all aspects of herpetology and published by the Austrian Herpetological Society, Natural History Museum, Vienna.

HerpNet. A collaborative effort by natural history museums to establish a global network of data for herpetological collections. Currently 53 institutions are participating in this project, and there is an open-ended invitation for other institutions to join. The mission is to make available the accumulated knowledge from more than 4 million specimens in worldwide museum collections by creating a distributed database with access from various portals. Current URL: http://herpnet.org

herptile. *n.* A vernacular and illegitimate term that is widely used as shorthand reference to amphibians and reptiles. The usage should be discouraged. See *herp.*

herptiliary. *n.* An outside enclosure that is used for maintaining captive amphibians or reptiles in near-natural conditions.

Hertwig's sheath. Part of the oral epithelium that is attached to the labial side of amphibian teeth.

hertz. (Hz) *n.* Cycles per second (cps), a derived *SI* unit of frequency (of a periodic phenomenon).

Hertz index. See *"effectiveness" of thermoregulation.*

Hesticobatrachia. *n.* A new monophyletic taxon, sister to *Ceratophryidae* within the new taxon *Chthonobatrachia*, and containing *Cycloramphidae* and the new taxon

Agastorophrynia (D.R. Frost et al., *Bull. Amer. Mus. Nat. Hist.* 297:1–370, 2006).

hetero-. Prefix meaning "different," "other," "variable," or "other than usual."

heterocannabilism. *n.* Cannabilism of nonkin. Cf. *filial cannibalism.*

heterochromatism. *n.* Reference to changes of color.

heterochronic mutation. A mutation that alters the timing of developmental events. See *heterochrony.*

heterochrony. *n.* Evolutionary changes in the timing or rate of development of specific organs or features. Specifically, the condition of having a different onset and cessation of growth, or a different rate of growth for a given feature, relative to that of this same feature in an ancestor.

heterocyclic. *adj.* Reference to any organic compound forming a ring consisting of carbon atoms and at least one atom other than carbon.

heterodont. *adj.* Having teeth that vary from one another in size and shape. Cf. *homodont.*

heterogametic sex. The sex that produces gametes containing unlike sex chromosomes. Cf. *homogametic sex.*

heterogamety. *n.* (*adj.* –ic) A condition of having different sex chromosomes in the same animal. Both male and female heterogamety occur in amphibians and reptiles. See *genetic sex determination.*

heterogamy. *n.* Union of gametes having different size or shape.

heterogeneity. *n.* The state of being dissimilar.

heterogeneous. *adj.* Characterized by elements or parts that are different or unrelated. Variable structure or nonuniform composition. Cf. *homogeneous.*

heterograft. *n.* A graft of tissue transplanted between animals of different species. (Syn. *heterotransplant, xenograft*)

heterologous. *adj.* Derived from a different source.

heterometabolism. *n.* A condition of having significant variation in the metabolic rate.

heteromorphic. *adj.* Having different forms, either as a function of the life history of a species or often in reference to appearances that are different from the "normal" form of a species. *Heteromorphic chromosomes* are homologous chromosomes that differ morphologically. This term should not be used as a synonym for *polymorphic.*

heteromorphosis. *n.* See *homeosis*

heterophagy. *n.* (*adj.* –ous) Feeding on a wide variety of food items.

heterophilic granuloma. See *pyogranuloma.*

heteroplastic. *adj.* Reference to *heterograft* or *chimera.*

heteroselection. *n.* Selection favoring *heterozygotes.* Cf. *homoselection.*

heterosis. *n.* Increased vigor of hybrid populations resulting from multiple characters associated with interspecific crosses. This term is used usually in reference to domesticated plants and animals. (Syn. *hybrid vigor*)

heterospecific. *adj.* Reference to members of different species. Cf. *conspecific.*

heterosynaptic modulation. A change in the efficacy of synaptic transmission at one synapse that is attributable to activity at another, separate synapse.

heterotherm. *n.* (*adj.* -ic) Basically ectothermic animals that are capable of varying degrees of endothermic heat production but do not regulate body temperature within a narrow range. Distinctions are sometimes made between *regional heterothermy* (variation of temperature in different parts of the body) and *temporal heterothermy* (variation of body temperature over time, due, e.g., to variable metabolic heat production attributable to activity).

heterothermy. *n.* 1. Generally, variation of body temperature. 2. A pattern of temperature regulation in which daily or seasonal variation in body temperature exceeds an arbitrary range of ±2 °C that defines homeothermy. Some physiologists prefer to restrict usage of this term to endotherms and not extend it to ectotherms.

heterotopic. *adj.* 1. Reference to presence or tranplantation of tissue in an abnormal location on the same organism. 2. Occurring in a wide variety of habitats.

heterotopic sulcus. A turtle shell seam that does not occur in the normal position for a species.

heterotransplant. *n.* See *heterograft* or *xenograft.*

heterotroph. *n.* (*adj.* –ic) An organism that depends for its nourishment on complex organic molecules derived from the ingestion of other organisms.

heterotypic. *adj.* Belonging to, or derived from, a different type. Cf. *homotypic.*

heterozygosity, heterozygous. *n., adj.* A condition of having one or more pairs of dissimilar alleles at one or more loci.

heterozygote. *n.* An individual that has inherited different alleles at one or more loci, and therefore does not breed true. Cf. *homozygote.*

heterozygote advantage. Reference to the circumstance of greater fitness associated with a heterozygote relative to either homozygote condition.

heuristic. *adj.* Providing direction in pursuit of a problem; a rule of thumb.

hexa-. Prefix meaning "six" or "sixfold."

hibernaculum. *n.* (*pl.* -a) The winter retreat where an animal hibernates or overwinters.

hibernal. *adj.* Of or pertaining to the winter. (Syn. *hiemal*) Cf. *aestival, autumnal, vernal.*

hibernation. *n.* A period of winter dormancy, involving prolonged rest and often suppressed metabolism in response to declining ambient temperatures. See *brumation.*

hide box. Any object used to provide a place for seclusion in the cage housing a captive amphibian or reptile.

hiemal. *adj.* See *hibernal.*

hierarchy. *n.* Generally, an integrated system comprising two or more levels wherein the higher levels control to some extent the activities of the lower levels. See *classification, dominance hierarchy*.

high energy phosphate bond or **compound**. A phosphorylated molecule that yields a large amount of free energy upon hydrolysis (e.g., ATP).

high-performance liquid chromatography. (HPLC) A chromatographic procedure that permits precise and rapid separation of molecular mixtures based on size, charge, and solubility.

hilus. *n.* A depression or pit on an organ, allowing entrance or exit of vessels and nerves.

Himalayan Pitviper. Vernacular name for the Asian pitviper *Gloydius himalayanus*.

hindbrain. *n.* See *rhombencephalon*.

hindgut. *n.* The terminal region of the gut, where unabsorbed parts of digested food are stored and eventually eliminated. Also refers to the embryonic structure from which the large intestine is formed. Cf. *foregut*.

hindgut fermentation. See *fermentation*.

hindlimb. *n.* Either of posterior limbs, each comprised of thigh, *crus,* and *pes*.

hinge. *n.* 1. A band of articulation where one broad bony element may move relative to another. Commonly used to describe the flexible union between parts of the plastron or carapace of turtles, permitting closure of the shell and sealing of the turtle within whenever it is threatened. See Fig. 6. 2. The softer, pliable region of skin between overlapping scales of reptiles, especially squamates. See Fig. 29. 3. See *hinged teeth*.

hinged teeth. Teeth with a ligamentous attachment confined to a certain region of the attachment site. The tooth folds at the hinge in response to pressure. The hinge is usually correlated with a durophagous diet but may also be a pressure-adapted mechanism. See A.H. Savitzky, *Amer. Zool.* 23:397–409, 1983 and F.C. Patchell & R. Shine, *J. Zool. Lond.,* Series A, 208:269–275, 1986.

hip. *n.* The region of the body around the articulation of the femur and pelvis.

hispid. *adj.* Having a roughened or spiny surface.

histamine. *n.* A vasoactive molecule that causes dilation of blood vessels and is formed by decarboxylation of histidine. It has a widespread distribution, including amphibian skin.

histochemistry. *n.* A branch of *histology* dealing with identification of chemical components of cells and tissues.

histogenesis. *n.* Formation and development of tissue.

histogram. *n.* A bar graph that illustrates a relative frequency distribution. The area over each class interval is proportional to the relative frequency of data within this interval.

histology. *n.* (adj. –ical) The study of cells and tissues and their compositional features, usually by microscopy in conjunction with thin slices of tissues using a *microtome*.

histolysis. *n.* The destruction of tissues.

histones. *n.* Small proteins that bind with DNA.

histopathology. *n.* The microscopic study of diseased tissue.

historical range. The known general distribution of a species as reported in current scientific literature.

historical zoogeography. Study of the distribution of animal species or higher taxa in context of their past history and the geographical evolution of the region in which they occur.

histrionicotoxins. *n.* A class of alkaloids possibly unique to the skin in certain genera of dendrobatid frogs.

hitchhiking. *n.* See *genetic hitchhiking*.

HL. See *Herpetologists' League*.

hogback. *n.* In geology, sharp-topped ridge formed by the erosion of steeply dipping beds.

Hognose Snakes, Hog-nosed Snakes. Collective vernacular name for various species of colubrid snakes belonging to the American genus *Heterodon*, neotropical genus *Lystrophis*, and Madagascan genus *Leioheterodon*.

Hognose Pitvipers. Vernacular name for nine neotropical species of pitvipers belonging to the genus *Porthidium*.

holarctic. *adj.* Distributed throughout North America, Asia, and Europe.

Holarctic region. A large zoogeographical region encompassing the *Palaearctic* (Old World) and *Nearctic* (New World) regions to include most of the northern hemisphere.

holder. *n.* See *adhesive organ* and *balancer*.

holism. *n.* (*adj. holistic*) A philosophical commitment to the belief that the entirety of a system (e.g. organism) is greater than the sum of its individual parts. Therefore, an organism cannot be understood entirely by studying its component parts in isolation. See *gestalt, reductionism*.

holo-. Prefix meaning "whole," "complete," "entire," or "total."

holoblastic. *adj.* Of or pertaining to a pattern of early cleavage in which the cleavage planes of a fertilized egg pass completely through the dividing cell, as in the eggs of anurans and salamanders. Cf. *meroblastic, discoidal cleavage*

Holocene. *n.* The present Epoch of geological time, also called *Recent*. See Table 1.

holocrine. *adj.* Reference to a glandular secretory mechanism in which the entire secretory cell is cast off and breaks up to release its contents.

holochordal. *adj.* Reference to amphibian centrum in which the segment of notochord from which the centrum is derived becomes calcified, i.e. the centrum is solid without a hollow center.

holocelous, holocoelous. *adj.* A condition of a vertebra in which the centrum remains hollow, and there is no evidence of calcification of the notochordal canal. This type of vertebra is typical of many primitive salamanders.

holocrine secretion. A secretory mechanism wherein entire secretory cells are sloughed and break up to release their contents.

holoendemic. *adj.* A species that has a narrow geographic distribution but is not of recent origin.

hological methods. Reference to study of the totality of a system to understand the operation as a whole without emphasis on the contribution of component parts. Cf. *merological methods*.

holonephros. *n.* A hypothetical ancestral vertebrate kidney that forms from the entire length of the mesomere with a nephron corresponding to each body segment Cf. *archinephros, mesonephros, metanephros, opisthonephros, pronephros*.

holophyletic. *adj.* Reference to an evolutionary lineage consisting of a species and all of its descendants.

holospondylous. *adj.* Term used by E.D. Cope to designate a centrum composed of a single bone, characteristic of extant tetrapods. The elements are fused. Cf. *aspidospondylous*.

holotrophic. *adj.* Reference to a predator that preys on a single species of organism.

holotype. *n.* The single specimen that is designated as the type specimen and used for the formal description of a nominal species.

holozoic nutrition. Nutrition that requires complex organic molecules.

Homalopsinae. *n.* A clade (subfamily) of Colubridae consisting of rear-fanged estuarine, marine, or freshwater snakes distributed from Pakistan and China to Australia.

homalopsines. *n.* A collective name for all member species of the colubrid family Homalopsinae.

homeo-. A prefix meaning "similar" or "entire."

homeobox. *n.* A domain or sequence of 180 base pairs near the end of certain genes having prominent roles in development, regulating cell position and differentiation. *Hox* genes are a well-known subset of homeobox containing genes, which are numerous and diverse in animals. See *Hox* genes and *homeodomain*.

homeodomain. *n.* A homeobox encoded domain consisting of 60 amino acids in proteins that bind DNA and usually act as transcription factors controlling the activity of downstream target genes. See *homeobox*.

homeomorph, homeomorphy. *n.* A state of having the same shape, with reference to morphologically similar organisms.

homeoplastic graft. Tissue that is grafted from one individual to another of the same species. (Syn. *homograft*)

homeosis. *n.* (*adj.* **homeotic**) The formation or occurrence of an organ or structure that is atypical or inappropriate for its location in an individual or species. The term was coined in 1894 by W. Bateson in his book *Materials for the Study of Variation Treated with Especial Regard to Discontinuity in the Origin of Species*. (Syn. *heteromorphosis*)

homeostasis. *n.* A condition of relative internal stability maintained by physiological control systems (e.g., body temperature, blood pressure, blood sugar or ion concentration, etc.). The concept was developed by W.B. Cannon (*The Wisdom of the Body*, 1932), although often attributed to Claude Bernard.

homeotely. *n.* (*adj.* –**ic**) Evolutionary changes of homologous structures.

homeotherm. *n.* (*adj.* -**ic**) An animal that regulates its body temperature within a relatively narrow range, regardless of environmental conditions, by controlling metabolic heat production and heat loss. By convention this term has been used in reference to birds and mammals, but it may have application to non-avian reptiles in limited circumstances (e.g., brooding pythons, large sea turtles, possibly some dinosaurs). (Syn. *homoiotherm*) See also *endotherm*.

homeothermy. *n.* (*adj.* –**ic**) Maintenance of a constant body temperature, independent of the means or environmental temperature.

homeotic mutations. Mutations that result in one developmental pattern being replaced by a different, but homologous one.

homeotype, homoetype. *n.* A specimen determined as conspecific when compared with the type of a species, judged by an author other than the original author of the type specimen. (Syn. *homotype*)

homeoviscous adaptation. Reference to temperature-induced changes in membrane fluidity. Adjustments are reversible and reflect largely changes in lipid composition that maintain near-normal fluidity and function of membranes when temperature changes. Note this phenomenon really is a type of *acclimation* rather than *adaptation*, although the ability to acclimate is clearly adaptive.

home range. 1. The area in which an individual organism might be active, including routine activity, foraging, and occasional visits to more peripheral points. (Syn. *activity area, home realm*) 2. The area in which an individual is typically active, exclusive of occasional wanderings to peripheral points. See also *minimum home range* and *territory*.

home realm. Term proposed by W.W. Milstead (*Amer. Midl Nat.* 65:127–138, 1961) to distinguish the total area used by an organism from the area of routine or intensive activity. He thought that usage of the term "*home range*" should be restricted to the latter.

home site. *n.* A place within the *home range* to which an animal regularly returns, offering seclusion and a focal point for territorial defense.

homing. *n.* The ability of an animal to return to a specific locality or geographic point, following natural or artificial displacement.

homo-. Prefix meaning "similar," "alike," or "the same."

homochromic. *adj.* Of or pertaining to a *mimic* that imitates the color of a model.

homodont. *adj.* Possessing teeth of a single type, without regional differentiation along the jaw. Cf. *heterodont*.

homogametic sex. The sex that produces gametes which carry only one kind of sex chromosome.

homogeneous. *adj.* Characterized by same, similar, or related elements. Uniform. Cf. *heterogeneous*.

homograft. *n.* See *homeoplastic graft*.

homolog, homologue. *n.* 1. In systematics, a character that defines a clade. 2. In evolutionary biology, homologs are characteristics (including DNA and protein sequences) that are similar in different species because they have been inherited from a common ancestor. 3. In cytology, see *homologous chromosomes*.

homologous. *adj.* 1. Generally, reference to similar characters having a common origin or ancestry. Structural correspondence, or equivalence of structure, based in common ancestry. 2. In a strict sense used by phylogenetic systematists, this word refers to characters that are descended from the same derived structure. 3. In wider usage in vertebrate biology, this term refers to characters that are descended from the same ancestral structure. See *homologous chromosomes*. 4. A third usage that has been employed by molecular geneticists denotes the meaning "similar" without any implications of evolutionary descent. Usage in this sense should be discouraged.

homologous chromosomes. 1. Chromosomes that are structurally similar and pair together at meiosis. 2. Chromosomes in different taxa that are descended from a common ancestral chromosome.

homology. *n.* The state of being *homologous*. Similarity of characters due to common ancestry can be referred to as *phylogenetic homology*, contrasted with *serial homology*—a special case in which successively repeated parts in the same organism are similar. Cf. *analogy, homoplasy*.

homology blocks. Reference to genes that are on the same chromosome in two different species.

homonym. *n.* Reference to a name for two or more different taxa having the same spelling. *Primary homonyms* are species names having the same generic names when originally proposed. *Secondary homonyms* result when two or more species with the same specific epithets are combined into the same genus subsequent to their original descriptions in different genera.

homophylic. *adj.* Exhibiting resemblance due to common ancestry.

homoplastic. *adj.* 1. Reference to a *homograft*. 2. Reference to structures that perform the same function or have the same appearance without being *homologous* in an evolutionary sense.

homoplasy. *n.* (*adj.* *–tic*) 1. Parallel or convergent evolution, producing structural similarity in organisms that are not due directly to inheritance from common ancestry. Cf. *analogy, homology*. 2. In cladistics, a character present in at least two clades but is absent in the common ancestor of the two clades.

homoselection. *n.* Selection favoring *homozygotes*. Cf. *heteroselection*.

homosynaptic modulation. A change in the efficacy of a synapse as a result of activity at that synapse.

homotopic. *adj.* See *orthotopic*.

homotype. *n.* See *homeotype*.

homotypic. *adj.* Conforming to the type or normal condition. Cf. *heterotypic*.

homoxenous. *adj.* Living within a single host during a parasite's life cycle.

homozygote. *n.* An individual or cell that has identical alleles at one or more loci in homologous chromosome segments. Cf. *heterozygote*.

homozygous. *adj.* A condition of two copies of the same allele at a diploid locus.

honest communication or **signaling.** In ethology, this term has reference to communication or signaling by an animal that enables a potential predator, mate, or same-sex rival to assess individual quality using the same signal. As example, prey-to-predator displays in *Anolis* lizards can be categorized as "honest" signals if they communicate the prey's alertness and ability to escape an attack, potentially benefiting both animals if they avoid a costly encounter. Honest signals are ones that are costly enough to make them uneconomical to produce if the true level or quality is less than indicated. See A. Zahavi, *J. Theor. Biol.* 53:205–214, 1975; *J. Theor. Biol.* 67:603–605, 1977; M. Leal, *Animal Behaviour* 58:521–526, 1999. See also *signaling theory*.

hood. *n.* The dorsoventrally flattened expansion of the neck, produced by moving the ribs upward and outward. Characteristic of Cobras and Australian elapids, but also occurs in various other groups of colubroid snakes.

hook. *n.* 1. Antiquated term for the *anal spur* or vestigial hind limb of boid snakes. 2. Used by E.H. Taylor (*U. Kans. Sci. Bull.* 35:577–922, 1952) in reference to the recurved process on the humerus of males of the frog genus *Centrolene*.

Hook-nosed Snakes, Hooknose Snakes. Vernacular name for the North American colubrid species belonging to the genera *Ficimia, Gyalopion,* and *Pseudoficimia*.

Hoplocercidae. *n.* A clade (family) of Central and South American lizards previously treated as a subfamily of the Iguanidae. This group was raised to family status by D.E. Frost and R.E. Etheridge (*Univ. Kans. Mus. Nat. Hist. Misc. Publ.* 81, 1989), and this revision has been widely accepted.

Hoplocercinae. *n.* A clade (subfamily) of iguanid lizards, primarily terrestrial inhabitants of the rain or dry forests in Panama and South America. Three genera and 12+ species are recognized. See *Hoplocercidae*.

horizon. *n.* See *soil horizon*.

horizontal evolution. Simultaneous evolution of geographically distributed populations, producing geographical variation.

horizontal life table. A *life table* in which the survivorship schedule is calculated by directly following a cohort of individuals from birth to death. (Syn. *cohort life table*)

horizontal septum. An incomplete muscular septum in the ventricle of noncrocodilian reptiles that partially divides the ventricle by separating the *cavum pulmonale* from the *cavum venosum* and *cavum arteriosum*. The latter two are partially separated by the *vertical septum*. (Syn. *interventricular septum, membranous body, Muskelleiste, muscular ridge, septum ventriculorum*) See Fig. 36.

horizontal undulatory movement. Older term synonymous with *lateral undulation* or *serpentine* limbless locomotion.

hormone. *n.* A chemical messenger that is secreted into the blood circulation or intercellular space and affects target tissues.

horn. *n.* A term applied generally to various pointed projections from skin, bones, or other structures, e.g. the posterior projections of the tongue of certain anurans.

Horned Frogs. Vernacular name for South Asian megophryid frogs belonging to the genus *Megophrys,* and also applied to South American species of frogs belonging to the leptodactylid genera *Ceratophrys*, *Chacophyrs,* and *Lepidobatrachus*. The name is derived from the presence of projections of skin from the eyelids to form various so-called "horns." See also *Smooth Horned Frogs*.

Horned Lizards. Collective vernacular name for distinctive species of American (U.S. and Mexico) iguanid lizards belonging to the genus (or crown clade) *Phrynosoma*. These lizards are spiny with short, flat bodies and tails and hornlike appendages protruding from the head. For a recent discussion of phylogenetic taxonomy of this group, see A.D. Leaché & J.A. McGuire, *Mol. Phylogent. Evol.* 39:628–644, 2006.

Horned Toad. An antiquated vernacular name previously applied erroneously to *Horned Lizards*. See also *Horned Frogs*.

Horned Treefrog. Vernacular name for species of South American hylid frogs belonging to the genus *Hemiphractus*.

Horned Viper. Vernacular name for various species of terrestrial vipers belonging to the genera *Bitis* and *Cerastes*, found in Africa, southern Asia, and the Middle East. A number of these snakes are desert species with characteristics that are convergent with those of the American *Sidewinder Rattlesnake*. The name has reference to a prominent epidermal horn that is present over each eye, usually derived from a single scale. See also *Rhinoceros Viper*.

horny. *adj.* Reference to keratinized skin or fibrous, epidermal derivatives.

horny layer. See *stratum corneum*.

horotelic evolution, horotely. Evolving at a "normal" or average rate. Cf. *bradytelic, tachytelic evolution.*

horst. *n.* In geology, an upraised block bounded by normal faults.

host. *n.* 1. An organism that is infected by a parasite or other infectious agent. 2. The recipient of a graft from another individual.

host race. A genetic race of a parasitic species that occurs on a particular host species.

host specificity. Reference to the extent to which a parasite is restricted in its utilization of host species.

hot or **heat rock.** A commercially available stone or object with a thermostated heater, used for providing a heat source for captive reptiles.

hotspot. *n.* 1. See *biodiversity hotspot*. 2. In molecular biology, a region of a polynucleotide that experiences a relatively high frequency of mutation or transposition.

hour. (h) A unit of time equal to 60 min or 3600 s.

Hourglass Treefrog. Vernacular name for the Middle American hylid species *Hyla ebraccata*.

House Gecko, Common House Gecko. Vernacular name for the Asian gekkonid species *Hemidactylus frenatus*, inadvertently introduced to many tropical Pacific islands since the 1930s. See also *Asian House Gecko*.

Housesnakes or **House Snakes.** *n.* Vernacular name for lamprophiine snakes, especially African species belonging to the genus *Lamprophis*.

***Hox* genes.** Genes that contain *homeoboxes* and code for proteins that regulate genes controlling development by specifying embryonic cell division along the anterior-posterior axis of the body. *Hox* genes are physically clustered, and their expression has sharply defined boundaries that specify precise anatomical boundaries during development. Recent work has shown that *Hox* genes determine anteroposterior patterning in the vertebrate embryo and are involved in tetrapod limb formation. They are also candidates for important roles in morphological evolution. As example, changes in *Hox* gene expression are shown to be involved in combined developmental changes related to limb loss and trunk elongation during the evolution of snakes (M.J. Cohn & C. Tickle, *Nature* 399:474–479, 1999).

HPA. See *hypothalamic-pituitary-adrenal axis*.

HPLC. See *high-performance liquid chromatography*.

HR1. A *hemorrhagin* found in venom of the Habu.

HR2. A *hemorrhagin* found in venom of the Habu.

HSL. Hue, saturation, and luminance—the three color components that form a color image.

hue. *n.* A distinction of monochromatic light (of single wavelength) that permits it to be classified as a particular color. Monochromatic light has a characteristic hue.

hula dance. A behavior performed by male newts (*Notophthalmus viridescens*) involving undulating motions of the lower body and tail during courtship. The behavior was described by P.A. Verrell (*Anim. Behav.* 30:1224–1236, 1982) and resembles "*vent shuffling*" described in ambystomid salamanders.

humeral. *n.* The second pair of plastron laminae in turtles, located between the *gulars* anteriorly and the *pectorals* posteriorly. See Fig. 6.

humerus. *n.* The most proximal bone of the forelimb of vertebrates (especially tetrapods), which articulates with the shoulder girdle proximally and the *radius* and *ulna* distally.

humeral gland. *n.* Reference to a large, flat, glandular area on the side of *Rana musica*, immediately posterior to the arm insertion. Used by M.L.Y. Chang and H.F. Hsu (*Cont. Biol. Lab. Sci. Soc. China* 8:137–181, 1932).

humicolous. *adj.* Living in, or on, soil.

humidity. *n.* The concentration of water vapor in the atmosphere or ambient air. The *absolute humidity* is the mass of water vapor per unit volume of air, also known as the *vapor density*. The term *relative humidity* refers to the ratio of the moisture in the air to the moisture it would contain if it were saturated at the same temperature and pressure, expressed as a percentage.

humidity index. A measure of effective moisture supply used in classifying climates, calculated from the excess of precipitation over potential evapotranspiration.

humoral. *adj.* 1. Pertaining to the humerus. 2. Reference to the antigenic portion of the host's immune reaction to an antigenic stimulus.

hump. *n.* The rounded protuberance on the lower back of anurans, relating to articulation of the vertebral column with the pelvis and urostyle. (Syn. *iliac crest, sacral hump, sacral prominence*)

Humpnose Pitviper. Vernacular name for three species of Asian viperid snakes belonging to the genus *Hypnale*.

Hundred-Pace-Viper. *n.* Vernacular name for the Far Eastern pitviper *Deinagkistrodon acutus*.

husbandry. *n.* The care and management of captive animals.

hyalia. *n.* Reference to *principal horns* of the anuran hyoid. Used by C.F. Walker (*Occ. Papers. Univ. Mich. Mus. Zool.* 371:1–11, 1938).

hyaline cartilage. A widespread type of cartilage characterized by a homogeneous appearance of the matrix, which lacks accessory supportive fibers. Cf. *elastic cartilage, fibrocartilage*.

hyaluronic acid. A mucopolysaccharide polymer that is common in jelly coats of eggs and in connective tissues.

hyaluronidase. *n.* An enzyme and component of snake venoms that catalyzes the hydrolysis of hyaluronic acid. Its effect is to decrease the viscosity of connective tissue and facilitate the spreading of fluids throughout tissues. Also known as "*spreading factor*."

hybrid. *n., adj.* 1. An offspring resulting from sexual reproduction involving two genetically dissimilar races, subspecies, or species. (Syn. *intergrade*) 2. A *heterozygote*.

hybrid breakdown. A reduction of fitness associated with F_2 or backcross populations produced by intercrossing of genetically different populations or species. This phenomenon can create a postzygotic *reproductive isolating mechanism*. Cf. *heterosis*.

hybrid inviability. Failure of hybrids of genetically disparate populations to survive to reproductive age, thereby creating a postzygotic *reproductive isolating mechanism*.

hybridization. *n.* The process of interbreeding to form hybrids (intergrades), usually between related, but different, species.

hybridogenesis. *n.* Reproduction in which a hybrid female mates with a male of one of her parental species. Fertilization occurs, but none of the paternal genome is incorporated into the offspring's genome.

hybridoma. *n.* A cell culture consisting of a clone of a hybrid cell formed by fusing cells of different kinds, used to produce monoclonal antibodies.

hybrid speciation. The origin of a new species that results from interspecific hybridization.

hybrid sterility. Failure of hybrids between different species to produce viable offspring.

hybrid swarm. A large number of interspecific hybrids at the boundaries of two different species or populations.

hybrid vigor. *n.* See *heterosis*.

hybrid zone. A geographical region in which there occur hybrids between two geographical races.

hydatid of Morgagni. See *sessile hydatid*.

Hydatinosalamandroidei. *n.* A newly designated monophyletic taxon within *Diadectosalamandroidei*; a sister taxon with *Plethosalamandroidei* and composed of *Perennibranchia* and *Treptobranchia* (D.R. Frost et al., *Bull. Amer. Mus. Nat. Hist.* 297:1–370, 2006).

hydration. *n.* 1. A condition of combination with water. Solvation when the solvent is water. 2. Fluid replacement by an animal.

hydraulic. *adj.* Of or pertaining to the effects that result from the movement and force of a liquid.

hydraulic permeability. Sievelike properties of a membrane, permitting passage of fluid under pressure.

hydric. *adj.* Wet, or reference to water properties.

hydro-. Prefix meaning "water."

hydrobiology. *n.* The study of life in aquatic habitats.

hydrocarbon. *n.* An organic compound composed only of carbon and hydrogen atoms.

hydrocolous. *adj.* Living in an aquatic habitat.

hydrodynamic. *adj.* Pertaining to the action of water, including waves and tides.

hydrofoil. *n.* An object that produces lift when placed in a moving stream of water.

hydrogen bond. A weak electrostatic attraction between a hydrogen atom bound to a highly electronegative element in a molecule and another highly electronegative atom in the same or a different molecule.

hydrologic cycle. The global movement of water and water vapor from the sea to the atmosphere, to the land, and back to the sea and atmosphere again. (Syn. *water cycle*)

hydrology. *n.* Formal study of the properties, circulation, and distribution of water. Hydrology is concerned with the movement and distribution of water at or near the Earth's surface, whereas *meteorology* deals with water as it moves through the atmosphere.

hydrolysis. *n.* Fragmentation of a compound by the addition of water, whereupon a hydroxyl group joins one fragment and the hydrogen atom the other.

hydrophilic. *adj.* Having an affinity for water.

Hydrophiidae. *n.* Formerly, a family of "true," fully marine, Sea Snakes and distinguished from Sea Kraits. See *Sea Snakes*, *Elapidae*, and *Hydrophiinae*.

Hydrophiinae. *n.* A subfamily of *Elapidae,* generally including Sea Snakes and Australopapuan elapids. Marine elapids (*Sea Snakes*) were traditionally treated apart from terrestrial elapids and given familial status, but phylogenetic studies have proven this distinction to be artificial.

hydrophobic. *adj.* The property of repelling moisture. Non-wettable, as a lipid surface like butter.

hydrops. *n.* A condition of abnormal accumulation of serous fluid in tissues or body cavity, applied in particular to excess fluid accumulation in the lymph sacs of anurans.

hydrosphere. *n.* The global water mass, including atmospheric, surface, and subsurface waters.

hydrostatic organ. Reference to a structure that is used for *buoyancy* regulation. The lungs of some larval amphibians, aquatic amphibians, and sea snakes have been shown to function as hydrostatic organs (usually in addition to respiratory function).

hydrostatic pressure. Force exerted over an area due to pressure in a fluid. Pressure exerted by a column of water, increasing by about 1 atmosphere per 10 m depth of water column. See *gravitational pressure*.

hydrotaxis. *n.* Movement of an organism toward or away from water or a moisture stimulus.

hydrothermal. *adj.* Pertaining to hot water, as at a hydrothermal vent.

hydrotroph. *n.* A nutritive substance secreted by the oviduct and eaten by advanced embryos in some viviparous amphibians. See also *aplacental viviparity*.

hygric. *adj.* Reference to moisture or humid conditions.

hygro-. Prefix meaning "moist," "wet," "damp," or "humid."

hygrocolous. *adj.* Living in moist environment.

hygrometer. *n.* An instrument used to measure moisture or humidity.

hygroscopic. *adj.* Capable of absorbing moisture from air.

hygrotaxis. *n.* Movement of an organism toward or away from moisture.

Hylidae. *n.* A large family of anurans having diverse reproductive modes, unusual modifications of the skull in some species, and a majority of species with arboreal habits and well-developed toe discs. Generally known as Tree Frogs, there are four subfamilies (some classifications), about 40 genera, and approximately 870 species distributed in North, Central, and South America, the West Indies, Eurasia, eastern Asia, the Australopapuan region, and extreme northern Africa. The family is recently classified as sister taxon to the new *Leptodactyliformes* within the new taxon *Athesphatanura* and excludes *Hemiphractinae* to contain three subfamilies: *Hylinae, Phyllomedusinae,* and *Pelodryadinae.* See D.R. Frost et al., *Bull. Amer. Mus. Nat. Hist.* 297:1–370, 2006.

Hylinae. *n.* A speciose subfamily of the *Hylidae*.

Hyloidea. *n.* See *Hyloides* and *Bufonoidea*.

Hyloides. *n.* A new monophyletic taxon of anurans, sister with *Ranoides* within the *Phthanobatrachia* and composed of *Sooglossidae* and *Notogaeanura.* See D.R. Frost et al., *Bull. Amer. Mus. Nat. Hist.* 297:1–370, 2006. In larger part, this taxon is equivalent to *Hyloidea* of traditional usage, excluding *Heleophrynidae*.

Hynobiidae. *n.* A clade (family) of relatively small, mostly terrestrial Asian salamanders (100–250 mm), currently having 43 species.

hynobiid. *n, adj.* A member of, or pertaining to, Hynobiidae.

hyobranchial. *adj., n.* Reference to *hyoid* or *hyoid apparatus* or those components of the larval visceral skeleton that eventually form the hyoid. (Syn. *hyobranchial apparatus; hyobranchium*)

hyobranchial apparatus. 1. See *hyobranchial.* 2. Groupings of muscles and bones associated with the tongue.

hyobranchial sinus. A blood-filled sac or sinus in the floor of the throat of certain centrolenid tadpoles, first described by H. Hoffmann (*Rev. Biol. Trop.* 52:219–228, 2004). The structure originates from the transverse throat vein in pairs, conspicuously ventral of the hyobranchial apparatus, and has unknown function. (Syn. *sinus hyobranchialis*)

hyobranchium. *n.* 1. Reference to *hyoid* or *hyoid apparatus*. 2. Sometimes used in reference to all parts of the postcranial visceral skeleton of gill-breathing larval amphibians.

hyoglossum. *n.* Alternative term for *hyoid apparatus*, as used by H.M. Smith (*Evolution of Chordate Structure*, p. 171, Fig. 7.11, 1960).

hyoglossus muscle. A tongue muscle, which in chameleons is used to return a projected tongue (with prey) to the mouth. Because the chameleon's tongue is projected so far from the mouth, the hyoglossus muscles must simultaneously exert strong force (sufficient to reel in large prey) and undergo an extraordinary degree of contraction.

hyoid, hyoid apparatus. *n.* A complex of bones and cartilaginous structures that are derived from gill arch elements and provide a suspensory structure for the tongue and larynx. In larval and neotenic amphibians these arch elements support the gills, but in metamorphosed amphibians and reptiles the elements are transformed into the tongue skeleton with secondary functions related to movement of the larynx or mouth floor. (Syn. *hyoglossum*)

hyoid arch. The second gill arch of ancestral vertebrates and vertebrate embryos. Elements of this structure contribute to the *inner ear* and *hyoid apparatus*. Derivatives of this arch are usually called *ceratohyals*, but numerous synonymies exist in hyoid and branchial arch nomenclature.

hyoid bone. A bone or complex of bones situated at the base of the tongue and supporting the tongue and its muscles.

hyoid copula. See *copula*.

hyoid cornu. Alternative term for *hyoid horn*, as used by S. Barrows and H.M. Smith (*Univ. Kans. Sci. Bull.* 31:227–281, 1947).

hyoid horn. The body of the hyoid apparatus, derived from the second gill arch and modified in structural detail through evolution in various species. (Syn. *hyoid cornu*)

hyoid plate. The large, flattened medial ventral cartilage of the *hyoid apparatus*, from which various processes arise.

hyoid process. The *principal horn* of the anuran hyoid.

hyolingual projection. Rapid ballistic projection of the tongue, used in prey capture by some amphibians and reptiles.

hyomandibula. *n.* The major upper skeletal element of the second branchial (gill) arch involved in jaw suspension in many fishes. In tetrapods this element becomes modified to form the *columella* or stapes of the middle ear.

hyoplastron. *n.* The second pair of bones in the *plastron* of turtles, bordered anteriorly by the *epiplastra* and *entoplastron*, posteriorly by the *hypoplastra*, and articulating laterally in most species with the *peripherals* to form the anterior margin of the bridge. See Fig. 7.

hyostapes. *n.* That portion of the lizard *columella* embryonically derived from the hyoid arch.

hypapophysis. (*pl.* **-es**) *n.* (*adj.* **hypapophysial**) A ventral projection from the vertebral centrum in some squamate reptiles. These are especially prominent, and with enamel tips, in egg-eating snakes that employ these structures to break the shell of swallowed eggs.

hypaxial. *adj.* 1. Below, or ventral to, the vertebral or body axis. 2. Reference to trunk muscle that lies, or originates, ventral to the vertebral column or body axis. Cf. *epaxial*.

hyper-. A prefix meaning "greater than," "above," "beyond," "above." or "higher."

hypercalcemia. *n.* Excessive levels of calcium, usually in reference to concentrations in blood plasma.

hypercapnia. *n.* A condition of higher than normal carbon dioxide in the blood. Cf. *hypocapnia*.

hyperemia. *n.* Increased blood flow to a tissue or organ.

hyperextension. *n.* Unusual or excessive extension of a joint, limb, or other body part.

hyperglycaemin. *n.* A component of snake venom that produces marked increases in levels of blood glucose.

hyperglycemia. *n.* Excessive levels of glucose in blood plasma.

hypernatremia. *n.* Higher than normal, or excess, of blood sodium.

Hyperoliidae. *n.* A clade (family) of small to medium-size frogs, many of which are brightly colored and arboreal with toe discs. Four subfamilies include *Hyperoliinae* (*Reed Frogs*), *Kassininae*, *Leptopelinae*, and *Tachycneminae*. Revised recently to exclude the *Leptopelinae* (D.R. Frost et al., *Bull. Amer. Mus. Nat. Hist.* 297:1–370, 2006).

Hyperoliinae. *n.* A subfamily of *Hyperoliidae*.

hyperosmotic. *adj.* Containing a greater concentration of osmotically active constituents than a solution of reference. Cf. *isoosmotic, hypoosmotic*.

hyperparathyroidism. *n.* Abnormally high activity of the parathyroid gland.

hyperphosphatemia. *n.* Abnormally high levels of blood phosphorus.

hyperplasia. *n.* (*adj.* **-tic**) An increase in the mass of a tissue due to proliferation of cell numbers, as occurs in a regenerating body part. Cf. *hypertrophy*.

hyperpnea. *n.* Increased lung ventilation. (Syn. *hyperventilation*).

hypersaline. *adj.* Highly or excessively salty.

hypertely. *n.* Evolutionary overspecialization, as in progressive allometry and disproportionate size of a structure or organism.

hypertension. *n.* increased blood pressure above normal levels. Cf. *hypotension*.

hyperthermia. (*adj.* **-ic**) *n.* A condition in which body temperature is higher than the normal range. This term is normally used in reference to endotherms. See *behavioral fever*.

hyperthyroidsm. *n.* Abnormally high activity of the thyroid gland.

hypertonic. *adj.* Having a higher tonicity or osmotic pressure than a reference solution. Cf. *isotonic, hypotonic*.

hypertrophy. *n.* An increase in the size of an organ or tissue attributable to increased volume of the constituent cells. Cf. *hyperplasia*.

hyperventilation. *n.* Increased lung ventilation.

hypervitaminosis. *n.* A nutritional disorder of captive animals caused by excess vitamins in the diet. Cf. *hypovitaminosis*.

hypnotic reflex. *n.* See *death feint*.

hypo-. A prefix meaning "less than," "below," or "under."

hypobranchial. *n.* The basal part of a horn on the *hyoid*, usually articulating with a *ceratobranchial*.

hypocalcemia. *n.* A condition of lower than normal blood calcium or deficiency of body calcium.

hypocapnia. *n.* A condition of lower than normal carbon dioxide in the blood (Syn. *hypocarbia*) Cf. *hypercapnia*.

hypocarbia. *n.* See *hypocapnia*.

hypocentrum. *n.* That part of a vertebra derived from the posterior part of a myotome. (Syn. *intercentrum*)

hypocentrum pleurale. See *pleurocentrum*.

hypochordal. *adj.* Lying ventral to the notochord.

hypochordal rod. A band of embryonic cells between the dorsal wall and notochord in amphibians, from the level of pancreas to tail, which disappears before hatching. (Syn. *sub-notochordal rod*)

hypodermis. *n.* A transitional subcutaneous region comprised of loose connective tissue and adipose tissue between the integument and deep body musculature. (Syn. *superficial fascia*)

hypoglossal. *adj., n.* Reference to the twelfth cranial nerve.

hypoglossum. *n.* A cartilaginous element beneath the tongue of turtles, only loosely associated with the hyoid apparatus. (Syn. *entoglossum*)

hypoglycemia. *n.* Below normal concentrations of glucose in blood.

hypohyal. *n.* A derivative of the *hyoid arch*, attached to the *copula* and *ceratohyal* and forming the basal part of the *hyoid horn*. In some salamanders this element disappears at metamorphosis. Historically considered a *ceratohyal* or *entoglossal* by some.

hypoischial cartilage. See *hypoischium*.

hypoischium. *n.* A cartilaginous or bony rod, single or paired, directed posteriorly from the symphysis of the *ischium*. (Syn. *cloacal cartilage, hypoischial cartilage*)

hypokalemia. *n.* Below normal concentrations of potassium in blood.

hypolimnion. *n.* The bottom layer of cold water in a lake. Cf. *epilimnion, metalimnion*.

hypomelanistic. *adj.* Reference to an individual whose suite of pigment cells lacks melanocytes.

hypomere. *n.* The unsegmented ventral division of the embryonic lateral *mesoderm*. The ventrolateral portion of the *myotome*.

hypomeric muscles. Muscles derived from the hypomere; *hypaxial* muscles.

hyponatremia. *n.* Below normal concentrations of sodium in blood.

hypo-osmotic, hyposmotic. *adj.* Having a lower concentration of osmotically active constituents than a solution of reference. Cf. *isoosmotic, hyperosmotic*.

hypophosphatemia. *n.* Below normal concentrations of phosphorus in blood.

hypophysectomy. *n.* Excision of the *pituitary gland*.

hypophysis. *n.* See *pituitary gland*.

hypoplasia. *n.* Lack of normal growth or development of an organ or part. Arrested growth. Opposite of *hyperplasia*.

hypoplastron. *n.* The third pair of bones in the *plastron* of turtles, bordered anteriorly by the *hyoplastra*, posteriorly by the *xiphiplastra*, and articulating laterally with *peripherals* to form the posterior bridge in most species. See Fig. 7.

hyposternum. *n.* See *sternum, metasternum*.

hypotension. *n.* (*adj.* –**ive**) A condition of low blood pressure. Cf. *hypertension*.

hypothalamic-pituitary-adrenal axis. (HPA) Reference to a cascade of events related to activation of hypothalamic releasing factors for pituitary hormones that, in turn, activate the release of hormones from the adrenal gland. Hormones in this axis are involved in a number of physiological responses, including *stress response*. (Syn. *stress axis*)

hypothalamic release-inhibiting hormones. (RIH; release inhibiting factors, RIF) Neurohormones originating in the hypothalamus that are carried by portal vessels to the anterior pituitary, where they inhibit the release of specific hormones. Cf. *hypothalamic releasing hormones.*

hypothalamic releasing hormones. (RH, releasing factors) Neurohormones originating in the hypothalamus that are carried by portal vessels to the anterior pituitary, where they stimulate the release of specific hormones. (Syn. *releasing* hormones) Cf. *hypothalamic release-inhibiting hormones.*

hypothalamo-hypophyseal portal system. A system of portal veins connecting the capillaries of the hypothalamic *median eminence* with those of the *anterior pituitary*, which transport hypothalamic neurosecretions directly to the anterior pituitary.

hypothalamus. *n.* (*adj.* **-ic**) That part of the *diencephalon* that forms the floor of the median ventricle (3rd ventricle) of the brain. This structure is important to regulation of the autonomic nervous system and endocrine function.

hypothenar tubercle. The small tubercle projecting below the base of the fourth or outermost digit on the forefoot of anurans.

hypothermia. *n.* (*adj.* –**ic**) Condition of lower than normal body temperature. This term is normally used in reference to endotherms.

hypothesis. *n.* An assertion or proposition that is tested by scientific study or experimentation. An assumption proposed as an explanation of observed facts and leading to testable predictions. A testable statement concerning the cause of a natural phenomenon in question.

hypotonic. *adj.* Having a lower tonicity or osmotic pressure than a reference solution. Cf. *istotonic, hypertonic*.

hypotype. *n.* In taxonomy, a specimen that is described or figured to extend knowledge of a previously described species.

hypoventilation. *n.* Reduced lung ventilation.

hypovitaminosis. *n.* A deficiency of vitamins. Cf. *hypervitaminosis*.

hypoxia. *n.* (*adj.* **-ic**) Reduced or inadequate oxygen levels (below normal).

hysteresis. *n.* A nonlinear change in the physical state of a system, such that the state depends in part on the previous history of the system. A time lag in the association between two variables is different depending on the direction of change. Using *stress* vs. *strain* curves as an example, hysteresis can be quantified by the ratio of the area between curves divided by the area under the loading curve. See Fig. 27.

I

IACUC. (Institutional Animal Care and Use Committee) A local committee at any institution (e.g., government laboratory, university) that oversees the care and use of living animals in scientific research.

***ibid.* (*ib.*)** Abbreviation of the Latin *ibidem*, meaning "in the same place" (as in same book or journal).

iButton. *n.* Reference to DS1921 Thermochron iButtons manufactured by Dallas Semiconductor (Texas, USA), adopted as a small lightweight, real-time temperature recording device that can be attached externally or internally in small animals. Reconstruction of iButtons for use in studies of small amphibians or reptiles is described in K.A. Robert & M.B. Thompson (*Herpetol. Rev.* 34:130–132, 2003).

Ice Age. A general term for any or all of several Periods during Earth's history when there was a major extension of polar and continental ice masses accompanied by fall in sea level. (Syn. *glacial epoch*)

IC_{50}. The concentration of a drug that is required for 50% inhibition *in vitro*. Cf. EC_{50}.

ichno-. A prefix meaning "footprint" or "track."

ichnofossil. *n.* A fossilized mark or remnant formed in soft sediment by the movement of an animal (e.g., a footprint). A trace fossil.

ichnology, paleoichnology. n. The study of footprints and other marks preserved as fossil evidence of animal and plant activities.

ichthyophagous. *adj.* Feeding on fishes. (Syn. *piscivorous*)

Ichthyophiidae. *n.* A clade (family) of relatively primitive caecilians represented by two genera and 38 species distributed throughout parts of southern Asia.

Ichthyopterygia. *n.* A subclass or superorder of Mesozoic reptiles composed of predatory marine forms with short necks and porpoiselike body. The term (meaning "fish flippers") was introduced by Sir Richard Owen in 1840 to designate the Jurassic *ichthyosaurs* that were known at the time.

Ichthyosauria. *n.* A clade of extinct marine reptiles that had fish-like bodies, resembled dolphins, and gave birth to live young. See *Ichthyopterygia*.

ichthyosaur. *n.* A member of *Ichthyosauria*.

icon. *n.* A pictorial representation of a system or a component of a system.

iconotype. *n.* An illustration of a type specimen.

ICSH. See *interstitial cell-stimulating hormone*.

icterus. *n.* Jaundice or yellowish condition related to retention of biliary wastes.

ICZN. See *International Commission on Zoological Nomenclature*.

-idae. Suffix of a *family* name in zoological nomenclature (or subclass in botanical nomenclature).

ideotype. *n.* In taxonomy, a specimen examined by the author of a species but not collected from the type locality. (Syn. *idiotype*)

idiopathic. *adj.* Reference to disease condition of unknown cause.

idiotaxonmy. *n.* Traditional taxonomy related to individuals, populations, species, and higher taxa.

idiotype. *n.* See *ideotype*.

IGBP. See *International Geosphere-Biosphere Program*.

igneous rock. Rock formed by solidification of molten magma. Cf. *metamorphic rock, sedimentary rock*.

Iguana Iguana. A newsletter devoted to news and advisement about the *Green Iguana*.

Iguanas. *n.* Collective vernacular name for various species of iguanid, and especially iguanine, lizards.

Iguania. *n.* A basal clade of the order Squamata and sister taxon to *Scleroglossa*.

iguanid. *n, adj.* A member of, or pertaining to, *Iguanidae*.

Iguanidae. *n.* A robust clade (family) of lizards that includes eight subgroups. See *Iguaninae, Oplurinae, Phrynosomatinae* (elevated to family level in some classifications), *Tropidurinae, Polychrotinae, Hoplocercinae, Crotaphytinae,* and *Corytophaninae*.

Iguaninae. *n.* A clade of iguanid lizards including "Iguanas" (e.g., genus *Iguana*) constituting eight genera and about 34 species distributed from North to tropical South America, Galápagos Islands, the West Indies, and Fiji.

iliac crest. See *hump*.

ileum. *n.* The posterior section of the small intestine.

ilium. *n.* The dorsal (paired) element of the pelvic girdle that articulates with the sacrum.

illegitimate name. (*Nomen illegitimum*) A taxonomic name that is not in accordance with rules of the appropriate Code.

illuminance. *n.* The flux of light per unit area falling perpendicular to a surface.

imbibition. *n.* Passive uptake or absorption of water.

imbricate. *adj.* Overlapping, as shingles, and often used in reference to body scales of squamate reptiles.

immaculate. *adj.* Without mark. Reference to color patterns or portions of color patterns composed of a single color.

immature. *adj.* 1. Reference to any developmental stage(s) of an organism preceding attainment of sexual maturity. 2. Generally, reference to any incompletely differentiated system.

immigration. *n.* The process of individuals moving to, or entering, a population or region from another location. Cf. *emigration*.

immigration rate. 1. In *metapopulation models*, the immigration rate is the proportion of sites successfully colonized per unit time. 2. In the *equilibrium model* of island biogeography, the immigration rate is the number of new species arriving on an island per unit time.

immiscible. *adj.* Property of liquids that cannot be mixed to a homogeneous state.

immune. *adj.* 1. Being highly resistant to a disease because of humoral antibodies or cellular immunity. 2. Produced in response to an antigenic challenge, e.g. immune serum globulin.

immune response. A physiological response, or suite of physiological responses, to activation of the immune system by antigens, resulting in beneficial immunity to pathogenic microorganisms. The response is specific for the antigen involved, and is normally directed against foreign substances.

immune system. The organs, tissues, cells, and molecules responsible for immunity or protection from foreign substances.

immunity. *n.* The condition of being immune or secure against a disease. Nonsusceptibility to the invasive or pathogenic effects of foreign microorganisms or to the toxic effects of antigenic substances.

immunoassay, immunochemical assay. *n.* Any procedure that utilizes antigen-antibody reactions to localize or quantify the relative amounts of specific antibodies or antigenic substances.

immunoblotting. *n.* See *Western blotting*.

immunocompetent, immune competent. *adj.* A cell or organism that is capable of carrying out its immune function when properly stimulated.

immunoelectrophoresis. *n.* A combined methodology that identifies proteins by electrophoretic mobilities and their antigenic properties.

immunofluorescence. *n.* A method of determining the presence and location of antigen or antibody in tissue by a pattern of fluorescence, which results when the tissue specimen is exposed to the specific antigen or antibody labeled with a fluorescent material.

immunogen. *n.* A substance that causes an immune response.

immunohistochemistry. *n.* The application of antigen-antibody interactions and *immunoflurorescence* to *histological* techniques.

immunological distance. A measure of the difference between two antigens, estimated by immunological techniques.

immunological suppression. A condition in which the ability of the immune system to respond to most or all antigens is impaired.

immunological tolerance. A state of nonreactivity toward a substance that normally would be expected to elicit an immune response.

immunology. *n.* The science dealing with immunity, immunological reactions, and all of their related aspects.

immunosuppressive. *adj.* Reference to the action of various agents to inactivate a specific antibody to permit the acceptance of a foreign substance by an organism.

impacted. *adj.* Being wedged firmly. This term is used frequently in reference to impassable blockage of the gastrointestinal tract by hard feces, ingested stones, or undigested food items.

impaction. *n.* The condition of being impacted.

impact theory. The proposition that late Cretaceous (or Cretaceous-Tertiary boundary) mass extinctions of various groups of organisms resulted from collision of Earth with an asteroid or comet. This has been a popular theory of how dinosaurs went extinct. First published by L.W. Alvarez, W. Alvarez, F. Asaro & H. Michel (*Science* 208:1095–1108, 1980).

impedance. *n.* 1. The ratio of the pressure displacement to the volume displacement at a given surface. 2. The dynamic resistance to flow met by fluids or electric currents moving in a pulsatile manner.

impedance matching. Reference to adjustments of a sound-conducting system in which there is anatomical compensation for differences of impedance or resistance met by vibrational sound waves as they travel from air to the fluid of the inner ear.

imperial gallon. (gal) The *fps* unit of volume equal to 4.54609 dm^3. Cf. *US gallon*.

impervious. *adj.* Not permeable, usually with reference to water or air.

implant. *v., n.* 1. To insert or to graft tissue or other material into intact tissues or the body cavity. 2. Any material so grafted or inserted into the body.

impregnation. *n.* Fertilization, or union of egg and sperm.

imprinting. *n., v.* A form of learning in which individuals are exposed to key stimuli, usually during an early period of behavioral development, and form an association with an object or stimulus that elicits a stable behavior pattern. E.g., hatchling sea turtles imprint on natal beaches, and when they become adults they recognize appropriate cues and relocate the beach for nesting.

-inae. Suffix used to form names of subfamilies in zoological nomenclature.

inanition. *n.* A condition of starvation or multinutritional deficiency.

inappropriate name. In taxonomy, reference to a name that denotes a character, quality, or an origin not possessed by the taxon bearing that name.

inbred or **inbred strain.** *adj.* A condition of being genetically identical, or nearly so, due to high degree of *inbreeding*.

inbreed, inbreeding. *v.* Mating between closely related individuals. The (usually repeated) genetic crossing of closely related plants or animals.

inbreeding coefficient. 1. The probability that two alleles at the same locus in an individual are identical by descent. 2. A measure of the proportional loss of heterozygosity as a result of inbreeding.

inbreeding depression. Reference to decreased growth, survival, or fertility (fitness) following one or more generations of inbreeding.

inch. *n.* A derived *fps* unit of length equal to 2.54 cm.

incidence. *n.* The frequency of an event per period of time.

incidental catch. The capture of a species (e.g. sea turtle) while fishing for another species (such as shrimp).

incipient. *adj.* Nascent; used in reference to the initial stage of development of a structure or event. (Syn. *nascent*)

incipient species. Reference to populations that are in early process of speciation divergence, but may still interbreed in the absence of a specific barrier. (Syn. *semispecies*)

incised. *adj.* A synonym for *cuspate*, usually describing the shape of cutting edges of teeth or teeth-like structures.

incisor. *n.* A cutting tooth, normally with reference to the anterior teeth in the mammalian jaw.

incisura coracoidea. A deep notch separating the cartilaginous parts of the *coracoid* from the *procoracoid* in the shoulder girdle of salamanders.

incl. Abbreviation of the Latin *inclusus*, meaning "included."

inclination. *n.* The direction of a slope face. Aspect.

included niche. A niche of a species that occurs entirely within a larger niche of another species that is more generalized. The species having the included (smaller) niche survives by having competitive superiority in some limited part of the larger niche.

inclusion body. A minute foreign particle within a cell, e.g. a virus.

inclusive fitness. Reference to the combined fitness of an individual and its effects on the fitness of genetically related neighbors. See *kin selection*.

incompatibility. *n.* The failure or inability to cross-fertilize, graft, or otherwise form a viable association.

incomplete dominance. The failure of one allele to suppress completely the action of another. Cf. *complete dominance*.

incongruence. *n.* (*adj.* **incongruent**) Failure to match or fit. Cf. *congruence, congruent*.

incross. *n.* An individual that is mated with another from the same inbred line, often of the same genotype.

incubation. *n.* The period of egg development from oviposition to hatching of young. Reference can be to either artificial or natural conditions of the incubating environment. The term can also have reference to the act of incubating eggs.

incubation period. The time required for incubation or development of eggs.

indented key. A dichotomous key wherein the first part of a contrasting couplet is followed by all subsequent couplets. Each subordinate couplet is indented one step further to the right for clarity of presentation. Cf. *bracketed key*.

independence. *n.* (*adj.* **independent**) The condition of a variable when it is not influenced or controlled by another variable. Cf. *dependence*.

independent assortment. The random distribution of genes located on different chromosomes to various gametes during meiosis.

independent contrasts. See *phylogenetically independent contrasts*.

indeterminate. *adj.* Not definite, identified, or classified.

indeterminate growth. Growth that continues as long as an animal is living, albeit at slower rates in older animals. Thus, body size and age are correlated. Cf. *determinate growth*.

index fossil. A fossil that appears only in rock strata of a relatively limited span of geological age. A defining species indicator of a stratum. (Syn. *characteristic fossil, diagnostic fossil*)

index of dispersion. A measure of dispersion within a population based on a variance-to-mean ratio. A value of 1 indicates a random pattern, while values greater than 1 indicate an aggregated population, and values less than 1 indicate a uniform population.

index of similarity. The ratio of the number of species common to two communities to the total number of species present in both communities combined.

index species. See *indicator species*.

Indian Python. Vernacular name for the Asian boid species *Python molurus*.

indicator species. 1. A species that characteristically associates with a particular habitat or community. 2. A species used to characterize ecological response of a community to pollution exposure; or, more broadly, a species that characterizes or reflects habitat characteristics or the magnitude of a stress or stressor exposure. More recently, the term indicator species has been used in reference to an organism that accumulates pollutants in its tissues. Amphibians are widely considered to be useful indicator species. See V.H. Dale and S.C. Beyeler, *Ecological Indicators* 1:3–10, 2001. (Syn. *characteristic species, index species, ecological indicators*)

indigenous. *adj.* Native to a particular environment or region.

Indigo Snake. Vernacular name for the large and well-known New World colubrid snakes belonging to the genus *Drymarchon*.

indirect calorimetry. Sometimes used in reference to measurements of oxygen consumption by an animal. If certain parameters related to food composition are known, the oxygen consumption rates can be converted mathematically to units of heat production. Thus, the measures of heat production are "indirect."

indirect competition. Competitive interactions in which exploitation of a resource by one individual or species reduces its availability to other individuals or species. Cf. *direct competition.*

indirect development. Development that includes a free-living, feeding larval stage.

individual. *n.* A single organism or unit.

individual based model. An ecological model in which a computer is used to simulate the birth, growth, dispersal, and death of individuals in a population.

individual distance. The minimum distance between two conspecific individuals at which there is no overt aggression.

indicator species. A species that seems to be most indicative of a particular community, in the context of the surrounding landscape or in comparison with related communities.

Indo-Chinese Rat Snake. Vernacular name for the Asian colubrid species *Ptyas korros.*

indolealkylamines. *n.* Biogenic amines that are very widely distributed in various tissues of both plants and animals. They are present in serotoninergic neurons of the central and peripheral nervous system, and they constitute the most important and characteristic biogenic amines present in the skin secretions of amphibians.

Indomalaya, Indomalayan. *n.* One of eight *biogeographical realms* including the South Asian subcontinent and southeast Asia. Also a formal zoogeographical realm.

induced drag. In biomechanics, the drag or resistance to forward travel that is due to production of lift. See *drag.*

inducer. *n.* One of a class of effector molecules that stimulate a cell to increase the production of various metabolic enzymes.

inducible. *adj.* Property of a gene that is activated in response to a specific stimulus.

induction. *n.* From embryology, a morphogenetic effect wherein one cell mass determines the developmental fate of another cell mass through a chemical stimulus.

induction or **inductive method.** A method of reasoning that involves formulation of general rules or universal principles based on consideration of individual facts or singular observations. Cf. *deductive method.*

inert. *adj.* Inactive or physiologically quiescent.

inertia. *n.* The tendency of an object to remain at rest or in uniform motion in a straight line unless acted upon by external forces.

inertial endotherm. See *endotherm.*

inertial feeding. A mode of prey ingestion that involves a rapid release of prey by the jaws followed by rapid forward movement of the jaws to grasp the prey in a new position distal to the part entering the mouth. The inertia of the prey keeps it in a relatively stationary position while the head of the feeding animal progressively moves forward over the prey during a sequence of such movements. This mode of feeding is characteristic of Tegu and Monitor Lizards.

infarct, infarction. *n.* A localized area of *ischemic necrosis* caused when an artery is blocked by some obstruction.

inferior. *adj.* Situated below or directed downward. The lower of two or more anatomical parts.

inferior labial. Synonym for lower labial. See *labial.*

inferior *vena cava.* See *postcaval vein.*

infertile. *adj.* Not fertile; nonreproductive.

inflammation. *n.* A condition or reaction to injury, characterized by heat, swelling, redness, and pain.

influx. *n.* Movement of solute or solvent into a cell or tissue, across plasma membranes.

infra-. Prefix meaning "below," "beneath," or "smaller." Cf. *supra.*

infraciliary. (*pl.* –ies) *n.* Below the eye. See Fig. 16.

infracoracoid. *n.* F.B. Hanson (*Amer. J. Anat.* 26:41-115, 1919) pointed out the term *epicoracoid* was inappropriate and suggested use of *infracoracoid* in its place. However, few authors have followed this usage. See Fig. 3.

infradian rhythm. A biological rhythm with a period longer than 24 h (less than one cycle per day).

infrahumeral. *adj.* Of or pertaining to the ventral surface of the upper portion of the forelimb.

infralabial. *n.* Term used to denote *lower labial* or (less commonly) *sublabial* scale. See Figs. 11, 16.

infralabial glands. See *oral glands.*

infralabial prominence. Reference to a U- or V-shaped medial protrusion of the lower jaw, typical of many microhylid tadpoles. See D.G. Thibaudeau & R. Altig, *J. Morphol.* 197:63–69, 1988.

infralittoral. *adj.* See *sublittoral.*

inframarginal. *n.* Plate lying between the *marginals* of the *carapace* and the lateral margin of the *plastron*, hence bordering the *pectoral* and *abdominal* plates. In most species of turtle the *axillary* and *inguinal* plates are the only inframarginals.

inframarginal gland. A musk gland situated at the junction of the *carapace* and *plastron* in some species of turtles.

inframarginal pores. Pores located near the rear of the inframarginal scutes. These are found only in Ridleys, where they conduct secretory products of unknown function to the surface.

inframaxillary. *n.* See *chin shield.*

infraorbital vacuity. The opening in the skull below the orbit.

infraorder. *n.* A category of hierarchy classification below that of order.

infrared radiation. Electromagnetic radiation of wavelengths between 750 nm and 1 mm, between red visible light and radio waves. See also *far infrared, thermal radiation*.

infrared receptor. See *pit organ*.

infrared thermometer. A handheld instrument that focuses and detects the infrared radiation emitted by a distant object in order to measure its temperature. A lens is used to focus infrared energy onto a detector, which converts the energy to an electrical signal that can be displayed in units of temperature. Also called a *laser thermometer* if a laser is used to help aim the thermometer, or *noncontact thermometer* to describe the device's ability to measure temperature from a distance.

infra-rostral cartilage. See *mentomeckelian cartilage*.

infrasound. *n.* (*adj.* **infrasonic**) Sound or vibration of a frequency below the audible range of human hearing. Used especially in reference to low frequency body vibrations (1–10 Hz) of crocodilians that are used for acoustic communication during courtship in water.

infratympanic. *n.* Any one of the scales at the lower margin of the ear aperture of crocodilians.

infrazygapophysis. *n.* A lateroventral process on the anterior aspect of individual vertebra in caecilians, evidently a ventral arch element and extending forward alongside the intervertebral region and the posterior part of the next anterior centrum

infundibulum. *n.* 1. Generally, a funnel-shaped structure. 2. Used in reference to the *median eminence* of the hypothalamus, the *conus arteriosus*, and the opening of the oviduct near the ovary. See J.E. Girling, *J. Exp. Zool.* 293:141–170, 2002.

ingesta. *n.* The total intake of food, or gross energy ingested.

ingestion. *n.* Feeding, or consumption of food.

ingress. *n.* The act of entering or coming in. Cf. *egress*.

ingression. *n.* Inward movement of cells from an outer layer to the interior of an embryo.

ingroup. *n.* In systematics, any group of theoretically closely related species of interest to an investigator. The group being studied. Cf. *outgroup*.

inguinal. *adj., n.* 1. Of, or pertaining to, the groin. 2. The plate on the posterior margin of the bridge, between the *carapace* and *plastron*, in turtles. See Fig. 6.

inguinal amplexus. See *pelvic amplexus*.

inguinal fold. A fold of skin in the groin region of lizards of the Xantusiidae.

inguinal gland. Glandular tissue lying at or near the groin of certain turtles, crocodilians, and amphibians.

inguinal notch. The U-shaped indentation at either side of the rear of the turtle shell, through which the hind limb protrudes.

inheritance. *n.* Transmission of genetic information from parents or ancestors to offspring or descendants.

inhibin. *n.* A hormone released from gonads that inhibit hypothalamic-pituitary stimulation of gonadotropin release. There are two forms: inhibin A and inhibin B, differing in subunits.

inhibition. *n.* 1. Any action or process that acts to restrain reactions or behavior. 2. In neurophysiology, this term refers to interneuronal synaptic modulation that tends to reduce the activity or firing of action potentials in an affected neuron.

inhibitory. *adj.* Term used in neurophysiology to denote *neurons, neurotransmitters,* or actions that tend to reduce the probability of action potentials.

-ini. The ending of a name of a tribe in zoological nomenclature.

in litt. Abbreviation of the Latin *in litteris*, meaning in correspondence.

innate. *adj.* Inherited, instinctive behavior that is not learned.

innate releasing mechanism. An innate mechanism that releases a specific behavioral response to a *sign stimulus*.

inner ear. The internal part of the ear, comprising a sensory organ responsible for balance and equilibrium sense in addition to the sense of hearing. Pressure waves are transmitted to the inner ear by accessory structures including the *tympanic membrane* and *columella*. See *otic capsule*.

inner epidermal generation. See *epidermal generation*.

inner tarsal turbercle. A tubercle on the inner or medial side of the palm in some species of anurans.

innominate bone. The complex of bones making up each half of the pelvic girdle, becoming fused in birds.

inositol triphosphate. (IP3) An important intracellular second messenger released by the action of a specific phospholipase enzyme and binds to, and activates, a calcium channel in the endoplasmic reticulum.

inotropic. *adj.* Pertaining to the strength of contraction of the heart.

inscriptional rib formula. Notation for the number of pairs of attached and floating postxiphisternal *inscriptional ribs*. E.g., 3:1 indicates three attached, one floating rib.

inscriptional ribs. Slender calcified cartilages located ventral to the dorsal ribs in many lizards. These are tied to the dorsal ribs dorsally and the sternum and xiphisternal rods ventrally in the thoracic region. In the abdominal region they may or may not be attached to their corresponding rib.

insectivore. *n.* (*adj.* **–ous**) Feeding on insects; often inclusive of all arthropods.

insemination. *n.* Insertion of sperm into the genital tract of a female.

insertion. *n.* The relatively mobile site of attachment of a muscle. Cf. *origin*.

insessorial. *adj.* Adapted for perching.

in situ. Literally "in place," used in reference to a natural or original position (e.g., observation of heart activity while it functions naturally inside the animal).

insolation. *n.* Solar radiation reaching the Earth from the sun.

inspiration. *n.* The active process (ventilation) by which air is inhaled into the lung of an animal. Cf. *expiration.*

instantaneous birth rate. (*b*) The per capita birth rate of a population, equal to the number of births per individual measured over a short time interval. Calculated by dividing the birth rate by the current population size; units are births/(individual • time).

instantaneous death rate. (*d*) The per capita death rate of a population, equal to the number of deaths per individual measured over a short time period. Calculated by dividing the death rate by the current population size; units are deaths/(individual • time).

instantaneous rate of increase. (*r*) The potential rate of growth of a population in an infinite environment. The difference between the instantaneous birth rate (*b*) and the instantaneous death rate (*d*), equal to *b-d*. The instantaneous rate of increase also equals the per capita rate of population increase in a simple model of exponential population growth. The units of *r* are individuals/(individual • time). (Syn. *intrinsic rate of increase*)

instinct. *n.* An unlearned, characteristic pattern of innate behavior.

institutional resource collections. See *standard symbolic codes for institutional resource collections.*

insular. *adj.* Of or pertaining to islands.

insular endemic. An organism that has evolved in a restricted geographical area and has retained such limited range without invasion of a broader area.

insulin. *n.* A protein hormone synthesized and secreted by the beta cells of the pancreatic islets, functional with respect to control of cellular uptake of carbohydrates with secondary effects on lipid and amino acid metabolism. Mammalian insulin produces effects in amphibians similar to those observed in fish and mammals.

insulin-like growth factor-1. (IGF-1) A polypeptide hormone, originally called *somatomedin C*, that is an important growth stimulating hormone and part of a *growth factor system* regulating many cellular processes in virtually all tissues, including growth, differentiation, and apoptosis. IGF-1 is structurally similar to IGF-II and proinsulin and is primarily derived from the liver in response to growth hormone stimulation, but also is produced in the oviduct. This hormone is important for normal growth and function of reproductive and somatic tissues, and recently has been found to play important roles in growth and differentiation of reproductive tissues in reptiles. (Syn. *somatomedin*)

in syn. Abbreviation of the Latin *in synonymis*, meaning "in synonymy."

integral control. A control action that eliminates the offset inherent in *proportional control.*

integral proteins. Proteins that span the plasma membrane (of cells) and form selective filters and components of active transport mechanisms involved with input of nutrients and removal of cellular products and wastes.

integrated peripherals. See *peripheral device.*

integrins. *n.* A large family of receptor proteins that appear on cell surfaces of virtually every cell type in the animal kingdom. Structurally, these proteins interact with components of the extracellular matrix, and some participate in adhesion between cells or activate signal transduction pathways within cells.

integument. *n.* A covering. The skin and its derivatives.

integumental poison gland. See *nuchal venom gland.*

integumentary appendage. Any projection or outgrowth from the body, such as a limb, spine or crest, including *epidermal appendages* such as hair, nails, or feathers. See further comments under *epidermal appendage* and *scale.*

integumentary gland. 1. Small holocrine glands called dorsal integumentary glands lie below the scales one out from the midline on both sides along the dorsum of all crocodilians examined. In Crocodiles these glands are active in the first three months post-hatching and involute in time. The most like function is *pheromonal* possibly related to kin recognition, perhaps imprinting on fellow hatchlings or the mother during *creche* (*pod*) formation and parental guarding. The period of subsequent secretory activity of these glands is unclear. 2. This term could be used generically to include a variety of glands present in the skin or integument of all amphibians and reptiles.

integumentary sense organ. (ISO) Sensory pits present on the postcranial scales of crocodylids and gavialids, while absent from alligatorids. These structures are supposed to have a mechanosensory function, but alternative functions are possible and unresolved. The name was first used by P. Brazaitis (pp. 373–386 in G.J. Webb et al., *Wildlife Management: Crocodiles and Alligators*, 1987). See also K. Jackson et al., *J. Morphol.* 229:315–324, 1996. (Syn. *follicle pores, Poren, follicle glands, follicle pits*) Cf. *dermal pressure receptor* and *touch papillae*, alternative terms for dermal sensory receptors on the head of Alligators.

intense competition. Reference to interspecific competition involving a small number of species having considerable niche overlap. Cf. *diffuse competition.*

intention movement. A preparatory action exhibited prior to a complete behavioral response.

inter-. Prefix meaning "between."

interanal. *n.* An unpaired plate on the *plastron* in marine turtles, situated on the midline between the posterior ends of the anals.

interaortic foramen. An area of potential continuity between the left and right aorta of snakes, situated at the base of the caudal boundaries of the aortic valves and extending to the cranial point of fusion between the opposing aortic cusps. The foramen allows potential shunting of blood between the two aortae. See B.A. Young et al., *J. Morphol.* 216:141–159, 1993.

interarticular cartilage. See *intercalary cartilage*.

interbreed, interbreeding. *n.* Reference to mating or hybridization between different individuals, populations, races, or species.

intercalary. *n., adj.* Literally "inserted," but referring usually to growth that takes place neither at the base nor at the tip of an elongate structure. This term has been used in various terminologies in herpetological literature.

intercalary cartilage or **element.** A cartilaginous element between the ultimate and penultimate phalanges in the digits of various frog taxa. (Syn. *interarticular cartilage, intercalary ossicle, intercalary phalanx*)

intercalary ossicle. See *intercalary cartilage*.

intercalary phalanx. See *intercalary cartilage*.

intercalary replacement. A type of tooth replacement in certain lizards and amphisbaenians wherein new teeth lie in an oblique series between the functional teeth and move obliquely into vacated sockets. (Syn. *alternate replacement, interdental replacement, anguimorph replacement*)

intercalary vertebra. Synonym for *odontoid process* of salamanders, as used by W.K. Parker (*Trans. Linn. Soc. London*, Series 2, Zool. 2:171–214, 1882).

intercalated. *adj.* Inserted between adjacent structures or strata.

intercalated disk. The junctional region between two connected cardiac muscle cells.

intercalation. *n.* 1. A developmental process whereby cells from different layers lose contact with their neighbors and rearrange into a single layer, which thereby expands laterally and increases in surface area. 2. Reference to the generation of missing positional values during regeneration when cells of disparate positional values are brought together following amputation.

intercanthal. *n.* Any one of the small scales on the dorsal aspect of the snout and situated between the *canthal* scales in Rattlesnakes. (Syn. *prefrontal*)

intercellular. *adj.* Between cells.

intercellular clefts. Lateral intercellular spaces between adjacent cells of epithelia, open at the basal ends but restricted at the luminal ends by tight junctions.

intercentrum. *n.* The more anterior unit of the centrum of certain *labyrinthodonts*. Remnants of this element contribute to cervical vertebrae and possibly, in minor ways, to other vertebral elements; but for the most part, the intercentrum becomes the intervertebral cartilage of the vertebral column of amniotes. 2. Used by various authors in reference to the *intervertebral disk* or *hypocentrum* when referring to the so-called basiventral part of a vertebra.

interchinshield. *n.* See *intergenial*.

interclavicle. *n.* A single dermal bone that lies on the surface of the sternum in reptiles and in primitive, extinct amphibians. The *omosternum* of frogs possibly represents a remnant of this structure in extant amphibians. See Fig. 3.

intercostal fold. This term is used in reference to the *costal fold* of salamanders but is more appropriate as the structure lies between the ribs and is not part of them. See *costal groove* or *fold*.

intercostal space. See *costal fold*.

interdental replacement. See *intercalary replacement*.

interdependent association. See *mutualism*.

interdigital membrane. Reference to the *web* joining the digits of many amphibians and reptiles.

interdorsal. (pl. **-ia**) *n.* One of the pair of *arcualia* arising from the anterior part of a *sclerotome*, above the *notochord*, according to the Gadow theory of vertebra formation. It fuses with arcualia below it and in the posterior part of the next anterior body segment to form a vertebra. (Syn. the *pleurocentrum* of other authors)

interface. *n.* 1. The contact surface between two zones or contiguous substances. 2. In abstract, the term can also be used in contexts of disciplines, ideas, or concepts.

interfemoral. *n.* Any of the scales situated on the ventral surface of the body between the hind legs of lizards.

interference competition. An interaction in which individuals behave in a manner that limits the access of one population to a resource it shares in common with another (e.g., by territorial behavior). Cf. *exploitation competition*.

interference microscope. A microscope used for observing transparent structures, and permitting quantitative determinations of the relative retardation of light by various objects.

interfertile. *adj.* Capable of interbreeding.

intergeneric. *adj.* Between genera.

intergenial. *n.* Any scale situated between the paired *chin shields* of snakes. (Syn. *interchinshield*)

interglacial. *adj., n.* A warm period between two glaciations or ice ages.

intergenual extent. A measure defined by C.J. Goin and M.G. Netting (*Ann. Carnegie Mus.* 28:137–167, 1940) as the distance between the knees of frogs while the femora are extended at right angles to the body.

intergradation zone. Reference to a boundary area occupied by a population of intergrades separating two adjacent subspecies.

intergrade. *n., v.* 1. Reference to a *hybrid* or the interbreeding of closely related species or subspecies in the wild. 2. An individual from an intermediate population, usually possessing intermediate characters. 3. To gradually merge or blend into one another, pertaining to aspects of morphology or color pattern of geographically intermediate populations formed from differently patterned morphs where their distributions come into contact.

intergular. *n.* A single plate or pair of plates between the *gular* plates of turtles.

interhyal. *n.* A dorsal cartilage associated with the *hyoid horn* of some reptilian *hyoids.* From A.S. Romer (*Osteology of the Reptiles*, p. 422, 1956).

interleukins. *n.* Proteins secreted by leukocytes that promote the growth and differentiation of immune system cells.

interlimb length. See *axilla-groin.*

intermandibular. *adj.* Reference to the space between the mandibles.

intermandibular gland. See *mental gland.*

intermaxillary. *n.* See *premaxillary.*

intermaxillary gland. A mucus gland in the nasal septum between the premaxillary bone and the nasal capsule in many amphibians. One or more ducts open into the roof of the mouth, and the secretions render the tongue adhesive. (Syn. *internasal gland*) See also *oral glands.*

intermediary metabolism. Chemical reactions within cells related to transformations of nutrient molecules into structural and energy molecules required for maintenance and growth of the cell.

intermediate host. A host that is involved in the life cycle of a parasite, but in which the parasite does not become sexually mature. (Syn. *secondary host*)

intermedin. *n.* An early term for *melanocyte stimulating hormone.*

intermedium. *n.* The central of three small bones situated at the proximal end of the *carpus* and *tarsus*. Its occurrence in the forelimbs of lizards is erratic but is fairly consistent in other amphibians and reptiles (called *semilunar* in turtles). Fusion can occur, usually with the *ulnare* of the forelimb and, more consistently, with the *tibiale* (to form the *astragalus*) in the hindlimb. Most references to this bone are with respect to the forelimb.

intermittent breathing or **ventilation.** A ventilatory pattern characteristic of many reptiles in which ventilatory periods (one or more breaths) are interrupted by nonventilatory periods (cessation of breathing) of varying duration.

intermittent locomotion. Bursts of locomotion interspersed with periods of muscular inactivity.

internal. *adj.* Within, close to, or in the direction of the center of a structure.

internal carotid artery. A branch of the common carotid artery that supplies the upper jaw and cranium. Cf. *external carotid artery.* See also *carotid artery.*

internal carotid foramen. An opening in the roof of the mouth through which the internal carotid artery enters the cranial cavity.

internal colonization. In a metapopulation model, propagules originate only from occupied sites. Thus, if there is regional extinction, colonization ceases because there is no external source of propagules.

internal fertilization. Sexual union of sperm and egg, hence reproduction, within the female's body. Characteristic of all reptiles and a few amphibians.

internal gill. The organ of respiratory gas exchange developed in most fishes as outgrowths of the wall of the pharynx in association with gill arches and perfused directly by blood from the aortic arches. There are many references to the "internal gills" of tadpoles following development of the operculum, but this usage is inaccurate. The term is not to be confused with gill filters, which protrude from the internal side of the gill arches. The filamentous gills extending from the outer edges of the gill arches are "internal" only in the sense they are located behind the operculum. The gill filters are, in location, more "internal" than are the respiratory gill structures, and the latter can also be considered to be "external" gills. Cf. *opercular gill.*

internal naris. (*pl.* **-ae**) *n.* Nasal openings into the mouth or throat. (Syn. *choanae*)

internal pterygoid. Reference by some authors to the *pterygoid* in terminologies that include "*external pterygoid*" as a synonym for *ectopterygoid.*

internarial distance. A measure of distance between the medial margins of the external nares.

internasal. *n.* 1. Any of enlarged plates or scales on the dorsal surface of the head of lizards and snakes, situated between the *nasals* and directly behind the *rostral* scales; usually paired in snakes. (Syn. *anterior frontal, prefrontal*). See Figs. 10, 14. 2. A medial, cranial dermal bone present in some extinct amphibians.

internasal gland. See *intermaxillary gland.*

internasal plate. See *planum basale.*

internasal ridge. The sharp edge on the tip of the snout and *canthus rostralis* in *Crotalus willardi*, formed by upward bending of the outer edges of the *internasals* and *canthals.*

internasal space. The area of skin between the external nostrils of amphibians.

internasoloreal. *n.* The scale on the dorsal surface of the head in the legless lizard *Anniella* formed by fusion of the *internasal* and the *loreal.* Used by W.R. Coe and B.W. Kunkel (*Trans. Conn. Acad. Sci.* 12:349–403, 1906).

International Code of Zoological Nomenclature. A set of guidelines that guarantees consistency of application of the Linnean hierarchy in systematic nomenclature.

International Commission on Zoological Nomenclature. (ICZN) The body that is responsible for the International Code of Zoological Nomenclature. It compiles official lists and indeces and rules on matters of priority and validity related to taxonomic names in zoology.

International Council for Science. (ICSU) A nongovernmental organization representing a global membership that includes both national scientific bodies (107 members) and international scientific unions (29 members). The ICSU provides an international network and forum for discussion of issues relevant to policies for international science, planning, and coordinating interdisciplinary research, and support of conferences and exchange of data and ideas. Current URL: http://www.icsu.org

International Geosphere-Biosphere Program. (IGBP) A research program dedicated to studies of Global Change and providing scientific knowledge that improves the sustainability of the living Earth. The IGBP was started in 1987 by the International Council for Science. Current URL is http://www.igbp.kva.se

International Herpetological Society. A UK-based international herpetological organization founded in 1969. Current URL: http://www.international-herp society.co.uk/

International Society for the History and Bibliography of Herpetology. (ISHBH) A nonprofit organization founded in 1998 in Guelph, Canada, to join individuals who have interests in history as well as the bibliography of herpetology. The Society is dedicated to disseminating knowledge of these topics among its members and to the general public. Current URL: http://www.t-ad.net/ishbh/

International Species Inventory System. (ISIS) An international system for administration and dissemination of biological information on animal species held in captivity.

International Union of Biological Sciences. (IUBS) A nongovernmental, nonprofit organization, established in 1919 to promote the study of biological sciences through global communications and programs. The body is responsible for modifications to the International Code of Zoological Nomenclature, based on recommendations from the ICZN and conducted at a General Assembly or approved International Congress. Current URL: http://www.iubs.org

International Union for the Conservation of Nature and Natural Resources. (IUCN) An international organization currently known as the World Conservation Union, comprised of international states, government, and nongovernmental agencies working toward goals of sustainable development, preservation of biodiversity, and general lasting improvement in the quality of life. This is the world's largest and most important conservation network operating through a unique global partnership. Use of the name World Conservation Union began in 1990, although the organization was founded in 1948 as the International Union for the Protection of Nature (IUPN). Many people still know the Union as the IUCN. The Union monitors the state of the world's species through the IUCN Red List of threatened Species. Headquarters are in Gland, Switzerland. Current URL: http://www.iucn.org

international units. (Système Internationale) (SI) Units adopted internationally for conventional usage, based on six primary units: meter, *m* (length); kilogram, *kg* (mass); second, *s* (time); ampere, *A* (electric current); Kelvin, *K* (temperature); candela, *cd* (luminous intensity). All other SI units are derived from these six basic units. See Table 2.

International Zoological Yearbook. An annual publication of the Zoological Society of London, intended for workers in zoological and wildlife parks throughout the world.

interneuron. *n.* A neuron that transmits communication and lies between two neurons with which it has synaptic contact.

internode. *n.* A line connecting two branch points or nodes in a phylogenetic tree, representing at least one ancestral species from a speciation event. Cf. *node.*

interoccipital. *n.* 1. A medial scale or plate on the dorsal head of some lizards, lying immediately behind the *interparietal.* 2. The fourth plate on the dorsal midline behind the *rostral* in blind snakes. 3. A small centrally situated scale that separates the *occipitals* on the poserior head region of crocodilians.

interoccipital spine. A median bony spine lying between the occipital spines in Horned Lizards (*Phrynosoma*).

interoceptor, interoceptive receptor. A sensory receptor that responds to internal stimuli inside the body. Cf. *exteroceptor, proprioceptor.*

interocufrontal. *n.* Any scale lying between the *supraoculars* and the *frontal* in viperid snakes.

interoculabial. *n.* Any of small scales below the eye of Rattlesnakes, bordered by the *postfoveals* anteriorly and the anterior *temporals* posteriorly.

interocular. *adj.* Reference to the region of the dorsal surface of head between the orbits.

interocular width. The distance between closest points of eye margin to midline of head in snakes. Used by B.C. Mahendra (*Proc. Indian Acad. Sci.* 4:230–238, 1936).

interolecranal extent. The distance between the elbows of a frog when the humeri are extended at right angles to the body axis. This is a rather unreliable measurement and seldom has been used in taxonomy.

interorbital. *n., adj.* 1. Any of small and sometimes irregular scales situated between the orbits on the dorsal head surfaces of squamates. These are especially characteristic of rattlesnakes and boids. (Syn. *intersupraocular*) 2. Reference to the dorsal region of the skull between the eyes.

interorbital bar. A dark transverse band across the dorsal surface of the head that contacts the eyes in some snakes.

interorbital distance. The transverse distance between the orbits in frogs, usually measured from the midpoints of the eyes.

interparietal. *n.* 1. A scale on the dorsal midline of the head, lying between the parietals in lizards. When present, the external opening of the *parietal eye* lies in this scale. See Fig. 14. 2. This term may be used for any of the midline scales on the dorsal surface of the head of blind snakes. 3. Alternative term for the *frontoparietal* of lizards. 4. Alternative term for the *postparietal bone* in the reptilian skull. 5. An unpaired cephalic scale lying in the notch formed posteriorly by contiguous *parietals* in turtles.

interparietal foramen. *n.* See *parietal foramen.*

interpectoral seam. The narrow line between adjoining *pectoral* plates in the plastron of turtles.

interplastral. *n.* A scale on the ventral midline of the plastron of some turtles, of very erratic occurrence. Reported by H.H. Newman (*Biol. Bull.* 10:68–114, 1906).

interpolation. *n.* The process of determining the value of an intermediate function within a range of known values, without using the function equation.

interpreocular. *n.* Any small scale lying between the large upper *preocular* and the crescentic lower *preocular* in rattlesnakes.

interpterygoid vacuity. Any large opening on the midline of the *palate*, single or double, between the *pterygoid* bones.

interrenal cell. Any of cells that lie in folded cords comprising the major part of the adrenal gland of reptiles. These cords are interspersed between chromaffin cells and appear to be equivalent to the cortical tissue of mammals. Used by W. Fox (*J. Morphol.* 90:481–554, 1952).

interrenal tissue. Endocrine tissues that produce corticosteroids and become the cortex of the adrenal gland. Cf. *chromaffin tissue*.

interrugal space. The depressed area lying between the *frontal ridges* on the head of *Anolis* lizards.

interrupted. *adj.* 1. The condition of labial teeth in anuran larvae where there is a gap or space in the middle of a row, in contrast to *entire*. 2. This term also has more general application to interruptions or gaps in structures.

interscalar tissue. Reference to integument lying between, and adjoining, scales of reptiles. See *hinge* and Fig. 29.

Intersex. *n.* Reference to any deviation from the normal phenotypic state of the two sexes. An abnormal individual may have characteristics that are intermediate between the two sexes, but the individual is of one sex genetically.

interspace. *n.* Any space between similar structures or elements of a color pattern.

interspecific. *adj.* Reference to comparisons between or among different species, in contrast to *intraspecific*.

interspecific competition. Competition for limited resources among individuals belonging to different species.

interstapedial. *n.* See *stylus*.

intersternal. *n.* See *mesoplastron*.

interstitial. *adj.* (*n.* **interstitium**) 1. Pertaining to, or situated between, parts; or in the interspaces of a tissue. 2. Reference to the fluid-filled spaces, or fluid itself, between cells.

interstitial cells. Cells that are located between tubules of vertebrate testes and secrete testosterone. (Syn. *Leydig cells*)

interstitial fluid. Liquid within the spaces surrounding (outside of) cells, generally similar to blood plasma but with lower content of protein. The liquid is external to the blood and lymphatic vessels. (Syn. *tissue fluid*)

interstitial granules. Very small, round scales found between larger scales or tubercles in some lizards.

interstitial nephritis. Inflammation of the supporting (interstitial) tissues within the kidney.

interstitial skin. Reference to usually thinner skin situated between scales.

interstitium. *n.* Interstitial tissue.

interstitial cell-stimulating hormone. (ICSH) See *luteinizing hormone*.

intersupraocular. *n.* See *interorbital*.

intertidal. *adj. n.* The shoreline zone between the highest and lowest tidal reaches (equivalent to *eulittoral*).

interval estimate. In statistics, an estimate that establishes a range in which a population parameter might be considered to occur. Cf. *point estimate*.

interventral. (*pl.* **-lia**) *n.* One of a pair of *arcualia* arising from the anterior part of a *sclerotome*, below the *notochord*. It fuses with arcualia above it and in the posterior part of the next anterior body segment to form a vertebra. (Syn. *cranihaemal*) See also *pleurocentrum*.

interventricular canal. The connection between the *cavum venosum* and *cavum arteriosum* in the single ventricle of noncrocodilian reptilian hearts, recognized by F.N. White (*Anat. Rec.* 135:129-134, 1959).

interventricular septum. See *horizontal septum*.

intervertebral body. A pad of fibrocartilage situated between the articular ends of successive vertebral centra and forming a flexible joint. In some cases the cartilage or fibrous connective tissue fuses with the vertebral unit, either anteriorly or posteriorly. Note the term "intervertebral disk" is reserved for the structure characteristic of mammals. (Syn. *intervertebral cartilage*)

intervertebral cartilage. See *intervertebral body*.

interzygapophysial ridge. A horizontal ridge connecting the *pre-* and *postzygapophyses* in snakes, as used by W. Auffenberg (*Tulane Stud. Zool.* 10:131–216, 1963).

intestinal fermentation. See *fermentation*.

intestine. *n.* The posterior, elongated segment of the digestive tract, extending from the stomach to cloaca, where food is digested and the resulting nutrients are absorbed.

intra-. Prefix meaning "within."

intracardiac shunt. Because the ventricle of noncrocodilian reptiles and amphibians is undivided, anatomical features of the heart result in the potential for intracardiac shunting. Thus, a fraction of blood returning from body tissues and "intended" for distribution to the lung (for reoxygenation) instead recirculates to the body via the systemic outflow tracts; similarly, a fraction of oxygenated blood returning from the lung and normally "intended" for the body tissues instead recirculates to the lung via the pulmonary outflow tracts. These conditions are termed *intracardiac shunts* and result in either pulmonary bypass or systemic bypass (or both) of some fraction(s) of blood passing through the undivided ventricle. Partial or complete pulmonary bypass is called a *right-left shunt* (*R-L shunt*), and partial or complete systemic bypass is called a *left-right* (*L-R shunt*) shunt. The mechanisms controlling the magnitude and direction of shunting may involve nervous control of the relative resistance or pressure differences in the pulmonary and systemic outflow tracts. An increase in pulmonary vascular resistance will promote the translocation of a fraction of blood from the systemic circuit into the pulmonary circuit, and vice versa. These mechanisms have been termed *pressure shunting*. In contrast, a passive mechanism for shunting can involve blood that is located in a space of the ventricle common to both pulmonary and systemic circulations at different phases of the heart cycle and is subsequently "washed" into the "wrong" circulation by inflowing or outflowing blood moving in the "correct" direction. This latter mechanism has been termed *washout shunting*. See N. Heisler & M.L. Glass, pp. 334–353 in K. Johansen & W. Burggren, eds., *Cardiovascular Shunts: Phylogenetic, Ontogenetic and Clinical Aspects*, 1985; J.W. Hicks, pp. 425–483 in C. Gans & A.S. Gaunt, eds., *Biology of the Reptilia*, Vol. 19, Morphology G, 1998. (Syn. *cardiac shunt*)

intracellular. *adj.* Within a cell or cells.

intracellular space. The body fluid compartment located within cells.

intrademic selection. In genetics and evolutionary biology, selection that occurs within a local interbreeding population.

intrafusal fiber. A muscle fiber within a *muscle spindle* organ, forming neuromuscular spindles. See also *extrafusal fiber*.

intrageneric. *adj.* Within the same genus, or among members of the same genus.

intraguild predation. Reference to competitors that exploit common, limiting resources and also interact with one another as predator and prey. The phenomenon is common and can either reverse or reinforce the outcome of competitive interactions between species.

intramembranous ossification. The process of membrane bone formation, directly from mesenchyme without a cartilage precursor, giving rise to bones of the lower jaw, skull and pectoral girdle, bone in skin, and vertebrae in some urodeles and apodans. Cf. *endochondral ossification*.

intramuscular. *adj.* (*adv.* **intramuscularly**) Within muscle. Common usage is with reference to a mode of drug injection.

intranasal. *n.* See *septomaxilla* or *septomaxillary bone*.

intrasexual selection. See *sexual selection*.

intraspecific. *adj.* Reference to comparisons within a single species (e.g. different populations), in contrast with *interspecific*.

intraspecific competition. Competition for limiting resources among individuals of the same species.

intratarsal joint. Reference to an ankle in which the line of flexion passes between the calcaneum and the astragalus. Cf. *mesotarsal joint*.

intravascular. *adj.* Within a blood vessel.

intravenous. (**IV**) *adj.* (*adv.* **intravenously**) Within a vein. Common usage is with reference to a mode of drug injection.

intrinsic. *adj.* Belonging entirely to a structure, individual, or system; originating within it. Cf. *extrinsic*.

intrinsic diversity. See *Shannon-Weiner index*.

intrinsic isolating mechanism. Any genetically determined factor that prevents interbreeding between individuals of a sympatric species. Cf. *extrinsic isolating mechanism*.

intrinsic rate of increase. See *instantaneous rate of increase*.

intro-. A prefix meaning "inward."

introduced. *adj.* (*n.* **introduction**) Reference to individual animals or species that are brought from an area where they occur naturally to one where they are not indigenous. (Syn. *exotic, alien, invasive*)

introgression, introgressive hybridization. *n.* The incorporation of genes from one species into the gene pool of another by hybridization.

intromission. *n.* Literally the introduction of one part or object into another, applied in biological literature to sexual union.

intromittent organ. A male reproductive structure used for internal fertilization during copulation. This term includes the *penis* of crocodilians and turtles, the *hemipenes* of snakes and lizards, the *phallodeum* of caecilians, and the tail of *Ascaphus* frog.

intussusception. *n.* A condition in which one loop of the intestine invaginates into an adjoining intestinal segment.

in utero. Within the uterus.

invagination. *n, v.* Infolding of a surface or membrane.

invalid name. (*nomen invalidum*) In taxonomy, a name that is not validly published or is unavailable. Cf. *valid name*.

invasive, invasive species. *adj., n.* Reference to a nonnative species, *exotic* or *alien* to a region in which it occurs but was not historically present.

inverse density dependence. A condition in which the influence of an environmental factor tends to enhance population growth as population density increases, or to diminish population growth as population density decreases. Cf. *density dependence*.

inverse transformation. Transformation of a data set by taking the inverse of each number (x) by computing 1/x. This essentially makes very small numbers very large, and vice versa. Therefore, one must be careful to reflect or reverse the distribution prior to applying an inverse transformation so the ordering of the values will be identical to that of the original data. See *data transformation*.

inviability *n.* (*adj.* **inviable**) Generally, reduced vigor or ability to survive. In population biology, a measure of the percentage of individuals that fail to survive in a particular cohort when compared with another.

in vitro. *adv., adj.* In an artificial environment outside the body.

in vivo. *adv., adj.* Within a living organism or tissue.

involution. *n.* 1. Generally, rolling or turning inward. 2. In embryology, the migration of cells into the gastrula at the blastopore. 3. A process by which an expanding epithelium turns over on itself and continues to spread in the opposite direction along its basal margin. 4. The progressive degeneration that occurs naturally with advancing age.

ion. *n.* An atom or group of atoms carrying a positive (*cation*) or negative (*anion*) charge.

ionic bond. An electrostatic bond between atoms or molecules.

ionization. *n.* 1. The dissociation of a compound into ions in a solution. 2. Any process by which a neutral atom or molecule acquires a positive or negative charge.

iota link. A direct unbranching link in a food web involving a prey species and its unique predator or parasite. Cf. *gamma link, lambda link.*

ipsilateral. *adj.* Pertaining to the same side. Cf. *contralateral.*

iridescence. *n.* The condition of reflecting bright and changing colors attributable to structural properties of a surface in contrast to pigments.

iridic guanism. *n.* The development of *guanophores* in the iris of amphibians. Described by R.C. Stebbins (*Univ. Calif. Publ. Zool.* 48:377–512, 1949)

iridocyte. *n.* A specialized cell that is filled with crystals of guanine and various lipophores. (Syn. *guanophore, iridophore*)

iridophore. *n.* A chromatophore characterized by the presence of the purine *guanine*. Crystals of guanine impart a noniridescent bluish to silvery white color that varies depending on other surrounding pigments in the skin. (Syn. *guanophore*) See Fig. 33.

iris. *n.* A muscular disc with a contractile hole situated in front of the eye lens that acts as an adjustable aperture. The iris is functionally equivalent to the diaphragm of a camera lens, which controls light entry, clarity, and depth of field.

irradiance. *n.* Reference to the *radiant flux density* on a surface (e.g., W/m^2).

irritability. *n.* The ability to perceive and respond to stimuli. Sensitivity.

irruption. *n.* An abrupt and irregular increase in the size of a population, usually associated with some set of favorable changes in the environment and mass movement of animals. An example would be increased presence of amphibians following a wet year preceded by drought.

ischemia. *n.* (*adj.* **-ic**) The absence of blood flow to an organ or tissue.

ischial spine. A posteriorly projecting extension of the lateral part of the ischium in certain salamanders.

ischium. *n.* The posteriormost of the three bones that comprise each half of the pelvic girdle in tetrapods.

ISHBH. See *International Society for the History and Bibliography of Herpetology.*

ISIS. See *International Species Inventory System.*

island arc. A curved line of islands.

island biogeography, theory of. A principle developed by R.H. MacArthur and E.O. Wilson (*The Theory of Island Biogeography*, 1967) that the number of species existing on an island is a function of both island area and distance from the mainland, and is determined by the relationship between immigration (greater on nearer and larger islands) and extinction (greater on smaller islands). See also *subsidized island biogeography hypothesis.*

island-mainland model. A metapopulation model in which local extinctions are independent of one another, and colonization occurs via a continuous source of migrants. At a community level, this model is equivalent to the equilibrium model of *island biogeography.*

Island Night Lizard. Vernacular name for the largest xantusiid species, *Xantusia riversiana.* Formerly *Klauberina riversiana*; note that *Klauberina* has been synonymized with *Xantusia.*

islets of Langerhans. Aggregates of hormone-secreting cells found in the pancreas of vertebrates. Two types of cells secrete *insulin* and *glucagon*, respectively.

ISO. See *integumentary sense organ.*

iso-. Prefix meaning "equal."

isobar. *n.* A line on a chart or map connecting points of equal atmospheric or hydrostatic pressure.

isodont. *adj.* Condition in which teeth are of relatively uniform size and appearance.

isoelectric point. The pH at which the net charge on a protein is neutral (zero).

isoenzymes. *n.* See *isozymes.*

isoforms. *n.* Functionally related proteins that differ slightly in their sequences of amino acids.

isogamy. *n.* A mode of sexual reproduction in which sex cells of opposite mating types have similar size and morphology. Cf. *anisogamy.*

isogenetic. *adj.* Of the same origin. (Syn. *isogenous*)

isogenous. *adj.* See *isogenetic.*

isograft. *n.* A graft of tissue between two individuals having identical genotypes.

isogram. *n.* A line on a map joining points of equal numerical value with respect to a given parameter.

isolate. *n.* A geographically or ecologically isolated population or group of populations.

isolating mechanism. See *reproductive isolating mechanism.*

isolation. *n.* In biogeography and ecology, reference to condition of a population being separated from others such that interbreeding with others is prevented by extrinsic or intrinsic barriers.

isolecithal. *adj.* Reference to an egg in which the yolk is evenly distributed throughout the cytoplasm. Cf. *telolecithal*.

isomers. *n.* Reference to compounds having identical molecular formulas but with different three-dimensional molecular shapes or spatial orientation.

isometric contraction. Contraction and generation of force during which a muscle remains at fixed length and does not shorten significantly. Cf. *eccentric contraction*.

isometric growth. Growth with geometric similarity in which relative proportions of body parts remain constant with changes in total body size. Cf. *allometric growth*.

isometry. *n.* (*adj.* **-ic**) Proportionality of shape regardless of size. The relative proportions of the body parts remain constant as an individual grows. (Syn. *geometric similarity*) Cf. *allometry*.

iso-osmotic, isoosmotic. *adj.* Having a concentration of osmotically active constituents equivalent to that of a reference solution. Cf. *hyperosmotic, hypoosmotic*.

isophagous. *adj.* Reference to predators that are selective with respect to prey items but are not restricted to a single food type.

isopleth. *n.* A line on a chart or map connecting points of equal concentration of a given parameter.

isosmotic. *adj.* See *iso-osmotic*.

isotelic. *adj.* Producing or tending to produce the same effect. See also *homoplastic*.

isotherm. *n.* A line of equal temperature.

isotonic. *adj.* Having a *tonicity* equivalent to that of a reference solution. Cf. *hypertonic, hypotonic*.

isotonic contraction. A state of muscle contraction in which the generated force remains constant while the muscle changes length. See *concentric contraction* and *eccentric contraction*.

isotope. *n.* Any of two or more forms of an element having the same number of protons (atomic number), but a different number of neutrons (atomic weight). See also *stable isotope*.

isotropy. *n.* See *anisotropy*.

isotype. *n.* 1. In taxonomy, a duplicate of a holotype from the collection containing the holotype. 2. A type described from two congeneric species. 3. A form that occurs in a variety of locations.

isovolumetric, isovolumic. *adj.* Of equal volume.

isozymes. *n.* Multiple forms of a single enzyme found in the same animal species or even in the same cell. See *allozymes*. (Syn. *isoenzymes*)

isthmus. *n.* 1. A narrow strip of land connecting two larger land masses. 2. The aglandular portion of the reptilian oviduct. See J.E. Girling, *J. Exp. Zool.* 293:141–170, 2002.

iteration. *n.* A repetition procedure used in computation wherein operations are repeated until a good fit is obtained.

iteroparity. *n.* (*adj.* **–ous**) Reference to organisms having repeated periods of reproduction during their life history. Cf. *semelparity*.

IUBS. See *International Union of Biological Sciences*.

IUCN. See *International Union for the Conservation of Nature and Natural Resources*.

IUCN Red List. A list of animal species and subspecies considered to be threatened with extinction, including those thought to be already extinct in the wild. See also *International Union for the Conservation of Nature and Natural Resources*.

IV. See *intravenous*.

J

Jacare, Yacare. *n.* A Brazilian name for the *Caiman.*

Jackknife method. A method for estimating species richness in a community based on occurrence of species unique to a quadrat and sampled from a series of quadrats, calculated as

$$R = s + (n-1/n)^k$$

where *s* equals total number of species in *n* quadrats, *n* the number of quadrats, and *k* the number of unique species.

Jackson's Chameleon. Vernacular name for the popular African Chameleon *Chamaeleo jacksonii.* The head is adorned with three horn-like structures.

Jackson's method. An archaic method of treating venomous snake bite developed by Dr. Dudley Jackson. The method involves making a series of cross-shaped incisions at and around the fang punctures, extending these incisions above the bite as the swelling progresses, and applying suction (using rubber balls or mouth) at each incision.

Jacobson's gland. A large, compound tubular gland that empties into the *vomeronasal organ* of some amphibians.

Jacobson's organ. See *vomeronasal organ.*

Jacura. *n.* Vernacular name for the northern species of *Tegu.*

Japaluras. *n.* Collective vernacular name for Asian species of agamid lizards belonging to the genus *Japalura* (also called *Mountain Lizards*).

Japanese Journal of Herpetology. See *Current Herpetology.*

Japanese Rat Snake. Vernacular name for the colubrid species *Elaphe climacophora.* Also called *Aodaisho* locally.

Japanese Toad. Vernacular name for the bufonid species *Bufo japonicus.*

Japanese Treefrog. Vernacular name for the Japanese hylid species *Hyla japonica.*

Jararaca. *n.* Vernacular name for a large neotropical Lancehead pitviper *Bothrops jararaca.*

Jararacussu. *n.* Vernacular name for the South American pitviper *Bothrops jararacussu.*

Javan Mudsnake. Vernacular name for the Asian colubrid species *Xenodermus javanicus.*

jaw. *n.* Either of the two often complex parts of the mouth skeleton, lower jaw, and upper jaw. See also *mandible, maxilla, palatoquadrate.*

Jawclap. *n.* A behavior of crocodilians in which the jaws are suddenly and sharply clapped shut, either below or on the water surface, creating an acoustic signal employed during courtship. The behavior is often accompanied by a *headslap* and serves to deter rival males as well as attract females.

jaw sheath. Oral structures in tadpoles formed by the fusion of palisades of keratinized cells along the lateral margins overlying the infrarostral and suprarostral cartilages, serving as cutting or abrasive feeding surfaces, usually curved in shape, and typically with serrated edges. The term is used by some in preference to "beak" because it is unlikely these structures are homologous with the beaks of birds and turtles. See R. Altig, *Herpetologica* 26:180–207, 1970; H.C. Kaung, *Develop. Biol.* 11:25–49, 1975; H.C. Kaung & J.J. Kollros, *Anat. Rec.* 188:361–370, 1976. However, the term beak is, in most cases, analogous with those in birds and turtles, and the "beak" terminology is useful because it implies a particular action (unlike "jaw sheath"). See Fig. 1.

jelly. *n.* A mucoid, gelatinous substance secreted by the walls of the oviducts of amphibians and deposited on the eggs as they pass through the structure.

jelly coat. See *jelly envelope.*

jelly envelope. Reference to concentric layers of clear, gelatinous material deposited on the eggs of amphibians as they pass through the oviduct. The number of envelopes varies with species and may be deposited by glands in the cloaca as well as the oviduct. (Syn. *coat, egg capsule, egg envelope, envelope, gelatinous coat, jelly coat*)

jelly float. Reference to a wide, marginal ridge formed shortly after ovulation in the outer jelly envelope surrounding individual eggs of the frog *Kaloula.* Each egg in a clutch is separate from others and is kept on the surface of the water by the float.

jelly strand or **string.** The elongate strand of jelly that is deposited on the eggs of toads in the genus *Bufo.* The outer layers of the strand are common to the entire clutch, while two inner layers surround each individual egg.

jet wind effect. Reference to a wind that is intensified as a result of being funneled through a narrow canyon or mountain gap.

joint. *n.* Any junction or point of articulation of two skeletal elements.

Jolly-Seber method. A method used in mark-recapture studies to estimate population size of an open population.

The calculation involves a series of three or more periods of sampling in which marked individuals are counted and additional individuals are marked for individual recognition.

Jordan's rule. An ecological principle stating the nearest relatives of a given species are found not in the same area or in a remote one, but are generally in an adjacent region and separated by some type of barrier.

Journal of Herpetological Medicine and Surgery. A journal published by the Association of Reptilian and Amphibian Veterinarians.

Journal of Herpetology. A journal of herpetology published by the *SSAR*.

jowls. *n.* The fleshy parts beneath the lower jaw.

joule. (J) A derived SI unit of energy equivalent to 0.239 calories (cal).

jubal. *n.* Distinctly broadened or enlarged scale or scales situated immediately posterior to the head in some skinks of the genus *Eumeces*. A differentiated *nuchal*.

jubilee. *n.* Reference to an unusual aggregation or mass migration of aquatic animals, including sirens and other salamanders. Origin of this term is discussed in A. Carr, *A Naturalist in Florida*, 1994.

jugal. *n.* 1. A dermal bone of the cheek in vertebrates, found in the skull of reptiles and some modern amphibians. This bone borders the orbit ventrally and extends back behind the maxillary to form part of the lipline, forming the anterior half of the *zygomatic arch* when present. 2. Used by some authors in reference to scales situated below the eye in crocodilians.

jugal ligament. A ligamentous connection extending from the posterior extremity of the maxilla to the lateral extremity of the suspensorium in *Salamandra*.

jugular. *adj.*, *n.* 1. Pertaining to the throat or neck. 2. Used in reference to scales lying between the lower *labials* and the *chinshields* in snakes. See Fig. 12.

jugular body. Prominent lympho-myeloid structures morphologically similar in some respects to lymph nodes of mammals and situated lateral to the point of divergence of the pulmocutaneous artery and the aorta, described in several adult anuran amphibians. The function of these structures is not known, but they are believed to play a role in the production of lymphocytes and may be involved in some aspects of the differentiation and maintenance of the immune response capacity.

Jugular vein. Vessels that provide venous drainage from the brain, head, and neck, sometimes as *external jugular* and *internal jugular* vessels.

Jumping Pitvipers. Collective vernacular name for heavy-bodied, neotropical viperids belonging to the genus *Atropoides*.

jungle. *n.* Dense seral vegetation, particularly characteristic of tropical regions having high levels of precipitation.

Jungle Toads. Vernacular name for southeast Asian species of bombinatorid anurans belonging to the genus *Barbourula*.

junior homonym. In taxonomy, the later published of two homonyms. Cf. *senior homonym*.

junior synonym. In taxonomy, any of two or more synonyms other than the earliest published. Cf. *senior synonym*.

Jurassic. *n.* The second and middle Period of the Mesozoic Era, also known as the Age of Dinosaurs. Dinosaurs dominated the terrestrial vertebrate fauna, and the earliest birds appeared. See Table 1.

juvenile. *n.* A newborn, hatchling, or very young animal in an immature stage of development up to the time of attainment of sexual maturity. Juveniles are considered by some to comprise an age class distinct from that of *neonates* (D.J. Morafka et al., *Herpetol. Monogr.* 14:353–370, 2000).

Juxtaglomerular apparatus. A regulatory unit within the kidney consisting of *macula densa* cells in distal tubular segments and specialized cells at the glomerular afferent arterioles. Best studied in mammals, this unit participates in regulation of glomerular filtration and renal function.

juxtaposed. *v.* To place side-by-side or in contact. (Syn. *apposed*)

juxtaposed scales. Scales having edges that are touching but not overlapping.

juxtaposition. *n.* Placement of things side-by-side or in immediate contact. (Syn. *apposition*)

juxtasplenic body. In some snakes and varanid lizards the dorsal lobe of the *pancreas* is connected to the ventral lobe by an isthmus and is closely associated with the spleen. This anterior portion has been termed the juxtasplenic body and consists primarily of islet tissue.

K

K. See *carrying capacity*.

***K* and *r* selection theory.** See *r-K selection*.

kallikrein-kinin system. The enzyme *kallikrein* catalyzes *kinin* from plasma protein precursors. Kinins are involved in regulating inflammatory responses by mediating histamine release, promoting localized vasodilation of blood vessels, increasing capillary permeability, and activating pain receptors. The system has been demonstrated in crocodilians and is probably widespread among amphibian and reptilian taxa.

karstic. *adj.* Reference to irregular limestone strata permeated by streams and typically with sinks, caves and other subterranean passages.

karung. *n.* Snake skin, especially that of *Acrochordus* spp., having flattened scales and used in the commercial manufacture of clothes, bags, and other items.

karyotype. *n.* The chromosomal complement of a cell, individual, or species, often illustrated with a diagram showing the chromosomes placed in order of size.

karyogram. *n.* A photographic representation of the chromosome complement (*karyotype*) of a cell.

karyology. *n.* The branch of cytology that entails study of nuclei, especially the structure of chromosomes.

karyotic. *adj.* Nucleated.

karyotype. *n.* 1. The chromosome complement of a cell or individual. 2. Structural characteristics of the chromosome set. 3. Reference to those individuals having an identical complement of chromosomes. 4. A *karyogram*.

Kassinas. *n.* Collective vernacular name for African hyperoliid frogs belonging to the genus *Kassina*. Also called *Running Frogs*.

Kassininae. *n.* A subfamily of anurans belonging to *Hyperoliidae*.

kb. See *kilobase*.

keel. *n.* (*adj.* **-ed**) A prominent ridgelike process or raised line. Often used in reference to protruding midlines of scales, plates, or tail fins. (Syn. *carina*)

Keelbacks. *n.* Vernacular name for several Asian and Eurasian genera of natricine snakes, including *Amphiesma*, *Atretium*, *Macropisthodon*, *Rhabdophis*, *Tropidonophis*, and *Xenochrophis*. The name also has been applied to neotropical species in the genus *Helicops*.

keeled scale. Reference to a condition of scalation in squamate reptiles in which scales bear a prominent central keel or protruding midline. Cf. *smooth scale*.

Kelvin scale. (K) A temperature scale that specifies –273.16 °C as absolute zero, and each degree is equal to 1 degree Celsius. See *absolute zero*.

Kemp's Ridley Seaturtle. See *Cheloniidae*.

Kendall's rank correlation coefficient. A nonparametric statistic for determining the significance of association between two variables, using ranks for values within their respective samples.

keradont. *n.* See *labial teeth*.

keratin. *n.* (*adj.* **–ous**) A hard or tough fibrous, nonsoluble protein ("structural" protein) produced in the epidermis of both reptiles and amphibians and forming the basic material for scales, horns, claws, spines, etc., as well as the outermost layer of the epidermis, the *stratum corneum*. Generally, there are two types of keratin differentiated from keratinocytes: *alpha* (α), which is helical, hair-like keratin, and *beta* (β), which is pleated sheet, feather-like keratin. The latter is somewhat harder or stiffer due to a larger number of disulfide linkages within the protein matrix. β-keratins form the outer scale surfaces of lepidosaurs and crocodilians, and the shell scutes of chelonians. α-keratins occur vertically below β-keratins of scales and occur exclusively at hinge regions of crocodilians and the leg and neck skin of chelonians. For further discussion see H.B. Lillywhite & P.F.A. Maderson, *Biology of the Reptilia*, Vol. 12, Physiology C, pp. 397–442, 1982.

keratinization. *n.* The development of, or conversion into, keratin.

keratinocyte. *n.* An epidermal cell that has capacity to synthesize *keratins* and/or *lamellar bodies*.

keratinosome. *n.* See *lamellar bodies*.

keratitis. *n.* Inflammation of the cornea.

keratogenesis. *n.* (*adj.* **keratogenic**) Reference to the synthesis and formation of keratin.

keratohyalin-like granules. Structures resembling keratohyalin granules of the mammalian stratum granulosum, observed in the *clear layer* in epidermis of lizards and in the Tuatara during a limited period of the renewal phase of the shedding cycle, and questionably in the hinge region of crocodilian scales. These granules contain sulfur-rich and histidine-rich proteins and participate in the process of hardening of the clear layer that molds the spinulae and superficial microornamentation of the *Oberhautchen*.

keratophagy. *n.* Literally the eating of keratin or keratinous materials. Often used in reference to the act of eating one's own shed skin, common in many amphibians and reptiles during *ecdysis*.

ketone. *n.* A compound having a carbonyl group (CO) attached by a carbon to hydrocarbon groups.

ketone bodies. Products of fat and pyruvate metabolism that are formed from acetyl CoA in the liver and oxidized by muscle and the central nervous system during starvation.

key. *n.* An organized list of character states that facilitate identification of species, subspecies, or other taxa.

key character. A diagnostic character.

key stimulus. See *sign stimulus*.

key species. Any species that can be used to assess the degree of utilization or quality of a habitat or area.

keystone predator. The dominant predator having major influence on community structure. See *keystone species*.

keystone species. A concept originated by R.T. Paine (*Amer. Nat.* 100:65–75, 1966; *Amer. Nat.* 103:91–93, 1969) to identify species that are exceptionally important to the structure of communities and help to maintain their organization and diversity. This concept was extended by M.E. Powers and L.S. Mills (*Trends Ecol. Evol.* 10:182–184, 1995) and M.E. Powers et al. (*BioScience* 46:609–620, 1996) to interpret keystone status as only those species having a large, disproportionate effect, with respect to their biomass or abundance, on their community. See also L. Khanina, *Conservation Ecology* 2(2):R2, 1998 (Current URL: http://www.consecol.org/Journal/vol2/iss2/resp2).

kidney. *n.* One of two principal organs characteristic of vertebrates that functions in excretion and water balance.

kilo-. Prefix used to denote unit x 10^3.

kilobase. (kb) *n.* A unit of length of DNA or RNA strand equal to 10^3 bases or 10^3 base pairs.

kilocalorie. (Kcal; Cal) *n.* A derived cgs unit of energy equal to 1000 calories or 4.1855 x 10^3 joules at 15 °C.

kilodalton. (kdal, kDa) *n.* A unit of *molecular weight* equal to 1000 daltons.

kilopascal. (kPa) *n.* The SI unit of current common usage for measurements of pressure in physiology. See *Pascal*.

Kimura's neutral model of evolution. The hypothesis that neutral mutations outnumber advantageous mutations and, therefore, most variability between species is neutral with respect to natural selection (M. Kimura, *Nature* 217:624–626; also pp. 208–233 in M. Nei & R.K. Koehn, eds., *Evolution of Genes and Proteins*, 1983). See *neutral mutation-random drift theory of molecular evolution*.

kinase. *n.* A class of enzymes that catalyze the transfer of a phosphate group from ATP to a second substrate.

kinematics. *n.* The study of animal motion, its course and patterns.

kinematic viscosity. The *viscosity* of a fluid divided by its density.

-kinesis. Suffix meaning "movement."

kinesis. *n.* (*adj.* **kinetic**) 1. Physical movement, often used in reference to cellular phenomena. 2. A change in rate of random movement of an organism as a result of a stimulus. Also measured as the linear or angular velocity changes that occur in response to an alteration in the intensity of a stimulus. 3. Used in herpetology to denote the degree of movement between parts of the skull, or the plastron of some turtles. See *cranial kinesis*; also *akinesis, amphikinesis, mesokinesis*.

kinetic. *adj.* Dynamic, relating to motion. This word is descriptive of tetrapod skulls having several movable units.

kinetic energy. Energy inherent in the motion of a mass. The energy of motion.

kinetic skull. A skull in which certain elements, or groups of elements, have mobility with respect to other units. See *cranial kinesis* and Figs. 18, 19.

King Brown Snake. Vernacular name for the Australian elapid species *Pseudechis australis*.

King Cobra. Vernacular name for the largest species of Cobra, the Asian *Ophiophagus hannah*.

kingdom. *n.* The highest category in the hierarchy of classification, above *phylum*. Amphibians and reptiles are classified in the kingdom *Animalia*.

Kingsnakes, King Snakes. *n.* Vernacular name for numerous species of American snakes belonging to the colubrid genus *Lampropeltis*. See also *Milk Snakes*.

kinin. *n.* See *kallikrein-kinin system*.

kinocilium. *n.* The longest and most complex hairlike element of a neuromast cell.

Kinosternidae. *n.* A clade (family) of turtles including the smallest turtles of North America, commonly known as *Mud* and *Musk Turtles*. There are 3 genera and 22 species occurring in eastern North America, Central America, and South America.

kinosternid. *n.* A member of, or pertaining to, *Kinosternidae*.

Kinosternoidea. *n.* A clade (superfamily) of extant turtles that includes *Dermatemydidae* and *Kinosternidae*.

kin recognition. The ability to identify relatives. As example, certain amphibian larvae are cannibalistic and discriminate between nonkin and cousins.

kin selection. The proposition that a social act is favored if it increases the *inclusive fitness* of the individual performing the behavior, used as an explanation for how altruism can evolve when it increases the fitness of relatives. See W.D. Hamilton, *J. Theoretical Biol.* 7:1–52, 1964. Cf. *group selection*.

kipuka. *n.* An island of vegetation isolated by lava flows.

Kirtland's Snake. Vernacular name for the American terrestrial natricine *Clonophis kirtlandii*.

kite diagram. A two-dimensional representation of a frequency distribution having samples typically arranged symmetrically about a common vertical axis. Such distributions are used to depict the vertical stratification of a population or community.

knob. *n.* A protuberance or rounded lump. Used with particular reference to "knobbed anal keels," which are enlarged keels found on scales near the anal region in certain species of snakes.

Knob-scaled Lizards. Collective vernacular name for Middle American species of lizards belonging to the family Xenosauridae, especially the genus *Xenosaurus.*

knockout. *adj., n.* Informal reference to an animal in which a functional gene has been deleted or replaced. Therefore, the protein originally coded by the gene cannot be expressed.

knot. (kn) A unit of velocity equal to 1 *nautical mile* per hour.

Komodo Dragon. Vernacular name for the largest Monitor Lizard (*Varanus komodoensis*) endemic to the Komodo Islands.

kPa. See *kilopascal.*

kraal. *n.* An enclosure. E.g., a protected enclosure around the nest of a sea turtle on a beach.

Krait. *n.* Collective vernacular name for tropical Asian elapid snakes belonging to the genus *Bungarus.* See also *Sea Krait.*

Krebs cycle. See *citric acid cycle.*

krenal. *adj.* Reference to the upper region of a stream that originates as a spring.

K-r spectrum. A linear representation of reproductive strategies with *K*-selected species and *r*-selected species represented at the two extremes. See *r-K selection.*

krummholz. *n.* A discontinuous region of scrub or stunted forest typical of windswept alpine regions in close proximity to the tree line. Cf. *elfin forest.*

Kruskal-Wallis test. A nonparametric test that employs a one-way *analysis of variance* by rank, used to test differences in mean level between several samples where the number of counts per sample is different.

kryal. *adj.* Reference to the upper region of a stream that originates as meltwater from a glacier or permanent icefield.

K-selected species. See *r-K selection.*

K selection or **strategy.** See *r-k selection.*

K strategist. See *r-K selection.*

K-T boundary. 1. The transition in geologic time from the end of the *Cretaceous* (K) to the beginning of the *Tertiary* (T), approximately 65 million years ago. 2. A site or area showing evidence of rocks and other materials from both periods. See Table 1.

K-T extinction. A phenomenon that occurred at the end of the *Cretaceous Era* and transition to the *Tertiary* (about 65 million years ago), involving the mass extinction of numerous species including many dinosaurs. The phenomenon is regarded as a continuum of related events, but not established as having a single overriding cause. See Table 1.

Kufi. *n.* See *Levantine Viper.*

Kukri Snakes. Collective vernacular name for species of snakes belonging to the colubrid genus *Oligodon.*

Kulezinski coefficient. An index of similarity in species composition between two communities that is relatively little influenced by different sample sizes, calculated as

$$S_k = c/(a+b)$$

where *c* is the number of species common to both, and *a* and *b* are the numbers of species occurring in communities A and B respectively.

kurtosis. *n.* The departure of a frequency distribution from a normal distribution, measured according to its relative flatness or peakedness.

Ky. Abbreviation denoting 1000 years.

kyphoscoliosis. *n.* A condition of both *kyphosis* and *scoliosis.*

kyphosis. *n.* Dorsoventral curvatures of the vertebral column when viewed laterally. Commonly seen in reptiles suffering from metabolic bone disease. The usage in herpetology is largely in reference to abnormal shell conditions in turtles and curvature of the vertebral column. See also *kyphoscoliosis.*

L

label. *n.* In cellular and molecular biology, reference to the attachment of a substance to any cell or molecule of interest, thereby allowing these "targets" to be readily identified, counted, or isolated from other objects either *in vitro* or *in vivo*. (Syn. *tag*)

labial. *adj., n.* 1. Reference to the upper or lower lip. 2. As a noun, labial refers to any of the scales in rows that border the lips of reptiles. (Syn. *supralabial, infralabial* in reference to upper and lower labials, respectively) See Fig. 11.

labial flap. Fleshy flaps (various shapes and sizes) that overhang the mouth in microhylid tadpoles. (Syn. *oral flap, labial fold*)

labial fold. A flap of tissue that protrudes from the upper lip and covers all or part of the lower lip in many amphibian larvae. See *labial flap.*

labial formula. Reference to methods for representing the number and position of *labial* scales in snakes.

labial gland. 1. Mucous or muco-serous glands opening into the lips of various reptiles. 2. A glandular area in skin covering the jaws in *Salamandra*. Used by E.T. Francis (*The Anatomy of Salamanders*, p. 262, 1934).

labial papilla. (*pl.* **-ae**) Usually used in the plural with reference to any marginal or submarginal papillae (but not buccal papillae) of the *oral disc* of tadpoles. (Syn. *oral papilla*) See Fig. 1.

labial pit. Specialized *infrared receptors*, present as deep indentations on the margins of individual labial scales of various boid snakes. See *pit organ.*

labial ridge. See *tooth ridge.*

labial teeth. Small, keratinous rasps arranged like teeth of a comb in transverse rows on the lips of anuran tadpoles. These are not true teeth. (Syn. *denticles, keradonts*) See Fig. 1.

labial tooth formula. Reference to a method for indicating the number of rows of *labial teeth* in larval amphibians. (Syn. *dental formula, lip teeth formula, tooth row formula*)

labile. *adj.* Plastic or readily modified.

lability. *n.* Instability with respect to a physical or chemical stress.

labiomental. *n.* See *sublabial.*

labio-ocular. *n.* See *oculolabial.*

labium. (*pl.* **-ia**) *n.* Generally, a lip or liplike structure. Sometimes used in reference to the *oral disk* surrounding the mouth of tadpoles in herpetological literature. See Fig. 1.

labyrinth. *n.* The membranous structure of the inner ear.

Labyrinthodontia. *n.* A subclass of extinct amphibians, including those that gave rise to reptiles. The term was originally applied to a diverse group of Paleozoic tetrapods, but is no longer employed.

labyrinthodont. *n.* A member of the *Labyrinthodontia.*

labyrinthodont teeth. Teeth with complex infolding of the enamel, a diagnostic character of labyrinthodonts.

Lace Monitor. Vernacular name for the Australian varanid lizard species *Varanus varius.*

laceration. *n.* A cut or tear, usually in skin.

Lacertidae. *n.* A clade (family) of lizards that includes the well-known Wall Lizards (genus *Lacerta*) of Europe. Outside Europe, lacertids have diversified into many habitats and occur in Eurasia, Africa, and islands of the Sunda Shelf. Some 25 genera and 250 species are recognized among three subfamilies: *Gallotinae, Lacertinae,* and *Eremiainae.*

lacertid. *n.* Collective name for any species of lizard belonging to the family Lacertidae.

Lacertiformes. *n.* A clade of lizards that includes *Gymnophthalmidae, Teiidae,* and *Lacertidae.*

Lacertilia. *n.* A clade (suborder) of squamate reptiles, the lizards, containing about 4300 species. (Syn. *Sauria*)

Lacertinae. *n.* A subfamily of *Lacertidae.*

Lacertoidea. *n.* A sister clade to *Scincoidea* within the *Scincomorpha.*

lacrimal. *adj.* Of or pertaining to tears, tear glands, or ducts. *n.* 1. A dermal bone of the skull of amphibians and reptiles, quite common in primitive forms but often lost in extant species. It is situated near the lacrimal gland at the anterior margin of the orbit. 2. An enlarged scale bordering the anterior of the orbit in rattlesnakes and separating the preoculars from postoculars.

lacrimal gland. A gland connected with each eye and secreting fluid ("tears") that helps to moisten the cornea. In comparison with *Harderian glands*, lacrimal glands are usually less well developed, or absent, in amphibians and reptiles. Cf. *Harderian gland.* See also *salt glands,* Fig. 20.

lacrymal. *n.* 1. Alternate spelling of *lacrimal.* 2. An alternate name for the prefrontal bone of Rattlesnakes according to L.M. Klauber (*Occ. Papers San Diego Soc. Nat. Hist.* 5:1–61, 1939).

lachrymal pit. Older reference to the facial *pit organ* of pitvipers.

lactacidemia. *n.* An excess of *lactic acid* in the blood.

lactacid oxygen deficit. Excess consumption of oxygen above resting levels, utilized to dissipate lactic acid that accumulates during anaerobic activity. Cf. *alactacid oxygen deficit.*

lactate, lactic acid. *n.* A product of the incomplete oxidation of carbohydrate via the pathway of glycolysis in the absence of oxygen.

lactate dehydrogenase. (LDH) An important enzyme of glycolysis that catalyzes the conversion of pyruvic acid to *lactic acid.* It is important in adaptations to anaerobic conditions and has been the object of many evolutionary studies. This enzyme also has been isolated as a component of some elapid snake venoms.

lacuna. (*pl.* –ae) *n.* An interior opening or space, commonly used in reference to the minute internal cavities found within bone.

lacunal. *n.* Any one of relatively large scales forming the inner border of the pit in crotaline snakes. First used by C.H. Lowe & K.S. Norris (*Trans. San Diego Soc. Nat. Hist.* 12:47–64, 1954). L.M. Klauber (*Rattlesnakes*, 1956) suggests that if lacunals lie entirely exteriorly, they should be considered *foveals* rather than lacunals.

lacunar tissue. A cell type seen during the shedding cycle in the epidermis of lepidosaurian reptiles, present as enlarged cells with misshapen nuclei that appear to lie in a vacuole and associated with immigrant eosinophils prior to shedding. The condition arises in cells between α-tissue above and the *clear layer* below (innermost cells of an *epidermal generation*).

lacunolabial. *n.* A large scale in certain pitvipers, formed by a union of the prelacunal and second or third supralabial scales.

lacustrine. *adj.* Reference to, or living in, lakes or ponds.

laevogyrinid. *adj.* Reference to condition in many anuran tadpoles having a single spiracle on the left side of the body. Cf. *amphigyrinid, mediogyrinid, paragyrinid.*

lagena. *n.* An outgrowth of the ventral wall of the *sacculus* of the inner ear in amphibians, part of the membranous labyrinth having hair cells sensitive to sound. This structure evolves to form the *cochlear duct* in reptiles and other tetrapods.

lagg. *n.* A moat-like area of shallow water, surrounding a peat mat and often dominated by sedges.

lag growth phase. A period of little or no growth of a population, preceding the *exponential* or *logarithmic growth phase*. Cf. *logarithmic growth phase.*

lagoon. *n.* The body of water located in the center of an atoll, or which separates a barrier reef from land.

LAGs. See *lines of arrested growth.*

lake. *n.* A relatively large inland body of water, fresh or saline, having negligible current and a narrow peripheral beach.

Lalagobatrachia. *n.* A sister taxon with Leiopelmatidae and containing all other living frogs. A newly designated monophyletic group containing *Xenoanura* and *Sokolanura* (D.R. Frost et al., *Bull. Amer. Mus. Nat. Hist.* 297:1–370, 2006).

Laliostominae. *n.* A subfamily of anurans belonging to *Mantellidae.*

lambda link. A dichotomously branching link in a food web that involves one predator feeding on two prey species. Cf. *gamma link, iota link.*

lambert. *n.* A unit measure of surface luminance equal to 1 lumen/cm^2.

lamella. (*pl.* –ae) *n.* 1. Literally, a thin sheet, leaf, or plate. 2. Usage in herpetology is largely in reference to transverse plates or scales on the ventral surfaces of the digits in lizards. These are often soft and consist of numerous "bristles" or setae, each with a terminal spatula, hook, or knoblike structure. See *subdigital lamellae.* (Syn. *scansor*) See Fig. 32. 3. Also used in reference to thin sheets or plates of tissue, as found in *lamellar bone.*

lamellar bodies. Discreet lipid-enriched secretory organelles present in amniote epidermis generally (but questionably in amphibians) where they contribute to the transepidermal water barrier. In mammals and lepidosaurian reptiles, the lipid contents are exocytosed into the extracellular spaces where they become visible as distinct lamellae that fill the extracellular regions in the stratum corneum. In most lepidosaurs lamellar bodies are limited to the *mesos* layer of stratum corneum, but they are also present in α-*cells* of *Sphenodon* and certain snakes. In avian integument, the lipid contents of epidermis are derived from comparatively large lamellar bodies (called *multigranular bodies*) 3–5 times larger than those of mammals, in addition to large lipid droplets that resemble sebum or oil. (Syn. *Odland bodies, keratinosomes, cementosomes, membrane-coating granules, lamellar granules, multigranular bodies, mesos granules*)

lamellar bone. Bone structure represented by a lamellar arrangement of the calcified matrix, which results from variable orientation of fibrils in successive lamellae. Lamellar bone typically forms during periods or at ages characterized by relatively slow skeletal growth.

lamellar granules. See *lamellar bodies.*

lamina. *n.* Literally a thin, flat plate or layer. Used in herpetology in reference to *subdigital lamellae*, epidermal *plates* of turtles, and the upper part of the neural arch above the interzygapophysial ridges. All of these usages are currently not in favor.

lamina externa. A small rectangular cartilage lying against the lateral surface of the nasal sac in the chondrocranium of larval amphibians. (Syn. *ethmo-palatine, pars plana, planum terminale, prepalatine cartilage*)

lamina propria. The connective tissue layer of mucous membrane.

lamina terminalis. Little used synonym for *basement membrane.*

laminal spur. A sharp projection from the margin of several plastral plates (largely *humerals*) in juvenile Gopher Tortoises (*Gopherus*).

laminar. *adj.* Comprised of layers or laminae.

laminar bone. Bone with cortices arranged into one or more irregularly stratified zones. Layering of the cortex is a direct result of extensive rebuilding and remodeling that accompany skeletal growth.

laminar flow. A pattern of flow wherein fluid movement is streamlined and can be modeled as layers that move in parallel or slide past each other in parallel. Cf. *turbulent flow*.

laminar nucleus. The central, juvenile area of the lamina from which growth extends outward or peripherally in many turtles. See *areola*.

laminate. *adj.* Consisting of a series of thin, parallel, plate-like structures.

L-amino oxidase. An important digestive enzyme of some snake venoms that catalyzes the breakdown of many substances, deamination and oxidation of amino acids, and may serve to activate other venom components.

lampbrush chromosome. A chromosome characteristic of the primary oocytes of vertebrates, named for their fuzzy appearance when viewed at low magnification beneath a microscope. The largest known chromosomes are the lampbrush chromosomes of salamanders.

Lamprophiinae. *n.* A clade (subfamily) recognized in some classifications to include a large number of African and Madagascan colubrid snake genera and about 220 species known as *House Snakes*.

LAN. See *local area network*.

Lanceheads. *n.* Collective vernacular name for about 30 species of New World terrestrial pitvipers belonging to the genus *Bothrops*. See *Common Lancehead, Fer-de-Lance, Terciopelo, Barba Amarilla*.

lanceolate. *adj.* Resembling the sharp head of a lance or spear in shape.

land-bridge. *n.* 1. A more or less continuous connection between adjacent land masses, forming a potential route for dispersal and migration. 2. A young island (< 10,000 years old) whose origin was the result of a rise in sea level, erosion, or a marine inundation.

landform. *n.* A characteristically shaped feature of Earth's surface, e.g. a hill or valley.

Langley. (ly) *n.* A derived *cgs* unit of solar radiation equal to 1 cal/(cm^2 • min), or 697.8 W/(m^2 • s).

La Niña. This term refers to the extensive cooling of the central and eastern equatorial Pacific Ocean, which results in cooler than normal ocean temperatures across the central and eastern tropical Pacific Ocean, increased convection or cloudiness over tropical land masses, stronger than normal (easterly) trade winds across the Pacific Ocean. La Niña translates from Spanish and means "The Little Girl." (Syn. sometimes called *El Viejo*, or Old Man, *anti-El Niño*, or simply "*a cold event or episode*") Cf. *El Niño*.

Lanthanotidae. *n.* A varanoid clade (family) comprised of a single species of lizard found in Borneo and sometimes placed within Varanidae as a monotypic subfamily.

Laplace's Law. The principle that transmural pressure in a thin-walled tube is proportional to the wall tension divided by the inner radius of the tube.

Lapparentophis defrennei. The oldest undisputed snake, which lived between early and late Cretaceous in Algeria (about 100 million years ago). Other fossil snakes with well-developed hind limbs include *Pachyrachis problematicus* and *Haasiophis terrasanctus*, and their relationships to living snake species remains controversial.

lappet, lapette. *n.* A free, overhanging or overlapping flap or fold. Used in herptetological literature largely in reference to various skin folds in amphibians, flap-like structures on the tongue in some Geckos, and overlapping cartilage or bone.

lapse rate. The rate of decline in temperature of an air mass with increasing altitude without gain or loss of heat from an external source. The average global lapse rate is 0.64 °C/100 m. (Syn. *adiabatic lapse rate*)

large intestine. The posterior segment of the intestine, usually a comparatively straight tube of large diameter passing to the cloaca. *Villi* are generally lacking on its *mucosa*.

larva. (*pl.* -ae) *n.* An immature but free-living stage that is different in form from the adult and, unlike the embryo, is able to procure its own nourishment. See also *tadpole*.

larval stomach. Alternate term for *manicotto*, according to R.M. Savage (*Copeia* 1955:120–131, 1955). See *manicotto*.

larval transport. Reference to parental transport, usually on the back of a parent, that relocates exotrophic larvae from the site of oviposition of terrestrial eggs to an aquatic site.

laryngeal ventricle. A resonating chamber present as a depression in the wall of the *larynx*, posterior to the glottis, and used during vocal activity in certain anurans.

larynx. *n.* A complex of cartilaginous elements, fibers, and muscles at the pharyngeal opening of the trachea that functions to protect the opening and, in some species, permits vocalization.

laser thermometer. See *infrared thermometer*.

latent heat of vaporization. The amount of energy required to change a liquid to a gas of the same temperature.

latent period. 1. In physiology, the interval between activation of an action potential in a muscle fiber and the initiation of muscular contraction. 2. More generally, *reaction time*.

lateral. *adj., n.* 1. Pertaining to a side or position away from the median plane or midline of the body. The opposite of medial. 2. Describes a category of eye position in tadpoles in which the eyes protrude further laterally than the surrounding body surface and thus are included in a dorsal silhouette. 3. *n.* Any of the scales on the sides of a lizard other than *dorsals* or *ventrals*. 4. See *costal,* Fig. 4.

lateral dermal fold. See *costal fold.*

lateral field. The area between the paravertebral and lateral body stripes in some lizards.

lateral fold. A term used variously in reference to *costal fold* of salamanders, *longitudinal fold* of certain lizards (e.g., Anguidae), and *dorsolateral fold* of anurans.

lateral inhibition. In sensory processing, reciprocal suppression of excitation by neighboring neurons in a sensory network, the function of which enhances contrast at boundaries and increases the dynamic range of a stimulus. Neurons excited by a stimulus inhibit their neighboring neurons.

lateralis organ. Synonym for *neuromast* (R.W. Murray, *J. Exp. Biol.* 33:798–805, 1956).

lateral lamina. Synonym for *costal lamina* in turtles.

lateral line organ. Used by many authors in reference to *neuromast.*

lateral line system. A collection of epidermal sense organs distributed over the head and along the body of aquatic amphibians (and fish). Lateral line organs (*neuromasts*) function as mechanoreceptors or electroreceptors (*ampullary organs*) in some larval caecilians and aquatic salamanders. The lateral line system of amphibians differs from those of fishes by not having "pores" or channels. Thus, terminologies applied to amphibians should not include the term "pore." (Syn. *acoustico-lateralis system*)

laterally compressed. Flattened from side to side and thus deeper than broad.

lateral marginal. Equivalent to *marginal* lamina of turtles. See Fig. 4.

lateral nuchal fold or **pocket.** See *nuchal pocket.*

lateral occipital. *n.* See *exoccipital.*

lateral profile. A side view of a body or structure.

lateral process. 1. A prominent projection arising from the *hyoid plate* of the anuran *hyoid apparatus,* lateral to the *thyrohyal* and posterior to the *alary process.* 2. A poorly defined lateral portion of the upper jaw sheath of a tadpole, beyond which serrations are small to absent. See R. Altig, *Herpetologica* 26:180–207, 1970.

lateral sequence gait. Locomotion characterized by having only one foot off the ground at any one time, and the forefoot strikes the ground after the hind foot on the same side, followed by the hind foot on the opposite side.

lateral thyroid. See *ultimobranchial body.*

lateral undulation. The most commonly used mode of limbless locomotion in snakes. Horizontal waves travel down alternate sides of the body axis and generate a reaction force at fixed points in the animal's physical environment, usually surface irregularities in the substrate. There are no static points of contact with the substrate, and the body moves continuously against multiple (three or more) fixed points of force application. (Syn. *serpentine locomotion, sinusoidal locomotion, undulatory movement*)

laterosphenoid. *n.* A bone in the skull of reptiles, usually paired and located on the midline above the palate, at or behind the orbit. This element is derived from the chondrocranium and is present in crocodilians and possibly some lizards. (Syn. *orbitosphenoid, alisphenoid, otosphenoid*)

Laticaudidae. *n.* A former family-level taxon that included *Sea Kraits* as a separate lineage of marine (amphibious) snakes, distinct from the true Sea Snakes or *Hydrophiidae.* See *Sea Snakes, Elapidae,* and *Laticaudinae.*

Laticaudinae. *n.* A subfamily name used by some to designate a separate lineage of Sea Snakes, commonly known as *Sea Kraits.* Others regard this separation of taxa to be misleading (see *Elapidae*).

latichoanal fissure. A cleft extending from the lateral margin of the internal naris, which follows the deep cleft in the *prevomer* to the lateral extremity of the prevomerine tooth row. Described in *Pseudoeurycea* by I.L. Baird (*Univ. Kans. Sci. Bull.* 34:221–266, 1951).

Latin square. A set of symbols arranged in a checkerboard such that no symbol appears twice in any row or column. The arrangement is used to design the layout of replicate plots in field experiments.

latitude. *n.* In geography, lines of latitude measure north-south position between the poles. One degree of latitude = 60 nautical miles or 111 km. One minute of latitude = 1 nautical mile or 1.85 km. Cf. *longitude.*

latolent. *adj.* A type of movement observed in tortoises in association with olfactory sensing of the environment. The head is moved from side to side in a swinging motion, as a pendulum. Term coined by A. Eglis (*Herpetologica,* 18:1–8, 1962). Cf. *dirolent.*

Laurasia. *n.* A northern supercontinent or land mass that fragmented about 150 millions years BP during the Mesozoic to form the present continents of the Northern Hemisphere, including North America, Greenland, and Eurasia, excluding India. Cf. *Gondwana.*

Laurentobatrachia. *n.* A new monophyletic taxon of anurans, sister to the new taxon *Xenosyneunitanura* within the new taxon *Afrobatrachia,* and containing the families *Hyperoliidae* and *Arthroleptidae.* See D.R. Frost et al., *Bull. Amer. Mus. Nat. Hist.* 297:1–370, 2006.

lava. *n.* Magma on Earth's surface.

lavage. *n.* The irrigation or washing out of an organ.

Lava Lizards. Vernacular name for about seven species of tropidurine iguanid lizards belonging to the iguanid genus *Tropidurus* and *Microlophus* (formerly *Tropidurus,* endemic to the Galapagos Islands). Also called *Pacific Iguanas.*

lavender cell. See *violet cell*.

law. *n.* An empirical generalization of a biological principle that seems to be without exception and is confirmed by repeated successful testing.

Law of Laplace. See *Laplace's Law*.

l.c. See *loc. cit*.

LC_{50}. *n.* See *lethal concentration*.

L-C Treatment. See *ligature-cryotherapy treatment*.

LD, LD_{50}. *n.* See *lethal dose*.

LDH. See *lactate dehydrogenase*.

Leaf Chameleons. Vernacular name for Madagascan species of Chameleons belonging to the genus *Brookesia*. Also called *Dwarf Chameleons*.

Leaf-folding Frogs. Vernacular name for African species of hyperoliid frogs belonging to the genus *Afrixalus*. Eggs of these species are deposited on a leaf, which is then folded, and its opposite faces are glued together by a very adhesive oviductal secretion.

Leaf Frogs. Vernacular name for species of Middle and South American frogs belonging to the hylid genera *Agalychnis*, *Hylomantis*, *Pachymedusa*, *Phrynomedusa*, and especially *Phyllomedusa*.

Leafnose Snakes. Vernacular name for two species of Madagascan colubrid snakes in the genus *Langaha*, named for elaborate scaled appendages on the snout, which is leaf-like in females. The name also applies to snakes in the colubrid genus *Lytorhynchus*, distributed in North Africa and the Middle East, and to American colubrids in the genus *Phyllorhynchus*.

Leaftail Gecko, Leaf-tailed Gecko. Vernacular name for geckos belonging to the genus *Phyllurus* and also the genus *Uroplatus* from Madagascar.

Leaf-toed Geckos. Collective vernacular name for numerous and largely Asian and South American species of Geckos belonging to the genera *Hemidactylus* and *Phyllodactylus*.

learning. *n.* The accumulation of adaptive changes in behavior of individuals as a result of experience.

Least Geckos. Collective vernacular name for species of Middle American Geckos belonging to the genus *Sphaerodactylus*.

least squares, method of. A statistical method of estimating parameters of curve fitting based on minimizing sums of squared differences between the observed values and their respective expected values. Estimates obtained by this method are used in regression analysis and analysis of variance. See *line of best fit*.

Leatherback Turtle, Leatherback Seaturtle. An ancient species of Sea Turtle, *Dermochelys coriacea* (Dermochelyidae), presently under threat of extinction.

lebetase. *n.* A novel anticoagulant protease isolated from venom of the Levantine Viper (*Vipera lebetina*) by J. Sligur & E.P. Sligur (*J. Toxicol.* 11:91–113, 1992).

Lebetine Viper. See *Levantine Viper*.

lecithinase. *n.* A phospholipase enzyme that catalyzes the hydrolysis of phospholipids.

lecithotrophy. *n.* (*adj.* –**ic**) A fetal nutritional pattern in which nutrients are supplied to the embryo primarily by yolk supplied to the egg prior to ovulation. Terminology is from J.P. Wourms (*Amer. Zool.* 21:473–515, 1981). Cf. *matrotrophy*. See also *post-paritive lecithotrophy*.

lectotype. *n.* One of a series of syntypes designated by an author to serve as the type specimen subsequent to the publication of the original description.

leeward. *adj.* Reference to the side of a mountain, dune, or ridge that faces away from a wind or water current. Cf. *windward*.

left aorta. See *aorta* and Figs. 36, 37.

left-right (L-R) shunt. See *intracardiac shunts*.

leg index. A measure of the length of a frog's leg relative to the body, expressed as a percentage.

legitimate name. *Nomen legitimatum*. In taxonomy, a legitimate name published according to relevant rules of the Code.

Legless Lizards. Vernacular name for species of lizards belonging to the North American anguid genus *Anniella*, and the Australian gekkonid genus *Aprasia*.

Leguaan. *n.* A common name for Monitor Lizards (*Varanidae*) in South Africa.

Leiopelmatanura. *n.* A clade of anurans that includes all extant forms except Ascaphidae.

Leiopelmatidae. *n.* A family including *Ascaphus* (tailed frog) and *Leiopelma*—a single genus and four species of frogs representing the only amphibians native to New Zealand. They are unique among frogs in having inscriptional ribs embedded in musculature of the ventral body wall. A sister taxon to the newly designated *Lalagobatrachia*. See D.R. Frost et al., *Bull. Amer. Mus. Nat. Hist.* 297:1–370, 2006. See also *Ascaphidae*.

lek. *n.* A cluster of male territories, together with their occupants, visited by females for courtship and mating. The term is derived from a Scandinavian word meaning "play" and refers to both the males and the site or arena where they display. The Galapagos marine iguana is an example of a "lekking" species of reptile.

lemmanura. *n.* A synonym for Orton's Type 3 tadpole (G.L. Orton, *Syst. Zool.* 2:63–75, 1953) or discoglossid of O.M. Sokol (*Copeia* 1975:1–23, 1975). See P.H. Starrett, pp. 251–271 in J.L. Vial (ed.) *Evolutionary Biology of the Anurans. Contemporary Research on Major Problems*, 1973.

Lemuria. *n.* A continental land bridge throught to have connected South Africa, Madagascar, and India up to the end of the Cretaceous.

lens. *n.* The principal light-focusing structure in the eyes of vertebrates.

lens cell. A long cylindrical cell type that makes up the lens of the *parietal eye* in lizards and in *Sphenodon*.

lens size. A measure of the greatest diameter of the lens extracted from the right eye, used as a taxonomic character in salamanders.

lentic. *adj.* Pertaining to slow-moving or standing aquatic habitats, such as a pond. Cf. *lotic.*

lenticle. *n.* A type of *scale organ* that bulges out of the depression in which it resides.

lenticular. *adj.* Shaped like a lense. Biconvex.

lenticulare. *n.* A small bone in the carpus of crocodilians, lying between the cubitus and metacarpals at the level of the third and fourth digits (described by G. Cüvier, 1808).

Leopard Frog. Vernacular name for ranid frogs belonging to the North American species complex that includes *Lithobates pipiens, L. berlandieri, L. chiricahuensis, L. sphenocephalus,* and *L. yavapaiensis* (all formerly *Rana*). For recent taxonomic revision see D.R. Frost et al., *Bull. Amer. Mus. Nat. Hist.* 297:1–370, 2006.

Leopard Gecko. Vernacular name for the south Asian gekkonid species *Eublepharis macularius*, popular in herpetoculture and the pet trade.

Leopard Lizard. Vernacular name for species of iguanid lizards belonging to the American genus *Gambelia*.

lepidic. *adj.* Pertaining to scales.

lepidosaur. *n.* Any reptile inclusive of snakes, lizards, amphisbaenians, and Tuatara.

Lepidosauria. *n.* A clade (subclass) of reptiles that includes *Tuatara* and squamates. This group is a sister taxon to *Archosauria* (birds + crocodilians). See also *Eosuchia.*

Lepidosauromorpha. *n.* One of two primary divisions of *Diapsida* that includes extant squamates, Tuatara, and several fossil groups.

lepidosis. *n.* The arrangement and pattern of scalation.

lepospondylous. *adj.* 1. Reference to amphibian vertebrae which have centra formed by deposition of bone around the embryonic notochord. Present in salamanders and caecilians. 2. Used also in reference to vertebrae with a single spool-like centrum, derived from a *pleurocentrum* rather than a *hypocentrum.*

leptin. *n.* A hormone that regulates energy expenditure and body mass in mammals, also suggested to be expressed in lizards (P.H. Niewiarowski, M.L. Balk & R.L. Londraville, *J. Exp. Biol.* 203:295–300, 2000). Recent work suggests reptilian leptins have physiological functions similar to those in endotherms, thus aiding in the regulation of metabolic rate, fat stores, appetite, and reproductive cycles (S. Spanovich et al., *Comp. Biochem. Physiol.* B 143:507–513, 2006).

lepto-. Word element meaning "slender" or "delicate."

Leptodactylidae. *n.* A clade (family) of anurans with considerable variation in morphology, habits, and life histories. Formerly with 50 genera and approximately 1100 species distributed in southern North America, South America, and the West Indies. For taxonomic revisions see D.R. Frost et al., *Bull. Amer. Mus. Nat. Hist.* 297:1–370, 2006.

leptodactylid. *n.* A member of, or pertaining to, *Leptodactylidae.*

Leptodactyliformes. *n.* A new monophyletic taxon, sister to *Hylidae* within the new taxon *Athesphatanura*, and containing the new taxa *Diphyabatrachia* and *Chthonobatrachia* (D.R. Frost et al., *Bull. Amer. Mus. Nat. Hist.* 297:1–370, 2006).

leptodactylin. *n.* A toxic compound extracted from leptodactylid frogs.

Leptopelinae. *n.* A former subfamily of *Hyperoliidae*, placed recently within *Arthroleptidae* (D.R. Frost et al., *Bull. Amer. Mus. Nat. Hist.* 297:1–370, 2006).

leptopheny. *n.* (*adj.* –**ic**) *adj.* Reference to a population that exhibits a narrow range of continuous phenotypic variation. Cf. *monopheny.*

Leptotyphlopidae. *n.* A family of *Blind Snakes*, generally with burrowing habits and distributed from Africa, the Middle East, southwestern Asia, northern South America, Central America, and southern United States. About 105 species in this family are often referred to as *Worm Snakes* or *Thread Snakes*.

leptotyphlopid. *n.* A member of, or pertaining to, *Leptotyphlopidae.*

Leptotyphlopoidea. *n.* A superfamily containing the family Leptotyphlopidae, *Worm Snakes* or *Thread Snakes*.

leristas. *n.* Vernacular name for species of scincid lizards belonging to the Australian genus *Lerista*.

lesion. *n.* Localized tissue that has been damaged or altered by a pathological process.

Leslie matrix. A matrix representation of the birth and death parameters in an age-structured population growth model, developed by the population ecologist P.H. Leslie (*Biometrika* 35:183–212, 1945).

lethal. *adj.* Causing or pertaining to death by direct action.

lethal concentration. (LC_{50}) The concentration of a toxic substance that kills half of the organisms in an exposed test group. Cf. *lethal dose, LD_{50}* and *lethal time, LT_{50}.*

lethal dose. (**LD, LD_{50}**) A measure of the quantity of venom required to kill an animal. The maximum amount of venom that will permit survival of all animals in a test group is the LD_0. The amount of venom that will kill 50% of the individuals in a test group is the LD_{50}. And the minimum amount of venom that will kill all the animals in a series is the LD_{100}. The *minimum lethal dose* (MLD) is the minimum amount of venom required to kill an individual (just in excess of the LD_0). The *mean lethal dose* is determined by integrating the curve that correlates dosage with percentage mortality and is usually very close, or equal, to the LD_{50}. Cf. LC_{50} and LT_{50}.

lethal factor 1. A toxin isolated from *Heloderma* venom that causes hemorrhaging.

lethality. *n.* The frequency of deaths caused by a given condition. In herpetology, the percentage of deaths due to envenomation by snake bite out of the total number of snake bites.

lethal maximum and **minimum.** See *lethal temperature*.

lethal mutation. A genetic mutation that causes death.

lethal temperature. Either a measure of the high temperature that produces irreversible damage and death in an individual upon exposure (*lethal maximum*), or the low temperature that causes irreversible damage and death (*lethal minimum*). Cf. *critical thermal minimum* and *maximum*.

lethal time. (LT_{50}) The time required to kill 50% of test organisms on exposure to a given test concentration of a substance. Cf. *lethal concentration, LC_{50}* and *lethal dose, LD_{50}*.

lethargy. *n.* (*adj.* –**ic**) A state of drowsiness or suppressed activity that is pathologic.

leucistic. *adj.* White or colorless; lacking pigment.

leucophore. *n.* A type of *guanophore* (or synonym) characterized by presence of guanine in the form of fine granules.

leucolysin. *n.* A *hemotoxin* that degrades white blood cells.

leukocyte. *n.* A *white blood cell*, usually having amoeboid movement and functioning to protect the body from microorganisms that cause disease.

Levantine Viper. Vernacular name for the Asian viperine species *Macrovipera lebetina*. (Syn. *Kufi, Lebetine Viper, Mountain Adder*)

levator. *n.* A muscle that lifts or elevates a part.

level of significance. See *significance*.

lever. *n.* A rigid structure that transmits forces by turning, or tending to turn, at a pivot. See Fig. 23.

lever arm. The perpendicular distance between the line of action of an applied force (or component of such a force) and the associated pivot of a lever. Cf. *moment, moment arm*. See Fig. 23.

Leydig cells. 1. Interstitial cells that lie outside the seminiferous tubules of the testis of all vertebrates and synthesize androgenic steroid hormones. Leydig cells are the major source of testosterone in the circulation. (Syn. *interstitial cells*) 2. A large, gland-like cell in the epidermis of adult urodeles and apodans, probably derived from a unicellular gland and specialized to secrete mucus. They may also be present in tadpoles, but are more abundant in salamander larvae. They disappear during metamorphosis and are absent in adults. (Syn. *club cells, Leydig's gland*)

Leydig's duct. Equivalent to *ductus deferens* according to H. Gadow (*Amphibia and Reptiles*, p. 48, 1901).

Leydig's gland. Synonym for *Leydig cell* of the skin.

LH. See *luteinizing hormone*.

library. *n.* See *genomic library*.

lichenoid, lichenous. *adj.* Rounded, with irregular protrusions and peripheral branchings, as a lichen. This descriptor has been used in reference to blotching patterns of salamanders.

liebesspiel. *n.* Specifically "love play" (German), used in reference to the courtship dance of various salamander species. (Syn. *nuptial dance*)

Liebig's law. See *minimum, law of*.

life cycle. Reference to the series of developmental stages characteristic of organisms from fertilization through reproduction and death.

life history. The history of an individual organism, from the fertilization of the egg to its death. A large body of literature relates to significant features of the life cycle, emphasizing strategies that influence survival and reproduction. See *life history strategies, life history theory*.

life history strategies. Reference to a selected set of adaptations related to local environments and an organism's life cycle, such as fecundity, the timing of reproduction, longevity, etc. See *r-k selection*. See E.R. Pianka, *Amer. Nat.* 104:592–597, 1970.

life history theory. Theory that analyzes or shows how natural selection acts on life history traits including how long an animal lives, at what age it matures, and how many offspring it has. The theory is essentially mathematical in character and is used to explain why particular life history strategies are adaptive in particular environments. See E.R. Pianka, *Amer. Nat.* 104:592–597, 1970; S.C. Stearns, *The Evolution of Life Histories*, 1992.

life span. The maximum or mean duration of life (longevity) of an individual or group.

life table. A complete tabulation of mortality data for a population or cohort, organized with respect to age. See *horizontal life table*.

life zone. A biogeographical region having a characteristic biota.

lift. *n.* Descriptive of a force that is at right angles to oncoming fluid, air or water.

ligament. *n.* Tough connective tissue that forms attachment between two or more skeletal elements (bone or cartilage).

Ligament of Botallus. A connective tissue vestige of the *ductus Botalli* that remains in the adult after the embryonic pulmonary shunt has atrophied.

ligand. *n.* Any molecule that will bind to a complementary site such as a receptor, membrane, etc.

ligand-gated ion channel. A channel that opens when a specific molecule binds to the extracellular domain of the channel protein.

ligature-cryotherapy treatment. (L-C treatment) A technique for treatment of venomous snake bite developed by Herbert Stahnke. The method involves immediate ligature between the site of the bite and the body, followed by immersion of the envenomated part in ice water. The method is not widely accepted as an appropriate treatment.

light. *n.* Electromagnetic radiation with wavelengths between those of X-rays and those of heat (*infrared radiation*).

light adapted. Term used to describe an animal that has been maintained under fairly strong illumination for a period of time, with usual reference to effects on physiological color change.

lightness. *n.* With respect to color, lightness is the perceived level of emitted light relative to light from a region that appears white. The Munsell scale for lightness is called *value*.

likelihood. *n.* In statistics, an estimate of the relative support for each alternative hypothesis. The chance that a given event will occur.

limb bud. Developmental condition of a limb in which it appears as a bud-like outgrowth without any differentiated or recognizable external features. See Fig. 1.

limb bud stage. The developmental stage in the life of anuran larvae when the limbs appear as limb buds.

limbic lip. A lateral extension of the upper part of the *limbus* in certain lizards. This structure bends to form a large lip invaginated into the cochlear duct and encloses a relatively large volume of endolymphatic fluid. First described by C.C.D. Shute & A.d'A Bellairs (*Proc. Zool. Soc. Lond.* 123:695–709, 1953) and rejected as legitimate terminology by some.

limbic system. A portion of the forebrain, including the hippocampus and preoptic areas, and forming a higher correlative center in the brain.

limbus. *n.* 1. Literally an edge, fringe, or border. 2. Used in herpetology as synonymous with *cochlear cartilage*, a ring-like structure of modified connective tissue that supports the basilar membrane of the ear in lizards. Introduced by C.C.D. Shute & A.d'A Bellairs (*Proc. Zool. Soc. Lond.* 123:695–709, 1953). Such usage has been judged to be confusing and not acceptable.

limestone. *n.* A sedimentary rock composed mostly of calcite.

limited-area searching. Reference to behavior of male anurans that remain near one site during the breeding period and attempt to clasp only those individuals moving into the immediate area.

limiting basal layer or **zone.** See *basement membrane*.

limiting factor. An environmental factor that limits a population's abundance or distribution.

limiting membrane. See *basement membrane*.

limiting resource. A resource in critically short supply such that increases in the resource increase the size of a population that utilizes the resource.

limn-. Prefix meaning "lake." Cf. *limno-*.

limnetic. *adj.* Pertaining to open water of a lake or other bodies of fresh water, away from the bottom.

limnicolous. *adj.* Living in lakes.

limno-. Prefix meaning "freshwater" or "marshy." Cf. *limn-*.

Limnodynastidae. *n.* A clade (family) of anurans, sister with *Myobatrachidae* within the *Myobatrachoidea* and containing 8 or 9 genera distributed in Australia and New Guinea. See D.R. Frost et al., *Bull. Amer. Mus. Nat. Hist.* 297:1–370, 2006.

Limnodynastinae. *n.* A subfamily of *Myobatrachidae* in some classifications.

limnology. *n.* The study of freshwater systems, especially lakes, ponds, and other standing waters, and their associated biota.

Lincoln index. A method for estimating the size of a population based on sampling of animals among which some proportion have been previously marked and are recaptured, according to the formula

$$N = N_m \cdot N_s / N_{ms}$$

where N_m = the number of marked animals released, N_s = the number captured in a sample, and N_{ms} = the number of marked individuals in the sample. (Syn. *catch-mark-recatch method, mark-recapture method*)

linea alba. Literally a white line. Used in reference to the median ventral line of connective tissue extending the length of the abdomen between the rectus abdominis muscles.

lineage. *n.* A linear evolutionary sequence of descent from an ancestral species to a descendant species.

linea masculina. (*pl.* –ae) A band of fibrous connective tissue extending the entire length of each layer of the obliquus muscles at both dorsal and ventral borders and constituting a secondary sexual character in various diverse species of male frogs. See C.C. Liu, *J. Morphol.* 57:131–145, 1935; D.D. Davis and C.R. Law, *Science* 81:562–564, 1935. *Linea masculinae* have been shown to play a role in exhalation and sound production in the Tungara frog, *Physalaemus pustulosus* (C. Jaramillo et al., *J. Morphol.* 233:287–295, 1997).

linear distance methods. In morphometrics, reference to metrics based on linear measurements of anatomical features.

linearity. *n.* The adherence of a response to the equation R = kS, where R = response, S = stimulus, and k = a constant.

linear regression. A technique for fitting a line that defines the expectation of how much one factor increases or decreases from a unit increase in a second factor. Regression analysis is applicable when the relationship between two variables is one of functional dependence of one on the other. That is, the magnitude of the dependent variable is assumed to be determined by (is a function of) the magnitude of the independent variable, whereas the reverse is not true. See *line of best fit, regression coefficient*.

Lined Skinks. Vernacular name for various species of Skinks belonging to the genera *Eumeces* and *Hemiergis*. See also *Five-lined Skink*.

Lined Snake. Vernacular name for the American colubrid species *Tropidoclonion lineatum*.

line of best fit. A straight line that represents the best moving average for a linear group of observed points associated with bivariate axes. The fit requires that the sum of squares of the deviations of the observed points from the moving average be minimum.

lines of arrested growth. (LAGs) Reference to developmental patterns in which there are pauses in bone deposition resulting from periods of slower or arrested rates of growth. These commonly appear as a thin line that appears darker than the surrounding tissue. A less common pattern is a broader and less distinct line that also stains darker, referred to as an *annulus*. Alternating with LAGs or annuli are broad zones that stain homogeneously light and represent areas of active bone formation. Together, a broad zone followed by either a LAG or annulus represents a *skeletal growth mark*, useful in studies of *skeletalchronology*.

line transect. *n.* A method used to sample the distribution of organisms in a given habitat, based on counts of species and individuals in association with a straight line through the habitat. Cf. *belt transect*.

lingual. *adj.* Pertaining to the tongue.

lingual cartilage. See *otoglossal*.

lingual feeding. Feeding by capturing prey with the tongue, as in many amphibians and lizards.

lingual fossa. The small, semicircular indentation in the ventral side of the *rostral* scale of snakes, through which the tongue protrudes from the closed mouth. (Syn. *rostral crease*)

lingual glands. See *oral glands, salt glands* and Fig 20.

lingual papilla. Reference to those buccal papillae that rest on the lingual *anlage* (R. Wassersug, *Misc. Publ. Mus. Nat. Hist. Univ. Kansas* 68:1–146, 1980).

lingual process. Synonym for *entoglossal*, as used by many authors.

lingual salt glands. See *salt glands*.

lingual shelf. A ledge or ridge on the inner (lingual) side of the lizard *dentary*.

linkage. *n.* See *genetic linkage*.

linkage disequilibrium. The nonrandom distribution of the alleles of genes at two or more loci, not necessarily on the same chromosome, into the gametes of a population. Linkage is said to generate a disequilibrium relative to a random distribution to gametes.

Linnaean classification. Reference to the traditional system of classification of organisms established by Carolus Linnaeus (Latinized name of Carl von Linné) and based on similarity of form and structure.

lipase. *n.* An enzyme that breaks down lipids to their constituent glycerol and fatty acids.

lipid. *n.* A general class of hydrophobic molecules that include fatty acids, neutral fats, waxes, steroids, and phosphatides.

lipid bilayer. The continuous double layer of lipid molecules that forms the basic structure of a biological membrane.

lipogenesis. *n.* The formation of fat from nonlipid sources.

lipophilic. *adj.* Having an affinity for lipids.

lipophore. *n.* A chromatophore characterized by the presence of lipid inclusions in which naturally occurring pigments are dissolved. These cells occur usually at the outer layers of the dermis and generally appear either yellow (*xanthophores*) or red (*erythrophores*).

lipoprotein. *n.* A lipid-protein complex, usually in reference to components of the plasma membrane.

lipovitellins. *n.* Egg yolk proteins derived from *vitellogenin*.

Lissamphibia. *n.* A clade that includes the common ancestor of frogs, salamanders, and caecilians, and its descendants. Some have proposed using the term Amphibia as the formal name for this clade and Temnospondyli for a larger clade that includes extant amphibians and all fossils more closely related to them than to amniotes. For discussion see M. Laurin (*Annales des Sciences Naturelles* 1998:1–42, 1998).

liter. (L) *n.* See Table 2.

lithic. *adj.* Pertaining to rock.

lithophagy. *n.* Eating of stones or rock; suspected in some crocodilians as a means of controlling buoyancy or aiding digestion.

litorin. *n.* See *bombesin*.

little horn. Reference to the paired structures on the anterior end of the copula of the hyoid in Salamandridae, which extend into and form a support for the tongue. (Syn. *anterior radial, hypohyal*)

litter. *n.* 1. Collective animals that are produced at a multiple birth. 2. Fallen plant material that is recent and only partially decomposed, forming a surface layer on some soils.

Littersnakes. *n.* Vernacular name for species of mostly Middle American colubrid snakes belonging to the genus *Rhadinea* (also known as *Brown Snakes*).

littoral, littoral zone. *adj., n.* Pertaining to the shore or intertidal; the shallow open-water region lakeward or seaward of the shoreline

live-bearing. Giving birth to young that have developed beyond the egg stage.

living fossil. Reference to an organism that is known from the fossil record and subsequently found to be extant.

living tags. Grafts of tissue that are transferred from one part of the body to another. This is an experimental technique of marking hatchling animals in a manner by which they might be recognized years later when they are juveniles or adults.

lizard. *n.* A squamate reptile other than snakes and amphisbaenians, including iguanian lizards—the most basal or "primitive" living squamates—gekkotan, scincomorph, and anguimorph lizards.

llanura. *n.* Spanish for an area having little relief; a plain.

load. *n.* A force applied to a solid object.

loam. *n.* A rich soil containing less clay than silt and sand.

lobe. *n.* 1. Any of the enlarged, outward curved parts (usually three) belonging to a single segment of the rattlesnake rattle. (Syn. *ring*) The lobes are separated by strong constrictions, and the anteriormost one is always larger than the next behind it. 2. The parts of the turtle plastron meeting at the bridge, one being called the "*anterior lobe*" and the other the "*posterior lobe.*" 3. The individual enlargements of a *multiple testis* in salamanders. 4. See *occipital lobe*.

lobule. *n.* 1. A small segment or lobe. 2. Term used by various authors in reference to the structural unit of the testis of urodeles, also called cyst, capsule, or tubule by various authors.

local area network. (LAN) A network of communication among computers that are linked within a relatively small area such as a building or operational division.

local extinction. The disappearance of a single population within a metapopulation. See *extirpation*.

local population. See *deme*.

loc. cit. (*l.c.*) Abbreviation of the Latin *loco citato*, meaning "at the place cited."

locomotion. n. (*adj.* **locomotory**) The ability, or act, of an organism to move from place to place.

locus. *n.* A specific site on a given chromosome; the physical region of the chromosome at which a gene resides.

loess. *n.* A fine-grained deposit of wind-blown silt and clay, characterized by a high porosity often near 60%. Such deposits may blanket hills and valleys downwind of a source of fine sediment, such as a desert or a region of glacial outwash.

logarithmic growth phase. A period in which maximum population growth occurs, with number of individuals doubling per unit time. (Syn. *exponential growth phase*) Cf. *lag growth phase*.

Loggerhead Turtle, Loggerhead Seaturtle. Vernacular name for the marine turtle *Caretta caretta* (Cheloniidae).

logistic growth model. A model of population growth based in the concept that resources are limiting and there is density dependence in the instantaneous birth rate and/or instantaneous death rate. The model generates a characteristic S-shaped curve, and population growth decelerates to a constant carrying capacity that reflects the resources available in the environment.

logistic curve. A sigmoid curve fitted to an exponential function such as the logistic equation.

logistic equation. A simple mathematical model of a population growing from a small number of individuals to a limited population in a finite environment, first published by P. Verhulst (1845, 1847). The differential form is

$$dN/dt = rN[(K-N)]/K$$

where r is the rate of maximum population growth (Malthusian parameter), N is number of individuals, and K is the limiting number of individuals (so-called *carrying capacity*). (Syn. *Verhulst logistic equation* or *logistic growth curve*)

logistic growth. Reference to population growth that follows a sigmoid curve.

logistic growth curve. See *logistic equation*.

log-normal distribution. Descriptive of a frequency distribution for a variable that, when converted to its logarithm, is normally distributed.

logotype. *n.* In taxonomy, a type determined from a written description in the absence of both a specimen and an illustration.

log transformation. Logarithmic transformations are a class of transformations in which numbers are expressed as y to the x power in a variety of possible ways. Log transformations are undertaken to improve the normality of variables. Note if natural logarithms are taken, the log of numbers less than 1 are undefined and a constant must be added to the minimum value of the distribution, preferably to 1, prior to transformation. See *data transformation*. Cf. *square root transformation*.

longevity. *n.* The average life span of the individuals in a population under a given set of conditions.

Long-fingered Frogs. Collective vernacular name for species of frogs belonging to the African arthroleptid genus *Cardioglossa*.

longitude. *n.* Lines of longitude, or meridians, run between the North and South Poles and measure east-west position. The prime meridian is assigned the value of 0 degrees and runs through Greenwich, England. Meridians to the west or east of the prime meridian are measured in degrees. Cf. *latitude*.

longitudinal. *adj.* Reference to length. Running along, or parallel to, the long axis, in contrast to *transverse*.

longitudinal dorsal scale row. A row of scales running parallel to the long axis of the body.

longitudinal dune. A large, symmetric ridge of sand parallel to the wind direction. (Syn. *seif*)

longitudinal smooth muscle. The outer layer of smooth muscle oriented along the long axis of the small intestine.

longitudinal type. An arrangement of scale rows in which the scales form a straight, rather than oblique, series. See *scale row*. (Syn. *rectiform*)

longitudinal valve. See *spiral fold*.

Longnose or **Long-nosed Snake.** Vernacular name for species of snakes belonging to the colubrid genus *Rhinocheilus* (especially *R. lecontei*).

longshore. *adj.* Reference to currents or movements that are parallel to a coastline.

lophate. *adj.* 1. Having smooth longitudinal ridges. 2. Reference to a pattern of *epidermal microsculpturing*. See R.M. Price *J. Herpetol.* 16:294–306, 1982.

Lora. *n.* Vernacular name for the arboreal Green Parrotsnake, *Leptophis ahaetulla*, native to Middle and South America.

lore or loreum. (*pl.* **-res, -ra**) *n.* The side of the head between the eye and nostril.

loreal. *n., adj.* 1. A scale, or reference to region, between the *nasals* and *preoculars* on each side of the head. The scale is usually single in snakes but several in lizards, and it has taxonomic importance. It is typically present in colubrids, but always absent in elapids. See Figs. 11, 16. 2. Of or pertaining to the *lore*.

loreal pit. Older term for the *sensory pit* or *pit organ* (opening to infrared sensory receptors) on the side of the head of *pitvipers*.

lorica. *n.* See *cuirass*.

loricate. *adj.* Possessing a *cuirass*.

lorilabial. *n.* Any of a number of scales situated in groups or longitudinal rows between the *loreals* and *supralabials*, sometimes extending between the eye and *labials*. This term has been variously used by authors in relation to scalation of both lizards and snakes.

lost year. In reference to sea turtles, the period of time between hatching and the attainment of a carapace length of 20–30 cm, during which young turtles are epipelagic in habits and rarely encountered. The period may not correspond, in actuality, to a single year.

lotic. *adj.* Of, relating to, or living in actively moving water such as a stream. See *fluvial*. Cf. *lentic*.

Lotka-Volterra equations. Simple models for predator-prey systems or two competing populations, based on logistic equations.

lottery hypothesis. The assumption that unpredictability plays some role in the development of biological communities.

lower labial. See *labial*.

lower rostral. See *mental*.

Loxocemidae. *n.* A clade (family) containing a single species of snake ("*New World Python*") inhabiting tropical dry forests of Mexico and Central America.

LT_{50}. See *lethal time*.

Ludwig effect. A generalization attributable to W. Ludwig (1950) that a species tends to be more diversified, or polymorphic, in the center of a long established range rather than at its margins.

Ludwig's theory. A proposition that new genotypes can be added to a gene pool if they can utilize new components of the environment and thereby occupy a new niche, even if they are competitively inferior in the ancestral niche.

lumbar. *adj.* Pertaining to the region between the pelvic girdle and the true ribs, when present, or the posterior half of the dorsal and dorsolateral areas of the body, when ribs are absent.

lumbar gland. A prominent glandular area in the loin of certain frogs. (Syn. *lumbar organ*)

lumbar organ. See *lumbar gland*.

lumbar vertebra. A presacral vertebra situated between the pelvic girdle and the thorax. Specific definitions of this term were developed for mammals and are generally inapplicable to reptiles and amphibians. Substitute terms include *poststernal vertebra, dorsal vertebra, trunk vertebra, prescral vertebra*, and *thoracicolumbar vertebra*.

lumen. (*pl.* **lumina**) *n.* 1. The principal internal cavity or space of an organ. 2. The interior channel within a tube or duct, as in a fang of a venomous snake. 3. A derived SI unit of light flux (*lm*) defined as the light energy emitted per second within a unit solid angle of 1 steradian by a uniform point source of 1 candela luminous intensity.

luminance. *n.* A measure of light emitted per unit of projected surface area, perpendicular to the direction of view and measured as candelas/m^2.

luminescence. *n.* Light emission from a body that is not attributable to temperature. The source of light energy may originate from a chemical reaction or from an external source that is directed into the body from outside. See *bioluminescence, fluorescence*.

luminosity. *n.* The relative quantity of light emitted or reflected. Brightness.

lumpiness. *n.* In ecology, reference to patterns of discontinuous distributions of adult body sizes of species comprising a particular community. That is, body mass distributions are fundamentally discontinuous with "*clumps*" and "*gaps*" in the distribution, and such patterns are inferred to reflect important processes that structure ecosystems across scales in time and space. See C.S. Holling, *Ecol. Monogr.* 62:447–502, 1992.

lumping. *n.* In taxonomy, the practice of ignoring minor variation in the definition or recognition of taxa, relative to "splitters" who utilize relatively more minor variation in defining taxonomic units. Cf. *splitting*.

lunar. *adj.* Pertaining to the moon.

lunar day. See *day*.

lunate. *adj.* Crescent-shaped or bent with a broad and rather shallow fork.

lung. *n.* An air sac, usually paired, derived from endoderm of the floor of the pharynx and used for aerial respiratory gas exchange. In amphibians the lungs are relatively simple and saccular, but in reptiles they are more complexly subdivided and approach the condition seen in mammals. See also *saccular lung, tracheal lung, vascular lung* (Fig. 41).

lungless salamanders. Collective vernacular name for salamanders belonging to the family *Plethodontidae*.

lungworm. *n.* Any of a number of parasitic nematodes (especially the genus *Rhabdias*) present in the lungs of amphibians and various squamate reptiles.

lunula. *n.* An ossification developed in the semilunar cartilages and femoro-fibular disc of lizards.

lure. *n.* Reference to any appendage or body part that is displayed and used to attract potential prey items. Examples are the tongue of Alligator Snapping Turtles and the colorful tail tips of juveniles in many snake species. See *caudal luring.*

luring. *n., v.* Use of an anatomical lure to attract prey. See *caudal luring.*

luteinizing hormone. *(LH)* A gonadotropin secreted by the anterior pituitary that acts with *follicle-stimulating hormone* (FSH) to induce ovulation and release of estrogens and progestogens (e.g., progesterone) from the ovary in females and stimulates testicular Leydig cell steroidogenesis in males. Found in all vertebrates studied to date except for squamates. (Syn. *interstitial cell-stimulating hormone, ICSH*)

lux. (lx) *n.* A derived SI unit of illumination, defined as 1 lumen/m^2.

Lygosominae. *n.* A subfamily of lizards belonging to the *Scincidae.*

lymph, lymphatic fluid. *n.* Clear plasma-like fluid carried in lymphatic vessels.

lymphadenitis. *n.* Inflammation of the *lymph nodes.*

lymphadenopathy. *n.* Enlargement or swelling of the lymph nodes, usually due to acute inflammation.

lymphatics, lymphatic system, lymphatic vessels. *n.* A collection of blind-ended vessels that drain filtered extracellular fluid from tissues and return it to the blood circulation. The term sometimes is applied to the system of *lymph glands* that function in immune responses.

lymph gland. An organ on the course of a lymphatic vessel, containing lymph and white blood cells that remove foreign bodies from the lymph.

lymph heart. Accessory "hearts" that impart movement to lymph along the lymphatic vessels and assist the return of lymphatic fluid to the cardiovascular system. These are present in fishes, some amphibians, some reptiles, and embryonic birds, often where lymphatic vessels enter veins. These are not true hearts but consist of striated muscle in the walls of lymphatic vessels that slowly contract and develop pressure to drive the lymph.

lymph nodes. Aggregations of lymphoid tissue along the course of lymphatic system that produce lymphocytes and filter the lymphatic fluid.

lymphocytes. *n.* Leukocytes produced in lymphoid tissue that lack cytoplasmic granules and have a large, rounded nucleus and participate in immune responses.

lymphoid tissue. Sites of blood cell differentiation outside of the bone marrow (e.g., spleen).

lymph sacs. Lymphatic spaces that are especially prominent in anuran amphibians and possibly function to store fluids in terrestrial species.

lyophylize. *v.* To create a stable preparation of a biological substance by freeze-drying.

Lyre Snake. *n.* Vernacular name for two species of rear-fanged colubrid snakes belonging to the American genus *Trimorphodon.*

lysin. *n.* An agent of *lysis.*

lysis. *n.* Destruction, dissolution, or decomposition of cells or other structures.

M

m. 1. Mole (expressed in Daltons or kilodaltons) 2. Molar.

Ma. An abbreviation for "million (years) ago."

Mabuyas. *n.* Vernacular name for species of scincid lizards belonging to the genus *Mabuya*.

maceration. *n.* Process of breaking into smaller pieces.

macro-. A prefix meaning "very large" or "very large scale."

macrocephaly. *n.* The condition of a grossly enlarged skull or head, usually associated with certain age classes of some turtles.

macroclimate. *n.* The climate of a major geographical region, usually based on meterological factors measured about 1.5 m above the ground to avoid topographical, edaphic, and vegetational influences. Cf. *microclimate*.

macroevolution. *n.* Reference to evolutionary patterns that involve changes in taxonomic categories above the species level and events that originate new higher taxa. The term also is used to describe large-scale changes in gene frequencies that occur over a geological time period. This term is used particularly in paleontology. (Syn. *macrophylogenesis*) Cf. *microevolution*.

macrolecithal. *adj.* Reference to an egg with a large amount of yolk, regardless of its distribution. Characteristic of reptiles.

macrogland. *n.* Prominent aggregations of multiple glands, as in clusters of mucous or granular glands of amphibians (W.E. Duellman & L. Trueb, *Biology of Amphibians*, p. 370, 1986).

macromolecule. *n.* A molecule of relatively large molecular mass, generally greater than several thousand and including proteins, nucleic acids, polysaccharides, etc.

macromorphology. *n.* Gross external morphology. Cf. *micromorphology*.

macromutationism. *n.* The idea or theory that new taxa evolve rapidly by means of a major genetic mutation that establishes reproductive isolation and new adaptations all at once.

macrophage. *n.* Any large, mononuclear phagocytic cell derived from monocytes in the walls of blood vessels and in loose connective tissue. See *phagocyte*.

macrophagy. *n.* (*adj.* **-ous**) Reference to feeding on relatively large prey or food particles. Cf. *microphagous*.

macrophylogenesis. *n.* See *macroevolution*.

macroscopic. *adj.* Relatively large size, visible to the unaided eye. (Syn. *megascopic*) Cf. *microscopic*.

Macrostomata. *n.* A clade that includes most advanced lineages of snakes except for basal *alethinophidians*.

macrotaxonomy. *n.* See *beta taxonomy*. Cf. *microtaxonomy*.

macroteiid. *n.* A member of the lizard family Teiidae. Cf. *microteiid*.

macula. (*pl. –ae*) *n.* 1. A spot, stain, or thickening. 2. Typical usage is in reference to clusters of hair cells forming organs of equilibrium in the *vestibular apparatus* of the inner ear. The macula is the major sensory epithelium of the *utriculus*. Cf. *crista*.

macula adherens. See *desmosome*.

macula densa. A region of specialized cells in the glomerular apparatus that senses sodium chloride concentration and flow of distal tubular fluid in the kidney.

macula lagena. (*pl. –ae*) A sensory area of the reptilian *cochlear duct*.

macula neglecta. A small sensory spot in the *utriculus* of fishes and amphibians.

Madagascan. A zoogeographical realm including Madagascar.

Madagascan Hog-nosed Snake. Vernacular name for a few species of Madagascan snakes belonging to the genus *Leioheterodon*. (Also called *Brown Snakes*.)

Madagascar Tree Boa. Vernacular name for the Madagascan boid species belonging to the genus *Sanzinia*.

magainins. *n.* A Hebrew word meaning "shield," first applied to a family of peptides isolated from the skin of *Xenopus laevis* and having broad-spectrum antimicrobial activity (M. Zasloff, *Proc. Nat. Acad. Sci.* 84:5449–5453, 1987).

magnetic compass orientation or **taxis.** See *geomagnetotaxis*.

magnetic field. Region of magnetic force that surrounds the Earth.

magnetite. *n.* A naturally occurring magnetic oxide of iron.

magnetoreception. *n.* The ability to detect fluctuations in magnetic fields. Use of directional ("compass") information derived from the Earth's magnetic field has been demonstrated in a wide variety of animals including amphibians and sea turtles. See *geomagnetotaxis*.

magnum. See *uterine tube*.

maintenance energy. Energy used by a cell or organism for "idling" metabolic processes other than growth, storage, or reproduction.

Maja. *n.* Vernacular name for the Cuban boid *Epicrates angulifer*.

major histocompatibility complex. (MHC) One or more gene clusters (depending on taxon) and encoded cell surface antigens that affect resistance to disease and parasites, having a central role in immune responses and rejection of transplanted tissues.

malachite green. A chemical used for treatment of various fungal infections on the shells of captive aquatic turtles.

malar. *adj.* Reference to the cheek, cheekbone, or side of head. *n.* 1. Synonym for the *jugal* bone of the skull. 2. Used by some authors in reference to a scale on the lower jaw of some lizards and amphisbaenians.

Malayan Gharial. See *False Gharial.*

Malayan Pitviper. Vernacular name for the viperid species *Calloselasma rhodostoma.*

male. *n.* The sperm-producing form of a bisexual species, symbolized ♂.

male release call. The vocalization emitted by either a male or female frog that is in *amplexus* and is grasped by another male. The call informs the clasping male that pairing has already taken place and usually results in release. The call is accompanied by a *warning vibration* in some species. (Syn. *alarm call, sex-warning vibration, warning call, warning croak, warning chirp, warning vibration*)

malignant. *adj.* In medicine, a clinical course that progresses to death. Cf. *benign.*

malnutrition. *n.* Undernourishment; insufficient nutrients for normal growth and maintenance.

Malthusian model. An expression of exponential population increase, where the rate of increase in numbers $dN/dt = rN$, where r is a constant rate of increase and N is the original population size.

Mamba. *n.* Collective vernacular name for various species of African elapid snakes belonging to the genus *Dendroaspis.* These are generally arboreal, slender, swift-moving, and potentially very dangerous snakes.

mamba intestinal protein 1. See *AVIT.*

mambin. *n.* A protein component of snake venom, isolated from Mamba venom, that acts on blood platelets or thrombocytes. (Syn. *dendroaspin*)

mammal-like reptiles. See *Therapsida.*

mammilary process. A lateral projection on the anterior zygapophysis in snakes, directed outward and forward as a site of muscle attachment.

Mamushi. *n.* Vernacular name for the Japanese pitvipers *Gloydius halys* and *Gloydius blomhoffi.*

management. *n.* The process of working with the characteristics and interactions of habitats, wild animal populations, and humans to achieve specific goals.

mandible. *n.* 1. The lower jaw. 2. Keratinized labial parts of the oral disk in tadpoles, lying inside the labial teeth rows and bordering the mouth opening. The term has been applied to the lower keratinized structure alone as well as to both upper and lower structures.

mandibular. *adj.* Of, or pertaining to, the lower jaw.

mandibular arch. The first visceral arch.

mandibular fossa. See *adductor fossa.*

mandibular gland. See *musk gland, gland of Gabe.*

mandibular suspensorium. See *suspensorium.*

mandibular raking. An unusual feeding mechanism or prey transport in which the tooth-bearing elements of the lower jaw rotate synchronously in and out of the mouth, dragging prey into the esophagus. This mechanism is reported in small, burrowing Thread Snakes (family Leptotyphlopidae) and is the only vertebrate feeding mechanism known in which prey are transported exclusively by movements of the lower jaw. Reported by N.J. Kley & E.L. Brainerd (*Nature* 402:369, 1999). Cf. *pterygoid walk.*

mangrove. *n.* 1. A tidal salt marsh community found in tropical and subtropical regions dominated by salt-tolerant trees and shrubs, especially those of the genus *Rhizophora.* 2. Common name for salt-tolerant plants of the genus *Rhizophora.*

Mangrove Frog. See *Crab-eating Frog.*

Mangrove Snake, Mangrove Cat Snake. Collective vernacular name for colubrid species of snakes belonging to the genus *Boiga,* especially the species *B. dendrophila.* See also *Crab-eating Water Snake.*

Mangrove Pitviper or **Mangrove Viper.** Vernacular name for the Asian pitviper *Trimeresurus purpureomaculatus.*

manicotto, *manicotto glandulare. n.* A glandular, stomach-like swelling of the foregut immediately posterior to the esophagus in anuran tadpoles. The structure is similar to a stomach in storage function, but it differs in its range of pH, being part of the larval pancreatic complex. (Syn. *larval stomach*)

Mann-Whitney U-Test. A nonparametric test using sequential ranks of sample counts to determine whether two independent random samples are drawn from populations having the same distributions.

Mantella. *n.* Collective vernacular name for species of Madagascan frogs belonging to the family Mantellidae, and especially the genus *Mantella.*

Mantellidae. *n.* A clade (family) of anurans that has undergone extraordinary radiation in habits and morphology, with species of *Mantella* evolving bright coloration and toxic skin secretions that are convergent with some Neotropical dendrobatid frogs. Five genera and approximately 140 species occur in Madagascar. Three subfamilies include *Boophinae, Laliostominae,* and *Mantellinae.* For taxonomy see D.R. Frost et al., *Bull. Amer. Mus. Nat. Hist.* 297:1–370, 2006.

Mantellinae. *n.* A subfamily of *Mantellidae.*

Mantel test. A test of *spatial statistics* that computes a correlation between two distance matrices to assess the extent of spatial autocorrelation among measurements that are made in the field. See also *Partial Mantel test.*

manubrium. *n.* Alternative term for *omosternum* or *sternum*. See Fig. 3.

manubrium sternii. See *omosternum*.

manus. *n.* The hand or forefoot, comprised of wrist-palm-fingers.

manuscript. (MS) *n.* A typewritten or handwritten text, typically the copy submitted for publication and not reproduced as multiple copies.

manuscript name. In taxonomy, an unpublished scientific name.

Many-banded Coral Snake. Vernacular name for the Coral Snake species *Micrurus multifasciatus*.

Many-lined Salamander. Vernacular name for the North American plethdontid species *Stereochilus marginatus*.

Map Turtle. Vernacular name for American species of emydid turtles belonging to the genus *Graptemys*.

Marbled Salamander. Vernacular name for the American salamander *Ambystoma opacum*.

Margalef's index (SR) A univariate index of the biodiversity of a sample or assemblage of organisms, weighted in favor of species richness. Calculated as

$$SR = (S-1) \log N$$

where S is the number of species and N the number of individuals in the sample or assemblage.

marginal. *adj.* Reference to a margin or border. *n.* 1. Any of the plates forming a sharp angle at the margins of the carapace of turtles. See Fig. 4. 2. Any of the scales lying immediately below and bordering the *dorsals* in crocodilians.

marginal belt. A ring of presumptive mesoderm in the amphibian blastula, similar to the grey crescent of the zygote before division begins and intermediate in pigmentation between the pigmented animal hemisphere and the unpigmented vegetal hemisphere. (Syn. *marginal zone*)

marginal canal. See *Bidder's canal* or *duct*.

marginal habitat. Reference to habitat with few species due to adverse physical or other conditions.

marginal height. A height measurement for turtle shells used by F.G. Benedict (*Carnegie Inst. Wash.*, Publ. 425:1–539, 1932) and equal to the straight-line distance from the surface on which a shell is resting to the ventral edge of the marginal plates, or to the bend in the marginals (edge of shell).

marginal lamina. One of the outer epidermal plates along the edge of the carapace in turtles.

marginal papillae. Any papilla or papillae on the margin of the *oral disc* of a tadpole, commonly as a circumoral marginal series or with dorsal or ventral gaps. See Fig. 1.

marginal zone. See *marginal belt*.

marginobrachial. *n.* See Fig. 4.

marginofemoral. *n.* See Fig. 4.

marginolateral. *n.* See Fig. 4.

marginocollar. *n.* See Fig. 4.

marine. *adj.* Reference to the sea or ocean.

Marine Iguana. See *Galapagos Marine Iguana*.

marine terrace. A broad, gently sloping platform that may be exposed at low tide.

Marine Toad. See *Giant Toad*.

marinobufagin. *n.* A *bufagen* isolated from *Bufo* (*Chaunus*) *marinus*.

marinobufotoxin. *n.* A bufotoxin isolated from *Bufo* (*Chaunus*) *marinus* and formed by conjugation of *marinobufagin* with suberylarginine.

marker. *n.* In molecular biology, reference to a gene, antigen or fragments of DNA, RNA, or proteins used in various contexts to map chromosomes, distinguish cell types, or calibrate electropohoretic gels.

mark-recapture. A technique used for decades by ecologists and wildlife biologists in which animals are captured and then physically marked or tagged in some way with a personal identification number or feature before being released. Recapture of the individual at a later time can provide valuable information about the animal, its population or species (growth, estimates of population size, movements, etc.). Amphibians and reptiles have been marked using various techniques including use of metal tags, PIT tags, scale clippings, paint, notching of turtle shells, freeze or hot-wire branding, tattooing, fluorescent dyes, and staining the rattles of rattlesnakes. See *Lincoln index*.

marsh. *n.* An ecosystem having more or less continuously waterlogged soil dominated by emersed herbaceous plants, but without a surface accumulation of peat.

Marsh Crocodile. See *Mugger*.

marsupial cartilage. See *ypsiloid bone or cartilage*.

Marsupial Frog. Collective vernacular name for numerous species of South American hylid frogs belonging to the genus *Gastrotheca*. These frogs are named for reproductive specializations involving brooding of tadpoles within a dorsal pouch, with direct development in most species and a lengthy period of larval development in water after tadpoles leave the pouch in some others.

marsupial pouch. A sac-like cavity formed by skin on the back of females of the anuran genus *Gastrotheca* and used for brooding fertilized eggs that are received through an opening above the cloaca.

marsupium. *n.* See *pecten*.

mask. *n.* 1. Reference to a broad, dark stripe that extends horizontally across the orbits, usually from nostril to eye, or to tympanum, or beyond. The term is often used in descriptions of frog color patterns. 2. Reference to face-like markings on the back of the *hood* of various cobras.

Massasauga. *n.* Vernacular name for the Pigmy Rattlesnake species *Sistrurus catenatus*.

mass. *n.* The numerical measure of an object's inertia, or resistance to being accelerated. The quantity of matter contained in an object.

mass effect. In biogeography and ecology, reference to spatial dynamics in which there is a net flow of individuals created by differences in population size (or density) in different patches (A. Schmida & M.V. Wilson, *J. Biogeography* 12:1–20, 1985).

mass extinctions. Relatively large extinctions that occur during geologically brief periods of time in the fossil record, generally involving between 25 and 50% of all fossil families and up to 90% of species. Such extinctions involve diverse taxonomic groups due to global ecological circumstances that suggest a common or related cause. Patterns of extinction can be catastrophic (abrupt), stepwise, or gradual. Mass extinctions at the end of the Cretaceous are the most famous because they involved the loss of dinosaurs, flying reptiles, marine reptiles, and numerous invertebrates including ammonites.

mass-specific metabolic rate. The metabolic rate of a unit mass of tissue, usually determined by dividing the metabolic rate of an animal by its mass.

mass spectrometry. A technique that can identify the gases in a gaseous mixture based on their characteristic ratios of mass to charge.

mast cell disrupting and chemotactic peptides. A relatively new family of peptides from amphibian skin, having potent chemotactic and disruptive effect on both rat and human mast cells. These are potential messengers in hypersensitivity, inflammatory, and other immunological processes and may be subject to manipulation for therapeutic value. These peptides include *granuliberin* from *Glandirana* (formerly *Rana*) *rugosa* and *pipinins* from *Lithobates* (formerly *Rana*) *pipiens*.

mast cells. Cells that are abundant in lymph nodes, spleen, bone marrow, connective tissue, and skin, thought to be counterparts of blood basophils.

mast crop. Production of offspring at one time to reduce effective predation.

mastication. *n.* Chewing or grinding of food with teeth.

Mastigure. *n.* Vernacular name for species of African and Asian agamid lizards belonging to the genus *Uromastyx*.

mastoid. *n.* A term sometimes used for *squamosal* or *supratemporal* by early authors.

Matamata, Mata Mata. *n.* An unusual and well-known chelid turtle (*Chelus fimbriatus*) of South America, characterized by a broad flattened shell with three prominent keels, a long neck with sensory cutaneous flaps, a fleshy proboscis, and suction feeding on fish.

maternal. *adj.* Pertaining to, or derived from, the female parent. Cf. *paternal*.

maternal inheritance. Inheritance derived soley from the maternal parent, controlled by extranuclear or cytoplasmic genetic factors. See *extranuclear inheritance*.

maternal mRNA. Messenger RNA found in oocytes and early embryos, derived from the maternal genome during oogenesis.

mathematical model. The symbolic representation of hypotheses or assumptions about a system in the form of an equation or set of equations, used to describe or predict the behavior of a system.

mating. *n.* Sexual union between two individuals of opposite sex in order to reproduce.

mating ball. A mating phenomenon seen in some snake species, especially Garter Snakes, in which males outnumber females in breeding aggregations and from 10 to 100 males surround and simultaneously court a single receptive female. See J.B. Gardner, *Copeia* 1955:310, 1955 and *Copeia* 1957:48, 1957. This term should not be confused with *balling posture* or *balling*.

mating system. Reference to the pattern of matings between individuals of a population, including such factors as extent of inbreeding, pair-bonding, and number of simultaneous mates.

matriarchy. *n.* A social group led by a female. Cf. *patriarchy*.

matric potential. The force with which water is held in a matrix such as a soil by capillary action and adsorption to colloids. Adhesive and cohesive forces bind the water and reduce its potential energy compared to that of free water. (Syn. *matric pressure*) See *water potential*.

matric pressure. *Matric potential* expressed in units of pressure.

matrilinear inheritance. Transmission of cytoplasmic particles only in the female line.

matrix. (*pl.* **matrices**) *n.* 1. The intercellular substance of a tissue. 2. The material in which a fossil or other object is imbedded. 3. Used in reference to the fleshy tail tip in rattlesnakes, from which the rattle segments develop. 4. A rectangular array of *x* rows, each containing *n* numbers or elements arranged in columns.

matrotrophy. *n.* A fetal nutritional pattern in which nutrients are supplied by extravitelline means alternative to yolk during fetal development. The term is used sometimes to describe post-paritive matrotrophy as well as placental provision. Terminology is from J.P. Wourms (*Amer. Zool.* 21:473–515, 1981. See also D.G. Blackburn, *Copeia* 1994:925–935, 1994.) Cf. *lecithotrophy*.

maturation. *n.* (*adj.* **mature**) The attainment of sexual maturity, with individuals having differentiated gametes.

Mauthner's cell. A giant neuron present, usually as a pair, in teleost fishes and larval amphibians (absent in amniotes). Each cell is located in the medulla at the base of the VIII cranial nerve and used in control of trunk and tail movements, especially in rapid escape responses. Each Mauthner cell sends a single, large-diameter and fast-conducting axon down the medulla and spinal cord where it synapses with motor neurons innervating axial muscles at each spinal segment. Most amphibians have Mauthner cells as larvae, and these have been shown to persist in adults where the structures may be reduced and of unknown function.

maxilla, maxillary. *n., adj.* Either of the two, usually tooth-bearing, principal dermal bones in the upper jaw, separated anteriorly by the *premaxillaries*. In turtles the bone is without teeth and covered with a horny beak. In viperid snakes species, the maxilla is strongly modified as the support for the fang (Fig. 21). (Syn. *supermaxillary*)

maxillary arch. Reference to the dermal bones forming the upper jaw, consisting of *premaxillary, septomaxillary, maxillary, jugal,* and *quadratojugal*. Some of these elements may be absent or fused with other bones in some taxa.

maxillary formula. A method first used by E.R. Dunn (*Caldasia* 3:155–224, 1944) to indicate the total number of teeth, their relative sizes and position on the maxillary of a snake.

maxillary groove. A well defined channel in the midline of the palate in some species of trionychid turtles.

maxillary segment. One of three parts of the *kinetic skull* of lizards, used by J. Verluys (1912) and refined by T.H. Frazzetta (*J. Morphol.* 111:287–319, 1962).

maxillary teeth. Teeth located on the two maxillae of the upper jaw in amphibians and reptiles.

maxillopalatal arch. See *palatomaxillary arch*.

maxillopalatine bone. A complex dermal bone forming the lateral teeth-bearing portion of the maxillary arch in caecilians.

maximum aerobic speed or **velocity.** The locomotor speed at which the rate of aerobic respiration is maximal.

maximum entropy inference. A quantitative generalization of *Occam's razor*.

maximum extinction rate. The maximum rate at which resident species on an island go extinct, occurring at conditions when all of the source pool species occur on the island.

maximum feeding rate. The asymptotic per capita feeding rate of predators in a *Type II* and *Type III* functional response. The units are victims/(time • predator).

maximum homology hypthesis. A proposition that the optimum reconstruction of a phylogenetic tree is one that maximizes identity due to common ancestry, as indicated by homologous genetic coding sites. (Syn. *Red King Hypothesis*)

maximum immigration rate. The maximum rate at which new species arrive on an island, usually achieved when the island has no resident species.

maximum likelihood methods. In phylogenetic inference, these methods are mathematically consistent and derived from the general statistical approach of the same name by R.A. Fisher. The methods use statistical criteria to arrive at character-based scores and a preferred phylogenetic tree. The method combines probabilities of observing patterns of character states such that a single overall probability is maximized. The method selects the ancestral trait value with highest likelihood on a given phylogenetic hypothesis, given a model of trait evolution (defined by the user). Maximum likelihood methods have found greatest application to molecular data for gene sequences and proteins. Cf. *distance methods, parsimony methods, Bayesian analysis*. See also *optimization*.

maximum parsimony. A hypothesis that optimum reconstruction of ancestral character states is the one that requires the fewest mutations in the phylogenetic tree to account for contemporary character states. Maximum parsimony minimizes trait evolution on a given phylogenetic tree. See *optimization*.

MCH. See *melanocyte concentrating hormone*.

meadow. *n.* An area of closed herbaceous vegetation dominated by grasses, often near a stream.

mean. ($\bar{x}$) *n.* The arithmetic mean is the average, or central tendency, of a distribution or series of numbers, calculated as $\sum x/n$ where x is an individual value and n is the sample size. The mean is denoted as a statistic of a sample as $\bar{x}$, and as a population parameter by μ.

mean activity temperature. The arithmetic mean or average of the body temperatures of animals within their *activity temperature range*. (Syn. *mean body temperature*)

mean clasp time. The average length of time spent in clasping spells, calculated by dividing the total response time by the number of clasping spells. Used by W.M.S. Russell (*Behaviour* 7:113–188, 1954).

mean deviation. A measure of variation within a set of data given by the arithmetic mean of the deviation from the mean.

mean lethal dose. See *lethal dose*.

mean selected temperature. The arithmetic mean or average of body temperatures measured from animals in a thermal gradient. See *selected body temperature range*.

mean square. In analysis of variance, the sum of squares divided by the corresponding number of *degrees of freedom*.

meatus. *n.* A canal or opening.

mechanical isolation. Reproductive isolation attributable to incompatibility of male and female genitalia.

mechanistic modeling. A mathematical description of relationships between variables with emphasis on biological, physical, or chemical principles linking the variables to explain the underlying mechanism of relationships.

mechanistic theory. The general proposition that natural processes are determined mechanistically and can be explained by the laws of chemistry and physics.

mechanoreceptor. *n.* A sense organ that responds to small changes in mechanical force (distortion or pressure).

Meckelian groove. An elongated depression left in the lower jaw of many animals, attributable to *Meckel's cartilage*, which is absorbed in adults having served as a core for bony deposits.

Meckelian orifice. See *adductor fossa*.

Meckelian vacuity. See *adductor fossa*.

Meckel's canal. The internal cavity of the *dentary* bone in certain reptiles, which can be subdivided anteriorly into a ventral space housing *Meckel's cartilage* and a dorsal space containing the inferior alveolar nerve.

Meckel's cartilage, Meckelian cartilage. The cartilage that forms the entire lower jaw of early vertebrates and forms the core around which the dermal bones of the lower jaw are deposited in amphibians and reptiles. The cartilage is distinguishable in early embryos, but is lost in practically all adults. The posterior end forms the ossified *articular*. The anterior end of this cartilage may ossify as an independent element, the *symphysial,* in anurans.

mediad. *adv.* In the direction of the midline.

medial. *adj.* 1. Situated toward or at the midline or medial plane of a body or structure. 2. Describes a *vent tube* of tadpoles that opens parallel with the plane of the ventral fin or tail muscle.

medial lower labial. See *mental*.

median. *adj.* Situated in the middle or in the plane that divides an animal or structure into two equal halves. *n.* 1. The large scale on the top of the head of xantusiid lizards, anterior to the paired frontals and between the anterior ends of the supraocular series. 2. The middle value in a group of numbers arranged in order of size. The median can be estimated from a histogram by finding the smallest number such that the area under the histogram to the left of that number is 50%. Cf. *mean, mode*.

median eminence. A structure at the base of the hypothalamus, continuous with the hypophyseal stalk and containing the primary capillary plexus of the *hypothalamo-hypophyseal portal system*.

median eye. See *parietal eye*.

median gular. A scale situated between the chin shields or on the midventral line posterior to the chin shields but anterior to the first ventral.

median labial. See *mental*.

median ridge. A transverse flaplike ridge of variable shapes suspended from the roof of the buccal cavity of tadpoles (R. Wassersug, *Misc. Publ. Mus. Nat. Hist. Univ. Kansas* 68:1-146, 1980).

mediogyrinid. *adj.* Reference to anuran tadpoles that have a single, median *spiracle*. Cf. *amphigyrinid, laevogyrinid, paragyrinid*.

mediostapedial bar. The central part of the columella in lizards, connecting the *footplate* at the proximal end with the *extracolumella* at the distal end.

mediostapes. *n.* Synonym for the *stylus* of anurans.

Mediterranean climate. Reference to a climate that is characterized by hot, dry summers and mild, wet winters. Such regions include the coastal areas of Mediterranean countries, southern California, central Chile, and the Cape region of South Africa.

Mediterranean Gecko. Vernacular name for the gekkonid species *Hemidactylus turcicus*, which occurs naturally in the Middle East and has been introduced widely elsewhere due to its propensity to travel readily with humans.

medium. *n.* 1. The nutritive substance provided for the growth of a given organism in the laboratory. 2. The fluid environment, air or water, immediately surrounding an organism.

medulla. *n.* 1. The inner portion or core of an organ. 2. See *medulla oblongata*.

medulla oblongata. The posterior part of the brain stem continuous with the *pons* above and the *spinal cord* below. Sometimes called simply the *medulla*.

medullarin. *n.* A paracrine, diffusible secretion released by the medulla of the undifferentiated early embryonic gonad. It stimulates differentiation of the medulla, inhibits the cortex, which results in testis formation, and also stimulates development of male secondary characteristics. E. Witschi (*Biol. Rev.* 9:460–488, 1934) first theorized from amphibian experiments that corticomedullary inductors, *cortexin* and *medullarin,* induced sex differentiation in amphibians. Cf. *corticin*.

medullary. *adj.* Pertaining to the innermost portions of an organ. Cf. *cortical*.

medullary bone. Reference to ossified tissue that lines the inner core of bone. Cf. *cortical bone*.

medullary cord. See *sex cord*.

medullary cardiovascular center. Groups of neurons located in the medulla that are involved in the integration of information used in control and regulation of blood circulation.

medullary respiratory center. Groups of neurons located in the medulla that control the activity of motor neurons associated with breathing or ventilation.

mega-. (M) Prefix meaning "large," "great" or "greater than normal." Also used to denote unit x 10^6.

megadont. *adj.* Possessing teeth that are irregular in size and appearance, often in reference to strongly enlarged teeth on some part of the maxillary bone.

megafauna. *n.* Larger forms of animal life, or the largest forms within a given community.

Megagaea. *n.* See *Arctogaea*.

megascopic. *adj.* Reference to a structure that is observable with the unaided eye, or with use of a hand lens. (Syn. *macroscopic*) Cf. *microscopic*.

Megophryidae. *n.* A clade (family) of anurans, sister with *Pelobatidae* within the *Pelobatoidea*, having very glandular skin and unusual clusters of granular glands in both sexes of various species. Largely inhabitants of tropical stream edges and forests, 11 genera and approximately 130 species are distributed in central and southern Asia. Sometimes called "*Toadfrogs*."

meio-. Prefix meaning "smaller" or "less than."

meiolecithal. *adj.* See *microlecithal.*

meiosis. *n.* A process of cell division in which gametes are formed by rearrangement of genetic material such that the number of chromosomes in each cell is halved (*reduction division*). Cf. *mitosis.*

Melanesia. *n.* A geographical area and an ethnological unit including islands of the West Pacific south of the equator and some islands of volcanic origin (Papua New Guinea, New Caledonia, Fiji, New Hebrides, Bismarck archipelago, and the Solomon islands). These are often considered to be a distinct biogeographical unit separate from Polynesia.

melanin. *n.* The dark pigment, derived from oxidation products of tyrosine and dihydroxyphenol compounds, present in skin and internal peritoneum. The color varies from dark reddish to dark brown or black. See *melanophore.*

melanin concentrating hormone. See *melanocyte concentrating hormone.*

melanism. *n.* (*adj* **melanistic**) Excessive pigmentation or blackening of the skin or other tissues, usually of genetic origin.

Melanobatrachinae. *n.* A subfamily of *Microhylidae.*

melanoblast. *n.* A cell originating from the neural crest that develops into a *melanocyte.*

melanocyte. *n.* A dendritic clear cell that synthesizes tyrosinase and, within their *melanosomes*, the pigment *melanin.*

melanocyte concentrating factor. See *melanocyte concentrating hormone.*

melanocyte concentrating hormone. (MCH) A neuropeptide hormone that causes fish to lighten, but induces dispersion of melanin and darkening in some tested amphibians and reptiles. The effect is 1/600 as potent as MSH, and the role of MCH in the control of skin color is not clear. MCH also has antagonistic actions in the brain affecting feeding behaviors aggression, arousal, and reproductive function in mammals and fishes. (Syn. *MCH, melanin concentrating hormone, melanocyte concentrating factor*)

melanocyte stimulating hormone. (MSH, α-MSH) A neuropeptide hormone formed in and secreted from the intermediate lobe of the pituitary of amphibians and reptiles (and also the anterior lobe of mammals). The hormone acts on melanocytes to stimulate the dispersion of melanin granules within these cells and thereby cause darkening of the skin. Physical stress elevates plasma levels of MSH, and in reptiles this hormone exhibits trophic properties that can stimulate fetal growth, protein synthesis, wound healing, liver regeneration, and neural regeneration. The earlier term for this melanotropic hormone was *intermedin.* (Syn. *α-stimulating hormone, MSH, intermedin, melanophore expanding hormone, melanophore stimulating hormone, melanotropin*)

melanomacrophage. *n.* See *melanophage.*

melanophage. *n.* Macrophage cell that phagocytizes the dermal melanophores in the larval skin of urodeles during metamorphosis. These cells are also numerous in the liver of amhibians and reptiles, and are extremely numerous in some species. These cells function as macrophages and free radical scavengers. The brown pigment seen in melanophages consists of both iron and melanin granules. In many diseases, and especially chronic illnesses, they will increase in both size and number. (Syn. *melanomacrophage*)

melanophore. *n.* A *chromatophore* containing *melanin*, which is generally dark brown to black in color. See Figs. 33, 34.

melanophore gap. A spot of white on the abdomen and lower flanks of plethodontid salamanders, contrasting with the darker areas of same regions.

melanophore index. (MI) A semiquantitative method of indicating the degree of dispersion of melanin granules in a melanophore, based on an arbitrary scale of pigment morphology or darkening. Originally developed by L. Hogben & D. Slome (*Proc. Roy. Soc. Lond.* B 108:10–52, 1931).

melanophore stimulating hormone. See *melanocyte stimulating hormone.*

melanosome. *n.* Any of the granules within the melanocytes that contain tyrosinase and synthesize *melanin.*

melanotropin. *n.* See *melanocyte stimulating hormone.*

melatonin. *n.* A predominant hormone associated with the *pineal* organ, but also found in the brain and eyes of ectothermic vertebrates. Melatonin synthesis is largely under photic control, and levels fluctuate rhythmically with the light cycle. This substance influences ectotherm thermoregulation, voluntary exposure to light intensities, and also contracts melanophores thereby causing lightening of the skin. Melatonin inhibits maturation of gonads in ranid frogs and is also associated with "blanching" of tadpoles and larval salamanders held in complete darkness.

melting point. The lowest temperature at which a solid will begin to liquefy.

membrana basilaris. See *basement membrane.*

membrana nictitans. See *nictitating membrane.*

membrana prima. See *basement membrane.*

membrana propria. See *basement membrane.*

membrana terminans. See *basement membrane.*

membrane. *n.* A thin layer of tissue that covers a surface, lines a cavity, or divides a space or organ.

membrane-coating granules. See *lamellar bodies.*

membrane filament. A filamentous structure that runs longitudinally along the free edge of the *undulating membrane* in the spermatozoon of amphibians.

membrane potential. A transmembrane electric potential, characteristic of all cells. The electric potential is measured from within a cell relative to the potential of the extracellular fluid, which by convention is considered to be zero.

membrane transport proteins. *Integral proteins* that transport specific classes of molecules across membranes.

membranous body. See *horizontal septum*.

membranous labyrinth. A series of fluid-filled sacs and canals forming the inner ear within the otic region on either side of the head in vertebrates. These house the organs of equilibrium and the receptors for vibrational or sound waves.

Mendelian genetics or **inheritance.** Reference to the inheritance of genes according to the laws governing the transmission of chromosomes to successive generations. Named in honor of G. Mendel, the Austrian monk who studied inheritance of traits in garden peas during the period 1856–1866. (Syn. *chromosomal inheritance, Mendelism*)

Mendelian population. An interbreeding group of organisms that share a common gene pool.

Mendelism. *n.* See *Mendelian genetics*.

Mendel's laws. Principles relating to the segregation and independent assortment of chromosomal genes, first elucidated by the Austrian monk G. Mendel.

meningoencephalitis. *n.* Simultaneous inflammation of the membranes that cover the brain and the spinal cord.

meninx. *n.* (*pl.* **meninges**) Membranous envelope surrounding the central nervous system.

meniscus. *n.* 1. A crescent-shaped fibro-cartilage situated between two articulating surfaces. 2. The curved upper surface of a column of liquid, caused by capillarity. The orientation of curvature depends on the nature of the fluid and the wall of the container.

meniscus pterygoideus. A *meniscus* situated between the *pterygoid* and the *basipterygoid process* in various squamates and turtles.

mental. *n., adj.* 1. The single median scale situated on the front edge of the lower jaw in snakes and lizards, bordered on both sides by the first *lower labials* and directly in front of the *chinshield* of lizards. (Syn. *lower rostral, median labial, medial lower labial, mentum, symphyseal, symphysal, symphysial*) See Figs. 12, 15. 2. Sometimes used in reference to the *symphysial*, a bone of the lower jaw. 3. Of or pertaining to the chin or chin region; lower jaw.

mental cartilage. See *mentomeckelian cartilage*.

mental gland. 1. A gland beneath the skin of the intermandibular area in certain male salamanders, usually distinguished by slight bulging and different coloration of the skin. The secretions of this gland are sexually stimulating to females. (Syn. *chin gland, chin pad, intermandibular gland, mental hedonic gland cluster*) 2. Paired epidermal invaginations situated in the throat region of more than 20 genera of turtles. Secretions from glandular tissue associated with these structures evidently function in intraspecific communication related to intermale combat or mating behaviors.

mental groove. The median furrow or depression on the midline between the *chin shield* scales in the lower jaw of snakes. (Syn. *submental groove, symphysial groove*) See Fig. 12.

mental hedonic gland cluster. See *mental gland*.

mentomeckelian bone. A term used by some herpetologists in reference to the *symphysial bone*.

mentomeckelian cartilage. A paired cartilage at the anterior end of the Meckelian cartilages in the lower jaw of amphibian larvae, differentiating to form the support for the horny mandibles. The two cartilages may ossify separately in transformed individuals, forming the *symphysial* bone (also called the *mentomeckelian bone*), or they may fuse with the *dentary* bone. (Syn. *infra-rostral cartilage, mental cartilage*)

mentum. *n.* See *mental*.

mercy call or **cry.** See *fright cry*.

meridian. *n.* A north-south reference line passing through the geographical poles of Earth. See *longitude*.

Meridianura. *n.* A newly designated taxon of anurans, sister with *Hemiphractidae* within the *Nobleobatrachia*, and containing *Brachycephalidae* and *Cladophrynia*. See D.R. Frost et al., *Bull. Amer. Mus. Nat. Hist.* 297:1–370, 2006.

meridional. *adj.* Southern. Located in, or pertaining to, the south.

meristic. *adj.* Reference to number of parts, or discrete, discontinuously variable characters (e.g. scales of a snake). Cf. *ameristic*.

Merkel cell. An epidermal cell type of amphibians, generally found in the basal layer joining neighboring epithelial cells by desmosomes. Merkel cells are associated with nerve endings and together constitute a Merkel cell-axon complex. Similar cells or complexes occur at various locations in skin of the three Orders of extant amphibians, and are also present in most groups of vertebrates including mammals. Merkel cell complexes in amphibians are considered to be rapidly adapting mechanoreceptors and possibly have neuroendocrine functions as well.

mero-. Prefix meaning "incomplete" or "part."

meroblastic. *adj.* Reference to a type of early embryonic egg cleavage in which there is only partial or incomplete division as a consequence of large amount of yolk present, as in reptilian and avian eggs. Cf. *holoblastic, discoidal cleavage*.

merological methods. Reference to study of the component parts of a system in order to derive a concept of the operation of the whole. Cf. *hological methods*.

Mertensian mimicry. A type of mimicry in which one species is mildly poisonous and thus serves as a model for a fatally poisonous species. Cf. *Batesian mimicry*.

mesa. *n.* A broad, flat-topped hill bounded by cliffs and capped with a resistant rock layer.

mesangium. *n.* Cells and matrix between the glomerular capillaries.

mesencephalon. *n.* The central major division of the brain, which includes paired dorsal swellings of the *tectum*. (Syn. *midbrain*)

mesenchyme. *n.* Embryonic connective tissue composed of branched, loosely organized cells, often with the capacity to migrate.

mesentery. *n.* A membranous sheet that attaches various organs to the body wall.

mesial spine. Used by L.M. Klauber (*Rattlesnakes*, p. 694, 1956) in reference to spines in the crotch of the *hemipenis* of Rattlesnakes. Cf. *shoulder spine*.

mesic. *adj.* Reference to a habitat or a condition that is moderately moist. The term is used to describe organisms that occupy a moist habitat.

meso-. Prefix meaning "intermediate" or "middle."

Mesobatrachia. *n.* A clade of anurans that includes the extant *Pipidae, Rhinophrynidae, Megophryidae, Pelodytidae,* and *Pelobatidae*.

mesocavernous. *adj.* Reference to cracks and voids in shallow subterranean habitat often located between bedrock and overlying soil on volcanic islands.

mesoderm. *n.* The layer of embryonic cells situated between ectoderm and endoderm and giving rise to muscle, bone, connective tissue, blood, linings of all body cavities, the mesenteries, and epithelia of various organs.

mesoglyph. *n.* Reference to a fang condition in some snakes in which the fang lies near the center of a comparatively short maxillary bone, with smaller teeth anteriorly, and the rear third of the bone is edentulous.

mesokinesis. *n.* (*adj.* **-etic**) A type of *kinesis* in which the point of movement or rotation is between the *frontal* and *parietal* bones of the *dermatocranium*. See also *metakinesis, prokinesis,* and Fig. 18.

mesolecithal egg. An egg with a moderate amount of granular yolk throughout, but with a heavy concentration of it in one hemisphere. This is typical of amphibian eggs. Cf. *telolecithal egg.*

mesomegadont. *adj.* Reference to a condition of dentition in which the maxillary bone bears anterior teeth that decrease gradually in size, followed by a middle region of distinctly enlarged, aglyphous teeth, followed posteriorly either by a diastema and then a series of more or less uniform smaller teeth or by a continuing series of decreasing teeth with no diastema. Term introduced by H.M. Smith (*Turtox News,* 30:214–218, 1952).

mesomere. *n.* The small middle division of the embryonic lateral mesoderm.

mesomorphic. *adj.* Reference to intermediate characters or traits.

mesonephric duct. The collecting duct of the mesonephric kidney, which functions in most adult male amphibians as a urogenital duct and in male amniotes as a sperm duct. (Syn. *ductus deferens; Wolffian duct*)

mesonephros. *n.* The middle of three laterally paired embryonic kidneys of vertebrates, and the functional kidney of fetal amniotes as well as the adult kidney of fishes and amphibians. (Syn. *opisthonephros*) Cf. *archinephros, holonephros, metanephros, pronephros.*

mesopelagic. *adj.* Reference to the midwater depth zone of oceans (usually between 200 and 1000 m).

mesophilic. *adj.* Of or pertaining to organisms that thrive under intermediate or moderate environmental conditions, such as moderate temperature or moisture. The term is sometimes restricted to conditions of moderate moisture levels.

mesoplastral. *adj.* Situated in the middle of the plastron of turtles.

mesoplastron. *n.* A dermal ossification in the bony *plastron* of certain turtles, situated usually as a pair of bones between the *hyoplastron* and *hypoplastron* bones of the plastron. These elements are characteristic of primitive turtles, but are lost in most modern species except *Sternotherus* and the Pelomedusidae.

mesopodial. *n.* Any of a series of small bones lying between the *epipodials* and *metapodials* in the limb of a tetrapod, comprising the *carpus* of the forelimb and the *tarsus* of the hindlimb.

mesopterygoid. *n.* 1. A separate bone in *Anguis*, derived from the large, flat, anterior part of the pterygoid as seen in most reptiles. Used by W.K. Parker (*Trans. Linn. Soc. London,* 2:171–214, 1882).

mesoptychial. *n.* Reference to any of the scales situated on the external surfaces of the *gular fold* of lizards.

mesoptychium, mesoptychis. *n.* The middle and posterior throat region in lizards, delimited by the pregular and gular folds.

mesorchium. *n.* A mesentery that supports the testis.

mesos granules. See *lamellar bodies.*

mesos layer. A stratum of differentiated cells within the lepidosaurian *stratum corneum*, consisting of keratinized cells derived from α-cells and alternately sandwiched between lamellar lipids in many species. The layer separates the outer β-keratin from the α-layer beneath. See Fig. 31.

mesosaurs. *n.* Extinct stem amniotes from the early Permian and the first *sauropsid* to become specialized for an aquatic existence.

mesostapedial. *n.* A cartilage or bone in the middle ear of anurans, forming part of the *columella*, extending from the *stylus* to the *extracolumella* (as used by E.D. Cope, *Bull. U.S. Natl. Mus.* 34:1–525, 1889). This series of cartilage or bony elements may become fused in some species, thereby losing their distinctiveness.

mesosternum. *n.* 1. The bar formed by fusion of xiphisternal rods in some lizards. 2. Synonym for *metasternum* in frogs. See Fig. 3.

mesotarsal joint. Reference to an ankle in which the calcaneum and astragalus fuse, and the line of flexion passes between them and the distal tarsals. Cf. *intratarsal joint.*

mesothelium. *n.* A single-celled lining of body cavities.

mesotic cavity. A term used in reference to the chamber of the middle ear in reptiles (S.B. McDowell Jr & C.M. Bogert, *Bull. Amer. Mus. Nat. Hist.* 105:1–142, 1954).

mesotocin. *n.* See *antidiuretic hormone.*

mesovarium. *n.* A mesentery that supports the ovary.

Mesozoic. *n.* The middle Era of the Phanerozoic Eon, characterized by the development and extinction of dinosaurs. See Table 1.

messenger RNA. (mRNA) A type of RNA molecule that functions during translation to specify the sequence of amino acids in a nascent polypeptide.

meta-. Prefix meaning "after," "change," "between," or "among."

metabolic acidosis. A decrease in body fluid pH at constant P_{CO_2}, usually a result of metabolic production of H^+ and insufficient H^+ excretion or excessive OH^- excretion by the kidney.

metabolic alkalosis. An increase in body fluid pH at constant P_{CO_2}, usually a result of gastrointestinal absorption of OH^- or metabolic production of OH^- or excessive H^+ excretion by the kidney.

metabolic bone disorder. A progressive, debilitating, and potentially lethal disease resulting from insufficient dietary calcium (or exposure to ultraviolet light).

metabolic ecology, metabolic theory of ecology. A new, informal name for a recent body of theory based in mathematical application of generalizations related to energy metabolism as underpinning for understanding large-scale patterns in ecology. Quantitative models based in first principles of physics, chemistry, and biology are used to understand how processes of growth, survival, and reproduction are constrained in individual organisms, and how these in turn constrain processes at the level of populations, communities, and ecosystems. Developed by J.H. Brown and colleagues, the use of energetics as a unifying principle in ecology is controversial. For presentation of theory, see J.H. Brown et al., *Ecology* 85:1771–1789, 2004.

metabolic depression. Reference to a significant depression of the standard rate of metabolism, usually triggered by some environmental condition or stress and regulated internally by biochemical/physiological mechanisms.

metabolic intensity. See *metabolic rate, mass-specific metabolic rate.*

metabolic oxygen conformer. See *oxygen conformer.*

metabolic oxygen regulator. See *oxygen regulator.*

metabolic pathway. A specific sequence of enzymatic reactions that converts one molecule or substance into another.

metabolic rate. The rate of conversion of chemical energy into heat, which is a byproduct of all energy-requiring biochemical reactions necessary to sustain life. It reflects the "intensity" of the sum total of energy utilization in the body. It is commonly measured as heat energy released per unit time, measured either directly in a calorimeter or indirectly as rate of oxygen consumption (which can be converted using appropriate energy equivalents). In formal writing it is more correct to say "rate of metabolism" than to say "metabolic rate." See also *mass-specific metabolic rate, standard metabolic rate.*

metabolic scope. See *aerobic metabolic scope.*

metabolic theory of ecology. See *metabolic ecology.*

metabolic water. Water that results from oxidative cellular metabolism.

metabolism. *n.* The sum total of all chemical and physical processes that are used for anabolic synthesis of macromolecules in cellular assembly and catabolic degradation of macromolecules for energy production. See *metabolic rate.*

metabolizable energy. (ME) That part of consumed energy that is made available to cellular metabolism and is not excreted either in urine or in feces. See also *assimilation.*

metacarpal. *n.* Any bone that connects fingers (*phalanges*) with bones of the wrist (*carpals*).

metacarpus. *n.* The collection of *metacarpal* bones that comprise part of the forefoot between the wrist and fingers.

metachromatic dye. A dye that stains tissues two or more colors.

metachrosis. *n.* Change of color in animals. (This term has been used with variants of meaning by herpetologists, with reference to abnormal coloration, whether it changes or not, and the ability to change color.)

metacommunity. *n.* A set of local communities that are linked by dispersal of multiple and potentially interacting species. Terminology from D.S. Wilson, 1992; Review of concepts in M.A. Leibold et al. (*Ecology Letters* 7:601–613, 2004). See also M. Holyoak, M.A. Leibold & R.D. Holt, eds., *Metacommunities. Spatial Dynamics and Ecological Communities*, 2005.

metadata. *n.* Data on data and databases.

metagamic sex determination. The condition in which sex of offspring is not genetically fixed but is largely the result of external environmental factors.

metakinesis. *n.* (*adj.* **-etic**) A type of *cranial kinesis* in which the point of movement or rotation is between the *parietal* and *supraoccipital* bones, or between *dermatocranium* and *neurocranium*. See also *mesokinesis, prokinesis,* and Fig. 18.

metalimnion. *n.* A vertical zone of a lake consisting of a midlevel layer of rapidly decreasing temperature. (Syn. *thermocline*) Cf. *epilimnion, hypolimnion.*

metamere. *n.* One of the serially repeated structural units along the body axis. See *metamerism.*

metamerism. *n.* (*adj.* **-ic**) Serial repetition of more or less similar body parts or units along an anteroposterior axis (e.g., ribs, costal grooves). Often considered synonymous with *segmentation* in literature, but the latter term is sometimes restricted to embryonic conditions.

metamorph. *n.* An immature anuran that has recently transformed from the larval stage of an *exotrophic* tadpole.

metamorphic climax. Reference to the period when metamorphic changes are most rapid and profound. (*Gosner stages* 42–46).

metamorphic rock. Rock formed by restructuring of preexisting rock influenced by high temperature and pressure. Cf. *igneous rock, sedimentary rock*.

metamorphosis. *n.* (*adj.* **metamorphic**) The marked structural transformation of an animal from one stage of its life history to another. Strictly, a change in structure or shape, but used most often in reference to the transformation from larval to adult life stages in amphibians. See also *secondary metamorphosis.* (Syn. *transformation*) Cf. *minor metamorphosis*.

metanephric duct. See *ureter.*

metanephros. *n.* The posterior of three laterally paired embryonic kidneys of vertebrates, becoming the adult kidney of amniotes. Cf. *archinephros, holonephros, mesonephros, opisthonephros, pronephros*.

metaneural. *n.* See *suprapygal.*

metaphysis. *n.* The region that is transitional between the *diaphysis* (shaft) and *epiphyses* of long bones.

metaplasia. *n.* (*adj.* –**ic**) Changing of one type of tissue into another. Cf. *hypertrophy*.

metapodial. *n.* A collective term for any of the bones of the limb lying between the mesopodials and the phalanges, viz. *metacarpals* and *metatarsals*.

metapopulation. *n.* A group of several local populations that are linked by immigration and emigration; a group of conspecific populations coexisting in time but not in space. The original meaning was used in a stricter sense as a collection of populations that experience local extinction and recolonization (D. Levins, *Bull. Entomol. Soc. Amer.* 15:237–240, 1969).

metapterygoid. *n.* An endochondral bone of the *palatoquadrate*, present in fishes, absent in amphibians, but possibly represented in fossil reptiles by a small bone lying in the same position as the *meniscus pterygoideus*. Used by E.S. Goodrich (*Studies in the Structure and Development of Vertebrates*, p. 430, 1930).

metarteriole. *n.* An arterial capillary

metastasis. *n.* Spreading of malignant neoplastic cells from the original site to another part of the body.

metasternum. *n.* The most anterior part of the sternal apparatus lying posterior to the *epicoracoids* and *coracoids*, when parts of the sternum are distinguished. (Syn. *hyposternum, manubrium, mesosternum, pars ossea, sternum, xiphisternum*) See Fig. 3.

metatarsal. *n.* Any of the *metapodials* in the hind foot.

metatarsal tubercle. An elevated epidermal thickening on the posterior portion of the bottom of the foot in anurans.

metatarsus. *n.* The part of the hind limb occupied by the metatarsals.

metatype. *n.* 1. In taxonomy, a specimen from the type locality determined by the original author subsequent to the original description of the species. 2. A specimen subsequently determined by the author after comparison with the type.

metautostyly. *n.* See *autostyly.*

metazoan. *n.* Reference to an organism that possesses more than a single cell. A multicellular animal.

metencephalon. *n.* The anterior segment of the *rhombencephalon*, which includes the *cerebellum* and part of the *medulla oblongata*.

meteorology. *n.* The study of weather, emphasis being on atmospheric conditions and their short- or long-term changes.

methemoglobin. *n.* Hemoglobin in which the iron atom of the heme moiety has been oxidized (Fe^{2+} to Fe^{3+}). It is thereby nonfunctional and does not bind oxygen.

methyl ketones. Compound molecules present in the epidermis of some snakes and lizards, known to act as *pheromones* in some squamate species.

metischial process. A projection arising from the ischium of New Zeland Geckos. Described by N.G. Stephenson (*J. Linn. Soc. London, Zool.* 44:278–299, 1960). (Syn. *spina ischii*)

metonym. *n.* A taxonomic synonym (name given to a taxon that already has a valid name).

meter, metre. (m) *n.* The standard *SI* unit of lenth, equal to 39.37 inches.

-metric. Suffix meaning "measure."

metric scaling. A technique for ordination. (Syn. *principal coordinate analysis*)

metric system. A decimal system of weights and measures in which the gram, meter, and liter are basic units of weight, length, and capacity. The *SI units* system is an extension of the metric system. See Tables 2 and 3.

metric traits. See *quantitative character.*

Mexican Beaded Lizard. Vernacular name for the Middle American helodermatid species *Heloderma horridum*. See also *Beaded Lizard.*

Mexican Garter Snake. Vernacular name for the Middle American colubrid species *Thamnophis eques.*

Mexican Leaf Frog. Vernacular name for the hylid species *Pachymedusa dacnicolor.*

Mexican Treefrog. Vernacular name for Middle American hylid species belonging to the genus *Smilisca*, especially *Smilisca baudinii.*

Mexican West-coast Rattlesnake. Vernacular name for the American pitviper *Crotalus basiliscus.*

MHC. See *major histocompatibility complex.*

MHD. See *minimum hemorrhagic dose.*

MI. See *melanophore index*.

micelle. *n.* A microscopic particle made from an aggregation of amphipathic molecules in solution. Common usage of this word is usually related to digestion and absorption of lipids.

Michaelis constant. (K_m) The substrate concentration at which the reaction rate of an enzyme is half maximal.

Micrixalidae. *n.* A monotypic taxon of anurans, sister to the new taxon *Ametrobatrachia* within the new taxon *Telmatobatrachia*, and containing the Indian genus *Micrixalus*. See D.R. Frost et al., *Bull. Amer. Mus. Nat. Hist.* 297:1–370, 2006.

micro- (μ) 1. Prefix meaning "very small" or "on a small scale." 2. Used to denote unit x 10^{-6}.

microarray. *n.* See *DNA chip*.

microcirculation. *n.* Reference to blood circulation through *capillaries*, usually including the capillary system and the associated *arterioles* and *venules*.

microclimate. *n.* The local climatic conditions (radiation, wind, humidity, etc.) immediately around an animal or at any of its places of potential habitation (near the ground). (Syn. *bioclimate*)

microcomplement fixation. A technique used to estimate immunological distances.

microcosm. *n.* A small community that is considered representative of a much larger ecological system.

microdermatoglyphics. *n.* See *epidermal microsculpturing*.

microelectrode. *n.* A tiny hollow glass needle that can be inserted into tissues or cells for recording of electrophysiological phenomena.

microenvironment. *n.* The local conditions that immediately surround an organism. (Syn. *microhabitat*)

microevolution. *n.* A pattern of evolutionary change that involves changes in gene frequencies within a population, involving relatively few generations over a relatively short period of time. Generally used to describe genetic changes within the lifetimes of populations at or below the species level. Cf. *macroevolution*.

microfauna. *n.* A very small form of animal life, or the smaller forms within a given community.

Microfil®. *n.* A compound or series of compounds that opacify microvascular and other spaces of nonsurviving animals and postmortem tissues under physiological injection pressure. The compound series is available in multiple radiopaque colors, as well as clear. Commonly used in *cast-corrosion* techniques and visualization processes for microcirculation and microanatomy.

microfilaments. *n.* 1. Generally, filaments within cytoplasm or cellular organelles that have a diameter less than 10 nm (e.g., actin). 2. Contractile cytoskeletal actin filaments of 6-nm diameter.

microfilaria. *n.* A first-stage juvenile of any filariid nematode that is viviparous and usually found in blood or tissue fluids of the definitive host.

microflora. *n.* 1. The flora of a small geographic area. 2. In physiology, the microorganisms of the gastrointestinal tract.

Micro Frog. Vernacular name for the the African ranid frog, *Microbatrachella capensis*.

microgeographic race. Reference to local races restricted to very limited areas, usually having considerable migration distances in between. Such races are considered to be components of a subspecies.

microhabitat. *n.* The immediatae surroundings in which an organism lives. (Syn. *microenvironment*)

Microhylidae. *n.* A clade (family) of anurans characterized by two or three palatal folds in adults, and a suite of larval characters including absence of cornified denticles, imperforate nares, and a glottis fully exposed on the floor of the mouth. Most microhylids are relatively small and have extremely variable life histories. There are 69 genera and approximately 350 species distributed in the eastern United States throughout Central and South America, sub-Saharan Africa, southern and eastern Asia, and northern Australia. Known as *Narrow-mouthed Frogs*, there are nine subfamilies: *Scaphiophryninae, Cophylinae, Asterophryinae, Genyophryninae, Brevicipitinae, Phrynomerinae, Microhylinae, Discophinae,* and *Melanobatrachinae*. This family was recently partly revised as a sister taxon to Afrobatrachia, and composed of 8 subfamilies (*Asterophryinae, Cophylinae, Dyscophinae, Gastrophryninae, Melanobatrachinae, Microhylinae, Scaphiophryninae*) in addition to several nominal genera not assigned to subfamily (D.R. Frost et al., *Bull. Amer. Mus. Nat. Hist.* 297:1–370, 2006).

Microhylinae. *n.* A subfamily of *Microhylidae*.

microlecithal. *adj.* Reference to an egg that contains very little yolk for nourishment of the developing embryo. This type of egg usually occurs only in animals that have some mechanism for exchange of nutrients between maternal and embryonic fluids. (Syn. *meiolecithal, oligolecithal*)

micrometeorology. *n.* Study of very small and local scale aspects of microclimate such as turbulence, diffusion, and heat transfer.

micrometer. (μm) *n.* A derived metric unit of length, equal to 10^{-6} meter. This word replaces the use of *micron* (μ), a unit of equivalent length found in earlier literature.

micromorphology. *n.* Microscopic surface morphology, usually studied with aid of the scanning electron microscope. Cf. *macromorphology*.

micron. *n.* See *micrometer*.

Micronesia. *n.* Reference to numerous smaller oceanic islands within the warmer part of the western Pacific Ocean, recognized as a geographical and ethnological area sometimes considered to be a distinct biogeographical unit, distinct from *Polynesia*.

micronutrients. *n.* Trace elements required in minute quantities for optimal growth.

microornamentation. *n.* 1. Reference to the fine structure on the surfaces of the hemipenis of snakes. 2. See *epidermal microsculpturing.*

microphagy. *n.* (*adj.* **-ous.**) Reference to feeding on relatively small prey. Cf. *macrophagy.*

microphthalmic. *adj.* Possessing smaller than normal eyes.

microplicae. *n.* See microridges.

microreticulate. *adj.* Having a delicate network of anastomosing ridges. Reference to a pattern of *epidermal microsculpturing*. See R.M. Price *J. Herpetol.* 16:294–306, 1982.

microridges. *n.* Reference to extensive surface folds on epithelial cells in a variety of vertebrate species, surmised to function in facilitating the even spreading and gross retention of a mucous coat. (Syn. *microvillar ridges, microplicae, cytoplasmic folds*)

microsatellite DNA. *n.* Tandem, repeated sequences of noncoding DNA referred to as *microsatellite, minisatellite,* and *satellite DNA*, depending on the length of the repeat (2 to more than 2000 base pairs). The repeat units tend to be highly polymorphic, and several loci taken together produce a genetic fingerprint for a species that is consistent and unchanging from one individual to the next. Most microsatellite loci have many alleles and can be useful in analyzing population or phylogenetic relatedness. Microsatellites are a powerful class of genetic markers used in a variety of molecular studies such as determining fine scale population structure, parentage, and individual identification.

microscopic. *adj.* Reference to objects that cannot be observed without the use of a microscope.

microsculpturing. *n.* See *epidermal microsculpturing.*

microsite. *n.* 1. A specific and precise location. 2. In ecology, a spatial term that refers to a site that is capable of holding a single individual. 3. An assemblage of microscopic fossils.

microspectrophotometer. *n.* An optical system that combines a spectrophotometer with a microscope.

microspecies. *n.* 1. A *microgeographic race* or other infrasubspecific group. 2. A component of a *species aggregate.*

microspheres. *n.* Microscopic plastic spheres that are injected usually into vascular blood streams where they are transported downstream in blood and subsequently lodge in capillaries. These are available in various sizes, which can be matched to the vascular dimensions of the animal in question. Their presence in tissues allows quantification of bloodflow patterns according to various methods, depending on whether the microspheres are fluorescent or radiolabeled. Microspheres can also be applied to other uses, such as following movement or location of items in the gut.

microsurgery. *n.* Surgery that is carried out while viewing the object through a microscope.

microteiid. *n.* A member of the lizard family *Gymnophthalmidae*. Cf. *macroteiid.*

microtome. *n.* A hand or motor-driven machine used for cutting thin slices of paraffin-embedded tissue in preparation for staining and viewing by microscopy.

microtopography. *n.* Topographic features considered on a small scale.

microtubule. *n.* A cylindrical cytoplasmic structure made of polymerized tubulin and found in many cells as a constituent of motile structures such as mitotic spindles, cilia, and flagella.

microvillar ridges. *n.* See *microridges.*

microvillus. (*pl.* **-i**) *n.* Minute cylindrical projection from a single cell surface that greatly increase its surface area, usually on absorptive epithelia. Cf. *villus.*

microzoogeography. *n.* Study of the distribution of geographical races, species or genera over relatively limited areas.

midbrain. *n.* See *mesencephalon.*

middle ear. Collectively, the middle ear cavity together with cartilages and/or bony ossicles involved in transmission of vibrational stimuli from airborne waves to the inner ear.

middle ear cavity. See *tympanum.*

middle height. A measurement of turtle shell dimension, taken by measuring the distance from a flat surface on which the shell rests to the middle of the third vertebral plate. Used by J. Van Denburgh (*Proc. Cal. Acad. Sci.* 1:1–6, 1907). See Fig. 8.

middorsal. *adj.* Reference to the portion of the dorsal surface of a body or structure that lies along the midline.

midgut. *n.* The region of the alimentary canal where chemical digestion of food takes place.

midrib. *n.* The cartilaginous central portion of the vocal cord in certain species of anurans, e.g. some bufonids. First used by W.H. McAlister (*Copeia* 1961:86–95, 1961).

midsagittal fold. A medial fold of skin in the gular region of some lizards.

midsagittal plane. A median parallel plane passing dorsoventrally through the long central axis of the body.

midventral. *adj.* Reference to the portion of the ventral surface of a body or structure that lies along the midline.

Midwife Toad. Vernacular name for species of European anurans belonging to the discoglossid genus *Alytes*, and also African anurans belonging to the hyperoliid genus *Alexteroon.*

Mie scattering. In atmospheric science, reference to large-particle scattering of radiation due to dust, smoke, and other aerosols in the atmosphere. Cf. *Rayleigh scattering.*

migrant selection. Selection based on individuals of differing genotype having variable migratory abilities (and therefore founding new colonies with differential success).

migration. *n.* (*adj.* **migratory**) The directed movement of individuals, hence gene flow, between different regions or populations. Reference may also be to a behavior pattern that involves a regular or recurring movement from one location to another by a group of animals of a given species in response to changes in weather, season, food availability, or other environmental factors that stimulate the behavior.

migration coefficient. The proportion of a gene pool represented by migrant genes per generation.

migratory range. The maximum range over which non-reproducing organisms can move, either actively or passively.

mile. (mi) *n.* A secondary *fps* unit of length equal to 1.609344×10^3 meters.

milieu. *n.* The characteristic surroundings or environment of an organism or, internally, a cell, tissue, organ, or other structure.

milk. *v.* Reference to extracing venom from a venomous snake or lizard by manipulating means.

Milksnakes, Milk Snakes. *n.* Vernacular name for some New World colubrid Kingsnakes, especially *Lampropeltis triangulum*. There is controversy whether populations of this widely distributed snake represent a single species.

Miller-Weber thermometer. See *cloacal quick-reading thermometer*.

milli-. A word element meaning "thousand" or "thousandth," used to denote unit $\times 10^{-3}$.

milliliter. (ml) A metric measure equal to liter $\times 10^{-3}$.

millimeter. (mm) A metric measure equal to meter $\times 10^{-3}$.

millimicron. (mμ) See *nanometer*.

milt. *n.* The seminal fluids and sperm that are released by a male anuran during clasping, usually during near-simultaneous expulsion of eggs by the female.

mimesis. *n.* (*adj.* **mimetic**) See *mimicry*.

mimetic. *adj.* Reference to simulation of a function, process, etc.

mimetic polymorphism. Mimicry in which the mimic is polymorphic, and the different morphs resemble models belonging to more than one species.

mimic. *n., v.* 1. An organism that closely resembles another, usually unrelated, organism. The organism which it resembles is known as the *model* and is unpalatable or in some way adapted to deter predators that also prey on the mimic. 2. To imitate.

Mimic Chameleon. See *False Chameleon*.

mimicry. *n.* The phenomenon where organisms resemble one another in context of gaining an advantage in predator avoidance. See *Batesian, Mertensian, Müllerian, and Peckhammian mimicry*. (Syn. *mimesis*)

mineralcorticoids. *n.* Steroid hormones that are synthesized and secreted by the adrenal cortex and function to regulate the volume and ionic composition of body fluids.

mineral cycles. Reference to cycling of minerals within the biosphere. The cycles of mineral elements of biological importance are referred to as elemental cycles, mineral cycles or nutrient cycles. Energy flows through the biosphere, but nutrients and other chemical elements of biological importance cycle within ecological systems. The scientific discipline that deals specifically with these phenomena is called *biogeochemistry*.

mineralization. *n.* The process by which a mineral constituent is formed from other softer tissue or organic material (e.g., calcium phosphate in bone). The deposition of metallic ions in the organic matrix of connective tissue.

minimum activity area. See *minimum home range*.

minimum hemolytic dose. See *minimum indicating dose*.

minimum hemorrhagic dose. (MHD) The minimal quantity of venom that will produce a 10 mm diameter hemorrhagic lesion within 24 h after intracutaneous injection into a rabbit. Used by H. Kondo et al. (*Jap. J. Med. Sci. Biol.* 13:43–51, 1960).

minimum home range. The area delimited by plotting all points of capture of an individual organism and then connecting the outermost points with straight lines. (Syn. *minimum activity area*)

minimum indicating dose. A generic term in reference to the minimum amount of venom that is necessary to produce any specified reaction, determined from serial dilutions produced by halving the amount of venom mixed with a fixed amount of carrier test substance. (Syn. *MSD, minimum skin reaction; MHD, minimum hemolytic dose*, etc.)

minimum, law of. A generalization that the minimal requirement with respect to any specified factor is the ultimate determinate in controlling the distribution or survival of a species. It is based in the idea that growth of an organism is limited by the essential element that is present in the smallest proportion to the requirement of the organism. (Syn. *Liebig's law*)

minimum lethal dose. (MLD) See *lethal dose*.

minimus, minimum. *n.* An earlier term for the outermost and usually smallest digit on a foot.

minisatellite DNA. See *microsatellite DNA*.

Mink Frog. Vernacular name for the American ranid species *Lithobates* (formerly *Rana*) *septentrionalis*. For recent taxonomic revision see D.R. Frost et al., *Bull. Amer. Mus. Nat. Hist.* 297:1–370, 2006.

minor metamorphosis. Reference to metamorphic changes in morphology and ecology involving minimal variation between larval and postlarval stages, e.g. as occurs in some salamanders in contrast to anurans.

MLD. See *minimum lethal dose, lethal dose*.

Miocene. *n.* A geologic Epoch within the Tertiary Period. See Table 1.

mire. *n.* A collective term for various *bogs* and *fens*. Used to describe a waterlogged ecosystem in which vegetation is rooted in wet peat.

miscible. *adj.* Descriptive of liquids that can be mixed together in any ratio without separation.

misogamy. *n.* See *reproductive isolation*.

Mississippian. *n.* The fifth Period of the Paleozoic Era beginning about 350 million years ago and ending about 320 million years ago. This division of the geologic table is employed in North America as the lower and older subdivision of the Carboniferous, as used on other continents.

Mississippi Green Water Snake. Vernacular name for the American natricine species *Nerodia cyclopion*.

mitigation. *n.* Compensation required for human alteration of natural resources or habitat pivotal to the survival or well-being of listed species.

mitochondrion. (*pl.* –ia) *n.* A self-reproducing, membrane-enclosed organelle where ATP is produced during aerobic metabolism. The density of mitochondria usually is correlated with the metabolic intensity of a tissue.

Mitochondrial DNA. (mtDNA) A circular duplex of DNA, having a genetic code slightly different from the "universal" genetic code and residing within mitochondria, encoding a small subset of mitochondrial functions. Because only the egg cell contributes significant numbers of mitochondria to the zygote, mtDNA is maternally inherited. Mitochondrial genomes experience higher mutation rates than those in nuclear genomes and thus have relatively more power to reveal evolutionary differences between closely related species.

mitosis. *n.* A process of cell division achieved by simple separation of the contents to create two genetically identical daughter cells, each having a full set of chromosomes. Cf. *meiosis*.

mitotic index. The proportion of cells within a tissue that are undergoing mitosis at any given time.

mixed layer. A layer of water that is well mixed by wave action or thermocline convection.

mixed nerve. A nerve that contains axons of both sensory and motor neurons.

mixing ratio. In atmospheric science, this term refers to the mass of water vapor per unit mass of dry air.

mixis. *n.* Biparental sexual reproduction.

mks system. The system of scientific units based on the *meter-kilogram-second* as in the *SI* units.

ml. See *milliliter*.

MLD. See *lethal dose*.

mm. See *millimeter*.

mM. Millimolar concentration. See *molarity*.

mobility. *n.* The ability or tendency for an organism or population to change its location or distribution with time.

Moccasin. *n.* Vernacular name for North American pitvipers belonging to the genus *Agkistrodon*, particularly Cottonmouths (*Lowland* or *Water Moccasin*) and Copperheads (*Highland Moccasin*). See *Water Moccasin*.

Mock Viper. Vernacular name for Asian colubrine snakes belonging to the genus *Psammodynastes*. As the name implies, these snakes exhibit some features that resemble vipers that are found in the same regions.

modality. *n.* 1. The state or quality of a stimulus or sensation (e.g., sound, light, etc.). 2. In statistics, the state of having one (*unimodal*), two (*bimodal*) or more (*polymodal*) modes or peaks in a frequency distribution.

mode. *n.* The value that occurs most frequently in a given set or distribution of data. See also *modality*. Cf. *mode*.

model. *n.* 1. A physical or mathematical representation of an object (animal), or an interacting set of processes that determine properties of the object or its relationship to other entities. A conceptual (nonquantitative) or mathematical simulation of reality for purposes of describing, analyzing, or understanding nature. Modeling involves clearly stating a set of assumptions, then using rules of logic to deduce consequences. 2. In behavioral ecology, an organism that is imitated by a mimic.

mode of reproduction. See *reproductive mode*.

modern. *adj.* Living at the present time, or during recent historic time. (Syn. *extant*)

modern synthesis. See *neo-Darwinism*.

modification. *n.* Any noninherited, environmentally induced change in a phenotype.

modifier. *n.* In genetics, reference to a gene that modifies the phenotypic expression of a nonallelic gene.

modulus of elasticity. See *elastic modulus*.

modulatory agent. A substance that either increases or decreases the response of a tissue to a physical or chemical signal.

Mohave Rattlesnake. See *Mojave Rattlesnake*.

moisture index, moisture deficit index. (I_m) The relationship of potential evapotranspiration (*PE*) to total precipitation (*P*), expressed as a percentage: $I_m = [(P - PE)/PE] \bullet 100$. Other indices are also used in agricultural and geographical literature to quantify aridity or to indicate the portion of total precipitation used to support plant growth, but all are not included in this *Dictionary*.

moisture stress. The tension or force by which water is retained in soil.

Mojave Rattlesnake. Vernacular name for the American pitviper *Crotalus scutulatus*.

molality. *n.* The number of moles of a solute in a kilogram of a pure solvent. Cf. *molarity*.

molarity. *n.* The number of moles of a solute in a liter of solution. Cf. *molality*.

molchpest. *n.* A little-understood disease affecting captive salamanders, often in epidemic proportions, characterized by a variety of symptoms including inflammation of skin and the development of extensive abscesses. (Syn. *newt plague*)

mole. (mol) *n.* Avodagro's number (6.023×10^{23}) of molecules of an element or compound, equal to the molecular weight in grams. The standard *SI unit* for amount of a substance.

molecular biology. A branch of modern biology that attempts to understand biological phenomena in terms of molecular mechanisms. Biochemical and physical techniques are often used to investigate problems related to genetics. See *molecular genetics.*

molecular chaperone. See *chaperone.*

molecular clock. A postulation that, when averaged across the entire genome of a species, the rate of nucleotide substitutions in DNA remains constant. Hence, the degree of divergence in nucleotide sequences between two species can be used to estimate their divergence node. Similarly, the degree of divergence in the amino acid sequences of a protein can be used to estimate the length of time that has elapsed since the divergence of two species from a common ancestor, assuming a constant rate of amino acid substitutions for a given family of proteins. (Idea originally proposed by E. Zuckerkandl & L. Pauling, pp. 97–166 in *Evolving Genes and Proteins*, ed. V. Bryson & H.J. Vogel, 1965.) A phylogenetic tree constructed on the assumption of a molecular clock has all terminal taxa equidistant from the root.

molecular genetics. A subdivision of genetics that investigates the structure and function of genes at the molecular level.

molecular markers. Reference to genetic markers (DNA sequences) that identify a species and are used in a variety of molecular studies to examine relatedness in various contexts. See *microsatellite DNA.*

molecular phylogeny. Phylogenetic relations inferred from genetic similarities and differences in the nucleic acid sequences coding for identified proteins.

molecular systematics. The use of nucleic acid and protein data to investigate variation and evolutionary relationships among organisms.

Mole Kingsnake. Vernacular name for the American colubrid snake *Lampropeltis calligaster*, or its southern subspecies. See also *Prairie Kingsnake.*

Mole Salamanders. Collective vernacular name for salamanders belonging to the family Ambystomatidae, especially the genus *Ambystoma*. The name also is applied specifically to the species *Ambystoma talpoideum*.

Mole Skink. Vernacular name for the American scincid lizard *Eumeces egregius*. Note that H.M. Smith (*J. Kansas Herpetol.* 14:15, 2005) recently proposed that *Plestiodon* be a replacement name for the genus *Eumeces* in North America.

Mole Vipers. Collective vernacular name for species of African snakes belonging to the Atractaspididae, also called *Burrowing Adders, Burrowing Asps* or *Stiletto Snakes.*

molluscivore. *n.* An organism that eats mollusks.

molt, moult. *v.* To shed or slough, usually used in reference to the outermost skin or epidermis.

molting. *n.* The process of shedding the skin; *ecdysis*, exfoliation.

moment. *n.* The measure of the tendency of a force to rotate a body, quantified as the product of force times the perpendicular distance from the point at which a force is applied to the point of rotation.

moment arm. A *lever arm.*

moment of inertia. The tendency of an object to resist movement.

momentum. *n.* The capacity of an object moving in a straight line to overcome resistance. The product of mass times velocity.

monandry. *n.* (*adj.* –**ous**) Reference to a female that mates with a single male.

monarchy, monarchistic dominance. Dominance of an intraspecific group of individuals by a single animal that represses interactions of the other members of the group, both male and female.

monimostyly. *n.* A condition of jaw suspension in which the quadrate bone is firmly fixed to the cranium and cannot be moved, characteristic of turtles and crocodilians.

Monitors, Monitor Lizards. A collective common name for numerous species of the *Varanidae*, also known as *Goannas* in Australia and *Leguaans* in South Africa.

Monkey Frog. Vernacular name for neotropical arboreal hylid frogs belonging to the genus *Phyllomedusa.*

mono-. Prefix meaning "one," "single."

monoallochronic ovulation. See *ovulation.*

monoautochronic ovulation. See *ovulation.*

monobasic. *adj.* See *monotypic.*

monocellate. *adj.* Possessing a single eye-like spot or *ocellus*.

monochromatic light. Light of a single wavelength.

monochronic. *adj.* Of single occurrence.

Monocled Cobra. Vernacular name for the Asian species *Naja kaouthia*.

monoclonal antibody. A homogeneous antibody that is produced by a clone of antibody-forming cells and binds to a single type of antigen, used as specific markers in immunoassays.

monocondylar, monocondylous. *adj.* Reference to possessing a single median *occipital condyle* on the skull. Characteristic of reptiles, in contrast to the double condyle of amphibians.

monoculture. *n.* The culture or cultivation of a single species to the exclusion of others.

monodactyl. *adj.* Having only a single digit on a foot, either finger or toe, as in the hind foot of certain skinks.

monogamy. *n.* (*adj.* –**ous**) A reproductive system in which individuals have only a single mate during a breeding season. Cf. *polygamy.*

monogeneric. *adj.* Reference to a taxon having only one genus.

monogenic character. A character that is determined by expression of a single gene.

monograph. *n.* A comprehensive work published on a single subject or group of organisms.

monokinesis. *n.* A type of kinesis in which there is only one line of movement in the skull, as distinct from *akinesis* and *amphikinesis*. Coined by T.H. Frazetta (*J. Morphol.* 111:287–319, 1962).

monomer. *n.* A simple compound that constitutes a repeated unit of a polymer, the latter formed by repetition of a single reaction.

monomorphic population. A population in which a trait or character of interest is invariable due to fixation of one allelic form of the gene that expresses the trait.

mononuclear. *adj.* Possessing a single nucleus.

monophagy. *n.* (*adj.* –ous) Utilizing only one type, or single species, of food. (Syn. *monotrophy*) Cf. *polyphagy*.

monopheny. *n.* (*adj.* –ic) The occurrence of a single phenotype in a population. Cf. *leptopheny*.

monophyletic group. A natural taxon composed of two or more species, including the known or hypothesized ancestral species and all of its descendants sharing a single common ancestor. Members of a monophyletic group are *sister taxa*, and a monophyletic group is also called a *clade*. Cf. *polyphyletic group*.

monophyly. *n.* (*adj.* -etic) A condition where a group of species have all been derived from a single common ancestor. Having one origin. Monophyletic groups are identified by shared, uniquely derived character states. Cf. *paraphyly, polyphyly*.

monospecific. *adj.* 1. Reference to a genus that contains a single species. 2. A synonym for *monovalent* with reference to antivenins.

monospondylous vertebra. A condition of vertebrae in which a vertebral segment is composed of a single centrum. Cf. *dispondylous vertebra*.

monosynaptic. *adj.* Pertaining to, or passing through, a single synapse.

monotaxic. *adj.* Belonging to the same taxonomic group.

monothetic key. A dichotomous key wherein each couplet has single contrasting statements that require a simple "either/or" answer. Cf. *polythetic key*.

monothetic group. In systematics, a group of organisms that are classified together on the criterion they possess a unique set of characters that is both sufficient and necessary for inclusion in the group thus defined.

monothetic taxon. A taxon defined by a unique combination of diagnostic characters that comprise both necessary and sufficient criteria for identification. Cf. *polythetic taxon*.

monotopic. *adj.* Reference to occurrence or origin in a single locality or geographical area. Cf. *polytopic*.

monotrophy. *n.* See *monophagy*.

monotypic. *adj.* A taxonomic group that contains only a single taxon of the next lower level (i.e., a monotypic genus contains but a single species). (Syn. *monobasic*) Cf. *bitypic, polytypic*.

monovalent. *adj.* 1. Capable of combining with only one antigenic specificity or with only one antibody specificity. 1. Containing antibodies for, or antigens of, one strain of a given species. 3. Having a valency of one.

monsoon. *n.* A seasonal summer storm characterized by a high amount of precipitation in a relatively short period.

monsoon forest. A type of deciduous forest of eastern and southeastern Asia, having a monsoon-dependent climate.

montane. *adj.* Reference to mountains, or dwelling in mountains.

Montane Pitvipers. Vernacular name for four species of Middle American pitvipers belonging to the genus *Cerrophidion*.

Montpellier Snake. Vernacular name for the European colubrid species *Malpolon monspessulanus*.

moor. *n.* An open elevated region of wet and acidic peat, characteristically dominated by heathers, sedges, and grasses.

moraine. *n.* A body of till either being carried on a glacier or left behind after a glacier has receded.

morbidity. *n.* 1. A condition of being diseased or morbid. 2. The frequency of a disease per period of time (generally one year).

moribund. *adj.* In a dying state.

Morisita's index of dispersion. An index of dispersion that is independent of the sample mean. The index has a value equal to, greater than, or less than unity for random, contagious, and uniform distributions, respectively.

Morisita's index of similarity. A measure of similarity between two communities, ranging from 0 to about 1 for no similarity and complete similarity, respectively.

-morph. Word element meaning "form" or "shape."

morph. *n.* A genetically determined variant in a population, usually a unique body form or coloration. (Syn. *form*) See *morphotype*.

morphine. *n.* See *alkaloids*.

morphocline. *n.* A graded, geographic series of homologous character states. (Syn. *morphological transformation series*) See *cline*.

morphogen. *n.* Any compound that is produced in a localized region of a developing organism and diffuses to adjacent cells, which are influenced to enter a specific developmental pathway.

morphogenesis. *n.* (*adj.* –tic) Reference to developmental processes that transform a developing cell, tissue, or organism from an immature to more mature state. (Syn. *differentiation*)

morphology. *n.* (*adj.* **morphological**) 1. The study of form and structure. A science dealing with the structures of organisms, including their developmental and evolutionary histories. (See also *functional morphology*) 2. The form of an organism, especially considered as a whole.

morphological color change. Reference to color changes that occur over days, weeks, or longer due to developmental, ontogenetic, or seasonal changes in chromatophore patterns. Such changes may be permanent, seasonal, or reversible. Cf. *physiological color change*.

morphological cross section. A cut through the area of a muscle perpendicular to its longitudinal axis at its thickest part. Cf. *physiological cross section*.

morphological index. The ratio of the hind limb length divided by the tail length, which is the reciprocal of the *allometric index*. Used by J.L. Dolphin & E. Frieden (*J. Biol. Chem.* 217:735–744, 1955).

morphological series. A graded series of fossils that show continuous variation in a given character or suite of characters.

morphological species concept. (morphospecies) See *species*.

morphological transformation series. See *morphocline*.

morphometry. *n.* (*adj.* –**ic**) 1. Reference to the quantitative parameters of structures. A quantitative measurement of a form or structure. 2. The scientific measurement or analysis of the shape or form of organisms. The word *morphometrics* refers to the study of covariances with organismal shape. Historically, morphometrics has been less concerned with the geometric analysis of shape per se and more concerned with how biological processes affect shape. Contemporary morphometrics includes a synthesis of approaches to the quantitative analysis of variation in shape.

morphometrics. *n.* See *morphometry*.

morphoplasy. *n.* (*adj.* **morphoplastic**) The morphological potential or plasticity of a developing organ or organism.

morphospecies. *n.* See *species*.

morphotype. *n.* 1. In taxonomy, an example or specimen selected to represent the differentiation in form or structure of a population that exhibits a given variant (morph). 2. A list of morphological characters presumed to be present in an ancestral species. 3. Reference to any distinct, more or less stable morphological form of a specified organism.

morrinas. See *arribada*.

mortality. *n.* 1. Generally used in reference to death. 2. In ecology, reference to death rate expressed as a percentage or as a fraction of the population. This is the current medical index for snake bite and is expressed as the number of annual deaths due to snake bite in relation to the number of inhabitants of a region. Cf. *natality*.

morula. *n.* An embryonic stage prior to the *blastula*, characterized by a cluster of cleaving *blastomeres*.

mosaic evolution. Reference to evolutionary change in one or more body parts without simultaneous changes in other parts, resulting in uneven rates by which characters of an organism undergo modification within a phylogenetic lineage. Nonharmonious character transformation.

mosasaurs. *n.* Extinct, predatory marine reptiles (not dinosaurs) having a body shape similar to that of living varanid lizards. Some researchers believe that snakes and mosasaurs had a common ancestor, a theory first suggested by E.D. Cope in 1869 and revived in the 1990s.

motile. *adj.* Having capacity for movement.

motility. *n.* 1. Movement. 2. In digestive physiology, this refers to contractions of the alimentary canal which mix and transport ingested material along its length.

motivation. *n.* A behavioral drive directed toward a specific need or goal.

motoneuron or **motor neuron.** n. An efferent neuron that carries impulses to an effector organ.

motor endplate or **end plate.** The neuromuscular junction (synapse) where neurons innervating muscle activate its fibers.

motor unit. A unit of motor activity, defined by a collective number of muscle fibers that are innervated by the same neuron.

mottled. *adj.* Marked with spots or blotches of different colors, or with cloudy spots of different shapes and sizes, often as haphazard arrays.

Mountain Adder. See *Levantine Viper*.

Mountain King Snakes. Vernacular name for species of colubrid snakes *Lampropeltis pyromelana* and *L. zonata*.

Mountain Rat Snake. See *Green Rat Snake*.

Mountain Yellow-legged Frog. See *Yellow-legged Frog*.

mountain range. A group of closely spaced mountains or parallel ridges.

mountain-stream type. One of the seven types of tadpoles distinguished by G.L. Orton (*Syst. Zool.* 2:63–75, 1953).

mouse unit. The lethal dose of a snake venom for an individual mouse.

mouth. *n.* The opening formed by the rupture of the oropharyngeal membrane to connect the buccopharynx with the exterior. Proper use should be without reference to mouthparts.

mouth brooding. See *Rhinodermatidae*.

Mouth-brooding Frog. See *Rhinodermatidae*.

mouth-disc, mouth disk. See *oral disc*.

mouth gaping. See *gaping*.

mouthparts. *n.* Collective reference to all soft or keratinized feeding structures (as in tadpoles) largely external to the *mouth*. See *oral disk*. (Syn. *oral apparatus*) See Fig. 1.

mouth rot. See *stomatitis, Pseudomonas infection*.

mRNA. See *Messenger RNA*.

MS. (*pl.* **MSS**) 1. Abbreviation of the Latin *manuscripta*, meaning "manuscript." 2. Used as an abbreviation for Master of Science degree.

MSD. See *minimum indicating dose*.

MSH. See *melanocyte stimulating hormone*.

MSH potency. A measure of the ability of a substance to stimulate melanin dispersion, quantified by changes in light reflection following immersion of frog skin in a solution containing the test substance.

MS-222. See *tricaine methanesulfonate*.

mtDNA. See *mitochondrial DNA*.

mtDNA lineages. Lineages or phylogenies derived from data on mitochondrial DNAs.

muciferous crypt. Reference to the *neuromast* of the lateral line system in *Ascaphus* tadpoles. Used by H.T. Gaige (*Occ. Pap. Univ. Mich. Mus. Zool.* 84:1–9, 1920).

mucin. *n.* 1. The mucopolysaccharide that forms the chief lubricant of *mucus*. 2. Also a synonym for *mucus*.

mucocyte. *n.* A cell that produces *mucus*.

mucopolysaccharide. *n.* A class of polysaccharides that contain hexoseamine, may or may not be combined with protein, and forms various mucins when dispersed in water.

mucoprotein. *n.* See *glycoprotein*.

mucormycosis. *n.* A highly infectious disease of anurans, prevalent in Australia and caused by the fungus *Mucor amphibiorum*. The disease affects the skin and internal organs.

mucosa. *n.* A *mucous membrane*, or tissue that contains or secretes *mucus*.

mucosal. *adj.* 1. Pertaining to the side of an epithelial tissue facing the lumen of a body cavity or the exterior of the body. Cf. *serosal*. 2. Pertaining to a mucous membrane.

mucous gland. Any gland that produces and secretes *mucus*, usually as a thick, colorless liquid. This is one of the principal gland types in the integument of amphibians. Fig. 28. Cf. *serous gland*.

mucous membrane. A membrane that secretes *mucus*. (Syn. *mucosa, tunica mucosa*)

mucro. *n.* (*pl.* **mucrones, mucronations**) An abrupt, terminal point or spine. (H.M. Smith, *Handbook of Lizards*, p. 29, 1946, restricts this term to the spine lying on the median line of a single scale and refers to any lateral spines on either side of the mucro as *denticules*.)

mucronate. *adj.* Terminating in a sharp point, a projecting spine at the free, posterior median tip. Spines to the side of the median posterior edge of a scale are termed *denticules*. See *mucro*.

mucus. *n.* (*adj.*, **-ous**) A viscous, protein-containing mixture of mucopolysaccharides that lines the mucous membranes and is secreted from mucous glands, serving to moisten and lubricate. (Syn. *mucin*)

mud. *n.* Fine sediment particles having diameters less than 0.0625mm (sometimes defined as less than 0.002 mm) forming fine grained, slimy, marine, or lacustrine detrital sediment.

Mudpuppy. Vernacular name for certain proteid salamanders belonging to the genus *Necturus*. See also *Waterdogs*.

Mud Salamanders. Vernacular name for several species of North American plethodontid salamanders belonging to the genus *Pseudotriton*. See also *Red Salamanders*.

Mudsnake. *n.* Vernacular name for the North American colubrid *Farancia abacura* and the Javan colubrid *Xenodermis javanicus*.

Mud and **Musk Turtles.** Collective vernacular name for species of turtles belonging to the family Kinosternidae (*Kinosternon* and *Sternotherus* spp.). The name *Mud Turtle* also applies to various species of African pelomedusid turtles belonging to the genus *Pelusios*.

Mugger. *n.* Vernacular name for the Asian crocodilian species *Crocodylus palustris*. (Syn. *Marsh Crocodile*)

Mulga Snake. Vernacular name for the Australian elapid genus *Pseudechis*.

Müllerian duct. 1. The oviduct. 2. The paired embryonic ducts originating from the peritoneum that connect with the urogenital sinus to develop into the uterus and oviduct. (Syn. *Muller's duct, paramesonephric duct, oviduct*)

Müllerian mimicry. Resemblance between species when both have characteristics that deter or are unpleasant to predators. Cf. *Batesian mimicry*.

Müller's duct. See *Müllerian duct*.

multi-. Prefix meaning "multiple" or "many."

multicameral. *adj.* Reference to multiple chambers, usually used with reference to lung structure.

multicarinate. *adj.* With more than a single keel.

multicellular. *adj.* Composed of many cells.

multicuspid. *adj.* Having more than one cusp, as on the occlusal margin of a tooth.

multidimensional species concept. See *species*.

multifactorial. *adj.* 1. Reference to many factors. 2. A status of phenotypic characters when they are controlled by the integrated action of multiple, independent genes. (Syn. *multigenic, polygenic*)

multifid. *adj.* Cleft into many parts.

multifocal. *adj.* Found in more than one location.

multigenic. *adj.* See *multifactorial*.

multigranular bodies. See *lamellar bodies*.

multimodal distribution. A distribution having more than one mode.

multiparous. *adj.* (*n.*, **-ity**) Reference to production of more than one, and usually several, offspring at birth. (Syn. *pluriparity*) See *parity*. Cf. *uniparity, biparity*.

multiple-entry key. An identification key that allows the operator freedom to select any character in any sequence from a given list, often utilizing some form of punch card system.

multiple regression. A biometric method of relating one dependent variable to more than one (several) independent variables.

multiple testes. 1. A male gonad having a series of enlargements separated by constricted regions often greater in length than the enlargements. This type of gonad has been described in several diverse genera of salamanders

(R.R. Humphrey, *Biol. Bull.* 43:45–67, 1922). 2. Abnormal, multiple organs reported in frogs exposed to pesticides such as atrozine. Multiple ovaries are observed similarly in female frogs.

multipotent. *n.* See *pluripotent.*

multiserial. *adj.* Describes structures that occur in three or more series at a site, e.g. tooth rows on a tooth ridge in tadpoles. Cf. *biserial, uniserial.*

multistate character. A character having three or more different states.

multivalent. *adj.* See *polyvalent.*

multivariate analysis. In statistics, generic reference to a number of techniques for the simultaneous analysis of more than one independent variable, including analysis of variance and covariance, regression, and correlation methods. Cf. *univariate analysis.*

multivoltine. *adj.* Producing more than one clutch or brood of offspring during a year. (Syn. polyvoltine) Cf. *univoltine.*

Munsell color ordering system. A color order system originally conceived by A.H. Munsell in 1905 to provide a system for specifying colors. The basic idea of the system is that color appearance can be described in terms of three attributes: *hue, chroma,* and *lightness.* See A.H. Munsell, *A Color Notation,* 17th ed., 1992.

mus. Abbreviation of the Latin *museum,* meaning "museum."

muscarinic toxin. A toxin and common component of African *Mamba* venom that causes *hypotension* and increased *peristalsis.*

muschelwulst. *n.* A German term adopted by T.S. Parsons (*Bull. Mus. Comp. Zool.* 120:103–277, 1959) for a lateral outpocketing of the wall of the olfactory region of turtles. (Syn. *concha*)

muscle, muscle organ. *n.* Collectively, an anatomical and functional group of *muscle fibers* together with supportive noncontractile tissues including connective tissue sheath, blood vessels, and nerves.

muscle fiber. A single syncytial cell of skeletal muscle, functioning as a contractile unit within a muscle. Muscle fibers produce force or tension, which may or may not move a body part.

muscle process. See *tendon process.*

muscular process. See (1) *pharyngeal process* and (2) *tendon process.*

muscular ridge. See *horizontal septum.* See Fig. 36.

muscle spindle. A fusiform bundle in striated muscle that houses specialized *intrafusal fibers,* specialized for proprioception and monitoring the degree of muscle stretch.

musculature. *n.* The characteristic arrangement or system of muscles of a given organism or its parts.

museum, museum of natural history. An institution or building dedicated to preservation and housing a collection of scientific and/or historical objects. Herpetological collections of preserved whole specimens are maintained in such museums throughout the world. See Table 4 for a list of acronyms associated with various museums.

museum collection codes. See *standard symbolic codes for institutional resource collections.*

mushroom body. A bony column projecting inward from the floor or wall of the vomeronasal organ and reducing the space to a dome-shaped chamber.

mushroom tongue. A *boletoid* tongue.

musk. *n.* A substance with a persistent and penetrating odor, secreted by special glands in various reptiles.

musk gland. A gland secreting a substance with a very strong, penetrating odor, usually of hedonic or scent marking function. These occur as paired glands located on the ventral aspect of the lower jaw in skin folds near the angle of the jaw (*angular gland, chin gland, gular gland, mandibular gland, throat gland*) and at the opening of the cloaca (*cloacal gland, paracloacal gland*) in crocodilians, as *anal glands* in snakes, and as an *angular gland* or *inframarginal gland* in turtles.

Musk Turtle. See *Mud Turtles.*

muskelleiste. *n.* See *horizontal septum.*

Mussurana, Musurana. (Portuguese **Muçurana**) *n.* Vernacular name for the colubrid snake species *Clelia clelia,* native to Central and South America. Also called *Zopilota* in Central America and *Cribo* on some Caribbean islands.

mutagen. *n.* A compound that produces mutations in genetic material.

mutagenic. *adj.* Causing mutations.

mutant. *n.* An organism that bears a mutant gene that is expressed in the phenotype.

mutation. *n.* A heritable, unprogrammed alteration in genetic material.

mutational load. Reference to genetic disability affecting a population due to accumulation of deleterious genes generated by recurrent mutation.

mutation rate. The number of mutation events per gene per unit time.

mutilous. *adj.* Harmless, without defensive structures.

mutualism. *n.* A relationship in which two species or groups of organisms have higher growth, survival and/or reproduction rates due directly to the presence of the other. Interactions benefit each of two species. Many studies indicate that more than two species are often involved in mutualistic associations. (Syn. *interdependent association*)

muzzle. *n.* Occasionally used as a name for the *snout.*

my, mya, myr. An abbreviation for "million years (ago)."

Myall Snake. See *Curl Snake.*

mybp. An abbreviation for "million years before present."

mycobacteriosis. *n.* Disease caused by members of the genus *Mycobacterium.*

mycobacterial. *adj.* Reference to infection, etc., caused by rod-shaped, aerobic, gram-positive bacteria of the family Mycobacteriaceae.

mycotic dermatitis. Inflammation of the skin caused by a fungal infection.

myelencephalon. *n.* The posterior segment of the *rhombencephalon*, which includes most of the *medulla oblongata*.

myelinated axon. An axon that has a *myelin sheath*, which increases the rate of signal conduction.

myelin sheath. A membranous, fatty sheath formed by cells that are wrapped around segments of various vertebrate axons, functioning to electrically insulate the axon and thereby increase the rate of signal conduction.

myeloblast. *n.* A cell that differentiates by aggregation to form multinucleated, striated muscle cells.

myeloid. *adj.* Reference to bone marrow.

myelomalacia. *n.* Degeneration or softening of the spinal cord.

myelotoxin. *n.* A toxin that destroys bone marrow and bone marrow functions. Said to be a component of venom from the lizard *Heloderma*.

myiasis. *n.* Infection or invasion of the body of an animal by diperan larvae.

Myobatrachidae. *n.* A monophyletic clade (family) of anurans with reduced or absent digital discs, and varied life histories including two unusual forms of egg brooding. There are 23 genera and approximately 122 species distributed in Australia, New Guinea and Tasmania, if limnodynastine species are included. Revised recently with 13 genera and sister to *Limnodynastidae*. See D.R. Frost et al., *Bull. Amer. Mus. Nat. Hist.* 297:1–370, 2006.

Myobatrachinae. *n.* A subfamily of *Myobatrachidae*.

Myobatrachoidea. *n.* A taxon (superfamily) of anurans, sister with *Batrachophrynidae* within the *Australobatrachia*, and composed of *Limnodynastidae* and *Myobatrachidae*. See D.R. Frost et al., *Bull. Amer. Mus. Nat. Hist.* 297:1–370, 2006.

myoblast. *n.* An embryonic precursor of skeletal muscle fibers.

myocarditis. *n.* Inflammation of cardiac muscle.

myocardium. *n.* (*adj.* **–al**) Heart muscle.

myocoel. *n.* The transitory cavity of the myotome.

myoepithelial cell. A cell with contractile capability that lines a channel or gland.

myofascitis. *n.* Inflammation of a muscle and its fascia.

myofibril. *n.* A longitudinal contractile unit of a muscle fiber, consisting of serially repeated *sarcomeres* and surrounded by *sarcoplasmic reticulum*.

myofilament. *n.* The contractile protein filaments that comprise the *myofibrils* of a muscle.

myogenic. *adj.* A property of muscle activation by rhythmic electrical activity inherent within a muscle cell or cells in the absence of neuronal input. Vertebrate hearts are myogenic and capable of beating without activation by neurons (although neurons are important in modulating the rhythm). Cf. *neurogenic*.

myoglobin. *n.* A monomeric iron-globin complex resembling a single unit of the hemoglobin molecule, found in vertebrate muscle and functioning to provide a reservoir of oxygen and possibly to enhance its diffusion into the mitochondrial regions of the fibers.

myoglobinuria. *n.* The presence of myoglobin in the urine.

myokymia. *n.* A brief, benign condition of spontaneous tetanic contractions of motor units or groups of muscle fibers, often involving adjacent or antagonistic units. This term may apply to involuntary twitching of the muscles of the tail, a response to stress seen in many species of snakes.

myolitic, myolytic. *adj.* Causing disruption of the muscle cell membrane with loss of function related to leakage of intracellular enzymes and potassium ions.

myomere. *n.* A segment or block of trunk muscle.

myometrium. *n.* The muscular part of the *uterus*.

myonecrotic. *adj.* Causing damage or destruction of muscle tissue.

myopathy. *n.* (*adj.* **–ic**) Reference to a pathological alteration of muscle.

myoplasm. *n.* The cytoplasm of a muscle cell. (Syn. *sarcoplasm*)

myoseptum (*pl.* **myosepta**) *n.* The connective tissue septum between the main tissue masses (*myotomes*) that form the striated axial musculature in vertebrates, usually visible in the tail of anuran larvae as chevron-shaped divisions between the myotomes.

myosin. *n.* A protein that constitutes the thick filaments and cross-bridges in muscle fibers. It also occurs in other types of cells where it is involved in various aspects of cellular motility.

myotome. *n.* The portion of embryonic *somite* that develops into striated skeletal muscle.

myotoxin. *n.* (*adj.* **–ic**) 1. A substance that is poisonous or destructive to muscle. 2. A class of toxins in viperid and colubrid venoms that inhibit the potassium or calcium channels without altering the acetylcholine receptors. Used largely in reference to components of rattlesnake venoms in herpetological literature.

myr. Abbreviation of "million years."

myrmecophagus. *adj.* Feeding on ants.

N

N, n. The total number of observations in a population, abbreviated *n*. 1. The number of observations in a sample from a population, or the number of observations in a data set, abbreviated *N*. 2. Abbreviation of the prefix *nanno*, denoting unit x 10^{-9}. 3. Abbreviation of the Latin *nomen*, meaning name. 4. Abbreviation of the Latin *novus*, meaning "new." 5. Abbreviation of the Latin *nudum*, meaning "naked."

NAAMP. See *North American Amphibian Monitoring Program*.

NAD. See *nicotinamide adenine dinucleotide*.

NAD nucleotidase. A component of some snake venoms that catalyzes the breakdown of nicotinamide N-ribosidic linkage of NAD.

Na⁺/K⁺ pump. See *sodium-potassium pump*.

nail. *n.* The enlarged scale forming a sharp spine at the tip of the tail in some turtles.

naïve. *adj.* Reference to an animal with no previous experience with a particular situation or stimulus.

najine. *n.* A term used in reference to any or all venoms from various types of cobra.

Nakamura's gland. See *nuchal gland*. This name coined by T.P. Maslin (*Proc. Calif. Acad. Sci.* 26:419–482, 1950) to honor the person who did much work on this structure.

naked. *adj.* Reference to a condition of snake hemipenis in which the basal region is devoid of ornamentation such as spines or papillae. Contrast with *nude*, which refers to the tip of hemipenis being devoid of ornamentation. The restricted definitions are from H.G. Dowling and J.M. Savage (*Zoologica*, 45:17–31, 1960).

nano, nanno-. (n) Prefix denoting unit x 10^{-9}.

nanometer. (nm) One-billionth (10^{-9}) of a meter. This is a preferred term for describing ultrastructural dimensions and replaces the *millimicron* (mµ), a unit of equivalent length, in earlier literature.

nape. *n.* The back of the neck.

Naped Snakes. Vernacular name for Australian elapid snakes belonging to the genus *Furina*.

nares. *n.* See *naris*.

narial. *adj., n.* 1. Reference to nares. 2. Used as another word for the *septomaxillary*.

narial flap. See *choanal papilla*.

narial plug or **valvule.** A small, hemispherical, and thickened swelling on the tongue of many salamanders and caecilians, usually paired and functions to close the internal nares.

naris. (*pl.* -es). *n.* The nostril or nasal opening, usually paired. Nares that open to the outside are termed *external nares* or *nostrils*, whereas those that open into the mouth or pharynx are termed *internal nares* or *choanae*. See also *narial*.

Narrow-headed Garter Snake. Vernacular name for the American colubrid species *Thamnophis rufipunctatus*.

narrow heritability. See *heritability*.

Narrow-mouthed Frog, Narrowmouth Toad. Collective vernacular name for anuran species belonging to the family *Microhylidae*. The vernacular Narrowmouth Toad applies especially to the American genus *Gastrophryne* and the Asian genera *Gastrophrynoides* and *Kaloula*.

nasal. *n.* 1. A scale or plate that occurs usually on the side of the head and borders or contains a naris. This scale may be entire or divided, with a suture separating what is usually called a *prenasal* and *postnasal*. See Figs. 11, 16. 2. A paired dermal bone lying on the middorsal surface of the skull, lying immediately behind the *premaxillary* and entering the external naris anterior to the *frontals*. This bone is absent in turtles. 3. Has been used in reference to the *septomaxillary*.

nasal capsule. Cartilaginous structures that form around the sensory epithelium of the *nasal cavity*.

nasal cavity. See *nasal sac*.

nasal chamber. The main portion of the nasal sac.

nasal cleft. The *nasal groove* in Blind Snakes.

nasal gland, nasal salt gland. 1. A form of *salt gland* that eliminates excess salts as secretions through a duct in the nasal cavity of some lizards. See *oral glands, salt glands*. See Fig. 20. 2. A paired gland situated lateral to the nasal cavity in the area between the eye and external naris in the Montpellier Snake (*Malpolon monspessalanus*). The glands discharge a watery secretion that contains variable amounts of proteins, salts, and lipids, which are discharged through ducts on the nasal valves. The secretions are rubbed onto the body with the head and are hypothesized to retard evaporative water loss.

nasal groove. A deep furrow or cleft running from the lip line across the nasal scale and through the nostril to the prefrontal scale in blind snakes of the family Typhlopidae. The groove does not divide the nasal into two parts, but merely lies on its surface. (Syn. *nasal cleft*)

nasal knob. See *ghara*.

nasal obstruction. A blockage of the nostril formed of dead epidermis, shaped like a short funnel and usually formed just prior to ecdysis.

nasal pore. An isolated pit or opening lying on the inner posterior wall of the nostril (external naris) in certain crotaline snakes. Original description by T.P. Maslin (*Copeia* 1942:18–24, 1942).

nasal sac. The sensory portion of the nasal apparatus lying between the external and internal nares. (Syn. *nasal cavity, olfactory sac*) See *nasal capsule*.

nasal septum. The bony or cartilaginous vertical, medial partition separating the nasal passages.

nasal turbinates. *n.* 1. Chambers of the nasal passages with olfactory receptors in the surface epithelium. 2. Small, scroll-like ossified or cartilaginous structures present in the anterior nasal passages of 99% of living mammals and birds. These structures increase the surface area of the nasal passages and act as intermittent countercurrent heat exchanges during inhalation and exhalation. Turbinates are covered with moist tissues that warm and humidify inspired air and, because of evaporative lowering of temperature, condense moisture from expired air. By this process, they function to reduce rates of evaporative heat and water loss that otherwise would accompany the high lung ventilation rates characteristic of endothermic taxa. The presence/absence of turbinate bones in extinct dinosaurs has been used to infer conclusions about their metabolic physiology (J. Ruben, *Annu. Rev. Physiol.* 57:69–95, 1995). (Syn. *conchae, respiratory conchae, respiratory turbinates*)

nasal valve. Any structure in the nostril that obstructs passage of air in one direction while permitting it in the opposite direction. Various valvular structures have been described in various amphibians and reptiles, including simple flaps and erectile tissue.

nascent. *adj.* Reference to origin or beginning. Coming into being. (Syn. *incipient*)

nasolabial. *n.* A scale on the head of blind snakes that enters the lip line and contains the external naris. The scale evidently results from the fusion of a *labial* scale with a *nasal* scale.

nasolabial gland. Any of a group of branched and often tubular glands with the orifice at the *nasolabial groove* of plethodontid salamanders. These glands function to irrigate the grooves.

nasolabial groove. A furrow or groove running from the external naris to the edge of the upper lip in plethodontid salamanders. The groove permits quick drainage of the nares and, in some species, extends well onto the *cirrus* of the lip.

nasolacrimal duct. A duct connecting the the corner of the eye with the *naris* (absent in turtles).

nasopremaxillary bone. A complex of dermal bone forming the anterior tooth-bearing portion of the maxillary arch in advanced caecilians.

nasorostral. *n.* Any of one or more small scales tha lie between the *nasal* and the *rostral* in some species of *Bothrops*.

naso-vomeral organ. See *vomeronasal organ*.

natal. *adj.* Of or pertaining to birth.

natality. *n.* Birth rate, usually expressed as the number of offspring produced per female or head of a population per unit time. Cf. *mortality*.

Natatanura. *n.* A new monophyletic taxon of anurans, sister to the new taxon *Allodapanura* within the new taxon *Ranoides*, and containing *Ptychadenidae* and a new taxon *Victoranura* (D.R. Frost et al., *Bull. Amer. Mus. Nat. Hist.* 297:1–370, 2006).

natatory membrane. See *web*.

native. *adj.* Indigenous; of natural occurrence in a region and not introduced by humans.

Natricidae. *n.* See *Natricinae*.

Natricinae. *n.* A subfamily of Colubridae including more than half the genera in the family, including familiar *Gartersnakes, Watersnakes*, and *Keelbacks*. This group is given familial status in alternate classification schemes (e.g., J.T. Collins, *J. Kans. Herpetol.* 19:18–20, 2006; N. Vidal et al., *Comptes Rendus Biologies* 330:182–187, 2007). Roughly 250 species are widely distributed in North America, Eurasia, and the East Indies.

Natterjack Toad. Vernacular name for the European bufonid species *Bufo calamita*.

natural classification. A hierarchical classification based on hypothetical phylogenetic relationships such that the members of each category in the classification share a single common ancestor. (Syn. *phylogenetic classification*)

natural group. A group of monophyletic taxa.

natural history. The study of nature, natural objects, and natural phenomena. When the term was coined, "history" meant "description," and in this context natural history is a description of nature, and naturalists are those who study nature. See H.W. Greene, pp. 99–108 in M.E. Feder & G.V. Lauder, eds., *Predator-prey Relationships: Perspectives and Approaches from the Study of Lower Vertebrates*, 1986; H.W. Greene, *Amer. Zool.*, 34:48–56, 1994; D.J. Schmidly, *J. Mammal.* 86:449–456, 2005.

natural key. An identification key constructed from a natural classification and having branching sequences that reflect presumptive evolutionary relationships of the group.

natural selection. Differential reproduction of phenotypes through time, expressed through differential survival of genotypes due to variable advantage in relation to biotic and physical forces in the environment. The "selection" results from nonrandom elimination of individuals (therefore genes) from a population. This process in evolution was identified by Charles Darwin.

natural taxon. A group of organisms that represent a unit actually derived by evolutionary events. Cf. *artificial taxon.*

naturalized. *adj.* Descriptive of an introduced species that has become successfully established.

nautical mile. 1. A secondary *SI* unit of length equal to 1852 meters (*International nautical mile*). 2. A secondary *fps* unit of length equal to 6080 feet (*UK nautical mile*).

neap tide. A tide of minimum range that is coincident with the first and third quarters of the moon. Cf. *spring tide.*

Nearctic. *n.* A *biogeographical (and zoogeographical) realm* of the northern hemisphere, consisting of North America from the Central Mexican Plateau northward to the Aleutian Islands and Greenland. This region is part of the *Holarctic.*

nearest neighbor method. A method of sampling populations and estimating population density based on the distance from individuals to nearest neighbors.

neck. *n.* The constriction between the crown and root of teeth in many species of iguanid lizards. (Syn. *shaft*)

necropsy. *n.* A postmortem examination performed to ascertain the cause of death. (Syn. *autopsy*)

necrosis. *n.* (*adj.* **necrotic**) A state of tissue wherein morphological changes are indicative of cellular death caused by progressive enzymatic degradation. The condition may be localized or involve a tissue or organ more generally and is typically due to infection or injury.

necrotic or **necrotizing dermatitis.** Inflammation of the skin that results in the death of the tissue.

necrotic or **necrotizing stomatitis.** Inflammation of the oral cavity that results in the death of the tissue.

negative Bohr shift. See *Bohr shift.*

negative feedback. See *feedback.*

negative selection. The removal of deleterious mutations from a population. (Syn. *purifying selection*) Cf. *positive selection.*

nektonic. *adj.* 1. Reference to the middepths of a body of water. 2. A type of free-swimming tadpole inhabiting middle depths of open water and rasping food from submerged surfaces, distinguished by G.L. Orton (*Syst. Zool.*, 2:63–75, 1953).

neo-. Prefix meaning "new" or "newer."

Neobatrachia. *n.* A monophyletic clade of advanced anurans, which, as originally conceived (O.A. Reig, 1958), includes the *Bufonoidea* and *Ranoidea*. These include the most speciose families and greatest diversity of extant frogs. All frogs excluding the *Archaeobatrachia*. More recently, this clade was classified as sister with *Anomocoela* within the *Acosmanura*, containing *Heleophrynidae* and *Phthanobatrachia* (D.R. Frost et al., *Bull. Amer. Mus. Nat. Hist.* 297:1–370, 2006).

Neocaudata. *n.* A clade that includes all extant salamanders except the *Sirenidae.*

neochoanate. *adj.* A condition in some lizards where the openings of the internal choana and the vomeronasal organ are not continuous, rather, being separated from each other by an overlap or contact between the *vomer* and *maxillary bones.* Cf. *paleochoanate.*

neo-Darwinism. A post-Darwin concept that species evolve by natural selection, resulting in adaptive phenotypes arising from changes attributable to mutations. The term is based in neo-Darwinian synthesis (also called *evolutionary synthesis* or *modern synthesis*) that developed roughly between 1920 and 1950. The synthesis represents a union between Darwinian theory of natural selection and Mendelian genetics, resulting in a reformation of principles in a new language of mathematical population genetics and subsequent extension to other fields within what is commonly termed "evolutionary biology." See also *synthetic evolution.*

neodrosopterin. *n.* See *drosopterin.*

neoendemic. *adj.* An endemic species that has a limited geographical range attributable only to its recency of origin.

Neogaea. (Neogea) *n.* A zoogeographical area originally consisting of both *Nearctic* and *Neotropical* regions, but which is now used generally to refer to the Neotropical region only. Cf. *Arctogaea, Notogaea, Palaeogaea.*

Neogene Period. A unit of geological time consisting of the Miocene, Pliocene, Pleistocene, and Holocene Epochs (see Table 1). Traditionally, the Neogene ended at the end of the Pliocene Epoch, just before the beginning of the Quaternary Period. However, there is a movement among some geologists to draw different hierarchical boundaries, producing somewhat confusing terminology and disagreement.

neoholotype or **neotype.** In taxonomy, a new type specimen that is selected in the absence of the original type.

neoichnology. *n.* The study of tracks, burrows, and other trace structures made by living organisms.

neomorph. *n.* A new morphology or structure in a derived species that has no ancestral equivalent.

neonate. *n.* (*adj.* **neonatal**) A newly born individual, also used by many authors in reference to *hatchlings*. Neonates comprise an age class distinct from *juveniles.* D.J. Morafka et al. (*Herpetol. Monogr.* 14:353–370, 2000) define neonate as the first post-paritive age class, which expresses the effects of pre-paritive parental investment, influences of pre-paritive environments (egg, nest, etc.), and in which transitory aptive features particularly enhance the success of parition and, or, first dispersion. The maximal extent of the neonatal stage is 10% of ontogeny of a young reptile prior to sexual maturity.

neonatology. *n.* The study of *neonatal* biology. See D.J. Morafka et al., *Herpetol. Monogr.* 14:353–370, 2000.

neontology. *n.* The study of living or extant life, as contrasted with *paleontology* (study of extinct life forms).

neonychium. *n.* A fetal claw pad in some lizards and crocodilians that surrounds the sharp claw and appears to prevent ripping of embryonic membranes.

neopalatines. *n.* Paired toothless bones lying posterior to the vomers in some neobatrachian anurans and usually articulating with the maxilla and sphenethmoid. These elements are not homologous with the palatine bone of other vertebrates.

neoplasm. *n.* A localized proliferation of cells, not subject to the usual limitations for growth. Such growths are either *benign* or *malignant*. See *tumor*.

neopodium. *n.* Part of the tetrapod limb containing structures that have arisen in evolution strictly as part of the limb and are not derived from ancestral fin elements. These include the digits and their supporting bones. Term was proposed by Westoll, according to G.L. Orton (*Science* 120:1042–1043, 1954). Cf. *archepodium*.

neorbital. *adj.* A condition of turtle skull in which the frontal bone enters the rim of the orbit, typical of most modern taxa of turtles. Cf. *paleorbital*.

neorhachitomous. *adj.* A *rhacitomous* vertebra having a double centrum consisting of both a pleurocentrum and a dominant hypocentrum. Used by E.E. Williams (*Quar. Rev. Biol.* 34:1–32, 1959).

neotene. *n.* A *neotenic* organism.

neotenic, neotenous. *adj.* Reference to the condition of *neoteny*.

neoteny. *n.* A developmental process or state in which a larva grows and becomes sexually mature while retaining an otherwise larval body form or features. Metamorphosis and development of the adult body form is either delayed or absent. Historically, several terms have been introduced to distinguish different patterns of neoteny. *Total obligatory neoteny* or *absolute neoteny* refers to the absence of metamorphosis. *Partial obligatory neoteny* or *semi-larval neoteny* refers to retention of some larval characters while others metamorphose. *Total facultative neoteny* refers to sexual maturity in larval conditions, but total metamorphosis can be induced experimentally. *Accidental neoteny* refers to a neotenic condition induced by exceptional circumstances in animals that normally metamorphose. *Partial neoteny* refers to simple retardation or delay of metamorphosis, as in overwintering tadpoles. Cf. *paedogenesis, paedomorphosis, progenesis*.

Neotropic. *n.* One of eight *biogeographic realms*, as well as a formal *zoogeographical realm*, comprised of the New World tropics.

Neotropical Rattlesnake. Vernacular name for the American pitviper *Crotalus durissus*.

Neotropical Vine Snakes. Vernacular name for species belonging to the American colubrid species *Oxybelis*.

neotropics. *n.* The tropical part of the New World, including South America, Central America, parts of Mexico, and the West Indies.

neotype. *n.* In taxonomy, this refers to a newly designated type specimen that is selected in the absence of existing type material.

nephric. *adj.* Reference to the kidney.

nephric duct. See *ductus deferens*.

nephric ridge. The posterior region of intermediate mesoderm.

nephric tubule. See *nephron*.

nephritis. *n.* Inflammation of the kidneys.

nephrocoel. *n.* The cavity of the embryonic *mesomere* comprising a coelomic chamber of the *nephrotome*.

nephron. *n.* The morphologic and functional unit of the vertebrate kidney, comprised of the various segments of an individual tubule that forms urine. (Syn. *nephric tubule, renal tubule*) See also *uriniferous tubule* and *renal corpuscle*.

nephrosis. *n.* A degenerative process involving the kidneys, sometimes reversible.

nephrostome. *n.* A ciliated, funnel-shaped opening leading from the coelom into an excretory tubule where fluid or particles are carried from the coelomic cavity into either the veins or the nephrons. Found only in primitive vertebrates and some extant amphibians. In larval amphibians, connections with the kidney tubules are often lost before metamorphosis, but in a few species (*Alytes, Discoglossus*) the connection with nephrons is retained throughout life. This term is confusing, as it is most commonly used in connection with excretory structures of invertebrates. (Syn. *coelomoduct, peritoneal funnel*)

nephrotome. *n.* The nephrogenic part of the *mesomere*, and the segmental forerunner of the nephron in the urinary structure of the embryo.

nephrotoxic. *adj.* Pertaining to toxicity that destroys cells of the kidneys.

nepionotype. *n.* The specimen on which the first description of the larval stage of a species was based.

neritic. *adj.* Reference to the area of seas near the coastline and extending to a level of 200 m.

nerve. *n.* A bundle of neurons, axons, or processes that are enveloped by connective tissue. These have specific connections and transmit neural information to and from structures within the body. Cf. *neuron*.

nerve cord. The principal bundle of neurons and associated cells that develops dorsal to the notochord. Also refers to the primitive or embryonic primordia of spinal cord and brain.

nerve sheath or **nerve cord sheath.** The fibrous connective tissue that envelops the *nerve cord*.

nerve fiber. See *axon*.

nerve impulse. See *action potential*.

nerve processes. Fibers that emanate from the soma of neurons, including *dendrites* and *axons*. (Syn. *neurites*)

nervous system. The collection of neurons and their connectivity in an animal's body.

nest. *n.* A protective structure or location where eggs are deposited for incubation and, in some cases, attendance. See also *foam nest*.

nested hierarchy. A system of consecutively subordinate categories.

nesting. *vt, n.* The process of depositing eggs in a nest or nest cavity. Cf. *nesting behavior*.

nesting behavior. Reference to the activity of forming a receptacle or shelter to hold and hatch eggs. Found in some modern amphibians and reptiles, and also described in dinosaurs.

nestling. *n.* A neonate or young animal residing entirely within the nest from the time of birth or hatching until the time of emergence. Definition from J.K. Tucker et al. (*Copeia* 1998:488–492, 1998).

net flux. The sum of *influx* and *efflux* through a membrane or other material.

net precipitation. Reference to the proportion of total precipitation that actually reaches the ground, calculated as total precipitation less that component that is intercepted by plant surfaces.

net primary production. (NPP) *Gross primary production* less that consumed by catabolic processes of respiration, often referred to simply as *production* or *net production*.

net reproductive rate. (R_0) The mean number of female offspring produced by a female over her lifetime, equal to the gross number of offspring produced, discounted by the chances of female survivorship through different ages. Units are numbers of offspring.

net secondary production or **productivity.** Reference to the total assimilation of organic matter by a primary consumer, individual, or population per unit time per unit area or volume, less that consumed by catabolic processes of respiration.

neural. *adj.* Reference to *neurons*, nerves, or nervous tissue. *n.* 1. Any of the bony plates on the median row of the carapace of turtles, corresponding to the *central* vertebral laminae. There are usually eight, each having attachment of a neural spine of an underlying dorsal vertebra to its ventral surface. See Fig. 5. Bounded on the midline by the nuchal anteriorly and the pygals posteriorly. (Syn. *central, vertebral*) 2. The mid-dorsal row of bony tubercles and scales on the tail-trunk of *Chelydra*.

neural arch. The arch of bone or cartilage between the *centrum* and *neural spine* of a vertebra, enclosing the *spinal cord*. (Syn. *neurapophysis*) Cf. *hemal arch*.

neural canal. The canal that forms beneath the *neural arches* to contain the *spinal cord*.

neural crest. Embryonic ectoderm that forms at the raised edges (neural folds) or crest of the upturning *neural plate*, eventually differentiating to form peripheral neurons, pigment cells, the adrenal medulla, and other tissues.

neural fold. Either of the two upturned edges of the neural plate, arching together during *neurulation*.

neural groove. The longitudinal depression on the upfolding *neural plate* between neural folds.

neural lobe. The *neurohypophysis* of the pituitary.

neural network. An interconnecting group of neurons, forming the basis for communication in the nervous system. (Syn. *neuronal network*)

neural plate. 1. The thickened plate of *ectoderm* that develops into the brain and spinal cord. 2. Either of two forks produced by the bifurcation of *neural spine* in caudal vertebrae of Sand Boas, genus *Eryx* (M.S. Sood, *Proc. Indian Acad. Sci.* 14:390–394, 1941).

neural spine. The dorsal process that arises from a neural arch. (Syn. *processus spinosus*)

neural tube. The embryonic central nervous system. The tubular structure becomes the brain and spinal cord, formed by closure of the upfolding margins of the *neural plate*.

neurapophysis. *n.* See *neural arch*.

neurectoderm. *n.* Ectodermal tissue formed from the neural plate, which differentiates into the brain and spinal cord. (Syn. *neural ectoderm*)

neurilemma. *n.* The thin sheath that surrounds the axon cylinder, or if it is myelinated, the myelin of a nerve fiber.

neurites. *n.* See *nerve processes*.

neurocoel. *n.* The cavity of the neural tube.

neurocostal. *n.* The paravertebral row of tubercles and scales on the tail and trunk of *Chelydra*. Used by H.H. Newman (*Biol. Bull.* 10:68–114, 1906).

neurocranium. *n.* Part of the skull that is derived from the *chondrocranium* and encloses the brain and special sense organs such as nose, eyes, and ears. (Syn. *brain case*)

neuroepithelium. *n.* 1. Epithelium comprised of cells specialized to function in reception and transduction of external stimuli. 2. Ectodermal epithelium from which the central nervous system is derived.

neuroethology. *n.* A subdiscipline of behavior in which neuronal activity is studied during a behavior to determine how activity in neuronal circuits generates and controls the behavior.

neurogenic. *adj.* Activation by activity in motor neurons. Cf. *myogenic*.

neuroglia. *n.* Supportive tissue of the central nervous sytem. Cells that support, nourish, and insulate neurons. See *glial cells*.

neuroglobin. *n.* A vertebrate molecule related to *hemoglobin*, expressed intracellulary in the central and peripheral nervous system, cerebrospinal fluid, retina, and endocrine tissues. Neuroglobin is a monomer that reversibly binds oxygen with an affinity that is greater than that of hemoglobin and functions to increase oxygen availability to neural tissue, providing protection during hypoxic or ischemic conditions. Identified by T. Burmester et al., *Nature* 407:520–523, 2000.

neurohormone. *n.* A chemical that is released by neurons and exerts hormonal effects outside the nervous system.

neurohypophyseal hormones. Hormones of the neurohypophysis: *vasotocin*, *mesotocin*, and *oxytocin* (mammalian).

neurohypophysis. *n.* The posterior, or neural, lobe of the pituitary gland, which develops from the floor of the diencephalon and is derived embryonically from an outpocketing of the hypothalamus. This is a neuronally derived reservoir for hormones and consists of a neural lobe, which makes up its bulk, and a neural stalk connected to the hypothalamus from which neurosecretions pass. (Syn. *pars nervosa, posterior pituitary*)

neurological. *adj.* Pertaining to the nervous system.

neurolytic. *adj.* Destructive of nervous tissue.

neuromast. *n.* A lateral line organ consisting of a pear-shaped group of compactly arranged cells, both supporting and sensory, embedded in the epidermis. A single kinocilium and numerous stereocilia emerge at the top of each sensory cell, which has synaptic contact with neurons and a ribbon-like *cupula* projecting from its apical surface. These function as *mechanoreceptors*. See also *lateral line system*, which in amphibians does not include "pores" or channels that occur in fishes. See Fig. 35.

neuromodulation. *n.* A change in function of neurons attributable to chemical messengers (*neuromodulators*) that are released from axon terminals but diffuse more widely than do typical neurotransmitters and often produce longer-lasting effects.

neuromodulator. *n.* See *neuromodulation*.

neuromuscular junction. See *motor endplate*.

neuron. *n.* (*adj.* **-al**) An individual nerve cell, including dendrites, cell body, and axon(s). (Syn. *neurone, nerve cell*) Cf. *nerve*.

neuronal circuit. A set of interconnected, functionally related neurons. (Syn. *neuronal network*)

neuronal integration. Ongoing summation of all synaptic inputs to a postsynaptic cell, which collectively determine the signaling pattern of the cell.

neuronal network. See *neural network, neuronal circuit*.

neuroparalytic. *adj.* Causing paralysis by means of action primarily on neural mechanisms controlling skeletal muscle.

neuropathy. *n.* Reference to any disease of the nervous system.

neuropeptide. *n.* A peptide molecule that functions as a neurotransmitter.

neurophysins. *n.* Proteins associated with neurohypophyseal hormones stored in granules in the neurosecretory terminals. These proteins are cleaved from the hormones prior to their secretion.

neurophysiology. *n.* A subdiscipline of physiology involving study of the structure and function of neurons and the nervous system.

neuropil. *n.* A dense mass of interwoven and synapsing neuronal processes and glial cells.

neurosecretory cell or **neuron.** A neuron that releases *neurohormones*.

neurotensin. (NT) *n.* A tridecapeptide which is present in nervous and/or intestinal tissue from all classes of vertebrates studied, including reptiles and amphibians. Studies have suggested that NT is a hypophysiotropic factor in frogs and may also be involved in visual processing and release of MSH from the pars intermedia.

neurotoxin. *n.* (*adj.* **-ic**) A substance that is poisonous or destructive to nervous tissue or neuromuscular junctions. Used largely in reference to components of snake venoms in herpetological literature. Numerous pollutants are also neurotoxins, including heavy metals, organic solvents, and polychlorinated biphenyls (PCBs). Cf. *hemotoxin*.

neurotransmitter. *n.* A chemical messenger that is released by a presynaptic nerve terminal and interacts with receptor molecules on the postsynaptic membrane. Neurotransmitters act to either activate or inhibit the postsynaptic cell with which they communicate. (Syn. *transmitter*)

neurula. *n.* A developmental stage of the vertebrate embryo in which the neural axis is fully formed and histogenesis proceeds rapidly.

neurulation. *n.* The process that converts the neural plate into a neural tube.

neustonic. *adj.* An ecomorphological guild of surface-feeding tadpoles, including lentic or lotic species that filter particles in or near the surface film using an upturned oral disc.

neustophagia. *n.* Skimming of particulate matter from the surface of water, first applied to feeding behaviors in some river turtles by D. Belkin & C. Gans (*Ecology* 49:768–769, 1968).

neutral equilibrium. See *passive equilibrium*.

neutralize. *v.* To reverse or eliminate the effects of a substance or condition.

neutral model of evolution. See *neutral mutation-random drift theory of molecular evolution*.

neutral mutation. A mutation that either has no measurable phenotypic effect, or results in phenotypic expression that does not change fitness in relation to the present environment.

neutral mutation-random drift theory of molecular evolution. A theory which emphasizes that nucleotide substitutions during the course of evolution are largely the result of random fixation of neutral or nearly neutral mutations, rather than attributable to positive Darwinian selection. Many protein polymorphisms appear to fit this paradigm of molecular evolution. Neutral mutations are not without function. They can be equally effective as ancestral alleles in promoting survival and reproduction of organisms that carry them, and they are maintained in the population by a balance between mutational input and random extinction. Neutral theory attributes a large role to *genetic drift*, but does not deny the role of natural selection in determining the course of adaptive evolution. Theory proposed by M. Kimura (*Nature* 217:624–626, 1968; J.L. King and T.H. Jukes, *Science* 164:788–798, 1969). (Syn. *Kimura's neutral model of evolution, neutral theory of molecular evolution,* or simply *neutral theory of evolution*)

neutral theory of evolution. See *neutral mutation-random drift theory of molecular evolution.*

neutral theory of molecular evolution. See *neutral mutation-random drift theory of molecular evolution.*

new generation. See *epidermal generation.*

New Guinea Plateless Turtle. See *Pig-nosed Turtle.*

new name. See *nomen novum.*

Newt. *n.* A colloquial term for any of various small, semi-aquatic salamanders, especially those of the family Salamandridae.

newton. (N) A derived *SI* unit of force, defined as the force required to accelerate 1 kg of mass 1 m/s per second.

newt plague. See *Molchpest.*

New World. Reference to the Americas.

New World Python. Vernacular name for the loxocemid species *Loxocemus bicolor.*

New Zealand. A *zoogeographical realm* including New Zealand.

New Zealand Frog. Vernacular name for species of frogs belonging to the genus *Leiopelma* and family Leiopelmatidae.

NGO. A generic acronym for a nongovernmental organization.

niche. *n.* See *ecological niche.*

niche breadth. The range for any factor within which a species can function. The concept is used in context of a niche as a multidimensional space and indicates upper and lower limits for a given parameter representing one axis of the hyperspace.

niche diversification. Divergence of structure, function, or behavior between species that compete for an identical resource. (Syn. *character displacement*)

niche overlap. Overlap of niche dimension(s) with implication there is direct competition for a given resource by two or more species.

nicotinamide adenine dinucleotide. (NAD) A coenzyme that is widely distributed in organisms and participates in many enzymatic reactions. It plays a key role in cellular metabolism.

nictitating membrane or **nictitating eyelid.** A thin, transparent fold of skin which can be drawn across the cornea from the inner angle of the eye. It functions as a *third eyelid*, situated beneath the movable upper and lower eyelids to clean and protect the corneal surfaces without obstructing the animal's vision. Present in certain amphibians and reptiles. (Syn. *membrana nictitans, third eyelid*)

nidicolous. *adj.* 1. Remaining in the nest for some while after hatching. Cf. *nidifugous.* 2. A developmental mode whereby an anuran tadpole of some form develops, but the individual does not feed. The morphology varies from a fully formed tadpole to developmental variations that deviate from the developmental pattern of a typical tadpole. See also R. Altig & G.F. Johnston, *Smithsonian Herpetol. Info. Serv.* 67:1–75, 1989.

nidifugous. *adj.* Leaving the nest almost immediately after hatching. Cf. *nidicolous.*

nidus. (*pl. –i*) n. (*adj. –al*) Literally, a nest or place where eggs or young are deposited. Also used in reference to a focus of infection, cellular growth, etc.

Night Adders. Vernacular name for about six species of African vipers belonging to the genus *Causus*, unlike other viperids in having round pupils, smooth scales, and large head plates.

Night Frogs. Collective vernacular name for African species of arthroleptid frogs belonging to the genus *Astylosternus.*

Night Lizards. Vernacular name for various species of lizards belonging to the *Xantusiidae*, especially the genus *Xantusia.*

Night Snakes. Vernacular name for colubrid snakes belonging to the genus *Hypsiglena*, inhabiting drier parts of Mexico and the United States.

Nile b., A., Nile b. sulfate. An oxazin dye that stains fatty acids blue.

Nile Crocodile. Vernacular name for the crocodilian *Crocodylus niloticus.*

Nile Monitor. Vernacular name for the African Monitor Lizard *Varanus niloticus.*

nitric oxide. (NO) A free radical gas that acts as a neurotransmitter and mediates a wide range of physiological effects including regulation of blood pressure, macrophage defense systems, and intercellular communication. Endogenous NO is produced by the conversion of L-arginine to L-citrulline and NO, a reaction catalyzed by the enzyme NO synthase. In amphibians and reptiles, NO has been shown to play a role in modulation of breathing and regulation of systemic vascular tone. See N. Skovgaard et al., *Comp. Biochem. Physiol.* 142A:205–214, 2005.

n.n. or ***n. nov.*** Abbreviation of the Latin *nomen novum*, meaning "new name."

no. Abbreviation of the Latin *numero*, meaning "number."

NO. See *nitric oxide.*

nobelian bone. Either of a pair of subpelvic, posterior skeletal rods, associated with the "tail" or intromittent organ of the frog *Ascaphus.*

Nobleobatrachia. *n.* A new monophyletic taxon sister to *Australobatrachia* within the *Notogaeanura* and composed of *Hemiphractidae* and *Meridianura* (D.R. Frost et al., *Bull. Amer. Mus. Nat. Hist.* 297:1–370, 2006).

nociception. *n.* The perception of pain.

nociceptor. *n.* A receptor for pain or stimulation by injury.

nocturnal. *adj.* Active primarily or exclusively at night.

nod. *n.* A repeated raising and lowering of the head, in rapid succession and in a vertical plane, when a male lizard is in the presence of a conspecific female. This behavior is characteristic of the males of many species of lizards, especially iguanids, and is a display of sex recognition by the male.

node. *n.* In systematics, a point in a cladogram where one branch splits off from another. Each node represents a common ancestor, and the branches are the lineages derived from it by speciation events.

node-based, node-defined. Describing a taxonomic group that is defined as the descendants of the most recent common ancestor of two other groups. Cf. *stem-based*.

nodose. *adj.* Having nodes or projections.

nodule. *n.* A small node or compact mass of tissue that is hard and can be detected by touch.

noise. *n.* 1. In colloquial usage, this refers to variation in experimental data attributable to uncontrolled effects comprising a component of variance called experimental error. 2. In physiological data acquisition systems, any unwanted electrical signal that interferes with the specific parameter of interest that is being recorded or monitored.

nom. Abbreviation of the Latin *nomen*, meaning "name."

nomen. **(nom.)** (*pl. nomina*) Latin, meaning "name."

nomenclature. *n.* (*adj.* -al) The application of distinctive names (*scientific names*) to organisms and taxa, in accordance with international rules. By convention, this involves assignment of a binomial scientific name comprised of genus and species, or in the case of subspecies, three names (genus, species, followed by subspecies). The species name is capitalized, followed by species and subspecies names in lower case; all are printed in italics (e.g., *Crotalus viridis helleri*). Scientific names are based in Latin or Greek, scholarly languages of the 18th century and carryover from the work of the Swedish botanist Carolus Linnaeus who produced the system of binomial names. Formal biological nomenclature attempts to avoid communication problems associated with multiple, regionally based names through the use of single scientific names for each species and taxon.

nomen dubium. Latin for "dubious name." In scientific nomenclature, a name originally or historically proposed for a given organism or group, but subsequently considered inaccurate or inadequate for various reasons, e.g., an incomplete description.

nomen illegitimum. See *illegitimate name*.

nomen invalidum. See *invalid name*.

nomen legitimatum. See *legitimate name*.

nomen novum. Latin for "new name." In scientific nomenclature, a name proposed as a replacement for an existing name.

nomen nudum. Latin for "naked name." In scientific nomenclature, a name that is considered invalid because of inadequate or lost specimens on which the description is based.

nomen validum. See *valid name*.

nominalistic species concept. See *species*.

nominate race. The first described race or subspecies of a species, after which other subspecies are based or compared. The subspecific name of the nominate race or subspecies is the same as that of the specific name.

non. Latin, meaning "not" or "not of."

nona-. Prefix meaning "nine," "ninefold."

non-adiabatic. Reference to a thermodynamic process in which heat enters or leaves the system. Cf. *adiabatic*.

nonadditive genetic variance. See *dominance genetic variance*.

nonaptation. *n.* A neutral character currently not subject to natural selection (S.J. Gould & E.S. Vrba, *Paleobiology* 8:4–15, 1982).

non-contact thermometer. See *infrared thermometer*.

non-Darwinian evolution. Reference to genetic changes in a population that are produced by forces other than natural selection. This term is usually associated with the neutralist view of evolution. See *neutral mutation-random drift theory of molecular evolution*.

nondimensional species concept. See *species*.

nonessential amino acids. See *essential amino acids*.

nonhomologous chromosomes. Chromosomes that do not synapse during meiosis.

non-overlapping generations. Condition of a population in which parents reproduce once and then die, such that successive generations do not overlap.

nonparametric statistics. See *statistics*.

nonpathogenic. *adj.* Not causing disease.

nonpolar. *adj.* Reference to chemical groups or components of molecules that are water insoluble.

non-poisonous. Reference in herpetology to a nonvenomous animal, incapable of inflicting serious or lethal injury.

nonrandom. *adj.* In statistics, having an *a priori* probability of occurrence of zero or unity. Cf. *random*.

nonrandom mating. See *assortative mating*.

non-random sample. A sample drawn with reference to some specific attribute of a population. E.g., females in a population of lizards.

nonspecialized pseudautotomy. See *pseudautotomy*.

nonspiking neuron. A neuron that transmits information without action potentials.

non-venomous. Reference in herpetology to an animal that lacks, and cannot inflict, venom.

non-viable. Incapable of normal development or survival.

noose, noosing. *n., v.* Reference to a technique of capturing lizards by means of employing a small noose or loop of thread, usually with a slipknot, which is attached to the end of a pole and is placed around an animal's head or neck. The lizard is usually prodded to run forward, or the noose is pulled abruptly back, thereby ensnaring the animal and tightening the noose at its neck or shoulders. The same principle is used in methods for capturing crocodilians and sometimes snakes. (Syn. *snaring*)

noradrenaline. *n.* See *norepinephrine*.

norepinephrine. *n.* A neurotransmitter secreted by many postganglionic sympathetic neurons, some cells of the central nervous system, and (as a hormone) the *adrenal medulla*. (Syn. *noradrenaline*)

normal activity range. The body temperatures at which a free-ranging animal engages in its normal routine of behaviors. First introduced by R.B. Cowles & C.M. Bogert (*Bull. Amer. Mus. Nat. Hist.* 83:263–296, 1944). See *activity temperature range*.

normal curve. In statistics, a typically symmetrical bell-shaped graph of a normal distribution.

normal distribution. A commonly used probability distribution in statistics, the graph of which is bell shaped and matches a vast number of continuous distributions in nature. (Syn. *Gaussian distribution*)

normalizing selection. See *stabilizing selection*.

normal solution, normality. A solution that contains one gram-equivalent weight of solute dissolved in sufficient water to make a liter of solution.

norm of reaction. See *reaction norm*.

normoxic. *adj.* Pertaining to a normal level of oxygen, equivalent to atmospheric if reference is to the environment.

North American Amphibian Monitoring Program. (NAAMP) A long-term monitoring program to track the status and trends of anuran populations in the eastern United States, based on roadside surveys of calling amphibians.

northern blotting. In molecular biology, a hybridization technique used to identify specific RNA by probing with radiolabeled cloned genes an electrophoretically resolved array of RNAs. See *Southern blotting, western blotting*.

Northern Leopard Frog. Vernacular name for the American ranid frog species *Lithobates* (formerly *Rana*) *pipiens*. See also *Leopard Frog*.

Northern Pacific Rattlesnake. See *Pacific Rattlesnake*.

Northern Water Snake. Vernacular name for the American natricine species *Nerodia sipedon*.

Northwestern Garter Snake. Vernacular name for the American colubrid species *Thamnophis ordinoides*.

nose. *v.* To rub the dorsal head surface along the ventral side of a female, as in the courtship of males in many species of salamanders. The snout may rub the entire ventral surface of a female's body during this behavior, but often the attention is focused on the throat region.

nostril. *n.* Either one of a pair of external openings of air passages associated with the snout and leading to the lungs internally. See *naris*.

notch. *n.* (*adj.* **-ed**) A sharp, V-shaped indentation. Commonly used in reference to structures on the posterior margin of scales, the margins of toe pads in certain arboreal frogs, or between laminae of the carapace and plastron of turtles.

note repetition rate. In reference to anuran vocalization, the number of repetitions of a note within a call per unit time.

Nothopsinae. *n.* In some classifications, a clade (subfamily) of Central and South American colubrid snakes containing about 20 species.

notocentrous. *adj.* A type of vertebra in which the centrum is formed entirely above the notochord, from the dorsal arches.

notochord. *n.* The rod-shaped fibrocellular cord defining the primitive skeletal axis of all embryonic vertebrates. In most vertebrates it is replaced entirely or nearly so by the vertebral column during development.

notochordal. *adj.* Reference to a type of vertebra in which the centrum is hollow, with a funnel-shaped cavity at either end connected by a small aperture. Used by A.S. Romer (*Osteology of Reptiles*, p. 223, 1956) and found in adult Tuataras, geckos, trunk vertebrae of turtles, and embryos of all modern reptiles.

Notogaea. (Notogea) *n.* Name given to designate the combined *Australasian* and *Neotropical* zoogeographical regions of the world. Cf. *Arctogaea, Neogaea, Palaeogaea*.

Notogaeanura. *n.* A newly designated monophyletic taxon of anurans, sister with *Sooglosidae* within the *Hyloides* and composed of all hyloid taxa except *Sooglossidae* (D.R. Frost et al., *Bull. Amer. Mus. Nat. Hist.* 297:1–370, 2006).

Nothosauria. *n.* A relatively generalized stem group of Mesozoic marine reptiles giving rise to *plesiosaurs*. See *Sauropterygia*.

novelty. *n.* In evolutionary biology, this term is used generally and indiscriminately for evolutionary changes that represent departure from earlier states. In correct usage, this term should apply to qualitative as well as quantitative departure from ancestral condition. The evolution of novelty involves *key innovations* of molecular, cellular, or developmental mechanisms. For example, key innovations in the mobility of the jawbones of snakes affected a cascade of other specializations in prey capture and led, in part, to the successful radiation and diversification of snakes.

NPP. See *net primary production*.

n. sp. Abbreviation of the Latin *species nova*, meaning "new species."

NT. See *neurotensin*.

nucha. (*pl. –ae*) *n.* (*adj.* **nuchal**) The nape, or back, of the neck.

nuchal. *adj.* Relating to the neck. See Figs. 4, 5.

nuchal bone. The bone lying on the anterior margin of the turtle carapace, at the midline. (Syn. *prenuchal, proneural*) See Fig. 5.

nuchal crest. A series of central enlarged scales, or fold of skin, running longitudinally on the back of the neck of many lizards.

nuchal endolymphatic sacs. Sacs in the nuchal region of some lizards that are connected to the inner ear by a duct and contain calcium carbonate deposits.

nuchal gland. Any of a series of cutaneous glands located in the paravertebral region of the neck in several species of Asian natricine snakes belonging to the genera *Rhabdophis, Macropisthodon,* and *Balanophis*. First described by K. Nakamura (*Mem. Coll. Sci., Kyoto Imperial Univ.,* Series B 10:229–240, 1935). Unlike other vertebrate integumentary glands, nuchal glands develop exclusively from mesodermal tissue without involvement of ectoderm. Nuchal gland secretions contain cardiotonic steroids identified as bufadienolides and are discharged when the overlying skin is broken or damaged. The secretions are irritable to mucous membranes and presumably are protective against predators. Behavioral studies have shown that *Rhabdophis tigrinus* direct the nuchal glands toward a predator and ventroflex the neck to expel the glandular products. (Syn. *integumental poison gland, Nakamura's gland, nuchal gland, nuchal poison gland, nuchal venom gland, nuchodorsal gland*)

nuchal lamina. The medial, anteriormost lamina of the turtle carapace, overlying the bone of the same name. (Syn. *nuchal shield, precentral*) See Fig. 4.

nuchal pocket. A pouchlike infolding of skin on the side of the neck in various lizards. (Syn. *lateral nuchal fold, lateral nuchal pocket*)

nuchal poison gland. See *nuchal gland.*

nuchal rosette. Several semi-fused nuchal scales forming an oval structure on the neck of some crocodilians.

nuchal scale. 1. Any of the scales on the nape of the neck of crocodilians, usually forming two rows and often reinforced with osteoderms. 2. Any one of the scales situated on the neck of lizards immediately behind the head (Fig. 14). (Syn. *post-occipitals, cervicals*) See also *jubal.*

nuchal shield. See *nuchal lamina.*

nuchal venom gland. See *nuchal gland.*

nuchodorsal gland. See *nuchal gland.*

nuchomarginal. *n.* Any of a series of smaller scales arranged in a ring around the nuchal scales on the neck of crocodilians. See also *postnuchal.*

nuclear envelope. A double membrane surrounding the nucleus of a cell. The outermost membrane is studded with ribosomes.

nuclease. *n.* An enzyme that hydrolyzes nucleic acids and their residues.

nucleic acids. Nucleotide polymers of high molecular weight. See also *DNA, RNA.*

nucleosidase. *n.* An enzyme that hydrolyzes nucleic acids.

nucleotide. *n.* A monomeric unit of nucleic acids, consisting of a purine or pyrimidine base, a ribose or deoxyribose sugar, and a phosphate group.

nucleus. *n.* 1. A membrane-bound, central organelle within a eukaryotic cell that contains its principal genetic material. 2. An anatomically and functionally distinct group of neuronal somata in the central nervous system. 3. An atomic nucleus is the central, positively charged mass surrounded by a cloud of electons.

nude. *adj.* Reference to absence of ornamentation on the tip of the snake hemipenis. Contrast with *naked*, which refers to lack of ornamentation at the basal region of hemipenis. Restricted definitions by H. Dowling and J. Savage (*Zoologica*, p. 217–228, 1960).

null hypothesis. A standard hypothesis of statistical testing for differences between sample means, which states there is no difference between the populations from which the samples are drawn. One then determines the probability of a difference equal to, or greater than, the one actually observed. If this probability is 0.05 or less, the null hypothesis is rejected, and the difference is said to be significant.

numerical character. A *quantitative character.*

numerical response. In population ecology, this term has reference to the per capita growth rate of a predator population as a function of prey abundance. Cf. *functional response.*

numerical taxonomy. See *phenetic method.*

nuptial brush. A synonym for *glove* in male *Xenopus*. Used by H. Gadow (*Amphibia & Reptiles*, p. 146, 1901).

nuptial dance. See *liebespiel*. Used by S. Bishop (*Handbook of Salamanders*, p. 7, 1943).

nuptial excrescence. See *nuptial pad*. Used by H. Gadow (*Amphibia & Reptiles*, p. 32, 1920) and many other authors.

nuptial pad. A cyclic-appearing region of roughened and darkened skin associated with breeding activities of male anurans and formed by glands with granular cytoplasm. These are usually evident on the thumb or other digit, but can also occur as patches on the underside of limbs, chest or chin in various species. Such "pads" are thought to function for clasping the female during amplexus and may possibly be used in combat with other males. (Syn. *nuptial excrescence*)

nuptial tubercle. Reference to any of numerous small, white and rigid pustules on the chin of male *Batrachylodes trossulus*, a frog native to Solomon Islands. Used by W.C. Brown & G.S. Myers (*J. Wash. Acad. Sci.* 39:379–380, 1949).

nurse. *v.* To provide some manner of *parental care* for the developing embryo or newborn of a species.

nutrient. *n.* A substance ingested as food, which promotes growth or provides energy for metabolic processes.

nutrition. *n.* Of, or pertaining to, the combined processes or functions of ingestion, digestion, and assimilation of food.

nutritional secondary hyperparathyroidism. A condition of excess phosphorus or insufficient calcium, causing the bones to weaken in captive reptiles. (Syn. *rickets* when observed in juveniles; *osteomalacia* when observed in adult animals)

nutritive foramen. See *pedicular foramen*.

Nyctibatrachidae. *n.* A taxon (family) of anurans, sister to *Ranidae* within the *Ranoidea*, and containing the genera *Nyctibatrachus* and *Lankanectes* distributed in Sri Lanka and India. For taxonomy see D.R. Frost et al., *Bull. Amer. Mus. Nat. Hist.* 297:1–370, 2006.

O

Oak Toad. Vernacular name for the American bufonid species *Anaxyrus* (formerly *Bufo*) *quercicus*. For taxonomic revision see D.R. Frost et al., *Bull. Amer. Mus. Nat. Hist.* 297:1–370, 2006.

oar. *n.* The vertically compressed tail of a *sea snake*.

obdurator foramen. Synonym for *cordiform foramen*. This is not strictly homologous with the structure of the same name in mammals, as the obdurator nerve does not pass through it, except in turtles.

obdurator process. A small process on the median, anterior *bronchial process* of the amphibian cricoid cartilage.

Oberhautchen. *n.* The outermost layer of beta keratin that forms external *microornamentation* in the epidermis of lepidosaurs. The layer is a unique constituent of the lepidosaurian epidermis. See Fig. 31.

-obin. Suffix meaning "fibrinogen clotting," used in a nomenclatural scheme for describing exogenous hemostatic factors in snake venoms. By adding a portion or designated abbreviation of a snake's scientific name to the suffix, one obtains a designation for the fraction being identified. For example, using the name "*gabonica*" (species name for the Gaboon Viper) one obtains *gabonobin*.

objective. *adj.* Factual, or having real status. Cf. *subjective*.

obligate, obligatory *adj.* Essential or necessary. Restricted or limited to a specific condition of life or environment. Not flexible with respect to a specified condition of environment. Cf. *facultative*.

oblique septum. A membrane separating the heart, pericardial cavity, lungs, and pleural cavities from the peritoneal cavity. Present in some lizards and all snakes and crocodilians.

oblique type. A type of scale row arrangement in which the transverse rows are curved with various degrees of oblique angle present, usually 30° to 45°.

obliterative coloration or **shading.** Graded coloration on a body that neutralizes relief and thereby gives the appearance of a flat surface.

obtuse. *adj.* Blunt or rounded at an extremity.

Occam's razor, Ockham's razor. The principle of favoring the simplest sufficient hypothesis even though others are possible. Parsimony as a philosophical principle in systematics.

occidental. *adj.* Western or westerly.

Occidozyginae. *n.* A subfamily of the anuran family *Dicroglossidae*.

occipital. *adj.* 1. Of, or pertaining to, the back of the head or skull (*occiput*). 2. See *occipital bones*. 3. See *occipital scale*.

occipital arch. The collective *occipitals* that grow upward and around the nerve cord, forming a posterior part of the chondrocranium.

occipital bones. Endochondral bones at the posterior end of the cranium, including unpaired basioccipital, interoccipital and supraoccipital, and paired exoccipitals. All may not be present in any given skull. (Syn. *occipitals*)

occipital condyle. A rounded, bony protrusion from the back of the tetrapod skull, presenting a single or double surface that articulates with the first cervical vertebra (the *atlas*). All reptiles have a single occipital condyle that articulates with the atlas, while amphibians have two.

occipital flap. See *occipital knob*.

occipital groove. A furrow or channel wherein occurs the occipital artery, located between the otic capsule and the occiput in toads of the genus *Anaxyrus* (= *Bufo*). First used by O. Sanders (*Herpetologica* 9:25–48, 1953).

occipital knob. A rounded swelling of dense fibrous tissue, adherent to the skin, beginning at the anterior margin of the frontoparietals in the interorbital region and extending back to the occiput in frogs allied to *Rana doriae*. In some species the posterior margin is free and forms an *occipital flap*. Used by M.A. Smith (*J. Nat. Hist. Soc. Siam* 4:215–229, 1922).

occipital lobe. A flap of skin at the posterior dorsal border of the head in *Chameleo*, either separate on either side of the head or joined medially. Erectile in some species. Defined by A.S. Rand (*Copeia* 1961:411–414, 1961).

occipital recess. A deeply excavated area in the lateral process of the *exoccipital bone*, forming the floor of the tympanic cavity.

occipital segment. One of three segments of the *kinetic* lizard skull, comprised of most cartilaginous bones of the braincase.

occipital scale. Used variously to denote the *parietal* scale of squamates in older literature, the *interparietal* of lizards, the scales posterior to and bordering on the parietals and (if present) *interparietal*, and the *post-parietal* of blind snakes.

occipito-petrosal. *n.* The composite ossification formed by fusion of the deeper, ossifying parts of the occipital ring with the ear capsules in salamanders of the genus *Salamandra*.

occiput. *n.* The posterior part of the vertebrate skull, where it joins the vertebral column.

occlude. *v.* 1. To close or bring together, as in closing the jaws. 2. To prevent passage, as in a blood vessel.

occlusion. *n.* The contact pattern of a bite, relating the manner in which teeth close together.

occult. *adj.* Concealed from observation, or without apparent symptoms.

Oceania. *n.* 1. One of the eight *biogeographic realms* including Polynesia, Fiji, and Micronesia. 2. The islands of the Pacific and surrounding seas.

oceanic. *adj.* Of, or pertaining to, the ocean.

oceanic climate. A climate characteristic of islands and continental margins where the annual temperature range is less than the average for that latitude due to proximity of an ocean or sea.

oceanic island. An island that has never been connected to a land mass, originating directly from beneath the sea by volcanic activity or other geologic event. Cf. *continental island*.

oceanography. *n.* Study of the oceans.

ocellated. *adj.* Possessing ocelli.

ocellus. (*pl.* **-i**) *n.* A circular, eye-like spot consisting of concentric rings of contrasting color.

octa-. Prefix meaning "eight" or "eightfold."

ocular. *adj.* Pertaining to the eye. *n.* 1. Reference to any scale bordering the eye, including *preoculars, postoculars, supraoculars,* and *suboculars.* 2. An *eye shield* or scale covering a rudimentary eye, as in blind snakes. 3. An enlarged and irregular cephalic scale of crocodilians, situated between the *supraciliaries* and *supraoculars*.

ocular elygium. See *elygium*.

ocular window. A wide, unpigmented oval field of skin situated laterally of the eyes in some centrolenid tadpoles, thought to enable laterally-directed vision from dorsally projecting eyes.

oculolabial. *n.* A scale in *blind snakes* formed by fusion of the eye shield and the labial beneath it.

oculomotor. *adj., n.* Reference to the third cranial nerve.

OD. See *Optical density*.

Odland bodies. See *lamellar bodies*.

odontoblast. *n.* A cell that produces dentine.

odontoid. *n.* A bony process that resembles and functions as a tooth. Such structures occur on several bones in the mouth of various frog species.

odontoid peg. See *odontoid process*.

odontoid process. *n.* A toothlike process, especially that on the *axis*, or second cervical vertebra, of reptiles (and also birds and mammals) that articulates with the *atlas* and enables lateral movement of the head. (Syn. *odontoid peg*)

odontophores. *n.* Teeth-bearing processes of the vomer and palatine bone in amphibians.

odontostichos. *n.* A generation of teeth that belong to any particular *zahnreihe*.

odor. *n.* A sensation related to detection of chemicals by sensory, nasal olfactory epithelium. Cf. *vomodor*.

oesophageal. *adj.* Alternative spelling for *esophageal*.

offprint. *n.* A separately printed copy of an individual article or paper that has also appeared in a larger publication.

ohm. *n.* (Ω) A unit of electrical resistance, equivalent to the resistance of a column of mercury 1 mm^2 in cross-section and 106 cm long.

-oidea. Suffix used to form names of animal superfamilies.

oikos. *n.* Habitat.

oil immersion. Reference to an objective lens that gives the highest resolution with the light microscope when the space between the lens and glass coverslip (over the object to be examined) is filled with a small drop of oil having the same refractive index as the glass.

oinodont. *adj.* Possessing a single gap anywhere in the maxillary series teeth. Term proposed by F. Wall (*Snakes of Ceylon*, p. xviii, 1921).

old generation. See *epidermal generation*.

old growth forest. A forest that has undergone successional stages and has culminated in a relatively stable and complex ecosystem characterized by presence of old and (where environmental conditions are favorable) large trees in addition to an associated biotic community and distinctive functional attributes.

Old World. Reference to the collective continents of Europe, Africa, and Asia.

olfaction. *n.* (*adj.* **-ory**) The sense of smell, or sensory perception of molecules released at some distance from an animal. Chemoreception by the nasal and vomernasal organs. Cf. *gustation*.

olfactory. *adj.* Reference to the sense of smell.

olfactory bulbs or **lobes.** The most anterior outposts of the brain, in which nerve fibers from the olfactory receptors are received and relayed back through the olfactory tract to the cerebral hemispheres.

olfactory sac. See *nasal sac*.

oligo-. Prefix meaning "few," "little."

Oligocene. *n.* The third Epoch of the Tertiary Period and the third oldest of seven geological Epochs of the Cenozoic Era. Recent geological time scales assign an age of 34 to 24 million years before present. See Table 1.

olecranon. *n.* A projection on the proximal end of the ulna.

oligodont. *adj.* 1. Possessing only a few small, and widely spaced, teeth. 2. Having few teeth functional at any one time. Cf. *polydont*.

oligohaline. *adj.* Reference to organisms that are tolerant of only a moderate range of salinities.

oligolecithal. *adj.* See *microlecithal*.

oligomer. *n.* A molecule comprised of relatively few monomeric subunits.

oligonucleotide. *n.* A linear sequence of up to 20 joined nucleotides. See *oligomer*.

oligopeptide. *n.* See *peptide*.

oligophyodont. *adj.* Reference to condition in crocodilians and several taxa of lizards having several sets of deciduous teeth and an eventual complete set of permanent teeth. Used by H.M. Smith (*Evolution of Chordate Structure*, p. 271, 1960).

oliguria. *n.* Production of a small volume of urine.

olivary enlargement. Part of the esophagus of alligators, which may contain food or small stones.

Olive Python. Vernacular name for the boid species *Liasis olivaceus*.

Olive Ridley Seaturtle. See *Cheloniidae*.

Olive Sea Snake. Vernacular name for the elapid sea snake species *Aipysurus laevis*.

Olm. *n.* Vernacular name for the proteid cave-dwelling salamander *Proteus anguinus*.

Olympic Salamanders. Vernacular name for salamanders belonging to the American family Rhyacotritonidae and the genus *Rhyacotriton*.

omentum. *n.* A membrane that joins one internal organ to another.

omni-. Prefix meaning "all," "universally."

omnispective method. A method of classification that uses weighted phenotypic similarities as criteria of relationships, taking evolutionary history into account but without full phylogenetic analysis. The method also incorporates pragmatic and intuitive considerations. Cf. *cladistics, evolutionary method, phenetic method*.

omnivore. *n.* (*adj.* –ous) An animal that feeds on both animal prey and plant materials.

omolitum. *n.* See *suprascapula*.

omoplate. *n.* See *scapula*.

omosternum. *n.* Anterior, midventral ossification or cartilaginous element in the shoulder girdle of frogs. The expanded cartilaginous tip may be called the *episternum*, when present. (Syn. *manubrium, manubrium sternii, pars ossea, presternum*) See Fig. 3.

omphalallantoic placenta. A placenta in which the allantois invades the cleft between the isolated and main yolk masses seen in omphaloplacentas and fuses with the isolated yolk mass to form the embryonic portion of the placenta. Unique to squamates and found in all viviparous snakes.

omphaloplacenta. *n.* A yolk sac placenta at the abembryonic pole from the outer wall of a yolk mass that is completely isolated from the main yolk mass. An isolated yolk mass in apposition to the uterine epithelium. The omphaloplacenta is likely an avenue for water uptake and is probably involved in histotrophic nutrient transport from the mother to the embryo via apocrine secretion (inferred from poor blood supply; see M.B. Thompson et al., *Internatl. Congr. Series* 1275:218–225, 2004). Unique to squamate reptiles. Cf. *chorioallantoic placenta*. For terminology discussion see J.R. Stewart & D.G. Blackburn, *Copeia* 1988:839–852, 1988.

oncogene. *n.* A gene that induces uncontrolled cellular proliferation.

oncotic pressure. Colloid *osmotic pressure*, attributable largely to protein in blood plasma.

one-tailed test. In statistics, a test of a hypothesis in which the critical region comprises values on only one side of the mean (the left or right hand tail) of the sampling distribution. A test for directional differences. Cf. *two-tailed test*.

onomatophore. *n.* A name-bearing type specimen deposited as a voucher in a museum collection, important for the proper and valid creation of new names in zoological nomenclature.

ontogeny. *n.* (*adj.* –ous or -etic) Reference to complete development of an organism, from fertilized egg to adult. The life history of an individual, both embryonic and postnatal.

oo-. Prefix meaning "egg."

oocyte. *n.* A cell that produces eggs by meiotic division. A developing *ovum*.

oogenesis. *n.* The formation of eggs, including *vitellogenesis* and formation of egg membranes.

oogonium. (*pl.* **oogonia**) *n.* A female germ cell that gives rise to an *oocyte* (developing egg cell).

oology. *n.* The scientific study of eggs.

oophagous. *adj.* Feeding on eggs.

oophorectomy. *n.* See *ovariectomy*.

oospecies, oogenus, oofamily. An identification of organisms by species, genus, or family according to the classification of their eggs or fossilized egg remains.

opacity. *n.* The reciprocal of *transmittance*.

op. cit. Abbreviation of the Latin *opere citato*, meaning in the publication cited. Usage in the text of a work avoids repetition of a bibliographic reference.

open population. A population of organisms in which there is exchange of individuals, and therefore gene exchange, with other populations (i.e. *immigration* and *emigration*). Cf. *closed population*.

open system respirometry. An experimental technique in which the oxygen consumption and carbon dioxide production of an animal are determined by measuring the gas content of air before and after it flows through a chamber in which the animal is confined. Cf. *closed system respirometry*.

operational definition. A definition having to do with the operation of a process or system and stated in terms of properties relevant to a given experimental situation, rather than more fundamental characteristics.

operational taxonomic unit. (OTU) In numerical taxonomy, used to designate any item, individual, or group that is used for comparison or analysis.

operative temperature. **(T_e)** Essentially, the equilibrium temperature of an inanimate object, or of an animal having minimal metabolism and lacking physiological controls, in relation to a given thermal environment. It is the product of interactions between biophysical and morphological factors that influence an ectotherm's body temperature. Operative temperature can be measured using a suitable physical model of an animal, or a taxidermic mount, as a "T_e thermometer" that is equilibrated with its surrounding environment such that its temperature is not changing. T_e can also be computed using heat transfer equations. The concept of operative temperature was originally developed by C.E. Winslow et al. (1937) for human studies, but was later applied to biophysical ecology of animals by G.S. Bakken & D.M. Gates (pp. 255–290 in D.M. Gates & R.B. Schmerl, eds., *Perspectives of Biophysical Ecology*, 1975). See also G.S. Bakken, *Amer. Zool.* 32:194–216, 1992. *Standard operative temperature* is defined in the same way as operative temperature, except that a reference environment has a fixed condition of convection, usually *free convection* (zero wind speed).

operator. *n.* 1. In molecular biology, a gene that regulates the synthetic activity of closely linked structural genes via its association with a regulator gene. 2. In behavioral ecology, the organism that mimicry acts to deceive.

opercle. *n.* The *ear flap* of a crocodilian.

opercular. *n.* See *splenial*.

opercular cavity or **chamber.** See *gill chamber, peribranchial chamber.*

operculare. *n.* A term variously applied to the *splenial* of the lower jaw, the *columella* of amphibians, and the *prearticular* of turtles. Because of the variation of usage and potential confusion, usage of this term is not recommended.

opercular fold. A fold of integument that grows posteriorly over the gill chamber and forms the *operculum* in anuran larvae. Not homologous with the operculum of fishes.

opercular gill. A term sometimes used to describe the gill structure of anuran larvae, as the entire gill chamber is enclosed by a fleshy *operculum*. These are true external gills, and the term *internal gill* is not accurate. See *internal gill.*

opercular or **opercularis muscle.** A muscle of the middle ear of amphibians that adjusts hearing sensitivity, originating on the suprascapula and inserting on the cartilaginous or bony operculum.

opercular sac. See *gill chamber.*

opercular canal or **tube.** Reference to single or paired structures that connect the branchial chamber to the external opening of the *spiracle* in tadpoles with midventral or ventrolateral spiracles.

operculum. *n.* 1. A flaplike lid or cover, used in reference to the fleshy fold covering of a gill chamber in larval amphibians and, especially, bony fishes. The structures in the two taxa are not homologous. 2. A bony element derived from the *otic capsule* and present in the *oval window* as a distinct element, fused with the *columella*, or absent in amphibians. It functions as a sound conducting apparatus in adults of many species in which it is connected to the suprascapular by the opercular muscle. This element is absent in amniotes. 3. The *earlid* of crocodilians.

operon. *n.* A segment of DNA consisting of an operator and its associated structural genes.

Ophidia. *n.* An alternative term for *Serpentes*, a suborder of the *Squamata* comprised of the snakes.

ophidian. *adj., n.* A snake, or having snake-like properties.

ophidiasis (or –ism). *n.* See *envenomation.*

ophioadenosinetriphosphatase. *n.* An enzyme of snake venom that destroys ATP and ADP.

ophiocholinesterase. *n.* See *colubercholinesterase.*

ophio-L-amino acid oxidase. Any component of snake venom capable of catalyzing the oxidation of amino acids, thereby accelerating proteolytic enzyme activity and aiding digestive function.

ophiology. *n.* The study of snakes.

ophio-oxidase. *n.* A component of many snake venoms that activate proteolytic enzymes and thereby hasten lysis.

ophiophagy. *n.* (*adj.* –ous) The eating of snakes as prey items. Of or pertaining to a predator that eats snakes.

ophiotoxemia. *n.* See *envenomation.*

ophiotoxicology. *n.* The study of snake venoms and their actions.

ophiotoxin. *n.* An older term for a neurotoxin isolated from Cobra venoms.

ophiotoxismus. *n.* See *envenomation.*

ophis. *n.* Greek name for snake. See D.C. Cannatella (*J. Herpetol.* 24:322–323, 1990) for discussion of the uses of this name in systematics in relation to variation of ancient Greek languages.

ophisthotonic spasm. A condition in which the head and lower limbs of animals (except snakes) are bent backwards while the body is arched forward.

ophitoxaemia. *n.* See *envenomation.*

ophthalmia. *n.* Inflammation of deeper structures of the eye. (Syn. *ophthalmitis*)

ophthalmic. *adj.* Reference to the eye.

ophthalmitis. *n.* See *ophthalmia.*

opioid peptides. A class of peptides unique to the skin of South American and neotropical hylid frogs of the subfamily *Phyllomedusinae*. Frog opioid peptides include *deltorphins* and *dermorphin*.

opisthocoelous. *adj.* Condition of a vertebral unit in which the anterior face of the centrum is convex, while the posterior aspect is concave, as in most salamanders and many chelonians. Cf. *amphicoelous, procoelous*.

opisthodontous. *adj.* Reference to presence of teeth at the back of the maxilla and separated from the preceding ones by a *diastema*.

opisthodont pathway. An evolutionary sequence of fang development in which the fang of venomous snakes is found on the posteriormost point of support on the maxillary, as in ophisthoglyphs. See also *proterodont pathway*.

opisthoglyph. *n.* (*adj.* –ous) A condition in venomous snakes where the posterior pair of teeth on each maxilla are enlarged or grooved to function as fangs (some colubroids). (Syn. *rear-fanged, ectoglyph*) Cf. *aglyph, proteroglyph, solenoglyph*.

opisthoglyphodont. *adj.* See *ectoglyph*.

opisthomegadont. *adj.* Having enlarged, aglyphous teeth on the posterior end of the maxillary bone, sometimes separated from smaller anterior teeth by a *diastema* (*diaopisthomegadont*) or continuous with smaller teeth and sometimes grading into them (*synopisthomegadont*).

opisthonephros. *n.* The adult kidney of anamniotes, which develops from all or most of the nephrotome posterior to the pronephros. See *mesonephros*.

opisthotic. *n.* 1. The posterior part of the otic capsule, derived from the chondrocranium. Present in most reptiles where it may be fused with the exoccipital, but apparently lost in most amphibians. (Syn. *epiotic, paraoccipitale*) 2. This word was once applied to the *supratemporal* of lizards.

Oplurinae. *n.* A clade (subfamily) of iguanid lizards that are terrestrial, rock-dwelling, or arboreal and occur in subhumid to arid areas in Madagascar. Two genera and seven species are recognized.

opportunism. *n.* A theory that organisms evolve as historical conditions permit and eventually occupy all potential niches.

opportunistic species. A species that is specialized to exploit newly opened habitats, usually related to dispersal ability and rapid reproduction rates (*r-selected*).

opposable. *adj.* Capable of being placed opposite, as the fused digits of chameleons and the thumb and other digits of some frogs.

opsin. *n.* The protein portion of photosensitive molecules contained in photoreceptors of the retina.

optic. *adj.* Reference to the eye(s).

optic chiasm, optic chiasma. A swelling beneath the hypothalamus of the vertebrate brain where the two optic nerves meet and, in some species, cross the midline and project to the contralateral side of the brain.

optical density. (OD) A measure of how readily light traverses a solution, inversely related to *transmittance*.

optical isomers. Molecular isomers that, due to asymmetry, rotate the plane of a beam of plane-polarized light when passed through a solution of the molecules. Such molecules are given the prefix *D* or *L* depending on whether the plane of light is rotated to the right (*dextro*) or to the left (*levo*).

optic lobe. Either of two lobes comprising primary visual centers at the dorsal part of the midbrain.

optimal. *adj.* Most favorable; often used to suggest optimality of characters, processes or relationships without knowledge that this is actually the case. See *optimality theory*.

optimal foraging theory. See *optimality theory*.

optimality theory. Ecological theory concerned with how organisms perform well and acquire resources (e.g., *optimal foraging theory*). Optimal "solutions" of organisms in various performance situations (feeding, moving, etc.) are those that maximize some measure of performance (known as "currency") and, ultimately, fitness. The use of optimality models does not carry the assumption that all organisms are optimal.

optimization. *n.* In taxonomy, the determination of the polarity of character states by inspection of the structure of particular trees. Optimization methods estimate the ancestral trait values on a phylogenetic tree. Commonly used optimization criteria are: *Maximum Parsimony* (MP) which minimizes the amount of trait change, and *Maximum Likelihood* (ML) which maximizes the likelihood of a trait at a node given likelihood values for trait evolution.

optimum. *n.* Reference to a condition that is most favorable to the maintenance or function of a system.

optomotor reaction. A response to rotation of amphibians and reptiles, in which the individual turns first the eyes, then the head, and finally the head and trunk in the direction opposite to that of rotation.

Orajel®. *n.* A topical anesthetic that has been used to immobilize amphibians for photography (H. Kaiser & D.M. Green, *Herpetol. Rev.* 32:93–94, 2001). The product is an over-the-counter medication normally sold for alleviating pain in teething infants. See also *Anbesol®*.

oral. *adj.* Reference to the mouth.

oral apparatus. See *mouthparts*.

oral disc or **disk.** The mouthparts surrounding the mouth opening of larval anurans, including jaws or beaks, labia (lips), and arrangement of horny, rasping cusps, tooth ridges, and marginal papillae. Considerable adaptive variation is related to environment and feeding habits. This term does not refer to the *mouth*. See R. Altig, *Herpetologica*, 26:180–207, 1970. (Syn. *buccal apparatus, labium, mouth-disc* or *disk, mouthparts*) See Fig.1.

oral festoon. A fringe of papillae forming the outer border of the *oral disc* in anuran larvae.

oral flap. See *labial flap*.

oral glands. A variety of glands present in the buccal cavity of tetrapods. In various amphibians, mucous glands are located on the tongue, and a large *intermaxillary gland* may be associated with the palate. In reptiles, strips of glandular tissue are present along the upper (*supralabial* *glands*) and lower (*infralabial glands*) lips, within the tongue (*lingual glands*) or below it (*sublingual glands*), in association with the snout (*premaxillary* and *nasal glands*), and along the roof of the mouth (*palatine gland*). These glands release mucus to lubricate prey during handling and transport. See also *choanal gland, dental gland, lacrimal gland, Harderian gland, Duvernoy's gland, rictal glands, salt glands, gland of Gabe, taste buds, venom gland,* and Fig. 20.

oral hitching. Reference to a means of movement by extension and retraction cycles of the oral disc in *rheophilous* tadpoles, which thereby maintain attachment to the substrate (R. Altig & E.D. Brodie Jr., *J. Herpetol.* 6:21–24, 1972).

oral papilla. See *labial papilla.*

oral sucker. See *adhesive organ.*

oral tube. Reference to the the cylinder-shaped mouth or protruding mouthparts of tadpoles.

oral valve. Paired structures that separate the mouth from the buccal cavity in anuran larvae.

orbit. *n.* The bony cavity containing the eye.

orbital. *adj., n.* 1. Pertaining to the orbit. 2. An alternative term for *ocular* and combined forms of the word, such as "postocular."

orbital appendage. Reference to dermal projection from the edge of the upper eyelid in certain frogs (e.g., the genus *Megophrys*). Used by R.F. Inger (*Fieldiana: Zoology*, 33: 183–531, 1954).

orbitosphenoid. *n.* 1. A paired bone in the skull of reptiles, derived from the chondrocranium and usually located on the midline above the palate, at or anterior to the orbit. Cf. *pleurosphenoids.* 2. Various authors have used this word for the *sphenethmoid* or *presphenoid.* (Syn. *sphenolateral*)

orbitotympanic. *n.* See *pretympanic.*

order. *n.* A taxonomic category ranking below *class* and above *family.*

ordered transformation series. A hypothesis and structure of the pathway followed in an evolutionary modification of a set of homologous characters. Cf. *unordered transformation series.*

ordinate. *n.* The vertical or y-axis of a graph. Cf. *abscissa.*

ordination. *n.* A method used to summarize patterns in data from ecological communites on a single graph. Points representing communities or taxonomic units in a multi-dimensional space are represented in numerical analyses such that distances between the points are inversely related to the similarities.

Ordovician. *n.* The second Period of the Paleozoic Era, extending from roughly 495–430 million years ago. See Table 1.

organ. *n.* A distinct part of an organism consisting of one or more tissues that perform a particular function.

organ culture. The maintenance and growth of organs or organ parts, including primordia, *in vitro.*

organelle. *n.* A specialized structure within a cell, such as mitochondrion, lysosome, cilium, etc.

organic. *adj.* 1. Pertaining to living or dead organisms, or to chemicals synthesized by them. 2. Chemical compounds based on carbon rings or chains.

organism. *n.* An individual living thing, whether microbe, plant, or animal.

organizer. *n.* A specialized region of the embryo that is capable of determining the differentiation of tissue into defined structures. Grades of *primary, secondary,* and *tertiary* organizers are recognized.

organo-. Greek word element meaning "organ."

organochlorine. *adj., n.* An organic compound containing chlorine. Some of these compounds are used as pesticides, are persistent in the environment, and are highly toxic to vertebrates.

organ of Leydig. Term used in reference to the *parietal eye* of reptiles.

organogenesis, organogeny. *n.* The origin or development of organs.

oriental. *adj.* Eastern or easterly.

Oriental region. A *biogeographical realm* of the world, including southern Asian countries separated from the Australasian region at *Wallace's Line.* See *Indomalaya.*

orientation. *n.* Active positioning of, or by, an organism in relation to a given axis, stimulus, or direction of a stimulus.

orientation circle. A circular pattern seen in the tracks made by a Sea Turtle, either when a female is crawling between a nesting site and the sea or when a hatchling crawls to the sea. The pattern is thought to be related to direction-finding behavior.

origin. *n.* 1. The source or beginning of anything. 2. In anatomy, the more fixed end or attachment of a muscle. Cf. *insertion.*

original count system. A method used to identify scale rows. See *scale row formulas.*

original designation. In taxonomy, the designation of the type of a taxon when it is first published. Cf. *subsequent designation.*

Origin of Species. An abbreviated name for the celebrated book by Charles Darwin, first published in 1859, documenting the phenomenon of evolution and proposing a theory to explain its process.

Ornamental Snake. Vernacular name for Australian elapid snakes belonging to the genus *Denisonia*, especially *Denisonia maculata* that is known only from certain river drainage systems of Queensland.

ornamentation. *n.* The condition of being decorated or embellished, used usually in reference to surface structures such as spines, horns, crests, etc. A visible body feature of an organism that functions primarily in social behavior. See *micro-ornamentation*.

ornith-. A prefix meaning "bird" or "birdlike."

ornithine cycle. A cyclic series of reactions which convert potentially toxic ammonium from nitrogen metabolism to urea.

Ornithischia. *n.* One of two stem subgroups of *Dinosauria*, regarded as monophyletic and a sister taxon to *Saurischia*. The group can be defined as all Dinosauria closer to *Triceratops* than to birds (K. Padian and C.L. May, *New Mexico Mus. Nat. Hist. Sci. Bull.* 3:379–381, 1993) and was particularly abundant in the Cretaceous.

ornithischians, ornithischian dinosaurs. *n.* Reference to any or all members of the *Ornithischia*.

ornithophagous. *adj.* Feeding on birds. Exclusive *ornithophagy* is characteristic of a number of snake species or populations that occur on islands.

ornithopods. *n.* A clade of herbivorous *ornithischians* that includes Duck-billed Dinosaurs.

orobatic. *adj.* Associated with montane regions.

orogeny. *n.* (*adj.* **-ic**) The process of uplifting of mountains.

orographic desert. A *rain shadow desert* on the leeward side of a mountain range.

orographic precipitation or rain. Rainfall resulting from supersaturation of an air mass when it is forced upward over an orographic barrier such as a mountain range. On the *windward* side cooling leads to condensation; thus, very often, the *leeward* side is drier (*orographic desert, rain shadow*). Cf. *convective precipitation, frontal precipitation*.

oropharyngeal cavity. A collective chamber of combined *mouth* and *pharynx*.

ortho-. Prefix meaning "straight," "upright" or "at right angles."

orthogenesis. *n.* A concept of unidirectional change during the evolution of a group of related organisms. A diagram of such phylogeny shows a straight line with no side branches because ancestors evolve into new species without temporal overlap of ancestors and descendants.

orthologs. *n.* 1. Homologous genes (or any DNA sequences) that became separated because of a speciation event. They are derived from the same gene in the last common ancestor. 2. Genes or proteins in different species that are so similar they are assumed to have originated from a single ancestral gene. Cf. *paralogs*.

orthoselection. *n.* Continuous selection on members of a lineage such that evolution continues in a given direction over time.

orthotopic. *adj.* 1. Transplantation involving movement of tissue to a location that is its normal or natural site on another individual. (Syn. homotopic) Cf. *heterotopic*. 2. Reference to an individual or group in its normal habitat.

orthotype. *n.* In taxonomy, a type by original designation.

os. *n.* Latin word meaning "bone."

os antebrachium. See *radio-ulna*.

oscillatory tongue protrusion. Up-and-down movement of a protruded tongue.

oscillogram. *n.* A record made from an *oscilloscope* showing variations in the amplitude of a signal with time. See *oscilloscope*.

oscilloscope. *n.* An instrument that amplifies and plots the amplitude of an electric signal in the vertical displacement of a beam of electrons emitted by a cathode ray tube. Time is indicated as the beam of electrons is driven from left to right by a time base generator, and its position is visualized on a phosphor screen. Thus, the size of a signal fed into the oscilloscope is plotted on the screen as a function of time. Today a computer frequently supplies the display functions of an oscilloscope.

os coccygeum. The elongate, unsegmented bone lying posterior to the sacrum in frogs, composed of a number of fused vertebrae. The terms *urostyle* and *coccyx* have been used inappropriately for this structure.

os cruris. Synonym for the *tibiofibula* in frogs.

os en ceinture. Term used to identify the *sphenethmoid*.

os innominatum. Either of the two lateral halves of the single fused bone of the pelvis in adult reptiles.

os magnum. The distal *carpal bone* located at the base of the third *metacarpal bone*. Present in the forelimb of most turtles, lizards, and Tuatara. It is fused with the *unciform* in crocodilians.

osmium tetroxide. (OsO_4) A common fixative used in electron microscopy.

osmoconformer. *n.* An animal that has body fluids of the same osmotic concentration as that in the environment. The term is usually applied to aquatic or amphibious species. Cf. *osmoregulator*.

osmolality. *adj.* Reference to concentration of osmotically active solutes in osmoles of solutes per kg of solvent.

osmolar. *adj.* Reference to the concentration of osmotically active particles or solutes in solution. See *osmolarity*.

osmolarity. *n.* Reference to the concentration of osmotically active solutes, or effective osmotic pressure, in osmoles of solutes per liter of solution.

osmole. (Osm) The standard unit of osmotic concentration, equivalent to the amount of solute substances that dissociates in solution to form one mole (Avogadro's number) of particles (molecules or ions).

osmolyte. *n.* A substance that functions to raise the osmotic pressure (and lower the freezing point) of a body fluid.

osmometer. *n.* An instrument that measures the *osmotic pressure* or concentration of a solution.

osmoregulation. *n.* Maintenance of internal *osmolarity* with respect to that of the environment. This regulatory process includes adjustments of solute content in relation to the volume of water.

osmoregulator. *n.* An organism that regulates its internal osmolarity in the face of changes in external or environmental osmolarity. Cf. *osmoconformer*.

osmosis. *n.* The movement of pure solvent from an area of low solute concentration to an area of high solute concentration through a semipermeable membrane.

osmotic flow. The osmotic flux attributable to osmotic pressure.

osmotic pressure. The pressure that potentially can be created by osmosis between two solutions separated by a semipermeable membrane. The force necessary to prevent osmotic flow between the two solutions, measured in terms of pressure.

osmotic water. Water that is held in close contact by clay particles and is less available to soil organisms than *capillary water*.

ossa praeplastralia. See *preplastron*.

osseous. *adj.* Of or pertaining to bone.

ossicle. *n.* Any small, usually irregularly shaped bone. Used especially with reference to such structures in the middle ear.

ossicula auditus. See *auditory ossicles*.

ossification. *n.* The process of bone formation. The process is unique to vertebrates and involves deposition of hydroxyapatite (calcium phosphate) on a collagenous matrix.

ossified. *adj.* Transformed into bone.

osteitis. *n.* Inflammation of the bone.

osteo-. Prefix meaning "bone" or "relating to bones."

osteoblast. *n.* (*adj.* **–ic**) A cell that is associated with bone production and secretes the matrix as it matures.

osteoclast. *n.* A cell that is associated with bone resorption.

osteocyte. *n.* A mature osteoblast that has become entrapped within bone matrix and maintains the mineral content of bone.

osteoderm. *n.* An osseous plate or deposit in the dermal layer of the skin, including the large, thickened structures of crocodilians and the *epithecals* of Leatherback Turtle. There is often a one-to-one correspondence between epidermal scales and osteoderms in reptiles. Large or "compound" osteoderms may be primitive or derived structures (as in skinks). The *gastralia* of certain reptiles are supposedly derived from osteoderms, and bony deposits in the skin of some frogs have also been termed osteoderms. Osteoderms are nonhomologous structures that have evolved independently a number of times. Cf. *dermal scale*. (Syn. *dermal bone, dermal ossicle*)

osteofascial tunnel. A structure formed by snake vertebrae and intermuscular septa, extending over numerous vertebrae and functioning to enclose and link individual muscle segments to form long chains without immediate attachment. Intermuscular septa, tendons, and other connective tissues enclose the muscle segments in the tunnels or longitudinal columns. The vertebrae are probably involved only in some of the tunnels but not directly in others. The tunnels are best developed around the three largest epaxial muscles, which are ones with long bellies and long tendons, hence needing to be held near the axis during contraction and axial bending. The tunnels retain muscles in their proper positions during contraction. Used by A.d'A Bellairs & G. Underwood (*Biol. Rev.*, p. 193–237, 1951).

osteogenic. *adj.* Reference to formation of bone.

osteoid. *n.* The uncalcified organic matrix of bone.

osteolith. *n.* A fossil bone that has become completely mineralized.

osteological. *adj.* Of or pertaining to bones or the skeleton.

osteology. *n.* The scientific study of bones, their development, and structure.

osteomalacia. *n.* A disease characterized by softening or decomposition of bone due to impaired mineralization, with excess osteoid accumulation. See *nutritional secondary hyperparathyroidism*.

osteomyelitis. *n.* Inflammation of the bone marrow cavity, usually attributable to an infection by bacteria.

osteon. *n.* An arrangement of bone cells into highly ordered concentric rings or layers, with bone matrix surrounding blood vessels and nerves running through a central canal. These form a *Haversian* system.

osteopathy or **osteopathology.** *n.* Disease or injury of the bones.

osteopenia. *n.* Below normal bone mineral density.

osteoporosis. *n.* A disease of bone characterized by decreased bone mineral density, altered microarchitecture of bone, and altered content of noncollagenous proteins.

ostium. *n.* A small opening into a duct or space.

ostium abdominale tubae. The fimbriated opening into the infundibulum of the oviduct in female vertebrates.

ostium pharyngeum. (*pl.* *ostia pharyngea*) Opening of the Eustachian tube(s) into the pharyngeal region of the mouth.

otic. *adj.* Reference to the ear(s).

otic apparatus. The collective bony structures of the middle ear.

otic capsule. The embryonic case of cartilage, and in some adults endochondral bone, that surrounds the inner ear. (Syn. *auditory capsule, inner ear*)

otoconium. (*pl.* **-ia**) *n.* An extremely minute *otolith*, found, for example, in the large endolymphatic bags on the side of the neck of certain geckos.

otoglossal, otoglossal cartilage. The anterior cartilaginous extension of the basibranchial, which supports the tongue in some salamanders (e.g., ambystomids). (Syn. *lingual cartilage*)

otolith. *n.* Any calcareous concretion that lies on hair cells of equilibrium organs in the inner ear of vertebrates. See also *otoconium*.

oto-occipital. See *exoccipital*. Also used in reference to the bone formed by fusion between the *exoccipital* and the *opisthotic*.

otoparietal plate. Term proposed by O. Sanders (*Herpetologica*

9:25–47, 1953) for the postorbital process of the frontoparietal bone in *Anaxyrus* (= *Bufo*). (Syn. *tectum supraorbitale*)

otoparietal ridge. An elevated part of the *otoparietal plate*, located on the posterior margin of the orbit; actually a *cranial crest* and making up part of the postorbital crest.

otosphenoid. *n.* 1. Used by various authors in reference to the *prootic* or *laterosphenoid*.

otostapes. *n.* The proximal portion of the *columella*, typically including the footplate, that is embryologically derived from the otic apparatus.

OTU. See *operational taxonomic unit*.

ouchterlony diffusion plates. Gel diffusion plates used in immunological testing, often with snake venom.

outbreeding. *n.* The crossing of genetically unrelated plants or animals. (Syn. *crossbreeding*)

outcrop. *n.* A surface exposure of bare rock, not covered by soil or vegetation.

outcross. *n.* See *outbreeding*.

outer epidermal generation. See *epidermal generation*.

outgroup. *n.* In systematics, a group of organisms that is related to, but removed from, the group under study. One or more outgroups are included in comparisons of taxa to determine which of two homologous character states may be inferred to be *apomorphic*. Cf. *ingroup, sister group*.

overall similarity. In systematics, reference to the proportion of shared character states.

oval window. The membranous area on the lateral or ventral wall of the *otic capsule*, separating the inner ear from the middle ear and adjoining the *footplate* of the middle ear apparatus. See *fenestra ovalis*.

ovarian. *adj.* Of or pertaining to the ovary.

ovarian cycle. See *ovulation*.

ovariectomy. *n.* Excission or removal of an *ovary*. (Syn. *oophorectomy*)

ovariosalpingectomy. *n.* The surgical removal of the *ovary* and *oviduct*.

ovary. *n.* The female reproductive organ (gonad) that produces ova or eggs.

ovary cord. See *secondary sex-cord*.

oviduct. *n.* The tube that conveys eggs from the ovary to the *uterus* or to the exterior via the cloaca. Called *Müllerian duct* in embryos, but the term is often carried over to adults. It arises embryonically as a derivative from the nephric duct in some amphibians, but in others, and in reptiles, it originates independently. The oviduct synthesizes and secretes proteins and other substances that nourish and encapsulate the ova and also aid in fertilization. (Syn. *Müllerian duct, Müller's duct, paramesonephic duct*)

oviductal. *adj.* Pertaining to the *oviduct*. Note this word is often misspelled as *oviducal*.

oviductal egg. An egg within the oviduct.

oviductal sinus. A region of the oviduct formed distal to the medial junction of the two ovisacs anterior to the cloaca in the Tailed Frog *Ascaphus truei* and the African bufonid *Nimbaphrynoides occidentalis*. This structure is not "urogenital" because the Wolffian ducts and urinary bladder empty into the cloaca. See D.M. Sever et al., *J. Morphol.* 248:1–21, 2001.

oviform. *adj.* Shaped like an egg.

ovigerous. *adj.* 1. Carrying or bearing eggs. 2. Reference to modification or differentiation for egg transportation (e.g., the *brood pouch* of *Gastrotheca*).

oviparity. *n.* (*adj.* **oviparous**) A mode of reproduction in which underdeveloped eggs are laid or spawned by the female, encased within membranes sometimes overlain by a shell. Embryonic development takes place outside the female, and nutrition is *lecithotrophic*. Cf. *ovoviviparity, viviparity*.

oviposition. *n.* (*v.* **oviposit**) The act of laying or depositing eggs.

ovipositor. *n.* A term applied to the extended cloaca that is protruded from the body and used to deposit eggs in or on the epidermal pouches of certain female anurans such as *Pipa pipa*.

ovisac. *n.* 1. That portion of the amphibian oviduct distal to the *ampulla*, called "uterus" by some authors. This and the ampulla are the only regions of the oviduct that possess intrinsic exocrine glands. See *ovotheca*. 2. A thin, vascular capsule into which eggs fall when expelled from the ovary, and which must then rupture to permit eggs to escape into the coelom.

ovophagous. *adj.* Reference to eating of eggs.

ovotheca. (*pl.* **-ae**) *n.* The expanded part of the oviduct in which unfertilized eggs are temporarily retained just before amplexus by female anurans. Ovothecae may be discrete or fused into either partitioned or unpartitioned chambers, but sharing a common opening into the cloaca. (Syn. *ovisac, uterine dilation, uterine chamber, uterus*)

ovoviviparity. *n.* (*adj.* **ovoviviparous**) A mode of reproduction in which fertilized eggs remain in the female reproductive tract, where they develop without substantial nourishment from the mother and there is no placenta. The young hatch or break through the egg membranes immediately before, during, or just after, the eggs are laid. Cf. *oviparity, viviparity, semioviparity*. See D.G. Blackburn (*Herpetol. J.* 4:65–72, 1994) for opinions on terminology. Currently, ovoviviparity is a term that is no longer used by persons studying reptiles.

ovulation. *n.* The release of an *ovum* or egg from the *ovary*. Ovulation patterns have been classified by S.S. Guraya (*Ovarian Follicles in Reptiles and Birds*, 1989) and R.J. Etches & J.N. Petitte (*J. Exp. Zool.* (Suppl) 4:112–122, 1990). *Polyautochronic* refers to simultaneous ovulation of many ova from both ovaries. *Monoautochronic* species, which include most Geckos, ovulate a single ovum simultaneously from each ovary, whereas *monoal-*

lochronic species, e.g. *Anolis* spp., ovulate one ovum from either the right or left ovary.

ovum. *n.* A unfertilized gamete of female animals, produced by the ovary and capable of developing into a new individual of the same species following fertilization by a sperm or male gamete.

oxbow lake. A crescent-shaped lake formed in an abandoned stream channel that is isolated from the present channel by a meander cutoff and sedimentation.

oxidant. *n.* The electron acceptor in a redox reaction.

oxidation. *n.* Reference to a loss of electrons or an increase in the net positivity of an atom or molecule, classically defined by the combination of a molecule with oxygen or the removal of hydrogen from it. Biological oxidations usually are achieved by removal of hydrogen from a molecule.

oxidation-reduction reactions. Chemical reactions in which electrons are transferred from one molecule to another. Because electrons are transferred to an oxidizing agent, which becomes reduced, oxidation and reduction are always coupled. See *oxidation*.

oxidative phosphorylation. The formation of high energy phosphate bonds in the phosphorylation of ADP to ATP, accompanied by transport of electrons from the substrate to oxygen. (Syn. *respiratory chain phosphorylation*)

oxidative stress. Reference to overwhelming of antioxidant defenses by *prooxidants*, sometimes exacerbated by nutritional deficiency, high temperature, hypoxia, and exposure to xenobiotics.

oxygenated. *adj.* Having adequate or normal levels of oxygen. Cf. *deoxygenated*.

oxygen association curve. See *oxygen dissociation curve*.

oxygen conformer. Reference to an animal in which the rate of metabolism is dependent on the ambient oxygen such that the rate of oxygen consumption is dependent on the partial pressure of ambient oxygen (P_{O2}). Cf. *oxygen regulator*.

oxygen debt. Reference to the extra oxygen required to replenish oxygen stores and to oxidize the products of anaerobic metabolism that accumulate in muscle tissue during periods of intense physical activity. The term refers to a condition in which oxygen demand is greater than oxygen supply to a tissue. The "debt" is repaid by elevated levels of aerobic metabolism following exercise in which there is an anaerobic component. See *excess post exercise oxygen consumption*.

oxygen dissociation curve. A graphic plot (curve) that describes the relationship between the extent of combination of oxygen with hemoglobin (saturation) and the partial pressure of oxygen in the gas phase. (Syn. *oxygen equilibrium curve, oxygen association curve*)

oxygen equilibrium curve. See *oxygen dissociation curve*.

oxygen isotope ratios. The ratio of ^{18}O to ^{16}O. Such ratios in bones of dinosaurs have been used to infer physiological attributes of these extinct animals, as the ratio in bone phosphate depends on body temperature and the $^{18}O/^{16}O$ ratio of body water at the time of bone deposition. Interpretations of these ratios is, however, controversial.

oxygen minimum layer. A depth zone in a water column of lake or ocean where the level of dissolved oxygen is depleted. This is usually below the thermocline at about 500 m.

oxygen regulator. Reference to an animal in which the rate of metabolism is independent of the ambient oxygen, and the standard rate of oxygen consumption is independent of the partial pressure of ambient oxygen (P_{O2}). Cf. *oxygen conformer*.

oxyhemoglobin. *n.* Hemoglobin that is combined with oxygen bound to the iron atoms of the heme groups.

oxyntic cell. A granular, acidophilic cell in the gastric mucosa of frogs, which elaborates both hydrochloric acid and a precursor of pepsin.

oxytocin. *n.* A mammalian peptide hormone secreted by the posterior pituitary; experimentally functions as a diuretic in a few investigated amphibians and induces oviposition in reptiles.

Oxyuraninae. *n.* A subfamily of elapid snakes in some classifications, including about 110 species distributed in Australia and New Guinea. Alternatively classified within the elapid subfamily *Hydrophiinae*.

P

p. 1. Probability. 2. Abbreviation of *pico*, denoting unit x 10^{-12}.

P_{50}. In physiology, the partial pressure of oxygen at which hemoglobin or a solution of whole blood is half saturated with oxygen.

P_i. Inorganic phosphate.

P_1. Parental generation.

P_2. Second parental generation (grandparents).

pacemaker, pacemaker tissue. *n.* A cell or tissue that is depolarized to produce action potentials spontaneously and rhythmically, usually to "drive" electrically coupled cells or tissue.

pacemaker potential. The spontaneous and rhythmic depolarization produced by pacemaker tissue.

pachy-. Prefix meaning "thick."

pachycephalosaurs. *n.* Bipedal, head-butting *ornithischian* dinosaurs.

Pacific. *n.* A *zoogeographical realm*. See *Oceania*.

Pacific Boas. Collective vernacular name for various boid species belonging to the genus *Candoia*, distributed on island chains of the Pacific.

Pacific Chorus Frog. Vernacular name for the North American hylid species *Pseudacris regilla* (formerly *Hyla regilla*, Pacific Treefrog). Recently three species have been recognized within the *Pseudacris regilla* complex: *P. hypochondriaca*, Baja California Chorus Frog; *P. regilla*, Pacific Chorus Frog; and *P. sierra*, Sierra Chorus Frog. See E. Recuero et al., *Mol. Phylogenetics Evol.* 41:511, 2006.

Pacific Geckos. Vernacular name for various species of geckos belonging to the genus *Gekko*.

Pacific Iguanas. See *Lava Lizards*.

Pacific Newt. Collective vernacular name for North American salamandrid species belonging to the genus *Taricha*. See also *newt*.

Pacific Rattlesnake. Vernacular name for *Crotalus oreganus oreganus* (Northern Pacific Rattlesnake) and *C. o. helleri* (Southern Pacific Rattlesnake). See also *Western Rattlesnake*.

Pacific Treefrog. See *Pacific Chorus Frog*.

Pacinian corpuscle. Pressure receptors found in the deep skin, muscles, joints, and connective tissues of vertebrates, consisting of a nerve ending surrounded by a laminated capsule of connective tissue. These receptors are stimulated by mechanical distortion of the connective tissue laminae, which are arranged like the leaves of an onion.

pad. *n.* Reference to any thickened tissue that is resilient and resembles a cushion. Used in reference to *subdigital pads* of lizards comprised of numerous *lamellae*, the expanded tips of digits in many frogs (especially arboreal species), *basiventrals* between gastrocentrous vertebrae, the tail tips of certain lizards, and the subarticular or palmar tubercles in certain bufonids. See *adhesive pad, digital pad, subdigital pad, friction disk, friction pad*.

paddle. *n.* The flattened limb of marine turtles, specialized for swimming.

paddle stage. The phase of life cycle in which larval anurans develop limb buds that become paddle-like in form, ending with the differentiation of the foot.

paedogenesis. *n.* Precocious sexual reproduction by young individuals or larvae. Stephen J. Gould (*Ontogeny and Phylogeny*, 1977) rejects this term in favor of its synonym *progenesis*. Cf. *neoteny, paedomorphosis, progenesis*.

paedomorphosis. *n.* (*adj.* **paedomorphic**) An evolutionary change that results in truncated development in descendant species compared with ancestral species, and therefore adults retain immature or juvenile characteristics of ancestors. Paedomorphosis was defined originally by W. Garstang (*Zool. J. Linnean Soc.* 35:81–101, 1922) to mean "shaped like a child." Cf. *neoteny, paedogenesis, progenesis*.

pain cry. See *fright cry*.

Painted Belly Monkey Frog. Vernacular name for the South American hylid frog *Phyllomedusa sauvagii*.

Painted Frogs. Collective vernacular name for species of Palearctic anurans belonging to the family Discoglossidae, and especially the genus *Discoglossus*. See also *Midwife Toad*.

Painted Turtle. Common name for *Chrysemys picta*, a well-known member of *Emydidae*.

pair bonding. An intimate and long-lasting association between male and female animals of the same species.

palaeo-, paleo-. Prefix meaning "old," "ancient."

Palaeobatrachidae. *n.* A clade of extinct anurans comprising at least 17 species in five genera, evidently the sister group of the Pipidae.

Palaeogaea. (Palaeogea) *n.* A biogeographical area originally comprising the Palearctic, Ethiopian, Oriental, and Australian regions. Cf. *Arctogaea, Neogaea, Notogaea*.

palaeorbital. *adj.* A condition in which the frontal bone of the skull of turtles is shut off from the orbital rim. Commonly seen in extinct turtles, but rare in modern turtles (e.g., *Dermochelys*).

palatal. *adj.* Of or pertaining to the roof of the mouth.

palatal complex. Reference to the bones and soft tissues that comprise the *palate.*

palatal foramen. A large opening in the secondary bony palate of reptiles.

palatal notch. A V-shaped feature in the roof of the mouth, formed by divergence of the palatal bones from the midline posterior to a line of common contact. This feature has been used in classification of some lizard genera.

palatal teeth. The innermost row of teeth on either side of the palate and positioned on the *palatine* and *parashphenoid* bones in snakes.

palatal valve. A soft but muscular extension at the rear of the mouth, just behind the tongue and secondary palate. This structure can be raised or lowered to keep water out of the throat. Common to all crocodilians, the palatal valve enables them to struggle with, and drown, prey that are held in the mouth underwater without themselves taking in water. (Syn. *gular valve*)

palate. *n.* The roof of the mouth, or partition separating the nasal from oral cavities. In some reptiles a bony *secondary palate* may divide the buccal cavity into an upper respiratory chamber and a lower alimentary chamber. A secondary palate is partially developed in chelonians and is complete in crocodilians. It evolved as an adaptation for feeding in water. See *secondary palate.*

palatine. *adj., n.* 1. Relating to or located in the palate (roof of the mouth). 2. Either of a pair of dermal bones in the roof of the mouth, situated immediately behind the *vomers* (or *prevomers*) and usually separated by contact with intervening *vomer* and *pterygoid*. This is a very distinctive element in snakes, bearing teeth and forming the anterior part of the pterygoid-palatine arch.

palatine erection. Reference to an upward rotation of the *palatine* with respect to the *pterygoid*, which functions to assist prey transport during the *pterygoid walk* of snakes. See G. Haas, *Zool. Jahrb. Anat.* 52:347–403, 1930; A. Deufel & D. Cundall, *J. Morphol.* 258:358–375, 2003.

palatine gland. See *oral glands* and Fig. 20.

palatine teeth. 1. Teeth situated on the *palatine* bone of snakes, contributing the anterior part of inner tooth rows. 2. *Vomerine* or *paravomerine* teeth in amphibians. 3. Generally, any teeth located on bones of the palate.

palatomaxillary arch. A complex of mobile bones in the roof of the mouth in snakes, which work together during deglutition. Elements include the *maxillary, palatine, pterygoid,* and the *ectopterygoid*. (Syn. *maxillopalatal arch* by some authors)

palatopterygoid. *n.* A unit of bone in the roof of the mouth of snakes, formed by fusion of the *palatine* and *pterygoid* bones to form a continuous arch, usually bearing teeth.

palatoquadrate or **palatoquadrate cartilage.** Embryonic cartilage of tetrapods, associated with ossification of dermal bones of the upper jaw. The posterior aspect ossifies as the *quadrate.* (Syn. *pterygoquadrate cartilage*)

Palearctic. *n.* 1. One of eight *biogeographical realms* of the world, comprised of Europe, the former USSR, Africa north of the Sahara, and most of Asia north of the Himalayas. 2. Also a formal *zoogeographical realm.*

Pale-headed Snakes. See *Broad-headed Snakes.*

paleo- or **palaeo-.** A prefix meaning "past" or "ancient."

paleobiogeography. *n.* The study of the geographic distribution of animals and plants in the geologic past.

paleobiology. *n.* The study of the living forms and life processes of the geologic past. Or, specifically, the study of existing fossil organisms.

Paleocene. *n.* The first geologic Epoch in the modern *Cenozoic Era.* The Paleocene followed the mass extinction event at the end of the *Cretaceous Period*, which marked the demise of the dinosaurs. Due to climatic conditions of the Paleocene, reptiles were more widely distributed than they are at present. See Table 1.

paleochoanate. *n.* A condition in which opening of the internal *choana* and the *vomeronasal organ* are continuous, without separation by overlap or contact between the vomer and maxillary bones. Described in lizards; see also *neochoanate.*

paleoclimate. *n.* The climate during a specified interval of geologic time.

paleoclimatology. *n.* The study of climate in a designated interval of the geologic past.

paleocortex. *n.* The more lateral part of the cerebral cortex.

paleoecology. *n.* The study of the relationships between organisms and their environments in the geologic past, especially with reference to fossil communities.

paleoenvironment. *n.* Reference to environmental conditions of the geologic past.

Paleogene, Palaeogene. *n.* A unit of geologic time consisting of the *Paleocene, Eocene,* and *Oligocene Epochs.* Lasting about 42 million years, mammals and birds evolved into diverse faunas following the mass extinction that ended the preceding *Cretaceous Period.* See Table 1.

paleogenic. *adj.* A distribution pattern in which a taxon is represented by remnant populations living in one or more disjunct regions, which carries the implication that such populations are remnants of a group previously more widely and continuously distributed.

paleogeography. *n.* Study of the distribution and form of Earth's landmasses in the geologic past.

paleoichnology. *n.* See *ichnology.*

paleontology. *n.* The study of extinct forms of life, usually by means of examining their fossils.

paleospecies. *n.* Reference to successive species in a phyletic lineage that are assigned ancestor and descendant status based on the geological strata in which they appear. See *chronospecies*.

Paleotropic. *n.* A *zoogeographical realm* equivalent to *Ethiopian* or *Afrotropical* biogeographical realm.

Paleozoic. *n.* The earliest of the three geologic Eras, spanning approximately 300 million years and subdivided into six geologic Periods. Multicelled animals underwent a dramatic radiation and diversification at its beginning. See Table 1.

palette (pallet, pallette). *n.* A flattened, expanded, and rounded area resembling an artist's palette in shape. Used rarely in herpetology in reference to *subdigital disks* and *subarticular tubercles*.

pallor. *n.* Paleness of skin and mucous membranes, usually indicative of anemia or low blood pressure. 2. Lightening of the skin due to contraction of melanins.

palmar. *adj.* Reference to the ventral or cupping surface of the forefoot, situated between the base of the fingers and wrist, upon which the digits close when flexed.

palmar tubercle. A *tubercle*, or rounded protuberance, on the palmar surface of the forefoot of amphibians.

palmate. *adj.* 1. Having the distal portion broad and flat. 2. Webbed, thus possessing an interdigital membrane. Rarely used, but common in older amphibian literature.

palmation. *n.* A web.

Palm Pitvipers. Collective vernacular name for several species of *Bothriechis* inhabiting Latin America.

palpate. *v.* (*n.* **palpation**) To examine by feeling or touching. Typically a method used to locate either ovarian follicles, ova, heart, or ingested prey in an animal, feeling for such by hand.

palpebra. *n.* An eyelid.

palpebral. *adj, n.* 1. Generally, relating to the eyelid. 2. A bone in the dermis of the upper eyelid of certain crocodiles and lizards. 3. The *supraocular* or small scales on the upper eyelids of lizards.

palpebral appendage. The orbital appendage or horn on the eyelid of certain anurans.

palpebral disk. 1. A transparent area in the center of the lower eyelid of certain lizards, and in some species of amphibians. (Syn. *palpebral spectacle, window*) 2. Reference to *supraocular disk*.

palpebral membrane. The transparent eyelid that lies beneath a true eyelid and can extend from the inner to outer corner of the eye, characteristic of many anuran species.

palpebral spectacle. See *palpebral disk*.

paludicolous. *adj.* Inhabiting marshy habitats.

palustrine. *adj.* Of, or inhabiting, swamps or marshes; a ponded body of water smaller than 8 ha.

pan. *n.* A soil horizon of clay or highly compacted material.

pan-. Prefix meaning "all," "whole," "completely."

pancake coil. Reference to the *resting coil*, or a flattened, basking coil, of snakes. Used by L.M. Klauber in reference to Rattlesnakes (*Rattlesnakes*, Vol. 1, p. 378, 1956).

Pancake Tortoise. Vernacular name for the African tortoise *Malacochersus tornieri*.

pancreas. *n.* An organ closely associated with the vertebrate intestine that secretes digestive enzymes and produces endocrine secretions, including the hormones insulin and glucagon. In most species the pancreas is closely associated with the *spleen*.

pancreatic duct. A duct that carries secretions from the pancreas to the small intestine.

pancreatic islets. See *islets of Langerhans*.

pancreozymin. *n.* See *cholecystokinin*.

pandemic. *adj.* 1. Ubiquitous or widely distributed. 2. Reference to a disease that reaches epidemic proportions simultaneously in many parts of the world.

Pangaea. (Pangea) *n.* The single supercontinent that existed during the midPaleozoic, comprised of all modern continents in direct physical contact with a minimum of isolating physical barriers. This supercontinent began breaking apart 200 million years ago to form the supercontinents of Laurasia and Gondwana, from which the present continents derived.

panmictic index. (P) A coefficient of inbreeding that measures a population's system of mating as a deviation from the heterozygosity frequencies expected under random mating. A measure of relative heterozygosity. ($1 - P = F =$ Wright's inbreeding coefficient.)

panmictic population. A local population in which mating is completely random.

panmixis. *n.* See *random mating*. Cf. *assortative mating*.

panophthalmitis. *n.* Inflammation of all tissues and structures of the eye.

panting. *n.* Increased breathing frequency that increases the rate of ventilatory air movement, hence evaporation of water and removal of heat from membranes associated with the upper respiratory airways, throat, and mouth. Generally, ectothermic reptiles such as lizards avoid hyperthermia by behavioral means, and thus conserve water. Panting is a last line of defense in animals that might be faced with overheating for various reasons.

pantropic, pantropical. *adj.* Occurrence throughout the tropics and subtropics.

paper chromatography. *Chromatography* in which filter paper serves as the stationary phase.

papilla. (*pl.* **–ae**) *n.* A small, nipple-shaped, and usually pliable projection or elevation. This term has been used in reference to fleshy projections on the hemipenis of snakes, the septal ridge of turtles, pustules lining the vent of some male salamanders, and the labial teeth of tadpoles. See also *oral papilla, filiform papilla*. See Fig 1.

papilla amphibiorum. A papilla in the wall of the *sacculus* of the inner ear of Lissamphibians, near the junction with the *utriculus*. This special sensory structure is larger than the *basilar papilla* and seems to play the main role of hearing in amphibians, which may or may not possess a *basilar papilla*. (Syn. *amphibian papilla*)

papilla basilaris. An auditory receptor organ represented by a sensory area within a short tube (*cochlear duct*) which projects caudally from the *sacculus* of the inner ear of many amniotes lacking a more elongated projection (such as the cochlea of mammals). (Syn. *basilar papilla*)

papillary bar. Reference to a hyaline inclusion in the *basilar membrane* of some reptiles. Term suggested by M.R. Miller according to I.L. Baird (*Amer. Zoologist* 6:431–436, 1966).

papillary cone. See *conus papillaris*.

papillary fringe. A border of minute, fingerlike projections, especially in reference to the edge of the *oral disk* in larval amphibians.

papillate, papilliferous, papillose. *adj.* Having or bearing papillae, or nipple-like protuberances.

papilloma. *n.* A benign lobulated or arborizing tumor of the epidermis. Can be caused by several different viruses including herpesvirus, papillomavirus, and poxvirus.

para-. Prefix meaning "beside," "by," "along," or "beyond."

parabasal. *n.* See *parasphenoid*.

parabolic dune. A deeply curved dune in a region of abundant sand, often anchored by vegetation and formed with the horns pointing upwind.

parabronchial lung. The lung unique to birds, in which respiratory gas exchange takes place in *air capillaries* distributed in numerous small pockets along the walls of relatively stiff tubular *parabronchi* that run in parallel and connect with a system of *air sacs*. Air flows through the *parabronchi* in one direction throughout the ventilatory cycle, and respiratory gases are exchanged between *air capillaries* and blood capillaries comprisng a system of *crosscurrent exchange*.

parabronchus. (*pl. –i*) See *parabronchial lung*.

paracapsular sinus. A unique, fluid-filled structure in the middle part of the *tympanum* or middle ear cavity of chelonians.

paracellular pathway. A route for potential water and solute movement that passes between, rather than through, adjacent cells of an epithelium.

parachordal. *n., adj.* 1. Either of a pair of cartilaginous elements lying laterally to the notochord in the cranial region of early embryos. These form the foundation for development of the braincase. 2. Lying laterally to the notochord.

parachute. *n.* Free, membranous extensions on various parts of the body and appendages of certain lizards and anurans, enabling a slowing of descent during a free fall when leaping or gliding from one object to another or to the ground. Examples occur on the head, body, limbs, and tail of gekkonid lizards of the genus *Ptychozoon* and on the toe webbing in certain rhacophorid frogs.

parachuting. *n.* A mode of free fall or gliding in which the rate of descent is slowed by use of flattened, often membranous extensions of body parts. This term is used in preference to *gliding* when the angle of descent is less than 45° from vertical. See J.A. Oliver, *Amer. Nat.* 85:171–176, 1951.

paracloacal. *adj.* Beside the cloaca.

paracloacal gland. See *musk gland*.

paracolumella. *n.* The *columella* in reptiles and most fossil amphibians, distinguished from the *eucolumella* in being derived partly from the quadrate according to H.M. Smith (*Evolution of Chordate Structure*, p. 214, 1960). Other authors, however, consider the columella to be strictly a hyoid derivative.

paracrine secretion. Reference to secretions that affect neighboring cells adjacent to the origin of the secretion. Cf. *autocrine secretion, endocrine secretion, exocrine secretion*.

paradaptation. *n.* The concept that some features of a character may not be adaptive or owe their properties to natural selection.

paradiapophysis. *n.* A term preferable to *synapophysis* according to W. Auffenberg, who credits A. Remane, 1936 (*Tulane Stud. Zool.* 10:131–216, 1963).

paradigm. *n.* A term used generally as a synonym for model, hypothesis, or theory. The term is used variously in reference to a ruling model that has replaced others, or to a known example that serves as a model or pattern for a more general phenomenon. The term may also be defined as a distinctive pattern of major assumptions, concepts, and propositions in a substantive area.

Paradox Frog. Vernacular name for species of anurans belonging to the family Pseudidae, especially the species *Pseudis paradoxa*.

paraffin section. In classical histology, a section of tissue cut by a microtome after it has been imbedded in a paraffin wax.

parafibula. *n.* A sesamoid cartilage in the femoro-fibular ligament of lizards.

parafollicular cells. See *ultimobranchial glands*.

paraganglia. *n.* Chromaffin tissue that is adjacent to sympathetic nerve ganglia.

paragyrinid. *adj.* Reference to a tadpole having a single spiracle on the left side but well below the body axis, sometimes nearly midventral in position (G.F. Johnston & R. Altig, *Herpetol. Rev.* 17:36-37, 1986). Cf. *amphigyrinid, laevogyrinid, mediogyrinid*.

parahyoid. *n.* An ossification located on the ventral *hyoid plate* in certain anurans.

parallax. *n.* An apparent displacement of an object due to a change in the observer's position. *Binocular parallax* is the seeming difference in position of an object as seen separately by one eye and then by the other, the head remaining stationary.

parallel evolution. 1. Reference to the same or a similar trend that evolves independently in two or more lineages, usually, but not always, related to one another. 2. The maintenance of constant differences in the evolution of characters in two unrelated lines. (Syn. *parallelism*)

parallelism. *n.* 1. Similarity of structures or features that evolve independently in different lineages, usually of common ancestry. 2. See *parallelism.*

parallel key. See *bracketed key.*

parallel muscle. A muscle organ in which all the muscle fibers lie in the same direction and are aligned with its long axis. Cf. *pinnate muscle.*

parallel processing. Reference to information processing in which multiple pathways in the nervous system simultaneously carry information about a particular input or output.

paralogs. *n.* 1. Homologous genes that have separated because of gene duplication events. 2. Two or more different genes in the same species that are so similar in nucleotide sequence that they are assumed to have originated from a single ancestral gene. Cf. *orthologs.*

paralysis. *n.* Loss or impairment of motor function due to lesion of the neural or muscular mechanism. By analogy, *sensory paralysis* refers to impairment of sensory function.

paramesonephric. *adj.* Reference to the primordium of the female reproductive tract.

paramesonephric duct. See *Müllerian duct.*

parameter. *n.* (*adj.* **parametric**) 1. A true measurable property of a population, in contrast to an observed or sample value. In statistics, a value that specifies one of the various members of a family of probability distributions, e.g. the mean of a normal distribution. 2. A constant in a mathematical expression that distinguishes specific cases, having a definite, fixed value in one case but different values in other cases. 3. A variable whose measure is indicative of a quantity or function that cannot itself be precisely determined by direct methods, e.g. blood pressure as a measure of cardiovascular function.

parametric. *adj.* Pertaining to or defined in terms of a parameter.

parametric statistics. See *statistics.*

paramyxovirus. *n.* Any number of RNA viruses that are members of the family Paramyxoviridae. Several paramyxoviruses have been identified in snakes (and fewer isolates from lizards) where they can cause pneumonia and pancreatic necrosis.

paraoccipitale. *n.* An older term used for the *opisthotic* bone in turtles.

parapancreas. *n.* A secretory strucure of unknown origin and function, lying within or beside the ventral lobe of the pancreas of some sea snakes and having a separate duct into the duodenum. (Syn. *glandula parapancreatica*)

parapatry. *n.* (*adj.* **–ic**) Reference to populations or species that occupy areas that are contiguous or juxtaposed, sometimes with a narrow zone of overlap within which hybridization commonly occurs. Cf. *allopatry, sympatry.*

parapatric speciation. The gradual origin of new species from populations that maintain genetic contact during the entire process by a narrow zone of overlap. Cf. *allopatric speciation, alloparapatric speciation, peripatric speciation.* (Syn. *semigeographic speciation*)

paraphalangeal elements, paraphalanges. Cartilaginous structures associated with interphalangeal joints of gekkonid lizards. These elements seem to be associated with grasping abilities of the foot in contexts of digging and climbing modifications. See A.P. Russell & A.M. Bauer, *J. Morphol.* 197:221–240, 2005.

paraphalangeal rod. An accessory and usually paired supportive structure on either side of the phalanx of some lizards with pads on the digits. Described in the gekkonids *Hemidactylus* and *Peropus* by N.G. Stephenson (*J. Linn. Soc. Lond., Zool.*, 44:278–299, 1960).

paraphyletic group. A natural taxon composed of two or more species, including the known or hypothesized shared common ancestral species, but not all of its descendants. See *paraphyly.*

paraphyly. *n.* (*adj.* **-etic**) A condition in which several species of a group are derived from a hypothetical common ancestor, but not all sister species are in the group. An incomplete clade. Cf. *monophyly, polyphyly.*

paraphyseal organ. An alternate term for *paraphysis.*

paraphysis. *n.* An evagination of the brain, originating from the roof of the telencephalon and lying anterior to the *epiphyseal complex.* This structure is prominent in amphibians but less so in reptiles where it is more obvious in embryos. The dorsal sac of the paraphysis contributes to development of choroid plexuses in most vertebrates. (Syn. *paraphyseal organ, paraphysis cerebri*)

paraphysis cerebri. See *paraphysis.*

parapineal eye. See *parietal eye.*

parapineal organ. The anterior of two processes projecting dorsally from the diencephalon, which may be modified distally to form a light sensitive organ, usually called a *parietal eye.* The parapineal is the functional median eye in *Sphenodon* and in lizards. (Syn. *anterior epiphysis, parietal organ*)

parapophysial articulation. An articulation between caudal vertebrae in snakes of the genus *Eryx*, involving a process immediately below the neural spine and extending forward to fit a groove on the preceding vertebra. See *accessory lateral plate.*

parapophysis. *n.* A *transverse process* originating from the centrum of a vertebra and serving as an articulating surface for the lower head of a bicipital rib. If a rib has a single head, the *parapophysis* may be fused with the *diapophysis.* See *synapophysis.*

parapsid. *adj.* Reference to a reptilian skull in which there is a single fenestra or temporal opening, situated dorsally with the postorbital and squamosal bones meeting below. (Syn. *euryapsid*)

paraquadrate. *n.* Used by various authors in reference to the *quadratojugal* of crocodilians and the *squamosal* of amphibians, lizards, and *Sphenodon*.

Parareptilia. *n.* One of two major *sauropsid* lineages, including turtles and an assortment of fossil groups including mesosaurs. Cf. *Eureptilia*.

parasagittal. *adj.* Located on a plane that parallels the sagittal plane.

parasagittal fold. A longitudinal fold of skin lateral to the midgular region in some lizards.

parasagittal plane. A plane parallel to the median sagittal plane of the body. Cf. *sagittal plane, frontal plane, longitudinal*.

parasite. *n.* Any organism that associates with, and is metabolically dependent on, another living organism (host) for completion of its life cycle. Typically, the parasite association is to some extent detrimental to the host. See *parasitism*.

parasite or **parasitic drag.** In biomechanics, drag that is not due to the production of lift but results from a body's surface friction and adverse backflow in the wake when moving through a fluid. See *drag*.

parasitic male. See *satellite male*.

parasitism. *n.* A symbiotic, intimate association between two different kinds of organism that benefits one member, the *parasite*, but is harmful to the other, termed the *host*.

parasitology. *n.* The study of *parasites* and *parasitism*.

paraspecific. *adj.* Refers to the property of some antivenins to be *polyvalent* or *polyspecific*.

parasphenoid. *n.* An unpaired dermal bone that forms along the midline of the palate, forming immediately beneath the *chondrocranium*.

parasphenoid rostrum. Used by A. Romer (*Osteology of Reptiles*, p. 476, 1956) in reference to the *cultriform process*.

parasphenoid teeth. Teeth that are borne by the *parasphenoid* bones of the skull in certain amphibians.

parasternal. (*pl.* **-lia**) *n.* A single V-shaped bone, preformed in cartilage, and located on the ventral internal surface of the body between the *sternum* and *pubis* in lizards. This structure has been variously interpreted in past writings, but A.S. Romer (*Osteology of Reptiles*, p. 431, 1956) argued these are not homologous with *gastralia* of other reptiles. (Syn. *post-sternal rib*; see also *parasternum*)

parasternum. *n.* The entire structure that occupies the area between the sternum and pubis in certain lizards, comprised of a series of *parasternalia*.

parasympathetic nervous system. A division of the autonomic nervous system which controls visceral functions and acts usually antagonistically to the sympathetic division of the autonomic system. Its peripheral ganglia lie generally close to the target organs and are associated with craniosacral outflow from the spinal cord.

paratelic. *adj.* Reference to superficially similar characters or traits derived by convergent evolution.

paratenic host. A host in which a parasite survives without undergoing further development. The host in this case may serve the simple function of transport. (Syn. *transport host*)

parathyoid, parathyroid gland. *n.* An endocrine gland near the thyroid gland, functional in controlling calcium and phosphate metabolism.

parathyroid hormone. (**PTH**) A peptide hormone secreted by the parathyroid gland in response to low calcium levels, and which mobilizes calcium stores while reducing calcium excretion by the kidneys.

paratoid gland. *n.* Paired dorsal rows of glands present in some salamanders. Cf. *parotid gland, parotoid gland*. For discussion of terminology, see M.J. Tyler et al., *Herpetol. Rev.* 32:79–81, 2001.

paratympanic organ. A small structure of presumed sensory function found in the middle ear of tetrapods, thought to be homologous with the spiracular sense organs of fishes. It is found in the embryos of amphibians and reptiles, but appears reduced or lost in adults except for *Sphenodon* in which it may be 2 mm long with stratified epithelium but without innervation of sensory cells. The structure also is present in juvenile alligators. (Syn. *Vitali organ*)

paratype. *n.* Any specimen or specimens of a type series not designated as the holotype.

paraventral. *n.* Any of scales in the series that run longitudinally on either side of the ventral scales of snakes.

paravertebral. *adj.* Anatomical reference to either side of the mid-dorsal vertebral line.

paravertebral field. The area between the vertebral and paravertebral body stripes in some lizards.

paravertebral fold. A fold of skin extending along the paravertebral region of the body in some lizards.

paraviviparous. *adj.* Reference to an embryo completing a modified development using oogenic energy sources in a site other than the reproductive tract of the mother. See R. Altig & G.F. Johnston, *Smithsonian Herpetol. Info. Serv.* 67:1–75, 1989.

paravomer. *n.* One of two bony plates bearing teeth in the roof of the mouth of salamanders, derived from the vomer, and, in some species, attached to it.

paravomerine teeth. Teeth located on the *vomer*, which have been called both *palatine teeth* and *parasphenoid teeth*. However, most literature reporting teeth on the parasphenoid actually refers to paravomerine teeth.

paraxial mesoderm. Paired strips of mesoderm that form along the neural tube of an embryo, becoming arranged as segmentally arranged *somites* in the trunk.

PARC. See *Partners in Amphibian and Reptile Conservation*.

Pareatinae. *n.* A subfamily of semi-arboreal colubrid snakes specialized for feeding on snails and slugs, distributed in southern Asia and the East Indies.

parenchyma. *n.* The essential or functional elements of an organ.

parental behavior. The activity of caring for newly born offspring by either or both parents. The term refers to all behavioral contributions by the parent(s) to offspring after oviposition or parturition. For further discussion of terminology see L.A. Somma, *Parental Behavior in Lepidosaurian and Testudinian Reptiles*, 2001.

parental care. This phrase typically has reference to all nongametic and postfertilization contributions of parents to the survival of their offspring, including viviparity and other physiological or structural contributions. For further discussion of terminology see L.A. Somma, *Parental Behavior in Lepidosaurian and Testudinian Reptiles*, 2001. See also *brooding*.

parental generation. (P) The generation consisting of the immediate parents of the F_1 generation.

parenteral. *adj.* Reference to bodily entry by means other than the digestive tract (e.g., injection).

parent species. An existing species that gives rise to one or more recognizably new species and then continues relatively unchanged or may disappear.

paresis. *n.* Incomplete paralysis, sometimes an effect of envenomation.

parietal. *n., adj.* 1. Either of a pair of dermal bones forming a principal part of the braincase at the roof of the skull, located between the *frontals* and the *occipitals*. The parietals are usually paired, but they may fuse to form a single bone or fuse with the frontal to form a single frontoparietal bone. 2. Either of a pair of large scales on the head of a snake, located immediately behind the *frontals* and forming the last pair of scales usually distinguishable from the body scales. (Syn. *occipital*) See Fig. 10. 3. Used to denote any of the large scales on the head of lizards located posterior to the *frontoparietals* and sometimes bordered posteriorly by *occipitals* or scales distinguishable from the rest of the body. See Fig. 14. 4. Generally used to denote the head scale of Blind Snakes, genus *Typhlops*. 5. Generally used in reference to the dorsal part of the head where small scales might occur in the region otherwise occupied by larger *parietal* plates. See also Fig. 14. 6. Relating to the outer wall of an organ or cavity, or to the outermost surface.

parietal boss or **knob.** A rounded posterior extension of the *parietal* cranial crest of bufonid anurans. See Fig. 2.

parietal cells. The source of secreted hydrochloric acid in the *fundus* of the stomach.

parietalectomy. *n.* The surgery used to remove the parietal eye.

parietal eye. A median photosensory structure (not a true "eye") with an aperture that is surrounded by the parietal bone. This structure develops from the *parapineal organ* as an outgrowth of the forebrain. It responds to light and dark but does not form images. The photoreceptors of this structure resemble rods and cones in morphology, but they show chromatic antagonism consisting of a hyperpolarizing light response most sensitive to blue light and a depolarizing light response most sensitive to green light. This condition is unique among all known photoreceptors, and comparisons of the opsins and signaling molecules involved suggests evolutionary divergence of parietal eyes from visual eyes. (Syn. *frontal organ, median eye, organ of Leydig, parapineal eye, third eye*) (The term *pineal eye* should not be used, as the *pineal organ* does not participate in the formation of the parietal eye in amphibians or reptiles.)

parietal foramen. An unpaired, midline opening in the skull for a third, or *parietal eye*, usually on the middorsal line between the parietal bones, in the parietal bone, frontal bone, or near the frontoparietal suture. This structure was present in the primitive vertebrate skull, but is absent in modern amphibians, turtles, snakes, and crocodilians. (Syn. *interparietal foramen, pineal foramen*)

parietal organ. See *parapineal organ*.

parietal peritoneum. See *peritoneum*.

parietal-pineal complex. Reference to morphologically and functionally related structures located on the roof of the diencephalon. The parietal-pineal complex may produce catecholamines, serotonin, and melatonin, and may affect behavior, gonadal activity, thermoregulation, and color change. See *parietal eye, pineal, parapineal organ*.

parietal spot. See *brow spot*.

parietal spur. Used historically to designate the posterior part of the *cranial crest* or the frontoparietal ridge in *Bufo* (=*Anaxyus*). See Fig. 2.

parietal-squamosal arch. A slender bar of bone far back on the skull of certain *pleurodire* turtles, evidently the remaining connection between extensions of the *parietal* and *squamosal* bones following extensive emargination of the cheek region.

parietal sulcus. A narrow but prominent depression or furrow in the parietal region of some lizards.

parition. *n.* A general term for *parturition* and *oviposition*.

parity. *n.* The number of times a female has produced offspring. A parity of 0 indicates an animal has not given birth to offspring. A parity of 1, 2, 3, etc. indicates the number of times an animal has produced offspring, but the designation does not indicate the number of offspring per clutch or litter. See *uniparous, biparous, multiparous*.

-parous. Word element or suffix meaning "producing."

paroccipital. *adj., n.* 1. Near the occipital bone. 2. This term has been variously used by authors to designate the *opisthotic, exoccipital* (when bone is produced by fusion of *paroccipital* and *exoccipital*), and the *supratemporal* of squamates.

paroccipital process. A lateral extension of the opisthotic bone, extending laterally to the roofing bones, either the *tabular* or the *squamosal*. This element may be formed by bones in addition to, or instead of, the opishotic. It is present in various amphibians and reptiles, but especially prominent in lizards and in *Sphenodon*. See *parotic crest*.

parotic. *n.* Previously used as a synonym for *exoccipital*.

parotic crest. A term used interchangeably with *paroccipital process* and *parotic process* by E.S. Goodrich (*Studies on the Structure and Development of Vertebrates*, 1930).

parotic process. See *parotic crest*.

parotid gland. A cephalic, salivary gland located laterally behind the eye. In rear-fanged colubrid snakes this gland is known as *Duvernoy's gland* and is associated with enlarged and grooved teeth (Fig. 20). Delivery of its secretions immobilize prey. It has been suggested that this term should be used only in reference to the salivary glands of mammals (M.J. Tyler et al., *Herpetol. Rev.* 32:79–81, 2001). Cf. *paratoid gland, parotoid gland*.

parotoid gland. Hypertrophied cutaneous glands characteristic of most bufonid and some hylid anurans, located behind the eye and sometimes extending onto the neck and shoulder region. Cf. *paratoid gland, parotid gland*. For discussion of terminology, see M.J. Tyler et al., *Herpetol. Rev.* 32:79–81, 2001. See Fig. 2.

Parrotsnake, Parrot Snake. Vernacular name for neotropical colubrid snake species belonging to the genus *Leptophis*.

pars distalis. The anterior (= principal) lobe or distal part of the *adenohypophysis*.

pars endocardalis. A unique structure in the interventricular septum of the heart in crocodilians, evidently derived from the free margin of the embryonic incomplete interventricular septum.

parsimony methods. In phylogenetic inference, out of all the evolutionary trees that could possibly describe the relationships among a group of organisms, parsimony methods seek the tree that implies the fewest evolutionary changes in the characters being examined. Parsimony methods are closely linked to cladism. *Linear parsimony* permits reconstructions at ancestral nodes with no change, and permits ambiguous reconstructions. *Squared-change parsimony* minimizes large changes, spreading evolution over the internal tree branches. Cf. *distance methods, maximum likelihood methods, Bayesian analysis*.

parsimony, principle of. 1. The principle that preference is given to the simplest sufficient explanation or hypothesis to explain a phenomenon, even when other explanations are possible. See *Occam's razor*. 2. In systematics, the principle of invoking a minimal number of evolutionary changes to infer phylogenetic relationships.

pars cartilaginea. A term used for the cartilaginous parts of the sternal apparatus comprising the *episternum* anteriorly and the *xiphisternum* posteriorly. See Fig. 3.

pars intermedia. The intermediate lobe of the *adenohypophysis*, situated between the *pars distalis* and *pars tuberalis*.

pars juxtaneuralis. A term previously used to designate the reptilian *pars intermedia*.

Parsley Frogs. Vernacular name for three species of pelodytid frogs belonging to the genus *Pelodytes*, distributed in the Caucasus Mountains and in western Europe.

pars muscularis. The posteroventral segment of the crocodilian interventricular septum, probably homologous with the incomplete interventricular septum of other reptiles. See also *pars endocardalis*.

pars nervosa. See *neurohypophysis*.

pars ossea. Term formerly used to designate the bony parts of the sternal apparatus lying anterior to the clavicles and posterior to the epicoracoids (= *omosternum* and *metasternum*).

pars plana. See *lamina externa*.

pars pylorica. The posterior region of the stomach from which food passes into the intestine. Cf. *fundus*.

pars tuberalis. A structurally distinct region of the *adenohypophysis*, with unclear function, and forming a vascularized tube of cells that surround the hypophyseal stalk.

part. Abbreviation of Latin *partim*, meaning "part."

parthenogen. *n.* A *parthenogenetic* species.

parthenogenesis. *n.* (*adj.* **-tic**) Reproduction in which an egg develops into an embryo without fertilization by a sperm. Parthenogenesis has been reported for a few amphibians (ambystomatid salamanders and *Rana*) and about 30 species of squamates including at least one species of snake (anecdotally in others). Asexual amphibians and reptiles originate from hybridization of two species; their populations are all female, and reproduction is via clonal inheritance. Asexual reptiles reproduce without interactions with other individuals and do not require sperm to initiate or sustain development.

Partial Mantel test. A test for spatial autocorrelation, similar to the *Mantel test* except that three or more variables can be used. See *Mantel test*.

partial pressure (of a gas, x). (P_x) The pressure exerted by any specific gas that is part of a mixture; the pressure that one gas contributes to the total pressure in a mixture of gases. Thus, the partial pressure of oxygen in atmospheric air (symbolized P_{O_2}) is equal to the atmospheric pressure multiplied by the fractional concentration of oxygen in atmospheric air.

partition coefficient. In physiology, the ratio of the distribution of a substance between two different liquid phases (usually oil and water).

Partners in Amphibian and Reptile Conservation. (PARC) An inclusive and diverse partnership between public and private organizations dedicated to the conservation of herpetofauna and their habitats. PARC is a national (U.S.A.) and international conservation network and resource. Current URL: http://www.parcplace.org

parturition. *n.* The process of giving birth by *viviparity*. Cf. *oviposition*. See *parition*.

PAS. See *periodic acid Schiff procedure*.

Pascal. (Pa) A derived *SI* unit of pressure or stress, defined as a pressure of 1 newton/m^2. In physiological measurements, this unit (usually as *kilopascal, kPa*) is now used in addition to, or in place of, the more familiar *mm Hg*. 1 Pascal = 7.52 x 10^{-3} mm Hg (torr).

Pascal, laws of. The principles that: 1. In any point in a fluid, there is a pressure exerted in all directions from the point. 2. The pressure that occurs at any particular point in a fluid is equal to pressure anywhere else in the fluid at the same level. 3. Pressure increases with depth in any column of fluid. Such "laws" are named in honor of B. Pascal, a French mathematician, physicist, and philosopher (1623–62).

passage time. Reference to the time elapsed between ingestion of food and the elimination of feces that are derived from that food. The measurement can vary depending on whether the investigator notes the time of first appearance of feces, a mean defecation interval, or the appearance of plastic beads, dyes, or other markers present in the ingested food. (Syn. *clearance time*)

passive electrotonic transmission. Reference to passage of a graded change in membrane potential over or along a cell membrane, conducted with decay and without the activation of an action potential.

passive equilibrium. Reference to an unstable equilibrium that results from selective neutrality of alleles at a genetic locus, as occurs in Hardy-Weinberg equilibrium. (Syn. *neutral equilibrium*)

passive immunity. A condition of immunity created by inoculation with serum antibodies or lymphocytes from an immune animal rather than by exposure to the antigen.

passive integrated transponder. (PIT or **PIT tag)** A small, coded, electronic microchip enclosed in biocompatible glass that varies in size, usually between 10–14 mm long and about 2 mm in diameter. PIT tags are injected or inserted by surgical incision beneath an animal's skin or into muscle or body cavity. The term "passive" is derived from the fact that the tag is dormant until activated by a handheld reader that scans an animal and generates a close-range, electromagnetic field that activates the tag. When so activated, the tag transmits a unique number and thereby identifies the animal. The use of PIT tags in biological research began with studies first published in 1983 to determine the usefulness of the method for measuring fish movements.

passive process. Used to denote a process that does not require the expenditure of metabolic energy. Cf. *active process*.

passive transport. In cell biology, reference to transport across a membrane that does not require the direct input of energy. See also *facilitated diffusion*.

passive uptake. Reference to absorption of ions by diffusion or other physical processes that do not involve expenditure of metabolic energy.

patagial rib. Any of several pairs of false ribs used to support the fan-like wing, or *patagium*, of Flying Lizards (genus *Draco*).

patagium. *n.* (*pl.* –**ia**) The "wing" or airfoil of gliding animals (e.g. *Draco*), consisting of a web of skin between the body and limbs. (Syn. *axillar wing, axillary web, axillary wing*)

patch. *n.* 1. Generally, reference to an area of skin that differs from that surrounding it in color, texture, or vascularization. The "*seat patch*" or "*pelvic patch*" of anurans refers to the thin and highly vascular ventral integument in the region of the groin that is adapted for rapid water uptake. See also *verrucae hydrophilicae*. 2. In ecology, a spatial term that refers to a discrete area of habitat, capable of holding populations or communities.

patch clamping. In cellular physiology, a recording technique in which a microelectrode is sealed tightly against a cell membrane and the transmembrane potential is held constant, enabling the experimenter to measure ionic currents through single ion channels or across the membrane of an entire cell.

patch dynamics. A concept of a community as a mosaic of patches in which biotic interactions and abiotic disturbances occur. See *patch*.

patchiness. *n.* Reference to a pattern of distribution in which organisms occur in aggregations or clusters.

Patchnose Snake. Vernacular name for species of colubrid snakes belonging to the American genus *Salvadora*.

patella. *n.* Any flat, cup-shaped bone, usually a *sesamoid* element and widely known as the knee cap in mammals. A.S. Romer (*Osteology of Reptiles*, p. 376, 1956) has used this term to designate a sesamoid element in the tendon of extensor muscle of the thigh, proximal to its attachment with the tibia. He also refers to a sesamoid element in the triceps tendon proximal to its ulnar attachment as the *patella ulnaris* (ibid., p. 372).

paternal. *adj.* Reference to the male parent.

paternal inheritance. Reference to inheritance of characters that are governed by genes or self-reproducing organelles contributed to the offspring solely by the male parent. E.g., in animal embryos the centrioles are derived from a progenitor centriole carried by the fertilizing sperm.

patho-. Prefix meaning "suffering," "feeling," or "associated with disease."

pathogenesis. *n.* The development or course of a disease.

pathogenic. *adj.* That which produces disease or toxic symptoms. Infectious.

pathology. *n.* 1. The structural and functional manifestations of disease. 2. The study of the characteristics of disease, especially body changes that manifest responses or causes of disease states.

patristic character. A character or trait that is inherited by all members of a group from their most recent common ancestor.

pattern class. A putatively genetically nonexclusive population of individual organisms from the same geographic region with similar distinguishing characteristics.

PAUP. See *Phylogenetic Analysis Using Parsimony.*

pavimentous. *adj.* Large, flat, frequently quadrangular, and juxtaposed rather than overlapping.

PBT. See *preferred body temperature.*

P$_c$. See *critical oxygen tension.*

PCA. See *principal components analysis.*

PCBs. See *polychlorinated biphenyls.*

PCR. See *polymerase chain reaction.*

P$_{CO2}$. Symbol used in physiology to denote the partial pressure of carbon dioxide in blood, tissue, or medium.

peat. *n.* A brown, lightweight, unconsolidated or semiconsolidated deposit of plant remains.

peck. *n.* A unit of volume equal to 2 gallons or 9.09218 dm^3.

Peckhammian mimicry. A form of mimicry in which the predator mimics a feature of its prey, thus providing advantages in attracting prey.

pecten. *n.* See *conus papillaris.*

pectinate, pectiniform. *adj.* Comb-shaped, or having projections like the teeth of a comb.

pectineal process. See *prepubic process.*

pectoral. *adj., n.* 1. Pertaining to the chest or breast, or forelimbs. 2. Scales of the ventral chest in lizards. 3. The third pair of laminae on the *plastron* of turtles. (Syn. *thoracic*) See Fig. 6.

pectoral amplexus. A sexual embrace of anurans wherein the forelimbs of the male clasp the female from behind in the region of the chest. (Syn. *axillary amplexus, axillary embrace*)

pectoral crest. See *deltopectoral crest.*

pectoral fold. A fold of skin across the chest of some amphibians.

pectoral girdle. The skeletal elements that form the base of support for the anterior appendages and actions of its muscles, consisting of the *scapula, clavicle* or *coracoid, sternum,* and other bones. See Fig. 3.

pectoral gland. A small gland on the chest of some species of *Scaphiopus*, reported by W.F. Blair et al. (*Vertebrates of the United States,* p. 245, 1957).

pectoralis muscle. The muscles that connect the anterior chest with the shoulder girdle and humerus.

pedestal. *n.* That part of an endoglyph fang above the entrance to the lumen, forming a union with the maxillary bone.

pedicel, pedicle. *n.* 1. Generally, a slender stalk or stem, or a basal part of a structure. 2. Used in reference to a thin stem used to attach eggs of some salamanders to rocks or other objects. 3. The basal part of the amhibian tooth. 4. the fang of a snake. 5. The area of the hemipenis between the proximal truncus and apical region. (Syn. *peduncle*)

pedicellate. *adj.* Possessing a *pedicel*, or attached by one.

pedicellate teeth. The bipartite teeth characteristic of *lissamphibians*. Each tooth consists of a basal pedicel and a distal crown, usually capped by enamel. The latter is connected to the former by a hinge. The pedicel is composed of dentine covered by cementum (bone), while the hinge consists of uncalcified dentine or a ring of fibrous connective tissue.

pedicular foramen. A variable opening in the neural arch below the interzygapophysial ridge. Used by W. Auffenberg (*Tulane Stud. Zool.* 10:131–216, 1963).

pediment. *n.* A gently sloping erosional surface cut into the solid rock of a mountain range in a dry region, usually covered with a veneer of gravel.

pedomorphism, paedomorphism, paedomorphosis *n.* (*adj.* **pedomorphic, paedomorphic**) Condition in which a larva becomes sexually mature without attaining an adult body form. The condition can be achieved by either *neotony* or by *progenesis.*

peduncle. *n.* See *pedicel.*

peel. *v.* To "roll" the foot off the ground while maintaining a contact with the surface until the last phalanx of the longest digit has turned plantar-side up. Described by R.W. Haines (*J. Anat.* 86:412–422, 1952), who suggests this is a typical movement of the forefoot of lizards and gives maximum propulsive force.

peep order. The call or chorus sequence established by individuals of calling frogs, first described for the hylid *Hyla crucifer.* See also *precentor.*

pelagic. *adj.* Reference to living in the upper waters of the open ocean. Living in open water. Cf. *benthic.*

pelobatid. *n, adj.* A member of, or pertaining to, *Pelobatidae.*

Pelobatidae. *n.* A clade (family) of anurans, sister with Megophryidae within the Pelobatoidea and containing the genus *Pelobates* according to D.R. Frost et al. (*Bull. Amer. Mus. Nat. Hist.* 297:1–370, 2006). These frogs are characterized by having a well-developed, spadelike metatarsal tubercle with a well-ossified internally supporting prehallux on the hind feet, used for burrowing. Some classifications recognize three genera and 11 species distributed in North America, western Eurasia, and northwestern Africa, collectively known as "Spadefoot Toads." See also *Scaphiopodidae.*

Pelobatoidea. *n.* A clade (superfamily) of anurans, sister with *Pelodytoidea* within the *Anomoceola* and containing *Megophryidae* and *Pelobatidae* (and *Pelodytidae* of some authors).

Pelodryadinae. *n.* A subfamily of the *Hylidae* with species representing three genera inhabiting the Australopapuan region.

Pelodytidae. *n.* A clade (family) of anurans consisting of a single genus (*Pelydytes*) and three species distributed in western Europe and the Caucasus Mountains of western Asia. These frogs were previously included as a subfamily within Pelobatidae, but they are distinguished by skeletal characters from other pelobatids. They comprise a sister taxon with *Scaphiopodidae* within the *Pelodytoidea* according to D.R. Frost et al. (*Bull. Amer. Mus. Nat. Hist.* 297:1–370, 2006).

Pelodytoidea. *n.* A clade (superfamily) of anurans, sister to *Pelobatoidea* within the *Anomocoela* and containing *Pelodytidae* and *Scaphiopodidae* according to D.R. Frost et al. (*Bull. Amer. Mus. Nat. Hist.* 297:1–370, 2006).

Pelomedusidae. *n.* A clade (family) comprised of African Side-necked Turtles, including two genera and 18 species.

Pelomedusoidea. *n.* A clade (superfamily) of Side-necked Turtles, including the families *Pelomedusidae* and *Podocnemididae.*

pelvic. *adj.* Of or pertaining to the pelvis or hip, or to the hind limb girdle.

pelvic amplexus. A sexual embrace of anurans in which forelimbs of a male clasp the female from behind in the region directly in front of the hind limbs. (Syn. *inguinal amplexus*)

pelvic girdle. The bones and cartilages that form the base of support for the posterior appendages and the actions of their muscles.

pelvic gland. Part of a large, tubiform gland in the roof of the cloaca of male salamanders. This gland opens into the cloaca and functions in relation to production of spermatophores. It is homologous with the *spermatheca* in female salamanders. See also *abdominal gland, cloacal gland.*

pelvic patch. A specialized region of ventral skin in the vicinity of the groin of anurans. It is thin, highly vascular, and adapted for rapid uptake of water from moist surfaces or soil. Antidiuretic hormone (ADH) modifies both the blood flow to the pelvic patch and its permeability to water. (Syn. *seat patch, ventral seat patch, ventral patch*)

pelvic spur. Either of a pair of vestigial limbs, visible on either side of the anus in boid snakes and the Oriental Pipe Snake (*Cylindrophis*). These are generally relatively more enlarged in males than in females and may be used as organs of stimulation during courtship. (Syn. *anal claw, anal spur*)

Pelycosauria, pelycosaurs. *n.* One of two major groups of *synapsid* reptiles, sometimes called "Sail-backed Reptiles," originating from *cotylosaurs*, enjoying extensive radiation through the early Permian, and giving rise to *therapsids.*

penetration path. See *copulation path.*

penis. *n.* Male copulatory organ of turtles, crocodilians, primitive birds, and mammals, consisting of single or paired masses of erectile tissue called *corpora cavernosa.* See also *hemipenis, intromittent organ, phallodeum.*

penis bone. A *baculum*, or bony rod reinforcing the penis.

Pennsylvanian. *n.* An Epoch of the Carboniferous Period lasting roughly 25 million years. The Pennsylvanian is named after the state of Pennsylvania where rocks from this age are widespread.

penta-. Prefix meaning "five" or "fivefold."

pentadactyl, pentadactylous. *adj.* Possessing five digits (per limb).

pentadactylin. *n.* An antimicrobial peptide from the skin secretions of the South American Bullfrog, *Leptodactylus pentadactylus.* See J.D. King et al., *Comp. Biochem Physiol.* 141C:393–397, 2005.

pentose. *n.* A five-carbon monosaccharide sugar.

pepsin. *n.* A proteolytic enzyme secreted by the gastric mucosa.

peptidase. *n.* Any of a subclass of proteolytic enzymes that catalyze the hydrolysis of peptide linkages.

peptide. *n.* A molecule consisting of a linear array of amino acid residues. A short peptide chain is called an *oligopeptide*, and long chains are called *polypeptides.* One or more peptides comprise proteins. Bioactive peptides play important roles as neurotransmitters, neuromodulators and/or neuroendocrine agents and are present in amphibian skin secretions.

peptide bond. A covalent bond linking amino and carboxyl groups between two amino acids in a protein or peptide.

per-. Prefix meaning "through."

peramorphosis. *n.* An evolutionary change in which the juveniles of a descendant species display some of the characteristics of adult forms of the ancestral species. Cf. *paedomorphosis.*

per capita. Per individual.

perennial. *adj., n.* 1. Occurring throughout the year. 2. Reference to plants that persist for several years. Cf. *annual, biennial.*

Perennibranchia. *n.* A monophyletic group of salamanders and sister taxon with *Treptobranchia* within the *Hydatinosalamandroidei*. It contains the families *Proteidae* and *Sirenidae.* See D.R. Frost et al., *Bull. Amer. Mus. Nat. Hist.* 297:1–370, 2006.

perennibranchiate. *adj.* Retention of gills throughout life (characteristic of certain groups of salamanders).

Perentie. *n.* Vernacular name for the Australian varanid lizard species *Varanus giganteus.*

performance. *n.* Key tasks or functions such as feeding, digesting, moving, growing, regulating water, etc., that are crucial for survival and reproduction. This term and concept has become widely used in relation to studies of comparative and organismal biology linking phenotypic traits, "performance," and fitness.

perfusion. *n.* The passage of fluid over or through an organ, tissue, or cell. Commonly, the reference is to blood flow through capillaries of an organ.

peri-. Prefix meaning "around."

peribranchial chamber. The space in a tadpole containing the gills, anterior to the *diaphram* separating the *buccopharynx* from abdominal cavities and delimited

peribranchial wall — **peristome**

by the *operculum*. (Syn. *gill chamber, opercular chamber*)

peribranchial wall. See *diaphragm*.

pericapsular recess. That portion of the perilymphatic system that surrounds the otic capsule and proximal portion of the *extracolumella* in turtles.

pericardial. *adj.* Surrounding the heart.

pericardial chamber or **cavity.** The separation that occurs between the inner *pericardium* and *epicardium*.

pericardium. *n.* The fibroserous membrane that envelops the heart and secretes fluids to surround and lubricate it.

perichondral bone. Bone that is formed over underlying cartilage. Cf. *periosteal bone*.

perichondrium. *n.* The fibrous membrane that covers cartilage.

perichordal. *adj.* Reference to developmental events associated with the centrum. Goodrich (1930) refers to perichordal development as the appearance of cartilaginous *arcualia*, which join and fuse completely surrounding and eventually constricting the notochord, whereas Williams (1959) describes chondrification of the *perichordal tube*, eventually surrounding and constricting the notochord and persisting until it is replaced by bone.

perichordal tube. A continuous axial tube of sclerotome origin and surrounding the notochord early in embryonic development. The *autocentrous* vertebrae of tetrapods develop from these layers of cells according to E.E. Williams (*Quart. Rev. Biol.* 34:1–32, 1959).

periderm. *n.* The outer layer of cells in the embryonic skin.

perilymph, perilymphatic fluid. *n.* An aqueous fluid surrounding the labyrinth of the inner ear.

perinatal. *adj.* Relating to the period shortly before and after birth.

periocular. *adj.* Reference to areas or structures surrounding the eye.

period, Period. *n.* 1. The period of a cyclic phenomenon is the amount of time it takes for the variable to go through one complete cycle and return to its current or reference level. The period is measured in units of time. 2. A division of geological time (capitalized). See *geological Period*.

periodic acid Schiff procedure. (PAS) A procedure that stains for polysaccharides.

periodicity. *n.* 1. A regularly recurring change in an organism that is dependent on an external stimulus from the environment. Cf. *rhythm*. 2. In molecular biology, the number of base pairs per turn of the DNA double helix. 3. The periodic recurrence of an event in any system.

periodontal ligament. Fibrous tissue supporting a tooth in its socket, extending between the cement and alveolar bone. A region of organized collagen fiber matrix not calcified by osteoblast products.

periosteal bone. Bone that is formed over underlying bone that is completely ossified. Cf. *perichondral bone, endosteal bone*.

periosteum. *n.* (*adj.* **periosteal**) The sheet of fibrous connective tissue covering all bones, and having bone-forming potential. Cf. *endosteum*.

periotic, periotic element. *adj., n.* 1. Surrounding the internal ear. 2. A collective term for any one or more of the bones comprising the otic capsule. Includes the *prootics, opisthootics*, and, according to H. Gadow (*Amphibia and Reptiles*, p. 95, 1901), also includes the *epiotic*.

peripatric. *adj.* Living in a region peripheral to that of the main population of the species.

peripatric speciation. A mode of speciation that occurs in localized descendants of a founder population such as those isolated on the periphery of the distribution of the parental population or on islands, in contrast to *parapatric speciation*. The concept was proposed by E. Mayr who proposed that a bottleneck population, or reduction in population size, initiates genetic change (pp. 157–180 in *Evolution as a Process*, ed. J. Huxley, A.C. Hardy & E.B. Ford, 1954). (Syn. *founder effect speciation*) Cf. *dichopatric speciation*.

peripheral. *n.* 1. Any one of the marginal bones that form the outer border or margin of the turtle carapace, each one lying beneath a *marginal* lamina. See Fig. 5. 2. See *peripheral device*.

peripheral circulation. Reference to blood circulation through arteries, capillaries, and veins, distal to the heart.

peripheral device. A computer device, such as a printer, keyboard, monitor, etc., that is not part of the essential computer (i.e. memory and microprocessor). Peripheral devices may be external or internal, the latter often referred to as *integrated peripherals*.

peripheral isolate. A population isolated at or beyond the outer boundary of a species' range.

peripheral nervous system. All neurons or parts of neurons that lie outside the brain and spinal cord, including autonomic nerves, somatic motor nerves, sensory receptors, and sensory afferent neurons. The cranial and spinal nerves and their associated ganglia. Cf. *central nervous system*.

peripheral resistance unit. (PRU) A unit measure of the drop of pressure (ΔP, in mm Hg) along a vascular bed divided by the mean flow in ml/sec. The concept of *peripheral resistance* expresses the resistance to blood flow offered largely by the smaller peripheral vessels and is quite analogous to the concept of electrical resistance in an electric circuit.

peristalsis. *n.* A traveling wave of constriction (followed by relaxation) in tubular tissue such as the gut, produced by constriction of circular muscle. The movements function to move materials through the gut. (Syn. *peristaltic action*)

peristome. *n.* Literally "around the mouth" and used occasionally (e.g. T. Barbour, *Reptiles and Amphibians, Their Habits and Adaptations*, 1926) in reference to the mouth

peritoneal canal

of certain tadpoles having an oral *umbrella*. This term is usually used in reference to mouth structures of invertebrates and is not recommended for usage in herpetology.

peritoneal canal. A term given to small canals associated with the cloacal wall or penis and observed with either abdominal openings or blind ends in various turtles and crocodilians.

peritoneal cavity. The major body cavity that occupies much of the trunk and contains most of the digestive tract, organs of excretion and reproduction, and is bounded by the *peritoneum* lining the visceral organs and the inner aspect of the body wall.

peritoneal funnel. 1. An embryonic ciliated opening from the *nephrocoel* to the *coelom*. 2. See *nephrostome*.

peritoneal sheath. The white portion of the *parietal peritoneum* of some frogs.

peritoneum. *n.* A serous membrane of mesodermal epithelium that lines the *peritoneal cavity*. The portion of the peritoneum that covers the organs and forms their mesenteries is the *splanchnic* or *visceral peritoneum*. Whereas that portion lining the inside of the body wall is the *somatic* or *parietal peritoneum*.

peritonitis. *n.* An infection of the *peritoneum*.

perivascular lymphatics. Lymphatic vessels that surround blood vessels, especially common in snakes.

perivitelline membrane. A membrane formed beneath the *vitelline membrane* immediately after fertilization.

perivitelline space. The space between the *vitelline* and *perivitelline membranes*. (Syn. *capsular cavity*)

permeability. *n.* In general usage, the ease with which substances can pass through a membrane. More specifically, the *permeability coefficient* expresses the rate of flux of a substance passing through a membrane per unit of concentration difference (or driving force) across the membrane (units in cm/s).

permeability coefficient. See *permeability*.

permeable. *adj.* Allowing passage of a substance (e.g., ion or water) through a membrane.

Permian. *n.* The last geologic Period of the *Paleozoic Era*. Permian tetrapods included amphibians, sauropods, and synapsid reptiles. A fully terrestrial fauna evolved with the appearance of the first large herbivores and carnivores. See Table 1.

permissive. *adj.* In context of physiology, hormonal actions permitting target tissues to respond to hormonal, neuronal, or environmental stimuli.

peroneal. *adj.* Pertaining to the fibula or to the outer side of the leg.

Peron's Sea Snake. Vernacular name for the elapid species *Acalyptophis peronii*. Also called *Eyelash Sea Snake*.

per se. As such; of or by itself.

pers. comm. Abbreviation of "personal communication."

persistence. *n.* 1. In context of ecosystem dynamics, the survival time of the system or part of the system. 2. The length of time a toxic substance or other compound remains in the environment or some part of an ecosystem.

perturbation. *n.* A disturbance, used in reference to any departure of a living system from a steady state.

pes. (*pl.* **pedes**) *n.* The foot or hindfoot, comprised of ankle-sole-toes.

petalum. (*pl.* **-a**) *n.* One of a number of transverse, overlapping laminate structures on the hemipenes of some lizards.

petechiae. (*pl.*) *n.* (*adj.* **-ial**) Small hemorrhages, solitary or multiple.

Petersen method. A method of mark-and-recapture used for estimating the population size (N) of a closed population, calculated as

$$N = CM/R$$

where M is the number of individuals marked in first sample, C the number of individuals captured in a second sample, and R the number of marked individuals in the second sample.

petrocolous. *adj.* Living in rocky habitats.

Petropedetidae. *n.* A clade (family) of small frogs native to forests and savannas of sub-Saharan Africa. Formerly containing 13 genera and about 100 species, with 70% being species of *Phrynobatrachus*. The latter are now included in *Phrynobatrachidae*, with *Petropedetidae* containing four genera (*Conraua, Indirana, Arthroleptides, Petropedetes*) native to south India, tropical West and East Africa. For taxonomic revision see D.R. Frost et al., *Bull. Amer. Mus. Nat. Hist.* 297:1–370, 2006.

petrosal. *n.* A synonym for *prootic* according to S.W. Williston (*The Osteology of the Reptiles*, p. 27, 1925).

pg. Picogram.

P generation. See *parental generation*.

pH. A measure of acidity, specifically the negative logarithm (base 10) scale of hydrogen ion concentration of a solution: $pH = -\log [H^+]$. The measure of acidity is on a scale from 0 (acid) through 7 (neutral) to 14 (alkaline), and each whole unit represents a tenfold change in acidity/alkalinity.

-phagous. Suffix meaning "feeding on," "eating."

phagocyte. *n.* (*adj.* **–ic**) A cell that engulfs other cells or foreign particles, including microorganisms.

phagocytosis. *n.* The process of a cell engulfing other particles of matter.

phalangeal formula. Listing of the number of bones in each digit, beginning with the inner digit and proceeding in sequence to the outermost digit of the foot.

phalanx. (*pl.* **phalanges**) *n.* (*adj.* **phalangeal**) Any one of the bones of a digit.

phallodeum. *n.* An eversible portion of the male cloaca of caecilians, which is inserted into the female's cloaca and used as an organ of intromission during copulation. See *penis*.

phaneric coloration. Conspicuous coloration. Cf. *cryptic*.

phaneroglossal. *adj.* Possessing a tongue. Cf. *aglossal*.

Phanerozoic. (occasionally **Phanaerozoic**) *n.* The Eon covering roughly 545 million years during which abundant animal life has existed, divided into three Eras: *Paleozoic, Mesozoic,* and *Cenozoic*. See Table 1.

pharotaxis. *n.* Movement toward a place or location in response to a learned or conditioned stimulus.

pharyngeal. *adj.* Pertaining to the *pharynx*.

pharyngeal fold. A transverse membranous fold extending across the palate in front of the pharynx in microhylid frogs. Two or three such folds might be present, with the posterior one often bearing teeth.

pharyngeal gland. A gland lying near the internal nares of anurans. The secretions are thought to pass into the nares.

pharyngeal groove. A ciliated groove used for food collection and located along the outer margin of the branchial chamber in tadpoles of *Xenopus*. See also *pharyngobranchial tract*.

pharyngeal process. Any of several processes on the *cricoid cartilage* of the amphibian larynx. (Syn. *spina oesophagea*)

pharyngeal pump. Reference to movements of *ceratobranchials* that cause changes in the volume of the pharyngeal cavity during irrigation of the gills of tadpoles.

pharyngeal respiration. Reference to exchange of respiratory gases across the highly vascularized membranes of the pharynx, characteristic of many aquatic turtles, salamanders and anurans. See *buccopharyngeal respiration*.

pharyngeal slit. An elongated opening in the lateral wall of the pharynx.

pharyngobranchial tract. The mid-region of the pharyngeal floor in *Xenopus* tadpoles, which lies between the gill filters and is covered with ridges that entrap fine particles in secreted mucus. The particles trapped by mucus are carried by cilia in pharyngeal grooves at the sides of the pharynx into the esophagus. The pharyngobranchial tract functions similarly to *branchial food traps*, which in microphagous tadpoles of other species are covered by a flap or valve (ventral velum). For further clarification see R.J. Wassersug, *J. Morphol.* 137:279–288, 1972; *Occas. Pap. Mus. Nat. Hist. Univ. Kansas* 48:1–23, 1976; R.J. Wassersug & K. Rosenberg, *J. Morphol.* 159:393–426, 1979.

pharynx. *n.* The region of the digestive tract immediately posterior to the mouth, in which functional paired gill slits are present in adult and larval fish-like vertebrates, in all larval and some aquatic adult amphibians, and embryonically in other vertebrates. The throat.

phase contrast microscopy. A technique in microscopy that uses differential light refraction by different components of the specimen to enhance viewed images. The phase contrast microscope is an optical system that converts phase variations in the light focused on the image plane into visible variations in light intensity or contrast.

phasic. *adj.* Transient or recurring cycles of activity in nerve or muscle.

phasic receptor. A sensory receptor that responds only briefly to a stimulus, typically producing action potentials only at the onset, or during part, of a sustained stimulus.

phene. *n.* Any phenotypic character that is genetically determined.

phenetic. *adj.* Reference to overall similarity based on many characters selected without regard to evolutionary history, including character states arising from common ancestry, parallel evolution, and convergent evolution.

phenetics, phenetic method, phenetic taxonomy. A system of classification based on numerical comparison of large numbers of equally-weighted phenotypic characters without regard to phylogenetic relationships. (Syn. *numerical taxonomy, taximetrics*) Cf. *cladistics, evolutionary method, omnispective method*.

phenocline. *n.* A graded series of phenotype frequencies within the geographical range of a species. See *cline*.

phenogram. *n.* A branching diagram that links taxa according to estimates of overall similarity based on evidence from a sample of characters that are not judged as to whether primitive or derived. Numerical algorithms are used to create diagrams of overall similarity among species.

phenology. *n.* (*adj.* **-ical**) Reference to natural phenomena that recur periodically, e.g. migration, reproduction.

phenon. *n.* A group of phenotypically similar organisms, used in numerical taxonomy to replace taxon.

phenotype. *n.* The physical characteristics of an organism, which comprise the manifestation of genetic expression. Any property of an organism that can be measured. See also *character*. Cf. *genotype*.

phenotypic evolution. Change in the developmental program descendants inherit from their ancestors.

phenotypic plasticity. Phenotypic variation due to environmental influence on the phenotype associated with a particular genotype, representing genotype-environment interaction. See also *reaction norm, developmental plasticity*.

phenotypic sex determination. Reference to control of gonad development by nongenetic stimuli. See also *environmental sex determination, temperature dependent sex determination*.

phenotypic variance. The total variance observed in a character. See *genetic variance*.

phenotypic variation. The total variation among the individuals in a population.

pheromone. *n.* A chemical substance produced by an individual and released into the environment for the purpose of signaling between (among) individuals of the same species. Examples include substances that are used to mark territories, attract mates, or provide scents used in trailing.

philopatric. *adj.* Reference to organisms that tend to remain in the location where they were born.

pholidosis. *n.* The arrangement and pattern of scales, including shields and laminae, of reptiles. Cf. *scutation*.

pholidate. *adj.* Possessing scales; scaly.

-philous. Suffix meaning "loving" or "thriving in."

-phobic. Suffix meaning "intolerant of," or "lacking affinity for."

phonon. *n.* A quantum of sound energy.

phonotaxis. *n.* Movement of an animal relative to a sound source.

phosphagens. *n.* High energy phosphate compounds that donate phosphate groups used for rapid rephosphorylation of ATP from ADP (e.g. *creatine phosphate*).

phosphatase. *n.* Any of a group of enzymes that catalyze the hydrolysis of esterified phosphoric acid, with liberation of inorganic phosphate. Several different phosphatases occur in snake venoms.

phosphate bond energy. The energy liberated when a phosphorylated compound undergoes hydrolysis to form free phosphoric acid. See *adenosine triphosphate*.

phosphatidase. *n.* See *phosphatidase A*.

phosphatidase A. A *hemolysin* enzyme and component of snake venoms that generates lysolecithin from tissue phosphatides, the former being capable of weakening cell walls. (Syn. *phosphotidase, phospholipase, phospholipase A*)

phosphocreatine. *n.* See *creatine phosphate*.

phosphodiesterase, phosphoesterase. *n.* An enzyme component of snake venoms that hydrolyzes nucleic acids (DNA, RNA) and breaks down phosphorus containing substances in cell nuclei.

phosphodiester bonds. Bonds that link individual nucleotides in nucleic acid.

phosphoglycerides. *n.* Glycerol-based lipids found in biological membranes.

phospholipase. *n.* See *phosphatidase A*.

phospholipase A. See *phosphatidase A*.

phospholipid. *n.* A lipid containing phosphate esters that hydrolyze to fatty acids, glycerol, and a nitrogenous compound.

phosphorescence. *n.* Emission of light in darkness due to release of absorbed radiation. Sometimes this term is applied incorrectly to *bioluminescence*.

phosphorylation. *n.* The incorporation of a phosphate group into an organic molecule.

phosvitins. *n.* Egg-yolk proteins derived from *vitellogenin*.

phot. The *cgs* unit of surface illumination, equal to 1 lumen/cm^2.

photodissociation. *n.* Chemical dissociation induced by light.

photolysis. *n.* The decomposition of compounds by radiant energy, especially light.

photomicrography. *n.* The process of making photographs through a microscope.

photon. *n.* A quantum of electromagnetic energy. The smallest amount of light energy that can exist at any given wavelength.

photoperiod. *n.* See *photoperiodism*.

photoperiodism or **photoperiodicity.** *n.* (*adj.* **photoperiodic**) A physiological response to daylength. The *photoperiod*, or period of light in 24 hours, triggers such activities as breeding, migration, brumation, and other aspects of behavior.

photophase. *n.* Reference to the light phase in a 24-h period. Cf. *scotophase*.

photophore. *n.* A light-emitting cell, organ, or luminous spot, unknown among amphibians and reptiles. See *bioluminescence*.

phototopic vision. Color vision in bright light. Cf. *scotopic vision*.

photopigment. *n.* A pigment molecule that changes its energy state when it absorbs one or more photons of light.

photoreactivation. *n.* Reference to reversal of injury to cells caused by ultraviolet light, accomplished by subsequent exposure of affected cells to visible light.

photoreceptor. *n.* A biological light receptor.

phototaxis. *n.* Movement of an animal relative to light. Cf. *heliotaxis*.

phototransduction. *n.* The conversion of light energy incident on a receptor into a cellular response.

phragma. *n.* A lid.

phragmosis. *n.* Reference to blocking the entrance to a hole or burrow by use of a part of the body, usually roughened or hardened for this purpose. Examples are the heads of certain anurans and tails of burrowing snakes. The term was coined by W.M. Wheeler (*Quar. Rev. Biol.* 2:1–36, 1927) in reference to insect behaviors, but he also extended the usage to other groups of animals. The behavior was first observed in amphibians by T. Barbour (*Mem. Mus. Comp. Zool.* 44:242–243, 1914; also *Reptiles and Amphibians. Their Habits and Adaptations*, 1926).

Phrynobatrachidae. *n.* A taxon (family) of anurans, sister to *Pyxicephaloidea* within the new taxon *Africanura*, containing two genera with species restricted to sub-Saharan Africa. For taxonomic revision see D.R. Frost et al., *Bull. Amer. Mus. Nat. Hist.* 297:1–370, 2006.

Phrynomerinae. *n.* A subfamily of anurans belonging to *Microhylidae*.

Phrynosomatidae. *n.* A clade (family) of lizards previously treated as a subfamily of the Iguanidae: *Sceloporinae* or *Phrynosomatinae*. This group—including the genera *Callisaurus, Cophosaurus, Holbrookia, Petrosaurus, Phrynosoma, Sator, Sceloporus, Uma, Urosaurus,* and *Uta* —was revised to family status by D.E. Frost and R.E.

Etheridge (*Univ. Kans. Mus. Nat. Hist. Misc. Publ.* 81, 1989), which is widely accepted. Some systematists, however, remain in disagreement (e.g., J.R. Macey et al., *J. Mol. Evol.* 44:660–674, 1997).

Phrynosomatinae. *n.* A clade (subfamily) of iguanid lizards that are morphologically and ecologically diverse, including "Horned Lizards" (*Phrynosoma* spp.) and the speciose genus *Sceloporus*. There are nine genera and approximately 125 species distributed from southern Canada to Panama. See *Phrynosomatidae*.

Phthanobatrachia. *n.* A monophyletic clade of anurans, sister with *Heleophrynidae* within the *Neobatrachia*, and containing newly designated *Hyloides* and *Ranoides* (D.R. Frost et al., *Bull. Amer. Mus. Nat. Hist.* 297:1–370, 2006).

phyletic. *adj.* Reference to course of evolution or a direct line of descent.

phyletic evolution. The gradual transformation of one species into another without branching. (Syn. *anagenesis, successional speciation, vertical evolution*) Cf. *cladistic evolution*.

phyletic gradualism. A model of evolution in which species change gradually through time as a result of slow, direct transformation within a lineage, thereby producing a graded series of differing forms. Cf. *punctuated evolution*.

phyletic line or **lineage.** 1. A sequence of two or more successive *chronospecies* between two successive branching points of a phylogenetic tree. 2. A lineage that is relatively continuous and complete in the fossil record.

phyletic speciation. See *phyletic evolution*.

phyletic tree. See *phylogenetic tree*.

phyletic extinction. See *pseudoextinction*.

phyllolitorin. *n* See *bombesin*.

Phyllomedusinae. *n.* A subfamily of the *Hylidae* with member species representing seven genera inhabiting Central and South America. Some members of this family have evolved cutaneous lipid glands and wiping behaviors that "wax" the skin and thereby reduce evaporative water loss.

phyllospondylous. *adj.* A type of vertebra in which the centrum is incomplete above, with the spinal cord and the notochord lying in a single cavity and not separated by a bony area. The vertebra appears to be comprised entirely of the *hypocentrum* and *neural arches*, while the *pleurocentrum* is completely absent.

phylochronology. *n.* The study of populations in space and time using phylogenetic and population genetic methods. The approach attempts to determine variation in genetic diversity experienced by populations in context of a changing environment through time.

PhyloCode. *n.* A newly proposed system for naming organisms based on evolutionary relationships (*phylogeny*) rather than grouping them by shared characteristics. Proponents would denote groupings as nested "clades" descended from a common ancestor, rather than arranging groups in a traditional descending hierarchy of phyla, classes, orders, families, etc. The idea "to complete the Darwinian revolution" with a nomenclature based in evolutionary theory was proposed in a series of papers by J. Gauthier and K. de Queiroz, published between 1987 and 1994. Phylogenetic taxonomy provides a useful system for defining clade names based on evolutionary relationships among organisms. Acceptance of such a system of phylogenetic nomenclature has been controversial to present. (Syn. *phylogenetic taxonomy*) See K. de Queiroz & J. Gauthier, *Annu. Rev. Ecol. Syst.* 23:449–480, 1992.

phylogenetic. *adj.* Reference to evolutionary relationships within or between groups.

phylogenetically independent contrasts. A phylogenetically-based statistical method now widely used for testing hypotheses about character states in comparative data sets. The method compares values of sister taxa and extracts variation correlated with phylogeny, such that residual "corrected" variation can be analyzed for its relationship to present-day ecology. The method therefore gives priority to phylogeny and considers only the residual variation as potentially attributable to ecology. Independent contrasts was originally developed and applied for testing bivariate correlated evolution of continuous-valued characters. It can also be applied to multiple regression, ANOVA, ANCOVA, principle components analysis, and other statistical approaches including comparisons of rates of evolution across clades. See J. Felsenstein, *Am. Naturalist* 125:1–15, 1985, for first presentation of the method.

Phylogenetic Analysis Using Parsimony. (PAUP) A computer program that generates the most parsimonious dendrogram of taxa from a set of data.

phylogenetic bracket. See *extant phylogenetic bracket*.

phylogenetic character. A homologous feature, phenotype or trait of an organism or group of organisms.

phylogenetic classification. See *natural classification*.

phylogenetic distance. A measure of the degree of separation between two organisms or their genomes, expressed in various ways such as the number of accumulated sequence changes, number of years, or number of generations. The distances are often mapped onto phylogenetic trees, which depict deduced relationships among the organisms.

phylogenetic effect. See *phylogenetic inertia*.

phylogenetic homology. See *homology, homologous*.

phylogenetic inertia. 1. Reference to preadaptations and underlying genetic mechanisms that determine the direction and rate of evolutionary change. 2. Due to shared ancestry, closely related species are more likely to be similar to each other than are distantly related species. Thus, character data for different species may not be independent. This phenomenon also is called "*phylogenetic effect*" and could have a number of different causes.

See P.H. Harvey & M.D. Pagel, *The Comparative Method in Evolutionary Biology*, 1991.

phylogenetic noise. Patterns of character states due to *convergent evolution* that causes potential misinterpretation of phylogenetic relationships.

phylogenetic species. See *species*.

phylogenetic systematics. A method for reconstructing evolutionary trees in which taxa are grouped exclusively on the presence of shared derived characters. See *Hennigian systematics*.

phylogenetic taxonomy. See *PhyloCode*.

phylogenetic tree. A diagram that depicts the hypothesized geneological ties and sequence of historical ancestor-descendent relationships linking individual organisms, populations or taxa. When species are considered, they are represented by line segments, and points of branching correspond to speciation events, with a measure of relative or absolute time on one axis. (Syn. *evolutionary tree, phyletic tree, phylogram, Tree of Life*)

phylogeny. (also **phylogenetics, phylogenesis**) *n*. The relationships of groups of organisms according to their evolutionary history. Evolutionary history of a lineage or organism.

phylogeography. *n*. Investigations based in molecular biology that attempt to evaluate the spatial distribution of evolutionary lineages and their phylogenetic relationships. This discipline documents the genetic diversity within and among geographical areas and evaluates the phylogenetic distinctiveness of populations, providing insight into how organisms have historically responded to local landscape changes. See E. Bermingham & C. Moritz, *Mol. Ecol.* 7:367–369, 1998; J.C. Avise, *Phylogeography: The History and Formation of Species*, 2000. *Comparative phylogeographic* analyses contrast gene trees of codistributed species, and these methods facilitate searches for common historical events that might have affected multiple taxa in a given region.

phylogram. *n*. A *phylogenetic tree* wherein branch lengths are proportional to amount of "time" separating taxa. See also *phylogenetic tree*. Cf. *cladogram*.

phylum. *n*. A taxonomic category ranking above *class* and below *kingdom*.

physical environment. All structural and chemical factors of the environment, comprising the abiotic component of an ecosystem.

physiognomy. *n*. (*adj.* –**ic**) Characteristic features or appearance of vegetation or a plant community

physiograhy. *n*. (*adj.* –**ic**) The description of surface features of Earth.

physiological color change. Color changes that occur quickly in relation to nervous, hormonal, or local control mechanisms. Cf. *morphological color change*.

physiological cross section. A cut through the area of all muscle fibers perpendicular to their long axes. Cf. *morphological cross section*.

physiological dead space. See *dead space*.

physiological race. A race characterized by physiology or physiological characters.

physiological saline. An isotonic aqueous solution of NaCl for temporary maintenance of living cells or tissues.

physiology. *n*. The study of function in living organisms; dynamic processes and their underlying mechanisms in physical and chemical terms.

phytogeography. *n*. Study of the geographical distribution and relationships of plants.

phytolith. *n*. A gastrointestinal mass consisting of plant fibers.

phytoplankton. *n*. See *plankton*.

phytotelm. (pl. **phytotelmata**). *n*. A water-holding cavity in some part of a plant or plant product (e.g., leaf or nut shell); sometimes extended to include other small, isolated cavities of standing water (e.g., snail shell).

pia. *n*. The innermost covering of the brain and spinal cord.

Pickerel Frog. Vernacular name for the American ranid frog species *Lithobates* (formerly *Rana*) *palustris*. For recent taxonomic revision see D.R. Frost et al., *Bull. Amer. Mus. Nat. Hist.* 297:1–370, 2006.

pico-. (p) Prefix used to denote unit x 10^{-12}.

pie chart. A circle divided into radial sectors having sizes proportional to quantities they represent, used to represent data arranged in percentage categories.

Pielou's evenness. (J´) See *Equitability index*.

piezo-. Prefix meaning "pressure."

piezoelectric. *adj*. Pertaining to the electrical property of a crystalline substance in relation to changes in pressure. Load stress on bones may produce surface electric charges, for example.

Pig Frog. Vernacular name for the American ranid frog *Lithobates* (formerly *Rana*) *grylio*. For recent taxonomic revision see D.R. Frost et al., *Bull. Amer. Mus. Nat. Hist.* 297:1–370, 2006.

pigment. *n*. In biological usage, a molecular substance in cells or tissues that imparts color.

pigment spot. An unusual accumulation of melanophores in Axolotls, which can in some cases be transformed into actively growing tumors.

Pigmy Rattlesnake. Collective vernacular name for small Rattlesnakes belonging to the genus *Sistrurus*, especially the species *S. miliarius*. See also *Massasauga*.

Pig-nosed Turtle, Pignose Turtle. Common name for a large freshwater species *Carettochelys insculpta* (also called *New Guinea Plateless Turtle* and *Fly River Turtles*). This turtle has flippers shaped much like those of sea turtles, and the nares open at the end of a fleshy proboscis, resembling a pig's nose.

pileus. *n*. A black, cap-like marking on the top of the head.

piliform appendage. Elongated, hairlike, and transparent extensions of the shed skin, described in *bufonid* and *leptodactylid* anurans.

pillar. *n.* A buttress.

pineal, pineal body. An organ formed from the roof of the forebrain that acts as a light dosimeter, produces melatonin, and regulates the hormonal output of other endocrine organs. (Syn. *pineal organ, pineal gland, epiphysis, epiphysis cerebri*)

pineal end organ. An extraoptic, photosensitive medial structure in many anurans that develops as a dorsal outgrowth of the diencephalon, pinches off, and comes to lie in the dermis. The external terminus is marked by the *brow spot*. Receptor cells are present with ultrastructure that resembles photoreceptors in the retina of the eye, and the end organ is innervated by the pineal nerve. The structure is homologous with the parietal eye and has been implicated to function with respect to pigmentary adaptation, synchronization of circadian locomotor rhythms, sun-compass orientation, and polar-taxis. (Syn. *frontal organ, Steida's organ, stirnorgan*)

pineal eye. See *parietal eye*.

pineal foramen. See *parietal foramen*.

pineal gland. See *pineal, pineal body*.

pineal organ. See *pineal, pineal body*.

pineal stigma. An external indication of the location of the *pineal body*, usually in the center of a middorsal scale on dorsum of the head.

Pine Barrens Treefrog. Vernacular name for the North American hylid species *Hyla andersonii*.

Pine Snake. Vernacular name for several American colubrid snakes including *Pituophis melanoleucus* and *Pituophis ruthveni*.

Pine Woods Snake. Vernacular name for the American colubrid species *Rhadinaea flavilata*.

piniform. *adj.* Conical or cone-shaped.

pinky. *n.* Vernacular and slang name for a newborn, hairless mouse, which is a common food item used in the husbandry of captive snakes and lizards.

pinnate, pinnate muscle. *adj.* 1. Generally, feather-like, having a series of similar structures arranged in one or more series and branching from a central axis. 2. Descriptive of muscles having their fibers sloping inward toward one or more central tendons, an arrangement that confers greater force per unit cross-section or volume of muscle. Cf. *parallel muscle*. 3. Used in some literature to describe the gills of certain larval salamanders and the oblique flouncing on the snake hemipenis.

pinnigrade. *adj.* A form of swimming locomotion by means of using flippers as paddles (e.g., Sea Turtles).

pinocytosis. *n.* (*adj.* *pinocytotic*) Active ingestion of fluid into a cell by invagination of the cell membrane to form vesicles.

pint. A unit of volume equal to 1/8 gallon (= 0.56826 dm^3).

pioneer species. The first species to appear in an environment after a disturbance.

Pipanura. *n.* A clade of anurans that includes all extant forms except for the basal clades of *Ascaphidae, Leiopelmatidae, Bombinatoridae,* and *Discoglossidae*.

Pipe Snakes. *n.* Collective vernacular name for various species of snakes in the families *Anomochilidae, Aniliidae, Uropeltidae,* and *Cylindrophiidae* (or subfamily *Cylindrophiinae*). See also *Asian Pipe Snakes, Dwarf Pipe Snakes*.

Pipidae. *n.* A clade (family) of aquatic anurans, unique among frogs in lacking tongues. Five genera include the well-known *Pipa* (Surinam Toad) and *Xenopus* (Clawed Frogs, widely used in experimental biology) and represent approximately 30 species distributed in Panama, tropical South America, and sub-Saharan Africa. A sister taxon with *Rhinophrynidae* within the *Xenoanura*.

pipinin. *n.* A peptide isolated from *Rana pipiens* having potent chemotactic and disruptive actions on mammalian mast cells, which liberate histamine during inflammatory reactions. See also *mast cell disrupting peptides*.

Pipoidea. *n.* A clade of anurans that includes *Pipidae* and *Rhinopohrynidae*.

pipping. *n., v.* The act of slitting the eggshell just prior to emergence of a hatchling reptile.

piscivore. *n.* (*adj.* *–ous*) Reference to feeding on fish. (Syn. *ichthyophagous*)

pisiform. *n.* A sesamoid bone in the reptilian carpus, usually quite small and located on the outer margin, in a notch between the *ulna* and the *cuneiform*.

PIT. See *passive integrated transponder*

pit. *n.* Any pit or depression, often in a scale. Sometimes used in reference to femoral pores of lizards. See *apical pit, facial pit, scale pit, fossette, pit organ*.

pitch. *n.* Rotation of a swimmer or flyer around its transverse axis.

pitfall trap. A structure such as a can or bucket that is sunken into the earth with the top generally level with the substrate. Smaller animals accidentally fall into these structures and thereby become trapped. Pitfall traps are often used in conjunction with *drift fences* as an effective means to sample the diversity and abundance of amphibians and reptiles in field studies.

pith. *v.* A means of rapidly killing an amphibian or reptile by inserting a sharp instrument such as a dissecting needle between the skull and first vertebra and wiggling it to destroy the spinal cord and brain tissue near the point of entry.

pit organ. Specialized *infrared receptors* having evolved independently several times within the Boidae and once within Colubroidea (Crotalinae). Boid pit organs are located in upper and lower labial scales (and/or rostral scales), whereas crotaline pit organs are present as single structures between the eye and nostril on either side of the head (Fig. 21). A pit organ consists of a thin, innervated membrane stretched across an open cavity, a design that permits rapid and sensitive detection of changes in infrared radiation. The design permits very precise discrimination of the directionality and distance of an infrared source. Many studies have demonstrated or assumed that pit organs function in detecting and accurately striking warm-bodied prey, but recent investigations demonstrate these organs also are functional in behavioral thermoregulation (A.R. Krochmal & G.S. Bakken, *J. Exp. Biol.* 206: 2539–2545, 2003). (Syn. *lachrymal pit, loreal pit, facial pit, infrared receptor, labial pit, pit receptor*)

pit scale. Any scale lying in or on the border of the *pit organ* of pit vipers. See *lacunal* and *foveal*.

pit receptor. See *pit organ*.

pituicyte. *n.* A nonendocrine cell of the *neurohypophysis*.

pituitary gland. A central and complex endocrine organ situated at the base of the brain and connected to the hypothalamus by a stalk. It consists of anterior (*adenohypophysis*) and posterior (*neurohypophysis*) lobes, the former being of non-neural origin. The structure forms embryonically as a fusion of an outpocketing of the roof of the mouth (Rathke's pouch) and the hypothalamus. (Syn. *hypophysis*)

pituitary gonadotropins. Gonadotropic hormones released from the anterior pituitary. Principal pituitary gonadotropins are *luteinizing hormone* (LH) and *follicle stimulating hormone* (FSH).

pitviper, pit viper. Collective common name for numerous species of snakes belonging to the *Crotalinae,* all of which characteristically possess a *pit organ.* See also *Palm Pitvipers, Rattlesnakes.*

pivotal temperature. The narrow range of temperature during the incubation of eggs at which there is an abrupt change in the sex ratio of hatchlings from nearly all males to females, or vice versa, in species that have *temperature-dependent sex determination.*

pK. The negative log (base 10) of an ionization or dissociation constant, K. $pK = -\log_{10}K$.

pK′. The negative log (base 10) of the apparent ionization or dissociation constant, K′, used for a CO_2/HCO_3^- system in the Henderson-Hasselbalch equation. $pK' = -\log_{10}K'$.

placebo. *n.* An inactive substance or preparation used as a control in studies in which the effects of an active substance are being tested.

placenta. *n.* A vascular, nutritive connection between fetal and maternal tissues, through which the developing fetus exchanges nutrients, respiratory gases, and metabolic waste products. The wall of the uterus is typically the maternal contribution in placental amphibians and reptiles. Some embryonic amphibians have a pharyngeal (or gill) placenta that provides gas exchange between the embryo and mother. In viviparous reptiles, the embryonic portion is provided by the extraembryonic membranes. Definitions of this term vary in literature depending on the taxonomic usage. See J.R. Stewart & D.G. Blackburn, *Copeia* 1988:839–852, 1988 and references therein. Most viviparous species have simple placentae and are essentially *lecithotrophic.*

placental viviparity. Viviparity in which the embryo receives nutrition from the mother by means of the placental connection between them.

placentome. *n.* A regionally specialized structure of the *chorioallantoic placenta* in most highly placentotrophic lizards. On the maternal side, the uterine epithelium consists of enlarged cuboidal cells so that the maternal blood supply is no longer closely apposed to the embryonic vessels. Enlarged uterine villi protrude and invaginate in an interlocking manner against the external surface of the *chorion.* On the embryonic side, the placentome lies on the mesometrial axis of the egg and has an epithelium of large cuboidal cells containing large nuclei overlying a highly vascularized *allantois.* The structure of the placentome suggests a nutrient transport function.

placentotrophy. *n.* Provision of extravitelline nutrients (maternal-to-fetal nutrient transfer) accomplished by means of a placenta in viviparous squamates. Term suggested by D.G. Blackburn (*Copeia* 1994:925–935, 1994). See *matrotrophy.*

placode. *n.* A distinct, thickened plate of embryonic *ectoderm.*

Placodontia. *n.* A distinctive but highly varied group of Mesozoic marine reptiles characterized by short bodies and paddlelike limbs. See *Sauropterygia.*

plagiodont. *adj.* Having the palatal teeth of snakes set obliquely or in two convergent series.

plagiotreme. *adj.* Possessing a transversely oriented anal slit or vent, associated with possession of paired *hemipenes* in snakes and lizards. See also *cyclotreme.*

Plainbelly Water Snake. Vernacular name for the American natricine species *Nerodia erythrogaster.*

Plains Garter Snake. Vernacular name for the American colubrid species *Thamnophis radix.*

plane. *n.* A flat surface along which any two points can be connected by a straight line. These are useful with respect to anatomical description and dissection. **Dorsal plane:** A division at right angles to the median and transverse planes of the body or structure, dividing it into dorsal and ventral parts. **Median plane:** A longitudinal division along the midline of the body or part that divides the structure into equal left and right halves. **Parasagittal plane:** One parallel and to one side of the sagittal plane through a body or part. **Sagittal plane:** A vertical plane that divides the body or structure into left and right parts. **Transverse plane:** A division across the body or structure, at right angles to its longitudinal axis.

plane polarized light. Light that is vibrating in only one plane.

plankton. *n.* (*adj.* –**ic**) Microscopic and usually photosynthetic organisms that drift suspended in water of rivers, lakes, and oceans. The community is often described as composed of *zooplankton* (animals) and *phytoplankton* (plants), but this terminology is based on an obsolete division of living things into two kingdoms, plants and animals.

plantar. *adj.* Reference to the sole or ventral side of the hind foot. Cf. *palmar*.

plantar pad or **tubercle.** Regions of thickened skin on the sole of the hind feet in anurans.

plantigrade. *adj.* Reference to locomotion with the entire sole of the foot in contact with the ground.

planum basale. Part of the amphibian chondrocranium consisting of a broad plate of cartilage connecting the trabecular processes anterior to the internal nares. (Syn. *internasal plate, ethmoid plate*)

planum tectale. See *tectal plate*.

planum terminale. See *lamina externa*.

plasma. *n.* See *blood plasma*.

plasmalemma. *n.* See *plasma membrane*.

plasma membrane. The surface or outer membrane that surrounds cells. (Syn. *plasmalemma*)

plasma membrane transformation. Reference to a series of changes in the plasma membrane of uterine epithelial cells, stimulated by cyclic changes in hormonal concentration. This phenomenon is well known in mammals and was recently demonstrated in the uterus of a viviparous Australian Skink (M.J. Hosie et al., *J. Morphol.* 258:346–357, 2003).

plasticity. *n.* Generally, the capacity for change in response to an external influence. See *phenotypic plasticity*.

plastic strain. Strain in which a body is molded or bent under stress and does not return to its original shape after the stress is released.

plastral. *adj.* Pertaining to, or part of, the *plastron* of turtles. Also used as a general grouping term for any one of the series of laminae covering the plastron.

plastral formula. A method of indicating length relationships between the laminae of the plastron, first introduced by A. Loveridge and E.E. Williams (*Bull. Mus. Comp. Zool.* 115:163-557, 1957). In formulation, individual laminae are listed in order from the longest to the shortest.

plastron. (*pl.* **-ra**) *n.* The ventral turtle shell, consisting of paired bones overlain by a matching series of laminae or scutes. Cf. *carapace*.

plastron length. A turtle shell measurement usually corresponding to the median straight-line length of the plastron, but occasionally measured along the curve of the plastron.

plastrum. *n.* An archaic spelling of *plastron*.

Platannas. *n.* Vernacular name for pipid frogs belonging to the genus *Xenopus*, also known as Clawed Frogs.

plate. *n.* 1. A term that denotes a large scale, used variously in reference to ventral *scutes*, the bony elements of the turtle shell, *osteoderms* of Crocodiles, and the head *shields* of snakes. 2. A geological plate or major section of Earth's crust. See *plate tectonics*.

plateau. *n.* A broad, flat-topped area elevated above the surrounding land and bounded, at least in part, by cliffs.

Plated Lizards. Collective vernacular name for various species belonging to the *Gerrhosauridae*, African lizards closely related to *Girdled Lizards* (*Cordylidae*).

plate tectonics. A theory that Earth's surface is divided into a few large, thick plates that are moving slowly and changing in size. Intense geological activity occurs at the plate boundaries.

platyan. *adj.* Reference to a centrum that is more or less flat at each end. (Syn. *acoelous*)

platybasic. *n.* See *platyrabic*.

platycoelous (also **-celous, -coelian, -celian**) *adj.* Reference to a vertebra that has solid centra with slight or no hollowing or excavation of the articulating faces.

Platysternidae. *n.* A single species family of turtle, sometimes included as a subfamily of *Chelydridae*.

platytrabic. *adj.* Reference to a skull having no interorbital septum and lack of fusion between the embryonic trabeculae, giving a broad, low braincase as seen in snakes and modern amphibians. (Syn. *platybasic*) Cf. *tropitrabic*.

playa. *n.* A very flat, dry lake bed of hard, mud-cracked clay.

playa lake. A temporary, shallow lake formed following a rainstorm on a flat valley floor in an arid region.

plectrum. *n.* The anterior ossification in the *oval window* of anurans. The stapes, or *columella*, is sometimes called the plectrum in anurans. (Syn. *pseudoperculum, pseudooperculum*)

pleio-. Prefix meaning "more."

pleiomorphism. *n.* (*adj.* –**ic**) The occurrence of variable phenotypes in a genetically uniform group of organisms. See *phenotypic plasticity*.

pleiotropy, pleitropy. *n.* (*adj.* –**ic**) A condition in which a single gene influences multiple phenotypic traits. (Syn. *polypheny*)

pleisiotypy. *n.* (*adj.* –**ic**) Reference to the ancestral character state in a transformation series.

Pleistocene. *n.* The sixth geologic Epoch of the *Cenozoic Era*. See Table 1.

pleomorphic. *adj.* Having more than a single form or shape.

plesiomorph, plesiomorphy. (*adj.* –**ic**) 1. Reference, in systematics, to a character state that occurs in an outside group as well as the group of organisms being considered. A character that is present in an ancestral form and also retained in a descendant or descendants. 2. In evolution, this term refers to an original primitive feature thought to have arisen in an ancestor of all the taxa being considered. Cf. *apomorphy*. 3. Having similar shape or structure.

Plesiosauria. *n.* A clade of Mesozoic marine reptiles distinguished by unusual morphological modifications of the neck. See *Sauropterygia*.

plesiosaur. *n.* A member of *Plesiosauria*.

Plethodontidae. *n.* A diverse, speciose family of salamanders that have radiated extensively in the neotropics. Lungs are absent in all species (approx. 360), and fully metamorphosed individuals possess a nasolabial groove that aids in chemoreception. Collectively known as "lungless salamanders," distribution includes North and South America (majority of species) and Europe. A sister taxon with *Amphiumidae* and recently assigned within the more inclusive taxon *Xenosalamandroidei* (D.R. Frost et al., *Bull. Amer. Mus. Nat. Hist.* 297:1–370, 2006).

plethodontid. *n, adj.* A member of, or pertaining to, *Plethodontidae*.

Plethodontinae. *n.* A clade of *plethodontid* salamanders that includes *Hemidactyliini, Plethdontini, and Bolitoglossini*.

Plethodontini. *n.* A clade of *plethodontid* salamanders that includes the genera *Aneides, Plethodon,* and *Ensatina*.

Plethosalamandroidei. *n.* A newly designated monophyletic taxon containing *Rhyacotritonidae* and *Xenosalamadroidei*. A sister taxon with *Hydatinosalamandroidei* within the *Diadectosalamandroidei* (D.R. Frost et al., *Bull. Amer. Mus. Nat. Hist.* 297:1–370, 2006).

pleural. *adj., n.* 1. Relating to the wall of the thoracic cavity. 2. Any one of the bony plates covering the ribs in the turtle carapace, corresponding to the *costal* laminae. See Fig. 5.

pleural cavity. The cavity between the lungs and the wall of the thorax. In amphibians and the majority of reptiles the anterior and dorsal recesses that house the lungs are part of the general body cavity, but in crocodilians and some squamates the lung enclosures are completely closed from the remainder of the celom and are considered to possess a true *pleural cavity*.

pleural rib. A true primary rib articulating with a vertebra.

pleuroccipital. *n.* See *exoccipital*.

pleurocentrum. *n.* A lateral component of the vertebral centrum, derived from the anterior part of a sclerotome. Earlier called *interventralia* by Gadow. This element is present in extinct amphibians and predominates to become the body of each vertebral segment, or *centrum,* of amniotes.

Pleurodeira, Pleurodira. *n.* A clade (suborder) of extant turtles characterized by bending of the neck to the side ("side-necked turtles") rather than in a vertical plane. This suborder includes the families *Chelidae, Pelomedusidae,* and *Podocnemididae*. Cf. *Cryptodeira*.

pleurodire, pleurodiran. *n.* (*adj.* **-irous**) Reference to a turtle that withdraws the head beneath the carapace by bending the neck in a horizontal plane.

pleurodont ankylosis. Condition where rootless teeth are ankylosed to the inner side of the jawbone, set in a shelf-like inner, lingual wall. This is a common mode of implantation in lizards and snakes. (J.W. Osborn, *Symp. Zool. Soc. Lond.* 52:549–574, 1984) Cf. *acrodont, protothecodont, thecodont ankyloses*.

pleurodontous. *adj.* Dentition containing different types of teeth.

pleuroectoglyph, pleuroglyph. *n.* (*adj.* **pleuroglyphous**) Terms coined by H.M. Smith (*Turtox News*, 30:214–218, 1952) and W.H. Stickel (*Proc. Biol. Soc. Wash.,* 56:109–128, 1943), respectively, to indicate those *ectoglyphs* that have the groove or channel for passage of venom on the lateral margin of the tooth, as contrasted with *proectoglyph*. See *pleuro-ophisthoglyph, pseudo-opisthoglyph*.

pleuro-ophisthoglyph. *n.* A synonym for *pleuroglyph*, proposed by H.M. Smith (1947).

pleuroperitoneum. *n.* A term sometimes used in reference to the common membranous lining of the chest cavity and abdominal region in those vertebrates lacking a diaphragm.

pleurosphenoid. *n.* A paired skull bone in reptiles, derived from the chondrocranium and located on the midline above the palate, at or behind the orbit. H.M. Smith (*Evolution of Chordate Structure*, p. 202 and Fig. 8.4, 1960) distinguishes two sets of sphenoid bones, with *orbitosphenoids* as an anterior set and *pleurosphenoids* as a posterior set. These words are used almost interchangeably, however, in much herpetological literature. Also, the term *laterosphenoid,* described from crocodilians, is often substituted for pleurosphenoid.

pleurothecodont. *adj.* A type of tooth attachment intermediate between *pleurodont* and *thecodont*.

plexiform bone. Bone characterized by a regularly arranged, three-dimensional plexus of primary vascular channels. This represents a tissue adapted to massive deposition of bone in a relatively short time. Plexiform bone is not a prevalent type of bone in modern reptiles, although it is quite common in several extant mammals and occurs in limited amounts in young crocodilians.

plexus. *n.* A network of intermingling blood vessels or nerves.

plica. (*pl.* **-ae**) *n.* A fold or folded part.

plicate. *adj.* Pleated or folded.

plinth. *n.* (*adj.* **plinthlike**) A brick or rectangular block. Used in herpetology with reference to amphibian egg masses that are flattened and somewhat rectangular rather than spherical.

Pliocene. The last Epoch of the Tertiary Period. It is spelled **Pleiocene** in some older texts. See Table 1.

-ploid. A suffix used in genetics and cytology to designate a specified multiple of the chromosome set of the nucleus of an organism.

ploidy. *n.* See *polyploidy*.

plot. *n.* 1. In ecology, a circumscribed sampling area. 2. Reference to a graphic display of data.

plug-flow reactor. A digestive system in which a bolus of food is progressively digested as it moves through a long, tubelike digestive reactor. This concept was developed in chemical engineering and subsequently applied to animal digestive systems (D.L. Penry and P.A. Jumars, *Amer. Nat.* 129:69–96, 1987). Typically, the midgut of vertebrates functions as a plug-flow reactor, although there are variations of patterns, and other terminologies apply to animals other than amphibians and reptiles (see reference).

plumbeous. *adj.* Resembling the color of lead.

plume. *n.* The path taken by a continuous discharge of a fluid or product from an outlet.

pluriparous. *adj.* See *multiparous*.

pluripotent. *adj.* Descriptive of any cell or early embryonic tissue that has a number of possible developmental fates (but not all of the fates possessed by the zygote). (Syn. *multipotent*)

plurivorous. *adj.* Feeding on a variety of different prey items or food sources.

pluvial. *adj.* Reference to the action of rain or precipitation.

pluvial lake. A lake formed during an earlier time of abundant rainfall.

pneumatic. *adj.* Relating to or involved in the passage of air.

pneumonia. *n.* Acute inflammation of the lungs, a result usually of bacterial or viral infections in reptiles.

P_{O2}. Symbol used in physiology to denote partial pressure of oxygen in blood, tissue, or medium.

pocket. *n.* A bag-like, blind cavity. See also *acaridomatrium, nuchal pocket, postfemoral dermal pocket*.

pod. *n.* A collective name for a group of juvenile crocodilians.

Podarcis. *n.* A new electronic journal presented through the Internet in English and in Dutch. This periodical reports articles on various aspects of herpetology including reproduction and technology related to herpetoculture. Current URL: http://www.podarcis.nl/

Podocnemidae. *n.* A clade (family) of side-necked turtles that inhabit lakes and rivers and are active swimmers with flattened shells. Three genera and eight species are recognized inhabiting northern South America and Madagascar.

poecilo-. Prefix meaning "various" or "variable."

poikilo- Prefix meaning "various" or "variable."

poikilotherm. *n.* (*adj.* **-ic**) An animal in which body temperature tends to fluctuate with the surrounding ambient temperature. Cf. *ectotherm, homeotherm*.

Pointed Snakes. Vernacular name for West Indian colubrid snakes belonging to the genus *Uromacer* (also called *Vinesnakes* and *Tree Snakes*).

point estimate. In statistics, a single number used to estimate a population parameter. Cf. *interval estimate*.

point mutation. In molecular genetics, a mutation caused by the substitution of one nucleotide for another.

poison. *n.* A substance that is toxic when ingested or absorbed. Not to be confused with *venom*. A venomous animal actively delivers a toxic secretion, whereas a poisonous animal relies on passive means of harming another individual. Both poisons and venoms contain *toxins* and are mixtures of biologically active and inert substances. Cf. *toxin, venom*.

Poison Frogs. Collective vernacular name for species of anurans belonging to the American family Dendrobatidae, and especially the genus *Dendrobates*. See also *Rocket Frogs, Golden Poison Frogs*.

poison gland. See *venom gland, granular gland*.

poison lake. A venom chamber delimited by fibrous septa arising from the fibrous walls that surround the ophidian venom gland, each emptying separately into the venom duct.

poisonous. *adj.* Reference to any living organism having secretions or substances that, by means of ingestion, inhalation, absorption, injection, or other application, are capable of causing physiological disturbance or structural damage. It is becoming commonplace to restrict this term to organisms that have a harmful effect when partially or completely eaten by another animal and to use the term *venomous* for organisms that introduce venom into the body of another animal by means of fangs (as in reptiles) or stingers.

poisson distribution. A theoretical description of the probability that random, independent events, based on a unitary event of a particular size, will occur.

polar. *adj.* 1. Reference to areas within the Arctic and Antarctic circles. Cf. *temperate, tropical*. 2. Reference to chemical groups that are water-soluble, e.g. the hydrophilic side chain of an amino acid.

polarity. *n.* 1. Reference to a directional trend within a system. 2. Direction of modification of a character within a transformation series. 3. In phylogenetic systematics, the *plesiomorphic* or *apomorphic* status of a character. 4. The direction of magnetism.

polarization microscope. A compound light microscope that permits study of anisotropic properties of objects and renders the objects visible due to their optical property of anisotropy.

polarized transformation series. In systematics, reference to a *transformation series* in which the ancestral and derived conditions of the character set have been determined. Cf. *unpolarized transformation series*.

pole. *n.* Either extremity of a structure.

pollex, pollyx. *n.* The first toe on the forefoot, or innermost digit on the hand (a thumb when present). See also *prepollex*.

pollical rudiment. A rudimentary indicator of a *prepollex*, usually a small bump or tubercle on the inner side of the first digit of the hand in females of certain frog species in which males have a spiny prepollex.

polliwog. *n.* Reference to a tadpole or larval amphibian. In herpetology, this term is usually applied to a generalized type of tadpole that possesses a narrow fin used to produce undulatory movements for swimming.

pollutant. *n.* An agent of pollution. Any introduced substance that adversely affects a resource or ecosystem.

pollution. *n.* Contamination of a natural ecosystem, usually in reference to anthropogenic sources or causes.

poly-. Prefix meaning "many."

polyandry. *n.* (*adj.* **–ous**) A reproductive system in which a female mates with more than one male during a breeding season. Cf. *polygyny*.

polyautochronic ovulation. See *ovulation*.

polychlorinated biphenyls. (PCBs) A class of synthetic organic compounds widely recognized as toxic contaminants that are highly persistent in the environment. The toxicity of these compounds is related to the number of chlorine atoms in the molecule.

polychromatism. *n.* A condition where a population or species is composed of individuals representing several different colors or color patterns.

Polychrotinae. *n.* A clade of iguanid lizards that includes the familiar and well-studied *Anolis* species, in addition to other lesser known forms. There are eight genera and more than 450 species distributed from North to South America and throughout the West Indies. Nearly 400 recognized species belong to the genus *Anolis*.

polycythemia. *n.* Overproduction of red blood cells.

polydactyly. *n.* A condition of having more than the usual number of fingers or toes.

polydont, polyodont. *adj.* Possessing numerous teeth that are functional simultaneously. Cf. *oligodont*.

polygamy. *n.* (*adj.* **–ous**) A reproductive system in which animals have more than one mate during a breeding season (*polyandry* and/or *polygyny*). Cf. *monogamy*.

polygenic. *adj.* See *multifactorial*.

polygyny. *n.* (*adj.* **–ous**) A mating system in which the male may mate with more than one female during a breeding season. Cf. *polyandry*.

polymely. *n.* A congenital state of deformation in which a vertebrate has more than four limbs.

polymer. *n.* A compound composed of a linear sequence of simple molecules or residues, i.e. repeating subunits or monomers.

polymerase chain reaction. (PCR) In molecular genetics, an amplification technique for copying the complementary strands of a target DNA molecule simultaneously for a series of cycles until the desired amount is obtained. The technique uses specialized heat-stable enzymes of bacteria that exist at extremely high temperatures and results in a large quantity of the DNA sequence. Thus, it permits detection and retrieval of a specific sequency that was only a tiny fragment of the original DNA material.

polymodal. *adj.* A state of having more than two modes or peaks in a frequency distribution. Cf. *bimodal, unimodal*.

polymorph. *n.* One of two or more forms (morphs) of a polymorphic species.

polymorphic behavior hypothesis. A hypothesis that spacing behaviors of individuals limit the density of a population—a form of population self-regulation. Further, individual differences in spacing behavior are proposed to have a genetic basis and respond to natural selection. (Syn. *Chitty hypothesis*)

polymorphism. *n.* (*adj.* **-ic**) The sustained occurrence of two or more distinct and genetically determined forms (color pattern, morphologies, etc.) in a single species, usually representing two or more genetically distinct classes in the same interbreeding population. Such polymorphism may be transient or persistent, in which case the phenomenon is referred to as a *balanced polymorphism*. If the classes are located in different geographic regions, *geographic polymorphism* occurs.

Polynesia. *n.* A geographical and ethnological assemblage of numerous small oceanic islands scattered over the eastern Pacific Ocean, extending north to include Hawaii and east to Easter Island, sometimes also including New Zealand. A broader usage of the term includes both *Melanesia* and *Micronesia*.

polyopisthocoelous. *adj.* A vertebra in which more than one concavity of the *opisthocoelous* type occur on a single end.

polypeptide. *n.* See *peptide*.

polyphagia. *n.* Reference to eating an abnormal amount of food.

polyphagy. *n.* (*adj.* **–ous**) Feeding on more than one type, or multiple species, of food. Cf. *monophagy*.

polyphalangy. *n.* An increase in the number of phalanges in each digit.

polyphenism. *n.* The occurrence of several phenotypes in a population, the variation not due to genetic differences. The term is usually used in reference to the production of distinct morphological forms by a single species, representing adaptations in which a genome is associated with discrete alternative phenotypes in different environments.

polypheny. *n.* 1. See *pleiotropy*. 2. Extensive phenotypic variation that lacks genetic fixation.

polyphyletic group. A group of species classified together, but including some members that are descended from different ancestral populations and thus do not share a common ancestor. Cf. *monophyletic group*.

polyphyly. *n.* (*adj.* **-etic**) Reference to a condition in which a group of organisms includes species derived from more than one ancestral form. Cf. *monophyly, paraphyly*.

polyphyodont. *adj.* Having new teeth develop throughout life, involving cycles of tooth loss and replacement (e.g., as in snakes).

polyploidy. *n.* (*adj.* **polyploid**) A condition wherein the number of chromosome sets is greater than two. (Syn. *ploidy*)

polyprocoelous. *adj.* A vertebra on which there is more than a single concavity of the procoelous type on a single end.

polysaccharide. *n.* A long-chain carbohydrate such as glycogen, formed by polymerization of numerous monosaccharide units.

polyspecific. *adj.* A little used synonym for *polyvalent*.

polyspermy. *n.* Penetration of an egg by more than a single sperm.

polytene. *adj.* Possessing many duplicate chromatin strands.

polythetic key. A key wherein multiple contrasting characters are used simultaneously in at least some of the couplets. Cf. *monothetic key*.

polythetic taxon. A taxon defined by a large number of characters, none of which is diagnostic alone. Cf. *monothetic taxon*.

polytopic. *adj.* Descriptive of a distribution of subspecies or races in two or more geographically discontinuous areas. Cf. *monotopic*.

polytrophic. *adj.* Feeding on a variety of different prey species or food sources.

polytypic. *adj.* Reference to a taxon comprised of more than two subordinate principal taxa. Cf. *bitypic, monotypic*.

polytypic species. A species that is subdivided into a number of specialized races or geographical variants.

polytypy. *n.* In taxonomy, reference to practice of establishing a taxon based on more than one type.

polyuria. *n.* The production and voiding of an abnormal volume of urine.

polyvalent. *adj.* 1. In bacteriology, reference to a vaccine effective against two or more strains of the same species of microorganism. 2. Reference to an antibody molecule that will combine with more than a single molecule of its antigen. 3. In herpetology, used to denote an *antivenin* effective against the venoms of more than one species. (Syn. *multivalent, polyspecific*)

polyvoltine. *adj.* See *multivoltine*.

pond. *n.* A small body of standing fresh water, intermediate in size between a pool and a lake.

pond larva. One of three types of larva described in salamanders, characterized by possession of a dorsal fin extending forward to the forelimb, elongate featherlike gills, slender toes, and balancers during early stages. See also *stream larva* and *terrestrial larva*.

Pond Turtles. Collective vernacular name for species of turtles belonging to the family Emydidae, especially the genera *Actinemys, Chinemys, Emys, Clemmys,* and *Geoclemys, Heosemys, Mauremys.* See also *Painted Turtle, Scooter, Slider*.

pons. *n.* That part of the vertebrate brain that lies between the medulla oblongata and the midbrain, a ventral derivative of the metencephalon.

pontine. *adj.* Relating to the pons of the brain.

Pope process. A method for refining horse blood plasma to separate antibody molecules from nonfunctional protein fractions of antitoxins. Protocols in P.A. Christensen (*S. Afr. Snake Venoms*, p. 58, 1955) and A.J. Harms (*Biochem. J.*, 42:390–397, 1948). (Syn. *Pope's method*)

Pope's method. See *Pope process*.

Pope's Pit Viper, Pope's Tree Viper. Vernacular name for the Asian pitviper *Trimeresurus popeorum*.

popliteal. *adj.* Pertaining to the back of the knee.

Popperian method. See *deductive method*. Hypotheses can be falsified or corroborated, but not proven.

popping. *n.* 1. A method of sexing juvenile snakes, involving application of gentle pressure to evert, or "pop," the hemipenes from the base of tail, using the ball of one's thumb. 2. See also *cloacal popping*.

population. *n.* 1. A local, geographically defined group of conspecific organisms sharing a common gene pool and exhibiting reproductive continuity from generation to generation. (Syn. *deme*) Cf. *metapopulation*. 2. Generally, the entire set of individuals about which inferences are to be made. Also referred to as the *sampling universe*.

population biology. A subdiscipline of biology and ecology that involves study of the numbers of organisms, their life histories, reproductive and growth rates, and the interactions of populations in time and space.

population bottleneck. Reference to a sharp decrease in the size of a population with consequent reduction in size of the gene pool. See *genetic bottleneck*.

population density. In ecology, the number of individuals of a population per unit of living space (usually expressed as area).

population genetics. Investigation of the genetic composition of populations and the selective influences that determine gene frequencies. This field involves considerable mathematical modeling to elucidate the interaction of factors such as mutation, selection, migration, and population size with the fixation or loss of linked and unlinked genes. It provides the mathematical structure for investigating the process of *microevolution*.

population growth rate. The rate of change of population size, measured as the change in population size, (dN), during a very small interval of time (dt).

population pyramid. Depiction of the age-class structure of a population wherein the youngest class is at the base and successive classes are stacked above.

population size. The number of individuals in a finite population, denoted by the letter N.

population structure. Reference to the manner in which a population is structured with respect to life history variables (age class distribution, age at first reproduction, etc.) and subdivision into local demes, their size, and the extent of migration or gene flow between demes.

pore. *n.* A small opening or passage in the skin.

Poren. *n.* See *integumentary sense organ*.

porosity. *n.* The percentage of volume that is taken up by openings in rock or soil.

portal system or **vessels.** A set of venous blood vessels beginning and ending in capillary beds or sinuses of the liver.

positive feedback. See *feedback*.

positive selection. The retention of mutations that benefit an organism. (Syn. *Darwinian selection*) Cf. *negative selection*.

post- . A prefix that indicates behind, or the back part of a structure. Closer to the rear or tail than some other comparable part.

postabdominal. *n.* Alternate term for the *anal* scale or plate of snakes, used by various authors in herpetological literature. Also called *postabdominal scutella*. See Fig. 13.

postanal. *n.* 1. A large plate, usually paired, immediately posterior of the cloaca in males of some species of lizards. 2. Used in reference to scales lying immediately posterior of the cloaca in the snake genus *Xenopeltis*.

postanal bone. Either of a pair of bones lying just beneath the skin in the *postanal sac* on either side of the tail and immediately posterior to the cloaca in several families of lizards. (Syn. *cloacal bone, postcloacal bone, postcloacal ossicle*)

postanal gland. See *anal gland*.

postanal sac. Either of a pair of cavities located at either side of the tail immediately posterior to the cloaca in several families of lizards (Anguidae, Eublepharidae, Gekkonidae, Pygopodidae), thought to be homologous and similar in function to the *cloacal gland* of crocodilians and *anal glands* of snakes. (Syn. *cloacal sac, postcloacal sac, subcaudal sac*)

postano-coccyx. *n.* Part of the *coccygeum* that lies posterior to the anus, equivalent to the *ura*.

postauricular. *adj.* Reference to the neck area behind the external ear.

postaxial. *adj.* See *ulnar*.

postbranchial body. Antiquated term used for *ultimobranchial body* (or *gland*) by earlier authors.

postcaval vein. The adult vessel posterior of the heart that collects common venous return from the posterior systemic veins. (Syn. *posterior vena cava, inferior vena cava*)

postcentral. *n.* Single or paired *supracaudal* laminae located at the posterior edge of the chelonian *carapace*, just posterior to the centrals. (Syn. *anal, caudal, caudal marginal, marginal, postmarginal, pygal, supracaudal*) See Fig. 4.

postcephalic. *n.* Any of a series of scales located in a row behind and in contact with, the *parietals* and *posttemporals* of snakes.

postcloacal bone. See *postanal bone*.

postcloacal distance, postcloacal length. A measurement of distance from the median posterior edge of the cloaca to the tip of the tail in turtles. Used by J.E. Mosimann & J.R. Bider (*Can. J. Zool.* 38:19–38, 1960).

postcloacal gland. See *anal gland*.

postcloacal ossicle. See *postanal bone*.

postcloacal sac. See *postanal sac*.

postcloacal sheath. The chamber which holds the retracted *hemipenis* of lizards and snakes.

post coitum. After mating.

postconcha. *n.* A prominent outpocket of the lateral wall of the nasal cavity of crocodilians, located posterior and partly ventral to the posterior part of the *concha*.

postdisplacement. *n.* An evolutionary change in which a descendant species has a delayed onset of development or growth in certain structures, in comparison with the ancestral species. Cf. *predisplacement*.

posterior. *adj.* Directed toward or situated at the back. Opposite to *anterior*.

posterior cornu. Alternate term for *ceratobranchial*, used by several authors in herpetological literature.

posterior lobe. 1. The part of the chelonian *plastron* that lies behind the *inguinal notch*, according to A. Carr (1952). 2. Sometimes used in reference to the *posterior pituitary* or *neurohypophysis*.

posterior medial process. This term has been used to denote the *thyrohyal* of anuran hyoid.

posterior pituitary. See *neurohypophysis*.

posterior vena cava. See *postcaval vein*.

posterior vidian canal. An enclosed canal on the underside of the posterior braincase containing the internal carotid artery and palatine nerve.

posterolateral, posterolateral process. This term has been used to denote the *lateral process* of anuran hyoid.

posteromedial. *adj.* Pertaining to the midline or middle region of the rear of a body or structure.

postfemoral. *adj.* Pertaining to the posterior side of the femoral region.

postfemoral dermal pocket. The infolding formed by the skin behind the hind limb at its insertion point on the body in several species of lizards. (Syn. *acaridomatium, postfemoral pocket, postfemoral mite pocket*)

postfemoral mite pocket, postfemoral pocket. See *postfemoral dermal pocket*.

postfoveal. *n.* Any of small scales bordering the facial pit of crotaline snakes, located between the *lacunal* and *interoculabial* scales. See also *foveal, prefoveal*.

postfrontal. *n.* 1. A dermal bone in the skull roof of primitive tetrapods and in *Sphenodon*, where it is located between the *frontal* and *postorbital*. 2. Previously used to denote the *postorbital* bone in turtles, and the *prefrontal* scale in lizards and snakes.

postfronto-squamosal arch. A *supratemporal bridge* made up of the *postfrontal* and *squamosal* bones in most diapsid reptiles.

postganglionic neuron. Reference to an autonomic neuron that has its cell body in a peripheral ganglion, receives synaptic input from preganglionic neurons, and synapses with target organs. Cf. *preganglionic neuron*.

postgeneial, postgenial. *n.* Previously used in reference to the posterior *chin shield* of snakes.

postgular. *n.* An older term for the *humeral* lamina of the turtle plastron. See Fig. 6.

postgular fold. A transverse fold of skin posterior to the gular fold and immediately anterior to the points of forelimb insertion on the body. In many lizards, this fold may be incomplete medially.

post hoc. After this, or following. Used similarly to *a posteriori* in reference to biometric or statistical tests in which the results or comparisons of interest become evident only after experimental results are obtained. See *a posteriori*.

post-iliac gland. Reference to multiple small, light-colored subcutaneous glands located posterior and dorsal to the base of the thigh in the plethodontid salamander *Pseudoeurycea bellii*. First described by I.L. Baird (*Univ. Kansas Sci. Bull* 34:221–265, 1951).

postlabial. *n.* 1. Any of several enlarged scales situated posterior to, and in line with, the *labial* scales of lizards. 2. See also *postmandibular*.

postloreal. *n.* Reference to small, irregularly occurring scales located between the *loreal* and *preocular* scales of Rattlesnakes. Used by C.H. Lowe and K.S. Norris (*Trans. San Diego Soc. Nat. Hist.* 12:49–64, 1954).

postmalar. *n.* A scale on the chin of amphisbaenians, lying behind the *malar* and the *postgenials*, forming a series across the midline.

postmandibular. *n.* Any scale of a series lying immediately behind and below the lower *labials* in turtles, generally in two rows when used on sea turtles. (Syn. *postlabial*)

postmarginal. *n.* See *postcentral*, and Fig. 4.

postmating isolating mechanism. See *reproductive isolating mechanism*.

postmental. *n.* Paired or single scales lying immediately behind the *mental* in some snakes and lizard species. See Fig. 15.

postminimus. *n.* A cartilaginous digit external to the *minimus* in certain primitive salamanders.

post mortem. After death.

postnasal. *n.* A scale or scales located behind the nasal and anterior to the loreal, sometimes called a posterior *nasal*. (Syn. *frenonasal, preloreal*) See Fig. 11.

postnatal. *adj.* Occurring after birth, with reference to the newborn.

postneural. *n.* See *suprapygal*.

postnodal. *adj.* Used by F. Wall (*Snakes of Ceylon*, p. 322, 1921) to denote teeth that are posterior to a *diastema*.

postnuchal. *n.* The pair of crocodilian *nuchomarginals* that lie on the midline behind the nuchals, usually in contact with each other.

postnuptial. *adj.* Reference to events or conditions following completion of a breeding period, usually in reference to reproduction.

postoccipital. *n., adj.* 1. The anterior *nuchals* of crocodilians. 2. The *postparietal* of snakes. 3. The *parietal* or dorsal head scale behind the *occipital* in certain lizards. 4. Of or pertaining to the area immediately posterior to the occipital region.

postocular. *n., adj.* 1. A scale or scales at the posterior boundary of the orbit. (Syn. *postorbital*) See Fig. 11. 2. Of or pertaining to the region immediately behind the eye. (Syn. *postorbital*) See Fig. 2.

postocular spine. See *superciliary spine*.

postorbital. *n.* 1. A dermal bone in the skull of reptiles, located behind the orbit and generally separating the *jugal* and *frontal*. This element may be lost or fused with the *postfrontal* and is absent from modern amphibians. 2. Used for the *postocular* scale of snakes. See Fig. 11.

postorbital arch. An osseous bar extending from the roof of the skull to the upper lipline in those taxa such as lizards which have deeply excavated temporal openings and the cheek region is not solid bone. It is comprised of the *jugal* and *postfrontal*, and often an extension of the *frontal*.

postorbital fenestrae. (temporal fenestrae) Postorbital openings that allow space for attachment and contraction of muscles. The *Lepidosauria* and *Archosauria* have two paired postorbital fenestrae (*diapsids*); mammals (*synapsids*) have a single paired postorbital fenestra; *anapsids* lack postorbital fenestrae. These are important characters in separating major groups of terrestrial vertebrates.

postorbital furrow. A groove beginning at the posterior corner of the eye and extending posteriorly for variable distances in salamanders.

postorbital shelf. Reference to the skeletal area posterior of the eye in the skull of bufonids.

postorbito-squamosal arch. A supratemporal bridge in *Sphenodon*, comprised of the *postorbital* and *squamosal* bones.

postpalatine. *n.* The posterior part of the *pterygopalatine* in certain larval salamanders, which differentiates to form a distinct bone in the adult.

postparietal. *n.* 1. The scale located immediately posterior of the *preparietal* and below the *interparietal* in typhlopid and leptotyphlopid snakes. In others, reference is to one or more scales immediately posterior to the *parietal* plates. (Syn. *postoccipital*) 2. Reference to dermal cranial bone located posterior to the *parietal* in primitive forms. Typically a single median element in reptiles, which together with the *tabulars* may form a transverse row of bones that has been called the *supratemporals*. The bone may extend down over the margin of the occiput. (Syn. *interparietal, dermal supraoccipital, dermosupraoccipital*) 3. Distinct nape scales, separated from the *interparietal* by several smaller scales in certain lizards.

post-paritive lecithotrophy. Reference to obtaining nutrition from residual yolk following birth, sometimes sustaining overwintering of neonates and nestlings. See *lecithotrophy.*

post-partum, postpartum. After parturition. (Syn. *post-parturient*)

post-parturient. See *post-partum.*

postprandial. *adj.* Following a meal.

postprandial calorigenesis. Reference to increased metabolic rate following ingestion of a meal, related to the mechanical and biochemical costs of digestion and assimilation of digestive products. (Syn. *diet-induced thermogenesis, heat increment of feeding, heat of nutrient metabolism, specific dynamic action, specific dynamic effect*)

post-prezygapophysial length. Measurement of the distance from the posterior edge of the post-zygapophysial facet to the anterior edge of the prezygapophysial facet of a snake vertebra.

post-pubal rod. A paired structure of connective tissue in the frog *Ascaphus,* originating below the pubis and extending back along the base of the cloaca, one going around each side of the aperture. During sexual activity, changes in the shape of these structures bring the cloaca into position for internal fertilization. First described by R.M. Ritland (*J. Morphol.* 97:119–174, 1955).

postpuberal. *adj.* Reference to characters that first appear at sexual maturity and undergo cyclic changes, such as the swollen cloaca and nuptial pads of male *Triturus.* Cf. *prepuberal.*

postrictal. *adj.* Of or pertaining to the area just posterior to the corner of the mouth.

postrictal spine. An epidermal spine immediately posterior to the angle of the mouth in horned lizards (*Phrynosoma*).

postrostral. *n.* 1. A small scale bounded by the *rostral, anterior nasal,* and first *labial* scales in rattlesnakes. More generally, any scale lying between the *rostral* and *internasal* scales, or an odd *internasal* scale. 2. Collective term for the series of three middorsal scales behind the *rostral* in blind snakes. 3. A small scale or scales bordering the *rostral* posteriorly, anterior to the internasals in many species of lizards. See Fig. 14.

postseminasal. *n.* A term that was used by H.M. Smith (*Handbook of Lizards,* p. 82, 1946) to denote the posterior part of a divided nasal scale in lizards.

poststernal vertebra. Any rib located anterior to the first *sacral* vertebra and posterior to the last vertebra having a rib connecting to the sternum. This includes vertebrae termed lumbar in mammals.

postsupraloreal. *n.* A term used to designate small scales of rattlesnakes that may be present behind the *supraloreal,* usually below the *canthus rostralis* and anterior to, and slightly above, the *preciliary,* when this scale is present.

postsynaptic. *adj.* Reference to the receiving ("downstream") side of a synaptic connection. Cf. *presynaptic.*

postsynaptic toxins or **neurotoxins.** Neurotoxic components of snake venom that disturb functions by binding to receptors on postsynaptic membranes. (Syn. a–*toxins*)

posttemporal. *n.* 1. The scale or scales comprising the second vertical row of *temporals.* See Fig. 11. 2. A skull bone present in fossil, but not modern, amphibians.

post-term. *adj.* After term or due date. Said of eggs that go beyond the date they are expected to hatch.

post-thoracic. *adj.* Reference to the area immediately behind the thoracic region.

postzygapophysis. (*pl.* **-es**) *n.* See *zygapophysis.*

postzygotic isolating mechanism. See *reproductive isolating mechanism.*

potamic. *adj.* Pertaining to rivers, or transport by river currents.

potamicolous. *adj.* Living in rivers.

potential energy. Stored energy that can be released to do work. The energy of position, related either to gravity or elastic loading.

potential evapotranspiration. (PE) The potential water loss from the ground or water body surfaces by evaporation and from vegetation by transpiration, given a level of available soil water.

POTZ. See *preferred optimum temperature zone.*

pouch. *n.* See *brood pouch.*

pound. The *fps* unit of mass defined as 0.45359237 kg.

power. *n.* 1. The rate of doing work, equal to work per time, or force x distance x time^{-1}. 2. See *statistical power.*

pox. *n.* Any eruptive or pustular disease of the skin. Can be caused by one of several viruses such as poxvirus and herpesvirus.

ppm. Parts per million, a unit of concentration commonly used in aquatic biology and literature on pollution.

ppt. Parts per thousand, $^0/_{00}$.

prairie. *n.* A plain dominated by grasses and forbs, with few or no trees, and found in moderately dry temperate regions.

Prairie Kingsnake. Vernacular name for the American colubrid snake *Lampropeltis calligaster,* and its subspecies *c. calligaster.* See also *Mole Kingsnake.*

Prairie Rattlesnake. Vernacular name for the Rattlesnake species *Crotalus viridis.*

Prairie Skink. Vernacular name for the American scincid lizard *Eumeces septentrionalis.* Note that H.M. Smith (*J. Kansas Herpetol.* 14:15, 2005) recently proposed that *Plestiodon* be a replacement name for the genus *Eumeces* in North America.

pratal. *adj.* Pertaining to grassland and meadowland.

pratinicolous. *adj.* Living in grassland or meadowland.

pre-. Prefix meaning "in front of," or "anterior to," a part, object or condition. Closer to the head than some other comparable part.

preadaptation. *n.* A feature of an organism that fortuitously serves a new function. S.J. Gould & E.S. Vrba (*Paleobiology* 8:4–15, 1982) suggest the term preadapta-

preanal

tion be replaced with *preaptation*. See *exaptation*. Cf. *aptation*.

preanal. *n., adj.* 1. Generally, any scale in the pelvic area, but usually denotes any one of the scales in a row immediately anterior to the vent of lizards. 2. Used by some authors in place of the *femoral* lamina in turtles (Fig. 6) 3. Used by some authors to denote the *anal* scale or scute in snakes (Fig. 13). 4. Reference to the area immediately in front of the vent.

preanal groove. A longitudinal groove located on the midventral body of certain lizards, posterior to the abdomen and anterior to the vent. (Syn. *pubic groove*)

preanal organ. *n.* A generic term suggested by A. G. Kluge (*Bull. Amer. Mus. Nat. Hist.* 135:1–60, 1967) for holocrine specializations of lizard epidermis, including *femoral glands* and other similar structures in the peri-cloacal region. In contrast to *generation glands*, secretory activity is independent of the shedding cycle. Cf. *generation gland*. (Syn. *femoral organ*)

preanal pore. A part of the *femoral pore* series, used to distinguish those on the body from those on the leg. These are present only as preanal pores in some species. Openings of well-developed *preanal organs*.

preano-coccyx. *n.* Part of the *coccygeum* that lies between the sacrum and the anus, an area that is large or well-defined in frogs and in turtles.

prearticular. *n.* A dermal bone of the lower jaw, located on the inner surface between the *splenial* and *articular*. Present in snakes, turtles and *Sphenodon*, but fused with the *articular* in lizards and with the *angular* in crocodilians (see *prearticulo-angular*). Usually considered to be the second bone in the jaw of salamanders (with the *dentary*), and is present in caecilians. In frogs, the element is lost or fused to form the *angulosplenial*. (Syn. *dermarticular, gonial, goniale*)

prearticulo-angular. *n.* Reference to the bone which articulates with the *coronoid* and the *splenial* forming the posterior border of the inframeckelian fenestra in the lower jaw of crocodilians. Usually called the *angular* by most authors, but it has characteristics of both *prearticular* and *angular*.

preatlas. *n.* See *proatlas*.

preaxial. *adj.* Reference to the side of the forelimb containing the *radius* in amphibians. This term has been used because the limb cannot be placed in classical anatomical positions for defining planes without serious distortion. In this sense the term substitutes for *medial*.

prebrachial. *n.* Any scale on the anterior surface of the *brachium*. (Syn. *antebrachial*)

prebutton. *n.* The small pre-rattle segment on the tip of the tail of Rattlesnakes, present at birth and lost at the first ecdysis. See also *button*.

Precambrian, Pre-Cambrian. *n.* An informal name for the Eons of geologic time that came before the current *Phanerozoic* Eon. It was briefly also called the *Cryptozoic* Eon. The term Precambrian is somewhat dated, but is still used commonly by paleontologists and geologists. See Table 1.

precapillary sphincters. Rings of vascular smooth muscle that variably restrict blood flow into a capillary network depending on the degree of activated *vasoconstriction*.

precaudal vertebra. 1. All vertebrae of snakes and limbless lizards anterior to the caudal vertebrae. 2. The first two vertebrae posteror to the *sacral vertebrae* of the salamander *Pseudoeurycea*, which are transitional and exhibit characteristics of trunk as well as *caudal vertebrae* and are thus distinguished from caudal vertebrae.

precaval vein. The adult vessel anterior of the heart that collects common venous return from the anterior systemic veins. (Syn. *anterior vena cava, superior vena cava*)

precentor. *n.* Reference to the individual male which holds the position of "choir master" in a breeding chorus of frogs. This term was used by H. Gadow (*Amphibia and Reptiles,* p. 269, 1901). See also *peep order*.

precentral. *n.* The anteriormost medial lamina of the chelonian *carapace*. A synonym for *nuchal* suggested by A. Carr (*Handbook of Turtles*, p. 35, 1952). See Fig. 4.

prechoanal sac. One of a pair of pouches, lined with sensory epithelium and opening into a groove in the roof of the mouth in certain species of anurans. (Syn. *blindsac*) See also *sinus laterlais nasi*.

preciliary. *n.* A small scale bordering the upper anterior margin of the eye in Rattlesnakes. The *preocular* and *supraocular* are usually in contact anterior to this scale. (Syn. *superciliary*)

precipitation. *n.* In meteorology, precipitation refers to any water that falls from the atmosphere to Earth. The water may be in a liquid or solid state as rainfall, snow, hail, or sleet. See also *acid precipitation, convective precipitation, orographic precipitation, frontal precipitation.* Cf. *rainfall.*

precise alternation. Reference to alternation of calls within an anuran chorus, leading to the formation of duets, trios, quartets, etc.

precision. *n.* In statistics, how closely repeated measurements of the same quantity are to one another. Cf. *accuracy.*

precloacal distance or **length.** A measurement distance from the median posterior margin of the *plastron* to the anterior edge of the *cloaca* in turtles.

precloacal gland. Narrow glands located anterior to the cloacal opening of most amphisbaenians, producing pheromones used for sexual or individual recognition. See also *femoral glands, preanal organ.*

precloacal pore. The opening of an exocrine gland situated in the precloacal region.

precocial. *adj.* A condition in which a newborn animal births or hatches in an advanced state of development and is capable of living wholly or partly independently of the parent at an early age. Cf. *altricial*.

preconcha. *n.* A pronounced convexity at the lateral wall of the nasal cavity in crocodilians, anterior to the middle or true concha.

precoracoid. *n.* 1. The anterior bar of cartilage located immediately above the *clavicle* in frogs. (Syn. *procoracoid*) 2. This term has been used in reference to the large bone forming the ventral margin of the *glenoid* in turtles. 3. Used to denote the large, flat, anteriorly directed process in the *coracoid* of salamanders.

precursor. *n.* 1. Generally, something that is present at an earlier time than another thing. 2. An earlier form from which a later form is descended.

predation. *n.* The consumption of one animal (*prey*) by another (*predator*). (Syn. *depredation*)

predation pressure. The effects of predation on the dynamics of a prey population.

predator. *n.* An organism that feeds on other living organisms, usually of other species.

predator-to-prey ratio. The ratio of predators relative to prey in a population or community, expressed numerically or in terms of biomass.

predictive classification. A classification with high information content, considered to be more natural than others. Structural and functional properties of a taxon can be inferred from such classifications.

predisplacement. *n.* An evolutionary change in which a descendant species has an earlier onset of development or growth in certain structures, in comparison with the ancestral species. Cf. *postdisplacement*.

preferred body temperature. (PBT) The mean selected temperature maintained by an ectothermic animal in a laboratory temperature gradient. F.H. Pough & C. Gans (*Biology of the Reptilia,* 12:17–23, 1982) suggested this term has an anthropomorphic connotation and should not be used. Cf. *selected body temperature*.

preferred optimum temperature zone. (POTZ) The range of temperatures within which a particular ectothermic species functions optimally overall. It equates with the temperature range in which a species is normally active, but is somewhat misleading because the terminology implies a conscious choice. Cf. *activity temperature range*.

preferred substratum temperature. Definition and comments similar to *preferred body temperature* except that this term has reference to temperatures of the substrate on which an animal rests.

prefoveal. *n.* Any of the small scales bordered by the *nasal, loreal, supralabial,* and *lacunal* scales in pit vipers. See *foveal, postfoveal*.

prefrontal. *n.* 1. One or more scales on the top of the head immediately anterior to the *frontal.* (Syn. *intercanthal, internasal, frontonasal*) See Figs. 10, 14. 2. A dermal bone of the cranium, located at the anterior margin of the orbit and usually bounded dorsally by the *frontal* and laterally by the *lacrymal.* Sometimes called an *ectethmoid* if lateral outgrowths invade the chondrocranium.

preganglionic neuron. An autonomic neuron that has its cell body located in the central nervous system, projects its axon into the periphery, and synapses onto postganglionic cells. Cf. *postganglionic neuron*.

pregeneial, pregenial. *n.* The anterior chin shield of snakes.

pregonad. *n.* A sexually neutral structure that later differentiates into a definitive gonad.

pregular. *n.* Any of several scales on the ventral surface of the head in certain species of lizards, anterior to a gular fold if present.

pregular fold. One or several transverse folds anterior to the *gular fold* in some species of lizards.

prehallux. *n.* An extra or vestigial digit on the inner margin of the hind foot in anurans. This often occurs as a process on the innermost side of the first toe (*hallux*). Cf. *prepollex*.

prehensile. *adj.* Capable of seizing or grasping an object by wrapping around it, as seen, for example, in grasping tails of arboreal snakes and lizards.

prehension. *n.* Rapid grasping and capturing of prey, usually with jaws or claws.

preloreal. *n.* See *postnasal*.

premating isolating mechanism. See *reproductive isolating mechanism*.

premaxilla, premaxillary. *n.* Either of two dermal bones at the anterior end of the skull, located between the maxillary bones and often bearing teeth. These elements meet at the midline and form the anterior angle of the upper jaw in amphibians and reptiles. (Seldom used syn. *intermaxillary*)

premaxillary gland. See *oral glands, salt glands*.

premetamorphic stage. The foot stage of a tadpole, used by A.C. Taylor & J.J. Kollros (*Anat. Rec.* 94:7–23, 1946).

premetamorphosis. *n.* Describes the primary period of body growth, developmental stages 25–35, in a tadpole (W. Etkin, pp. 313–348 in W. Etkin & L.I. Gilbert, eds., *Metamorphosis,* 1968. See also *premetamorphic stage*.

premortem. *adj.* Before death; occurring or existing prior to death.

prenasal. *n., adj.* 1. Either the anterior *nasal* scale, if divided, or a small scale anterior to the nasal of squamates. See Fig. 11. 2. Either of a pair of small scales between the *internasals* and *rostral* on the middorsal line of the head. 3. Of or pertaining to the area immediately in front of the nasal region.

preneural. *n.* A bony element lying between the *nuchal* and the *neurals,* middorsally, in the carapace of certain trionychid and some fossil turtle species.

prenodal. *adj.* Reference to teeth that are anterior to the *diastema* on the maxillary bone. Used by F. Wall (*Snakes of Ceylon,* p. 322, 1921).

prenuchal. *n.* See *nuchal bone*.

prenuptial. *adj.* Reference to a phenomenon that occurs before the beginning of a breeding period, and usually related to reproductive processes or behaviors.

preocular. *n., adj.* 1. Generally, any scale or scales bounding the anterior margin of the orbit. (Syn. *anteocular, anteorbital, anterior orbital, antocular, frenoorbital, preorbital*) See Fig. 11. 2. Of or pertaining to the area immediately in front of the nasal region.

preopercular. *n.* Antiquated term for *splenial* of turtle skull.

preorbital. *n.* See *preocular*.

prepalatine cartilage. See *lamina externa*.

preparator. *n.* A person whose work or skill is preparing models or reconstructions (e.g., dinosaurs) for display in a museum.

preparietal. *n.* A head scale of typhlopid snakes located laterally to the suture between the second and third scales on the midline behind the *rostral* and followed by the *postparietal*. Used by F. Wall (*Snakes of Ceylon*, p. 7, 1921).

preparotoid. *n.* The cranial region or crest bordering the *parotoid gland* anteriorly in some anurans. (Syn. *supratympanic*) See Fig. 2.

preplastron. n. Either of paired, anterior bony elements of the plastron of trionychid turtles. Formerly termed the *entoplastron* and considered homologous with that bone in other turtles. (Syn. *ossa praeplastralia*)

prepollex, prepollux. *n.* A term used interchangeably with *pollex* for the medial or vestigial digit on the inner side of the first digit of the forefoot in some species of anurans. Cf. *prehallux*.

pre-prezygapophysial width. The measured distance from the outer edge of one prezygapophysial facet to the outer edge of the opposite prezygapophysial facet on a single vertebra of a snake. Used by W. Auffenberg (*Breviora* 114:1–16, 1959).

preprint. *n.* A printed work that is distributed in advance of its being published as a book or periodical.

prepuberal. *adj.* Reference to characters that develop before sexual maturity and are maintained thereafter. Cf. *postpuberal*.

prepubic process. A projecting section of the anterolateral margin of the pubis in certain New Zealand Geckos. (Syn. *pectineal process*)

prepubis. *n.* 1. The anterior element of the pelvis in crocodilians and certain fossil reptiles. 2. Synonym for the *epipubis* of *Ascaphus* frogs.

presacral vertebrae. A collective term that includes all vertebrae anterior to the first that is connected to the pelvis. The term may include *cervical* and *dorsal* vertebrae in reptiles, or *cervical* and *trunk* vertebrae in amphibians.

presaturation dispersal. Emigration of individuals from a population that is at low density or in an early phase of growth. Cf. *saturation dispersal*.

prescapular process. A term for the *acromial process* in the pectoral girdle of turtles.

presphenoid. *n.* Variously used for ossifications of the chondrocranium. (Syn. *septosphenoid, orbitosphenoid*)

pressure. *n.* Force per unit area, typically measured as *mm Hg* = 0.133 *kilopascals* (*kPa*) = 1330 dynes/cm^2.

Presssure-Balance Hypothesis or **model.** A proposed mechanism of venom expulsion from fangs of venomous snakes, based largely on studies of Spitting Cobras. Pressure builds up within the venom chambers of the fang sheath until physical displacement of the sheath allows venom flow through the fang, as during fang penetration. The idea is supported by experimental evidence from studies by B.A. Young and coinvestigators (*Bioscience* 52:1121–1126, 2002; *Can. J. Zool.* 81:313–320, 2003). Recently, Young et al. (*J. Exp. Biol.* 207:3483–3494, 2004) suggest that Spitting Cobras utilize a pressure-balance system wherein displacement of the fang sheath is actively produced by contraction of the *M. protractor pterygoideus* and ensuing displacement of the palato-maxillary arch.

pressure drag. In biomechanics, drag on a moving object caused by displacement of the fluid medium, backflow, and the formation of pressure gradients. Thus, it is due to the pressure difference between the front and midsection of a moving object. See *drag*. (Syn. *form drag*)

pressure potential. A component of *water potential* that arises as a result of an applied hydrostatic or pneumatic pressure.

pressure pulse. The difference between *systolic* and *diastolic blood pressure*.

pressure ridge. Reference by R.C. Stebbins (1954) and others to the small ridges of sand or other loose soil that form as a result of the backward force of each individual loop of a snake's body as it acts on the substrate during forward locomotion on a shifting surface. These small ridges enable one to ascertain the direction of movement of a snake from its track.

pressure shunting. See *intracardiac shunt*.

presternal vertebrae. A term used by some authors to refer to trunk vertebrae anterior to the first vertebra with a rib connecting to the sternum.

presternum. *n.* A term used by various authors in reference to the *omosternum* of the frog girdle or the *metasternum* of reptiles. See Fig. 3.

presubocular. See *subocular*.

presumptive. *adj.* Descriptive of a presumed condition or fate.

presupraloreal. *n.* Any of small scales which lie below the *canthals*, anterior to and bordered by the *supraloreal*, and posterior to the *postnasal* of Rattlesnakes.

presynaptic. *adj.* Reference to the sending ("upstream") side of a synaptic connection.

presynaptic facilitating toxins. Components of snake venoms that facilitate the release of acetylcholine at neuromuscular junctions. Cf. *presynaptic paralyzing toxins.*

presynaptic paralyzing toxins. Components of venom of many elapids and certain viperids that exert a paralyzing action by inhibiting the release of acetylcholine at neuromuscular junctions. (Syn. *β–toxins*)

pretemporal. *n.* A term designating any scale belonging to the first vertical row of *temporal* scales. See Fig. 11.

preterminal segment. See *sexual segment.*

pre-term. *adj.* Before the term or due date, usually in reference to early birth or hatching.

pretestis. *n.* The pregonad of male amphibians.

prethecodont. *adj.* A condition of a deep dental groove bordered by high lingual and labial ridges, but without interdental partitions, as in young Crocodiles.

pretympanic. *adj, n.* 1. Located just anterior to the tympanum on the side of the head. (Syn. *orbitotympanic*) See Fig. 2. 2. Any of a number of scales located immediately anterior to the external ear in lizards.

prevalence. *n.* In contexts of medicine or toxicology, the number of cases diagnosed at a given time expressing the present frequency of an event.

prevomer. *n.* A term used by several authors for the bone called the *vomer* in the palate of amphibians and reptiles. Confusion arises concerning debate about homology with the vomer of mammals. See *vomero-palatine.*

prevomerine teeth. Teeth located on the prevomer.

prezonal episternum. Alternate term for *episternum.* See Fig. 3.

prezygapophysis. (*pl.* -es) *n.* See *zygapophysis.*

prey. *n.* An animal that is hunted or captured for food by another animal.

prey set. The total range of prey items that are utilized by a particular predator.

prey refugia. Areas occupied by prey that potentially minimize their rate of encounter with predators. See R.J. Taylor, *Predation*, 1984.

prey transport. A term used in reference to swallowing, including manipulation of the prey by the various active elements of the jaws and mouth.

prezygotic isolating mechanism. See *reproductive isolating mechanism.*

primary. *n.* Reference to grooves on caecilians that correspond with the ends of the ribs and therefore vertebrae, thus being identical to the costal grooves of salamanders.

primary bone. Bone in which collagen fibers are not regularly oriented, formed during the growth of an animal. Immature or woven bone. Cf. *secondary bone.*

primary consumer. A heterotrophic organism that feeds directly on a primary producer. Cf. *secondary consumer.*

primary egg membranes. See *egg membranes.*

primary forest. Pristine, unlogged climax forest.

primary girdle. The inclusive endochondral derivatives of the pectoral girdle, preformed in cartilage, that may remain cartilaginous or ossify. These include the *scapula, suprascapula*, and the *coracoid*, in addition to any derivatives or components of these elements.

primary homonym. In taxonomy, each of two or more identical names of different species or subspecies established in the same genus. Cf. *secondary homonym.*

primary host. The initial or principal host of a parasite.

primary neurulation. Formation of the embryonic neural tube by uplifting of folds in the edges of the neural plate, which subsequently fuse. Cf. *secondary neurulation.*

primary note. The longer of two or more notes in a call sequence of anurans.

primary producer. An organism that synthesizes complex organic molecules from simple inorganic substrates. An *autotroph.*

primary production or **productivity.** The synthesis of organic matter by autotrophs. See *gross primary production, net primary production.*

primary receptor. See *receptor.*

primary sex cord. See *sex cord.*

primary sex ratio. See *sex ratio.*

primary sexual character. Reference to the ovaries or testes, i.e. the organs that function to produce gametes. Cf. *secondary sexual character.*

primary speciation. The splitting of a single species into two species, usually as a result of natural selection favoring different gene complexes in geographically disjunct populations. Cf. *secondary speciation.*

primary type. The original specimen on which a species was based. (Syn. *type, prototype, basic type*)

prime mover. In evolutionary biology, reference to the ultimate factor that determines the direction and rate of evolutionary change, involving *phylogenetic intertia* and *ecological pressure.*

primer DNA. In molecular biology, a single-stranded DNA required for replication by DNA polymerase in addition to primer RNA. Also used in reference to oligonucleotides of single stranded DNA that are synthesized for use in a *polymerase chain reaction.*

primer pheromone. A pheromone that interacts with the physiology of a recipient organism, eventually eliciting a response. Cf. *releaser pheromone.*

primitive, primitive character. *adj.* Reference to a trait or condition that is *basal* or of ancient evolutionary origin in a phylogenetic lineage. Cf. *advanced, derived.*

primitive streak. The embryonic structure of amniotes that forms most of the dorsolateral mesoderm by means of migration and involution of cells.

primordial. *adj.* Reference to the earliest stage in ontogeny or development of an organ or system. Primitive or primary.

primordial germ cells. Cells that migrate into germinal epithelium of gonads and ultimately give rise to gametes.

primordium. *n.* See *anlage.*

principal components analysis. (PCA) A linear-transformation method of transforming data to a coordinate system such that the first axis explains the maximum amount of variance (called the first principal component), the second (orthogonal to the first) explains the maximum of the remaining variance, and so forth. Most of the variance is typically explained by the first two or three principal components (which are normalized eigenvectors). PCA can be used to reduce the dimensionality in a dataset while retaining those characteristics of the dataset that contribute most to its variance.

principal coordinate analysis. See *metric scaling*.

principal horn. A large process arising from the antero-lateral corner of the anuran *hyoid plate*, generally curving posteriorly along the remaining *hyoid apparatus*. This term is often substituted for *ceratohyal*. (Syn. *corniculum, hyalia, hyoid process*)

principal lobe. See *pars distalis*.

prions. *n.* Infectious pathogenic particles composed exclusively of special proteins and at least 100 times smaller than a virus.

priority. *n.* In taxonomy, a principle that seniority of a name is fixed by the date of valid publication or availability.

prism. *n.* See *enamel prisms*.

prismatic enamel. Enamel with prisms. See *enamel prisms*.

prismless enamel. A general term describing enamel that lacks enamel prisms. Cf. *enamel prisms*.

pro-. Prefix meaning "before," "in front of," or "forward."

proamnion. *n.* A term that was previously applied to bilaminar blastoderm of early embryos, consisting of ectoderm and endoderm and giving rise to the anterior fold of the amnion. In reptiles, this area is not completely homologous with the "proamnion" of birds.

proatlas. *n.* 1. Either of a pair of inverted V-shaped elements anterior to the neural arch elements of the atlas in some reptiles, including *Sphenodon* and Chameleons in living forms. (Syn. *preatlas, supradorsal*) 2. The *odontoid process* combined with the posterior of a pair of intercalated cartilaginous arches, which form cups fused to the arch of the atlas in salamanders.

probability. *n.* A quantitative measure of how likely something is to occur, relative to all possible outcomes. The probability of an event is the long-term frequency of an event relative to all alternative events, usually expressed as a decimal fraction.

probability distribution. See *frequency distribution*.

probe. *n.* In molecular biology, this term refers to an identifiable biomolecule that is used to identify or isolate a gene, a gene product, or a protein. The probe or molecule is labeled with a radioisotope or tagged in other ways that render it easily identified. A *DNA probe* is a chemically synthesized, radioactively labeled segment of nucleic acid used to find a gene of interest by hydrogen-bonding to a complementary sequence.

proboscis. *n.* An elongate and flexible extension of the snout. (Syn. *rostral appendage*)

procaudal. *n.* See *suprapygal*.

procely, procoely. *n.* (*adj.* **-ous**) Reference to the condition of a vertebral unit in which the anterior socket of the centrum is concave and the posterior face is convex. This is a consequence of fusion of the *intervertebral disk* with the vertebra lying anterior to it, which then fits into a hollowed area on the next posterior vertebra. This condition is characteristic of most living reptiles.

process. *n.* A projection or outgrowth of bone.

processus alares. See *alary process*.

processus articular. See *zygapophysis*.

processus spinosus. See *neural spine*.

processus thyreoideus. The *thyrohyal* of anuran hyoid.

procoagulant. *n.* A toxin that activates the coagulation cascade.

procoelous. *adj.* Condition of vertebrae in which elements are concave anteriorly and convex at the posterior end, characteristic of anurans and extant reptiles. Cf. *amphicoelous, opisthocoelous*.

procoracoid. *n.* 1. The anterior bar of cartilage lying immediately above the clavicle in many frog species (Fig 3). 2. The anterior element where two coracoid ossifications are present in some reptiles. This name is used for the sole *anterior coracoid* ossification in living reptiles, some authors believing the true coracoid has disappeared. 3. The anterior part of the ventral plate of cartilage in salamanders, being distinguished from the coracoid part of the plate by the *incisura coracoidea*. 4. A prominent, ventral, and mesially directed bone in the pectoral girdle of turtles, usually called the *coracoid*. (Syn. *anterior coracoid, coracoid process*) 5. Reference to the *acromial process* in turtles. 6. A third center of ossification in some primitive reptiles.

procricoid. *n.* A median, cartilaginous element of the laryngeal skeleton of turtles, dorsal and anterior to the *cricoid* itself.

procryptic. *adj.* (*n.* **procrypsis**) Having protective coloration and behavior.

proctodeum. *n.* 1. The embryonic invagination of surface ectoderm that contributes to the hindgut, usually giving rise to the cloaca. 2. The posterior third and final chamber of the reptilian *cloaca*, through which all products of excretion and defecation must pass, in addition to female reproductive products. See also *coprodeum* and *urodeum*.

production, ecological. (P) The amount of increase in organic matter, usually measured per unit area of land surface or per unit volume of water and divided into *gross* (that added before any use) and *net* (that which remains stored after use). *Primary production* is the production of organic matter by autotrophs. *Secondary production* is the production of organic matter by heterotrophs.

production efficiency. The percentage of energy consumed by an individual or population that is converted into new biomass.

productivity. *n.* 1. Ecological productivity is the amount of increase in organic matter per unit area or volume per unit time (note this is a rate function) Cf. *production.* 2. Fertility.

productivity coefficient. A dimensionless quantity derived as the ratio of gross community *production* to *respiration.*

proenzyme. *n.* The inactive form of an enzyme before it is activated by removal of a terminal segment. (Syn. *zymogen*)

proectoglyph. *n.* Reference to a venomous, rear-fanged snake in which the channel for passage of venom lies on the anterior margin of the fang. (Syn. *pro-opisthoglyph*) Cf. *pleuroglyph.*

profile. *n.* A graph depicting changes in one variable against another, e.g. temperature as a function of depth.

profile drag. All the drag on a flying or gliding object other than that associated with the ends of the wings.

progenesis. *n.* 1. Accelerated development of reproductive organs relative to somatic tissue, leading to retention of formerly juvenile characters by adult descendants (*paedomorphosis*). 2. An evolutionary change in which sexual maturity occurs earlier in the descendant species than in an ancestor.

progenitor. *n.* An ancestor.

progeny. *n.* The offspring that result from a given mating between two individuals. Siblings.

progesterone. *n.* A steroid hormone secreted by the *ovary* and *corpus luteum*, essential for egg retention and pregnancy maintenance in viviparous species. It has many effects that support egg and embryo development, including expression of egg-white proteins and suppression of myometrial activity. Progesterone (P_4) also is the steroid precursor for testosterone and estradiol 17-β.

progestins. *n.* See *progestogens.*

progestogens. *n.* Substances with progesterone-like activity, having roles in the maintenance of pregnancy or gravidity and an important role in germinal vesicle breakdown (final oocyte maturation) in amphibians. (Syn. *progestins*)

proglyph. *n.* See *endoglyph.*

proglyphodont. *adj.* See *endoglyphous.*

prognosis. *n.* A forecast that specifies the course and termination of a disease.

progonad. *n.* The anterior part of a gonad, which contains few germ cells and differentiates into fat body. Cf. *pregonad.*

programmed cell death. See *apoptosis.*

progressive selection. See *directional selection.*

proinvasin. *n.* A component in the venom of *Agkistrodon piscivorus* that destroys the ability of blood to inhibit the action of hyaluronidase, which then acts as a "spreading factor."

project. *v.* In the nervous system, reference to transmission of neural activity to an innervated region.

prokinesis. *n.* A type of *cranial kinesis* in which the line of movement is located between the *nasals* and *frontals,* or dermatocranium, anterior to the ocular orbit. Coined by T.H. Frazetta (*J. Morphol.* 111:287–319, 1962). See Fig. 18.

prolactin. *n.* A hormone secreted from the adenohypophys, having multiple roles in amphibians and reptiles. Prolactin enhances the influence of thyroid hormone in regulating molting in lizards, and it decreases epithelial keratinization in newts, aiding to produce a smooth skin for aquatic habitats. Concentrations of prolactin peak with metamorphosis in anuran amphibians and sensitize oocytes to stimulation by gonadotropins or progesterone. Prolactin concentrations tend to decline with stress.

prolapse. *n.* Protrusion or partial expulsion of an organ through an orifice, or its falling or slipping out of place.

proliferation. *n.* Growth or multiplication; cell division; reproduction.

prolonged amplexus. The habit of certain anuran species (e.g., *Atelopus*) to remain in *amplexus* for periods of days, weeks, or even months during migration to breeding sites, etc.

prolonged breeder. Reference to a species having a reproductive period lasting longer than one month.

promegadont. *adj.* Reference to presence of distinctly enlarged, aglyphous teeth at the anterior end of the maxillary bone, followed either by a *diastema* and then gradually smaller teeth (*diapromegadont*) or by a continuous series of gradually smaller teeth with no diastema (*synpromegadont*). Terms are from H.M. Smith (*Turtox News,* Vol. 30, No. 12, 1952). (Syn. *scaphiodont*)

prometamorphic period. *n.* A period of metamorphosis marked by the onset of accelerated limb growth, including pigment changes, dehydration of tissue, and other changes, and ending as the *climax period* begins. First used by W. Etkin (*Physiol. Zool.* 7:129–148, 1934).

prometamorphosis. *n.* See *prometamorphic period,* also W. Etkin, pp. 313–348 in W. Etkin & L.I. Gilbert, eds., *Metamorphosis,* 1968.

promiscuous. *adj.* Reference to a mating system in which neither sex is restricted to a single mate.

pronation. *n.* (*v.* **pronate**) Turning of a part such that the ventral side is directed downward.

pronator. *n.* A muscle that rotates or turns the ventral side downward.

pronephros. *n.* The anterior section of mesomere that develops into a functioning, usually transient, kidney in embryonic vertebrates. Cf. *archinephros, holonephros, mesonephros, metanephros, opisthonephros.*

proneural. *n.* See *nuchal bone.*

pro-opisthoglyph. *n.* See *proectoglyph,* which replaces this term.

prootic. *n.* An endochondral skull bone derived from the chondrocranium, which houses part of the labyrinth of the inner ear, forms the anterior part of the *otic capsule*, and contributes the anterior border of the *fenestra ovalis*.

prooxidant. *n.* See *reactive oxygen species*.

propagule. *n.* In ecology, the minimum number of individuals that are required for a species to colonize a new or isolated habitat.

propagule rain. A metapopulation model in which there is a large, external source of colonists, and the probability of habitat patch colonization does not depend on the fraction of sites occupied in the metapopulation.

property. *n.* Any character or attribute of an organism or system.

prophylaxis. *n.* (*adj.* **prophylactic**) Preventive treatments intended to protect from disease or contain its spread.

propodial. *n.* 1. The most proximal segment of a vertebrate limb, or the single proximal bone in either limb (humerus or femur). (Syn. *stylopodium*)

proportional control. A control action with an output that is proportional to the deviation of the controlled variable from a desired or established set point.

proportion of area occupied. (PAO) See *site occupancy*.

proprioception. *n.* The perception of "body awareness" concerning position and relative motion of body parts.

proprioceptors. *n.* Sensory receptors situated primarily in muscles, tendons, or joints that sense tension and provide information about the position and relative motion of body parts. Such receptors are specialized *interoceptors*. Cf. *exteroceptor, interoceptor*.

proptosis. *n.* The prolapse of the eyeball from its orbital boundaries.

pros-, pro- Prefix denoting a positive condition.

Prosauropoda. *n.* A subgroup of *Dinosauria* and one of two distinct clades of *sauropodomorph* dinosaurs.

proscapula. *n.* See *acromial process*.

proscapular process. See *acromial process*.

prosencephalon. *n.* A principal brain division, anterior in location and equivalent to the primitive forebrain. The prosencephalon includes both the *telencephalon* and *diencephalon*. (Syn. *forebrain*)

prospective. *adj.* Potential, possible, or probable. Cf. *realized*.

prospective ecospace. See *ecological niche*.

prostaglandins. *n.* A subgroup of *eicosanoids* (long-chain fatty acids) that originate in various tissues and have multiple effects, mostly related to modifying contractions of smooth muscle (vascular, pulmonary, and reproductive systems), vascular permeability, blood pressure, and the actions of some hormones. Prostaglandins are derived from arachidonic acid.

prosthetic group. A portion of a complex protein that is not a polypeptide. Usually this is the active site of a protein and is essential to the function of an enzyme.

protandry. *n.* The arrival of males before females at a breeding site. Cf. *protogyny*.

protease. *n.* Any proteolytic enzyme. There are many proteases in snake venoms, especially those of Viperidae, and these are involved in the tissue destruction seen in necrosis. See also *proteinase, peptidase*.

protective call. See *distress call*. (Also previously used as a synonym for *alarm call*.)

protective coloration. See *cryptic coloration*.

Proteidae. *n.* A clade (family) of aquatic, paedomorphic salamanders lacking maxillary bones and possessing two pairs of larval gill slits (three in other salamanders). Six species are recognized in Europe and eastern North America.

protein. *n.* A large molecule composed of one or more polypeptide chains, each consisting of a linear chain of amino acids (therefore containing nitrogen).

protein clock. See *molecular clock*.

proteinase. *n.* Any enzyme that catalyzes the splitting of interior peptide bonds of a protein. See also *protease*.

proteinuria. *n.* The occurrence of excess protein in urine.

proteolysis. *n.* (*adj.* **proteolytic**) The hydrolysis of proteins resulting in splitting of peptide bonds and formation of smaller polypeptides.

proteolytic enzyme. An enzyme that degrades proteins or polypeptides. A component of many snake venoms. See also *proteolysis*.

proteolytic principle. Earlier term for any *proteolytic* component of snake venoms.

proteomics. *n.* The study of all the proteins that are expressed in an organism; i.e., global protein analysis.

proterodontous. *adj.* Presence of a larger tooth at the front end of the maxilla.

proterodont pathway. An evolutionary sequence in which the fang of venomous snakes has developed on the anteriormost point of maxillary support, and the palate has only a slight mobility, as in the elapids. See also *opisthodont pathway*.

proteroglyph. *n.* (*adj.* **-ous**) A condition in which a relatively long maxilla bears a single, hollow, venom-conducting fang on its anterior end, followed posteriorly by non-venom-conducting teeth (elapids). The fang is fixed and not erected due to the relative immobility of the maxillary unit. Cf. *aglyph, opisthoglyph, solenoglyph*.

proterotype. *n.* 1. Primary type. 2. All the original specimens collected at one time from which the types were selected.

prothrombin. *n.* A coagulation factor in blood.

proto-. A prefix meaning "earliest," "first," or "original." Sometimes used to designate an ancestral form that was the evolutionary precursor of a taxonomic group (e.g., protoamphibian, etc.).

protoadaptation. *n.* A transient form or state of a character, which becomes functionally useful in response to future selection, although originally evolved in response to different selection forces acting on ancestors (C. Gans, *Evolution* 33:265–272, 1979).

protocement. *n.* See *cement*.

protogyny. *n.* The arrival of females before males at a breeding site. Cf. *protandry*.

protonym. *n.* In taxonomy, a name that is not validly published but is reused and validly published subsequent to first publication.

protothecodont ankylosis. A condition whereby a gutter is formed by labial bone, to which a tooth is ankylosed. The base of each tooth is ankylosed into a more or less deep socket by the deposition of cementum (J.W. Osborn, *Symp. Zool. Soc. Lond.* 52:549–574, 1984). Cf. *acrodont, pleurodont, thecodont ankyloses*.

protothrix. (*pl.* **-iches**) *n.* Epidermal thickenings that project above the skin of amphibians, having various forms in different species and considered by some to be a forerunner of hair (H. Elias and S. Bortner, *Amer. Mus. Novitates* 29, 1–15, 1957). May also comprise a simple sensory organ consisting of depressed epidermal thickenings scattered on the skin surfaces of fishes and various amphibians. These may or may not bear a central hairlike projection.

prototype. *n.* 1. See *primary type*. 2. A hypothetical ancestral stage.

protozoan. *n.* Any of numerous single-celled organisms living primarily by animal-like, heterotrophic nutrition. These include flagellates, amoebae, sporozoans, etc., which often occur as *endoparasites* in amphibians and reptiles.

protozoiasis. *n.* Any disease caused by *protozoa*.

protraction. *n.* (*v.* **protract**) Forward movement or projection of a limb or part. Reaching out from the body in a sagittal plane.

protractor. *n.* A muscle that draws a segment forward or out from the body.

protrusible. *adj.* Capable of being protruded, or thrust out and forward.

protuberance. *n.* A projecting part.

protuberant. *adj.* Swollen beyond the surrounding surface.

provenance. *n.* The place of origin.

proximal. *adj.* That part of a limb or structure that is closest to its point of attachment or origin, and usually medial. Closer to the center or central axis of the body.

proximate. *adj.* In immediate relation with something else. With respect to causality, this term refers to an immediate mechanism that explains a biological phenomenon. Cf. *ultimate*.

proximate factor. An environmental factor that acts as an immediate stimulus affecting biological activity.

psammonic. *adj.* An ecomorphological guild of tadpoles including lotic species that inhabit loose sand in rapidly flowing streams.

Psammophiini. *n.* A clade of colubrid snakes that are generally slender, diurnal, and swift-moving species, most represented by Afro-Asian *Sandsnakes* and *Grass Snakes* (*Psammophis*).

psammophilous. *adj.* Reference to sandy substrate or habitat, and usually in context of an animal's habits.

pseudautotomy, pseudoautotomy. *n.* Reference to nonspontaneous loss of the tail by intervertebral breakage, usually without tail regeneration in some lizards and snakes. There is no specialized caudal morphology in *unspecialized* or *nonspecialized pseudautotomy*, whereas the tail is modified to increase its fragility in *specialized pseudautotomy*. See J.M. Savage & J.B. Slowinski, *Biol. J. Linn. Soc.* 57:129–194, 1996.

Pseudidae. *n.* A clade (family) of aquatic anurans characterized by having elongate ossified intercalary elements, which enhance expansive webbing on the feet. Known as *Harlequin Frogs* or *Paradox Frogs*, two genera with nine species inhabit parts of South America. These frogs are very likely nested within the Hylidae.

pseudo-. Prefix meaning "false."

pseudoaposematic. *adj.* A false warning, applied to harmless organisms that evolve adaptive aposematic colors or behaviors similar to those of a dangerous species.

pseudoangular. *n.* A compound bone of certain caecilians, formed by fusion of the *angular, prearticular, articular,* and *complimentale*.

pseudoautotomy. *n.* See *pseudautotomy*.

pseudocentrous. *adj.* A type of vertebra in which the centrum is formed by fusion of the bases of the dorsal and ventral arches, according to *Gadow's hypothesis*. This is specifically denied by E.E. Williams (*Quart. Rev. Biol.* 34:1–32, 1959), who considers the centrum to be of indepedent formation and should be termed *autocentrous*.

pseudocelous, pseudocoelous. *adj.* One of two basic types of *ophisthocely* in which the notochordal canal is hollow but covered anteriorly with an ossified or cartilaginous cap. Externally it appears the same as the *hemicelous* condition.

pseudocolumella. *n.* The *columella* in anurans, differing from the *eucolumella* in that the *plectrum* is derived from the otic capsule and the *stylus* is from the quadrate (H.M. Smith, *Evolution of Chordate Structure*, p. 214, 1960).

pseudoconcha. *n.* Any projection of the lateral wall of the nasal cavity other than simple lamellar processes, which are termed *conchae*, and including the conchae of most squamates and the *preconcha* and *postconcha* of crocodilians. Discussed in detail by T.S. Parsons (*Bull. Mus. Comp. Zool.* 120:103–277, 1959).

pseudodentary. *n.* The compound bone in the lower jaw of *Hypogeophis* caecilians, formed by fusion of the *dentary, splenial, coronoid, supra-angular,* and *mentomeckelian*.

pseudoextinction. *n.* A condition in which all the individuals of a given taxon disappear, but this fact is accompanied by the appearance of one or more new taxa directly descended from the previous one. (Syn. *phyletic extinction*)

pseudohermaphrodite. *n.* See *hermaphrodite*.

Pseudomonas infection. Infectious disease caused by Gram-negative bacteria in the genus *Pseudomonas*. Commonly isolated from the oral cavity of snakes with *stomatitis* (*"mouth rot"*) and pneumonia.

pseudoneoplasm. *n.* A tumor-like growth that is not a true tumor. See *pseudotumor*.

pseudo-opisthoglyph. *n.* Synonym for *pleuroglyph*, proposed by C.M. Bogert & J.A. Oliver (*Bull. Amer. Mus. Nat. Hist.* 83:297–426, 1945).

pseudoperculum, pseudo-operculum. *n.* See *plectrum*.

pseudopisthocelous, pseudopisthocoelous. *adj.* Reference to an *opisthocelous* condition attained without fusion of the intervertebral body to the centrum.

pseudoplacenta. *n.* Any structure that functions similarly to a mammalian placenta, permitting respiratory and other exchange between the female parent and developing embryos that are carried by the mother.

pseudopreocular. *n.* Synonym for *presubocular* (see *subocular*).

pseudo-replication. *n.* In statistical sampling, a term to denote the use of a datum that is falsely assumed to represent an independent sample. Observations or duplication of experimental design that are not independent.

pseudotail. *n.* The presence of a few postcloacal vertebrae in an unsegmented terminal shield in some caecilians.

pseudotooth. *n.* A hypertrophied bony projection on the jaw of certain amphibians that resembles a true tooth.

pseudotumor. *n.* Reference to an abnormal swelling in the skin that is not a true tumor but attributable to encysting parasites or cellular hypertrophy associated with wound healing or regeneration. See also *pseudoneoplasm*.

pseudotype. *n.* In taxonomy, a type that is designated in error.

pseudoviviparity. *n.* See *viviparity*.

psicolous, psilicolous. *adj.* Living in prairie or savannah habitats.

psolidomeiomology. *n.* The study of scale row reductions in snakes. Coined by P.J. Clark & R.F. Inger (*Copeia* 1942:230–232, 1942).

psychric. *adj.* Reference to low temperatures or cold habitats.

psychro-. Prefix meaning "cold."

pterapophysis. *n.* A wing or spine-like process located posteriorly on the neural arch of fossil snakes.

pterorhodophore. *n.* A cell containing a eumelanin core surrounded by a mass of the red compound pterorhodin.

Pterosauria. *n.* An extinct group of *archosaurs* closely related to dinosaurs, and the first vertebrates to evolve powered flight. Flying reptiles.

pterosaurs. *n.* Reference to any or all members of the *Pterosauria*.

pterygoid. *n.* Paired dermal bones forming a major part of the palate, located behind the *palatines* in contact with the *quadrate*, and are generally separated from one another posteriorly by the *parasphenoid* in amphibians. These may bear teeth, as in some snakes where it forms an important part of the *palatopterygoid*. (Syn. *internal pterygoid*) See Fig. 21.

pterygoid canal. See *vidian canal*.

pterygoid walk. Reference to movements of the bilateral palato-maxillary apparatus of alethinophidian snakes, which are alternately lifted and protracted to secure a more forward grip on a prey item during *prey transport*. Each side alternately pulls forward on the braincase as the palatal teeth are embedded in the prey, with alternate actions effectively drawing the snake's head over the prey item. The snake moves over the prey item rather than drawing it in. Unilateral, asynchronous jaw ratcheting movements of the upper jaws. Used by R.E. Boltt & R.F. Ewer (*J. Morphol.* 114:83–106, 1964). Cf. *mandibular raking*.

pterygopalatine. *n.* A bony rod in the palate of larval salamanders, persisting in some non-metamorphic salamanders but lost in the non-neotenic forms. See *postpalatine*.

pterygoquadrate cartilage. See *palatoquadrate cartilage*.

pterygoquadrate joint. An articulation between the *pterygoid* and *quadrate* in a kinetic skull.

ptosis. *n.* Prolapse of an organ or part. 2. Paralytic drooping of the upper eyelid, symptomatic of neurotoxic envenomation.

Ptychadenidae. *n.* A taxon (family) of anurans, sister to the new taxon *Victoranura* within the new taxon *Natatanura*, and containing three genera with species distributed in sub-Saharan Africa, Seychelles, and Madagascar. See D.R. Frost et al., *Bull. Amer. Mus. Nat. Hist.* 297:1–370, 2006.

pubic. *adj.* Relating to or located in the region of the pubis.

pubic groove. See *preanal groove*.

pubic symphysis. The joint between fused pubic bones in the pelvic girdle of many reptiles.

pubis. (*pl.* -es) *n.* The anterior and ventral bones of the pelvic girdle, often fused in many reptiles.

pubo-ischiadic fenestra. See *cordiform foramen*.

puboischiadic plate. The ventral surface of the pelvis, formed by fusion of the *pubis* and the *ischium*.

puboischiotibialis muscle. See *gracilis muscle*.

Puddle Frogs. Collective vernacular name for African ranid frogs belonging to the genus *Phrynobatrachus*. Breeding takes place in shallow standing water.

Puerto Rican Coqui. See *Coqui*.

Puff Adder. Vernacular name for the African viperid species *Bitis arietans*, a snake that hisses loudly during defensive displays. This is the most widely distributed viper in Africa. The name also is used by some to apply, collectively, to all members of the genus *Bitis*.

Puff-faced Water Snake. Vernacular name for the Asian homalopsine species *Homalopsis buccata*.

Puffing Snakes. Vernacular name for neotropical species of snakes belonging to the colubrid genus *Pseustes,* named for impressive neck inflation during defensive displays. (Also known as *Bird Snakes.*)

pulmocutaneous artery. A major arterial outflow vessel in anurans, derived from the sixth aortic arch and giving rise to a *pulmonary artery* carrying blood to the lungs and a *cutaneous artery* carrying blood to the skin. Oxygenation of blood in amphibians occurs both at the lung and at the skin to varying degrees.

pulmonary. *adj.* Pertaining to the lungs.

pulmonary artery. The artery (sometimes divided into two or more arterial branches) that conveys relatively oxygen-poor blood to the lung(s).

pulmonary edema. Leakage of fluid into the respiratory (exchange) spaces of the lung, characteristically creating a diffusion barrier and compromising gas exchange.

pulmonary process. Historically used in reference either to the *bronchial process* or a small, posterolateral projection on this structure in amphibians.

pulsatile. *adj.* 1. Reference to periodic behavior in the trace of a variable through time (as in a chart record), which produces an up and down (waxing and waning) characteristic of the trace. 2. Pulsating, as in the beating of a drum.

pulse. *n.* 1. In bioacoustics, the distinct pulsations of sound that constitute a note, often apparent as vertical marks on an audiospectrogram. 2. In physiology, the *pulse pressure* is the difference between the *systolic* and *diastolic* pressure of blood in a major artery. 3. In ecology, a pulse is descriptive of a sudden increase in abundance of organisms or species, often occurring at regular intervals.

pulse-chase experiment. A technique in which cells are briefly exposed (the pulse) to a radioactively labeled precursor of some macromolecule, and the metabolic fate of the label is followed during subsequent incubation in a medium containing only the nonlabeled precursor (the chase).

pulse rate or **pulse repetition rate.** In bioacoustics, the number of pulses per unit time, usually expressed as pulses per second.

pulvinar rostrale. A low prominence on either side of the midline at the tip of the upper jaw that interrupts the *sulcus marginalis* in the frog.

pulvinar vocal. (*pl. pulvinaria vocale*) Fibrous structures that support the vocal chords in some anurans.

pumiliotoxins. *n.* Toxic alkaloids found in the skin of a various dendrobatid, bufonid, ranid, and myobatrachid frog species, acting to increase the rate of opening and closing of sodium channels and thereby produce spontaneous activity of nerves leading to motor, cardiovascular, and respiratory disturbances. See also *decahydroquinolines.*

punctate. *adj.* Spotted or marked with points or small depressions, usually with distinct margins.

punctuated equilibrium or **evolution.** In evolutionary biology, reference to relatively brief episodes of speciation followed by long periods of species stability. This concept is derived from observation of the fossil record (N. Eldredge & S.J. Gould, pp. 82–115 in T.J.M. Schopf, *ed., Models in Paleobiology,* Freeman, Cooper & Co., San Francisco, 1972). A phylogenetic tree constructed on assumption of punctuated evolution has all character evolution at branches (nodes). Cf. *gradualism, phyletic gradualism, molecular clock.*

pungent. *adj.* Sharp but not pointed. Sometimes used to describe tubercles on hands and feet of anurans.

Punnett square. A checkerboard method (matrix) commonly used to determine genotypic and phenotypic ratios.

pupil. *n.* The aperture in the center of the *iris* through which light enters the eye. The pupil in the eyes of amphibians and reptiles varies greatly in size and shape and can be adjusted by contraction of the iris muscles to protect sensitive cells of the retina.

pure line. Reference to descendants of a single homozygous parent produced by inbreeding.

purifying selection. See *negative selection.*

purgative. *adj.* See *cathartic.*

purinergic. *adj.* Reference to *purines* or their derivatives in their role as neurotransmitters.

purines. *n.* A class of heterocyclic nitrogenous compounds, of which derivatives (purine bases) are found in nucleotides.

purpura. *n.* 1. Accumulation(s) of blood outside the vascular system, producing the appearance of bruising. 2. Eruption of red blotches on the skin due to blood seeping out of capillaries.

purse-string suture. A running stitch that is placed in a circle and pulled together to close an opening.

purulent. *adj.* Relating to, or forming, pus.

pushup. *n.* A rapid elevation and lowering of the body by means of forelimb movements in many lizards. This is an important *display* that functions in sex recognition and territorial defense. See also *headbob.*

pustular, pustulate. *adj.* Pertaining to, or consisting of, *pustules.*

pustule. *n.* A small, elevated pus-containing lesion of the skin. In older herpetological literature, this term was sometimes used in reference to "warts" of amphibians.

putrefaction. *n.* See *decay.*

P value. Probability value, expressed as a decimal fraction. See *probability.*

pygal. *n.* 1. The posteriormost bone in the middorsal line of the turtle carapace. This term has also been used variously to refer to the *suprapygal* and *postcentral* elements of the carapace. See Fig. 5. 2. Reference to the basal caudal vertebrae of reptiles, including those without chevrons but having ribs. Crocodilians have one, and lizards have two, pygals.

Pygopodinae. *n.* A subfamily of *Gekkonidae,* inhabiting Australia, New Guinea, and some adjacent islands.

pylangium. *n.* The caudal part of the amphibian *conus arteriosus*, which connects to the ventricle proper. Differentiated from the posterior part of the embryonic *bulbus cordis*. Cf. *synangium*.

pyloric. *adj.* Reference to the part of the stomach adjacent to the intestine. See *pars pylorica*.

pyloric sphincter. A sphincter muscle located at the opening of the stomach into the *duodenum*.

pylorus. *n.* (*adj.* **-ic**) The distal aperture of the stomach, opening into the *duodenum*.

pyogranuloma. *n.* A chronic inflammatory lesion having a pus-filled center. In reptiles this is called a *heterophilic granuloma*.

pyramid. *n.* In ecology, a graphic representation in which a community parameter such as biomass, energy, or numbers of organisms is proportional to the width of the pyramid at appropriate heights representing successive trophic levels.

pyranometer. *n.* An instrument for measuring the amount of solar radiation reaching Earth's surface. (Syn. *solarimeter*)

pyrexial. *adj.* Feverish, or having an elevated body temperature. The term is normally applied to a mammal (but see *behavioral fever*).

pyric. *adj.* Pertaining to fire.

pyriform. *adj.* Pear-shaped.

pyrimidine. *n.* A class of heterocyclic nitrogenous compounds, of which derivatives (pyrimidine bases) are part of nucleotides.

pyrogen. *n.* Fever-producing substances, which act by resetting of the thermoregulatory set point ("thermostat") to a higher level. See *behavioral fever*.

pyruvate. *n.* A salt, ester, or anion of pyruvic acid. Pyruvate is the end product of *glycolysis* and may be metabolized to lactate or to acetyl CoA.

pythocholic acid. A principal acid characteristic of the bile salts of boid snakes.

Pythonidae. *n.* See *Pythoninae*.

Pythoninae. *n.* A subfamily of *Boidae,* comprising about 30 species found in Asia, Africa, and Australia. This group is given family status (*Pythonidae*) by some.

pythons. *n.* Collective vernacular name for numerous snake species belonging to the *Pythoninae* (or *Pythonidae*), and especially the genus *Python*. See also *Rock Python*.

Pyxicephalidae. *n.* A taxon (family) of anurans, sister to *Petropedetidae* within the *Pyxicephaloidea*, and containing 12 genera with species distributed in sub-Saharan Africa. For taxonomic revision see D.R. Frost et al., *Bull. Amer. Mus. Nat. Hist.* 297:1–370, 2006.

Pyxicephaloidea. *n.* A taxon (superfamily) of anurans, sister to *Phrynobatrachidae* within the new taxon *Africanura*, and containing the families *Petropedetidae* and *Pyxicephalidae* (D.R. Frost et al., *Bull. Amer. Mus. Nat. Hist.* 297:1–370, 2006).

Q

Q_{10}. A *temperature coefficient* defined as the ratio of the rate of a reaction at a given temperature to its rate at a temperature 10 °C lower. For any given temperature interval, Q_{10} = rate at T_2 divided by rate at T_1, raised to the power of $(10/T_2 - T_1)$. See *van't Hoff equation*.

QTL. See *quantitative trait loci*.

quadrat. *n.* A sampling plot, usually replicated and used in ecological studies where distributional properties of organisms are being investigated.

quadrate. *n.* 1. A bone derived from the posterior part of the *palatoquadrate cartilage* and which forms the articulating surface with the lower jaw in all vertebrates except mammals (Fig. 21). In amphibians this bone is small and may be absent; it is fused with the *pterygoid* in caecilians. 2. In past usage, a large cephalic scale forming the posterior end of the *jugal* series. 3. *adj.* Square or squared.

quadratojugal. *n.* A dermal skull bone, located on the cheek region below the *squamosal* and occupying the lower edge of the skull between the *quadrate* and the *jugal*. This bone is reduced or absent in most amphibians and reptiles, although characteristic of primitive tetrapods. The element is absent in salamanders (except *Tylotriton*), snakes, and lizards, although a rudiment is claimed to exist in geckos. See also *quadratomaxillary*.

quadratomaxillary. *n.* The bone situated between the *quadrate* and the *maxillary* in anurans, formerly called the *quadratojugal*.

quadri-. Prefix meaning "four" or "fourfold," "square," or "right angles."

quadricarinate. *adj.* Having four keels or ridges.

quadrifurcate. *adj.* Having four prongs or forks.

quadripenis. *n.* Either of two parts of a bifurcate *hemipenis*, distal to the point of bifurcation.

quadruped. *n.* (*adj.* –al) Any animal that has, or walks on, four legs; used synonymously with tetrapod.

quadrupedality. *n.* The habit of walking on four legs.

qualitative. *adj.* Descriptive and non-numerical.

qualitative character. Any character having discrete states that are neither numerical nor morphometric. Cf. *quantitative character*.

qualitative multistate character. A qualitative character that exhibits several states that cannot be arranged linearly along a single axis. Cf. *quantitative multistate character*.

quantitative. *adj.* Numerical; based on counts or measurements.

quantitative character. 1. A character based on counts, measurements, or other numerical values. 2. Any character or trait that exhibits *quantitative inheritance*, thus continuous variation. Cf. *qualitative character*. (Syn. *metric trait, numerical character*)

quantitative genetics. The study of the inheritance of quantitative, phenotypic traits.

quantitative inheritance. Reference to phenotypes that are quantitative in nature and exhibit a continuous distribution. Such characters depend on the cumulative action of multiple genes on separate chromosomes, thus there is not clear-cut segregation into readily recognizable classes showing typical Mendelian ratios.

quantitative multistate character. A quantitative character that exhibits several states that can be arranged linearly along a single axis. Cf. *qualitative multistate character*.

quantitative trait loci. (QTL) Reference is to genes that control the expression of traits showing *quantitative inheritance*.

quantum evolution. A rapid evolutionary shift to a new adaptive zone under strong selection pressure.

quantum speciation. Rapid evolution of new species, usually involving founder effects and genetic drift. (Syn. *saltation*)

quarantine. *n.* The isolation of individual animals for purposes of observation or examination. Generally, animals that are brought into a research or husbandry collection are routinely isolated for the duration of the incubation period of most diseases to which they might have been exposed, prior to transfer into a general collection.

quart. *n.* A unit of volume equal to ¼ Imperial gallon or 1.1365 liter.

Quaternary. *n.* 1. The second Period of the Cenozoic Era, spanning the last 2–3 million years and extending to the present time. It contains the Pleistocene and Holocene Epochs. See Table 1. 2. Fourth in order. 3. Containing four elements or groups.

Queen Snake. Vernacular name for the American colubrid species *Regina septemvittata*. See also *Crayfish Snake*.

Quenouille's test. A nonparametric method for testing differences in mean level between several samples, independent of any assumption about population variance.

quercicobufagin. *n.* A *bufogenin* isolated from *Bufo quercicus*.

quiescence. *n.* (*adj.* **quiescent**) A temporary state of reduced activity or cessation of development.

Quill-snouted Snakes. Vernacular name for several species of African atractaspidid snakes belonging to the genus *Xenocalamus.* The highly modified snout of these snakes is used to dig for amphisbaenians.

quin-. Prefix meaning "five" or "fivefold."

quinquecarinate. *adj.* Having five keels or ridges.

quintangular. *adj.* Having five angles.

R

r. 1. Reproductive potential. *Intrinsic rate of increase* of a population. 2. Correlation coefficient.

race. *n.* This is a colloquially used term referring to regional populations that share visible characteristics. Reference is to a phenotypically or geographically distinctive subspecific group, composed of individuals that inhabit a defined geographical or ecological region and possess characteristic gene and phenotypic frequencies that distinguish it from other groups. See also *ecotype, geographic race, subspecies*.

Racer. *n.* Collective vernacular name widely used for various species of slender, often striped, swift-moving snakes belonging to the colubrid genera *Coluber, Argyrogena, Elaphe, Pantherophis, Chironius, Leptodrymus, Mastigodryas, Alsophis, Psammophis, Drymobius, Drymoluber, Gonyosoma*, and *Dendrophidion*. This name is sometimes used also in reference to *Masticophis spp.* (e.g., Red Racer for western forms of *Masticophis flagellum*).

Racerunner. *n.* A collective common name for various species of teiid lizards, closely related to *Whiptails*. These occupy tropical and temperate habitats in the Americas and closely resemble Wall and Sand Lizards of the Old World. The name also applies to iguanid lizards of the genus *Plica*, and to lacertid lizards of the genera *Eremias, Ommateremias, Pseuderemias*, and *Rhabderemias*.

rachis. *n.* Generally, any axial structure. Reference in herpetology usually is to the central shaft of the external *gill* of salamanders, which bears the respiratory *filaments*. (Syn. *branchial process*)

rad. *n.* A derived SI unit for the amount of absorbed radiation by an organism or tissue, defined as the amount of radiation causing 1 kg of tissue to absorb 0.01 joules of energy.

radial. *n.* 1. Any of the small, horn-like processes on the *copula* in salamandrids. The *anterior radial* is synonymous with the *little horn*, whereas the *posterior radials* appear *de novo* during metamorphosis and are thus not derivatives of the visceral skeleton. Both sets of radials function to attach the *hyoid* to the tongue. 2. Any of the many scales lying along the anterior surface of the forearm in *Xantusia*. *adj.* Pertaining to the *radius* and thus equivalent to *preaxial*.

radiale. *n.* The more anterior of three small bones located at the distal end of the *epipodials* of the tetrapod forelimb, adjacent to the *intermedium* and at the end of the *radius*. (Syn. *scaphoid*)

radiance. *n.* The radiant flux density emanating from a surface per unit solid angle (W • m^{-2} • sr^{-1}).

radiant emittance. The radiant flux density emitted by a surface (W/m^2).

radiant flux. The amount of radiant energy emitted, transmitted, or received per unit time.

radiant flux density. The radiant flux per unit of area (e.g., W/m^2).

radiant intensity. Reference to the radiant flux emitted from a surface per unit solid angle (W/sr).

radiation. *n.* 1. The emission and propagation of electromagnetic energy through space or a medium in the form of waves, without direct contact between objects. By extension, the definition can also include ionizing particles. 2. In evolutionary biology, the process by which a group of species diverge from a common ancestral form, resulting in an overall increase in biological diversity. 3. The occurrence of this phenomenon over a relatively short time.

radioactive decay, radioactivity. *n.* (*adj.* –ive) The spontaneous emission of charged particles or photons that accompanies disintegration of an unstable nucleus of a radioisotope.

radioactive isotope. Any isotope that stabilizes an unstable nucleus by emitting ionizing radiation. (Syn. *radioisotope*)

radioautography. *n.* See *autoradiography*.

radiocarbon dating. A technique for estimating the age of ancient organic materials by measuring the loss over time of radioactive carbon (carbon-14), which has a precisely known rate of decay. The absorption of radiocarbon in living tissues ceases at death, thus the amount remaining gives an indication of the time elapsed since the death of the organism.

radiogenic. *adj.* Relating to, or produced by, radioactivity.

radiograph. *n.* See *radiography*.

radiography. *n.* (*adj.* -ic) The making of film records (*radiographs*) of internal structures of the body by means of exposing film sensitive to X-rays or gamma rays. These procedures are especially useful for visualizing bony tissues.

radioimmunoassay. (RIA) *n.* An immunological technique for measuring minute quantities of antigen or antibody, hormones, various drugs, or other substances with the use of radioactively labeled reagents. The technique is a competitive binding assay that depends on displacement of a radiolabelled ligand from an antibody by a non-labelled standard or test sample, followed by separation of the labeled ligand into bound and unbound fractions.

radioisotope. *n.* A *radioactive isotope.*

radioisotopic dating. See *radiometric dating.*

radioluscent. *adj.* A condition that allows passage of x-irradiation without forming an image on a photographic emulsion.

radiometric dating. A method of dating that involves the measurement of decay in certain naturally occurring radioactive isotopes that decay at a constant, known rate. (Syn. *radioisotopic dating*)

radionuclides. *n.* Atoms that disintegrate by emission of electromagnetic radiation.

radioopaque. *adj.* A condition that impedes the passage of x-irradiation, thereby causing an image to be formed on a photographic emulsion.

radiotelemetry. See *biotelemetry.*

radio tracking. See *biotelemetry.*

radioulna. *n.* A single bone in the lower forearm of frogs that arises from fusion of the *radius* and *ulna.* (Syn. *antebrachial bone, os antebrachium*)

radius. *n.* The medial bone of the lower forelimb in tetrapod vertebrates. This element lies parallel to the *ulna*, but these are fused in amphibians. See *radioulna.*

rafting. *n.* Reference to dispersal of terrestrial organisms across water by means of association with floating objects. Passive drifting, usually on another object.

Rainbow Boas. Vernacular name for Central and South American boid species belonging to the genus *Epicrates*, especially the terrestrial boid *Epicrates cenchria.*

Rainbow Snake. Vernacular name for a large, semiaquatic North American colubrid *Farancia erytrogramma.*

Rainbow Water Snakes. Collective vernacular name for Asian homalopsine snakes belonging to the genus *Enhydris.*

rainfall. *n.* Water that is precipitated from the atmosphere and falls earthward. Cf. *precipitation* and see *acid rain, convective rain, orographic rain, frontal rain.*

rain forest. Forest with abundant and year-round rainfall (at least 254 cm). See *tropical rain forest.*

Rain Frogs. Collective vernacular name for African species of microhylid frogs belonging to the genera *Breviceps, Scaphiophryne,* and American frogs belonging to the genus *Eleutherodactylus.* The common name "Rain Frog" is derived from the fact that these frogs seldom emerge from beneath ground except during periods of precipitation.

rainout. *n.* Deposition of solid material by rain.

rain shadow. A region on the downwind side of mountains that has little or no rainfall because of the loss of moisture on the upwind side of the mountains. This phenomenon creates and maintains arid regions and is dependent on the prevailing wind direction.

rain shadow desert. A desert or arid land in a *rain shadow* on the *leeward side* of a mountain range. (Syn. *orographic desert*)

raker. *n.* See *gill raker.*

rale. *n.* An abnormal respiratory sound, heard by auscultation and indicative of a pathologic condition. (See also *rhonchus*)

ramicolous. *adj.* Living on twigs or branches.

ramus. *n.* (*pl.* **-i**) 1. Generally, a branch or branching part. 2. Either of the two branches of the lower jaws.

ranatensin. *n.* See *bombesin.*

ranatuerins. *n.* Antimicrobial peptides isolated from the skin of the American Bullfrog, *Lithobates catesbeianus* (formerly *Rana catesbeiana*). See J. Goraya et al., *Biochem. Biophys. Res. Commun.* 250:589–592, 1998.

ranching. *n.* The process of raising amphibians or reptiles from eggs, hatchlings, or neonates to some set market size for commercial purposes. Unlike farming, this is not a closed-cycle system as it continuously relies on wild populations as a source of the eggs or newborns. Cf. *farming.*

random. *adj.* In statistics, having an *a priori* probability of occurrence other than zero or unity. Also used to indicate haphazard, without a recognizable pattern.

random distribution. A spatial distribution in which the presence of one individual or object has no influence on the distribution of other individuals or objects. Cf. *clumped distribution, uniform distribution.*

random error. A deviation having direction and magnitude that cannot be predicted.

randomization. *n.* A fair assignment of treatments to experimental units such that the latter are independent.

random mating. A mating system in which every male gamete in a population has equal opportunity to fertilize every female gamete. (Syn. *panmixis*)

random numbers. In statistics, a table of numbers wherein the probability of any number occurring at any one time is constant and independent of all preceding numbers.

random process. See *stochastic process.*

random sample. A sample drawn without reference to any particular attribute of a population. Every member of the sampled population has an equal and independent chance of being included in the sample. Cf. *non-random sample.*

random sampling error. See *experimental error.*

random variable. In statistics, a variable that can assume any of a given set of values with assigned probability. It is a function that associates a unique numerical value with every outcome of an experiment. A random variable can be *discrete* or *continuous.* A *discrete random variable* is one that may take on only a countable number of distinct values, such as 0, 1, 2, 3, . . . etc. A *continuous random variable* is one that takes an infinite number of possible

values, usually measurements such as height, weight, concentration, etc. Two random variables are said to be independent if the value of one has no influence on the value of the other, and vice versa.
random walk. An irregular path such as that followed by a particle exhibiting Brownian movement.
range. *n.* 1. The limits of geographical distribution of a species or taxon. 2. In statistics, reference to a measure of variation within a set of data, given by the difference between the maximum and minimum values. Note that the range is a single number, not many numbers.
ranid. *n.* A member of, or pertaining to, *Ranidae.*
Ranidae. *n.* A clade (family) of anurans including many common frogs and a diversity of habits and life histories. Sometimes known as *true frogs,* there are 38 genera and more than 600 species, with cosmopolitan distribution except for extreme southern South America, the West Indies, and most of Australia. Recent taxonomic reconstructions have resulted in a number of generic changes reported by D.R. Frost et al. (*Bull. Amer. Mus. Nat. Hist.* 297:1–370, 2006).
ranivorous. *adj.* Feeding on frogs.
rank. *n.* The position of a given level of classification in relation to the levels above or below.
ranoid. *n., adj.* See *Acosmanura.*
Ranoidea. *n.* A monophyletic clade (superfamily) of anurans including relatively derived families characterized by complete fusion of the epicoracoid cartilages. This clade is a sister taxon to *Rhacophoroidea* within the new taxon *Aglaioanura,* and it is composed of *Nyctibatrachidae* and *Ranidae.* See *Ranoides.*
Ranoides. *n.* A new monophyletic taxon of frogs having worldwide distribution except for New Zealand, most of Australia, and southern South America. A sister taxon to the new *Hyloides* within the new taxon *Phthanobatrachia* and composed of the new taxa *Allodapanura* and *Natatanura* (D.R. Frost et al., *Bull. Amer. Mus. Nat. Hist.* 297:1–370, 2006). This taxon is equivalent to Ranoidea with addition of *Microhylidae* and *Brevicipitidae.*
raphe. *n.* A seam or suture. The line of union of the halves of symmetrical parts.
rare. *adj.* 1. Of seldom occurrence. 2. Reference to taxa with small world populations that, while neither *endangered* nor *threatened,* are at risk.
rarefaction. *n.* In ecology, a method for standardizing data samples for measurements of species diversity in a community.
rasorial. *adj.* Adapted for scratching or scraping the ground.
rasp. *n.* 1. The collective multiple rows of teeth on the dentigerous bones of fetal viviparous caecilians. These have fused pedicels that form a plate attached to the underlying dermal bone and are thought to be used in wearing away the oviductal lining to obtain nourishment. 2. Reference to a feeding mode of tadpoles in which keratinized mouthparts are used to harvest particulate materials from submerged surfaces or sediment.
rasping organ. The multiple rows of horny teeth surrounding the mouth of most anuran tadpoles.
rate constant. A proportionality factor by which the concentration of a reactant in an enzymatic reaction is related to the reaction rate. (Syn. *specific reaction rate*)
rate modulation. In physiology, reference to incremental contractile force of a muscle in proportional response to increases in the rate of nerve stimulation.
rate of natural increase. The percentage increase in a population per unit time, calculated as the birth rate less the death rate.
Rathke's glands. One or more pairs of muscle-ensheathed glands situated outside the peritoneal cavity, adpressed to the internal lateral aspect of the shell in most taxa of extant turtles. These glands are thought to secrete predator repellents and/or blood-borne metabolites such as lactic acid through pores located on the shell bridge or on skin of the axillary or inguinal regions.
Rat Snakes. Collective vernacular name for various species of relatively large, diurnal snakes belonging to colubrid genera *Bogertophis, Elaphe, Gonyosoma, Pantherophis, Ptyas, Spilotes,* and *Spalerosophis.* Some of these species are also known as *Tree Snakes, Trinket Snakes,* or *Whip Snakes.*
rattle. *n.* The segmented, keratinized caudal appendage of Rattlesnakes that produces a buzzing or whirring sound when shaken by specialized caudal musculature. Each keratinized segment of the rattle fits loosely within another, and a new segment is added at each ecdysis. (Syn. *rattle string, string*) See also *button, prebutton.*
rattle formula. A combined letter and numeric formula that indicates the number of rattle segments present in a rattle. "B" indicates the black base element, a number indicates the total count of loose segments, and "b" indicates the button. Thus, a formula B7b would indicate the presence of a base, seven rattles, and a button. First used by F.L. Heyrend and A. Call (*Herpetologica* 7:28–40, 1951).
rattle fringe. See *fringe.*
rattle matrix. See *matrix.*
rattle number. A number that identifies a specific segment of a rattlesnake rattle, beginning with the button as "1," the next as "2," and so on. See also *rattle formula.*
Rattlesnake. *n.* Collective vernacular name for American pitvipers belonging to the genera *Crotalus* and *Sistrurus,* named for the presence of a *rattle* as a tail appendage (in most species).
rattle string. See *rattle.*
Rattling Frog. Vernacular name for an African hyperoliid frogs belonging to the monotypic genus *Semnodactylus* (Kassininae). The advertisement call is a coarse "rattle" that lacks strong frequency modulation characteristic of *kassinas.*

raw data. Data that has been collected but not organized numerically or analyzed statistically.

ray. *n.* An extension of dorsal vertebrae to form the inner bony structure of the fin-like crest on the body and tail of certain lizards (e.g. *Basiliscus*).

Rayleigh scattering. In atmospheric science, reference to small-particle scattering of radiation that is most pronounced at shorter wavelengths so the scattered radiation is blue. This is the source of blue color in the sky. Cf. *Mie scattering*.

R_b. Boundary layer resistance. See *resistance*.

RBC. *Red blood cell.*

re-. Prefix meaning "again," "repeat."

reaction. *n.* In contexts of physiology and behavior, a change in the activity of an organism in response to a stimulus.

reaction force. A force produced by an object that is equal and opposite in direction to an *action force* exerted upon it. See Fig. 25.

reaction norm. The variability of *phenotypic traits* exhibited by a given genotype under the range of natural conditions common to the habitat of the species, or under standard experimental conditions. See *phenotypic plasticity*. (Syn. *norm of reaction*)

reaction time. The time interval between a stimulus and the response to the stimulus. See *latent period*.

reactive oxygen species (ROS). Highly reactive byproducts of cellular metabolism that promote oxidation of biomolecules; characteristic of all aerobic organisms. These effects are neutralized by antioxidant defenses of cells. (Syn. *prooxidant*)

realized. *adj.* Actual. Cf. *prospective*.

realized niche. See *ecological niche*. Cf. *fundamental niche*.

realm. *n.* The largest of the biogeographical units, encompassing major climatic or physiographic zones.

real time. A property of an event or system in which data are processed as they are acquired instead of being accumulated and processed at a later time.

rear-fanged. See *opisthoglyph*.

recapitulation. *n.* The repetition of ancestral adult stages in embryonic or juvenile stages of descendants.

recent. *adj.* Extant, still in existence.

Recent. *n.* The present Epoch of geological time, or *Holocene*. See Table 1.

receptaculum seminis. See *spermatheca*.

receptive. *adj.* Responsive to stimuli.

receptive field. The area of external stimulation, or field of receptors, that provide input to a given neuron at a higher level in neural processing (e.g. within the brain).

receptor. *n.* 1. Any nerve or modified sensory cell that receives and transduces a stimulus. A *primary receptor* is a sensory neuron, whereas a *secondary receptor* is a modified nonneural cell that responds to a stimulus and transmits a signal to a connecting sensory neuron (e.g., rods and cones of the eye). 2. A molecule on the surface of, or within, a cell that recognizes and binds with specific molecules that produce some effect within the cell.

receptor potential. A change in membrane potential elicited by sensory stimulation of a sensory receptor cell, which changes the flow of ionic current across the plasma membrane.

recessive. *adj.* A genetically determined characteristic that is masked by its dominant counterpart. See also *dominance*.

recessive lethal. Reference to an allele that kills a cell or organism in which it is homozygous.

reciprocal altruism. See *altruism*.

reciprocal crosses. Genetic matings of the sort A × B and B × A, where the individuals symbolized by A and B are different in genotype or phenotype, or both. Such crosses are employed to detect *sex linkage, maternal inheritance*, or *cytoplasmic inheritance*.

reciprocal hybrids. Hybrid offspring that are derived from reciprocal crosses involving parents of different species.

reciprocal inhibition. In physiology, the inhibition of the motor neurons innervating a set of muscles during excitation of their antagonists.

reciprocation call. A call produced by a female anuran in response to a male's courtship call.

recognition species concept. See *species*.

recombinant DNA. An engineered molecule of DNA created *in vitro*.

recombination. *n.* See *genetic recombination*.

reconstruction. *n.* 1. The process of preparing complete bones, partial or full skeletons of extinct animals using available fossil materials. The term also refers to a bone or skeleton assembled in this manner. 2. A pattern of internal bone growth in which new bone replaces preexisting bone tissues.

recount system. See *scale row formula*.

recovery dose. The largest quantity of venom that can be received by an animal and still recover if given 1 ml. of antivenin.

recrudescence. *n.* 1. Generally, reappearance or regrowth. 2. Renewal of reproductive interest and readiness of reproductive structures, usually following a period of seasonal inactivity. In herpetological literature, this term is used largely in reference to the active part of male testicular cycles following a *refractory period*.

recruitment. *n.* 1. Generally, reference to a graded increase in participation of individual units to provide a collective response. The opposite, a reduction of participation of individual units, is termed *derecruitment*. Examples of the former are an increase in the number of open capillaries during a vasodilation response, an increase in the number of neurons (having different thresholds) that are activated during a response to a stimulus, and an increase in the number of motor units that are activated during increasing intensity of muscle contraction. 2. The influx of new members into a population by reproduction or immigration.

rectal chamber. Any distinctive compartment of the *coprodeum*.

rectiform. *adj.* Lying in a straight series, as in reference to scale rows. See *scale row*.

rectilinear. *adj.* Reference to a straight-line pattern of movement or growth.

rectilinear locomotion. A mode of limbless locomotion in which a snake moves forward in a straight line, with no lateral movement. In this type of movement, muscles on both sides of the body act synchronously, sequentially contracting and relaxing to move the body forward. Muscles pull the skin forward relative to the ribs, following which the ventral scales anchor the body to the substrate. Other muscles then pull the ribs and vertebral column forward relative to the stationary ventral skin. Several waves of such symmetrical contractions usually pass down the body simultaneously and make it appear that the ventral skin is crawling on its own while the dorsal skin moves at a nearly even rate. This mode of locomotion is characteristic of heavy-bodied snakes such as vipers and boids.

RDB. See *Red Data Book*.

recurved. *adj.* Form of a tooth that bows backward, a condition favorable for snaring or holding prey.

Red-bellied Blacksnake. Vernacular name for the Australian elapid snake *Pseudechis porphyriacus*.

Redbelly Newt. Vernacular name for the American species *Taricha rivularis*.

Redbelly Snake. Vernacular name for the American colubrid species *Storeria occipitomaculata*.

Redbelly Toads. Collective vernacular name for South American bufonid species of anurans belonging to the genus *Melanophryniscus*.

Redbelly Turtles. Vernacular name for certain American species of the emydid genus *Pseudemys*.

red blood cell. See *erythrocyte*.

Red Data Book. (RDB) A list of animal species and subspecies that are known or suspected to be threatened with extinction, or which are believed to have become extinct.

Red Diamond Rattlesnake. Vernacular name for the American pitviper *Crotalus ruber*.

Red-eared Slider. Vernacular name for the American emydid turtle species *Trachemys scripta*.

Red Hills Salamander. Vernacular name for the North American plethodontid species *Phaeognathus hubrichti*.

Red King Hypothesis. See *maximum homology hypothesis*.

red leg. A disease in frogs, indicated by a red flush over the ventral surfaces of the hind limbs and, in some cases, the abdomen. The condition is a hemolytic septicemia commonly caused by the bacterium *Aeromonas hydrophila*, but infection with *Pseudomonas, Proteus, Citrobacter, Salmonella,* and *Escherichia coli* can also occur. The condition is highly contagious to other frogs.

Red-legged Frog. Vernacular name for the North American anuran species *Rana aurora*.

Red List of Threatened Species. See *IUCN Red List* and *International Union for the Conservation of Nature and Natural Resources*.

red muscle. A classification of muscle based on color, which does not adequately consider the full complexity of muscle composition and function. Red muscle tends to be highly vascular and rich in myoglobin. Red muscle is relatively resistant to fatigue and contains a high proportion of aerobic fibers. Cf. *white muscle*.

redox pair. Two atoms, molecules, or compounds that are involved in mutual reduction and oxidation.

Red Queen hypothesis. A mathematical model that predicts continuing evolutionary change in communities during conditions of constancy in the physical environment. Evolution is predicted from consideration that sympatric species have important effects on each other, and not all species will be at their local adaptive peaks. Hence, these species are capable of further evolution even though the physical environment is stable. In this system, any evolutionary advance in one species will influence others through a network of interactions, so these other species may become subject to selective pressures for change simply to catch up. The name of this hypothesis is based on the Red Queen in *Through the Looking Glass*, who said: "Now here, you see, it takes all the running you can do to keep in the same place."

Red Rat Snake. See *Corn Snake*.

red rods. Photosensitive cells of amphibian retinas that absorb wavelengths maximally at 502 nm.

Red Salamanders. Collective vernacular name for several species of plethodontid salamanders belonging to the genus *Pseudotriton* and, in particular, the species *P. ruber*. See also *Mud Salamanders*.

Red-sided Garter Snake. Vernacular name for the American colubrid species *Thamnophis sirtalis*. (Syn. *Common Garter Snake*)

Red-Spotted Toad. Vernacular name for the American bufonid *Anaxyrus* (formerly *Bufo*) *punctatus*. For taxonomic revision see D.R. Frost et al., *Bull. Amer. Mus. Nat. Hist.* 297:1–370, 2006

Red-Tailed Rat Snake, Red-Tailed Green Rat Snake. Vernacular name for the Asian colubrid species *Gonyosoma oxycephalum*, a species that exhibits impressive neck displays during defensive behaviors.

reduced. *adj.* In context of anatomy, smaller in size than in ancestral forms.

reduction, reduction formula. *n.* 1. The addition of electrons or hydrogen to a substance. Most biological reductions involve hydrogenations. 2. See *scale row formula*.

reduction division. See *meiosis*.

reductionism. *n.* The study of complex systems by reducing such systems to their constituent parts, followed by detailed analysis of the components rather than the integrated whole.

redundant character. Any character that does not contribute useful information to an analysis.

Reed Frogs. Collective vernacular name for species of African frogs belonging to the family *Hyperoliidae,* especially the African genus *Hyperolius* and the Madagascan genus *Heterixalus.*

Reedsnakes. *n.* See *Calamarinae.*

reef. *n.* A resistant ridge of calcium carbonate formed on the sea floor by corals and coralline algae.

Reef Geckos. See *Dwarf Geckos.*

reference. *n.* A bibliographic reference is the citation of the name of an author (or authors) and date of publication of a work, usually including title of journal or book, volume, and page numbers.

reflectance. *n.* 1. The property of a body or surface to reflect, rather than absorb, radiation. 2. For any given wavelength or spectrum of wavelengths, the proportion of radiation reflected relative to a perfectly reflecting surface.

reflectivity. *n.* The fraction of incident *radiant flux* at a given wavelength reflected by a material or object.

reflex. *n.* An involuntary action of effectors mediated by the nervous system.

reflex arc. A pathway of neurons that are involved in a reflex connecting sensory input and motor output.

reflexic, reflexive, reflexogenic. *adj.* Reference to an autonomic response or nerve-mediated reflex.

refraction. *n.* The bending of light rays when they pass between two media of differing density (e.g., air to water).

refractive index. The refractive power of a medium compared with that of air, which, by convention, is assigned a value of 1.

refractory. *adj.* Not responsive.

refractory period. 1. In reproductive biology, a period of testicular regression and inhibition of sperm production in males, or suppressed ovulation in females, usually following the time of mating and lasting typically several months. 2. In neurophysiology, the period of increased membrane threshold during and immediately following the local activation of an action potential. This prevents fusion of action potentials and thereby insures that each one is a discrete unit of information.

refuge, refugium. (*pl.* **refugia**) *n.* An area in which prey may escape from, or avoid, a predator or deleterious effects of the environment.

reg. *n.* A stony desert. (Syn. *serir*)

regeneration. *n.* The natural regrowth or renewal of a structure.

region. *n.* Any major area that supports a characteristic biota.

regional extinction. The disappearance of all local populations of a metapopulation. Cf. *extirpation.*

regression. *n.* 1. In statistics, a dependent relationship between one variable and one or more other variables. The term *simple regression* refers to consideration of only two variables. See *linear regression.* 2. A reversal in the direction of evolution. 3. Retreat of the sea from a land area.

regression coefficient. The rate of change of the dependent variable with respect to that of the independent variable.

regression line. A line that is fit to data on a graph and depicts how much an increase or decrease in the dependent variable is expected from a unit increase in the dependent variable.

regressive evolution. A process of increasing simplification of structure, or a trend of decreasing specialization.

regulation. *n.* The active maintenance of a controlled variable at, or close to, a set value. Cf. *control.*

regulator. *n.* Reference to an animal that maintains a state of relative internal homeostasis by means of biochemistry, physiology, or behavior (thermoregulator, osmoregulator, etc.).

regulator gene, regulatory gene. A gene that produces a protein that influences the expression of other distant genes and, hence, their products. A regulator gene produces a repressor protein that shuts off the structural gene activity of an operon by an interaction with its operator gene.

regurgitation. *n.* 1. Reverse movement of contents in the stomach, resulting in expulsion from the digestive tract via the mouth. The return of partly digested food from the stomach to the mouth. See also *emesis.* 2. Backflow of blood due to imperfect action of a heart valve.

reinforcement ridge. See *cutting edge* (of fang in snakes).

reinforcing selection. The operation of selection pressures at two or more levels of organization such as individual, family, or population, such that certain genes are favored at all levels and their spread or maintenance in a population is accelerated or reinforced. Cf. *counteracting selection.*

reintroduction. *n.* Reference to a plant or animal being moved to a location where it occurred historically.

rejecta. (F + U) *n.* In ecological energetics, reference to that part of the total food or energy ingested by an individual, population, or trophic unit (consumption) that is not utilized for production and respiration, and is lost as *egesta* and *excreta,* per unit time per unit area or volume. *Consumption* minus *assimilation.*

relative abundance. In ecology, a measure of the relative number of individuals, often expressed as the total number of individuals of one taxon compared to the total number of individuals of other combined taxa in a given area, community, or other measure of space. Cf. *absolute abundance.*

relative age. A statement of the approximate age of an object or feature in comparison with some other object or feature, rather than in terms of its age in years. Cf. *absolute age.*

relative dating. A method of geological dating in which the relative age of a rock is used, rather than the absolute age.

relative fitness. See *fitness.*

relative humidity. The ratio of ambient water vapor pressure (or density) to saturation water vapor pressure (or density) at air temperature, expressed as a fraction or percentage. See *humidity.*

relative ranking. In systematics, the procedure of assigning a formal rank by combining monophyletic groups with their sister groups to form a series of consecutively subordinate categories. Cf. *absolute ranking.*

relative water content. The actual water content at a given condition relative to the water content under conditions of saturation, expressed as a percentage.

relaxation of selection. Cessation or release of selection during experimental situations.

release call. A simple anuran vocalization given by males or unreceptive females when amplexed by a male.

release inhibiting hormones. A hormone that depresses the responsiveness of target tissues. See *hypothalamic release inhibiting hormones.*

releaser. *n.* See *sign stimulus.*

releaser pheromone. A pheromone that elicits a more or less immediate response in a recipient organism.

releasing hormones. A hormone that initiates activity of target tissues. See *hypothalamic releasing hormones.*

relict. *adj.* Reference to a population or group of organisms that survives and persists in one region after becoming extinct elsewhere, or a surviving species of a group of which others are extinct. A *relict species* may also refer to the existence of an archaic form in an otherwise extinct taxon.

remigration. *n.* The return of animals to a particular breeding area in successive years.

remodeling. *n.* 1. Regrowth, repair, or reconfiguration of new tissue. Common reference is to the repair or replacement of older bone, and the deposition of new bone in relation to altered stress forces. The term also applies to growth of new blood vessels (*vascular remodeling*) in response to hypoxic or damaged tissues. 2. The term is applied by some to evolution.

remote sensing. The science of remotely sensing attributes of the environment, including its biota. The term has reference to using our senses and various recording instruments to gather information from a distance, including use of satellites, cameras, radio telemetry, data logging systems, etc.

renal. *adj.* Pertaining to the kidney.

renal clearance. The volume flow of plasma containing the quantity of a substance that is freely filtered and appears in the glomerular filtrate of the kidney. The clearance of a substance can be calculated from knowledge of the plasma and urine concentrations of the substance and the urine flow rate, and provides information about kidney function with respect to secretion or reabsorption of various plasma components (ions, drugs, etc.).

renal corpuscle. A functional filtering unit of the kidney, consisting of a *Bowman's capsule* and *glomerulus.* Tubular urine is formed initially in a *nephron* as fluid that is filtered from blood plasma flowing in the glomerular capillaries into the tubular space of the surrounding Bowman's capsule. Compared to those of amphibians, there is a definite reduction in the size of glomeruli in reptiles, which reflects adaptation for the conservation of water. Some lizards and snakes have aglomerular tubules.

renal-portal system. A circulatory arrangement in vertebrates from fishes to reptiles (including birds, but not mammals) wherein blood leaving the tail and hind limbs passes through a venous capillary system associated with the kidneys before returning to the heart.

renal sexual segment. (RSS) A hypertrophied region of the nephron of reptilian kidney, described originally by O. Gampert (*Zeit. Wiss. Zool.* 16:369–373, 1866) and found subsequently to occur as a sexually dimorphic structure unique to squamate reptiles. Development of the RSS is related to maturation and subsequent steroidogenic activity of the testes.

renal tubule. Any of the numerous individual urine-forming tubules that comprise the kidney, equivalent to a *nephron.*

renewal phase. A stage during the shedding cycle of lepidosaurs during which cell production and differentation produce much of a new *inner epidermal generation.* Cf. *resting stage.*

reniform. *adj.* Kidney-shaped.

renin. *n.* A proteolytic enzyme produced by specialized cells in renal vasculature and which converts *angiotensinogen* (angiotensin precursor) to *angiotensin.*

replacement bone. Bone that replaces cartilage as it ossifies. See *endochondral ossification.*

replacement fang. Any one of a series of teeth that develop posterior to a functional fang and, in sequential order, eventually replace the functional fang when it is lost. (Syn. *accessory fang, reserve fang*)

replacement tooth. 1. Any of a series of teeth that develop and replace another. 2. In tadpoles, any one of several fully formed *labial teeth* interdigitated sequentially beneath an erupted tooth.

replicate. *n., v.* 1. In statistics, any repeated experiment. See *replication.*

replication. *n.* 1. In molecular biology, a duplication process that requires copying from a template. DNA replication involves complementary nucleotides being added to replication strands of the duplex molecule that are first separated at beginning of the process. 2. In statistics, repetition of an experiment or experimental design to obtain more information for estimating variance or experimental error. Duplication of experimental design such that the resulting data are independent. See *pseudo-replication*. 3. Synthesis of new DNA strands that are complementary to existing template strands. 4. The production of identical copies of form, or information from existing units.

repressible enzyme. An enzyme that decreases in rate of production when intracellular concentrations of certain metabolites are increased.

repression. *n.* 1. The cessation of synthesis of an enzyme when the product of the reaction it catalyzes reaches a critical concentration. 2. Inhibition of transcription or translation when a repressor protein binds to an operator locus on DNA or to a specific site on a mRNA.

repressor or **repressor protein.** *n.* A protein synthesized by a regulator gene that binds to an operator locus and blocks transcription of that operon.

reprint. *n.* A printed reproduction of a work that is already printed, usually reissued using the same plates that were employed for the original printing.

reproduction. *n.* The process and act of producing offspring.

reproductive isolating (or **isolation**) **mechanism.** Reference to any factor that tends to reduce or prevent successful mating and interbreeding between members of related but genetically divergent populations or species of organisms. Such factors include geographical barriers, seasonal (temporal) separation, behavioral or ecological differences, gamete incompatibility, and cytological, morphological, or physiological differences. Isolating mechanisms are frequently identified as *premating* (*prezygotic*) or *postmating* (*postzygotic*) depending on whether the mechanism tends to prevent mating (includes ecological, temporal, and ethological isolating factors) or tends to reduce the viability of embryo or offspring following mating (includes hybrid sterility, inviability, and breakdown). (Syn. *isolating mechanism, Wallace effect*)

reproductive isolation. Genetic isolation due to absence of interbreeding between members of different species or populations. See *reproductive isolating mechanism*. (Syn. *misogamy*)

reproductive mode. *n.* Reference to a combination of traits that includes oviposition site, ovum and clutch characteristics, rate and duration of embryonic and larval development, stage and size of hatchling, and type of parental care characteristic of a particular species or lineage (S.N. Salthe and W.E. Duellman, pp. 229–249 in J.L. Vial, ed., *Evolutionary Biology of Anurans*, 1973). See *oviparity, ovoviviparity,* and *viviparity*. The diversity of reproductive modes among amphibians is much greater than that observed in other groups of vertebrates and has been summarized by S.N. Salthe (1969), S.N. Salthe & W.E. Duellman (1973), M.L. Crump (1974, 1982), C.F.B. Haddad & P.A. Prado (2005), and others. See C.F.B. Haddad & P.A. Prado (*BioScience* 55:202–217, 2005) who describe 39 reproductive modes for anurans (updated from Duellman and Trueb, 1986).

reproductive potential. (*r*) 1. The theoretical logarithmic rate of population growth when unimpeded by environmental limitations. (Syn. *biotic potential*) 2. The number of offspring a female of a given age can be expected to produce.

reproductive rate. The number of offspring produced by an individual per unit time over a given period.

reproductive strategy. Reference to patterns of reproduction that have resulted from natural selection. A combination of physiological, morphological, and behavioral attributes that act in concert to produce the optimal number of offspring under given environmental conditions (W.E. Duellman & L. Trueb, *Biology of Amphibians*, p. 13, 1986).

reproductive success. The number of offspring from a given individual that survive to reproduce.

reproductive value. The expected number of offspring that remain to be born to individuals of age x, relative to the number of individuals of age x.

reptile. *n.* Generally, any member of the class Reptilia including living and ancestral forms. Traditionally, reptiles include turtles, crocodilians, squamates, and Tuataras. The discovery that birds (Aves) are the closest living relative to the crocodilians, or the sister clade to crocodilians and turtles, reveals that the Reptilia is an unnatural grouping (*paraphyletic group*). Crocodiles are more closely related to birds than to lizards, so the birds should be part of the Reptilia. To avoid confusion, the term *Sauropsida* is now often used to refer to the monophyletic group consisting of reptiles and birds. For more detailed information on this problem, see the *CNAH* web page on vertebrate taxonomy.

Reptile Database. See *EMBL Reptile Database.*

Reptilia. *n.* 1. A clade (class) that includes the last common ancestor of turtles, squamates, and birds and all of its descendants. Birds are included in Reptilia because they are included in the archosaur branch of reptiles and are the closest living relatives of crocodilians among extant amniotes. 2. A bimonthly publication of high quality and dedicated to sharing information among professionals and amateurs to promote responsible care of captive animals and the conservation of amphibians and reptiles in the wild. The magazine is published in four editions: Spanish, German, English, and Italian.

Reptilomorpha. *n.* An early stem lineage of tetrapods containing a diverse array of both amniotes and nonamniotes, giving rise to the immediate ancestors of the major amniote groups. Cf. *Batrachomorpha*.

rescue effect. 1. In metapopulation models, this refers to the reduction in the probability of local extinction that re-

sults from an increase in immigration or the fraction of sites occupied. 2. In *island biogeography*, the rescue effect is the reduction in the species extinction rate on near versus far islands. See II.J. Brown & A. Kodric-Brown, *Ecology* 58:445–449, 1977.

reserve. *n.* A tract of land set aside to preserve its natural condition, such as a nature preserve, national park, refuge, natural reserve, or other conservation land.

reserve fang. See *replacement fang*.

reservoir host. An animal or species that harbors a parasite, thereby serving as a source of infection for other animals.

residual, residual variance. An observable estimate of the unobservable error. With respect to *analysis of variance* or a *regression* analysis, that part of the variation of a dependent variable that cannot be attributed to a specific tested source of variation. The sum of the residuals within a random sample is necessarily zero, and thus the residuals are not independent. In contrast, *errors* are independent random variables if a sample is chosen from a population independently, and the sum of the errors need not be zero. Residuals are observable; errors are not. Cf. *error*.

residue. *n.* A term in biochemistry for a small subunit that forms a component of a larger molecule.

resilience. *n.* Reference to the ability of a biological community to continue to function when disturbed, or to return to its former state following a perturbation. Cf. *durability*.

resistance. (R) *n.* 1. Generally, the property or aggregate properties that hinder(s) flow of something in a circuit or flow system (e.g. electric charge, air, blood). 2. In electric circuits, the unit of measure is the ohm (Ω), defined as the resistance that allows 1 ampere (A) of current to flow when a potential drop of 1 volt (V) exists across the resistance. 3. With respect to evaporative water loss from skin, the *skin resistance* to water vapor flux (R_s) is the ratio of the difference of water vapor density (or pressure) acting across the skin ($WVD_{in} - WVD_{out}$) to the rate of evaporative flux (*CEWL*), or in other words, the reciprocal of *conductance*, usually in units of cm/s in herpetological literature. Sometimes this resistance includes a *boundary layer resistance* (R_b), which must be subtracted from a total resistance (R_t) to yield R_s.

resistivity. *n.* The electrical resistance of a conductor 1 cm in length and 1 cm^2 in cross-sectional area.

resolution. *n.* The smallest signal increment that can be detected by a measurement system. Resolution can be expressed in bits, in proportions, or in percent of full scale.

resonance. *n.* A physical process whereby an oscillating phenomenon is amplified

resonating organ. See *vocal sac*.

resonating sac. See *vocal sac*.

resorption. *n.* Mobilization of a substance from a tissue that is about to senesce and its translocation to a perennial tissue for storage and reuse.

resorption pit. Reference to the basal region of an old tooth that is eroded as the new tooth replaces it.

resource. *n.* Any element of the environment that can be utilized by an organism.

resource management. Purposeful actions taken by people to mitigate the potentially negative effects of using natural resources beyond their sustainable capacity to yield benefits.

resource partitioning. Reference to differential utilization of resources such as space and food by the different organisms or taxa that coexist in a given area or habitat.

resource tracking. A hypothesis related to host-parasite evolution in which ectoparasites "track" a particular resource such as skin or scales.

respiration. *n.* 1. In restricted sense, the intracellular oxidation of organic substrates by molecular oxygen, resulting in the generation of ATP energy and byproducts of CO_2 and H_2O. In more general usage, this term is often used to denote the process of exchange of O_2 and CO_2 between an animal and its environment. The phrase *external respiration* is sometimes used in reference to breathing, which more accurately is termed *ventilation*. Cf. *external respiration, gas exchange*. 2. In ecological energetics, that part of the energy assimilated by an individual, population, or trophic unit that is dissipated as metabolic heat, per unit time. Calculated as $R = A - P$, or $R = C - P - FU$ where A is *assimilation*, P is *production*, C is *consumption*, and FU is *rejecta* (feces and urine).

respiratory acidosis. A decrease in blood or body fluid pH associated with an increase of blood P_{CO2} due to a reduction in the rate of lung ventilation (*hypoventilation*).

respiratory alkalosis. An increase in blood or body fluid pH associated with a decrease of blood P_{CO2} due to an increase in the rate of lung ventilation (*hyperventilation*).

respiratory blood circulation. Reference to the vascular system that transports blood through the lung or gas exchange surfaces, resulting in oxygenation of blood and release of CO_2 from blood prior to its recirculation via the *systemic circuit*.

respiratory conchae. See *nasal turbinates*.

respiratory pigment. A collective term for any colored substance that reversibly binds oxygen, usually in association with oxygen transport or storage in the body. *Hemoglobin* is the universal respiratory pigment in amphibians and reptiles, as well as other vertebrates. There is great variability in the structure and functional properties of hemoglobins, related to environment and oxygen requirements of species.

respiratory chain phosphorylation. See *oxidative phosphorylation*.

respiratory quotient. (RQ) The ratio of CO_2 production to O_2 consumption by an organism. The quotient varies according to the type of food substrate being oxidized.

respiratory turbinates. See *nasal turbinates*.

respirometry. *n.* Reference to methods used for measurement of an animal's respiratory gas exchange.

response. *n.* A change in an organism or tissue as a result of a stimulus. An organismal response includes behavior.

response metameter. A method for assay of venom potency by expressing the relationship between a quantitative dose of venom and the time of death in an experimental animal. For details see P.A. Christensen, *South African Snake Venoms*, p. 12, 1955.

response stage. Any one of several stages in the development of response to stimulus in larval *Ambystoma*. Term developed in a series of papers by G.E. Coghill.

resting coil. Term used by L.M. Klauber (*Rattlesnakes*, pp. 476–486, 1956) for the position of a Rattlesnake in which the body is at rest coiled flat against the substrate with the head resting on the outer edge of the coil. (Syn. *pancake coil*) See also *striking coil*.

rest or **resting potential.** The normal membrane potential of a nonactivated cell (at "rest").

resting phase. A stage during the shedding cycle of lepidosaurs, defined as a period of time during which the epidermis consists of an incomplete *outer epidermal generation* lying atop a stratum germinativum (P.F.A. Maderson). This time extends between shedding of the old epidermal material and the point of formation of cells (lacunar and clear layer) that complete the outer generation. During the early period of this phase, little if any cell production or differentiation takes place, and this is known as the *perfect resting phase*. Cf. *renewal phase*.

restoration. *n.* Management actions to return a community or ecosystem to its original, natural condition.

restoration ecology. The science of returning a damaged ecosystem to a close approximation of its condition prior to a disturbance or anthropogenic alteration of state.

restriction enzyme or **endonuclease**. Any of many enzymes that cleave foreign DNA molecules at specific recognition sites.

restriction fragment. A fragment of a longer DNA molecule that results from digestion by a restriction endonuclease.

restriction fragment length polymorphism. (RFLP) Reference to intraspecific variation of DNA fragment lengths generated by a specific endonuclease. Restriction enzyme analysis is a methodology useful for evaluating genetic variation between or among populations or species, and for detecting specific mutations. A small set of RFLP markers from an individual can provide a *DNA fingerprint*, or specific pattern of bands, useful for individual or species recognition.

restriction map. A diagram that represents a linear array of sites on a DNA segment at which one or more restriction enzymes cleave the molecule.

resultant. *n.* A single vector that is equivalent to a given set of vectors.

rete. *n.* A dense vascular network (usually capillaries) or network of small fibers.

rete-cord. *n.* The group of cells proliferated from embryonic kidney tissue, which migrates toward the germinal epithelium and establishes relationships with the *sex-cords*. The rete-cord for females is termed *rete-ovary*, and the rete-cord for males is termed the *rete-testis*.

retention index (RI) A measure of the amount of actual *homoplasy* that is present in a phylogenetic tree as a fraction of the maximum possible homoplasy, used to assess how well a given data set fits a proposed tree.

rete-ovary. *n.* See *rete-cord*.

rete-testis. *n.* See *rete-cord*.

reticular, reticulate, reticulated. *adj.* In the form or pattern of a net, or having a network of anastomosing ridges; often used in relation to scale patterns or skin color pattern. See R.M. Price *J. Herpetol.* 16:294–306, 1982.

reticular lamina. See *basement membrane*.

reticulate speciation. The formation of new species by means of hybridization between two ancestral species.

Reticulated Python. Vernacular name for the large terrestrial Asian boid *Python reticulatus*.

reticulocyte. *n.* A young erythrocyte showing a basophilic *reticulum* under vital staining.

reticuloendothelial system. A network of phagocytic cells that reside in the bone marrow, spleen, and liver of vertebrates, where they cleanse the blood or lymphatic fluids of foreign particles.

reticulum. *n.* A small network; especially reference to small protoplasmic network in cells.

retina. *n.* (*adj.* **-al**) The innermost, light-sensitive, multimembrane layer of the eyeball, containing the sensory elements (rods and cones) for reception and transmission of visual stimuli.

retractile. *adj.* Capable of being drawn back.

retraction. *n.* (*v.* **retract**) Movement of a limb backwards.

retractor. *n.* A muscle that draws a segment back to its original position, or pulls back toward the body in a sagittal plane.

retractor muscle. An elongated muscle in the tail of male squamates, used to invert the everted hemipenes. See *retractor*.

retraherence. *n.* A temporary retreat from adverse weather conditions, to be contrasted with more profound physiological changes in response to season, such as *brumation*. Term used by C.J. Goin and O.B. Goin (*Introduction to Herpetology*, p. 164, 1962)

retro-. Prefix meaning "backwards."

retroarcitular process. The projection from the posterior end of the jaw, to which the *depressor mandibulae* muscle attaches.

retrobasal process. A median process described on the *hyoid* of *Anolis* lizards, used to extend and contract the gular dewlap.

retro-cloacal formation. Reference to paired, elongate, cecal-like structures in the dorsolateral aspect of the body cavity, entering the proctodeum through confluent openings in the middorsal wall in *Typhlops*.

retrogression. *n.* (*adj.* **–ive**) Evolutionary reversal toward a less complex state.

retroversion. *n.* In anatomy, the tipping or turning backward of an organ or other body part.

revalidated name. In taxonomy, a name that was published prior to the beginning date of a group, but subsequently validly published in a post-beginning point work.

reversal. *n.* (*adj.* **retroverted**) In evolutionary biology, the return of a structure or feature to an original state from a modified one, or change from a derived character state back to a more primitive state. As example, in vertebrate evolution limbs are considered to be derived from a limbless state, while snakes have evolved from limbed tetrapod ancestors and have lost their limbs in the process.

reverse countershading. A pattern of coloration in which the back is lighter than the ventral body. Seen in animals which habitually rest upside-down. See *countershading*.

reverse mutation. A change in a mutant gene that restores its ability to produce a functional protein. (Syn. *reversion*)

reversion. *n.* See *reverse mutation*.

reverted. *adj.* Turned back on itself.

revision. *n.* In taxonomy, a critical reappraisal of a group or taxon in relation to analysis of new material or a new interpretation.

Reynolds number. (**Re**) A unitless number related to the tendency of a flowing gas or liquid to become turbulent, which increases in proportion to its velocity and density and decreases in proportion to its viscosity. The ratio of the inertia of a fluid medium to its viscosity.

RFLP. See *restriction fragment length polymorphisms*.

rhabdias infection. Disease caused by parasitic nematodes of the genus *Rhabdias*, which infect the gas exchange organs and other tissues of anurans.

rhachitomous vertebra. A specialized *aspidospondylous* vertebra in which wedge-shaped units alternate in alignment and interlock with their neighbors. Characteristic of early amphibians.

Rhacophoridae. *n.* A clade (family) of mostly arboreal frogs with enlarged toe discs, varied reproduction, and some having extensive webbing on the feet (used for gliding in *Rhacophorus nigromaculatus*). There are eight genera and more than 300 species distributed in tropical Africa, southern and eastern Asia. For taxonomy see D.R. Frost et al., *Bull. Amer. Mus. Nat. Hist.* 297:1–370, 2006.

Rhacophoroidea. *n.* A taxon (superfamily) of anurans, sister to *Ranoidea* within the new taxon *Aglaioanura*, and composed of the families *Mantellidae* and *Rhacophoridae*. See D.R. Frost et al., *Bull. Amer. Mus. Nat. Hist.* 297:1–370, 2006.

Rhacosaurinae. *n.* A lizard subfamily of the *Gymnophthalmidae*.

rhamphotheca. *n.* A term used for the horny covering of the beak of birds, used also by some authors to denote the keratinous sheath covering the jaws of turtles and some dinosaurs. (Syn. *beak*)

rheo-. Prefix meaning "current," "flowing."

rheocolous. *adj.* Of, or associated with, the flowing water or current of streams.

rheogenic. *adj.* Producing electric current.

rheology. *n.* The study of fluid movement.

rheophilous. *adj.* A collective term for tadpoles that are modified in various ways to inhabit flowing microhabitats of lotic systems.

rheotaxis. *n.* (*adj.* **–ic**) A directed response to a water or air current.

rhinal. *adj.* Reference to the nose.

Rhinatrematidae. *n.* A basal clade (family) of caecilians that retain several relatively primitive characters and hypothesized to be the sister taxon of remaining caecilians. Two genera and nine species occur in South America.

Rhineuridae. *n.* A family of amphisbaenian comprised of the single species, *Rhineura floridana*. See J.R. Macey et al., *Molecular Phylogenetics Evol.* 33:22–31, 2004.

Rhinoceros Viper (or **Adder**). Vernacular name for the African viperid species *Bitis nasicornis*, named for prominent epidermal appendages that project from the snout. Also called *River Jack*.

Rhinodermatidae. *n.* A clade (family) that includes two species of frogs belonging to the genus *Rhinoderma*, which inhabit riparian environments along cold streams in southern Argentina and Chile. These frogs are known as *Mouth-brooding Frogs* because of a unique type of parental care wherein the male of a species carries or broods tadpoles in its mouth.

Rhinophrynidae. *n.* An anuran clade (family) including a single species (*Rhinophrynus dorsalis*) ranging from southern Texas to Costa Rica. A sister taxon with *Pipidae* within the *Xenoanura*.

rhodopsin. *n.* A photosensitive pigment found in retinal rods, also called *visual purple*.

rhomb. *n.* (*adj.* **-ic, -oid**) Shaped like a diamond, commonly in reference to skin color patterns.

rhombencephalon. *n.* The posterior major division of the brain, including the *cerebellum* and *medulla oblongata*. (Syn. *hindbrain*)

rhonchus. (*pl.* **-i**) *n.* A rattling sound in the throat, bronchus, or lung, indicative of a pathologic condition such as fluid accumulation. See also *rale*.

Rhyacotritonidae. *n.* A clade of salamanders characterized by unique glands posterior to the vent in adult males, loss of the operculum, and having greatly reduced lungs. Four species are generally associated with cold seepages and streams in northwest North American conifer forests.

A sister taxon with *Xenosalamandroidei* within the *Plethosalamandroidei*. See D.R. Frost et al., *Bull. Amer. Mus. Nat. Hist.* 297:1–370, 2006.

Rhynchocephalia. *n.* A basal clade (order) of lepidosaurs comprised of two species of *Sphenodon*, the Tuatara. These animals were widely distributed in New Zealand during the past, but now occur only on a few of the small islands off the New Zealand coast. This is the only surviving *rhynchocephalian*, but the group was diverse and widespread during the Mesozoic.

rhynchocephalian. *n, adj.* A member of, or pertaining to, *Rhynchocephalia*.

rhynchosaurs. *n.* Primitive relatives of archosaurs, formerly grouped incorrectly with rhynchocephalians (which are lepidosaurs). Rhynchosaurs were dominant herbivores during the *Triassic* and possessed stout bodies, broad and short skulls, and crushing beaks instead of toothed jaws.

rhythm, rhythmicity. *n.* A recurring change in the physiology or behavior of an organism, entrained by physical cues in the environment but with inherent mechanisms ("internal clock") that persist when external stimuli are withdrawn. See *circadian rhythm, circannual cycle*. Cf. *periodicity*.

RIA. See *radioimmunoassay*.

rib. *n.* A long, curved bone that develops along the body wall from the dorsal (vertebral) to ventral aspect of the body. See *abdominal rib, caudal rib, pleural rib, sacral rib, ventral rib*.

Ribbon Snake. See *Eastern* and *Western Ribbon Snake*.

ribonuclease. (RNase) *n.* An enzyme that degrades RNA, found in small amounts in many snake venoms.

ribonucleic acid. See *RNA*.

ribose. *n.* A pentose monosaccharide constituent of RNA.

ribosome. *n.* A ribonucleoprotein particle that occurs within cytoplasm and interacts with mRNA, tRNA, and amino acids during synthesis of polypeptide chains. The sites of *translation*.

Rice Frogs. Vernacular name for species of Asian frogs belonging to the microhylid genus *Microhyla*.

richness index. (*d*) A measure of species diversity, calculated as

$$d = (S-1)/\log H$$

where *S* is the number of species in the habitat or community and *H* is the total number of individuals of all species.

rickets. *n.* See *nutritional secondary hyperparathyroidism*.

rictal glands. Oral glands or labial glands associated with the corner of the mouth of snakes (S.B. McDowell, *J. Herpetol.* 20:353–407, 1986). Various glandular structures are distinguished either as *superior* or *inferior rictal glands*. See *oral glands, gland of Gabe*.

rictus. *n.* The opening or gape of the mouth, used variously by some authors in reference to the angle of the jaw or mouth. (Syn. *rictus oris*)

rictus oris. See *rictus*.

Ridgenose Rattlesnake. Vernacular name for the American pitviper *Crotalus willardi*.

Ridleys, Ridley Turtles. *n.* Vernacular name for species of marine turtles belonging to the genus *Lepidochelys* (Cheloniidae).

riesenzellen. *n.* A distinctive, specialized cell type found uniquely in the larval epidermis of members of Bufonidae. The term was first used by W. Pfeiffer (*Z. vergl. Physiol.* 52:79–98, 1966) who considered the cells to be involved in release of a supposed alarm substance (*pheromone*) similar to that found in certain fishes. The cells differentiate from basal epidermal cells, open at the skin surface, and ultimately disappear near the end of metamorphosis. The cells are secretory, but their function remains unclear.

rift. *n.* A location where Earth's surface is cracking apart due to tectonic activity.

right-left (R-L) shunt. See *intracardiac shunts*.

right aorta. See *aorta*.

rigor mortis. A condition of rigidity attributable to a failure of cross-bridges to become detached in dying muscle due to depletion of ATP.

rim, rimmed. *n, adj.* A raised margin of an opening such as the nares.

rimrock. *n.* 1. A cliff or ledge overlooking lower ground and formed by the outcropping of an elevated horizontal layer of resistant rock. 2. A cliff or vertical face of an outcrop or rock in a canyon wall.

ring. *n.* 1. A solid color completely encircling the body of a snake (e.g., as in Coral Snakes). 2. Used commonly in reference to a *segment* or *lobe* of a rattlesnake rattle.

ring embryo. Reference to an amphibian embryo in which an equatorial neural fold ("ring") has been induced by involuted mesoderm lying over an unually large blastopore.

Ringer's (or **Ringer**) **solution.** *Physiological saline* solution. The composition varies depending on the organism and experimental requirements, but usually contains sodium, potassium, and calcium chlorides.

Ringhal. *n.* See *Rinkhal*.

Ringneck Snake. Vernacular name for American species of colubrid snakes belonging to the genus *Diadophis*.

ring species. A graded series of populations forming an extensive cline and having a geographic pattern that curves around on itself with populations at the extreme ends overlapping but unable to interbreed.

ringwulst. *n.* A term used to describe the *annular pad* in the eye lens of pygopodid lizards.

Rinkhal. *n.* Vernacular name for African elapid snakes belonging to the monotypic genus *Hemachatus* (also called *Spitting Cobras*).

riparian. *adj.* Pertaining to habitat at the edges of rivers, streams, or other waterways, sometimes including lakes.

risk assessment. A tool to estimate the probability of extinction (or persistence) for a particular species.

ritual combat. Intraspecific rivalry between two or more male amphibians or reptiles that attempt to overthrow the other in elaborate, but usually noninjurious, fights or aggressive acts. The combat is usually related to attempts to gain possession of a female. See *combat dance* and Fig. 39.

ritualization. *n.* Reference to evolutionary modification of a behavior pattern such that it serves a communicative function or enhances the efficiency of a signal.

River Frogs. Collective vernacular name for African species of ranid frogs belonging to the genus *Phrynobatrachus*. The name also is applied to the American ranid frog *Lithobates* (formerly *Rana*) *heckscheri*. For recent taxonomic revision see D.R. Frost et al., *Bull. Amer. Mus. Nat. Hist.* 297:1–370, 2006.

River Cooters. See *Cooters*.

River Jack. Vernacular name for the viperid *Bitis nasicornis*, more commonly known as the *Rhinoceros Viper*.

River Turtles. Collective vernacular name for various species of turtles belonging to the chelid genus *Emydura* and *Rheodytes*, the dermatemydid *Dermatemys mawii*, emydid species belonging to the genus *Hardella*, and pelomedusid species belonging to the genera *Peltocephalus* and *Podocnemis*.

riverine. *adj.* Living in, or pertaining to, rivers.

r-K selection. A formerly popular body of theory in which relative population density is thought to serve as an important selective force on life history traits. The theory derives its name from the two constants of the logistic growth equation. The term "*r* selection" relates to a population that is maintained at low population density, such that resources for growth are not limited. Under these circumstances, the best reproductive strategy is to maximize offspring production. Thus, the traits expected under condition of *r*-selection are early, semelparous reproduction, large *r*, many offspring with poor survivorship, a Type III survivorship curve, and small adult body size. In contrast, the term "*K* selection" relates to an organism that grows in an environment that is chronically crowded. Here the better strategy is one that leads to fewer, high-quality offspring that are superior competitors. With such resource limitation, *K*-selection should favor late, iteroparous reproduction, small *r*, few offspring with good survivorship, a Type I survivorship curve, and large adult body size. The terms "*r* strategist" and "*K* strategist" refers to either *r*-selected or *K*-selected species, respectively. In spite of its popularity in textbooks and earlier literature (see R.H. MacArthur and E.O. Wilson, *The Theory of Island Biogeography*, 1967; E.R. Pianka, *Amer. Nat.* 104:592–597, 1970), the theory of r-K selection is beset with a number of problems, and not all organisms have life history traits that fit the predictions.

r-K spectrum. A range or linear system of reproductive and life-history strategies with *r*-selected and *K*-selected species representing the two extremes.

RNA. (ribonucleic acid). *n.* A nucleic acid comprised of adenine, guanine, cytosine, uracil, ribose, and phosphoric acid, functioning in the transcription of DNA and its translation into protein.

RNA polymerase. An enzyme that transcribes an RNA molecule from the template strand of a DNA molecule.

roach. *n.* Reference to the crest or ridge that appears in the center of the back in *Dipsosaurus* lizards, apparently as consequence of the raising of vertebral scales (K.S. Norris, *Ecology* 34:265–285, 1953).

Road Guarder. Vernacular name for Middle American species of colubrid snakes belonging to the genus *Conophis*.

road kill, roadkill. Reference to amphibians and reptiles that are killed on roads or highways by passing motor vehicles. Although clearly jargon, this term is becoming increasingly used and accepted as a means of identifying road mortality in relation to transportation infrastructure and concerns for wildlife and biodiversity.

Robber Frog. Collective vernacular name for numerous species of neotropical frogs belonging to the genus *Eleutherodactylus*.

robust, robustness *adj., n.* 1. Strongly built or formed. 2. Reference to a tendency for a model's predictions to hold up even if one violates some of its assumptions. 3. Tendency for a statistical method to produce a fair result, despite deviations in the data from the premises on which the method is based.

Rock Agamas. Vernacular name for agamid lizards belonging to the genus *Stellio*.

Rock Geckos. Vernacular name for African and Middle Eastern species of Geckos belonging to the genera *Afroedura* and *Pristurus*.

Rocket Frogs. Collective vernacular name for species of dendrobatid anurans belonging to the genus *Colostethus*.

Rock Frogs. See *Splash Frogs*.

Rock Lizards. Collective vernacular name for various species of lizards belonging to the iguanid genus *Petrosaurus*, the agamid genus *Stellio* (*Rock Agamas*), Indian *Psammophilus* spp., Bermuda Rock Lizard (skink) *Eumeces longirostris*, Iberian *Iberolacerta* spp. (formerly *Lacerta monticola*), and various other lacertid species.

Rock Python. Vernacular name for boid species belonging to the Australasian genus *Liasis*, the Asian species *Python molurus*, and the African *Python sebae*.

Rock Rattlesnake. Vernacular name for the American pitviper *Crotalus lepidus*.

rod. *n.* A photoreceptor cell of the vertebrate retina that is sensitive to light due to cellular properties and high degree of convergence onto second-order cells. Thus, rods respond to relatively low light intensities. See also *cone*.

roentgen. (R) See *röntgen*.

röntgen (or **roentgen**). (R) A derived (non-SI) unit of ionizing radiation in air, defined as the quantity of X-rays or gamma radiation that liberates by ionization 8.38×10^{-3} joules per kilogram dry air at constant temperature and pressure. Named in honor of Wilhelm Conrad Röntgen, who was a German physicist who first observed and recorded X-rays in 1895. Continued use of this non-SI unit is "strongly discouraged" by the National Institute of Standards and Technology.

Rohon-Beard cell. Intramedullary sensory cells in the dorsal root of spinal ganglia, mediating mechanoreception from free nerve endings in the epidermis of larval amphibians. These structures are transient and disappear at metamorphosis, their function being assumed by extramedullary spinal ganglia. Hence, they are considered to represent a primitive condition.

roll. *n.* Rotation of a swimmer or flyer around its horizontal (anteroposterior) axis.

ROM. See *reactive oxygen species*.

roof, false. See *false roof*.

rookery. *n.* A breeding or nesting place, often used and reused by a population. The term is usually applied to some mammals and birds, but also to various reptiles, particularly Sea Turtles.

root. *n.* 1. The portion of an organ or structure that is buried in, or arises from, tissue, e.g. that part of a tooth that is connected or attached to the supporting bone. 2. Divisions of sensory or motor nerves which emerge from the spinal cord to extend peripherally

rosette. *n.* 1. A structure with a resemblance to a rose. Examples in herpetology include the arrangement of visual cells in some amphibians, and certain scale patterns in squamates.

rostral. *n., adj* 1. The scale at the tip of the snout, bordering the mouth and separating the labial rows. (Syn. *apical, beakshield*) See Figs. 10, 14. 2. Synonym for *internasal* bone. 3. A unique medial, terminal bone in the skull of ceratopsian dinosaurs. 4. Situated toward the *rostrum* (oral-nasal region), which may be anterior, dorsal, or ventral with respect to head structures, cephalad with respect to other regions of the body.

rostral appendage. A little-used alternative name for *proboscis*.

rostral crease. The ventral groove in the rostral scale that accommodates protrusion of the tongue in various species of snakes. (Syn. *lingual fossa*)

rostral hump. An enlarged globular protrusion on the snout of the lizard *Lyriocephalus scutatus*.

rostrum. *n.* The snout, or a beak-shaped process.

Rosy Boa. Vernacular name for a small species of boid snake, *Charina trivirgata* (formerly *Lichanura trivirgata*), common to southwestern United States and western Mexico.

rotational feeding. Reference to a feeding mode in certain caecilians, which generate substantial spinning force during long-axis body rotations when feeding underground. These amphibians have evolved a highly derived skull and cranial musculature in response to a specialized head-first burrowing lifestyle. Dietary studies indicate caecilians are capable of generating considerable bite force as well as rotational spin force, which may exceed the former. See G.J. Measey & A. Herrel, *Biology Letters* 2:485–487, 2006.

rotator. *n.* A muscle that turns a limb. See *pronator, supinator*.

Rouget cell. A cell with slender, fibrillar processes that branch to embrace the walls of capillaries in amphibians. A capillary pericyte. Named after Charles Marie Benjamin Rouget, a French physiologist.

Rough Green Snake. Vernacular name for the American colubrid species *Opheodrys aestivus*.

Roughskin Newt. Vernacular name for the salamandrid species *Taricha granulosa*. The name also is applied collectively to other members of the North American genus *Taricha*. See also *Pacific Newts*.

Round Island Boa. Vernacular name for two species of unusual snakes belonging to the basal macrostamatan clade of *Bolyeriidae*.

round window. 1. A thin contact membrane between the perilymphatic fluid of the inner ear and the canal (Eustachian tube) that communicates with the pharynx; the fenestra rotunda of the mammalian cochlea. The structure permits inward movement of inner ear fluids at the *oval window* to dissipate outward in air at the round window. Primitively, the round window is buried in tissues at the margin of the skull, but opens into the middle ear chamber in advanced forms. (Syn. *fenestra rotunda*) 2. This term has been used in older literature with reference to the *tympanic membrane* of amphibians and reptiles. Such usage is incorrect.

routine metabolic rate. Reference to rate of metabolism determined during routine activity, thus intermediate between *standard* and *maximal* rates of metabolism.

Royal Snakes. Vernacular name for the species of the Asian genus *Spalerosophis*.

RQ. See *respiratory quotient*.

R_s. Skin resistance to water vapor flux. See *resistance*.

r-selected species. See *r-K selection*.

r selection. See *r-K selection*.

r strategist. See *r-K selection*.

R_t. Total resistance. See *resistance*.

Rubber Boa. Vernacular name for small species of boid snake, *Charina bottae* and *C. umbratica*, endemic to western United States.

Rubber Frogs. Vernacular name for African micryhylid frogs belonging to the genus *Phrynomantis*. The skin is smooth and rubbery to the touch.

Rubner's hypothesis. A theory suggesting that species which are inactive during part of the annual cycle have longer life spans than do those that are active continuously throughout the year.

rubrification. *n.* The appearance of a red coloration in a soil due to oxidative weathering and the formation of hematite.

rudimentary. *adj.* Imperfectly developed. Sometimes used incorrectly as a synonym for *vestigial*.

ruga. (*pl. –ae*) *n.* 1. A ridge or fold. 2. Sometimes used in reference to the *frontal ridge* of *Anolis* lizards.

Rugh's method. A technique for inducing ovulation and fertilizing frog eggs artificially. The method originally involved injection of pituitary or pituitary extracts into male and female frogs in attempt to induce amplexus (R. Rugh, *Biol. Bull*. 66:22–29, 1934). Modification of the method involves stripping the eggs from the female and placing them directly into a suspension of sperm.

rugose. *adj.* Wrinkled, or marked by ridges or rough surface.

rugosity. *n.* A ridge or fold.

Running Frogs. Collective vernacular name for species of African frogs belonging to the hyperoliid genus *Kassina*. Also collectively referred to as *Kassinas*.

runoff. *n.* That part of precipitation that is not held in soil but drains freely away.

runt. *n.* An undersized or dwarfed newborn or hatchling reptile.

rupicolous. *adj.* Rock-dwelling; living on or among rocks.

rusconian plug. See *yolk plug*.

Russell's Viper. Vernacular name for the Asian viperid species *Daboia russelli*. (Syn. *Tic Polonga*)

Russian Journal of Herpetology. An international journal founded in 1993, publishing articles covering both basic and applied research on extant and fossil amphibians and reptiles. All aspects of herpetology are published in English. The content emphasizes the conservation of amphibians and reptiles, their habitats, and the promotion of cooperation between Russian and foreign scientific and commercial organizations.

Ruthven's Frog. See *Allophrynidae*.

S

s. 1. *Standard deviation*. 2. *Selection coefficient*. 3. SI time unit, second.

sabulicolous. *adj.* Living in sand. (Syn. *arenicolous*)

sac. *n.* Anatomical term for a pouch or baglike organ or structure, usually with a narrow opening and containing a secretion or other substance.

sacci anales. Early term used for *cloacal bursae* by Agassiz (1857).

sacculus, saccule. *n.* The smaller and more ventral of two sac-like divisions of the membranous labyrinth of the vertebrate inner ear.

saccus endolymphaticus. See *endolymphatic sac*.

sacral. *adj., n.* 1. Located in or relating to the *sacrum*. 2. One of the bones of the sacrum.

sacral diapophysis. (*pl.* -ses) Reference to a *transverse process* that occurs on sacral vertebrae and articulates with the *ilium*. (Syn. *sacral process*)

sacral hump. See *hump*.

sacral process. See *sacral diapophysis*.

sacral prominence. See *hump*.

sacral rib. A rib associated with sacral vertebrae, often fused with the vertebra and sacrum, rendering it difficult to define. Whether these structures are true ribs or transverse processes has been the subject of controversy.

sacral vertebra. (*pl.* -ae) Any of one or more vertebrae that articulate with the pelvic girdle. These vertebrae bear stout ribs that connect with the *ilium*. See also *sacrum* and *cloacal vertebra*.

sacrococcyx. *n.* The bone formed by fusion of the sacral (last) vertebra and the *os coccygeum* in pelobatid anurans.

sacrum. *n.* A single bony complex consisting usually of fused sacral vertebrae. This is the last single unit in the vertebral column of anurans, which articulates with the *ilium* and the *os coccygeum*. In turtles and in many fossil reptiles this structure is formed by fusion of several sacral vertebrae which provide support for the pelvic girdle.

saddle. *n.* A blotch within a color pattern, extending from a broader dorsal shape to a more narrow shape on the lateral body.

Saddleback Toads. Collective vernacular name for South American species of anurans belonging to the family Brachycephalidae.

Sagebrush Lizard. Vernacular name for the phrynosomatine iguanid species of American lizard *Sceloporus graciosus*.

sagittal plane. Any anteroposterior dorsoventral plane or section along the longitudinal axis, or parallel to the median plane, of the body. The plane divides the body into right and left parts. Cf. *midsagittal plane*.

saggital suture. The juncture between the *frontoparietal bones* on the skull roof of anurans.

sahel. *n.* A zone of *savanna* and *scrub* lying directly south of the Sahara Desert.

Sailfin Lizard. Vernacular name for Asian species of agamid lizards belonging to the genus *Hydrosaurus*.

Salamandra. *n.* A German journal of herpetology, founded in 1964 as a successor to *Salamander*, organized in 1918. Published by the German Society for Herpetology and Herpetoculture.

Salamandridae. *n.* A clade (family) of small to moderate-sized salamanders colloquially referred to as "Newts." Salamandrids have variable life histories, but adults of many species are terrestrial and migrate to ponds for breeding. Approximately 62 species are recognized in North America, Europe, Africa, and western Asia. A sister taxon with *Ambystomatidae*.

salamandridine, salamandrin. *n.* See *samandaridin*.

Salamandroidea. *n.* A clade (superfamily) that includes all salamanders known to have internal fertilization, including all extant families except for *Sirenidae* and *Cryptobranchoidea*. Relationships of taxa within this clade are controversial.

Salientia. *n.* A monophyletic clade (order) including living frogs (*Anura*), and various fossil taxa characterized by numerous derived characters. Sometimes used synonymously with *Anura*.

saline. *adj.* Salty or pertaining to salts, usually with respect to soil or water rich in soluble salts.

saline solution. See *physiological saline*.

salinity. *n.* A measure of the total concentration of dissolved salts in water, usually expressed as grams per kilogram or parts per thousand. The average salinity of seawater is 35 parts per thousand.

saliva. *n.* An aqueous secretion of the mouth that aids in mechanical and chemical digestion.

salivary glands. Glands that secrete saliva into the headgut.

Salmonella. *n.* A genus of nonlactose fermenting, Gram-negative bacteria that are commonly isolated from the intestinal tract of captive reptiles and are capable of being transmitted to humans and other animals.

salmonellosis. *n.* Infectious disease attributable to bacteria of the genus *Salmonella* which commonly infect various reptiles, especially turtles.

saltation or **saltatory locomotion.** *n.* 1. Locomotion by jumping. 2. A mode of locomotion in snakes that occurs when a rapid straightening of the body from anterior to posterior lifts the entire animal off the substrate. The behavior is limited to smaller individuals and usually extreme circumstances associated, for example, with escape behavior. 3. A speciation theory involving origin of new species from sudden mutations that produce large phenotypic effects ("hopeful monsters" of R. Goldschmidt). 4. *Quantum speciation.*

saltational evolution. Sudden origin of a new type of organism, radically different and giving rise to a new group. (Syn. *transformational evolution*)

saltatorial, saltatory. *adj.* Adapted for hopping or jumping.

salt glands. Glandular structures that have diverse origins from *nasal glands* (lizards), *lacrimal glands* (marine turtles and Diamondback Terrapin), *lingual salivary glands* (Crocodiles and Gharials), posterior *sublingual glands* (Sea Snakes and File Snakes), and the *premaxillary gland* (the marine homolopsid snake *Cerberus*). Each of these gland types functions as an osmoregulatory organ, important for salt excretion that is accessory to the kidney.

saltigrade. *adj.* Reference to a leaping or hopping mode of locomotion.

salt marsh. A poorly drained coastal swamp, inundated by most tides and relatively flat.

Salt Marsh Snake. Vernacular name for the American natricine species *Nerodia clarkii*, unusual because of its semi-marine habits.

Saltwater Crocodile. Vernacular name for the crocodilian species *Crocodylus porosus*. (Syn. "*salty*" in Australia)

salt wedge. With respect to an estuary, reference to an inward-flowing front of saline bottom water that penetrates beneath the main body of outflowing fresh water that is relatively less dense.

salvage. *n.* The retrieval of dead parts or whole animals for scientific information or record keeping.

samandaridin, samandaridine. *n.* A toxic alkaloid secretion of the cutaneous glands of salamanders belonging to the genus *Salamandra*. The physiological effects are similar to those of *samandarin*, but severalfold less potent. (Syn. *salamandridine, salamandrine, salamandrin*)

samandarin, samandarine. *n.* An alkaloid toxin secreted by skin glands of salamandrid salamanders, which can cause paralysis, hemolysis, and convulsions.

samandatrin. *n.* An alkaloid toxin secreted by *Salamandra atra*, differing slightly in properties from the alkaloids derived from *Salamandra salamandra*.

sample. *n.* A subset of a *population* from which inferences are to be made about the population. See *random sample.*

sample size. (*n*) The number of observations in a sample.

sample stastitic. See *estimate.*

sample variance. (s^2) In statistics, the square of the *standard deviation* of the sample.

sampling. *n.* The process of taking a sample.

sampling error. See *experimental error.*

sand. *n.* Sediment composed of cohesionless particles with a diameter range of 2.0–0.6 mm.

sandbar. *n.* A submerged ridge of alluvial sand in shallow water.

Sand Boas. Collective vernacular name for various African and Eurasian species of boid snakes belonging to the subfamily *Erycinae*.

sand dune. A mound of loose sand grains heaped up by the wind.

Sand Frogs. Collective vernacular name for African species of ranid frogs belonging to the genus *Tomopterna*.

Sand Lizards. A collective common name for various Old World lizard species belonging to the *Lacertidae* (sometimes called "Wall Lizards" or "true lizards"). The name applies particularly to members of the genera *Mesalina* and *Meroles, Pedioplanis,* and *Psammodromus*.

Sand Monitor. Vernacular name for the Australian varanid lizard species *Varanus gouldii*. Also called *Gould's Goanna*.

Sand Racers. Vernacular name for species of African colubrid snakes belonging to the genus *Psammophis*.

Sand Skinks. Vernacular name for scincid lizards belonging to the genera *Neoseps* and *Scincus*.

Sand Snakes. Collective vernacular name for various colubrid snakes belonging to the Old World genus *Psammophis* and the American *Chilomeniscus*.

sand-swimming. Reference to the rapid burial and burrowing movements of various desert sand-dwelling lizards.

saprolegniasis. *n.* A disease observed in captive aquatic amphibians and amphibian larvae, usually associated with injuries to the skin. The disease pathology is related to infection by fungus, especially species of the genus *Saprolegnia*.

saprophytic. *adj.* Growing on dead or decaying tissue, but nonpathogenic.

sarafotoxins. *n.* Peptides isolated from the venom of *Atractaspis* that are potent vasodilators, structurally and functionally similar to endothelins (hormones present in mammalian endothelium).

sarco-. Prefix meaning "flesh" or "fleshy."

sarcolemma. *n.* The plasma membrane of muscle fibers.

sarcomere. *n.* A repeating contractile unit of skeletal muscle *myofibrils*, containing parallel bundles of actin and myosin that, when cross-bridges are interacting, function to "pull" opposite morphological boundaries of the sarcomere (Z lines) toward one another, thereby creating tension within the muscle. The tension generated by numerous individual sarcomeres in series is additive.

sarcoplasm. *n.* See *myoplasm.*

sarcoplasmic reticulum. (**SR**) A network of membranes that create limiting spaces surrounding the contractile elements of myofibrils. Calcium is stored within the SR and is released as free Ca^{2+} during excitation-contraction coupling.

satellite DNA. See *microsatellite DNA*.

satellite male. A male anuran that is silent but locates itself near a calling male in attempt to intercept females attracted to the call. (Syn. *parasitic male*)

Sators. *n.* Vernacular name for iguanid lizards belonging to the genus *Sator*.

saturation. *n.* (*adj.* – **ed**) 1. The condition of fatty acid molecules having no carbon-carbon double bonds. 2. Presence of a factor at a concentration or level that is equal to, or in excess of, that required for maximum response or activity. 3. In ecology, the equilibrium condition of a community in which immigration is balanced by extinction. 4. With respect to color, saturation is the perceived difference between a color and white, regardless of brightness. Cf. *chroma*.

saturation deficit. A measure of atmospheric humidity, derived by subtracting the actual water vapor pressure or density from the maximum possible vapor pressure or density at a given temperature. Cf. *relative humidity, water saturation deficit*.

saturation dispersal. The emigration of surplus individuals from a population at or near its carrying capacity. Cf. *presaturation dispersal*.

saturation vapor pressure. The highest pressure of water vapor that can exist in equilibrium with a plane, free water surface at a given temperature. See *water vapor pressure*.

Saukrobatrachia. *n.* A new monophyletic taxon of anurans, sister to the new taxon *Africanura* within the new taxon *Ametrobatrachia*, and composed of *Dicroglossidae* and a new taxon *Aglaioanura*. See D.R. Frost et al., *Bull. Amer. Mus. Nat. Hist.* 297:1–370, 2006.

Sauria. *n.* See *Lacertilia*.

saurian or **sauroid.** *adj.* Relating to, or resembling, a lizard.

Saurischia. *n.* One of two stem subgroups of *Dinosauria*, regarded as monophyletic and a sister taxon to *Ornithischia*. The group can be defined as all *Dinosauria* closer to birds than to *Triceratops* (K. Padian and C.L. May, *New Mexico Mus. Nat. Hist. Sci. Bull.* 3:379–381, 1993).

saurochorous, saurophilus. *adj.* Dispersed or distributed by lizards or snakes.

sauropod. *n.* A member of the *Sauropoda*.

Sauropoda. *n.* A subgroup of *Dinosauria* including members that are, in general, very large quadrupedal animals with small head and very long neck and tail. Sauropods were particularly diverse and abundant in the late Jurassic and include the largest terrestrial animals that ever lived.

Sauropodomorpha. *n.* A subgroup of Dinosauria that includes *Sauropoda* + *Prosauropoda* and all saurischians closer to them than to birds. This group is a stem and sister taxon with *Theropoda* within the *Saurischia*. The relatively long necks of sauropodomorphs suggest they browsed on high vegetation, but this subject has been controversial.

sauropsid. *n.* A member of *Sauropsida*, reptile or bird.

Sauropsida. *n.* A clade of amniotes that includes dinosaurs and reptiles, including the birds.

Sauropterygia. n. An infraclass or superorder of Mesozoic reptiles, which like the *ichthyosaurs*, were specialized for an aquatic mode of life. The group includes the *nothosaurs, plesiosaurs,* and *placodonts*. These reptiles, along with the ichthyosaurs, played significant roles as predators in marine communities. Sauropterygians are considered *diapsid*, but they are sometimes classified with turtles.

sauvagine. *n.* A 40-amino acid polypeptide first isolated from the skin of *Phyllomedusa sauvagei*. Pharmacological actions include hypotension accompanied by tachycardia, antidiuresis, and numerous neuroendocrine effects.

savanna, savannah. *n.* Open grassland with scattered shrubs and trees, characteristic of much of tropical Africa, South America, and Australia.

Savannah Monitor. Vernacular name for the Australian Monitor Lizard *Varanus exanthematicus*.

saw. *v.* Reference to production of a continuous, low-frequency vocalization in response to clasping by a second male, described originally in *Xenopus* (W.M.S. Russell, *Behaviour* 7:113–188, 1954).

Saw-scaled Viper. Collective vernacular name for viperid snakes belonging to the genus *Echis,* especially *Echis carinatus,* distributed in parts of Asia, Africa, and the Middle East.

saxatile. *adj.* See *saxicolous*.

saxicoline. *adj.* See *saxicolous*.

saxicolous. *adj.* Living among rocks or in rocky habitats. Rock-dwelling. (Syn. *saxatile, saxicoline*)

scalar. *adj.* Reference to something that is completely described by a single number.

scalares. *n.* Specialized body musculature seen in advanced, burrowing lizards where muscle bundles run from ribs to skin and allow backward and forward locomotion, either beneath the ground or on the ground surface. (Syn. *skin slips*)

scalation. *n.* Reference to the pattern or arrangement, as well as the number, of scales on a reptile. (Syn. *squamation*)

scale. *n.* A term generally denoting the keratinized epidermal outgrowths or platelike structures that cover reptiles, especially those of squamates (Fig. 29). The structures incorporate specialized terms such as *plate, scute, shield*, and *lamina*. Developmentally, scales arise as folds in an otherwise flat integument without inductive influence of the underlying dermis. By contrast, *appendages* such as feathers, hairs, nails, etc. are "localized centers of specialized epidermal and/or dermal cell prolilferation and differentiation within an otherwise unspecialized integument." See P.F.A. Maderson, *Amer. Zool.* 12:159–171, 1972.

scale clipping. A method of marking snakes and other reptiles by cutting a part of the edge of one or more scales, usually ventral, in a predetermined pattern or combination. The scale clips leave scars on the marked scales that allow subsequent identification of the individual.

scale count formula. See *scale row formula*.

scale erection. The raising of scales such that spines or rough edges project outward from the body surface. This behavior is seen commonly among various lizards as a defense mechanism to injure predators or to secure the body within a hole or crevice.

scale formula. See *scale row formula*.

scale fossa. See *apical pit*.

scale organ. Minute sensory structures on the surfaces of lizard or snake scales, usually in the form of knobs, depressions, or hairlike projections. See also *apical pits, lenticle*.

scale patch. See *stridulation organ*.

scale pit. See *apical pit*.

scale rot. A jargonish name for a form of *dermatitis* in captive reptiles, especially snakes, characterized by appearance of small blisters or discoloration, which may become necrotic.

scale row. A continuous series of dorsal scales in squamate reptiles, counted in sequence according to the specific arrangement. Arrangements of scale rows may be of three types according to P.J. Clark & R.F. Inger (*Copeia* 1942:230–232, 1942). In the *longitudinal type*, the dorsal scales form straight rows in three directions including the length of body, a diagonal series up and back from ventrals, and a diagonal series up and forward from the ventrals. This pattern is characteristic of most snakes. (Syn. *rectiform*) In the *diagonal type* of scale row, the longitudinal rows run downward and backward in a long curve. The third pattern is the *oblique type*, in which the longitudinal rows run parallel to the body, while the transverse rows are curved and oblique to various degrees.

scale row count or **formula.** Reference to several methods of expressing scale row counts in snakes.

scaling. *n.* The dimensional change of a structure (size, etc.) in relation to changes in body size or mass, or in relation to changes in the scale of another structure.

scalloped. *adj.* Having a continuous but wavy border, such that the edge resembles a scallop shell.

Scalyfoot Gecko. Vernacular name for Australian gekkonid species belonging to the genera *Delma* and *Pygopus*.

scandent. *adj.* See *scansorial*.

scanning electron microscope. (SEM) See *electron microscope*.

scansor. *n.* A series of overlapping plates, used in reference to *subdigital lamellae* of Geckos and Anoles. See *lamella*. and Fig 32.

scansorial. *adj.* Adapted for, or frequently engaging in, climbing. (Syn. *scandent*)

scap. *v.* Reference to capture of turtles by means of using a dip net.

scaphiodont. *adj.* See *promegadont*.

Scaphiopodidae. *n.* A clade (family) of American anurans, sister with *Pelodytidae* within the *Pelodytoidea* and contaning the genera *Scaphiopus* and *Spea*.

Scaphiophryninae. *n.* A subfamily of the anuran *Microhylidae*.

scapula. *n.* 1. In amphibians, the principal dorsal component of the pectoral girdle, preformed in cartilage and forming the dorsolateral margin of the *glenoid fossa*. The dorsal edge may remain cartilaginous and is designated *suprascapular*. (Syn. *collum scapulae, omoplate, scapula minor*) See Fig. 3. 2. In reptiles, the principal bony component of the *scapular blade*, forming the dorsal component of the *glenoid fossa* as in amphibians. This element is present in all reptiles having a shoulder girdle, but is highly modified in turtles where it forms two prongs at right angles, one of which represents the *acromial process*. Note the scapula of turtles has been designated the *coracoid* in some older literature.

scapula major. *n.* See *suprascapulum*

scapula minor. See *scapula*.

scapular. *n., adj.* 1. See *gular*. 2. Of or pertaining to the *scapula*.

scapular blade. The dorsal part of the *scapulocoracoid plate* that represents the area occupied by the *scapula* when it is ossified separately.

scapulocoracoid, scapulocoracoid plate. A substitute term for the primary girdle when the *scapula* and *coracoid* are fused and indistinguishable in certain reptiles and amphibians. The dorsal part is called the *scapular blade*, and the ventral part is the *coracoid plate.*

Scarlet Kingsnake. Vernacular name for the American colubrid species *Lampropeltis triangulum elapsoides.*

Scarlet Snake. Vernacular name for American colubrid species *Cemophora coccinea*.

scat. Word element for dung or fecal matter.

scatology. *n.* The study of animal feces.

scattering. *n.* A change of direction of particles or waves that results from collisions or interactions.

scatter diagram. A graph plotting paired or multiple measurements as points in order to indicate a general picture of the relationship between two or more variables.

scavenger. *n.* An animal that feeds on dead or decaying animals, typically those that have died naturally and not as a result of predation.

scent gland. See *anal gland*.

scent transfer. Reference to a method of feeding captive snakes, in which a normally unacceptable food item (e.g., dead mouse) is scented with something that is acceptable (e.g., another snake in the case of feeding an ophiophagous species).

Scheltopusik. *n.* Common name for the anguid lizard *Ophisaurus apodus*. See also *Glass Lizards*.

Schiff reagent. A reagent, used in the *PAS* procedure, that attaches to and colors aldehyde-containing compounds.

schizocoely. *n.* Reference to formation of a coelom by cavitation or splitting of the hypomere.

Schnabel method. A mark-recapture method for estimating population size in a closed population, employing a series of samples in which marked individuals are counted and further individuals are marked using identical marks throughout. The method is an extension of the Petersen method to a series of samples, computing a weighted average of Petersen estimates as an estimate of population density (Z.E. Schnabel, *Amer. Math. Monthly* 45:348–352, 1938).

school. *v.* To swim together in large numbers, attributable to social behaviors rather than accidental aggregation.

Schultheis thermometer. See *cloacal quick-reading thermometer*.

Schwann cell. A type of cell that enwraps various peripheral axons to produce an insulating *myelin sheath*.

scientific method. A means of gaining knowledge through objective procedures that include testing of hypotheses based on observations.

scientific name. Formal nomenclature for any organism, designating the genus and species (e.g., *Crotalus viridis*), or genus, species, and subspecies (e.g., *Crotalus viridis oreganus*). The name is first assigned when a species is formally described, and the name is recognized universally by scientists in contrast with *common names* that may vary from region to region. See *binomial nomenclature*.

Scincidae. *n.* A speciose clade (family) of lizards characterized by smooth, shiny cycloid scales underlain by *osteoderms* and surrounding hard and cylindrical bodies often with reduction of limbs. Approximately 115 genera and 1260 species are virtually cosmopolitan except for high latitudes in the northern hemisphere. Four subfamilies include *Feyliniinae, Acontiinae, Scincinae,* and *Lygosominae*.

Scincinae. *n.* A subfamily of *Scincidae*.

Scincoidea. *n.* A sister clade to *Lacertoidea* within the *Scincomorpha*.

Scincomorpha. *n.* A basal clade of skink-like lizards within *Autarchoglossa* and sister taxon to *Anguimorpha*.

scintillation counter. An instrument that detects and counts minute flashes of light that are produced in scintillation fluid by particles emitted from radioisotopes.

sclera. *n.* The tough, fibrous outer coat of the eye, often strengthened in reptiles by the *sclerotic ring*.

scleral, scleroid, sclerotic. *adj.* Reference to the *sclera* of the eye.

scleral or **sclerotic ossicle.** Any of a series of bony plates that form the *scleral ring*.

scleral or **sclerotic ring.** 1. A circle of bony plates that form in the *sclera*, usually closely surrounding the cornea, and resisting inward or outward pressure. These plates were present in primitive vertebrates and persist primarily in some fishes, numerous reptiles, and birds. They are absent in modern amphibians, snakes, and crocodilians. 2. Used to designate a complete ring of *ocular* scales around the eye of a snake, including *suboculars*.

Scleroglossa. *n.* A basal clade of the order *Squamata* and sister taxon to *Iguania*.

scleroprotein. *n.* A generic term for fibrous proteins, including *keratin*.

sclerotome. *n.* The inner division of the embryonic *somite* that contributes to formation of vertebrae.

sclerotic ring. See *scleral* or *sclerotic ring*.

Scolecomorphidae. *n.* A clade (family) of caecilians characterized by several unusual characters including vestigial eyes attached to tentacles. Five species representing a single genus have disjunct distributions in central Africa. Recently considered to be a subfamily of Caeciliidae by D.R. Frost et al. (*Bull. Amer. Mus. Nat. Hist.* 297:1–370, 2006).

Scolecophidia. *n.* A basal clade (infraorder) of snakes that includes what are generally known as Blind Snakes, families *Leptotyphlopidae, Typhlopidae,* and *Anomalepididae*.

scoliosis. *n.* A permanent lateral curvature in the vertebral column.

scope for activity. See *aerobic metabolic scope*.

scoptanura. *n.* Synonym for Orton's (1953) Type 2 tadpole (microhylid of O.M. Sokol, 1975). See P.H. Starrett, pp. 251–271 in J.L. Vial, ed., *Evolutionary Biology of Anurans. Contemporary Research on Major Problems*, 1973.

scopulus. *adj.* Reference to crags and steep overhanging cliffs.

scotophase. *n.* The phase of darkness during a 24-hour period. Cf. *photophase*.

scotopic vision. Sensitivity to dim light. Cf. *photopic vision*.

scramble competition. Competition for a resource that is equally partitioned between competitors such that some or all of the individuals obtain an insufficient amount of the resource for reproduction or survival, and where scramble for the resource can result in injury to an individual. E.g., active searching for females with male-male physical interactions in some anurans during the breeding period. Cf. *contest competition*.

scraper. *n.* See *shovel*.

scream. *n.* See *fright cry*.

scree slope. *n.* An area of weathered rock fragments at the foot, or on lower slopes, of a mountain or hill. Used synonymously with *talus slope*.

Screeching Frogs. Collective vernacular name for numerous species of frogs belonging to the family Arthroleptidae, especially the genus *Arthroleptis*.

scrub or **scrubland.** *n.* See *shrubland*.

SCUBA. Acronym for "self-contained underwater breathing apparatus."

sculpturing. *n.* Reference to patterns of pitted or sculptured surfaces of osteoderms or skull bones of crocodilians. See also *epidermal microsculpturing*.

scute. *n.* 1. Any large, flat, horny scale such as a *plate, lamina,* or *shield* on the surface of a reptile. See *transverse ventral plate.* 2. This term has been used in reference to enlarged pads on the upper surface of each digit tip in certain species of frogs.

scute clipping. See *scale clipping*.

scutellation. *n.* 1. A collective term for the epidermal scales of reptiles. 2. Reference to the pattern of *laminae* on the *carapace* or *plastron* of turtles.

scutellum. (*pl. –a*) *n.* 1. A small scale. 2. See *subcaudal*.

scutum. *n.* See *ventral*.

SD. *Standard deviation.*

SDA. See *specific dynamic action*.

SE, S.E. See *standard error*.

sea. *n.* A geographical division of an ocean, or an inland body of salt water.

sea finding behavior. The process whereby hatchling sea turtles correctly orient toward the sea upon emergence from the nest. The cues involved in this behavior are not well understood, but light is clearly important.

Sea Kraits. Collective vernacular name for amphibious marine elapids belonging to the genus *Laticauda* (classifed by some as either a separate family, *Laticaudidae*, or subfamily *Laticaudinae*). Also called *Banded Sea Snakes, Common Sea Kraits, Yellow-lipped Sea Krait*.

seam. *n.* The boundary or juncture between adjacent scales or laminae.

search image. A predator's pictorial memory of the appearance of its prey. A model for visual comparison.

Sea Snakes. Collective vernacular name for marine elapid snakes classified as a subfamily *Hydrophiinae*, including both *Sea Kraits* and other entirely aquatic taxa. Earlier classification schemes suggested a dichotomy between true Sea Snakes, *Hydrophiidae*, and the Sea Kraits, *Laticaudidae*. See *Hydrophiinae*.

season. *n.* Any of the principal climatic periods of the year. Cf. *aestival, autumnal, hibernal, serotinal, vernal*.

seasonal. *adj.* Periodicity related to seasons.

seasonal breeder. A species that breeds during a specific time of the year.

seasonal climate. Reference to any climatic regime that results in marked extremes of precipitation or temperature as well as seasonal responses in the biotic communities associated with such regions.

seasonal dichromatism. Reference to seasonal changes in skin coloration.

seat patch. See *pelvic patch*.

seat patch down. A phrase coined for the behavior of dehydrated toads that involves touching the ventral skin to a moist substrate or surface (S.D. Hillyard et al., *Physiol. Zool.* 71:127–138, 1998). See *water absorption response*.

Sea Turtle. Common name for any of seven species of turtles belonging to the families *Cheloniidae* and *Dermochelyidae*.

sec. Abbreviation of the Latin *secundum*, meaning "according to."

Secchi disc technique. A method for estimating the transparency of water by submerging a white disc of standard size and recording the depth at which it disappears from view.

second. (*s*) *n.* The standard *SI* unit of time.

secondary. *adj., n.* Generally, second in a hierarchical order. In herpetology this word has been used to designate the partial or complete groove dividing a primary fold in caecilians. Thus, a secondary fold ("secondary") lies between two primary folds ("primaries").

secondary bone. Bone that is formed during internal reconstruction after preexisting bone tissue has been dissolved and reconstituted. Cf. *primary bone*. See also *Haversian bone*.

secondary cartilage. Cartilage that forms after initial bone ossification is complete, usually in response to mechanical stress.

secondary consumer. Any heterotrophic organism that feeds directly on a primary consumer. Cf. *primary consumer*.

secondary contact. Reference to reestablishment of contact between populations after a period of separation.

secondary discontinuity. Discontinuous distribution of a species because of fragmentation of the original habitat resulting from geological, biotic, or climatological changes.

secondary forest. Reference to secondary vegetation in a logged or otherwise disturbed forest.

secondary girdle. Reference to the dermal bone derivatives of the pectoral girdle, including the *clavicle, interclavicle,* and *cleithrum*.

secondary homonym. In taxonomy, one of two or more identical specific or subspecific names for different taxa that were established in different genera and subsequently transferred into the same genus. Cf. *primary homonym*.

secondary host. An *intermediate host*.

secondary metamorphosis. A major change in structure and habitat preference that occurs in post-larval amphibians. Examples would include newts that undergo changes in the color, condition, and appendages of the skin during the periodic return to an aquatic habitat.

secondary neurulation. Formation of the embryonic neural tube by cavitation within a previously solid cord. This is characteristic of most fishes and occurs in the tail region of tetrapods. Cf. *primary neurulation*.

secondary note. In anuran vocalization, one or more notes that follow a primary note and are usually shorter.

secondary palate. A shelf of bone within the buccal or oral cavity, serving to place the internal nares immediately anterior to the pharynx. The secondary palate separates the respired (ventilated) airstream from the mouth cavity and has evolved independently several times. See *palate*.

secondary plant compounds. Chemicals produced by plants that are toxic or unpalatable to herbivores.

secondary production. The assimilation of organic matter by a *primary consumer*.

secondary receptor. See *receptor*.

secondary rib. See *gastralium*.

secondary sex ratio. See *sex ratio*.

secondary sexual character. Reference to phenotypic differences between sexes other than the reproductive organs and gametes. Cf. *primary sexual character*.

secondary speciation. Interspecific hybridization between distinct evolutionary lineages. Cf. *primary speciation*.

secondary upper supratemporal. The uppermost scale of a series of supratemporal scales, located immediately posterior to the primary supratemporals in Skinks of the genus *Eumeces*.

second messenger. A small, intracellular regulatory molecule that is itself under control of an extracellular "first messenger," such as a hormone. Examples include cAMP, cGMP, and calcium.

second order neuron. A neuron that receives input from primary sensory neurons.

secretin. *n.* A peptide hormone secreted in the intestine to regulate pH of duodenal contents via control of gastric acid secretion and buffering with carbonate.

secretion. *n.* 1. A substance or product that is produced and released by a cell or gland to perform a specific function. 2. The process of elaborating and releasing secretions.

secretory vesicle. A membrane-bounded vesicle that contains a secretory product.

section. *n., v.* 1. In histology, a section of embedded specimen that is cut with a microtome and subsequently stained for observation. To cut such a section. 2. A neutral term employed as a subdivision of a taxon, frequently considered to be equivalent to a subgenus if used as a primary division of a genus. 3. A group of species separated by some distinction from others of the same genus. 4. A subgroup of individuals belonging to a scientific organization or society.

sedative. *n.* Medication to induce a state of calm.

sedentary. *adj.* Pertaining to an animal at rest or inactive.

sediment. *n.* Loose, solid particles that can originate due to: (a) weathering and erosion of pre-existing rocks, (b) chemical precipitation from solution, usually in water, and (c) secretion by, or decay of, organisms.

sedimentary rock. Rock formed by sediment deposition, or by aggregation of inorganic materials from skeletal remains. Cf. *igneous rock, metamorphic rock*.

segment. *n.* Any one of the individual, interlocking parts of a Rattlesnake rattle. Cf. *button, rattle, ring*.

segmental duct. See *ductus deferens*.

segmentation. *n.* 1. Serial repetition of structure or body parts. See *metamerism*. 2. Embryonic cleavage or the subdivision of an embryo into cells. 3. The separation of gut contents into localized masses.

seif. *n.* See *longitudinal dune*.

selected body temperature. The temperature selected by an individual (or mean for a species) when placed in a thermal gradient in the laboratory, often expressed as a central range of temperatures. Cf. *preferred body temperature*.

selection. *n.* A process in which different genotypes achieve relative allotment and propagation among individuals of a population. The selective effect on a gene or genotype is related to the probability that carriers will reproduce. See *balanced selection, directional selection, disruptive selection, frequency-dependent selection, group selection, kin selection, natural selection, normalizing selection, r and K selection, sexual selection, stabilizing selection*.

selection coefficient. (*s*) The decrement of fitness relative to a reference (individual, phenotype, or genotype). Usually, the proportionate reduction in the average contribution of gametes to the next generation made by individuals of one genotype relative to those of another (usually more fit) genotype.

selection force. See *selection pressure*.

selection pressure. 1. Reference to the effectiveness of natural selection in altering the genetic composition of a population over generations and time. 2. Any factor of the environment (physical or biological demand) that results in natural selection. (Syn. *selection force*)

selective advantage or **disadvantage.** Reference to increased or decreased *fitness*, respectively, of one genotype relative to that of a competing genotpe.

selective breeding. Selection by humans of particular genotypes in a population because they exhibit desired phenotypic characters. A common practice today by *herpetoculturists*.

selective neutrality. A condition expressed when phenotypic characters related to certain mutant alleles are equivalent to that of the wild-type allele in terms of their fitness values. See *neutral gene theory*.

self-limiting. *adj.* 1. Any process that becomes limited in scope or magnitude due to inherent constraints or negative feedback controls. 2. A disease that is limited by its own characteristics rather than by outside influence.

selva. *n.* Dense, tropical rainforest, especially that in South America.

SEM. *Scanning electron microscopy* or *microscope*. See *electron microscope*.

semasis. *n.* Advertisement coloration or other characters. Cf. *crypsis*.

sematic. *adj.* Reference to a warning or signalling function of characters, usually coloration.

semelparity. *n.* (*adj.* – **ous**) Reference to a life history strategy in which all reproduction is concentrated in a single age ("big bang" reproduction). Cf. *iteroparity*.

semen canal. See *sulcus spermaticus*.

semestrial. *adj.* Reference to periods of six months.

semi-. Prefix meaning "half" or "partly."

semi-aquatic. *adj.* Living part of the time in water. (Syn. *amphibious*)

semi-arid. *adj.* Reference to areas that are comparatively dry and characterized by sparse scrub or trees.

semicircular canals. One of the equilibrial organs in the inner ear of vertebrates, which function to sense acceleration of the head with respect to the gravitational field.

semidirect development. Development in which there is a nonfeeding larval stage that receives nutrition from its yolk stores.

semigeographic speciation. See *parapatric speciation*.

semiholoblastic. *adj.* Of or pertaining to a pattern of early embryonic cell divisions somewhat intermediate between *holoblastic* and *meroblastic* cleavage, resulting in a blastodisc and some division of the yolk.

semilunate. *adj.* Shaped like a half-moon.

seminal groove. In turtles, the groove that carries sperm from the cloaca to the glans and thence into the female's cloaca.

seminal plug. 1. Dried semen that adheres to withdrawn hemipenes, indicating that a snake is actively producing sperm and, by inference, reproductively active. 2. Dried semen that obstructs the female's reproductive tract or genitalia. (Syn. *copulatory plug*)

seminal receptacle. In general, a cavity or structure for sperm storage. The term is applicable to the *spermatheca* of salamanders ("*receptaculum seminis*" in older literature) and the branched glandular sacs lying in the lumen of the oviduct in some snakes and lizards.

seminal vesicle (= ***vesicula seminalis***). See *ampulla ductus deferens*.

seminatural. *adj.* Reference to habitats or communities that have been modified by human activities.

seminiferous. adj. Producing or carrying semen.

seminiferous tubule. Long, convoluted structures within the testis of amniotes that support developing sperm. See *spermatic ampullae* for comparable amphibian structures.

semiochemistry. *n.* The study of chemical signals that mediate interactions between members of different species. See *pheromone*.

semiotics. *n.* The study of communication.

semioviparity. *n.* (*adj.* **semioviparous**) Reference to a species or reproductive mode that is intermediate between *oviparous* and *viviparous*. See *ovoviviparous*.

semipermeable membrane. A membrane that selectively allows certain molecules, but not others, to pass through it. Commonly used to describe membranes that pass water but not other solute molecules.

semispecies. *n.* 1. Populations that are intermediate between races and biological species. (Syn. *incipient species*). Such populations typically exhibit weak reproductive isolation and partial intergradation. 2. A component species of a *superspecies*.

***semitendinosus* muscle.** A muscle in anurans having a double origin on the posterodorsal and posteroventral rims of the pelvis and extending along the ventral thigh to insert on the tibiofibula. It functions to abduct the femur and pulls it ventrally to flex the knee. It is united with the *sartorius* into a single muscle in most primitive frogs.

semi-terrestrial. *adj.* Living only part of the time on land or at ground level.

senescence. *n.* Physiological deterioration associated with aging of individuals in post-reproductive ages.

senior homonym. In taxonomy, the earlier published of two *homonyms*. Cf. *junior homonym*.

senior synonym. In taxonomy, the earlier published of two or more *synonyms*. Cf. *junior synonym*.

sens. Abbreviation of the Latin *sensu*, meaning "in the sense of."

sensation. *n.* The perception of a sensory stimulus.

sense organ. See *integumentary sense organ*.

sense-plate. *n.* See *stirnorgan*.

sensitivity. *n.* The capacity of an organism to respond to stimuli.

sensor. *n.* A mechanical, electrical, or biological device that responds to a physical stimulus, or change in the immediate environment, and produces a corresponding electrical signal. In animals, sensors are equivalent to receptors and usually involve neural signaling.

sensory. *adj.* Pertaining to a sense organ or sensation.

sensory adaptation. A process by which a sensory neuron or system becomes less sensitive to stimuli during prolonged or repeated stimulation. See also *accommodation*.

sensory cone. A papilla-like sensory structure of amphibian integument, often arising from the center of a pit and derived from *protothrix*.

sensory drive, sensory drive hypothesis. In animal behavior, this term has reference to natural selection for communication signals to stimulate the receiver sensory system effectively, such that the nature of the most effective signal design depends on habitat conditions. Under this assumption, signal diversity evolves because species or populations come to occupy different habitat conditions where selection for effective communication promotes divergence in signal designs.

sensory fiber. An axon that carries sensory information to the central nervous system.

sensory filter. A neuronal circuit that selectively transmits some features of a sensory input and ignores or eliminates other features.

sensory modality. The qualitative feature of a stimulus, or category of sensory receptors that are tuned to receive a particular class of energy (e.g., light, sound, olfaction).

sensory neuron. A neuron that carries responses from a sensory organ to the brain or spinal cord. Cf. *sensory fiber*.

sensory paralysis. See *paralysis*.

sensory pit. See *pit organ*.

sensory reception. Absorption and transduction of stimulus energy by a neuron that produces a receptor potential, leading to action potentials that travel to the central nervous system.

sensu. **(sens.)** Latin, meaning "in the sense of." Used in nomenclature for author citations of misapplied names, preceding the name of the author who misapplies the name.

sensu stricto. **(*sens. str.; s. str.; s.s.*)** Latin, meaning "in a strict or narrow sense."

sentinel organism or **species.** See *bioindicator*.

separate or *separatum*. Offprint.

sepsis. *n.* The presence in the blood or other tissues of pathogenic microorganisms or their toxins.

septal ridge. A ridge-like extension of the lateral walls of the *nasal septum* of some turtles (e.g., *Trionyx*).

septentrional. *adj.* Northern or northerly.

septicemia. *n.* Reference to a disease that is spread systemically via blood circulation by the presence and persistence of pathogenetic microorganisms and their toxic products.

septomaxilla, septomaxillary. *n.* A skull bone located at the posterior edge of the external nostril, usually quite small but may overlie and protect the *vomeronasal organ*. This element is of erratic occurrence. (Syn. *intranasal, narial, nasal, turbinal*)

septosphenoid. *n.* An ossification derived from the chondrocranium, of irregular occurrence and located at a median position in the orbital region anterior to the *basisphenoid*.

septum. (*pl.* **septa**) *n.* A dividing wall or partition, *e.g.* the vertical tissue separating the nasal passages.

septum bulbi. See *spiral fold*.

septum ventriculorum. Used in reference to the *horizontal septum* and the complete wall between the ventricles of crocodilians.

seq. Abbreviation for the Latin *sequens*, meaning "following."

sequela. *n.* (*pl.* **–ae**) A pathologic condition that follows or occurs as a consequence of another condition or event.

sequence affinity. Phylogenetic affinity based on similarities in amino acid sequences.

sequencer. *n.* An apparatus used to determine the sequence of amino acids or other monomers in a biological polymer.

sequencing. *n.* Reference to determination of the order of nucleotide residues of DNA or RNA molecules or fragments, or of the amino acid residues of a protein or polypeptide.

sequential key. An identification key that is constructed as a sequence of alternative choices. (Syn. *dichotomous key*) Cf. *simultaneous key*.

sequestered defensive compounds. **(SDCs)** Toxins, used in chemical defense, that are obtained from the diet or environment and are stored in the consumer's tissues with minimal chemical modification.

ser. Abbreviation for the Latin *series*, meaning "series."

sere. *n.* A stage in the succession of a community.

serial. *adj.* Repeated, as in repetition of body parts.

serial homology. 1. Reference to the resemblance among different members of a single, linearly ordered series of structures within an organism (e.g., vertebrae). Similarity of repetitive structures within the same organism due to common embryonic origin. Cf. *homology*. 2. The correspondence of structures that occupy different spatial positions in a series of like structures.

series. *n.* In taxonomy, the sample of specimens available for study. 2. A category of classification intermediate in rank between *section* and *species*.

serir. *n.* A stony desert. (Syn. *reg*)

seroallantoic placenta. See *chorioallantoic placenta*.

serology. *n.* The diagnostic identification of antibodies in serum, plasma, or other body fluids.

serosa. *n.* The outermost layer of the *alimentary canal*.

serosal. *adj.* Pertaining to the side of an epithelial tissue facing the blood. Cf. *mucosal*.

serotherapy. *n.* The use of *antivenin* (from *serum*) for the treatment of snake bite.

serotinal. *adj.* Pertaining to the late summer. Cf. *aestival, autumnal, hibernal, vernal*.

serotonin. **(5-HT)** *n.* 5-hydroxytryptamine, a neurotransmitter that causes certain smooth muscles to contract rapidly, increases capillary permeability, and plays a metabolic role in the central nervous system. Temperature appears to modulate the circannual variations of 5-HT in the brain of some reptiles (D.C. Wilhoft & W.B. Quay, *Comp. Biochem. Physiol.* 15:325–338, 1965; B. Vivien-Roels & A. Petit, *Gen. Comp. Endocrinol.* 34:77, 1978).

serous. *adj.* Pertaining to clear, watery fluid derived from blood plasma, without cells or formed elements.

serous gland. An organ secreting a thin, watery fluid.

serpent. *n.* A literary and common name for a snake.

serpentarium. *n.* Reference to a building or complex of structures used to house a collection of live snakes. These are usually held for the purpose of public display, breeding, extraction of venom, or other scientific purposes. (Syn. *snake farm*)

Serpentes. *n.* A clade (suborder) of the order *Squamata* that includes all extant snakes, some 2900 species represented in two infraorders. (Syn. *Ophidia*)

serpentiform. *adj.* Shaped like a snake.

serpentine locomotion. See *lateral undulation*.

serrate, serrated. *adj.* Notched or having a sawlike edge.

serration. *n.* The state of being *serrated*; or a serrated structure.

Sertoli cells. Elongated somatic cells within seminiferous tubules of the testes that function in nurturing developing sperm.

serum. *n.* The clear, liquid fraction of clotted blood after cells and coagulated elements have been removed. See also *antivenin*.

serum antivenenosum. A term for *antivenin* used by P.A. Christensen, *South African Snake Venoms*, p. v, 1955.

sesamoid, sesamoid bone. *n.* A small, nodular-shaped bone embedded in a tendon, fibrocartilage, or joint capsule where there is exposure to pressure. Reference to sesamoid elements in frogs are also known as "*heterotopic skeletal elements*" (R.A. Nussbaum, *Herpetologica* 38:312–320, 1982) or *extraskeletal bones* (from W.M. Olson, *Zoology* 103:15–24, 2000).

sessile. *adj.* Nonmotile.

sessile hydatid. Functionless vestige of the infundibular aspect of the oviduct, located near the testis of male amniotes. (Syn. *appendix testis, hydatid of Morgagni*)

seta. (*pl.* **setae**) *n.* 1. Generally, a bristle-like structure. 2. See *subdigital setae* and Fig.32.

set point. The state to which a negative feedback control system tends to return when perturbed by an environmental disturbance.

set point temperature, set point temperature range. *n.* 1. A specific level of body temperature at which a thermoregulatory behavior is elicited, either heat-seeking or shade-seeking. Effectively, the "thermostat" setting (or settings) within the thermoregulatory center of the brain (*hypothalamus*) 2. The central percentage (e.g. 50%, 68%) of all observations of body temperatures recorded from captive reptiles or amphibians held in a thermal gradient (P.E. Hertz et al., *Amer. Naturalist* 142:796–818, 1993)

Sewall Wright effect. See *genetic drift*.

sex. *n.* Generally, any process that recombines genes from more than one source in a single organism. Characteristically, sex among amphibians and reptiles involves two individuals that produce unlike haploid gametes whose union (fertilization) restores a diploid condition and produces a new individual.

sex bob. A bobbing display characteristic of male *Sceloporus* lizards, during which the lateral body is compressed to reveal blue ventral coloration while raising and lowering the body (*push-ups*) using the forelegs.

sex call. Vocalization employed by male anurans at breeding sites, used to attract and stimulate conspecific females.

sex cell. See *gamete*.

sex chromosome. Homologous chromosomes that are dissimilar in the heterogametic sex.

sex-conditioned character. A phenotypic character that is shaped or influenced by the sex of the individual.

sex cord. Aggregates of germinal epithelium that produce seminiferous tubules, ampullae of testes, and medullary cords of ovary. *Primary sex cords* grow inward into the gonad and contain germ cells as well as supporting elements. Primary cords become prominent in male testes and give rise to spermatogenic tissue, whereas in females the primary cords (also called *medullary cords*) tend to degenerate while eggs eventually develop in secondary sex cords (also called *cortical cords*).

sex determination. The genetic or developmental mechanism by which sex is determined in a given species. See *genetic sex determination* and *temperature-dependent sex determination*.

sex differentiation. The process of diverse development of characters in male and female individuals.

sex hormone. Reference to any hormone that is produced by, or influences, the gonads (includes *gonadotropins, estrogens,* and *androgens*).

sex index. A ratio of sexually dimorphic characters that allows sex recognition in frogs of the genus *Rana*. The ratio of body length to diameter of the tympanum is generally below 10 for mature males and above 10 for females and subadult frogs. First used by B. Martof (*Amer. Midl. Nat.* 55:101–117, 1956).

sex limited character. A phenotypic character that is expressed in only one sex.

sex linkage. Reference to a condition in which a gene that produces a certain phenotypic character is located on a sex chromosome.

sex ratio. The ratio of males to females within a population, sometimes expressed in percent at fertilization (*primary sex ratio*), at birth (*secondary sex ratio*), and at sexual maturity (*tertiary sex ratio*).

sexual behavior. All activity that promotes or leads to reproduction.

sexual dichromatism. A difference of color pattern between intraspecific males and females.

sexual dimorphism. A condition in which male and female members of a species exhibit distinct differences in one or more characters, usually size, color, or structure.

sexual embrace. See *amplexus*.

sexual heteromorphism. Different numbers or morphology in the chromosome complement of the two sexes.

sexual homology. The correspondence of male and female structures that develop from identical embryonic primordia.

sexual kidney. Reference to the anterior, ribbon-like part of the kidney of salamanders, derived from the *mesonephros* and distinct in development from the *definitive kidney*.

sexual reproduction. The fusion of nuclei from haploid *gametes* derived from *meiosis* in two different individuals.

sexual segment. Part of the individual renal tubule in the kidney of squamate reptiles, preceded by a thin tubular portion and connected to the collecting tubules by a short terminal segment. The structure may be encapsulated by connective tissue and smooth muscle and secretes mucus in addition to spherical granules during sexual activity in males. (Syn. *preterminal segment*)

sexual selection. A theory proposed by C. Darwin, which holds that the frequency of traits can increase or decrease depending on the attractiveness or fertility of the bearer. It proposes a selective force that arises when certain individuals gain an advantage in mating, as occurs most commonly in males, producing, for example, larger body size. This is sometimes referred to as *intrasexual selection*. In *epigamic selection*, the female is the active selective agent and chooses a male from among a field of genetically variable males.

sex-warning vibration. See male *release call*.

Seychelles Frogs. Vernacular name for species of anurans belonging to the family Sooglossidae, endemic to the Seychelles Islands.

shaft. *n.* See *neck* (of tooth).

shagreened. *adj.* Rough to the touch; covered with numerous, small tubercles.

shaker. *n.* A composite bone formed from coalesced vertebrae at the tip of the tail, situated within the *matrix* and *rattle* of Rattlesnakes. (Syn. *style, urostyle,* as used by H. Gadow, *Amphibians and Reptiles,* p. 23, 1901)

shaker muscle. Specialized tail muscles used to shake the rattle of Rattlesnakes. This muscle is characterized by having numerous mitochondria, rich supply of blood vessels, and biochemical properties that confer rapid twitch cycles. Some Rattlesnakes have been shown to rattle continuously for periods as long as three hours, and at frequencies up to 90 Hz while at warmer body temperatures (35 °C).

sham. *n.* A surgical procedure that mimics another but lacks some aspect or experimental manipulation of the other and thereby serves as a control for comparison of effects on an animal. E.g., if a drug-releasing pellet was implanted subcutaneously into an animal via a trocar, a sham would repeat the same invasive procedure using a trocar on another animal, but without actually implanting the pellet.

shank. *n.* See *tibial segment.*

Shannon-Weaver index. See *Shannon-Weiner index.*

Shannon-Weiner index (H') A univariate index of diversity based on information theory, used as an index of biodiversity in a sample or aggregation. Also called *Shannon-Weaver index.*

Shaw's Sea Snake. Vernacular name for the elapid Sea Snake species *Lapemis curtus.*

Sharptail Snakes. Vernacular name for American colubrid snakes belonging to the genus *Contia.*

shear, shearing. *n.* Stress in an elastic body that results from loads acting in opposite directions along parallel, closely adjacent lines. The term also refers to movement in which parts of a body slide relative to one another and parallel to the forces being exerted. See Fig. 24.

shear stress. Stress due to forces that tend to cause movement or strain parallel to the direction of the forces. See *stress* and Fig. 24.

sheath. *n.* 1. Generally, a protective and usually membranous covering. 2. The term has been used to denote the membrane surrounding the base of retractile fangs and with reference to the *anal flap* of hylid frogs. 3. The hollow, basal part of a *labial tooth* embedded in the tissue of the tooth ridge of a tadpole.

shed. *v., n.* 1. To slough or cast off a dead (*stratum corneum*) outer generation of epidermis. 2. Sometimes this word is used as a noun to refer to the cast skin (epidermis) after it has been shed. (Syn. *desquamate, exuviate, molt, slough*) See also *ecdysis*.

shedding. *n.* See *ecdysis.*

Sheep Frog. Vernacular name for the Middle American microhylid species *Hypopachus variolosus.*

shell. *n.* 1. The enclosing, rigid covering of a turtle, consisting of bone and scale and including the *carapace, bridge,* and *plastron*. 2. The outer covering of amniote eggs, either flexible and distensible or rigid with calcification.

shell-breaker. See *caruncle.*

shell length. See *straight length* and Fig.9.

shell pit. Reference to any small, rounded hole in the bony and epidermal shell of turtles. Usually found in terrestrial genera and possibly of mycotic origin.

shell width. Generally, a measurement of turtle shell width, taken as the distance between imaginary perpendiculars erected at the sides of a turtle opposite the suture between the second and third costals, or between perpendiculars touching the sides of the shell between front and hind legs, including a *front straight width* and a *rear straight width,* or the maximum straight line distance across the carapace regardless of where it occurs. See Fig. 9.

she-male. A male within a *mating ball* that releases a pheromone attracting other males, as though the snake was a female. Because this distracts other nearby males, the phenomenon gives the "she-male" a competitive advantage in courtship, and such males have been shown to mate successfully with females more often than do normal males.

shield. *n.* A large scale such as the head *plates* of snakes, or any of the *laminae* of turtles.

Shieldtail Snakes, Shield-tailed Snakes. Common name for various species of snakes in the family *Uropeltidae.*

shifting balance theory. A theory proposed by Sewall Wright that considers biological evolution to proceed most rapidly when subpopulations of a species remain isolated for sufficient time to evolve distinctive adaptations that broaden genetic diversity when gene flow is reestablished among the populations.

shimmy burial. Synonym for *sand-swimming.*

Shingleback. *n.* Common name for a well-known Australian Skink (*Tiliqua rugosa*) known to form pair bonds between males and females that last for many years.

Shinisauridae. *n.* A monotypic clade (family) having a single species that is known from China and placed by some within the family *Xenosauridae*.

short-circuit current. A technique developed by H.H. Ussing and K. Zerahn (*Acta Physiol. Scand.* 23:110–127, 1951) to demonstrate active ionic transport across amphibian skin. They placed isolated frog skin between two chambers with amphibian's Ringer's solution on both sides. An equal and opposite potential difference provided by an external circuit was used to null the inside positive transepithelial potential. The current required to bring the spontaneous potential difference to zero is known as the *short-circuit current* and equals the algebraic sum of the fluxes of all ions moving across the skin. Ussing and Zerahn demonstrated that the short-circuit current is equivalent to the net sodium flux in the inward direction.

Short-headed Frog. Vernacular name for species of African anurans belonging to the microhylid genus *Breviceps*.

Short-tailed Snake. Vernacular name for the American colubrid species *Stilosoma extenuatum*. Also a collective vernacular name for snakes belonging to the Uropeltidae.

shoulder spine. Reference to spines on the proximal outer margins of a crotalid *hemipenis*, in contrast to the *mesial spines* (L.M. Klauber, *Rattlesnakes*, p. 694, 1956).

shovel. *n.* A dark, keratinized protuberance on the margin of the hind foot, used in digging by burrowing anurans. (Syn. *digger, scraper, spade, spur*)

Shovel-headed Treefrogs. Vernacular name for species of frogs belonging to the Middle American genus *Triprion*.

Shovel-nosed Frogs. Collective vernacular name for frogs belonging to the African family *Hemisotidae*, and especially the genus *Hemisus*. Shovel-nosed Frogs are unmistakable, being squat, depressed, and somewhat bloated in appearance with a small head and prominently pointed snout that is hardened for digging. The lower jaw is recessed.

Shovelnose Snake. Vernacular name for North American colubrid snakes belonging to the genus *Chionactis*.

shrubland. *n.* Reference to habitats such as *chaparral, heath*, etc. where shrubs dominate the landscape.

Shumway stages. A descriptive scheme for developmental stages of tadpoles based on normal development of *Rana pipiens* larvae (W. Shumway, *Anat. Rec.* 78:139–147, 1940). See *stage*.

shunt. *n.* Generally, and literally, an alternate route. In physiological and herpetological literature, the term is usually used in connection with alternate routes for blood flow. Shunting of blood flow in amphibians and reptiles may involve alternate routes outside the heart, *extracardiac shunts*, or modification of blood flow from the heart to either pulmonary or systemic circulations, so-called *intracardiac shunts* made possible because of the incompletely divided ventricle in amphibians and non-crocodilian reptiles. See *intracardiac shunts*.

SI. See *international units*.

sib. *n.* A jargonish, shortened form of *sibling*.

sibling. *n.* Progeny of the same parents.

sibling species. Reference to species that are reproductively isolated but nearly identical in morphology. (Syn. *cryptic species*)

siccicolous. *adj.* Living in dry, sandy deserts.

SID. See *strong ion difference*.

Side-blotched Lizard. Vernacular name for American species of lizards belonging to the iguanid genus *Uta*.

side branch. See *collateral*.

Side-necked River Turtles. A collective term for American species belonging to the *Podocnemididae*.

Side-necked Turtles, Sideneck Turtles. Collective vernacular name for pleurodire turtles belonging to the families *Chelidae, Pelomedusidae*, and *Podocnemidae*.

side organ. See *lateral line system*.

sidereal day. See *day*.

sidereal time. Time measurement based on rotation of Earth relative to distant stars. Cf. *solar time*. See *day*.

Sidewinder or **Sidewinder Rattlesnake.** *n.* Vernacular name for the American pitviper *Crotalus cerastes,* named for the mode of locomotion over desert sands that are the primary habitat of this species.

sidewinding. *n.* A mode of limbless locomotion associated with low-friction or shifting substrates such as desert sands. Sections of the body are alternately lifted, moved forward, and then placed down on the substrate, producing a series of separate, parallel tracks, each oriented at an angle to the direction of travel. During these movements the snake is in static contact with the ground at two points. When a body segment touches down on the substrate, the remaining posterior part of the snake essentially "rolls out" to contact the substrate as the snake moves forward (see Fig. 26). Virtually all forces directed against the substrate act vertically, which avoids the slipping that would result if the body contacted the ground at an angle. Most or all snakes are probably capable of sidewinding, however the behavior is best known as the normal mode of movement in desert vipers such as the Sidewinder (*Crotalus cerastes*). Cf. *saltation, slide-pushing*.

siemen. (*S*) A unit of electrical conductance, equivalent to the reciprocal of the *ohm*.

Sierra Garter Snake. Vernacular name for the American colubrid species *Thamnophis couchii*.

sievert. (Sv) The *SI* unit that replaces the *rem* as the measure of radiation dose equivalence. 1 Sievert = 100 rem.

Sigma. 1. (Σ) The summation of all quantities following this symbol. 2. (σ) The symbol denoting the standard deviation of a population.

sigmoid curve. An S-shaped curve.

sigmoid notch. The articular surface on the medial head of the reptilian *ulna*, which fits onto the trochlear condyle of the humerus.

signaling theory. With reference to ethology, a body of theoretical work that examines communication between individual animals. See *honest communication* or *signaling*.

signal-to-noise ratio. The relation between a true biological signal (event or process) and the random background activity that arises as a result of incidental variation in kinetic energy or other irrelevant events.

signal transduction. See *cellular signal transduction*.

significance, level of. In statistics, the probability of erroneously rejecting a true null hypothesis.

sign stimulus. In ethology, any behavior or physical attribute that acts as a stimulus to elicit a specific behavior or behavioral sequence in another member of the same species (*fixed action pattern*). (Syn. *key stimulus, releaser*)

silent mutation. A gene mutation that has no observable consequence at the phenotypic level.

silt. *n.* Sediment composed of cohesionless particles with a diameter of 1/256 to 1/16 mm.

Silurian. *n.* The third Period of the Paleozoic Era, extending from the end of the Ordovician Period to the beginning of the Devonian Period. See Table 1.

silvicolous. *adj.* Inhabiting woodland.

similarity index. A measure of the similarity in species composition of two communities. Several methods have been developed to express such similarity.

simple tongue extension. Protrusion of the tongue, sometimes touching the substrate, without oscillatory tongue movements.

Simpson index, Simpson's diversity index. (D) An index of diversity based on the probability of picking two organisms at random that are different species, calculated as:

$$D = 1 - \sum (P_i)^2$$

where P_i is the proportion of individuals of species *i* in a community of *s* species. The index is often used to quantify the biodiversity of a habitat. It takes into account the number of species present as well as the relative abundance of each species. The Simpson index was first proposed by the British statistician Edward Hugh Simpson (*Nature* 163:688, 1949).

simulation. *n.* The imitation of the behavior of a system by a model.

simultaneous key. An identification key in which the unknown is compared simultaneously with all taxa, thereby securing an unambiguous identification in a single step. Cf. *sequential key*.

simultaneous parsimony analysis. See *combined parsimony analysis*.

sincipital, syncipital. *n.* Collective reference to any scale called a *frontal* or a *parietal*.

sinciput. *n.* The front part of the head, in contrast with *occiput*. In squamate reptiles, this term is defined as the area anterior to a line connecting the posterior border of the orbits.

sine. Latin, meaning "without."

sinistral. *adj.* Of or pertaining to the left side. This term is often used in relation to location of the spiracle of tadpoles. Cf. *dextral*.

sink. *n.* In ecological contexts, a buffering reservoir and repository for energy or matter in which the input is very large relative to the output.

sinkhole. *n.* A closed depression found on land surfaces underlain by limestone.

sink population. A population in which the local birth rate is less than the local death rate and the immigration rate is greater than zero. Such populations depend on external immigration for their persistence.

sinuate. *adj.* Curved; having a curved border.

sinus. *n.* 1. A recess, cavity, or channel in an organ or part of a body. 2. A dilated part of a blood vessel.

sinus hyobranchialis. See *hyobranchial sinus*.

sinus lateralis nasi. A lateral diverticulum of the nasal cavity of urodeles, bearing a sensory epithelium innervated by a branch of the olfactory nerve. This organ may function similarly to the vomeronasal organ, but the two structures are not homologous. (Syn. *recessus maxillaries*) See also *blindsack, prechoanal sac*.

sinusoid. *n.* An expanded capillary, as found in the liver and certain glands.

sinusoidal. *adj.* Wavy or tortuous.

sinusoidal locomotion. See *lateral undulation*.

sinus venosus. A membranous chamber that receives venous blood and transmits it to the atrium, to which it is attached. The most posterior of the primitive heart chambers.

siphonium. *n.* A term used by H. Gadow (*Amphib. Rept.*, p. 437, 1901) to denote a narrow tube of connective tissue running between the middle ear and the large empty space enclosed within the lower jaw of crocodilians.

sire. *n.* The male parent in animal breeding terminology. Cf. *dam*.

siren, sirenid. *n., adj.* A member of, or pertaining to, the Sirenidae.

Sirenidae. *n.* A clade (family) that includes four species of long, slender North American salamanders (*Pseudobranchus* and *Siren*) lacking pelvic girdles and hind limbs. These amphibians are highly aquatic, paedomorphic, and eel-like in appearance.

Sirenoidea. *n.* A clade (superfamily) of ancient salamanders that includes the *sirens*.

Sirens. *n.* Vernacular name for several species of salamanders belonging to the North American sirenid genus *Siren*. These are paedomorphic, eel-like, freshwater salamanders with prominent external gills and no hind legs. Cf. *Dwarf Sirens*.

sister groups or **sister taxa.** A species or higher monophyletic taxon hypothesized to be the closest genealogical relative of a given taxon. In systematics, this term has reference to two groups having the same immediate common ancestor. Sister taxa are derived from an ancestral species not shared by any other taxon, and they appear on a cladogram as lineages that arise as branches from a single divergence node. Cf. *outgroup*.

sister species. The species with which another species shares a most recent common ancestor.

sit-and-wait predator. See *ambush predator*.

site occupancy. A metric used in ecology and conservation biology based on collection of simple presence/absence data across a landscape to make inference regarding species status. Such models have become widely adopted in surveys of many taxa and allow for imperfect species detection or "false negative" observations. The method has been identified by the Amphibian Research and Monitoring Initiative as the only metric that so far meets the program criteria for being nationally interpretable and regionally adaptable. See D.I. MacKenzie et al., *Occupancy Estimation and Modeling: Inferring Patterns and Dynamics of Species Occurrence*, 2006. (Syn. *proportion of area occupied*)

SI Units (Système International d'Unités) An internationally agreed system of metric units comprising seven basic units: meter, kilogram, second, ampere, Kelvin, candela, and mole. Other metric units may be derived units or supplementary units. See Table 2.

Six-lined Racerunner. Vernacular name for the American teiid species *Cnemidophorus* (= *Aspidoscelis*) *sexlineatus*.

size refuge. The phenomenon whereby predation is avoided by attaining a size that reduces or eliminates threats from gape-limited predators.

skeletalchronology. *n.* Reference to the use of *annuli*, *lines of arrested growth*, or other features of bone to infer past changes through time (growth rates, age, etc.).

skeletal growth mark. See *lines of arrested growth (LAGs)*.

skeletal muscle. Striated muscle that is activated by nervous stimulation and moves the various body parts of an animal. Cf. *smooth muscle*.

skeletogenesis. *n.* The growth or formation of the skeleton.

skeleton, skeletal system. *n.* The hard or bone framework of the body of animals, consisting of *cranial, axial,* and *appendicular* elements.

skewness. *n.* In statistics, the departure from symmetry of a population. Positive or negative skewness indicates the presence of a long thin tail on the right or left of the distribution, respectively.

skin. *n.* The anatomical and functional outer covering of an animal, comprised of *dermis* and *epidermis* in vertebrates. (Syn. *integument*) See Figs. 28–31.

skin blister. General reference to various pathological conditions resulting in abnormal protrusions of the skin. See *bleb, bulla, dermatitis*.

skin friction. In biomechanics, the component of drag force due to interactions or shearing forces of viscous fluid acting within the boundary layer adjacent to an object having motion relative to the fluid. (Syn. *viscous drag*)

skin glands. General reference to any glandular structure of the integument.

Skink. *n.* Collective vernacular name for numerous lizard species belonging to the family *Scincidae*.

Skink Tegus. Vernacular name for gymnophthalmid lizards belonging to the genus *Leposoma*. Note these lizards are neither true Skinks nor true Tegus.

skin resistance. (R_s) See *resistance*.

skin retention. Retention of successive layers of the dead outer layer of the epidermis to form shields of the carapace and plastron in turtles and the dorsal scutes of crocodilians.

skin shedding. See *ecdysis*.

skin slips. See *scalares*.

skull. *n.* See *cranium*.

skull segments. See *maxillary segment, occipital segment, stapes segment*.

skull table. The flattened, dorsal roof between the cheek regions in the skull of reptiles.

Sleepy Lizard. See *Shingleback*.

Slender Blind Snake. Collective vernacular name for species of snakes belonging to the family Leptotyphlopidae, and especially the genus *Leptotyphlops*.

slenderness index. See *body condition index*.

Slender Salamanders. Collective vernacular name for plethodontid species of salamanders belonging to the genus *Batrachoseps*.

Slender Skinks. Vernacular name for Skinks belonging to the subfamily Lygosominae.

Slender Snakes. Vernacular name for species of Asian colubrid snakes belonging to the genus *Trachischium*.

slide-pushing. *n.* A mode of limbless locomotion employed on low-friction substrates, involving alternating waves of body motion similar to lateral undulation, but without use of fixed points of the substrate to generate forward reaction forces. Instead, backward moving body waves are propagated so rapidly that they generate a sliding friction sufficient to produce a reaction force that propels the snake forward. The head points in the direction of travel as the snake gradually moves forward, even though it presents an appearance of flailing and slippage. Term introduced by C. Gans (*Symp. Zool. Soc. Lond.*, No. 52, pp. 13–26, 1984).

Sliders, Slider Turtles. Vernacular name for species of Middle American emydid turtles belonging to the genus *Trachemys*.

sliding friction. See *friction*.

slime gland. See *mucous gland*.

slip face. The steep, downwind slope of a dune, formed from loose, cascading sand that generally keeps the slope at the angle of repose (about 34°).

slough. *v., n.* 1. The act of ecdysis or shedding the skin; also used to denote the shed skin that results following removal from the body. (Syn. *shed*) 2. A swamp, bog, or marsh, especially one that is part of a flow way, inlet, or backwater.

sloughing. *n.* See *ecdysis*.

slow twitch fiber. See *twitch fibers*.

Slow Worm. Vernacular name for the anguid lizard *Anguis fragilis*.

slug. *n.* Jargonish reference to an unfertilized ovum passed by a female reptile.

Slug-eaters, Slug-eating Snakes. Collective vernacular name for several species of neotropical colubrid snakes belonging to the dipsadine genera *Sibon* and *Sibynomorphus*, (also called *Tree Snakes*), and the African colubrid genus *Duberria*. These snakes are specialized predators on slugs and snails. See also *Snail-eating Snakes* and *Slug-snakes*.

Slugsnakes, Slug Snakes. *n.* Collective vernacular name for about 20 species of Asian colubrid snakes belonging to the pareatine genera *Pareas* and *Aplopeltura*. These snakes are specialized predators on mollusks, including snails, and have evolved some features that parallel those of the New World dipsadine *Snail-* and *Slug-eating Snakes*. Also called *Blunthead Slug Snakes, Snail-eaters*.

Small-toed Geckos. Vernacular name for various species of Old World Geckos belonging to the gekkonid genus *Cnemaspis*.

Smith's fluid. A fixative for amphibian eggs, highly recommended by R.S. Rugh (*Experimental Embryology*, p. 16, 1962).

Smoky Jungle Frog. See *South American Bullfrog*.

Smooth Green Snake. Vernacular name for the American colubrid species *Liochlorophis vernalis*.

Smooth Horned Frogs. Vernacular name for South American species of leptodactylid anurans belonging to the genus *Proceratophrys*.

smoothing. *n.* Reference to averaging of data in time or space so as to compensate for random errors or variation on a scale smaller than that presumed to be significant with respect to a given problem.

smooth muscle. Nonstriated muscle characterized by having heterogeneously distributed myofilaments and association with glands and internal organs. Cf. *skeletal muscle*.

smooth scale. Reference to a scale of squamate reptiles that lacks prominent *keels* or central ridges, detectable by touch or by eye. Cf. *keeled scale*.

Smooth-scaled Water Snake. Vernacular name for species belonging to the Asian homalopsine genus *Enhydris*.

Smooth Snake. Vernacular name for species of colubrid snakes belonging to the genus *Coronella*.

SMR. See *standard metabolic rate*.

Snail-eating Snakes, Snail-eaters. Vernacular name for several species of tropical colubrid snakes belonging to the Neotropical dipsadine genus *Dipsas* and the Asian pareatine genus *Pareas*. These snakes are specialized predators on slugs and snails. See also *Slug-eaters, Slug-eating Snakes*.

snake. *n.* A member of the *Serpentes* (~2900 living species) anatomically distinct from lizards and, like amphisbaenians, part of a lineage that has undergone body elongation and limb reduction. Snakes are one of the most rapidly diversifying groups of vertebrates, ranging from very small wormlike Blindsnakes to giant constrictors such as Boas and Pythons, and including fossorial, terrestrial, arboreal, and marine forms.

snake bite. Reference to a bite from a snake, usually in context of defensive or accidental bites to humans.

snake bite serum. See *antivenin*.

snake charmer. A person who carries one or more snakes in a basket and entertains by playing a musical instrument while swaying to focus a snake's attention as it rises from the basket. This phenomenon is known mainly from various parts of Asia, and the snake is usually a cobra, which maintains an upright posture with the hood spread. (Syn. *snake wallah*)

snake farm. See *serpentarium*.

snake hook or **stick.** An instrument consisting generally of a pole or long handle with a piece of L-shaped metal at the end, used for picking up, transferring, or pinning venomous or other snakes, which are thus held away from the body.

Snake Lizard. Vernacular name for gekkonid lizards belonging to the genus *Lialis*.

Snakeneck Turtles. Vernacular name for Asian and South American species of chelid turtles belonging to the genera *Chelodina* and *Hydromedusa*.

snakestone. *n.* Colloquial name for a porous object such as chalk or charred bone, used for the treatment of snake bite by application to the affected area of skin with the belief that the venom is drawn out of the wound. See also *black stone*.

snake venom antitoxin. See *antivenin*. (This term was used by the British Pharmaceutical Code.)

snake wallah. A name used for a *snake charmer* in India and Pakistan.

Snapping Turtle. Collective vernacular name for species of turtles belonging to the American chelydrid genera *Chelydra* and *Macroclemys*, and also Australian chelid species belonging to the genus *Elseya*. See also *Alligator Snapping Turtle*.

snaring. *n.* See *noosing*.

sneaker male. Nonterritorial males that mimic females and make periodic forays into territories of other, usually dominant, males and copulate with their female.

snorkel. *n.* Reference to an elongated snout used for breathing while the body is below the water surface in some species of aquatic turtles.

snout. *n.* The anterior part of the head which includes the nostrils. (Syn. *muzzle*)

Snouted Treefrogs. Vernacular name for numerous Middle and South American species of hylid frogs belonging to the genus *Scinax*. Also introduced in the West Indies.

snout-vent length. (SVL) A measure of body length representing the distance from the tip of the snout to the anal opening or some defined border of the vent. Usually synonymous with *body length*, *head-body length*, or *standard length*. See Fig. 17.

social behavior. Any association and interaction between two or more members of a given species for a period of time, in circumstances other than the usual male-female interaction for reproduction.

social cohesion hypothesis. A hypothesis relating behavior to dispersal patterns, proposing that social interactions prior to emigration constitute the principal factor that determines dispersal patterns rather than agonistic interactions at the time of emigration.

social dominance. A phenomenon in which an animal achieves a priority of access to a limited resource over another individual (or individuals), usually through winning aggressive interactions and attaining stability of submissive behaviors on the part of social subordinates.

sociality. *n.* The tendency to form social groups.

social subordination hypothesis. The proposition that as population density increases, increased intraspecific competition leads to increased aggression and results in subordinate individuals being forced to disperse to suboptimal habitats.

Society for the Study of Amphibians and Reptiles. (SSAR) A nonprofit society established in 1967 to advance research, education, and conservation related to amphibians and reptiles. The Society was founded by J.T. Collins, C.J. Hirschfeld and K. Adler as an outgrowth of the Ohio Herpetological Society and publishes the scholarly *Journal of Herpetology, Herpetological Review, Catalogue of American Amphibians and Reptiles,* and various other publications that are issued at irregular intervals. Current URL: http://www.ssarherps.org/

sociobiology. *n.* Study of the biological basis of social behavior, using evolutionary theory as a foundation for the study and interpretation of social behavior in humans and other animals. This relatively new discipline was catalyzed by the work of E.O. Wilson and especially publication of his book *Sociobiology: The New Synthesis* (1975).

sociogram. *n.* A catalogue of all the social behaviors of a species.

sodium-potassium pump. (Na^+/K^+ pump) A membrane protein that functions to maintain asymmetric concentrations of Na^+ and K^+ ions across the plasma membrane, actively extruding Na^+ from the cell and taking up K^+ from extracellular fluids at the expense of metabolic energy.

soft-shell. *n.* Reference to a condition of calcium deficiency in captive turtles, producing an abnormally soft or pliable shell.

Softshells, Softshell Turtles. Collective name for various species of trionychid turtles (e.g., genus *Apalone*). These are highly aquatic turtles characterized by a leathery skin and a flattened, reduced shell lacking peripheral bones.

software. *n.* Programs and programmed operating routines used with computers. Cf. *hardware*.

soil. *n.* A layer of weathered, unconsolidated material on top of bed rock, often defined as containing organic matter and being capable of supporting plant growth.

soil horizon. Any of the distinctive layers of soil that are distinguishable by characteristic physical or chemical properties.

Sokolanura. *n.* A newly designated monophyletic taxon sister with *Xenoanura* within the *Lalagobatrachia* and containing *Costata* and *Acosmanura*. See D.R. Frost et al., *Bull. Amer. Mus. Nat. Hist.* 297:1–370, 2006.

solar day. See *day*.

solarimeter. *n.* An instrument used to measure the amount of solar radiation reaching Earth's surface. (Syn. *pyranometer*)

solar radiation. Radiant energy emitted by the sun. Solar energy reaching Earth's surface ranges in wavelength from about 290–3000 nm.

solar time. Time measurement based on the rotation of Earth relative to the sun. Cf. *sidereal time*. See also *day*.

sole. *n.* The bottom of the foot.

solenoglyph. *n.* (*adj.* **-ous**) A snake or condition in which each fang is erected by rotation of a reduced maxilla on the prefrontal bone (*viperids* and *atractaspidids*). The maxilla bears no teeth other than a tubular hollow fang, and the nonerected fang can be folded against the roof of the mouth. Cf. *proteroglyph, opisthoglyph, aglyph*.

solitary. *adj.* Living alone or by itself.

solubility. *n.* The tendency for a substance (*solute*) to mix with a liquid (*solvent*) to produce a homogeneous mixture (*solution*).

solubility coefficient. See *Bunsen solubility coefficient*.

solute. *n.* A substance that is dissolved in a *solution*.

solution. *n.* A homogeneous liquid mixture of two or more substances.

solvation. *n.* The clustering of solvent molecules around a solute.

soma. *n.* Generally, the body. Often used in reference to the cell body of a *neuron*.

somatic. *adj.* Pertaining to the body's "framework," usually the skeleton, muscles, and skin, but not the viscera.

somatic cells. Cells of an organism's body other than the germ cells. Somatic cells are usually diploid (2N).

somatic nervous system. That part of the nervous system that mediates conscious perception and controls voluntary activity in the skeletal muscles. (Syn. *voluntary nervous system*)

somatic peritoneum. See *peritoneum*.

somatomedin. *n.* See *insulin-like growth factor-1*.

somatopleure. *n.* A membrane derived from *ectoderm* and the outer sheet of the *hypomere*.

somatosensory system. Collectively, all proprioceptive neurons and neurons receiving stimulation from the skin.

somatostatin. *n.* A hypothalamic neurohormone that inhibits growth hormone release from the pituitary. (Syn. *GH-inhibiting hormone, GIH*)

somatotropin. *n.* See *growth hormone*.

somite. *n.* One of many paired block-like masses of embryonic *mesoderm* arranged segmentally alongside the nerve tube of a vertebrate embryo and developing to form segmental musculature and the vertebral column. (Syn. *epimere*) See *dermatome, myotome, sclerotome*.

sonicate. *v., n.* To subject a biological sample to ultrasonic vibration in order to fragment cells, macromolecules, or membranes. Also a term given to a biological sample that has been subjected to such treatment.

sonogram. *n.* A graphic display of the frequency characteristics of a sound, plotting frequency as a function of time. An image of a structure that is produced by *ultrasonography*.

Sonoran Coral Snake. Vernacular name for American Coral Snake species *Micruroides euryxanthus*. Also called *Western Coral Snake*.

Sonoran Green Toad. Vernacular name for the American bufonic species *Bufo retiformis*.

Sonoran Herpetologist. A publication of the Tucson Herpetological Society.

Sonoran region. A zoogeographical region comprising northern Mexico and the southern parts of North America.

Sonoran Whipsnake. Vernacular name for the American colubrid species *Masticophis bilineatus*.

Sooglossidae. *n.* A clade (family) of anurans including small, terrestrial frogs having arciferal pectoral girdles and partially fused epicoracoid cartilages. The relationships of sooglossids have been controversial. Currently three genera and five species are recognized from the Seychelles Islands. In recent taxonomic revisions, sister with *Notogaeanura* within the *Hyloides* (D.R. Frost et al., *Bull. Amer. Mus. Nat. Hist.* 297:1–370, 2006).

Sørensen coefficient. (S) A measure of the similarity between the species composition of two communities using only binary data (presence or absence), usually expressed as a percentage and calculated as

$$S = 2c/(a + b)$$

where a and b are the numbers of species in communities A and B respectively, and c is the number in common to both.

sorption. *n.* The processes of *adsorption* and *absorption*, collectively.

source pool. See *source population*.

source population. A population in which the local birth rate exceeds the local death rate, and the emigration rate is greater than zero. A *source pool* refers to the number of species in a mainland or source area that can potentially colonize an island.

source-sink model or **system.** A system with habitat-specific demography, such that species diversity (or number of individuals), especially in tropical forests, increases when restricted localities favorable to certain species (or individuals in a population) allow them to produce a surplus of emigrants and, hence, become a source of new individuals that disperse to less favorable sites nearby, the sinks. Based on R.D. Holt (*Theor. Population Biol.* 28:181–208, 1985; H.R. Pulliam, *Am. Nat.* 132:652–661, 1988).

South American Bullfrog. Vernacular name for the leptodactyline frog *Leptodactylus pentadactylus*. Also known as the *Smoky Jungle Frog*.

South American Rattlesnake. Vernacular name for *Crotalus durissus terrificus*.

Southern blotting. In molecular biology, a method for transferring DNA segments resolved from electrophoresis from an agarose gel to a nitrocellulose filter. The DNA segment of interest is subsequently probed with a radioactive, complementary nucleic acid, and its position is determined by autoradiography. The technique was developed by E.M. Southern in 1975. Cf. *northern blotting, western blotting*.

Southern Coral Snake. Vernacular name for the Coral Snake species *Micrurus frontalis*.

Southern Pacific Rattlesnake. See *Pacific Rattlesnake*.

Southern Spadefoot. Collective vernacular name for American scaphiopodid species belonging to the genus *Scaphiopus*. Cf. *Western Spadefoot*. See also *Spadefoot Toad*.

Southern Toad. Vernacular name for the American bufonid species *Anaxyrus* (formerly *Bufo*) *terrestris*. For taxonomic revision see D.R. Frost et al., *Bull. Amer. Mus. Nat. Hist.* 297:1-370, 2006.

Southern Water Snake. Vernacular name for the American natricine species *Nerodia fasciata*. See also *Banded Water Snake*.

sp. (*pl.* **spp.**) Abbreviation of species.

spadate. *adj.* Spade-shaped; sometimes used to describe the shape of digital disk pads of anurans.

spade. *n.* The enlarged metatarsal tubercle on the hind foot of some burrowing toads, used for digging. (Syn. *digger, shovel, scraper, spur*)

Spadefoot Frog. Vernacular name for Asian microhylid frogs belonging to the genus *Calluella*. Cf. *Spadefoot Toad*.

Spadefoot Toad. Vernacular name for species of anurans belonging to the families Scaphiopodidae and Pelobatidae. The name also is used in reference to species of Asian anurans belonging to the megophyrid genera *Leptobrachium* and *Megophrys,* and the myobatrachid genus *Notaden.* Cf. *Spadefoot Frog.* See also *Southern Spadefoot* and *Western Spadefoot.*

spanning-canaliculus. n. Any one of the fine lines that run perpendicular to, and connect, successive *grid-canaliculi* on bones.

spasm. *n.* A sudden, involuntary muscular contraction.

spatial isolation. The spatial separation of populations, leading to reproductive isolation.

spatial statistics. A subdiscipline of statistics that includes numerous different tests and techniques related to spatial analysis. The majority of techniques are used to determine the extent to which data are spatially autocorrelated, or to perform tests of hypotheses after spatial autocorrelation is taken into account. Spatial autocorrelation occurs when observations are nonindependent because of their spatial arrangement.

spatulate. *adj.* Having a flat and rounded tip, like a spatula.

spawn. *v., n.* 1. To release eggs from the body, usually in aquatic animals. Also a term that denotes the resulting *egg mass* of amphibians. 2. Older colloquial term for the young of a newt or snake.

Spearman's rank correlation coefficient. In statistics, a nonparametric test for determining the significance of association between two variables that have been ranked within their respective samples.

special leucocyte. A large type of blood cell with a polymorphic nucleus, known only from reptiles. Described by S.F. Wood (*Univ. Calif. Publ. Zool.* 41:9–22, 1935).

specialist. *adj., n.* A species having a relatively narrow breadth of resource utilization. Cf. *generalist.*

specialized. *adj.* 1. Characterized by having a narrow range of tolerance for an ecological condition. 2. Also used to describe a species that has relatively low potential for further evolutionary change; opposite to generalized. 3. Any character or feature that is greatly modified from the original ancestral state, usually to perform a restricted function in response to specific environmental conditions.

specialized pseudautotomy. See *pseudautotomy.*

speciation. *n.* Generally, the formation or evolution of species. Two contexts may be distinguished. First, reference is to the splitting of an ancestral species into daughter species that coexist in time (*cladogenesis*). Second, reference is to a gradual transformation of one species into another without an increase in species number at any time within the lineage (*phyletic evolution*). See *alloparapatric speciation, allopatric speciation, parapatric speciation, peripatric speciation, sympatric speciation.*

species. *n.* Groups of interbreeding populations that are evolutionarily independent from other such populations. *Biological* (*genetic* or *isolation*) *species* are breeding populations that are reproductively isolated from other breeding systems (E.B. Poulton, *Proc. Entomol. Soc. Lond.* 1903:lxxvii–cxvi, 1904; E. Mayr, *Systematics and the Origin of Species from the Viewpoint of a Zoologist,* 1942; *Animal Species and Evolution,* 1963; *Populations, Species, and Evolution,* 1970). *Geneological species* concept recognizes species as components of a lineage or phylogeny (D. Baum & K. Shaw, pp. 289–303 in *Experimental and Molecular Approaches to Plant Biosystematics,* ed. P.C. Hoch & A.G. Stevenson, 1995). *Ecological species* is a set of organisms exploiting (or adapted to) a single niche (L. Van Valen, *Taxon* 25:233–239, 1976; L. Andersson, *Taxon* 39:375–382, 1990; M. Ridley, *Evolution,* 1993, 1996). The concept also recognizes species as a lineage or closely-related set of lineages which occupy an adaptive zone minimally different from that of other lineages in its range and which evolve(s) separately from all lineages outside its range (L. Van Valen, *Taxon* 25:233–239, 1976). *Evolutionary species* concept defines species as lineages over time, each having its own independent evolutionary fate and historical tendencies (A. Templeton, pp. 3–27 in *Speciation and Its Consequences,* ed. D. Otte & J. Endler, 1989; E.O. Wiley & R. Mayden, in *Species Concepts and Phylogenetic Theory,* ed. Q.D. Wheeler & R Meier, 2000). *Paleospecies* (*successional species*) are assemblages of organisms that appear distinctly different as a consequence of phenotypic (morphological) transformation (G.G. Simpson, *Principles of Animal Taxonomy,* 1961; E.O. Wiley, *Phylogenetics. The Theory and Practice of Phylogenetic Systematics,* 1981). *Phylogenetic species* has reference to the smallest aggregation of populations that possess unique combinations of character states in comparable individuals (J. Cracraft, pp. 28–59 in *Speciation and its Consequences,* ed. D. Ott & J. Endler, 1989; K.C. Nixon & Q.D. Wheeler, *Cladistics* 6:211–223, 1990; Q.D. Wheeler, *The Phylogenetic Species,* 1996; Q.D. Wheeler & R. Meier, eds., *Species Concepts and Phylogenetic Theory,* 2000). *Taxonomic* (*morphological, phenetic, typological*) *species* are groups of coexisting organisms that are phenotypically distinct from others (E. Mayr, *Principles of Systematic Zoology,* 1969; M. Ridley, *Evolution,* 1993, 1996). *Morphospecies* are based on morphological characters alone, without consideration of other biological factors (G.G. Simpson, *Principles of Animal Taxonomy,* 1961). The *genotypic species cluster* concept relates species to groups of individuals that have no intermediates when in contact and remain distinct outside areas of overlap (J. Mallet, *Trends Ecol. Evol.* 10:294–299, 1995). *Biosystematic species* (*ecospecies, ecological species*) are populations that are isolated by ecological factors rather than ethological isolation. The *recognition species concept* defines species in context of sharing a common fertilization system and specific mate recognition system (H. Paterson, pp. 21–29 in *Species and Speciation*, ed. E.S. Vrba, 1985). The *cohesion species concept* is based in phenotypic cohesion through genetic and/or demographic cohesion mechanisms (A. Templeton, pp. 3–27 in *Speciation and Its Consequences*, ed. D. Otte & J. Endler, 1989; pp. 32–43 in *Endless Forms. Species and Speciation*, ed. D.J. Howard & S.H. Berlocher, 1998). The *cladistic species concept* recognizes species as represented by the distance between two successive branching points on a cladogram, and thus an entity delimited in time by successive speciation events (J. Cracraft, pp. 28–59 in *Speciation and its Consequences*, ed. D. Ott & J. Endler, 1989; M. Ridley, *Evolution*, 1993, 1996). A *multidimensional species concept* considers species as a multi-population system wherein distinct morphospecies are members of a single dispersed species network in which morphological variants are replacing each other geographically (E. Mayr, *The Growth of Biological Thought*, 1982). A *nondimensional species concept* considers noninterbreeding sympatric populations as distinct species, whereas populations that interbreed and exhibit morphologically intermediate forms are regarded as belonging to the same species (here and now; i.e., no dimension in space or time) (E. Mayr, *The Growth of Biological Thought*, 1982). The *nominalistic species concept* holds that species are a human-devised abstraction formulated as a convenient way of referring to large numbers of individuals, but without any real existence in nature. A species is a fundamental level of classification used in the systematic identification of living organisms. Any given species is designated by a *scientific name*.

species aggregate. A collective group of species that are morphologically similar and therefore difficult to distinguish.

species-area curve. A graphic relationship illustrating number of species found relative to the size of the area surveyed. This graphic analysis is often used with reference to island biotas. (Syn. *area-species curve*)

species-area relationship. In island biogeography, a relationship in which species richness increases non-linearly with increase of island area.

species complex. A collective group of closely related species, usually having a recent common ancestor.

species diversity. In general, reference to the number of species. The term incorporates three concepts. *Species richness* refers to the total number of species. *Species evenness* refers to the relative abundance of species. *Species dominance* refers to the most abundant species. See also *alpha diversity, beta diversity,* and *gamma diversity*.

species epithet. The second word in a species name, e.g. *viridis* in the name *Crotalus viridis*.

species equilibrium. A condition in which the extinction rate equals the rate of arrival of new species by immigration.

species evenness. See *species diversity*.

species flock. A monophyletic group of ecologically diverse and closely related species that have evolved within a single macrohabitat and a geographically restricted area.

species group. See *superspecies*.

species indeterminate. (*sp. indet., sp. ind.*) Latin, meaning "indeterminate species." A species that cannot be identified from the original description.

species nova. (*sp. nov., sp. n.*) Latin, meaning "new species" and cited after the binomen in place of the authority.

species packing. An increase in species richness within a narrow range of resource variation.

species richness. See *species diversity*.

species swarm. Reference to a large number of closely related species that occur together in the same geographical area and are derived by multiple splitting of an ancestral stock.

specific activity. The ratio of radioactive to nonradioactive atoms or molecules of an identical kind.

specific dynamic action. (SDA) See *postprandial calorigenesis*.

specific dynamic effect. See *postprandial calorigenesis*.

specific gravity. The ratio of the mass of a substance to the mass of an equal volume of water, determined at a specified temperature.

specific heat. The amount of heat required to elevate 1 g of a sustance 1 °C.; i.e. cal or joules per (g • °C). See also *heat capacity*.

specific humidity. The mass of water vapor divided by the mass of moist air.

specificity. *n.* Selective reactivity between substances, e.g. an enzyme and its substrate.

specific name. The second name in a *scientific name*, indicating the species.

specimen. *n.* 1. An individual that is representative of a group. 2. Any sample or part of material used for scientific study.

speciose. *adj.* Characterized by numerous species, extant or during evolutionary history.

Speckled Kingsnake. Vernacular name for the American colubrid species *Lampropeltis getula holbrooki*.

Speckled Racer. Vernacular name for neotropical species of colubrid snakes belonging to the genus *Drymobius*.

Speckled Rattlesnake. Vernacular name for the American pitviper *Crotalus mitchelli*.

spectacle. *n.* Any fixed, transparent or semi-transparent, noncorneal covering of the eye. The skin and sclera may be separate, as in aquatic amphibians (*primary spectacle*); there may be partial fusion of the skin and sclera, as in some Cave Salamanders (*secondary spectacle*); or the skin and sclera may be completely fused, as in most snakes and some lizards (*tertiary spectacle*). The spectacle is protective and serves the same function as eyelids. (Syn. *brille, eye cap*)

Spectacled Caiman. Vernacular name for crocodilians belonging to the genus *Caiman*.

Spectacled Cobra. Vernacular name for the Asian Cobra species *Naja naja*.

Spectacled Tegus. Vernacular name for species of lizards belonging to the family Gymnophthalmidae, and especially the genus *Gymnophthalmus*.

spectral radiance. Radiance per unit wavelength interval (W • m^{-2} • sr^{-1} • μm^{-1}).

spectrogram. *n.* A photograph or representation of a spectrum, or plot of amplitude and frequency.

spectrophotometer. *n.* An instrument that passes a beam of visible or ultraviolet light through a fluid-filled vial and measures the intensity of emerging wavelengths, thus the absorbance or transmittance of light at specific wavelengths.

spectrum. *n.* The array of individual wavelengths of electromagnetic radiation produced by refraction or diffraction.

sperm. (*pl.* **sperm** or **sperms**) *n.* A mature haploid male gamete. (Syn. *spermatozoon*)

spermatheca. (*pl.* **–ae**) *n.* 1. A sac used for the storage of spermatozoa in the female. 2. Reference to a system of tubules, sometimes connected to a common tube, in the dorsal cloaca of female salamanders. These function to collect sperm from the *spermatophore* and to retain them until fertilization takes place. Spermathecae may occur as simple tubular glands or as more complex compound glands opening into the roof of the cloaca. (Syn. *receptaculum seminis, seminal receptacle, sperm sac*)

spermatic ampullae. Cavities of testes in which sperm cells are produced in amphibians. Cf. *seminiferous tubules*.

spermatic canal, duct, or **groove.** See *sulcus spermaticus*.

spermatid. *n.* One of four haploid cells formed during meiosis in the male. Spermatids transform into *spermatozoa* without further division.

spermatocyte. *n.* A diploid cell that undergoes meiosis and gives rise to four *spermatids*.

spermatogenesis. *n.* The development of mature spermatozoa in the testis. This is an inclusive term for both male meiosis and spermiogenesis.

spermatogenetic cycle. The sequence of events related to *spermatogenesis*, or production of spermatozoa from germ cells.

spermatogenetic wave. Reference to a caudocephalic movement of the spermatogenetic process in the testis of the salamander, as described by R.R. Humphrey (*Biol. Bull.* 43:45–67, 1922). The more posterior lobules begin sperm production initially, followed by a wave of developmental change proceeding toward the anterior end of the gonad.

spermatogonia. (*s.,* **-ium**) *n.* Mitotically active cells in the male gonad that are the progenitors of primary spermatocytes.

spermatophore. *n.* A structure containing sperm, usually a gelatinous mass with a *sperm cap* on top, secreted by cloacal glands of a male salamander and deposited on a moist surface from which it is picked up by the cloacal lips of a female. Usually only the cap is taken into the cloaca where the spermatozoa are held within the *spermatheca*.

spermatozoan. (*pl.* **-zoa**) *n.* A haploid male *gamete* produced by *meiosis*.

sperm cap. A globular mass of spermatozoa and jelly, which forms the upper part of the *spermatophore* of salamanders. (Syn. *sperm capsule*)

sperm capsule. See *sperm cap*.

sperm duct. Any duct (always urogenital) that conveys spermatozoa. See *ductus deferens, mesonephric duct*.

spermiation. *n.* The process of releasing spermatozoa from the testis and their passage into the kidney and Wolffian duct.

spermiogenesis. *n.* The formation of spermatozoa from the *spermatids* produced during the meiotic divisions of *spermatocytes*.

sperm sac. See *spermatheca*.

sphenethmoid. *n.* A single median skull bone, derived from the *chondrocranium* and located internally at the anterior of the orbit. In amphibians, this element is formed by fusion of the *orbitosphenoids*. In primitive reptiles this is a narrow, vertical bone lying anterior to the *basisphenoid* and hollowed to accommodate the cerebral hemispheres posteriorly and olfactory tracts anteriorly. It is absent in living reptiles.

Sphenodontidae. *n.* A basal clade (family) of lepidosaur that includes the two species of Tuatara (*Sphenodon*), presently restricted in distribution to some smaller islands off the New Zealand coast. See *Rhynchocephalia*.

sphenolateral. *n.* See *orbitosphenoid*.

sphenotic. *n.* Reference to an ingrowth arising from the *postfrontal* and invading the *chondrocranium*.

sphincter. *n.* A ring of smooth muscle capable of constricting an opening or passageway, typically associated with parts of the alimentary canal or arterioles.

spicule. *n.* A small, pointed structure such as a tubercle, often bearing a keratinized or calcified tip. This term was used interchangeably with *spinule* by earlier herpetologists.

Spikethumb Frogs. Vernacular name for species of hylid frogs belonging to the Middle American genus *Plectohyla*.

spina. *n.* A sharp and rigid projection of the skin, derived wholly or in part from epidermis.

spina ischii. See *metischial process*.

spinal. *adj., n.* 1. Reference to the spine or spinal cord. 2. Sometimes used in place of *vertebral*.

spinal canal. The cavity that runs longitudinally through the vertebrate *spinal cord*, containing *cerebrospinal fluid* that is confluent with the cerebral ventricles.

spinal cord. The portion of the vertebrate central nervous system that is encased in the vertebral column, extending from the caudal end of the medulla and comprised of nerve tracts.

spinal nerve. Any of numerous nerves entering or departing from the spinal cord. Cf. *cranial nerve*.

spinal vein or **spinal venous plexus.** A vessel or plexus of vessels that course near the spinal cord in most, if not all, extant vertebrates. The vessels may be robust and serve as a collateral route for venous return to the heart. In snakes there is an elaborate vertebral plexus extending from the base of the skull to the tip of the tail and is thought to enhance blood circulation to and from the head during upright posture (K.C. Zippel et al., *J. Morphol.* 250:173–184, 2001)

spina oesophagea. See *pharyngeal process*.

spindle cell. A type of blood cell found in amphibians, evidently analogous to mammalian platelets in producing fibrous clots upon rupture.

spine. *n.* Any firm or rigid process projecting from the body. See also *basal hook, mesial spine, shoulder spine*.

spinose. *adj.* Bearing spines or spine-like structures.

spinulate. *adj.* Covered with tiny calcified spines as on the calyces of some squamate hemipenes.

spinulate scale organ. A specialized microscopic scale organ characterized by a single median spine or filament (R. Etheridge & K. de Queiroz, pp. 283–367 in R. Estes & G. Pregill, eds., *Phylogenetic Relationships of the Lizard Families. Essays Commemorating Charles L. Camp*, 1988).

spinule. *n.* 1. A very small or microscopic spine or spine-like structure. 2. This term is used in reference to minute structures expressed in lizards as hypertrophied elaborations of the *Oberhaütchen* across the body surface and on the palmar and plantar surfaces of the feet, ranging in height from approximately 1.5 µm to about 2.5 µm on the feet of geckos. See *subdigital setae* and Fig 32.

Spiny Lizards. Collective vernacular name for North and Middle American species of iguanid lizards belonging to the speciose genus *Sceloporus*. Recently, the species *Sceloporus magister* has been split into three species (J.A. Schulte II et al., *Mol. Phylogenetics Evol.* 39:873–880, 2006).

Spinytail Gecko. Vernacular name for various species of Australian Geckos belonging to the genus *Diplodactylus*.

Spinytail Iguana. Vernacular name for iguanid lizards belonging to the American genus *Ctenosaura*.

Spinytail Lizards. Collective vernacular name for various species of lizards belonging to the family Cordylidae, especially the genus *Cordylus*.

Spinytail Skink. Vernacular name for species of scincid lizards belonging to the genus *Egernia*.

spiracle, spiraculum. *n.* The small aperture between the outside environment and gill chamber of anuran tadpoles, formed after the operculum has completed its growth over the chamber and fuses with integument of the abdomen. The spiracle may be single (either medial or left side) or bilaterally paired. The first gill cleft of caecilians also is said to form a spiracle. (Syn. *branchial aperture*) See Fig. 1.

spiracular tube. An elongated passageway connecting the gill chamber with the spiracle opening in tadpoles of certain species of anurans.

spiral fold. A wall projection forming a septum that partially dividies the *conus arteriosus* in amphibians. This structure assists the separation of oxygenated and deoxygenated blood flows from the single ventricle into the outflow tracts. Note this structure does not act as a true valve. (Syn. *longitudinal valve, septum bulbi, spiral septum, spiral valve*)

spiral groove. See *sulcus spermaticus*.

spiral septum. See *spiral fold*.

spiral valve. See *spiral fold*.

spit. *n.* A fingerlike ridge of sediment attached to land but extending out into open water.

spitter. *n.* Jargon for any of various elapid snakes that have the ability to expel venom from the mouth in a fine stream, often accurately aimed at the face or eyes of an aggressor.

Spitting Cobra. Collective vernacular name for several species of Afro-Asian elapids (genera *Naja* and *Hemachatus*) that expel or "spray" venom from the fangs, often aiming defensively at the face or eyes of an aggressor. See also *Cobra, Rinkhals,* and *Spitter*.

splanchnic peritoneum. See *peritoneum*.

splanchnocranium. *n.* The skeleton of the visceral arches; the branchial arches and its derivatives.

splanchnopleure. *n.* A membrane derived from endoderm and the inner sheet of the hypomere.

Splash Frogs. Vernacular name for south Asian ranid frogs belonging to the genus *Staurois*. These frogs occur along streams, often perch on rocks, and are also called *Rock Frogs*.

spleen. *n.* A large lymphoid organ that produces red and white blood cells and macrophages, usually located in dorsal mesentery. It is universal in tetrapod vertebrates and is an important site of blood cell production.

splenial. *n.* A dermal bone deposited over Meckel's cartilage and located on the outer, ventral, and inner surfaces, ventral to the dentary of the lower jaw. This element is present in snakes, crocodilians, and some species of lizards. It is absent in some species of lizards, most turtles, *Sphendon*, and all amphibians. (Syn. *opercular, operculare, preoperculare*)

splenial ridge. A ridge formed on the posterior lingual surface of the *dentary* in *Ambystoma*, possibly representing the remains of the *splenial*.

splint bone. Individual ossifications that make up the separate rows of the gastralia in *Sphenodon*. Used by H. Gadow (*Amphibia and Reptiles*, p. 298, 1901).

splinter scar. Evidence of a healing injury to the carapace of a turtle, usually evident as exposed bone at an opening in the overlying lamina.

Split-jawed Boas. Vernacular alternate name for *Round Island Boas*, family *Bolyeriidae*.

splitting. *n.* In taxonomy, the practice of using relatively minor variation in the definition or recognition of taxonomic units, compared with "lumpers" who ignore such minor variations of characters in defining taxonomic units. Cf. *lumping*.

spontaneous. *adj.* Occurring or developing without apparent external stimulus or cause.

sporadic. *adj.* Scattered or occasional.

sporozoan. *adj., n.* 1. Pertaining to sporozoa. 2. A sporozoon. Members of the spore-forming protozoan phylum Apicomplexa.

sporozoon. (*pl.* **-zoa**) *n.* Individual organism of the Apicomplexa, a phylum of unicellular protozoan organisms, responsible for various blood, intestinal, muscle, and skin diseases in reptiles and amphibians.

spot. *n.* A small area that differs in color from that of the ground color, with or without a border. See also *rhomb*.

spotlighting. *n.* A standard technique for locating amphibians and reptiles at night, using a flashlight such that its light reflects off an animal's retina. Use of a brighter light has little benefit because it may disturb animals and does not enhance eyeshine against the background.

Spotted Turtle. Common name for a well-known North American species of the family Emydidae, *Clemmys guttata*.

spp. Plural abbreviation for species.

spreading factor. A component of snake venom that facilitates its spread through tissues of an envenomated animal. The component has been identified as *hyaluronidase*, sometimes acting in concert with other factors such as *proinvasins*.

spring. *n.* A place where water flows naturally out of rock onto the land surface.

Spring Peeper. Vernacular name for American hylid species *Pseudacris crucifer*.

Spring Salamander. Collective vernacular name for several North American salamander species belonging to the genus *Gyrinophylus*.

spring tide. A tide of maximum range that coincides with the time of the new and full moon. Cf. *neap tide*.

sprint speed. A velocity of movement that is achieved when an animal runs rapidly, usually in experimental circumstances such as on a treadmill. In ectotherms such as amphibians and reptiles, the level of locomotion implies maximal, short-term, and exhaustive expenditure of energy requiring anaerobic energy production beyond the aerobic maximum.

spur. *n.* A stiff, sharp spine, usually projecting from the hindlimbs or on either side of the vent, as in the vestigial hind limbs of certain primitive species of snakes. See also *calcar, laminal spur, pelvic spur, spade*.

spur-flap. *n.* A series of several enlarged and connected scales forming a flap-like structure along the external margin of the forelimb in the Australian turtle *Pseudemydura umbrina*. Described by E.E. Williams (*Breviora* 84:1–8, 1958).

spurious. *adj.* False; neither true nor genuine.

sputter coating. A method for coating a mounted specimen with a thin layer of metal prior to examination in a *scanning electron microscope*.

sputum. *n.* Saliva or mucus that is spit out of the mouth.

squalene. *n.* A triterpene hydrocarbon that has been observed in the epidermis of snakes and posited to contribute to the chemosensory recognition of and by male Garter Snakes (R.T. Mason et al., *Science* 245:290, 1989).

squama. (*pl.* **-ae**) *n.* A scale or thin, platelike structure.

Squamata. *n.* A basal clade (order) of *lepidosaurs* that includes lizards, snakes, and amphisbaenians. Squamates are a diverse and successful radiation and share several anatomical characteristics, including *streptostyly*, paired eversible male copulatory organs (*hemipenes*), and a distinctive vertebral joint (*procoely*). There are more than 7200 species., including *Lacertilia* (Sauria), *Amphisbaenia*, and *Serpentes* (Ophidia).

squamate. *n., adj.* Colloquial term for any reptile belonging to the *Squamata* (lizards, amphisbaenians, and snakes).

squamation. *n.* The condition of possessing scales, and the arrangement of scales on the body. This term is used little in modern literature. (Syn. *scalation*)

squamid. *adj.* Scaly.

squamosal. *n.* A dermal cheek bone lying below the temporal series and articulating with the *pterygoid* and the *quadrate*. It also contacts the *jugal* and the *quadratojugal* when these are present. In snakes the squamosal is absent, and the bone lying between the *parietal* and *quadrate* is called either *tabular* or *supratemporal*. In lizards the term squamosal has been applied to either or one (outermost) of two elements between the parietal and quadrate. In other modern reptiles the squamosal is large and prominent, and is part of the temporal arch when it is present. The squamosal is fairly prominent in amphibians where it has been referenced by numerous names.

squamosal angle. The angle formed by the *squamosal bone* and the maxillary ramus in anurans.

squamosal antrum. An extension of the tympanic cavity posteriodorsal to the tympanic membrane and capped by the *squamosal* or *supratemporal*. This structure is present in most turtles.

squamous. (also **squamate, squamose**) *adj.* 1. Covered with or composed of scales. 2. Scaly or platelike.

square root transformation. In statistics and data analysis, the transformation of data by taking the square root of every value renders values of random variables having Poisson distributions amenable to assumptions of statistical analysis. Because one cannot take the square root of a negative number, a constant must be added to move the minimum value of the distribution above zero. See *data transformation*. Cf. *log transformation*.

Squeakers. *n.* Vernacular name for African arthroleptid frog species belonging to the genus *Arthroleptis*. These are small, squat frogs with short legs, and long digits. The males call incessantly during both day and night in rainy weather.

squeeze box. A box or container in which a snake is restrained between a soft bottom and a transparent, usually plexiglass, "lid" that can be used to apply pressure and immobilize the animal. This technique provides a means of restraining a venomous or dangerous animal while its length is measured by drawing or fitting a string over its outline. While this method can also permit handling of the tail end of an animal while its head is immobilized, a plastic tube is recommended for that purpose. See H. Quinn & J.P. Jones, *Herpetol. Rev.* 5:35, 1974.

Squirrel Treefrog. Vernacular name for the North American hylid species *Hyla squirella*.

squirt gland. A serous gland in the skin of *caecilians* that elaborates a fluid causing severe reaction of mucous membranes when expelled by muscular contraction.

SR. See *sarcoplasmic reticulum*.

SSAR. See *Society for the Study of Amphibians and Reptiles*.

ssp. Abbreviation of subspecies.

stability. *n.* In community ecology, the stability of populations to withstand perturbations without marked changes in composition.

stabilizer. *n.* See *balancer*.

stabilizing selection. Selection in which there is removal of alleles that produce deviations from the average population phenotype. Such selection removes deviant individuals and will reduce the variance in subsequent generations. Cf. *directional selection, disruptive selection.* (Syn. *centripetal selection; normalizing selection*)

stable. *n.* Remaining constant or relatively unaltered for extended period of time.

stable age distribution. The constant relative proportion of individuals represented in each age of an exponentially increasing (or decreasing) population. A stable age distribution implies that a population is growing with fixed survivorship and fecundity schedules.

stable equilibrium. In population ecology, an equilibrium is stable if a population always returns to it after a small perturbation. See *equilibrium*.

stable isotope. A nonradioactive isotope of an element. Currently, quantification of one or more stable isotopes in tissue is being used to determine prey of an animal and to evaluate food web connections, water sources, and other aspects of metabolic status of plants and animals. Evolving methodologies will undoubtedly continue to enhance the utility of stable isotopes in ecological and physiological research.

stable limit cycle. A population cycle that is stable and returns to a pattern of cycles with the same amplitude and period if it is perturbed.

stable state. See *equilibrium*.

staff of Aesculapius. *n.* See *caduceus*.

stage, staging. *n.* Reference to the recognition and description of specific morphological features that are useful in comparing the sequence of events in a developmental continuum. More than 45 staging tables have been produced for amphibian embryos and larvae, but several are cited, and thus evidently used, most frequently: K.L. Gosner, *Herpetologica* 16:183–190, 1960 (general); pp. 44–66 in R.G. Harrison (ed.), *Organization and Develoment of the Embryo*, 1969 (*Ambystoma*); P.D. Nieuwkoop & J. Faber, *Normal Tables of* Xenopus laevis *(Daudin)*, 1956; W. Shumway, *Anat. Rec.* 78:139–147, 1940 (*Rana*); A.C. Taylor & J.J. Kollros, *Anat. Rec.* 94:7–24, 1946 (*Rana*). See also discussion in R.W. McDiarmid & R. Altig, eds., *Tadpoles. The Biology of Anuran Larvae*, 1999.

staghorn gill. A large, multilobed gill of the terrestrial larvae of certain salamanders, named because of resemblance to the branched horns of a deer.

staining. *n., v.* The application of one or more pigments to tissue sections to enhance contrast of structures to be examined histologically.

stalk. *n.* A term sometimes used to denote the central shaft of the gill of tadpoles, arising from the gill arch and branching to form numerous secondary lamellae of the gill.

stalked hydatid. Reference to the tip of the *Wolffian duct*, anterior to the *epididymis*, once it becomes strictly a pathway for sperm transport in male metanephric animals. (Syn. *appendix epididymis*)

standard deviation. (σ, s or SD) A quantity equal to the positive square root of the variance. Therefore, it has the same units as the original measurements. See also *standard error of the mean, variance.* (Syn. on rare occasions *root mean square deviation* or *root mean square*)

standard distance. An older morphological measurement defined as the distance from the tip of the snout to the center of the eye.

standard error of the mean, standard error. ($s_{\bar{x}}$, SE, SEM) The standard deviation of the mean, calculated by dividing the standard deviation by the square root of the sample size (= number of observations in the sample). See also *standard deviation, variance*.

standard geological time scale. A worldwide relative scale of geologic time divisions. See Table 1.

standard length. A term used to suggest a standard of measurement with respect to the size of some character or body of an animal. The term has been used in reference to *snout-vent length, basicranial length*, and *standard distance* by various investigators.

standard metabolic rate. (SMR) A standardized measure of resting metabolic rate in an ectotherm, conducted at a specified temperature while an animal is at rest and postabsorptive, during the inactive phase of its daily cycle. See *metabolic rate*. (Syn. loosely, but not strictly, *resting metabolic rate*)

standard operative temperature. See *operative temperature*.

standard pressure. An ambient pressure of 1 atmosphere (101 kPa = 760 mm Hg).

standard symbolic codes for institutional resource collections. A listing of these can be found in A.E. Leviton et al., *Copeia* 1985:802–832, 1985.

standard temperature and pressure. (STP dry, STPD) Specified conditions for gas measurements: 0 °C, 1 atmosphere (atm) pressure, dry air.

standing crop. The total mass, volume, or energy of organisms (biomass) that comprises all or part of a population or other specified group.

standing water. Reference to a body of water that does not experience continuous flow in a particular direction.

stapedial artery. An artery that branches from the carotid artery and carries oxygenated blood to the outer part of the head and jaws in amphibians and reptiles.

stapedial footplate or **plate.** See *footplate*.

stapes. *n.* See *columella*, which is the preferred term in non-mammalian vertebrates.

stapes inferior. A separate chondrification in the wall of the otic capsule that fuses indistinguishably during embryonic life in turtles of the genus *Emys*. The structure lies very close to the *footplate* of the *columella*.

stapes segment. One of three segments of the kinetic skull of lizards, as defined by T.H. Frazzetta (*J. Morphol.* 111:293, 1962). In some modified burrowing forms, the *stapes* may provide a significant aspect of kinesis.

Starling curve. A curve that describes the relationship between work done by the heart and filling pressure.

Star Tortoise. Vernacular name for the well-known Indian species of tortoise *Geochelone radiata* (also the Radiated Tortoise, popular in the pet trade), *G. elegans*, South African tortoises belonging to the genus *Psammobates*, and the Burmese Star Tortoise, *Geochelone platynota*.

stasimorphic. *adj.* Reference to a character that retains the ancestral condition. Cf. *apomorphic*.

stasimorphic speciation. Formation of new species without morphological differentiation.

stasipatric speciation. Speciation involving adaptively superior individuals that arise in a particular part of the geographic range of an ancestral species due to dispersal of a favorable chromosomal rearrangement that yields favorable homozygotes.

stasis. *n.* (*adj.* **–ic**) In evolutionary studies, this term applies to the persistence of a species without significant change over a considerable span of geological time.

stasis tadpole. A larval anuran that has been starved during the period prior to the beginning of metamorphosis. Term from S.A. D'Angelo et al. (*J. Exp. Zool.* 87:259–277, 1941). Cf. *critical tadpole*.

stat. Abbreviation of the Latin *status*, meaning "rank."

state. *n.* Reference to a particular expression or condition of a character.

static friction. See *friction*.

static life table. See *vertical life table*.

-statin. Suffix meaning "inhibiting platelet aggregation," used in a nomenclatural scheme for describing exogenous hemostatic factors in snake venoms. By adding a portion or designated abbreviation of a snake's scientific name to the suffix, one obtains a designation for the fraction being identified. For example, using the name "*gabonica*" (species name for the Gaboon Viper) one obtains *gabonistatin*.

stationary age distribution. A special case of the *stable age distribution* in which the instantaneous rate of increase, r, is zero. Both the absolute and relative numbers of individuals represented in each age class remain constant.

statistic. *n.* A numerical value that characterizes some property of a sample from a population, usually used as an estimate of the corresponding parameter of the population from which the sample was drawn. Cf. *parameter*.

statistical error. 1. A *Type I error* refers to a random fluctuation being taken as evidence for a positive effect. The risk of making a "false positive" error of this sort is symbolized by the Greek letter alpha (α). 2. A *Type II error* results when one fails to detect an effect when there is one. The risk of making a "false negative" error of this sort is symbolized by the Greek letter beta (β).

statistical power. The probability that a test will reject a false null hypothesis (that is, make a *type II error*). In other words, the ability of a test to detect an effect, given that the effect actually exists. As power increases, the chances of a Type II error decrease, and vice versa. Any statistical result that has a p-value has an associated power. Power analysis can be either before (*a priori*) or after (*post hoc*) data are collected. *A priori* power analysis is conducted prior to the conduct of research and is typically used to determine an appropriate sample size to achieve adequate power.

statistical test. See *test, statistics*.

statistics. *n.* The application of probability theory and inference to the analysis of scientifically collected data. Staststistical inference involves the acceptance or rejection of one conclusion from one or more alternatives according to a computed result based on observations. *Parametric* methods in statistical analysis assume that the data conform to a defined probability distribution. By comparison, *non-parametric* methods in stastical analysis are relatively free from conventional assumptions about underlying probability distributions. However, non-parametric tests can also suffer when normality assumptions are violated (D.W. Zimmerman, *J. Exp. Educ.* 67:55–68, 1998).

statoacoustic. *adj., n.* Reference to the eighth cranial nerve.

stat. rev. Abbreviation of the Latin, *status revivisco*, meaning "status revived." Used in nomenclature for a name that has had its status revived following an earlier change.

status of stocks. The evaluation of the abundance of a particular harvestable species and its potential for harvest.

statute mile. A unit of length equal to 5280 feet or 1.6093 km.

steady state. A state of a living system in which the value of a variable remains constant (opposing forces are balanced), but continuous expenditure of energy is required to maintain the system. Cf. *equilibrium*. (Syn. *dynamic equilibrium, dynamic steady state*)

steatitis. *n.* Inflammation of adipose tissue.

Stefan-Boltzmann law, relation or **constant.** (*s* or *δ*) A statement or constant used in reference to radiation heat transfer, where the radiant energy (flux density, W/m^2) emitted by an object is proportional to the absolute temperature raised to the fourth power. For a blackbody, the heat flux emitted is given by the product of *s* or *δ* (Stefan-Boltzmann constant, 5.67×10^{-8} W $\cdot$ m^{-2} $\cdot$ K^{-4}) and the absolute temperature raised to the fourth power.

stegochordal. *adj.* A type of *centrum* in amphibians, characterized by ossification solely in the dorsal arc of the perichordal sheath, forming a flat band running along the ventral surface of a transversely flattened centrum. See also *ectochordal* and *holochordal*.

stegocrotaphy. *n.* See *stegokrotaphy*.

Stegokrotaphia. *n.* A subgroup of *Gymniophiona*, composed of the families *Ichthyophiidae* (including subfamily *Uraeotyphlidae*) and *Caeciliidae* (including subfamilies *Scolecomorphidae* and *Typhlonectidae*). See D.R. Frost et al., *Bull. Amer. Mus. Nat. Hist.* No. 297:1–370, 2006.

stegokrotaphy. *n.* 1. Reference to a skull condition in which the bony roof is complete, without a gap between the squamosal and parietal. Cf. *zygokrotaphy*. 2. A condition of skull in which the only gaps on the dorsal surface are the nares, the orbits, and the parietal foramen (spelled *stegocrotaphy*).

Stegosauria. *n.* A suborder comprised of herbivorous reptiles having a double row of upright bony plates along the back, long hind legs, and a relatively small head. These were prominent during the Jurassic to Cretaceous periods.

Steida's organ. See *pineal end organ*.

stellate. *adj.* Star-shaped, often used in reference to spots of pigmentation or chromatophores.

stem. *n.* In systematics, the single beginning point of a dendrogram or cladogram.

stem-based, stem-defined. Reference to a taxonomic group that is defined as all those entities that share a more recent common ancestor with one group than another. Cf. *node-based*.

stem-defined. *adj.* See *stem-based*.

stem group. A stem lineage, or an ancestral group, having relatively primitive characteristics, from which a *crown group* having relatively advanced characteristics has evolved. This term is used especially in discussion of taxa that have a fossil record. The stem group is usually extinct while the crown group is extant.

stem reptile. See *Cotylosauria*.

steno-. Prefix meaning "narrow." Cf. *eury-*.

stenobathic. *adj.* Tolerating only a narrow range of depth. Cf. *eurybathic*.

stenocoenose. *adj.* Having a restricted distribution. Cf. *eurycoenose*.

stenohaline. *adj.* Reference to an animal that has a relatively narrow tolerance for external salinity. Cf. *euryhaline*.

stenohydric. *adj.* Moisture sensitive, tolerating only a narrow range of moisture or humidity. Cf. *euryhydric*.

stenohygric. *adj.* Tolerant of a narrow range of atmospheric humidity.

stenophagic, stenophagous. *adj.* Reference to animals that accept a relatively narrow or select range of food items as prey. Cf. *euryphagic*.

stenoplastic. *adj.* Characterized by a narrow range of developmental response and phenotypic variation in relation to changes of environment. Cf. *euryplastic*.

stenosis. *n.* A narrowing or contraction of a body passage, blood vessel, or opening.

stenothermal, stenothermic. *adj.* Reference to an animal that has a relatively narrow tolerance for temperature change. Cf. *eurythermal*.

stenotopic. *adj.* Tolerating a narrow range of habitats or having a limited distribution. Cf. *eurytopic*.

steppe. *n.* Short-grass vegetation of plains and plateaus in the interior regions with semi-arid climate and extreme seasonal temperature variation, most extensive in central and southwest Asia.

stereochemical. *adj.* See *steric*.

stereocilia. *n.* The numerous hairlike sensors of a *neuromast* cell.

stereoscopic vision. The ability to see objects in three dimensions.

stereospondylous. *adj.* A *monospondylous* vertebra derived entirely from the *intercentrum*, wherein the body and neural arches are fused into a single bone (H. Gadow, *Amphibia and Reptiles*, p. 284, 1901); probably derived from a *rhachitomous* vertebra (E.S. Goodrich, *Studies in the Structure and Development of Vertebrates*, p. 49, 1930). According to E.E. Williams (*Quart. Rev. Biol.* 34:1–32, 1959), the centrum is derived entirely from the hypocentrum. Cf. *embolomerous*.

stereotaxis. *n.* A directed response of a motile organism to continuous contact with a solid surface. (Syn. *thigmotaxis*)

steric. *adj.* Pertaining to the spatial arrangement of atoms or molecules. (Syn. *stereochemical*)

sterile. *adj.* Unable to reproduce; without viable gametes.

sternal. *adj.* Relating to or involving the chest.

sternal fontanelles. Membrane-covered openings in the sternum.

sternal horn. Reference to either branch of the *sternum* when it is bifurcate posteriorly.

sternal shield. An earlier term for any one of the epidermal laminae on the *plastron* of turtles.

sternal vertebra. Any presacral vertebra that is connected with the sternum by ribs. See also *postsernal* and *presternal vertebra*.

sternum. *n.* The skeletal elements of bone or cartilage that form in tetrapods at the ventral midline articulating with the pectoral girdle and a variable number of ribs. Absent in early amphibians; this element occurs only in anurans among extant amphibians. In lizards and in frogs the sternum may be differentiated into two or three parts, variously called *presternum, metasternum, mesosternum,* and *xiphisternum.* See Fig. 3. (Syn. *hyposternum* in frogs. In earlier literature, sternum was sometimes used to denote the *plastron* of turtles.)

steroid. *n.* A lipid, derived from cholesterol, belonging to a family of saturated hydrocarbons having 17 carbon atoms arranged in a system of four fused rings.

steroid hormones. A class of cyclic hydrocarbon hormones synthesized from cholesterol, including estrogens, progestogens, androgens, glucocorticoids, and mineralcorticoids.

sterol. *n.* A compound with the general chemical ring structure of a steroid, but having a long side chain and an alcohol group. Cholesterol is a common biological sterol.

stiffness. *n.* In biomechanics, a property of materials quantified by the slope of the elastic region of a graphic plot of *stress* as a function of *strain*. Cf. *toughness*.

stiftchenzellen. *n.* Possibly chemoreceptor cells described in the epidermis of anuran larvae (*Rana* spp.), which appear to degenerate at metamorphosis.

Stiletto Snakes. Collective vernacular name for various species of cylindrical, usually burrowing snakes belonging to the family *Atractaspididae*, especially the genus *Atractaspis*. Also called *Burrowing Asps* or *Mole Vipers*.

stillbirth. *n.* Birth of dead, but usually well developed, young in viviparous reptiles.

stimulus. (*pl.* –i) *n.* A quality of the environment that stimulates a *receptor*.

Stinkpot. *n.* Common name for the "Musk Turtle" *Sternotherus odoratus* (family Kinosternidae).

stippled. *adj.* Bearing numerous dots.

stippling. *n.* The effect produced by presence of numerous dots having contrast with the background color on which they appear.

stirnorgan. *n.* See *pineal end organ*.

stochastic. *adj.* Random or expected by chance.

stochastic model. A model in which some of the parameters vary unpredictably with time, reflecting random or chance events in nature. Cf. *deterministic model*.

stochastic process. A process subject to chance or random mechanisms. (Syn. *random process*)

stock. *n.* 1. A race or lineage. 2. A management term that refers to a harvestable portion of a species living within a certain geographic area. The stock might include part of a population or several populations.

Stokes's Sea Snake. Vernacular name for the largest species of elapid Sea Snake, *Astrotia stokesii*.

stomach. *n.* The anterior digestive region of the alimentary canal, following the *esophagus* and preceding the *intestine*.

stomatitis. *n.* An infection of the oral cavity in snakes, lizards, and other reptiles. The condition usually results from a mouth injury coupled with stress and may be caused by bacterial and viral pathogens. The condition is infectious, necrotic, and ulcerative. (Syn. *canker, mouth rot, ulcerative gingivitis*)

stomodeum. *n.* Anterior invagination of the embryonic ectoderm that becomes the lining of the mouth.

stone. *n.* A sediment particle larger than about 20 mm in diameter, and a component of gravel.

STP, STPD. See *standard temperature and pressure*.

straddle amplexus. A mode of *amplexus* wherein males of certain Malagasy ranid frogs sit astride the shoulders of a female while both grasp a suspended leaf.

straight length. A measurement of the maximum straight line length of a turtle carapace, taken along the midline between the posterior border of the supracaudal and the anterior border of the nuchal. (Syn. *carapace length, shell length*) See Fig. 9.

Straightneck Turtles. Collective vernacular name for species of turtles in the *Cryptodeira*.

strain. *n.* 1. In biomechanics, the deformation of an elastic solid that is caused by a load. Such response to an applied stress is measured as the ratio of the change in size of a structure to some basic size. E.g., if L is the instantaneous length of a structure and L_o is the initial length, then strain = dL/L_o. The variable is dimensionless, but it is not unusual for it to be expressed as a percentage. 2. In genetics, a strain refers to an intraspecific group of organisms possessing usually only one or a few distinctive traits that are maintained by humans for genetic experimentation or domestication. A similar term is *variety*, which is usually used when the differences between the intraspecific groups are more substantial.

strain gauge. A sensor in which the resistance is a function of applied force.

strand. *n.* Land bordering a body of water; a beach.

strategy. *n.* A preprogrammed set of behavioral or life history characteristics, or the choices a competitor should make in all circumstances that might arise. The term is used in ecology and evolutionary biology and comes from evolutionary game theory. An *evolutionary stable strategy* (ESS) is a strategy that persists in a population. See *game theory* and *tactic*.

stratified. *adj.* Formed of layers.

stratigraphic time divisions. Synonym for *geologic time divisions*. See Table 1.

stratigraphy. *n.* Study of the origin, composition, and relative chronology of geologic strata.

stratocladistic. *adj.* Reference to phylogenetic relationships that are inferred from weighted, derived similarities between fossil taxa and selected stratigraphic data not contradicted by morphological analysis.

stratophenetic. *adj.* Reference to phylogenetic relationships inferred from overall similarity of fossil taxa and their stratigraphic sequence.

stratum. (*pl.* **strata**) *n.* 1. A horizontal layer of material, especially one of several parallel layers arranged in parallel one on top of another. 2. A layer of tissue. 3. A layer of rocks deposited about the same geological time. A bed or layer of sedimentary rock having approximately the same composition throughout. 4. Any of a number of levels or divisions in an organized system.

stratum basale. See *stratum germinativum*.

stratum compactum. That region of the *dermis* that contains distinct, ordered bundles of collagen fibers and underlies glands in the less dense *stratum spongiosum* above it.

stratum corneum. The outermost layer(s) of epidermis consisting of dead, keratinized cells (filled with keratin fibers), or partially cornified squamous cells, usually forming a syncytium. This layer consists of multiple cell layers in lepidosaurs but is only 1-2 cell layers thick in most amphibians. (Syn. *corneal layer, horny layer*) See Figs. 28–31.

stratum germinativum. The basal or innermost layer of living cuboidal or columnar cells of the epidermis that mitotically give rise to all cell layers of the epidermis above it. This cell layer rests on a basement membrane that marks the epidermal-dermal boundary. (Syn. *basal layer or lamina, basal generative layer, germinative layer, stratum basale, stratum Malpighii.*) See Figs. 28–31. The terms *stratum spinosum, stratum profundum,* and *stratum granulosum* are commonly used in descriptions of mammalian epidermis and were used somewhat ambiguously by A.A. Zimmerman & C.H. Pope (*Fieldiana Zool.* 32:357–413, 1948) in description of the keratinization of the rattle of Rattlesnakes. These terms have rather specific meanings in mammalian histology, and their usage in reference to lepidosaurian epidermis is discouraged.

stratum granulosum. Several replacement layers of cuboidal or polyhedral cells in amphibian skin immediately below the *stratum corneum*. These cells contain small granules and are involved in keratin formation. This and the *stratum corneum* have tight junctions. For comments regarding lepidosaurs see *stratum germinativum*.

stratum intermedium. In the skin of lepidosaurs, the region between the outer and inner *epidermal generations* that splits at shedding, consisting of the α-layer of the outer epidermal generation and the two cell layers beneath it (including *clear layer*). For further discussion of terminology see P.F.A. Maderson, *J. Zool.* 146:98–113, 1965.

stratum laxum. See *stratum spongiosum*.

stratum Malpighii. See *stratum germinativum*.

stratum spinosum. Several layers of cuboidal or polyhedral cells in adult amphibian epidermis immediately above the *stratum germinativum*. For comments regarding lepidosaurs see *stratum germinativum*.

stratum spongiosum. That region of the dermis overlying the *stratum compactum* and containing large glands (in amphibians), collagen and elastic fibers, smooth muscle, chromatophores, fibroblasts, capillaries with blood cells, and various nerves. (Syn. *stratum laxum*)

stream. *n.* A moving body of water, confined to a channel and running downhill under the influence of gravity.

stream channel. A long, narrow depression, shaped and more or less filled by a stream.

stream discharge. The volume of water in a stream that flows past a given point in a unit of time.

Stream Frogs. Collective vernacular name for African species of ranid frogs belonging to the genus *Strongylopus*.

stream headwaters. The upper part of a stream near the source.

stream larva. One of three types of larva described in salamanders, characterized by reduced fins, short bushy gills, absence of balancers, short toes, and usually occurrence in flowing water. See also *pond larva, terrestrial larva*. (Syn. *brook larva, mountain brook larva, mountain stream larva*)

stream mouth. The place where a stream enters the sea, a large lake, or a larger stream.

stream terrace. A steplike landform found above a stream and its flood plain.

strength. *n.* 1. In biomechanics, the capacity of supportive material to resist force without breakage or permanent deformation. 2. The capacity of muscle to produce much force.

streptostyly. *n.* A condition of *cranial kinesis* in which the *quadrate* forms a mobile joint with the *squamosal*, a result of the evolutionary loss of the lower temporal arch and characteristic of snakes and lizards. See Fig. 18.

stress. *n.* 1. The internal condition of an object that results from the transmission of a load. In biomechanics, the ratio of an applied force to the cross-sectional area over which the force acts. E.g., if F represents an applied force and A the initial area on which the force acts, the resulting stress = F/A, thus expressed in units of pressure. See *compression, tension, shear.* 2. Any force that is disturbing to an equilibrial condition or homeostasis in an organism. 3. The sum of the biological reactions to any adverse stimulus that tends to disturb an organism's homeostasis. A stress need not be synonymous with injury or damage, as normal functions such as muscle use can also act as a *stressor.* See *stress response.*

stress axis. Reference to the *hypothalamic-pituitary-adrenal (HPA) axis.* See also *stress response.*

stress response. A stereotypic, multifaceted response that occurs when an animal is exposed to any of a variety of *stressors*, regardless of the stimuli. The response includes physiological changes related to activation of the *hypothalamic-pituitary-adrenal axis* (HPA) and secretion of "stress hormones" that include catecholamines (epinephrine and norepinephrine), adrenocorticotropin (ACTH), and glucocorticoids (cortisol and corticosterone). The term "*biological stress response*" is used to describe the activation of the HPA axis and the resulting endocrinological actions that tend to promote the maintenance of homeostasis in response to challenge of a stressor. Understanding of the stress responses in amphibians and reptiles is at an immature stage, and assumptions based on mammalian studies need further testing.

stressor. *n.* Any aspect or element of the environment that stresses an animal or any of its physicochemical systems. Stressors are real or perceived challenges to an organism's ability to function and meet its needs.

stretch receptor. A sensory receptor that responds to a mechanical stimulus of stretch, typically associated with muscle, blood vessels, or the lung.

striate, striated. *adj.* A condition of having numerous parallel ridges or streaks. The term is applied to scales that bear minute and parallel grooves and ridges. See also *striated muscle.*

striated muscle. Skeletal and cardiac muscle that is characterized by sarcomeres arranged in register, producing microscopic *striations* in appearance.

striation. *n.* A streak or scratch, or a series of such structures.

strict. *adj.* Obligate.

stride. *n.* The distance traveled from beginning to end of a cycle of terrestrial locomotion involving limbs, beginning when one foot is lifted off the ground and ending when it again contacts the ground.

stridulating organ. See *stridulation.*

stridulation. *n.* Sound production by means of rubbing opposing patches of hardened epidermis together, as the rubbing of keeled scales of certain snakes (e.g. Saw-scaled Viper, *Echis*) or the opposing patches of hardened epidermis in the soft skin of the lower femur and the upper tibiofibula region of turtles of the family Kinosternidae. The latter have been termed *stridulating organ.*

strike. *n.* A term applied generally to sudden lunges of the head of reptiles, especially rapid prey-directed movements of snakes. Strikes directed at prey items are generally more accurate than are those employed in defense.

striking coil. A term introduced by L.M. Klauber (*Rattlesnakes*, pp. 476–486, 1956) to denote the position of the body of an agitated Rattlesnake, where the anterior aspect of the body is elevated off the substrate and drawn into an S-shaped loop of head and neck. The term can be applied to many snakes. Cf. *resting coil.*

string. *n.* A jargonish word for the *rattle* of a Rattlesnake.

striocristate. *adj.* Having both longitudinal parallel lines and comb-like ridges. Reference to a pattern of *epidermal microsculpturing.* See R.M. Price *J. Herpetol.* 16:294–306, 1982.

striolophate. *adj.* Having both longitudinal parallel lines and smooth longitudinal ridges. Reference to a pattern of *epidermal microsculpturing.* See R.M. Price *J. Herpetol.* 16:294–306, 1982.

strioreticulate. *adj.* Having longitudinal parallel lines, some of which anastomose into a network. Reference to a pattern of *epidermal microsculpturing.* See R.M. Price *J. Herpetol.* 16:294–306, 1982.

stripe. *n.* A narrow band of contrasting color that runs lengthwise on the body of a reptile or amphibian, most commonly seen in squamates.

Striped Frogs. Vernacular name for species of African hyperoliid frogs belonging to the genera *Paracassina* and *Phlyctimantis*

Striped Newt. Vernacular name for the American salamander species *Notophthalmus perstriatus.*

striped phase. Reference to a longitudinal stripe, or multiple stripes, of color running along the length of a snake that is normally blotched, ringed, or spotted, due in some cases to recessive alleles.

Striped Racer. Vernacular name for the American colubrid species *Masticophis lateralis.*

Striped Swamp Snake. Vernacular name for the American colubrid species *Regina alleni.*

Striped Whipsnake. Vernacular name for the American colubrid species *Masticophis taeniatus.*

Stripe-necked Snake. Vernacular name for several species of Asian snakes belonging to the genera *Liopeltis* and *Gongylosoma.*

Stripeneck Turtle. Vernacular name for largely Asian species of emydid turtles belonging to the genera *Mauremys* and *Ocadia.*

striping. *n.* The appearance of stripes, or partial, suggestive stripes, in color patterns of snakes due to abnormal developmental defects or possibly adaptive variation in species that move rapidly for escape from predators.

stroke volume. The volume of blood ejected from the ventricle of the heart in a single heart beat. In crocodilians having two ventricles, as in birds and mammals, the term is applied to the output of a single ventricle.

stroma. *n.* The connective tissue framework of an organ.

strong ion. Any of several ions that are essentially completely dissociated in physiological solutions or body fluids because their pK is much lower or higher than the biological pH. Examples are Na^+, K^+, and Cl^-.

strong ion difference. (SID) The difference between the total concentration of strong cations and the sum of strong anions in body fluids, which reflects the magnitude of the buffer base and is an important quantity in acid-base physiology. See *strong ion*.

structural color. A color seen on the surface of an animal attributable to physical or structural attributes that differentially reflect light in various ways, not due to the presence of pigment. (Syn. *static color*)

structural gene. Any gene that determines the amino acid sequence of a polypeptide.

structural protein. Any protein that contributes to the shape and structure of cells and tissues.

Stubby Frog. Vernacular name for species of Australian frogs belonging to the myobatrachid genus *Neobatrachus*.

Stubfoot Toad. Collective vernacular name for species of bufonid anurans belonging to the genus *Atelopus*, Also called *Harlequin Frog*.

Student's *t*-test. A computational test to determine the statistical significance of the difference between two sample means. The method was developed by W.S. Gosset, a British statistician who used the pseudonym "Student" in his publications.

Stump Heads. Vernacular name for snakes belonging to the family *Anomochilidae*, also known as *Dwarf Pipe Snakes*.

Stumptail Chameleon. Vernacular name for African species of Chameleons belonging to the genus *Rhampholeon*

style. *n.* 1. A synonym for the *shaker* of rattlesnakes. 2. Sometimes used as an alternate term for the *xiphisternum* element of the sternum. See Fig. 3.

styliform. *adj.* Slender and terminating in a point.

stylohyal, stylohyale. *n.* 1. A term used for the *extracolumella* in snakes. 2. A synonym for the *ceratohyal* of salamanders. 3. A process connecting the *principal horn* of the anuran hyoid to the otic region.

stylohyoid. *n.* See *columella-Meckelian cartilage*.

stylopodium. *n.* The proximal part of a limb, comprising the elements closest to the body. Cf. *autopodium, zeugopodium*. (Syn. *propodial*)

stylus. *n.* A bony rod extending from the *footplate* or *plectrum* to the *extracolumella*, or to the intermediate cartilages, in amphibians. Presumably derived from elements of the hyoid, the structure is functional through its attachment to the jaw in larval urodeles, while nonfunctional in most adults. (Syn. *interstapedial, mediostapes*)

sub-. Prefix indicating "below" or "beneath" (Syn. *infra-*)

subacrodont. *adj.* Reference to *acrodont* dentition when *polyphyodont*.

subadult. *n.* An animal approaching sexual maturity.

subangular gland. A rounded protuberance located below the jaw angle in male turtles of the genus *Gopherus*. A strongly scented secretion is released from the gland, which swells during the breeding season and is used to mark territories.

subapical lamella. Reference to any single or divided *subdigital lamella* extending across the underside of the digits in many Geckos, exclusive of the larger distal structures.

subaqueous. *adj.* In or under water.

subarachnoid. *adj.* Between the arachnoid and pia mater.

***subarcualis rectus.* (SAR)** See *ballistic tongue projection*.

subarticular tubercle. A *tubercle* lying at the base of the digits in anurans.

subcaudal. *n., adj.* 1. Any scale on the ventral side of the tail, either in a single or divided (two) series in snakes. (Syn. *caudal, scutellum, urostege*) See Fig. 13. 2. Of or pertaining to the ventral surface of the tail. 3. An older term for the *anal* lamina on the plastron of turtles. See Fig. 6.

subcaudal sac. See *postanal sac*.

subcentral foramen. Either of a pair of openings on the ventral aspect of the vertebral *centrum*, located on either side of the hemal keel or spine and variable in size and shape.

subclavian. *adj.* Beneath the clavicle; located in the shoulder.

subclavian artery. The artery that supplies the forelimbs with oxygenated blood in amphibians and reptiles.

subclimax. *adj.* A state in ecological succession that precedes the final or climax community.

subcutaneous. *adj.* Beneath the skin, but between the skin and muscle compartments.

subdigital. *adj.* Of or pertaining to the ventral surface of the digit.

subdigital adhesive pad. Reference to *subdigital lamellae* and *setae* used for adhesion during climbing in various species of geckos and anoles. See *subdigital lamella* and Fig. 32.

subdigital lamella. (*pl.* **-ae**) Any scale or pad on the ventral side of a digit in lizards. The term has been used especially with reference to the specialized climbing pads of geckos, skinks, and anoline lizards, which may be comprised of numerous *setae*. The subdigital pads of geckos has been the subject of functional speculation and study for at least 200 years. See *subdigital setae*. (Fig. 32)

subdigital pad. See *pad* and *subdigital lamella*.

subdigital setae. Reference to microscopic extensions of the corneous epidermis that form minute hair-like structures on the *subdigital lamellae* of geckos and anoline lizards, often with divided, spatulate terminations. These are adhesive foot hairs. In geckos each seta branches into hundreds of 200-nm spatulae that make intimate contact with a variety of surfaces (Fig. 32). Currently subdigital setae are thought to adhere to environmental surfaces during climbing by means of van der Waals forces (K. Autumn et al. *Nature* 405:681–685, 2000; *Proc. Nat. Acad. Sci.*

U.S.A. 99:12252–12256, 2002; A.P. Russell *Integr. Comp. Biol.* 42:1154–1163, 2002). See *subdigital lamella, subdigital adhesive pad*.

subdominant. *adj.* An animal that is subordinate to the dominant form or individual.

subduction. *n.* The sliding of the sea floor beneath a continent or island arc.

subechinate. *adj.* Having blunt spine-like protuberances. Reference to a pattern of *epidermal microsculpturing*. See R.M. Price *J. Herpetol.* 16:294–306, 1982.

subfamily. *n.* A classification of organisms that is more precise than a family but more inclusive than a genus. A subclade of a family.

subfoveal. *n.* A small *foveal* scale located between the *lacunals* and *supralabials*.

subgenus. *n.* A classification of organisms that is more precise than a family but more inclusive than a genus. A subclade of a genus.

subisodont. *adj.* Reference to relatively uniform teeth, but some being slightly enlarged. Used primarily in relation to maxillary teeth in snakes.

subjective. *adj.* Lacking objectivity and assessed as a matter of opinion or judgement. Cf. *objective*.

sublabial. *n.* Term used by various authors to denote *lower labial*, scales lying below the *lower labials* in lizards, or the scales lying between *chinshields* and *lower labials* in snakes. See Figs. 11, 16.

sublethal. *adj.* Reference to a dose of toxin that is less than the amount required to kill.

sublingual. *n.* 1. The area beneath the tongue. 2. See *chin shield*.

sublingual glands. See *oral glands, salt glands*.

sublittoral. *adj.* Reference to the shallow water zone of a lake or sea, generally from shoreline to the continental shelf or at depths between 6 and 200 m at sea, and between 6 and 10 m in freshwater lakes. (Syn. *infralittoral*)

submandibular gland. Each of a pair of circular glands opening on the throat near the inner posterior margins of the lower jaws in crocodilians and secreting an odoriferous mucus. The secretions are probably used in social communication.

submarginal. *n.* A term used variously with reference to the *inframarginal, supramarginal*, or supernumerary laminae internal to the ventral surface of the second *marginal* in certain turtles.

submarginal papillae. Fleshy projections on the face of the *oral disc* of tadpoles except the margin and elsewhere in species lacking jaw sheaths.

submaxillary. *n.* See *chin shield*.

submental. *n.* See *chin shield*.

submental groove. See *mental groove*.

submissive. *adj.* Reference to behavior of an animal that is defeated or deterred by aggressive contact with another individual.

submissive coloration. Coloration that is displayed by a subordinate individual.

subnasal. *n.* A term that has been applied to the scale in the *canthal* series that lies immediately below the *nasal* in sceloporine lizards.

subneural process. A term for *epapophysis*, as substituted by W. Auffenberg (*Tulane Stud. Zool.* 10:131–216, 1963).

subnivean. *adj.* Below snow.

sub-notochordal rod. See *hypochordal rod*.

subocular. *n., adj.* In general usage, reference to the region or any scale that lies immediately below the eye, between the orbit and *upper labial* scales. Various authors have distinguished *pre-* and *post-suboculars*, and the term has also been used to designate the labial lying directly below the eye in some forms. See Fig. 16.

suborbital bar. Reference to a dermal ridge beneath the eye, usually extending posteriorly to the angle of the jaw.

subordinate gesture. See *head nod*.

subordinate taxon. A taxon of lower rank than that with which it is compared.

suborder. *n.* A taxonomic category of classification, ranking below *order* and above *family*.

subpleurodont. *adj.* Reference to a *pleurodont* dentition in which *intercalary replacement* takes place, thought to be transitional to an *acrodont* condition.

subpopulation. *n.* A breeding group within a larger population or species range that experiences reduced migration relative to other groups.

subrictal. *n.* A term for posterior *postlabials* when these are clearly distinguishable from the anterior *postlabials*, used in the lizard genus *Phrynosoma* wherein one of these scales may be enlarged and of taxonomic significance.

subrictal spine. The spine located below the corner of the mouth and the rectal spine in Horned Lizards (*Phrynosoma*).

subsequent designation. In taxonomy, the designation of the type of a taxon in a work that is published subsequent to the establishment of the taxon. Cf. *original designation*.

subsidence. *n.* Sinking or downwarping of a part of Earth's surface.

subsidized island biogeography hypothesis. A new hypothesis for describing and predicting patterns of species diversity on small islands and habitat fragments. The idea is a modification of traditional *island biogeography equilibrium theory* that incorporates the influence of spatial subsidies from the surrounding environment, which vary among islands and habitat fragments, on species diversities. That is, the role of allochthonous resources is incorporated into analysis of recipient communities' patterns of diversity. See W.B. Anderson & D.A. Wait, *Ecology Letters* 4:289–291, 2004.

subsistence capture. The capture of wild animals by peoples living in close contact with the species, where such capture is customary, traditional, and necessary for the sustenance of such individuals and their families or immediate kin groups.

subspecies. *n.* A taxonomic category used to distinguished "races" within a species, i.e. geographically and/or ecologically defined subdivisions of a species with distinctive characteristics. This designation is the third name in a *trinomial name* designating a subspecies. See also *race*.

Subspecific epithet or **name.** *Subspecies*, or the third name in a *trinomial name*.

subspectacular. *adj.* Reference to the area immediately beneath the spectacle of the eye in squamates.

substrate. *n.* 1. A substance that is acted on by an enzyme. 2. Equivalent word for *substratum*.

substrate race. A local race having a characteristic coloration resembling that of its *substratum*.

substratum. (*pl.* **substrata**) *n.* The ground or surface on which an animal rests or moves.

subtaxon. (*pl.* **–a**) *n.* A subdivision of a taxon.

subterranean. *adj.* Reference to location, condition or living below the ground surface. (Syn. *subterrestrial*)

subterrestrial. *adj.* See *subterranean*.

subulate. *adj.* Elongate and tapering to a point.

succession. *n.* An orderly process of development or change in community structure through time. In its general form, succession is a process of community or ecosystem change starting from an unoccupied or incompletely occupied site and terminating in a more or less stable condition (known as *climax*).

successional speciation. See *phyletic evolution*.

successional species. Successive *morphospecies* within a distinct evolutionary lineage and sufficiently distinct to justify the assignment of species names.

sucker. *n.* See *adhesive organ*.

sucking disc or **disk.** See *adhesive organ*.

suctorial. *adj.* Reference to the large ventral adhesive organ of tadpoles that maintain position in fast moving water by adhering to rocks with the a sucking disc. The term *suctorial disc* was used originally in reference to the ventral structure of modified integument and musculature immediately behind the mouth of tadpoles of the Asian frog *Staurois*, permitting adhesion to rocks in torrential streams. The definition has now been extended to include any ventral-facing oral apparatus of tadpoles that maintain position by adhering to rocks.

sudd. *n.* A floating mass of plant material.

sulcate. *adj.* Furrowed, or marked with *sulci*.

sulcus. (*pl.* **–i**) *n.* 1. A shallow groove or narrow channel, usually used in reference to brain tissue. Cf. *gyrus*. 2. This term also has been used for the impressions of *seams* on surfaces of bone in the turtle carapace and has been employed as shorthand for *sulcus spermaticus*.

sulcus coronarius. A term used by E.T. Francis (*The Anatomy of the Salamander*, p. 187, 1934) for the deep groove that separates the atria from the ventricle in *Salamandra*.

sulcus marginalis. A furrow or groove extending along the inner margins of the upper lip of many frogs, which contacts the rim of the lower jaw when the mouth is closed.

sulcus spermaticus. A distinct groove extending along the outer surface of an everted hemipenis, or penis in turtles, functional in forming a canal through which spermatozoa move to the opening of the female oviduct during copulation. (Syn. *ductus spermaticus, semen canal, spermatic canal, spermatic duct, spermatic groove, spiral groove*)

sulphonamide. *n.* An organic bactericidal compound used in the treatment of a variety of bacterial infections in captive reptiles and amphibians.

sum of squares. In an analysis of variance, the sum of the squared deviations from the means.

summation. *n.* The addition of response. 1. In muscle physiology, the addition of muscle tension due to repeated and rapid stimulation of muscle. 2. In neurophysiology, *spatial summation* refers to the integration by a postsynaptic neuron of simultaneous synaptic inputs from terminals of different presynaptic neurons. *Temporal summation* refers to the additive effects of synaptic inputs that arrive close together in time or temporal sequence.

Sunbeam Snakes. Collective vernacular name for various species of snakes belonging to the families *Loxocemidae* and *Xenopeltidae* (or equivalent problematic taxa), especially the neotropical species *Loxocemus bicolor* (also called *Burrowing Python*) and the Asian genus *Xenopeltis*.

sunfish type. See *tadpole*.

sunglasses. *n.* A term used for the black-bordered, semi-transparent area in the lower eyelid of some lizards, thought to provide protection against bright sunshine.

super-, supra-, sur-. Prefix meaning "above," "over" or "greater."

supercanthal. *n.* Any of a series of scales lying anterior to the *superciliary* series between the *canthals, frontonasals*, and *prefrontals* in lizards (characteristic of *Ophisaurus*).

superciliar, superciliary, supercilium. *n.* Used for *supraocular, superorbital, ciliary,* or *preciliary* scales by many early authors. See Figs. 10, 16.

superciliary spine. An elongated, pointed scale situated in the middle (gekkonids) or posterior end (*Phrynosoma* spp.) of the *superciliary* series of scales. (Syn. *postocular spine*)

supercontinent. *n.* An earlier and larger land mass composed of currently existing continents. See *Pangea*.

supercooling. *n.* Reference to cooling below the physical freezing (= melting) point of body fluids without formation of ice crystals.

superfamily. *n.* A taxonomic division above *family* and below *order* or *infraorder.*

superficial. *adj.* Toward or at a structure's surface.

superficial fascia. See *hypodermis.*

superimposed immunization. A technique for preparation of antivenin involving an animal being immunized against whole venom and subsequently being reimmunized using increased doses of a specific venom fraction such as a neurotoxin fraction. The antivenin prepared in this manner provides more protection than does antivenin prepared from whole venom alone.

superior taxon. A taxon of higher rank than the one with which it is compared.

superior *vena cava.* See *precaval vein.*

supermaxillary. *n.* A term from L.M. Klauber (*Occ. Pap. San Diego Mus. Nat. Hist.* 5:1–61, 1939) for the *maxillary* of a Rattlesnake.

supernatant. *n.* The fluid that overlies a precipitate following the centrifugation of a suspension.

supernumerary tubercles. Tubercles located on the lower surface of the digits, beneath the phalanges and between the subarticular tubercles.

superposition. *n.* A principle that within a sequence of undisturbed sedimentary rocks, the oldest layers are on the bottom, and the youngest are on the top.

superspecies. *n.* A monophyletic group of essentially *allopatric species.* A cluster of incipient species or related *semispecies.* The superspecies name is enclosed in square brackets between the two words of the binomen. (Syn. *collective species, species group*)

supertramp. *n.* Jargonish reference to species having a wide geographical range, efficient means of dispersal, and usually excellent colonizing ability.

supination. *n.* (*v.* **supinate**) The act of turning a part such that the ventral side faces upward.

supinator. *n.* A muscle that rotates or turns the ventral side upward.

supra-. Prefix meaning "above," "over" or "superior." Cf. *infra-.*

supra-anal keel or **tubercle.** A prominent keel on the lateral scales of the posterior body near the vent in some species of Coral Snakes. These structures are characteristic of male snakes, but occasionally occur in very large adult females of certain species. More properly termed *supracloacal keel.*

supra-angular. *n.* See *surangular.*

supracaudal. *adj., n.* 1. Reference to the dorsal surface of the tail. 2. Used variously by authors to denote the *postcentral* lamina (Syn. *anal, caudal, postcentral, postmarginal*) or the *suprapygal* bone of the turtle carapace. See Figs. 4, 5.

supraciliary. *adj.* 1. Above the eye. See Figs. 10, 16. 2. See *supraocular.*

supra citato. (*supra cit.*) Latin, meaning "cited above."

supradorsal. *n.* See *proatlas.*

suprageneric. *adj.* Above the rank of genus.

supraglenoid buttress. A thickened region of the reptilian scapular blade above the glenoid cavity.

suprahumeral. *n.* Of or pertaining to the dorsal surface of the upper portion of the forelimb.

supralabial. *n.* One of the scales bordering the upper lip. See Figs. 11, 16.

supralabial gland. 1. A glandular dermal ridge extending along the upper lip of some ranid frogs, extending posteriorly and turning down behind the vocal sac in male animals. This gland secretes mucus and granular gland products in various species. 2. See also *oral glands.*

supralarvation. *n.* Reference to the phenomenon of a larval amphibian achieving a larger body size than that of the metamorphosed adult.

supralittoral. *adj.* Relating to habitat above the high water mark of the shore of a sea or lake, defined as the shoreline region that lies entirely above the water's edge but is affected by waves and sea spray.

supraloreal. *n.* A term used for a scale in the upper row of *loreals* when several scales are present in the loreal region, most often with reference to vipers.

supramarginal. *n.* Used variously by authors to refer to any of a row of *laminae* between the *marginals* and *laterals* on the carapace of some turtles, the dorsal surface of a *marginal,* or a row of tubercles or scales between *costal* and *marginal* rows of the tail-trunk of *Chelydra.* See Fig. 4.

supramastoid. *n.* Used as a synonym for *supratemporal* or to denote the *squamosal* in Crocodiles.

supranasal. *n.* Generally used with reference to any scale lying directly above the *nasal* and lateral to the *internasal* of squamate reptiles.

supranasal sac. A recess formed by invaginated skin beneath the supranasal scale, having an inconspicuous slit-like opening between the supranasal and nasal scales. This structure is present in many snake species of the subfamily Viperinae and the related Causinae (Night Adders). Studies of innervation suggest the supranasal sac functions as a heat detector like the labial pits of boas (D.S. York et al., *Anat. Rec.* 251:221–225, 1998).

supranumerary tubercle. Any tubercle on the bottom of the hand or foot of an anuran that is not a *subarticular tubercle.*

supraoccipital. *n.* A median cartilaginous bone, derived from chondrocranium and forming the dorsal border of the *foramen magnum* and extending dorsally to the *parietals.* Present in most reptiles and in caecilians.

supraocular. *adj., n.* 1. Reference to the area immediately over the eye. 2. Reference to a shield or scales lying dorsally above the orbit in snakes or lizards. (Syn. *superciliar,* *superciliary, supercilium, supraciliary, supraorbital, palpebral*) See Figs. 10, 14.

supraocular disk. The dorsal part of the head immediately above the eye in *Anolis* lizards, usually comprised of one or more series of scales. (Syn. *palpebral disk*)

supraocular ridge. The elongated margin of the *supraocular* scale which overlies the eye of a number of species of snakes, particulary various vipers. (Syn. *supraorbital ridge*)

supraorbital. *n.* 1. The region medial to, or overlying, the orbit. See Fig. 10. 2. Used as a synonym for the *supraocular* of snakes or the *palpebral* bone.

supraorbital ridge. See *supraocular ridge*.

supraorbital semicircle. A term used generally in reference to an arc of small scales separating the *supraocular* from the medial head scales. The term has also been applied to a semi-circular series formed by the *prefrontals, frontals, parietals,* and *interparietals* in certain genera of lizards. (Syn. *circumorbital, circumorbital semicircle*)

suprapericardial body. See *ultimobranchial glands*.

suprapterygoid fenestra. A conspicuous large space (but occluded with cartilage) present in the occipital aspect of the skull of bufonid anurans, bounded ventrally by the parasphenoid arm of the *pterygoid*, laterally by the *squamosal*, and dorsally and medially by the *prootic*.

suprapygal. *n.* Either of usually two bones independent of vertebrae and situated at the midline of the turtle carapace, between the last bone associated with the vertebrae and the *pygal*. (Syn. *epipygal, metaneural, postneural, procaudal, pygal, supracaudall*) See Fig. 5.

suprarenal gland. See *adrenal gland*.

supra-rostral cartilage. Reference to principal support of the mandibles and labial teeth of most tadpoles, missing in forms lacking mouth armament, such as microhylids.

suprascapula. *n.* Reference to the upper end of the *scapula*, which may be cartilaginous or ossified. (Syn. *adscapulum, episcapulum, omolitum, scapula major, suprascapularis*)

suprasquamosal. *n.* See *supratemporal*.

suprastapedial. *n.* See *extracolumella*.

supratemporal. *n.* 1. A dermal skull bone of primitive tetrapods, forming the central bone in a row extending from the *postfrontal* to the posterior end of the skull along the outer edge of the *parietal*. 2. The term also is used to denote large scales on the upper temporal region of the head in marine turtles and in some lizards.

supratemporal bridge. Reference to the arch of bone extending dorsolaterally and separating two distinct skull openings in *diapsid* skulls. The term may be applied to either *postfronto-squamosal* arch or *postorbito-squamosal arch*.

supratemporal foramen. The upper *temporal fossa* or opening.

supratibial. *adj.* Of or pertaining to the dorsal surface of the lower leg.

supratympanic. *n., adj.* 1. Above the eardrum. 2. A term used in reference to the enlarged scale immediately above the Tympanic membrane of turtles and also the two or three large scales that cover the *opercle* of crocodilians. 3. See *preparotoid*.

supratympanic fold or **ridge.** An extension or fold of skin above, and sometimes overlying, the tympanum of various anurans, occasionally extending posteriorly to overlie the shoulder in some species.

surangular. *n.* A dermal bone of the lower jaw, overlying the *Meckelian cartilage* between the *dentary* and *articular*, with which it may fuse in some reptiles. This element is present in all reptiles, and absent in most amphibians. (Syn. *supra-angular, suprangular*)

surf. *n.* Waves that break on the shore.

surface coat. See *coat*.

surface emissivity. See *emissivity*.

surface-feeding type. One of seven types of tadpole distinguished by G.L. Orton (*Syst. Zool.* 2:63–75, 1953).

surface tension. The elasticity of the surface of a fluid, which tends to minimize the surface area at each interface.

surfactant. *n.* A surface-active secretion by type II cells in the pulmonary epithelium of all vertebrates, consisting primarily of lipids and protein, and functioning to reduce surface tension, maintain osmotic balance, and prevent adhesion of collapsed pulmonary surfaces.

Surinam Toad. Vernacular name for species of Middle and South American anurans belonging to the pipid genus *Pipa*. The name applies, in particular, to the species *Pipa pipa*.

survival rate. The percentage of individuals surviving from one developmental stage, year class, or life stage to the next stage or succeeding period.

survival value. Reference to the effectiveness of a character in conferring fitness.

survivorship. *n.* The proportion of individuals from a given cohort that survive at a given time or age. See *survivorship curve, survivorship schedule*.

survivorship curve. A graph of the number of surviving individuals of a given cohort plotted against age. Three principal patterns of survivorship have been identified: Type 1, a survivorship curve in which the probability of surviving decreases with age; Type 2, a survivorship curve in which the probability of survival is constant with age; and Type 3 in which the probability of survival increases with age.

survivorship schedule. A schedule of probabilities that an individual survives from birth to the beginning of age *x*.

suspension. *n.* A dispersion of fine insoluble particles in a fluid.

suspension feeding. See *filter feeding*.

suspensorium. *n.* 1. Anything that suspends or holds up a part. 2. With reference to anatomy of lower vertebrates, the chain of bones or cranial elements that forms the side wall of the mouth cavity and connects the lower jaw with the skull. Sometimes called *mandibular suspensorium.*

sustainable developoment. Human development of an area or habitat that is economically viable without continual degradation of the environment or loss of renewable resources.

suture. *n.* A seam or groove formed between two adjacent parts of a structure, including the lines of articulation between bones of the turtle shell, boundaries between two abutting plates or scales, seams between interlocking bones of the skull, and the line of fusion of the two edges of the venom canal in snake fangs. In general, reference is to a union or seam between bones at an immovable joint. (Syn. *crease, seam, sulcus*)

SVL. See *snout-vent length.*

swamp. *n.* An area that is saturated or periodically inundated with standing water, typically dominated by woody plants and having wet, spongy ground.

Swamp Frog. See *African Swamp Frog* and *Australian Swamp Frogs.*

Swamp Snake. Vernacular name for American snakes belonging to the colubrid genus *Seminatrix.*

swell. *n.* In geology, an area that has subsided less than a surrounding area and is covered by a thinner sequence of sedimentary deposits.

swell mechanism. Reference to rapid expansion of parts of the head of lizards and presumably snakes, which facilitates loosening the old, outer generation of epidermis at the time of *ecdysis.* The mechanism involves a sphincter on the internal jugular vein, which upon contraction causes engorgement of erectile tissue around the nostrils, the sinus at the base of the nictitating membrane, and several other sinuses. This same mechanism also functions with respect to a fright reflex in Horned Lizards, *Phrynosoma,* resulting in ejection of blood from the eyes.

swimming frenzy. A period of heightened activity and rapid swimming of hatchling Sea Turtles when they set out to sea following emergence from the nest.

syllable. *n.* See *trill note.*

symbiont. *n.* Any organism involved in a symbiotic relationship with other organisms. See *symbiosis.*

symbiosis. *n.* An interactive relationship in which an organism lives with another organsim, to their mutual benefit. Usually the one organism lives inside, or is attached to, the other one.

symmetrical. *adj.* Characterized by equal and corresponding morphology or characters on opposite sides of a plane.

symmorphosis. *n.* The concept that functional capacity in each of a series of linked physiological components of a system is well matched to the typical demands of the system.

sympathetic nervous system. A division of the *autonomic nervous system* with peripheral ganglia located generally close to the spinal cord and derived from thoracolumbar outflow. Increased activity acts antagonistically to that of the *parasympathetic* system and generally activates responses associated with vigorous activity or the "fight-or-flight" response.

sympatric speciation. A relatively uncommon speciation process in which two or more populations inhabit common or overlapping geographical range and become reproductively isolated. The process may involve selection for ecological, behavioral, or phenological segregation of evolving populations. Morphological differentiation for use of resources is correlated with characters used in mate choice.

sympatry. *n.* (*adj.* **-ic**) Living in the same geographic location, with reference usually to the overlap in geographic range of two closely related species, which remain otherwise distinct. Cf. *allopatry, parapatry.*

symphygnathine. *adj.* Reference to a skull wherein the maxillae meet in front of the premaxillary bones. Used in reference to microhylid frogs by R.G. Zweifel (*Amer. Mus. Novit.* 1766:1–49, 1956).

symphysal, symphyseal, symphysial. *adj., n.* 1. Generally, reference to the symphysis of the lower jaw. 2. Used by many early authors to designate the *mental* of snakes. See Fig. 12. 3. Used to designate the cartilage and the bone ossified from it at the anterior end of *Meckel's cartilage.* The cartilages support the lower lip of larval amphibians. (Syn. *mental, mentomeckelian bone*)

symphysial groove. See *mental groove* and Fig. 12.

symphysis. (*pl.* **–es**) *n.* 1. The union or articulation of two opposing halves of a structure. 2. An *amphiarthrosis* having a pad of collagenous fibers or fibrous cartilage separating the bones.

symplesiomorphy. *n.* (*adj.* **-ic**) Reference to primitive characters that are found in the common ancestor of groups sharing the character, thus indicating common ancestry for these groups. Cf. *synapomorphy.*

sympleisiotypy. *n.* The common possession of a *pleisiotypic* character. Cf. *synapotypy.*

symporter. *n.* A carrier protein that transfers two solutes in the same direction across a biological membrane. Cf. *uniporter.*

symptomatic. *adj.* Exhibiting, or pertaining to the nature of, a symptom. Usually used in reference to disease.

syn. See *synonym.*

syn-. A prefix meaning "associated," "together," "joined," "fused," or "united."

synangium. *n.* The cranial aspect of the conus arteriosus, which exits the pericardium and divides into right and left arterial branches. (Syn. *truncus impar*; cf. *pylangium*)

synapomorphy. (*pl.* **-ies**) *n.* (*adj.* **–ic**) A derived character state or characters (*apomorphies*) shared by two or more taxa.

(Syn. *derived similarity*) See also *plesiomorphy*. Cf. *symplesiomorphy*.

synapophysis. *n.* A term used for the transverse process of a vertebra that articulates with a single-headed rib. The designation is used when the articulating surface is regarded as attributable to a fusion of the *diapophysis* and *parapophysis*, especially if the area is shared by neural arch and centrum. See *paradiapophysis*.

synapotypy. *n.* The common possession of a derived character state. Cf. *symplesiotypy*.

synapse. *n.* The functional junction between two neurons or between a neuron and an effector such as muscle. The presynaptic (transmitting) cell influences the activity in the postsynaptic (receiving) cell.

synapsid. *adj., n.* 1. Reference to a reptilian skull in which there is a single temporal opening or fossa, and the postorbital and squamosal meet above it. Characteristic of therapsid reptiles and mammals. 2. A member of the *Synapsida*. See also *postorbital fenestrae*.

Synapsida. *n.* A major group of tetrapods that includes *pelycosaurs* of the lower Permian and their mammal-like descendants, the *therapsids*, which gave rise to mammals at the end of the Triassic. These all have a single temporal opening with a bar of bone beneath it.

synaptic cleft. The small space separating the cells at a synapse.

synaptic efficacy. The effectiveness of a presynaptic action potential in producing a postsynaptic potential change, hence response.

synaptic facilitation. An increase in *synaptic efficacy*.

synaptic inhibition. Any effect acting at a synapse that reduces the probability of firing or level of excitation of the postsynaptic cell.

synaptic plasticity. The capacity for long-lasting or permanent changes in *synaptic efficacy* as a result of experience.

synaptic transmission. The transfer of a signal (via a neurotransmitter or passage of membrane potential change) between a presynaptic and postsynaptic neuron.

synaptic vesicles. Membrane-bound vesicles that contain neurotransmitter molecules and occur typically in terminal zones of axons.

synarthrosis. *n.* An immovable joint. If such a connection between elements is of bone, it is termed a *synostosis*; if between cartilage, a *synchondrosis*; if of fibrous connective tissue, a *syndesmosis*.

syncephalic. *adj.* A rib having a single head resulting from the fusion of the *capitulum* and the *tuberculum*.

synchondrosis. *n.* Union between two bones having a cartilaginous zone of contact, but with little or no movement. The cartilage may be lost through ossification. See *synarthrosis*.

synchronic. *adj.* Synchronous or contemporaneous. Cf. *allochronic*.

synchronous breeding. Reference to a circumstance when all females arrive at the breeding site roughly at the same time.

synchronous calls. Reference to breeding calls in which male anurans respond to one another rapidly such that calls overlap and produce rhythmic bursts of sound.

syncline. *n.* A geologic fold in which the layered rock usually dips toward an axis.

syncope. *n.* Faint, or temporary loss of consciousness, due to generalized cerebral *ischemia*.

syncranterian. *adj.* Possessing maxillary teeth with no gap separating the posteriormost members of the series, which are not conspicuously enlarged. Cf. *diacranterian*.

syncytium. *n.* (*adj.* –al) An aggregation of cells without cell boundaries, thus forming multinucleated cytoplasm.

syndactyl. *adj.* Possessing feet having two or more digits fused together into a single unit, as in Chameleons.

syndesmosis. *n.* (*adj.* -otic) 1. See *synarthrosis*. 2. Union between two bones having rough contiguous surfaces with an interosseus ligament between them, allowing moderate movement.

syndrome. *n.* A set of symptoms occurring together.

synecology. *n.* An older term for a branch of ecology that deals with the study of groups of organisms associated together as a unit.

synergism, synergy. *n.* Cooperative and reinforcing action between two or more agents such that the combined or total effect is greater than the sum of the individual, component actions.

synergist. *n.* In physiology, a muscle or agent which acts with another to amplify an effect. E.g. two or more muscles act together to produce motion in the same direction.

syngamic sex determination. Determination of sex as a result of fusion of gametes.

synonym. (*syn.*) *n.* In taxonomy, one of two or more scientific names applied to the same taxon. A *senior synonym* is the earliest available name for a taxon, whereas a *junior synonym* is any available name other than the senior synonym.

synonymize. *v.* To conclude that one described taxon is a junior synonym of another having an earlier name (the senior synonym).

synonymy. *n.* A chronological list or record of the scientific names applied to a species and its subdivisions.

synopsis. *n.* 1. A summary or synthesis of current knowledge. 2. A brief description of the essential features of a taxon.

synostosis. *n.* See *synarthrosis*.

synovial. *adj.* Relating to the viscous lubricating fluid that occurs within joint capsules and tendon sheaths.

synovial joint. A *diarthrosis*, defined by a synovial capsule consisting of dense fibrous connective tissue lined by a *synovial membrane* that secretes a lubricating *synovial fluid*.

synovial fluid. See *synovial joint*.

synovial membrane. See *synovial joint.*

synsacry. *n.* A condition in certain frogs characterized by fusion of the sacral and one presacral vertebrae, completely obscuring the location of the vertebral ball and socket between them.

synthesis, modern. See *neo-Darwinism.*

synteny. *n.* The property of being on the same chromosome. *Conserved synteny* indicates that genes on the same chromosome in one species are also on the same chromosome in the comparison species. (Syn. *homology blocks*)

synthetic evolution, theory of. A view that evolution is best explained by a synthesis of prior knowledge and theory, especially including mutation and selection. See *neo-Darwinism.*

syntopic. *adj.* Occurring at the same place or in the same microhabitat, thereby capable of interbreeding. Cf. *allotopic.*

syntype. *n.* In taxonomy, a member of the type series of a taxon whose author did not designate a holotype.

system. *n.* An organized assemblage of individual items into a single integrated whole.

Système Internationale (SI). See *international units.*

systematics. *n.* The study of phylogeny and classification of organisms based on evolutionary relationships. The field also may be described as the study of biological diversity and its origins. Refers to systems of classification. (Syn. *biosystematics*) See *taxonomy.*

systemic. *adj.* Pertaining to, or affecting, the entire body. A generalized, in contrast to local, response. See also *systemic blood circulation.*

systemic arch. See *aortic arch.*

systemic blood circulation or **circuit.** Reference to the vascular system that transports blood throughout the body, other than the lung or respiratory surfaces. Cf. *respiratory blood circulation.*

systole. *n.* (*adj.* –**ic**) That part of the heart cycle when the cardiac muscle is contracting and ejecting blood into the arterial system. Cf. *diastole.*

systolic pressure. The highest arterial blood pressure measured during the systolic phase of the cardiac cycle.

T

t. Student's *t* statistic used for testing the difference between two sample means.

tabular, tabulare. *n.* A dermal skull bone of primitive tetrapods, forming the last bone in a series between the *postfrontal* and the back end of the skull, extending along the outer edge of the parietal. It is located above the *otic capsule* and may extend downward over the occiput. In snakes, a single bone remains in this region, between the *parietal* and *quadrate*, and has been called *tabular*. Similarly, a bone in lizards between the *parietal* and *squamosal* has been called tabular.

tachy-. Prefix meaning "fast," "quick."

tachycardia. *n.* Reference to increased heart rate above the normal, resting level. Cf. *bradycardia*.

Tachycneminae. *n.* A subfamily of *Hyperoliidae*.

tachygen. *n.* A structure that appears suddenly during evolutionary history of a lineage.

tachygenesis. *n.* Accelerated ontogenetic or phylogenetic development.

tachykinins. *n.* A family of peptides named for a rapid ability to act on smooth muscle (in contrast to slower-acting kinins and bradykinins). These substances occur in amphibian skin as well as brain and gut tissues.

tachyphylaxis. *n.* Rapid immunization to a toxic substance following previous exposure to small doses of the same substance.

tachytelic evolution, tachytely. Relatively rapid evolution. Cf. *bradytelic, horotelic evolution*.

tactic. *n.* A term used in evolutionary biology to describe behavioral or morphological characters having expression that is contingent either on environmental conditions or on the status of individuals in which they appear. The term is useful for describing phenotypes that are flexible in their expression, in contrast to those that segregate strictly according to Mendelian genetics.

tactile. *adj.* Of, or relating to, touch.

tactile alignment. A behavior of courting snakes in which the male attempts to align his tail with that of the female such that the cloacas are adjacent. See *tactile chase*.

tactile bristle. A slender, hair-like projection from the center of the apical pits of certain lizards. (Syn. *apical bristle, bristle, hair, tactile hair*)

tactile chase. Reference to courtship behaviors in which a male snake follows and attempts to mate with a female. The male repeatedly tongue flicks and attempts to align his body or tail with that of the female, and the movements become spasmodic as he repeatedly attempts to locate her cloaca with his own (*tactile alignment*). See also *tail-search copulatory attempt*.

tactile hair. See *tactile bristle*.

tactition. *n.* The sense of pressure perception, usually involving the skin.

tad. *n.* Slang term for tadpole occasionally, and inappropriately, appearing in literature.

tadpole. *n.* Term for the nonreproductive larval life stage of any anuran. The term does not refer to immatures of other amphibians. See *tadpole type*.

tadpole type. Reference to descriptive stages in the evolution of tadpole structure and function, described by P.H. Starrett (pp. 251–271 in J.L. Vial, ed., *Evolutionary Biology of the Anurans: Contemporary Research on Major Problems*, 1973) following G.L. Orton (*Syst. Zool.* 2:63–75, 1953; *Syst. Zool.* 6:79–86, 1957). Type 1 is the most primitive and represented by pipoid larvae. Type 2 is derived, types 3 and 4 are more derived, and type 4 is more specialized than type 3.

tag. *n.* 1. A physical or chemical device that is attached to an animal and permits future identification of that particular individual. 2. See *label*.

taiga. *n.* The northern coniferous forest lying to the south of the Arctic tundra zone.

tail. *n.* 1. A caudal region posterior to the trunk, defined by various authors as either the body posterior to the anus or posterior to the sacrum, which are not necessarily the same. The terms *ura* and *coccygeum* have been proposed for these definitions, respectively, but have not been followed by most herpetologists. 2. The intromittent organ of the frog *Ascaphus*, which extends from the cloaca, has been referred to as a tail by some authors.

tail autotomy. See *autotomy*.

tail axis. An imaginary line that connects the body terminus and the tip of a straightened tail.

tail crest. The dorsal margin of the *tail fin* on a tadpole or certain lizards.

tail curling. A display behavior in which the tail is curled up and held high over the head by certain lizards and snakes (e.g. many coral snakes). The tail in this position may be held stationary or waved from side to side.

tail dropping. Colloquial substitute for *autotomy*.

Tailed Frog. Collective vernacular name for the North

American ascaphid frog species belonging to the genus *Ascaphus*.

tail filament. A fine bristle-like appendage of connective tissue extending from the tip of the tail in certain salamanders. In some species the structure is renewed at the breeding season.

tail fin. The thin, nonmuscular, and often colorless margins of the tail of tadpoles. See Fig. 1.

tail flipping. Reference to movement of tadpoles, especially hatchlings, by frantic flips of the tail (in contrast to regular undulations).

tail gland. See *anal gland*.

tail length. In general, the measured distance from the vent to the tip of tail, variously measured from either the anterior or posterior margin of the vent, or from the rear margin of the anal plate in snakes and lizards. In rattlesnakes, the measurement is from the vent to the proximal rattle.

tail ratio. Total length (body plus tail) divided by tail length. Cf. *body ratio*.

tail-search copulatory attempt. A courtship behavior in which a male snake attempts to locate a female's tail with his own, wrapping it around that of the female and moving the coils back and forth preceding *tactile alignment*.

tail sheath. Dense layer(s) of apparent connective tissue in the proximal third of the tail of some tadpoles, commonly seen in preserved specimens.

tail spine. Reference to the spiny or nail-like terminal of the tail in certain turtles; the terminal scale on the tail tip of snakes; and the vertical protuberance on the base of the tail of male spine-tailed salamanders (*Mertensiella* spp.), which is tactile and used to stimulate the female during mating.

tail tremor. Quivering or rapid wiggling of the tail in certain lizards during periods of excitement or prey stalking.

tail vibrating. Reference to thrashing or rapid shaking of the tail when a snake is disturbed or threatened. Characteristic of a number of different taxa, the tail movements often produce sounds similar to that of a rattlesnake when the animal is among dry leaves or other substrate debris.

tail walk. A courtship behavior exhibited by some plethodontid salamanders, involving a stimulated female following a male that is waving his sharply bent tail, culminating in deposition of a spermatophore. Cf. *waltz*.

tail waving. See *caudal luring*.

Taipan. *n.* Vernacular name for Australasian elapid snakes belonging to the genus *Oxyuranus*.

talus, talus slope. *n.* 1. See *astragalus*. 2. From geology, a pile of rock fragments or weathered debris at the base of a cliff, or a mantle of rock fragments on a slope below a rock face. Used synonymously with *scree slope*.

tapestry. *n.* A more or less continuous cover of trees on a steep slope.

tapetum. *n.* 1. A layer of cells. 2. See *tapetum lucidum*.

tapetum lucidum. A reflective choroid formed by accumulation of guanine crystals in the retinal pigment epithelial cells, found in the eyes of nocturnal species and producing, incidentally, the nocturnal "eye-shine" of animals when a light is directed at their eyes.

taphonomy. *n.* 1. The scientific study of the changes that have modified animal remains during the period between the death of an animal and its discovery and examination as part of the fossil record. Such changes may be biological or physical in nature. 2. The pathways and processes by which an assemblage of fossils is formed. The term was coined in 1940 by the Russian paleontologist I.A. Efremov.

TAR. An older abbreviation for "thermo-activity range." See *activity temperature range*.

target effect. In island biogeography, the effect of island area on immigration rate.

target organ or **tissue.** The organ or tissue on which a hormone has its effect and which possesses receptors for the specified hormone.

tarichatoxin, tarichotoxin. *n.* A potent neurotoxin found in the skin and internal organs of newts (*Taricha* spp.) and similar to tetrodotoxin from Puffer Fish. The toxin blocks nerve impulses and results in respiratory paralysis and death. The same toxin has also been isolated from some species of frogs.

tarsal. *n.* 1. Any bone located between the tibia or *fibula* and the *metatarsals*. 2. A scale in reptiles, located on the hind foot between the digits and the articulation with the *tibia* and *fibula*, often used with prefixes to indicate the position of the scale.

tarsal fold. A distinct ridge or fold extending along the margin of the tarsal region of the foot in many species of frogs. (Syn. *tarsal ridge*)

tarsal ridge. See *tarsal fold*.

tarsal spur. An outgrowth of the heel.

tarso-metatarsal articulation. The articulation between the *metatarsus* and the *tarsus*.

tarsus. *n.* The bones which form the articulation between the foot and leg (ankle).

taste buds. Chemosensory organs or receptors responsive to chemicals entering the mouth. The chemoreceptors of taste, present in the mouth and pharynx of amphibians, reptiles, and birds.

tautology. *n.* Unnecessary repetition of a word or statement.

taxic aggregation. Term used to describe an asocial aggregation caused by individuals responding to some sort of *taxis* rather than social interactions.

taximetrics, taxometrics, taxonometrics. *n.* See *phenetic method*.

-taxis, -taxy. Suffix meaning "arrangement" or "arrangement of."

taxis. *n.* A locomotor response associated with orientation to a stimulus direction or gradient.

taxon. (*pl.* **taxa**) *n.* A general term for a group of organisms recognized as a unit of classification. A hierarchical category in a classification. Specifically, a group of organisms classified together on the basis of sharing a single common ancestor.

taxonomic. *adj.* Relating to *taxonomy*.

taxonomic category. Reference to the rank of a taxon in a classification hierarchy.

taxonomic congruence. The degree to which different classification schemes postulate similar groupings of the same organisms.

taxonomic distance. Any measure of dissimilarity between or among taxa.

taxonomic group. Any taxon, including all subordinate taxa.

taxonomic hierarchy. A hierarchical system of taxonomic categories arranged in an ascending series of ranks. Eight principal ranks are recognized in Zoology: Domain, Kingdom, Phylum, Class, Order, Family, Genus, Species. Additional categories are present in many classification schemes, usually identified with the use of *super-* and *sub-* prefixes.

taxonomic position. The position of a taxon in a hierarchy of classification.

taxonomic species. See *species*.

taxonomy. *n.* (*adj.* –ic) The study of the naming and classification of living organisms according to an established set of principles based in theory and techniques. Classically, the description and classification of organisms was based on morphology, but increasingly considers genetic variation and interrelationships. The term was coined in 1813 by A.P. de Candolle (a botanist) for the process of classification. See *systematics*.

TCS Technique. An older treatment for envenomation by snake bite referred to as "Tourniquet, Cut, and Suck" method. The procedures involve application of a tourniquet proximal to the bite, multiple X-shaped incisions at and near the bite, and application of suction to the incisions. Formerly a commonly accepted method for treatment of snake bite, this is not currently recommended as an effective treatment. Cf. *AAA treatment*.

tecnophagy. *n.* (*adj.* –ic, -ous) The act of a female animal eating her own eggs, observed in skinks and a few snakes.

tectal plate. The posterior part of the amphibian chondrocranium, overlying the nasal capsule. Term coined by G.M. Higgins (*Illinois Biol. Monogr.* 6:1–91, 1920) to distinguish this from the mammalian cribiform plate. (Syn. *lamina cribosa, planum tectale*)

tectonic, tectonics. *adj., n.* Relating to the deformation of the structure of Earth's crust. *Tectonics* refers to the study of the structural features and processes of Earth's crust.

tectorial membrane. A membrane that covers a patch of sensory hairs in the inner ear.

tectum. *n.* The roof of the midbrain, representing a region of grey matter that is an important "center" for brain activity controlling the body in amphibians and, to a large degree, in reptiles.

tectum supraorbitale. See *otoparietal plate*.

TED. See *turtle excluder device*.

teeth. *n.* See *tooth*.

Tegus. *n.* Collective vernacular name for species of lizards belonging to the family Teiidae, especially comparatively large species of the genus *Tupinambis* distributed in many South American habitats and heavily exploited by humans. See also *macroteiid*.

T-1824. See *Evans blue*.

teiid. *n, adj.* A member of, or pertaining to, *Teiidae*.

Teiidae. *n.* A clade (family) of lizards often referred to as *macroteiids*, which are active, diurnal inhabitants of a range of habitats from deserts to tropical rain forests. Nine genera and 125 species are distributed in North America, Central America, South America, and the West Indies. Two subfamilies are recognized: *Teiinae* and *Tupinambinae*. This and the family *Gymnophthalmidae* comprise the *Teiioidea* within *Scincomorpha*.

Teiinae. *n.* A subfamily of *Teiidae*.

Teioidea. *n.* A clade of lizards comprised of the sister groups *Teiidae* and *Gymnophthalmidae* within *Lacertiformes*.

tela subcutanea. A term of fascia or connective tissue that underlies the *dermis* and joins skin to structures beneath it.

tele-. Prefix meaning "beyond."

telencephalon. *n.* The anterior terminal segment of the brain, derived from the embryonic prosencephalon and including the cerebral hemispheres. (Syn. *cerebrum*)

teleology. *n.* A doctrine that natural phenomena result from, or are shaped by, design or purpose.

teleonomy. *n.* A doctrine that structure and function of all characters have evolved by natural selection.

Telmatobatrachia. *n.* A new monophyletic taxon of anurans, sister to *Ceratobatrachidae* within the new taxon *Victoranura*, and containing *Micrixalidae* and the new taxon *Ametrobatrachia* (D.R. Frost et al., *Bull. Amer. Mus. Nat. Hist.* 297:1–370, 2006).

telobranchial body. See *ultimobranchial body*.

telocinobufagin. *n.* A *bufogenin*.

telolecithal egg. An egg, typical of reptiles and birds, which contains a large quantity of yolk that is concentrated toward one pole and does not participate in cleavage and blastula formation. The latter are limited to a clear protoplasmic cap. Cf. *mesolecithal egg*.

TEM. See *electron microscope*.

Temnospondyli. *n.* A diverse, long-lived, worldwide group of non-amniote tetrapods (*batrachomorphs*), characterized by flat, immobile skulls and a reduction of the hand to four fingers. The modern amphibians, or at least the frogs, likely originated within this group.

temnospondylous. *adj.* A type of vertebra characterized by the presence of a double-disk centrum composed of two distinct and unfused parts. The condition may be either *embolomerous* or *rhachitomous*.

temperate. *adj.* Reference to latitudes between the tropics and polar circles in each hemisphere. Cf. *polar, tropical.*

temperate rain forests. Forests between the tropics and the polar circles, from 35° to 55° north or south latitudes, where mild, wet winters and warm summers occur. These forests are cold and rare, and conifers are the dominant trees. The annual precipitation exceeds 250 cm in some areas, predominantly in the form of fog or snow.

temperature. *n.* An arbitrary scale of heat intensity. See *Fahrenheit* and *Celsius*. Cf. *heat.*

temperature acclimation. See *acclimation.*

temperature coefficient. See Q_{10}.

temperature-dependent sex determination. (TSD) A process of sex determination in which the sex of an individual is determined not by genotype but by the temperature experienced by embryos during critical periods of development. TSD is characteristic of some lepidosaurs, most turtles, and all crocodilians. For a recent book devoted entirely to this important subject, see N. Valenzuela & V. Lance, eds., *Temperature-dependent Sex Determination in Vertebrates*, 2004. Cf. *genetic sex determination.*

temperature profile. A plot of isotherms showing the distribution of temperatures throughout part of a habitat or microhabitat (such as a sand beach where turtles nest).

Temple Pitviper. Vernacular name for *Tropidolaemus wagleri*, a pitviper native to southeastern Asia. Also called *Wagler's Palm Viper.*

temporal. *adj. n.* 1. Reference to the region of the skull behind the eye. 2. Used to denote the *squamosal* bone of the skull. 3. A synonym for supratemporal, with reference to marine turtles. 4. Reference to scale or scales behind the *postoculars,* below the *parietal,* and above the upper *labials* in snakes and lizards. The term has been used with other adjectives or prefixes to describe relative position of scales, which are often arranged vertically in two or more rows. See Figs. 11, 16. 5. Of or relating to time.

temporal arches. The structural consequence of fenestration of the vertebrate skull, with reference to the upper and lower arches or bars formed by the defining skull bones. (Syn. *temporal bar*)

temporal bar. See *temporal arches.*

temporal fenestrae. See *temporal opening.*

temporal fossa. See *temporal opening.*

temporal fovea. A *fovea* that lies high on the posterior part of the retina.

temporal horn. Reference to any of the large, elongated hornlike spines on the sides or posterior margin of the head in certain Horned Lizards (*Phrynosoma* spp.).

temporal isolation. The absence of interbreeding between populations due to different reproductive seasons or timing of gamete production. A premating isolating mechanism. See *reproductive isolating mechanism.*

temporal opening. Used generally in reference to any gap between the bones of the cheek region of the reptilian skull. (Syn. *temporal fenestrae, temporal fossa, temporal vacuity*)

temporal vacuity. See *temporal opening.*

temperolabial. *n.* A large scale that projects downward between the fifth and sixth *supralabial* scales of some snakes, especially a number of Australian elapid species.

temporomastoid. *n.* Antiquated term that denoted the *squamosal* in salamanders.

tendon. *n.* A band or cord of fibrous connective tissue that is continuous with muscle fibers and attaches muscle to bone or cartilage. The tendon is noncontractile, but transfers force from muscle contraction to the element on which it attaches.

tendon process. A protuberance at the proximal end of certain ribs in some lizard genera, which attach to tendons of the dorsal musculature. (Syn. *muscle* or *muscular process*)

tensile force. 1. An applied force in a direction that tends to pull apart an object. 2. The force produced by muscle contraction.

tension. *n.* Stress in an elastic solid that results from a load directed away from the object and perpendicular to its surface. See *stress* and Fig. 24.

tentacle. *n.* 1. A protrusible structure lodged in a groove or canal of the maxillary bone at the posterior end of the naso-lacrimal duct of caecilians. The lumen opens to the vomeronasal organ, and the structure bears chemoreceptors that are strongly inferred to be functional in the identification and capture of prey. 2. A cutaneous projection in larvae of *Xenopus*, which has a cartilaginous core and has been compared with the *balancers* of salamanders.

Tentacled Snake. Vernacular name for the homalopsine species *Erpeton tentaculatus* of southeast Asia. This aquatic species is named for prominent, paired rostral protuberances that are probably sensory but have unknown function.

tentacular foramen. In caecilians, the depression or opening on the lateral skull containing the tentacular apparatus through which the tentacle is protruded.

ter-. Prefix meaning "three" or "threefold."

teratogen. *n.* (*adj.* **-ic**) Any agent or influence that causes physical defects or deformity in a developing embryo.

teratology. *n.* A branch of embryology dealing with studies of malformations.

teratoma. *n.* 1. A tumor that consists of a disorganized aggregation of different tissue types. 2. A term commonly used in place of *dysplasia* in the past, appropriate only if the implant of embryonic tissue into the adult host possesses neoplastic properties.

Terciopelo. *n.* Vernacular name for the Central American pitviper *Bothrops asper*. (Syn. *Fer-de-lance, Barba Amarilla, Yellow-jaw Tommygoff*)

terminal. *adj.* Reference to the termination of an anatomical structure.

terminal awn. A barbed process, or an elongate, terminal papilla at the apex of the hemipenis in some snakes (called "*terminal awn*" by H.G. Dowling and J.M. Savage, *Zoologica* 45:17–31, 1960).

terminal extinction. The loss of an entire species population without production of any descendant lineage.

terminal immigrant. An immigrant to a new area in which it can survive but cannot reproduce.

terminal phalanx. The distalmost phalanx of a digit.

terminal shield. The ventral unsegmented terminal area including the cloaca found in some caecilians.

terminal taxa. Reference to taxonomic groups that occur at the ends of branches in a *cladogram*. The tips of branches in a *phylogenetic tree*.

Terra. *n.* A herpetological publication of the Belgian Herpetological Society.

terraced. *adj.* A condition of unequal projection of a scaled surface, as at locations where nonoverlapping scales differing in size and shape meet each other.

terrain fear factor. A concept that prey species will alter their use of space and foraging patterns according to features of the terrain and the extent to which those features affect risk of predation. Proposed as a conceptual model for assessing the relative predation risk effects associated with encounter situations (W.J. Ripple & R.L. Beschta, *Forest Ecology and Management* 184:299–313, 2003).

terrapin. *n.* Informal term having reference to any of a number of freshwater turtles, especially semiaquatic species (*Pond Turtles, Diamondback Terrapins,* etc.).

terrarium. (*pl.* –ia) *n.* A container used to house captive amphibians or reptiles, usually of glass or plastic construction and closed but with ventilation. The internal space of such terraria usually contain a substrate material, water dish, hide box, plants, or other elements appropriate to the environment of the captive species.

terrestrial. *adj.* Living on land, predominantly at ground level.

terrestrial larva. One of three types of larva described for salamanders, characterized by large, flattened, three-lobed gills, absence of tailfins, and large amount of yolk. The entire larval life occurs in the egg without an aquatic phase in water. See also *pond larva, stream larva*.

terricolous. *adj.* Living on or in soil, spending most of its active life on, or below, the ground.

territorial behavior. Any behavior associated with the establishment, possession or defense of an area that is defended against incursion by other members of the species. Such behaviors typically involve ritualized displays or combat. (Syn. *territoriality*)

territoriality. *n.* 1. Defense of an area by an individual or group for its exclusive use. 2. An instinctive behavior related to the occupation and defense of a space referred to as a *territory*.

territory. *n.* (adj. *-ial*) An area of habitat that is defended against entry usually by other members of the same species. See also *home range*.

Tertiary. *n.* One of the major Periods of the geologic time scale, extending from the end of the Cretaceous Period about 65 million years ago to the beginning of the Quaternary Period about 1.8 million years ago. The Tertiary Period covers the time roughly between the demise of the dinosaurs and the beginning of the most recent ice age. See Table 1.

tertiary egg membranes. See *egg membranes*.

tertiary sex ratio. See *sex ratio*.

tessellate, tessellated. *adj.* Checkered, with squares or blocks of color abutting only at the corners.

test. *n.* In statistics, a process used to determine whether to accept or reject a statistical hypothesis, or whether observed values differ significantly from expected values.

testis. (*pl.* –es) *n.* The male gonad in which *spermatozoa* are produced. See Fig. 38.

testis cord. See *sex cords*.

testosterone. *n.* A steroid androgen synthesized in the testes, secreted primarily from interstitial cells (*Leydig cells*) of the testes, and responsible for appearance and maintenance of male secondary sex characteristics.

test statistic. A statistic on which a *statistical test* is based.

Testudines. *n.* The clade (order) that includes all extant turtles and tortoises, including 13 families and more than 285 species. (Syn. *Chelonia, Testudinata*)

Testudinidae. *n.* A clade (family) of turtles commonly called *tortoises*. These are terrestrial with highly domed shells, stout limbs, heavy bodies, and unwebbed feet. Most are herbivorous or omnivorous. Approximately 11 genera and about 40 species are recognized, ranging from southern North America, South America, southern Eurasia and Africa, Madagascar, and a number of oceanic islands.

Testudinoidea. *n.* A clade of extant turtles that includes *Testudinidae, Geoemydidae (Bataguridae),* and *Emididae*.

testudoid or **testudinal.** *adj.* Relating to or resembling a turtle or tortoise.

tetanus. *n.* An uninterrupted, sustained muscle contraction caused by high frequency motor impulse stimulation.

Tethys. *n.* The epicontinental sea separating *Laurasia* from *Gondwanaland* following the breakup of *Pangaea* during the Mesozoic.

tetra-. Prefix meaning "four" or "fourfold."

tetradactyl. *adj.* Having four digits.

tetraiodothyronine. *n.* See *thyroxine*.

tetramer. *n.* A structure that results from an association of four subunits.

tetrapod. *n.* A four-limbed vertebrate animal, including members of the *Tetrapoda*: amphibians, reptiles, birds, and mammals.

Tetrapoda. *n.* A superclass of vertebrates having four limbs with carpals, tarsals, and digits. This taxon incorporates amphibians, reptiles, birds, and mammals.

tetrodotoxin. *n.* See *alkaloids*.

TEWL. In physiology, used as an abbreviation for *transepidermal water loss* and, by others, for *total evaporative water loss*. The latter term would include evaporative loss from respiratory surfaces (during breathing) in addition to that from the skin surfaces.

Texas Coral Snake. Vernacular name for the American elapid species *Micrurus tener*.

textilotoxin. *n.* An extremely toxic component of venom from the Australian elapid *Pseudonaja textilis*.

thalamus. *n.* Either of paired masses of nerve cell bodies forming the lateral wall of the diencephalon, serving as an important relay center for sensory and motor nerve traffic.

thamnocolous. *adj.* Living on or in shrubs and bushes.

thamnophilic. *adj.* Thriving on or in shrubs and bushes.

thanatocoenosis. *n.* Reference to an assemblage of fossils that form after the death of the organisms, but does not necessarily indicate a corresponding behavioral association of these same individuals in life.

thanatophidia. *n.* A term to denote venomous snakes collectively; not a formal taxonomic unit.

thanatosis. *n.* See *death feint*.

theca. *n.* 1. A case or sheath. 2. An outer coat of egg follicles formed by connective tissue cells in the ovary. 3. Also used in herpetology as a name for the *brood pouch* of Marsupial Frogs (*Gastrotheca* spp.).

thecal. *adj., n.* 1. Reference to any bone or derivative of the primary dermal ossification in turtle shell. 2. A term for any of the independent membrane bones derived from the dermis and forming the bony shell of turtles, overlain by *epithecals*.

thecodont. *adj., n.* 1. A mode of tooth attachment in which a relatively long cylindrical base is set in a deep bony socket, usually affixed by uncalcified tissues such as collagen fibers. The term is best restricted to those cases, such as mammals, crocodilians, and archosaurs, in which there is a true periodontal membrane between the tooth and alveolus. In most true thecodont reptiles, cementum is deposited on the tooth and on the walls of the alveolus, but there is always soft tissue between the two (J.W. Osborn, *Symp. Zool. Soc. Lond.* 52:549–574, 1984). Cf. *acrodont, pleurodont, protothecodont ankyloses*. 2. See *Thecodontia*.

Thecodontia. *n.* An obsolete term no longer used by phylogenetic systematists because its content and meaning are synonymous with those of *Archosauria*. The term "*thecodont*" is still legitimately used in reference to socketed teeth. See *thecodont*.

thenar. *adj.* Pertaining to the palm or base of the thumb.

theory. *n.* An explanation for observed phenomena that has a high possibility of being true, often based on a large number of observations and tested hypotheses. No theory is absolute, however, and remains open to scrutiny and further testing.

therapeutic. *adj.* Serving to heal or cure.

therapeutic index. See *efficacy index*.

therapsid. *adj., n.* Reference to a member of the *Therapsida*, a clade of the *Synapsida*.

Therapsida. *n.* An order of *synapsids* that includes mammals and their extinct basal relatives, previously known as "mammal-like reptiles."

therm-. Prefix meaning "heat."

therm. *n.* A secondary *fps* unit equal to 1×10^5 BTU or 1.05506×10^8 joules.

thermal. *adj., n.* 1. Relating to heat. 2. A doughnut-shaped bubble of circulating warm air.

thermal balance. A state of equilibrial temperature. See *balance*.

thermal conductance. See *conductance, thermal*.

thermal conductivity. (*k*) A property of a material or object that describes the rate at which heat flows within the object for a given temperature difference (e.g., $W \cdot cm^{-1} \cdot {}^{\circ}C^{-1}$). Cf. *thermal conductance*.

thermal diffusivity. A measure of the rate at which a temperature disturbance, or change in heat flow, at one point in a body travels to another point. Thermal diffusivity is expressed by the relationship $K/\rho c_p$, where K is the coefficient of thermal conductivity, ρ is density, and c_p is the specific heat at constant pressure. Thermal diffusivity also is sometimes expressed as the rate of change of temperature in a transient heat transfer process. The higher the thermal diffusivity of a material, the higher the rate of temperature propagation.

thermal environment. A part of the physical environment relating to all elements and properties affecting heat exchange between a system (organism) and its environment.

thermal inertia. A tendency to slow the gain or loss of heat due to the contribution of an animal's mass to its *heat capacity*.

thermal radiation. *n.* Electromagnetic radiation emitted from a body due to its temperature, generally in the *far infrared* for animals at physiological temperatures and many terrestrial objects such as trees, rocks, etc. See *infrared radiation*.

thermistor. *n.* A semiconductor sensor that exhibits a repeatable change in electrical resistance as a function of temperature. Most thermistors exhibit a negative temperature coefficient. Cf. *thermocouple*.

thermo-. Prefix meaning "heat."

thermo-activity range. (TAR) See *activity temperature range*.

thermobiology. *n.* The study of the effects of heat on organisms and biological processes.

thermoception. *n.* The sense of heat (and the absence of heat, or cold), usually involving the skin and internal skin tissues.

thermocline. *n.* A mid-layer of water in a lake or ocean characterized by rapidly decreasing temperature (with depth). See *metalimnion*.

thermoconformer. *n.* A species or animal whose body temperature conforms to that of its environment and in which behavioral thermoregulation is weak or absent. Generally, tropical species of amphibians and reptiles tend to conform to ambient temperatures moreso than do temperate species. See *conformer*.

thermocouple. *n.* A temperature sensor created by joining two dissimilar metals and used to measure temperatures accurately. The junction produces a small voltage or potential difference that varies as a function of temperature. Cf. *thermistor*.

thermocycle. *n.* A cycle of temperature change.

thermodynamics. *n.* Formal designation of the study of energy exchanges between a system and its environment. The strength of this discipline lies in the generality of well-known first and second "laws" of thermodynamics that provide insight to understanding numerous problems related to systems (including organisms) and their environment.

thermogenesis. *n.* Literally the production of heat, used in relation to heat production by metabolic means (e.g. muscle contractions of brooding pythons).

thermometer. *n.* A device for measuring the temperature of an object or environment. See *cloacal quick-reading thermometer, infrared thermometer, thermistor, thermocouple*.

thermoperiodic. *adj.* Reference to responses of organisms to periodic changes of temperature.

thermophase. *n.* Reference to the part of a thermocycle during which higher temperatures prevail. Cf. *cryophase*.

thermophilic behavior. Heat-seeking behavior.

thermophobic behavior. Heat-avoiding behavior.

thermoreceptor. *n.* A sensory nerve ending specifically responsive to temperature changes.

thermoregulation. *n.* Regulation of body temperature. See *behavioral thermoregulation*.

thermotaxis. *n.* Movement of an organism toward or away from a heat stimulus.

Theropoda. n. A stem-based sister taxon to *Sauropodomorpha*, which together comprise the basal clades of the *Saurischia*. The name theropod means "beast foot" and was proposed by O.C. Marsh in 1881 for all the carnivorous dinosaurs known at that time. Birds are considered by many to be direct descendants of theropods.

theropods. *n.* Reference to any and all members of the Theropoda.

thigmotactic. *adj.* Crevice-dwelling, or reference to behaviors of wedging between unyielding objects.

thigmotherm. *n. (adj. -ic)* An animal that utilizes heat transfer by conduction from the environment, usually from contact with substrate, as a primary means of thermoregulation.

thigmothermy. *n.* Thermoregulation, or primary heat transfer, achieved by conduction from the environment, especially the substrate on which an animal rests.

thinic. *adj.* Reference to sand dunes.

thinicolous. *adj.* Living on sand dunes.

thinophilous. *adj.* Thriving on sand dunes.

thin layer chromatography. (TLC) Reference to *chromatography* in which the stationary phase is a thin layer of absorbent silica gel or alumina spread onto a flat glass plate.

thin shield. See *chin shield*.

third eye. See *parietal eye*.

third eyelid. See *nictitating membrane*.

thoracic. *adj., n.* Pertaining to the chest or thorax. In older literature this term was used to denote the *pectoral* lamina of turtles. See Fig. 6.

thoracicolumbar vertebra. See *dorsal vertebra*.

thoracic vertebra. This term has been used variously to denote a vertebra that possesses freely articulating ribs, a vertebra articulating with ribs that reach the sternum, and a vertebra situated in the region enclosed by elongated rotatable ribs and bounded posteriorly by the diaphragm. These definitions are based on mammalian distinctions and are inapplicable to amphibians and many reptiles. Substituted terms include *trunk vertebra* in amphibians, *sternal vertebra* in Sphenodon, *dorsal* or *thoracicolumbar vertebra* in reptiles, and *presacrals* in both amphibians and reptiles.

thoracolumbar vertebra. See *dorsal vertebra*.

thorn. *n.* A term that has been used to denote the terminal spine on the tail of typhlopid snakes.

Thorny Devil. An extraordinary species of Australian agamid lizard, *Moloch horridus*, covered with spiny, warty protuberances and narrow channels between scales that act to "wick" water droplets of dew or rain to the mouth. This lizard displays several traits that are convergent with similar appearances in American Horned Lizards.

Thornytail Iguana. Vernacular name for species of South American iguanid lizards belonging to the genus *Uracentron*.

Thoropidae. *n.* A new, monotypic taxon (family) of anurans, sister to *Dendrobatidae* within the *Dendrobatoidea*, and containing the Brazilian genus *Thoropa* (D.R. Frost et al., *Bull. Amer. Mus. Nat. Hist.* 297:1–370, 2006).

Thread Snakes. *n.* Vernacular name for various species of snakes in the family Leptotyphlopidae. See also *Worm Snakes*.

threat behavior. Any innate action or response of a species related to aggression or intimidation of potential predators.

threatened. *adj.* Reference to taxa that are likely to become endangered within the foreseeable future. This designation is essentially the same as the *"vulnerable"* category used by the IUCN.

3,5,3-triiodothyronine. (T_3) *n.* A tyrosine-derived hormone bearing iodine and secreted from the thyroid gland. This hormone has complex functions in metabolism, as does *thyroxine*.

threshold potential, threshold stimulus. Reference to the membrane potential at which a response, or action potential, is elicited. The *threshold stimulus* is the minimum stimulus energy required to produce a detectable response in a receptor or neuron.

throat fan. See *dewlap* and Fig. 40.

throat gland. See *musk gland*.

thrombin. *n.* A blood enzyme that is activated by prothrombin and converts fibrinogen to fibrin, which polymerizes into a fibrin clot.

thrombin-like enzymes. Glycoprotein enzyme components of viper and pit viper venoms that act as anticoagulants *in vivo*.

thrombocyte. *n.* A small, nucleated blood cell responsible for clotting of blood and other processes.

thrombocytopenia. *n.* A deficiency of *thrombocytes*.

thrombolectin. *n.* Protein component of some snake venoms that act on blood platelets or thrombocytes.

thrombosis. *n.* Formation of a blood clot or solid mass from blood constituents within a blood vessel or the heart.

thrombus. *n.* A blood clot that impedes circulation within a blood vessel. Cf. *embolism, embolus*.

throughfall. *n.* Water from precipitation that penetrates a tree canopy and reaches the forest floor without running down the trunks of trees.

thrust. *n.* 1. A force that produces forward motion. Thrust can result from displacement of a fluid (air or water). 2. In geology, a low-angle reverse fault.

thumb pad. A *nuptial pad* consisting of an enlarged, cornified area on the basal joint of the first digit in male anurans.

thymine. *n.* A pyramidine base and component of DNA.

thymus, thymus gland. *n.* A gland in or near the neck that produces lymphocytes and establishes the immunological potential of a young animal.

thyrohyal. *n.* 1. A prominent, ossified projection on the posterior medial edge of the *hyoid plate* of anurans, derived from the fourth *ceratobranchial*. (Syn. *posterior-medial process, processus thyreoideus, thyroid horn, thyroid process, thyrohyoid process*) 2. Antiquated synonym for *thyroid bone*. 3. Sometimes used in reference to the posterior bifurcation of the snake *hyoid*, but considered by some to be homologous with the *hypohyal* of other reptiles.

thyrohyoid process. See *thyrohyal*.

thyroid, thyroid gland. *n.* A gland located in the ventral region of the neck that produces iodinated hormones *thyroxine* and *triiodothyronine*.

thyroid bone. A small, median ossification of the *hyoid* skeleton in certain salamanders. The element is located anterior to the *pericardium* and posterior to the *hyoid plate*, to which it is connected during larval life by the *urobranchial shaft*.

thyroid cartilage. A cartilaginous element in the larynx of crocodilians (and mammals), in addition to the *cricoid* and *arytenoids*.

thyroid fenestra. See *cordiform foramen*.

thyroid hormones. See *thyroxine* and *triiodothyronine*.

thyroid horn. See *thyrohyal*.

thyroid membrane. A membrane associated with the processes of the hyoid and often surrounds the thyroid gland in anurans.

thyroid process. See *thyrohyal*.

thyroid-stimulating hormone. (TSH) A glycoprotein hormone that is secreted from the anterior pituitary and functions to stimulate secretory activity of the thyroid gland. (Syn. *thyrotropin*)

thyrotropin. *n.* See *thyroid stimulating hormone*.

thyrotropin-releasing hormone. A hormone released from hypothalamic secretory axon, which stimulates secretion of *TSH*.

thyroxine. (T_4) *n.* A hormone derived from thyrosine and secreted from the thyroid gland. It bears iodine and has complex functions in metabolism and a variety of physiological processes, including amphibian metamorphosis, dormancy, thermoregulation, and nutritional state, often acting in concert with catecholamines. (Syn. *tetraiodothyronine*) Cf. *triiodothyronine*.

tibia. *n.* The inner and larger bone of the lower part of the leg, between femur and foot, of tetrapod vertebrates.

tibial. *n.* Reference to any of the scales on the hindlimb of lizards, situated between the knee and ankle. The term is often supplemented with prefixes to indicate position.

tibiale. *n.* See *astragalus*.

tibia length. A morphometric measured variously as the distance between the convex surface of the knee to the convex surface of the heel in frogs, with both tibia and tarsus flexed; or the distance from center of knee to center of heel; or from heel to fold of skin at the knee.

tibial epiphysis. An accessory endochondral center of ossification at the end of the tibia in some reptiles.

tibial gland. A granular *macrogland* found on the lower leg of some species of bufonid and myobatrachid anurans.

tibial notch. A notch in the distal tibial epiphysis in some lizards.

tibial segment. The portion of leg containing the tibia and fibula. (Syn. *crus, shank*)

tibial spur. A bony ridge or projection on the tibia in some salamanders.

tibiofibula. *n.* The epipodial element in the hind limb of amphibians, formed by fusion of the *tibia* and *fibula*. (Syn. *os cruris*)

tick. *n., v.* 1. Vernacular name for various species of blood-sucking acarid parasites, a number of which infect reptiles—usually attaching to softer interscale tissue near hinge regions, limb pockets, and skin folds. 2. A term coined by W.M.S. Russell (*Behaviour*, 7:114–188, 1954) for the production of clicking sounds by a female *Xenopus* if not ready to spawn when clasped by a male.

Tic Polonga. Vernacular name for the venomous Russell's Viper, *Daboia russelli*, of south and southeast Asia.

tidal current. Alternating horizontal movement of water associated with the rise and fall of a tide.

tidal flat. An extensive area of land that is alternatively covered and uncovered by tidal water, consisting mostly of unconsolidated mud or sand.

tidal marsh. A marshy coastal area formed of mud and root mats of halophytic plants, regularly inundated by high tides.

tidal volume. In respiratory physiology, this refers to the volume of air that is moved in or out of the lungs with each individual breath.

tide. *n.* (*adj.* –**al**) The periodic rise and fall of the ocean water masses and atmosphere, produced by the gravitational effects of the moon and sun on Earth.

Tiger Rattlesnake. Vernacular name for the American pitviper *Crotalus tigris*.

Tiger Salamander. Vernacular name for *Ambystoma tigrinum* and its several subspecies.

Tiger Snake. Vernacular name for various Australian elapid snakes belonging to the genus *Notechis*, and Eurasian and African snakes belonging to the colubrid genus *Telescopus* (also called *Catsnakes*).

tight junction. An area of fused membranes between adjoining cells, which functions to impede or prevent the passage of extracellular material between the adjoining cells. See also *zonula occludens*.

till. *n.* Unsorted and unlayered rock debris carried by a glacier.

timber line. A line at higher altitudes above which trees do not grow. *Tree line*.

Timber Rattlesnake. Vernacular name for the North American pitviper *Crotalus horridus*. See also *Canebrake Rattlesnake*.

time constant. (τ) Generally, a measure of the rate of accumulation or decay in an exponential process, equivalent to the time required for an exponential process to reach 63% of completion.

time series. Data or samples taken at given time intervals.

Tinctanura. *n.* A new monophyletic taxon of anurans sister to *Cryptobatrachidae* within the *Cladophyrnia*, containing *Amphignathodontidae* and the new taxon *Athesphatanura*. See D.R. Frost et al., *Bull. Amer. Mus. Nat. Hist.* 297:1–370, 2006. Members are cosmopolitan in temperate and tropical areas of the continents, Madagascar, Seychelles and New Zealand.

tine. *n.* A prong, or slender projecting part, used in reference to the forked tips of the tongue of snakes and some lizards.

tiphic. *adj.* Pertaining to ponds.

tiphicolous. *adj.* Living in ponds.

tissue. *n.* A population of organized cells that perform the same orchestrated function (e.g., muscle).

tissue fluid. See *interstitial fluid*.

titer. *n.* The amount of a standard reagent necessary to produce a certain result in a titration.

titillate. *v.* A term to describe the act of a male turtle when, during courtship, it rapidly fans and taps the face of a female, often while swimming backward. In some species, long claws on the forefoot are rapidly vibrated during titillation. Term used by J.A. Oliver (*Natural History of North American Amphibians and Reptiles,* p. 332, 1955).

TL. Tail length or total length.

TLC. *Thin layer chromatography*.

toad. *n.* A categorical name for certain anurans, once used to differentiate terrestrial anurans (toads) from amphibious or aquatic ones (frogs). In the vernacular, members of the genus *Anaxyrus* (formerly *Bufo*) are still called toads, but the name also is used for various other anurans. Thus, lines are blurred and there is not strict differentiation of toads from frogs. The name is especially relevant to numerous species comprising the family Bufonidae, while members of the bufonid genus *Anaxyrus* are referred to as the "true toads." See also *Asian Toads*.

toadfly. *n.* A fly (*Bufolucilia bufonivora*) that deposits eggs within the nasal openings of toads. The larvae feed mainly on mucus but may penetrate the eyes and brain as they develop.

Toadfrog. *n.* See *Megophryidae*.

Toadheads, Toadhead Agamas. Collective vernacular name for various species of agamid lizards belonging to the genus *Phrynocephalus*.

toadlet. *n.* A term sometimes used for a recently transformed toad.

toe clip. *v.* A method for marking lizards or amphibians by clipping one or more toes in specified combinations to identify individuals. Most published reports of toe-clipping document few adverse effects, but there are exceptions especially in amphibians.

toe disc. *n.* See *disc*.

Tokay, Tokay Gecko *n.* A species of large Gecko (*Gekko gecko*) common to parts of Asia.

tolerance. *n.* Capability to withstand something, usually used in reference to some physical aspect of the environment (e.g., temperature) or to invasive chemicals affecting physiology (e.g., salts, contaminants, etc.).

toluidine blue. A metachromatic basic dye used in cytochemistry.

tomium. *n.* A term used by A. Carr (*Handbook of Turtles*, p. 36,

1925) to denote the cutting edge of the keratinized *beak* or jaw sheaths of turtles.

tomography. *n.* The recording of internal body images by means of an X-ray device that moves the X-ray source in one direction as the film is moved in the opposite direction, such as to show a predetermined feature in detail while blurring the details of other features.

ton. *n.* A secondary *fps* unit of mass equal to 2240 pounds or 1.016047 kg. Cf. *tonne.*

tongue flick. The periodic protrusion and sometimes up and down movement of the tongue in snakes and many species of lizards. Typically, the tongue is extended for a few seconds, then withdrawn and brought into close association with the vomeronasal organ. The behavior increases in frequency and intensity during arousal in contexts of trailing, prey detection, and defense. The principal function is assumed to be chemosensory, the tines sampling scent molecules and transferring them to the vomernasal organ.

tongue worm. See *worm.*

tonic. *adj.* Steady, or slowly adapting. Usually used in reference to muscle or neuronal activity.

tonic fibers. Muscle fibers that contract very slowly, producing prolonged, sustained contractions with low force. These fibers do not produce action potentials or "twitches," but exhibit graded states of contraction in postural muscles. Tonic fibers are common in reptilian muscle, whereas they are comparatively rare in mammals. Cf. *twitch fibers.*

tonicity. *n.* The relative osmotic pressure of a solution under a given set of conditions, defined by the osmotic response of cells when immersed in a solution. See *hypotonic, hypertonic, isotonic.*

tonic immobility. See *death feint.*

tonne. (t) A derived unit of mass equal to 1000 kg, or 2205 pounds. Cf. *ton.*

tonofilaments. *n.* Synonymous with *keratin* filaments.

tonus. *n.* Sustained, partial contraction of resting muscle with low force, produced by basal levels of neuromotor activity.

tooth. (*pl.* **teeth**) *n.* A small mass of dermal bone or epidermal hard structure, usually pointed or sharp and partially calcified, attached to the jaw and sometimes palatal structures, and used in feeding (seizing, holding, manipulating, and sometimes reducing prey) or defensive biting. Teeth are thought to have arisen phylogenetically from the bony armor of primitive fishes. See also *labial teeth.*

tooth formula. Reference to any of several methods of expressing the number and pattern of teeth present in an animal. See also *labial formula, maxillary formula.*

tooth replacement. Movement of new teeth into vacated sockets to take the place of teeth previously in the same position, sometimes as successive waves of replacement. See *intercalary replacement, vertical replacement, unimodal replacement.*

tooth ridge. In tadpoles, reference to serial, transverse fleshy ridges on the face of the *oral disc* surmounted by a row or rows of labial teeth generated in mitotic zones in the base of the ridges. These are present in some tadpoles that never have labial teeth. (Syn. *dental plate, labial ridge*)

tooth row. A single sequence of *labial teeth* in larval amphibians, arrayed in parallel series on the lips. See Fig. 1.

tooth row formula. See *labial formula.*

tooth series. Collective term for erupted teeth at a given site and each replacement tooth sequentially below it.

topical. *adj.* Confined to a small area, or of local occurrence.

topographic desert. A desert region of low rainfall located in the center of a continental land mass.

topographic map. A map on which elevations are shown by means of *contour lines*.

topography. *n.* (*adj.* –**ic**) Reference to all surface features of a geographical area.

topology. *n.* 1. Generally, the study of the properties of geometric figures that are subjected to deformations such as twisting or bending. 2. In systematic biology, reference to the branching order of a *phylogenetic tree*, based in morphological or molecular data.

toponym. *n.* In taxonomy, a scientific name based on the name of a locality, or the place-name itself.

topotype. *n.* 1. In taxonomy, reference to a specimen from the type locality but not necessarily part of the type series. 2. A population that has differentiated in response to local geographic factors.

top predator. The predator at the upper trophic level of a *food chain.*

torpor. *n.* (*adj.* **torpid**) A period or state of inactivity that is normally induced by cold or other adverse climatic condition. This term is used more often in contexts related to endotherms than to ectotherms. See *brumation.*

torque. *n.* A turning force, or moment. The product of a force times the perpendicular distance between its line of action and the pivot around which an object turns, or tends to turn, as a consequence of its application. See also *torsion.*

torr. *n.* A unit of pressure equal to 1 mm column of mercury. This unit was formerly used commonly in respiration physiology literature, but is now used rarely. See *Pascal, kilopascal.*

Torrent Salamanders. Vernacular name for North American species of rhyacotritonid salamanders belonging to the genus *Rhyacotriton.*

torsion. *n.* The tendency for opposing forces to twist an object, causing intermediate portions to deform or slip past each other. The stress or strain produced in a body as a result of forces of *torque*. See Fig. 24.

tortoise. *n.* Collective vernacular name for species of turtles belonging to the family Testudinidae.

tortoiseshell, tortoise shell. *n.* Colloquial term for the laminae of the Hawksbill Turtle, *Eretmochelys imbricata*, formerly widely used for ornamentation before the advent of plastics.

total. *n., v.* The overall sum of numbers, or determining such a quantity.

total body length. See *total length*.

total length. (TL) Straight line distance measured from the tip of the snout to the tip of the tail. Sometimes called, inappropriately, total body length. (Syn. *standard length, head-body length* in anurans)

totipalmate. *adj.* Having fully webbed feet.

totipotent. *adj.* Having potential for various specializations during development. Cf. *unipotent*.

touch papillae. Mechanosensory papillae confined to cranial scales of all crocodilians, distinguishable from postcranial sensory structures on the scales of crocodylids and gavialids. See M. Von During, *Z. Anat. Entwickl.-Gesch.* 143:81–94, 1973; *Abhandlungen Rhein.-Westfal. Akad.* 53:123–134, 1974. Cf. *integumentary sense organs*.

toughness. *n.* In biomechanics, a measure of the amount of energy required to break an object, quantified by the area under a stress vs. strain curve equivalent to the product of stress × strain, which equals work per volume. Thus, toughness is quantified by the area under the stress vs. strain curve at the breaking point of the object in question. Weak, stretchy materials like skin may take as much energy/volume to break as do strong, nonstretchy materials like bone. Thus, a material can be tough by being either strong or stretchy, or both. Cf. *stiffness*.

toxemia, toxaemia. *n.* A word variously misused by herpetologists to indicate *envenomation*.

toxic. *adj.* Poisonous or harmful.

toxic index, toxicity index. The number of minimum lethal doses of venom per injection or bite, arrived at by dividing the average amount of venom injected in a single bite by the MLD for that venom.

toxicity. *n.* The degree of virulence of a poison.

toxic metals. Pollutant metals that may be toxic when accumulated in biological tissue; unique in being neither created nor destroyed by human activity but can be transformed to more or less toxic forms and redistributed in the environment.

toxicology. *n.* The study of naturally occurring poisons (*toxins*) and their effects on organisms. More generally, this field is now considered to include study of the adverse effects of numerous substances on health and on the environment.

toxigenicity. *n.* The property of producing *toxins*.

toxin. *n.* A poison, especially a naturally occurring protein substance produced by living organisms. A chemical of biological origin that can adversely affect the physiology of another individual. Generally, venoms and poisons both contain toxins, but toxins are pure substances. The venoms of Elapidae are particularly rich in toxins. Cf. *poison, venom*.

toxinology. *n.* The science dealing with the toxins produced by pathogenic bacteria and certain higher plants and animals.

toxoid. *n.* See *anavenom*.

trabecula. *n.* (*adj.* –ar, *pl.* -ae) 1. Generally, a supporting or anchoring strand of connective tissue, often forming a fibrous or muscular inner partition, wall, bar, or column in an organ. 2. An elongate cartilage situated on either side of the ventral midline of the embryonic *chondrocranium*, running from the nasal capsule to the level of the hypophysis. These structures are fused anteriorly and the members fuse with the *parachordals* posteriorly. 3. Any of the spongy or fingerlike projections from the ventricular wall of the heart, alternating with deeper recesses. 4. Anastomosing spicules in cancellous bone forming a meshwork of intercommunicating spaces filled with bone marrow. 5. The structural walls of honeycomb units (*faveoli*) within the reptilian lung.

trace element. An element that is essential for normal growth and development of an organism, but required only in very small amounts.

trace fossil. Any fossil evidence of the movement or activity of an organism, e.g. a footprint or trackway.

tracer. *n.* A labeled compound or radioactive isotope.

trachea. (*pl.* –ae) *n.* (*adj.* –al) A membranous tube and air-conducting pathway between the pharynx and the bronchi of the lung. See also *tracheal lung*.

tracheal lung. A respiratory gas exchange structure of vascularized, respiratory tissue that arises along an expanded tracheal membrane in certain snakes and caecilians. It may consist of a vascularized dorsal surface of the trachea, or may be expanded into a baglike, elastic structure. It is a derived condition in snakes that lies craniad to the heart. (Syn. *anterior lung*) See Fig. 41.

tracheal process. The *bronchial process* of the amphibian cricoid cartilage, as used by earlier authors.

trackway. *n.* Term used to denote the impression of a pattern of locomotion across any surface capable of recording by means of imprint the footprints or body marks of an animal.

tract. *n.* A collection of nerve fibers coursing together in the brain or spinal cord.

tractable. *adj.* Tame or capable of being handled easily.

tradeoff. *n.* In ecological and life history theory, tradeoffs occur when large values for one trait lead to, or are compensated by, small values for another. The concept often assumes that energy apportioned among traits is limited, which is not always the case.

trail pheromone. A chemical laid down as a trail by one individual and followed by another member of the same species.

trait. *n.* See *character*.

trait evolution. The sequence of changes of a feature or phenotype on a *phylogeny*.

tramp species. *n.* A species having a wide geographic range and an efficient means of dispersal.

trans-. Prefix meaning "across."

trans. *adj.* In context of molecular configurations, reference to particular atoms or groups on opposite sides.

transcellular pathway. A route through an epithelium in which molecules or water cross through cells.

transcription. *n.* Formation of an RNA chain with a base sequence that is complementary to the informational base sequence in DNA.

transducer. *n.* A device that changes the form of energy of an input signal. A *sensor*.

transduction. *n.* The transformation of one form of energy into another. See *transducer*.

transect. *v., n.* 1. To cut across. 2. A line or narrow belt used to survey the distribution or abundance of organisms across a given area.

transepidermal water loss. (TEWL) Reference to water flux (via evaporation in terrestrial organisms) across the skin surfaces of the body. See also *evaporative water loss*.

transfaunation. *n.* Incorporation of species into the fauna of a new area when transferred beyond their natural geographical range by human means.

transfer RNA. (tRNA) One of 20 small molecules of RNA that functions to transfer amino acids from their activating enzymes to the ribosomes.

transformation. *n.* 1. Generally, a change. 2. A commonly used synonym for *metamorphosis*. 3. In statistics, a change of a variable used to simplify calculations or to transform the distribution of a random variable to a normal distribution. See *data transformation*. 4. In evolutionary biology, an implied evolutionary transition of a systematic character from one state to another state.

transformational evolution. See *saltational evolution*.

transformation series. A sequential series of character states that reflects the evolutionary trend, or transformation, of an homologous character.

transfusion. *n.* The transfer of blood or blood products from one individual to another.

transgenesis. *n.* Transfer of genes from one species to an unrelated species.

transgenic. *adj.* Descriptive of an organism that contains one or more genes artificially transferred from an individual of another species.

transient. *adj.* 1. Of short duration. 2. A stage in the phylogeny of a species.

transient epidermal gill. A term equivalent to *external gill* of other authors, developing from the branchial arches and undergoing atrophy before metamorphosis. See B. Viertel, pp 162–219 in P.G. Hepper, ed., *Kin Recognition*, 1991.

transient polymorphism. Reference to a *polymorphism* that exists in a population temporarily during a period when an allele is being replaced by one that is superior to it.

transilience. *n.* Speciation by a rapid shift of a population across an adaptive valley from its original adaptive peak to a new one.

transiliens bone. A sesamoid in the tendon of the *adductor mandibularis externus* of turtles of the genus *Gopherus*, mechanically aiding the vertical pull of muscle on the mandible.

transillumination. *n.* The passage of light through an object, with manipulation of the light source used as a technique for determining viability or (to some degree) contents of an egg, colloquially called "candling."

translation. *n.* The process of organizing amino acids on a polypeptide corresponding to the base sequence of mRNA.

translocation. *n.* The deliberate introduction or reintroduction of a species from another area by human intervention.

transmission electron microscope. (TEM) See *electron microscope*.

transmissivity. *n.* The fraction of incident *radiant flux* at a given wavelength transmitted by a material or object.

transmittance. *n.* The ratio of the radiation energy transmitted through a surface to the energy falling on it. The reciprocal of *opacity*.

transmitter. *n.* See *neurotransmitter*.

transmontane. *adj.* Reference to the other (relative to observer) side of the mountains. In California, reference is to geographic regions east of the Peninsular Ranges. Used extensively, for example, by G. Pickwell, *Amphibians and Reptiles of the Pacific States*, 1947. Cf. *cismontane*.

transmural pressure. Literally, pressure acting across a wall. In physiology this term refers to the difference in pressure across the wall of a blood vessel (= blood pressure – interstitial pressure).

transpalatine. *n.* Antiquated term for *ectopterygoid*, common in older literature.

Trans-Pecos Rat Snake. Vernacular name for the American colubrid species *Bogertophis subocularis*. See also *Desert Rat Snake*.

transpeninsular. *adj.* Ranging throughout the length of a peninsula.

transplantation. *n.* An experimental procedure, commonly used in embryology, involving excision of tissue from one organism (the donor) and insertion into another organism (the host).

transport host. A host in which a parasite survives without undergoing further development. (Syn. *paratenic host*)

transposons. *n.* Mobile genetic elements, which, as specific insertions, have been used as phylogenetic markers.

transverse. *adj.* Placed or running at right angles to the long axis; across. Cf. *longitudinal*.

transverse dune. A relatively straight, elongate dune oriented perpendicular to the prevailing wind.

transverse plane. A plane passing at right angles to the long axis of an animal, in essence producing a cross-section through the body.

transverse process. Reference to any of the lateral projections on a vertebra, including *diapophysis, parapophysis*, and *synapophysis*. These are a common type of process and separate the epaxial and hypaxial muscles.

transverse tubules. (T tubules) Tubules that are continuous with the plasma membrane of muscle fibers and extend deeply into the fiber, where they are closely apposed to the terminal cisternae of the sarcoplasmic reticulum. Depolarization of these membranes—attributable to action potentials conducted via the T tubules—stimulate the release of sequestered calcium to activate muscle contraction.

transverse ventral plates or **scale rows.** 1. Enlarged ventral scales or *scutes* on the ventral surfaces of most snakes, having the long dimension transverse to the longitudinal body axis. The posterior free edges of these scales are important in locomotion with respect to frictional interactions with the substrate. (Syn. *abdominal, gastrostege, ventral, scutum*) See Fig. 13. 2. Sometimes used to identify the arrangement of transverse ventral scales in crocodilians.

trapezium. *n.* The first *carpal* in the forefoot of turtles, lying at the base of the first phalanx.

trapezoid. *n.* The second *carpal* of the forefoot of turtles, situated at the base of the second phalanx.

traplining. *n.* A term used to describe a feeding strategy wherein an animal visits a sequence of widely dispersed food sources and obtains a small part of its food requirement at each feeding station.

treatment effect. In statistics, the change in response produced by a particular experimental treatment.

treatment sum of squares. In an *analysis of variance*, reference to that part of the total sum of squares attributed to differences between treatments.

TreeBoas, Tree Boas. *n.* Collective vernacular name for arboreal boas belonging to the genera *Boa, Candoia,* and *Corallus.*

Tree Cobras. Vernacular name for several species of arboreal cobras belonging to the genus *Pseudohaje.* These snakes occur in forested regions of tropical Africa.

Tree Frogs, Treefrogs. Collective vernacular name for numerous species of arboreal frogs worldwide, but usually applied to frogs of the family *Hylidae* (including, but not limited to, the genera *Hemiphractus, Hyla, Litoria, Osteocephalus, Osteopilus, Nyctimantis, Nyctimystes, Pelodryas, Phrynohyas, Scinax, Smilisca, Sphaenorhynchus, Stefania, Tepuihyla, Flectonotus, Anotheca, Trachycephalus,* and *Triprion*) and Rhacophoridae (*Chirixalus, Chiromantis, Nyctixalus, Polypedates, Rhacophorus*). The name also applies to the hyperoliid genus *Leptopelis* and the microhylid genera *Cophyla* and *Platypelis.*

Tree Iguanas. Vernacular name for South American species of iguanid lizards belonging to the genus *Liolaemus.*

tree line. A line which marks the northerly, southerly, or upper altitudinal limit of tree cover. (Syn. *timber line*)

Tree Lizards. Vernacular name for American species of iguanid lizards belonging to the genus *Urosaurus*, in particular *U. ornatus.* See also *Brush Lizards.*

Tree of Life. See *phylogenetic tree.*

Tree Snakes. A collective vernacular name for various species of arboreal snakes belonging to several taxa, including, but not limited to, the colubrid genera *Boiga, Aplopeltura, Dipsadoboa, Imantodes, Sibynomorphus, Tripanurgos, Uromacer.* See also *Boomslang.*

Tree Toads. Collective vernacular name for species of bufonid anurans belonging to several genera including *Dendrophryniscus, Laurentophryne, Leptophryne, Nectophryne, Nectophrynoides, Pedostibes.* The name also is used in reference to leptodactylid species belonging to the South American genus *Hylodes.*

trematode. *n.* A member of *Trematoda*, a class of parasitic flatworms.

Treptobranchia. *n.* A newly designated monophyletic taxon of salamanders and sister taxon with *Perennibranchia* within the *Hydatinosalamandroidei*. It contains the families *Ambystomatidae* and *Salamandridae.* See D.R. Frost et al., Bull. Amer. Mus. Nat. Hist. 297:1–370, 2006.

triad. *n.* Any grouping of three similar items.

Triassic. *n.* The first geological Period of the Mesozoic Era following the Permian and preceding the Jurasic Period. See Table 1.

tributary. *n.* A small stream that flows into a larger stream, thereby adding water to the stream.

tricaine methanesulfonate. (MS-222) A commonly used anesthetic for fishes and amphibians, convenient because it is absorbed across the skin. (Syn. *aminobenzoic acid ethyl ester*)

tricarboxylic acid cycle. See *citric acid cycle.*

tricarinate. *adj.* A condition in which a single scale or lamina bears three distinct and separate *keels*, usually applied to turtles.

tricolor. *n.* A term used colloquially among herpetoculturists for any King Snake of the genus *Lampropeltis* (and sometimes Coral Snakes of the genus *Micrurus*) that has a pattern of three alternating rings or bands of contrasting color, usually black, red and white, or yellow. This grouping includes most of the *L. triangulum* group, commonly known as "Milk Snakes."

triconodont. *adj.* Condition of an individual tooth that bears three similar points or cones on the crown.

tricolith. *n.* Reference to gastrointestinal mass composed of ingested hair or feathers.

tricuspid. *adj.* Having three points or cusps and used in reference to heart valves or tooth crowns.

Tridactyl or **tridactylous.** *adj.* Having feet that bear three digits.

trifid. *adj.* Partially divided into three parts.

trigeminal. *n., adj.* Denoting the fifth cranial nerve. The trigeminal nerve is the facial sensory nerve of most vertebrates and innervates the thermosensitive pits of boid snakes and pit vipers.

triglodytic. *adj.* Cave-dwelling.

triglyceride. *n.* A neutral molecule composed of three fatty acid residues esterified to glycerol. These are a fundamental constituent of many lipids.

triiodothyronine. (T_3) *n.* A second thyroid hormone initially isolated from mammals but now known to occur in all vertebrates. Cf. *thyroxine.*

trill. *v., n.* Reference to a call or vocalization characterized by rapid repetition of the same note.

trilobate. *adj.* Having three lobes.

Trinket Snakes. Vernacular name for various Asian snakes, including species in the genera *Elaphe, Gonyosoma, Coelognathus, Orthriophis,* and *Euprepiophis.*

trinomen. *n.* See *trinomial name.*

Trinomial or trinominal name. A scientific name, or subspecific name, consisting of genus, species, and subspecies designations (e.g. *Crotalus viridis oreganus*).

Trionychidae. *n.* A clade (family) of turtles with flattened bodies and reduced bony portions of the shell. These are fully aquatic turtles and range from North America to northern Mexico, sub-Saharan Africa to the Middle East, southern and southeast Asia, and islands of the Sunda Shelf. Fourteen genera and about 27 species are recognized in two subfamilies: *Cyclanorbinae* and *Trionychinae.*

Trionychinae. *n.* A subfamily of *Trionychidae.*

Trionychoidae. *n.* A clade (superfamily) of extant turtles that includes *Trionychidae* and *Carettochelidae.*

tripartite. *adj.* Having three parts.

triplo-. Prefix meaning "threefold."

triploid. (3N) *adj.* Having three chromosomes of each kind.

trivial name. The specific name or epithet.

triramous. *adj.* Having three principal branches.

tritiated water. See *tritium.*

tritium. (T or ^{3}H) *n.* A radioactive isotope of hydrogen with an atomic mass of three. Tritium combines with water to form tritiated water (T_2O) used as a tracer in metabolic studies.

triturating surface. An area for grinding, usually in reference to the jaws of certain turtles that have a broad mastication surface for smashing and grinding hard-shelled prey items.

tRNA. See *transfer RNA.*

trocar. *n.* A sharply pointed shaft used to penetrate skin or blood vessels (usually a vein) for purposes of introducing a cannula, catheter, transducer lead, or other tubular object.

trochanter. *n.* Either of two bony processes at the upper end of the femur.

trochlea. *n.* (*adj.* –**r**) A pulley-shaped part. In anatomic nomenclature, this term designates various bony or fibrous structures through, or over, which tendons pass, or with which other structures articulate.

trochlear. *n., adj.* 1. Reference to the fourth cranial nerve. 2. Reference to the bony grooves within certain joints.

troglodyte. *n.* (*adj.* –**ic**) A subterranean or cave-dwelling organism.

troglomorphy. *n.* Reference to evolutionary modification of the morphology of lineages or organisms associated with existence in caves.

troglophile. *n.* (*adj.* –**ic**) An animal frequently found in underground caverns or passages.

Trogonophidae. *n.* A family of *amphisbaenians.*

trophic. *adj.* Reference to feeding, nutrition, or predator-prey interactions. See *trophic hormones.*

trophic cascade. A progression of indirect effects by predators across successively lower trophic levels. Various uses of this term during the past 40 years have led to some controversy and ambiguity regarding the concept. The term was first used by R.T. Paine (*J. Anim. Ecol.* 49:667–685, 1980). See also J.A. Estes, K. Crooks & R. Holt, pp. 857–878 in S. Levin, ed., *Encyclopedia of Biodiversity,* Vol. 4, 2001.

trophic hormones. Any of a class of "nourishing" hormones involved in triggering cell growth and development. Not to be confused with *tropic.* Cf. *tropic hormones.*

trophic level. The position of an organism or species within a food chain. See *food chain.*

trophodynamics. *n.* Study of the energy relationships of feeding strategies and food webs.

trophotaxis. *n.* Movement of an animal toward or away from a food stimulus.

tropibasic, trypybasic. *adj.* See *tropitrabic.*

tropic. *adj.* See *tropic hormones.*

tropical. *adj.* Refer to geographic regions between the Tropic of Cancer and Tropic of Capricorn. Cf. *polar, temperate.*

Tropical Caecilians. Collective vernacular name for species of caecilians belonging to the family Scolecomorphidae.

tropical dry forest. Deciduous to semi-evergreen forest during a dry season, with canopy having few epiphytes and ranging from 2 to 40 m in height. There is a rain-free dry season 4–8 months long, and the beginning and duration of the rainy season is variable. Tropical dry forests once occupied about 60% of the forested tropics, but today less than 1% of this highly threatened vegetation remains in its original state.

Tropical Frogs. Collective vernacular name for numerous New World species of anurans belonging to the family Leptodactylidae, especially the genus *Eleutherodactylus.*

These frogs are highly variable in size, structure, and appearance.

tropical rain forest. A dense growth of tall, broadleaf evergreen trees and heavy vines, forming an almost complete shade cover over ground. Rain forest is indicative of abundant rain with no dry season and a fairly uniform high temperature. Tropical rain forests lie between the tropic of Cancer (23.5 °N. latitude) and the tropic of Capricorn (23.5 °S. latitude). However, only a small proportion of the vegetation in the tropics is strictly rain forest. (Syn. *selva*)

Tropical Water Snakes. See *Water Snakes*.

tropic hormones. Any of a class of hormones that have a primary function to act on other endocrine glands and regulate hormone secretion by such target glands. Tropic hormones stimulate and maintain their target endocrine tissues. Not to be confused with *trophic*. Cf. *trophic hormone*.

Tropidophiidae. *n.* A clade of relatively small snakes known as *Dwarf Boas*, having four genera and about 35 species. Distribution includes the West Indies, Central and South America. Known as *Dwarf Boas* or *Wood Snakes*.

Tropidophioidea. *n.* A clade (superfamily) of snakes that includes the families *Tropidophiidae* and *Xenophidiidae*.

Tropidurinae. *n.* A clade of iguanid lizards having variable morphology and natural history, and inhabiting a broad range of habitats. Twelve genera and approximately 275 species are recognized, distributed in South America, Galápagos Islands, and the West Indies.

-tropism. Suffix meaning "to turn."

tropitrabic. *adj.* Reference to a skull having an interorbital septum, derived from fusion of the embryonic trabeculae immediately anterior to the level of the hypophysis and continuous with the internasal septum. (Syn. *tropibasic, trypybasic*) Cf. *platytrabic*.

troposphere. *n.* The lowest major layer of the atmosphere, extending from Earth's surface to altitudes of 18–20 km at the equator and 8–10 km at the poles. The troposphere contains approximately 80% of the mass of the atmosphere and most of its water vapor, clouds, and carbon dioxide. Characterized by active convection, most weather events occur within this part of the atmosphere.

trot. *n., v.* A mode of locomotion where forelimbs and hind limbs on opposite sides of the body thrust backward at the same time.

True Chameleons. See *Chamaeleonidae*.

true extinction. The disappearance of all the individuals possessing a very similar but variable genome, without the appearance of any subsequent daughter species. Cf. *pseudoextinction*.

true frogs. See *Ranidae* and *frog*.

true toads. See *Bufonidae* and *toad*.

true vipers. Collective vernacular name for species of viperid snakes belonging to the subfamily *Viperinae*.

truncate, truncated. *adj.* Flattened or appearing to be cut off; often used in reference to the blunt appearance of extremities.

truncus. *n.* The proximal portion of the hemipenis.

truncus arteriosus. Literally arterial trunk, used usually as substitute name for the *conus arteriosus* of amphibians.

truncus impar. See *synangium*.

trunk vertebra. Any presacral vertebra, except for the cervical in salamanders. Used variously as interchangeable with *dorsal* and *presacral* of amphibians, and equivalent to *dorsal vertebra* in reptiles.

truss. *n.* A rigid framework consisting of an assemblage of solid parts.

tryptophyllins. *n.* A large family of nonhomogeneous peptides isolated from skin of phyllomedusine frogs. Little is known about the pharmacological actions and function(s) of these peptides.

TSD. See *temperature-dependent sex determination*.

TSH. See *thyroid stimulating hormone*.

t-test. See *Student's t test*.

Tuatara. *n.* Vernacular name for members of the genus *Sphenodon*. See *Rhynchocephalia*.

tuba. *n.* See *uterine tube*.

tube. *n.* A term used variously to designate certain parts of the oviduct.

tubercle. *n.* Any small, rounded discrete projection from the skin. (Syn. *tuberosity*)

tuberculate. *adj.* Bearing discrete, rounded bumps on the skin. Note this term should not be applied to spines or sharply pointed projections, only to projections that are rounded.

tuberculate keel. A term sometimes used for the median ridge on the carapace of certain species of turtle, wherein there appear isolated, elevated bumps.

tuberculum (of rib). *n.* A dorsal projection of bone from a rib that articulates with the *transverse process* of a vertebra. An accessory head that is present in primitive tetrapods, while many extant amphibians and reptiles have departed from this primitive pattern. Cf. *capitulum*.

tuberosity. *n.* See *turbercle*, which is the preferred descriptor.

tubocurarine. *n.* See *curare*.

tubule. *n.* A small tube.

tumor. *n.* 1. Swelling. 2. An uncontrolled and progressive growth of cells. (Syn. *neoplasm*)

tundra. *n.* A biome that prevails above tree line, distinguished by vegetation and climate associated with polar latitudes or high altitudes. The regions are generally without trees, but occasional dwarf tree species can survive when protected from the wind. Conditions may be harsh, with strong winds, low temperatures, and low rainfall year-round.

Tungara Frog. Vernacular name for the leptodactylid frog *Physalaemus pustulosus,* made known in particular by studies of reproductive behavior, sexual selection, and communication. See M.J. Ryan, *The Tungara Frog. A Study in Sexual Selection and Communication*, 1985. Also called Central American Mud-puddle Frog.

tunic, tunica. *adj., n.* A covering or coat, usually a membranous surrounding cover of an organ or a distinct layer of the wall of a hollow structure. Sometimes used in reference to the outer covering of the eyeball, or the scale on the skin that covers the margins of the brille of the eye.

tunica adventitia. The fibrous outer coat of various tubular structures, such as the outer layer of blood vessel walls.

tunica albuginea. A dense, white fibrous sheath enclosing a part or organ (e.g., gonads).

tunica intima. The inner lining of blood vessel walls.

tunica media. The middle layer of blood vessel walls, consisting of smooth muscle and elastic tissue.

tunica mucosa. Mucous membrane.

Tupinambinae. *n.* A lizard subfamily of the *Teiidae*.

turbitity. *n.* (*adj.* **turbid**) A quality of opaqueness in water due to suspended particulate matter or stirred-up sediment.

turbinal. *n.* 1. A term for the *septomaxillary*, used by various herpetologists. 2. This term is sometime used synonymously with *turbinates*.

turbinates. *n.* See *nasal turbinates*.

turbulence. *n.* Irregular motion or agitation of fluids inducing a condition of *turbulent flow*.

turbulent flow. A pattern of flow in fluids wherein there occur sharp gradients and inconsistencies in velocity and directionality; formation of eddy currents. Cf. *laminar flow*.

turgor. *n.* Swelling or distension.

turnover. *n.* 1. Extinction and replacement of species in a given area. 2. That fraction of a population that is exchanged per unit time (recruitment relative to mortality and emigration). 3. In ecological energetics, the ratio of productive energy flow to standing crop biomass in a community. 4. Reference to circulation of water in deep temperate lakes that mixes deeper water with surface water and tends to equalize temperature throughout the water column in spring and autumn.

turnover rate. 1. Reference to the dynamic replacement of atoms or markers in a tissue or organism without any net change in the total number of atoms or marker molecules. 2. In biogeography theory, reference to the number of species arriving or disappearing per unit time for an island community in equilibrium.

turtle. *n.* A member of the *Testudines*. About 300 species of living turtles include amphibious terrapins, terrestrial tortoises, and marine turtles. The phylogeny of turtles is controversial with respect to whether the clade is a sister group to the Diapsida (traditional view) or should be nested phylogenetically within the diapsid radiation (suggested by comprehensive molecular and morphological analyses).

Turtle and Tortoise Newsletter. A publication of the Chelonian Conservationists and Biologists.

turtle excluder device. (TED) A structure that is fitted to a trawl and designed specifically to reduce the incidental catch of turtles and other nontarget objects while maintaining normal capture levels of target organisms such as shrimp. The U.S. National Marine Fisheries Service has developed a TED that is effective in reducing mortality of turtles while maintaining, or actually increasing, the catch of shrimp.

Turtle-headed Sea Snakes. Vernacular name for the elapid sea snake species belonging to the genus *Emydocephalus*.

Twigsnakes. *n.* Vernacular name for specialized, arboreal colubrid snakes belonging to the African genus *Thelotornis*, and one of few colubrids capable of lethally envenomating humans although bites are rare. The snakes are also referred to by some as *Vine Snakes*.

twin species. A pair of *sibling species*.

Twin-spotted Rattlesnake. Vernacular name for the American pitviper *Crotalus pricei*.

twitch fibers. The most common striated muscle fiber type that produces all-or-none twitches in response to all-or-none action potentials. Contraction of these fibers is graded according to sequential, repetitive activation or an action potential train. Further classification of fiber types distinguishes *slow twitch* from *fast twitch* fibers. Cf. *tonic fibers*.

Two-legged Worm Lizards. Vernacular name for amphisbaenians belonging to the family Bipedidae.

Two-striped Garter Snake. Vernacular name for the American colubrid species *Thamnophis hammondii*.

two-tailed test. In statistics, a test for differences between means of population samples in which the critical region includes values on both sides of the mean of the sampling distribution (both left and right-hand tails). Cf. *one-tailed test*.

ty. Thousand years.

tylotose. *adj.* A term coined by F. Wall (*Snakes of Ceylon*, p. xix, 1921) for a knobbed condition; having a boss or swelling.

tympanic. *n., adj.* 1. Term used for the *squamosal* bone in the skull of amphibians. 2. Also used to denote the *quadrate* bone in the skull of snakes. (Syn. *infratympanic pretympanic*) 3. Pertaining to the *tympanum* or ear.

tympanic annulus. A cartilaginous outgrowth from the *quadrate* that forms a ring around the *tympanic membrane* of anurans, usually attached to the *squamosal* and the *crista parotica*. (Syn. *tympanic ring, tympanic cartilage*)

tympanic cartilage. See *tympanic annulus*.

tympanic cavity. See *tympanum*.

tympanic disc. See *tympanic membrane*.

tympanic membrane. *n.* The membrane that covers the external opening of the middle ear chamber or vestibule and functions in the mechanical reception of sound waves and their transmission to the middle ear. See *tympanum*.

(Syn. *eardrum, tympanic disc;* incorrectly *round window*) See Fig. 2.

tympanic ring. See *tympanic annulus*.

tympanorictal distance. A measure of the shortest distance between the anterior margin of the typanum and the corner of the mouth, used in bufonid taxonomy.

tympanum. *n.* 1. The middle ear cavity just medial to the tympanic membrane where sound is transmitted from the external membrane to the inner ear. Present only in anurans among the Amphibia. (Syn. *middle ear cavity, tympanic cavity*) 2. This word is often used, erroneously, with reference to the *tympanic membrane* or eardrum.

Tyndall scattering. The scattering and polarization of sky light by colloidal particles in a dispersed system, which produces bluish color.

typ. Abbreviation of the Latin *typus,* meaning "type."

type. *n.* See *type specimen, primary type.*

-type. Suffix meaning "image," "type."

Type I error. In statistics, the rejection of a null hypothesis when it is true and should be accepted. See *statistical error.*

Type I functional response. A linear increase in the number of prey consumed per predator per unit time as prey abundance increases. This is unrealistic because it assumes that predators can always increase their feeding rates. Cf. *Type II* and *Type III functional response.*

Type I survivorship curve. A survivorship curve in which survival probabilities are relatively high for young individuals and relatively low for old individuals.

Type II error. In statistics, the acceptance of a null hypothesis when it should be rejected. See *statistical error.*

Type II functional response. A response in which the number of prey consumed per predator per unit time is described by an asymptotic curve as victim abundance increases. A Type II functional response can occur because of predator satiation and constraints on handling time. Cf. *Type I* and *Type III functional response.*

Type II survivorship curve. A survivorship curve in which survival probabilities are relatively constant across different ages.

Type III functional response. A response for which an asymptotic curve describes the number of prey consumed per predator per unit time as prey abundance increases. The curve is S-shaped due to conditions that the feeding rate accelerates at low victim abundances, but then decelerates and approaches an asymptote at high victim abundance. This curve can arise because of search images and predator switching behavior. Cf. *Type I* and *Type II functional response.*

Type III survivorship curve. A survivorship curve in which survival probabilities are relatively low for young individuals and relatively high for old individuals.

type genus. A genus that might be chosen as a type reference for a family of organisms.

type locality. The location from where a *holotype* or *lectotype* specimen originated.

type material. A shared term designating all *type specimens* collected from the wild and retained in a museum.

type series. The specimens on which a description of a species is based.

type species. The species that is designated as the type of a genus or subgenus.

type specimen. The individual (original or ideal) animal selected by a taxonomist to serve as the basis for naming and describing a new species.

Typhlonectidae. *n.* A clade (family or subfamily) of highly aquatic, viviparous caecilians with five genera and 13 species in South America. Recently considered to be a subfamily of *Caeciliidae* by D.R. Frost et al. (*Bull. Amer. Mus. Nat. Hist.* 297:1–370, 2006).

Typhlopidae. *n.* A family of Blind Snakes having six genera and approximately 255 species distributed in Central and South America, the West Indies, southern Africa, Eurasia, Australasia, and Australia. Fossorial snakes with reduced eyes.

typhlopid. *n.* A member of, or pertaining to, *Typhlopidae.*

Typhlopoidea. *n.* A superfamily of snakes that includes the families *Typhlopidae* and *Anomalepididae.*

typical race. Reference to a subspecies that bears a subspecific name identical to that of the species (e.g., *Thamnophis elegans elegans*).

typogenesis. *n.* A period of explosive evolution during which new forms are produced rapidly.

typological species. A species defined according to characters of the *type specimen.*

typological thinking. Consideration of the members of a population of organisms as replicas of, or deviations from, a hypothetical type. A tendency to regard type specimens as "typical" of a species, allowing for little variation.

typology. *n.* A method of classification or study of natural groups based on the assumption that all members of a taxonomic unit conform to a given morphological plan without significant variation.

typostasis. n. A comparatively static phase of evolutionary history in which few new forms are produced.

U

ubiquitous. *adj.* Existing or seeming to exist everywhere. Widely ranging.

ulcer. *n.* (*v.* **ulcerate**) An open sore on the skin or mucous membrane, often discharging pus.

ulceration. *n.* The opening of a sore on the surface of the skin or a hollow organ.

ulcerative gingivitis. See *stomatitis*.

ulna. *n.* The larger of two bones of the forelimb in tetrapods, which articulates with the *humerus* and *radius*.

ulnar. *adj.* Reference to the *ulna* or the ulnar (medial) aspect of the arm. (Syn. *postaxial*)

ulnare. *n.* The most posterior of three small bones situated at the distal end of the *epipodials* of the tetrapod forelimb, at the end of the *ulna* and adjacent to the *intermedium* with which it is sometimes fused.

ultimate. *adj.* 1. Final point of a result; fundamental principle. 2. With respect to causality, this term refers to the evolutionary explanation for a biological phenomenon (i.e., the value or selective advantage for an organism). Cf. *proximate*.

ultimobranchial glands or **body.** A lobulated, glandular structure located lateral to the glottis and having a follicular structure similar to the thyroid gland. They are variously paired or single in amphibians and reptiles and absent following metamorphosis in certain anurans. The ultimobranchial gland of the Green Iguana consists of clumps of cells and epithelial-lined follicles. These structures secrete *calcitonin*, which is involved in calcium storage and metabolism, but the precise physiological role(s) of these glands has been difficult to establish with certainty. (Syn. *accessory thyroid gland, corpus Y, lateral thyroid, parafollicular cells, post-branchial body, suprapericardial body, telobranchial body*)

ultra-. Prefix meaning "beyond."

ultracentrifuge. *n.* A powerful centrifuge that can attain speeds as high as 60,000 rpm and generate sedimentation forces 500,000 times that of gravity, used to sediment macromolecules.

ultramicrotome. *n.* An instrument used to cut ultrathin sections, 50–100 nm thick, using knives made of polished diamond or broken plate glass and tissue specimens embedded in plastic.

ultradian rhythm. A biological rhythm having a periodicity greater than a day.

ultrasonic. *adj.* Pertaining to ultrasound.

ultrasonography, sonography. *n.* (*adj.* **-ic**) The use of high-frequency sound to determine density and characteristics of animal tissues or organs by analysis of reflected sound waves. An ultrasound-based diagnostic imaging technique used to visualize internal features of animals, most commonly in relation to human medical practice.

ultrasound. *n.* Sound waves having a frequency higher than the limit of audible frequencies in humans, ca. 20,000 Hz.

ultrastructure. *n.* Reference to minute structures that are visible only using a high level of magnification that can be obtained with the electron microscope.

ultraviolet radiation. (**UV**) Radiation emissions of wavelengths between 180 and 390 nm, between X-rays and visible light in the electromagnetic spectrum. This is an invisible component of sunlight and is also discharged from human-made sources such as fluorescent lamps and lasers. What is called "extreme ultraviolet" emissions extend to wavelengths of just a few nm. See also UVB.

µm. Micrometer (*micron*).

umbelliform. *adj.* Describes an upward-facing *oral disc* in tadpoles, used for harvesting particulate matter from a surface film.

umbo. *n.* A rounded elevation, used in reference to the central *areola* of plastral laminae in turtles.

umbraculum. *n.* A fleshy projection of the iris onto or over part of the pupil in some ranid tadpoles, thought to protect the eye from excessive light. See E.D. Van Dijk, *Ann. Natal Mus.* 18:231–286, 1966. Cf. *elygium*.

umbrella. *n.* A membranous, funnel-shaped extension of the lips of tadpoles of some anuran species, which act as a surface float when extended. In some species there are horny teeth present on the inside of the structure, which may thus serve as a rasping organ. It has also been suggested to function as a surface-feeding device. (Syn. *fimbriated peristome, float, funnel*)

umwelt. *n.* The total sensory input of an individual or species.

UN. Acronym for United Nations.

unavailable name. Any name proposed for a taxon that does not meet the requirements of the *International Code of Zoological Nomenclature*. Cf. *available name*.

unavailable work. In taxonomy, a work that is published prior to the beginning point of a group, or one that does not conform to the provisions of the Code and has been annulled by the *ICZN*.

UNCED. Acronym of the *United Nations Convention on Biological Diversity* signed in Rio de Janeiro in 1992.

unciform. *adj., n.* 1. Hooklike; used in reference to the tip of the upper jaw in certain turtles. 2. A term for the bone in the hand of anurans, and formed by fusion of the fourth and fifth carpals into a single element.

uncinate, uncinate process. *n.* A curved process on the lateral posterior margin of trunk ribs, overlapping the next posterior rib and adding strength to the thorax. This structure is typical of birds and is also found in crocodilians and *Sphenodon* among modern reptiles. (Spelled *uncinnate process* by Goodrich, 1927.)

undulate. *adj.* Having a wavy margin.

undulating membrane. A finlike membrane running the length of the flagellum-like tail of the sperm of amphibians. Wavelike movements of this membrane assist locomotion.

undulatory movement. See *lateral undulation*.

UNEP. Acronym of the *United Nations Environment Programme*.

UNESCO. Acronym of the *United Nations Educational, Scientific, and Cultural Organization*.

ungual. *adj.* Associated with or bearing a claw.

ungual canal. A narrow tube enclosed by the upper and lower *ungual plates* in the claw of *Hemidactylus*. First used by B.C. Mahendra (*Proc. Indian Acad. Sci.* 13:288–306, 1941).

ungual plate. The dorsolateral element of the claw in *Hemidactylus flaviviridis*, comprising an upper and lower piece and enclosing the *ungual canal*. First used by B.C. Mahendra (*Proc. Indian Acad. Sci.* 13:288–306, 1941).

unguiculate. *adj.* Bearing claws or nails.

unguis. *n.* (*pl.* **-ues**) Claw.

uni-. Prefix meaning "one," "single."

unicarinate. *adj.* Possessing a single *keel* (carina).

uniform convergence hypothesis. The proposition that convergent evolution in all parts of a phylogenetic tree occurs at a uniform rate.

uniform distribution. A spatial distribution pattern in which observations, points, or individuals are more uniformly spaced than if the distribution was random. The pattern suggests that the presence of one individual reduces the probability that another will occur immediately nearby. Cf. *clumped distribution, random distribution*.

uniform rate hypothesis. The proposition that any two lineages in a phylogenetic tree diverge at a constant rate with respect to each other.

unimodal. *adj.* A state of having a single mode or peak in a frequency distribution. Cf. *bimodal, polymodal*.

uniparental. *adj.* Having a single parent, as in parthenogenetic organisms.

uniparous. *adj.* (*n.* **–ity**) Producing a single offspring in each brood. See *parity*. Cf. *biparous, multiparous*.

uniporter. *n.* A carrier protein that transports a single solute from one side of a biological membrane to the other. Cf. *symporter*.

unipotent. *adj.* Capable of developing into only one type of cell. Cf. *totipotent*.

uniserial, uniseriate. *adj.* Describes structures that occur in a single set or series at a site, e.g. one tooth row on a tooth ridge in tadpoles. Cf. *biserial, multiserial*.

unisexual reproduction. Reproduction without a paternal contribution to the genetic makeup of the offspring.

unisexual species. An all-female species. All unisexual amphibians, and most of approximately 40 species of unisexual squamates, have formed from the "accidental" hybridization of two bisexual species in relatively recent evolutionary history.

United Nations Educational, Scientific and Cultural Organization. (UNESCO) An international organization founded in 1945 with the goal of promoting international cooperation in education, science, and culture among its member states.

United Nations Environment Programme. (UNEP) An international organization founded to promote cooperation in confronting problems and promoting understanding of the environment.

unit evolutionary period. The time in millions of years during which a divergence of 1% occurs between the initially identical nucleotide sequence present in two branches of a lineage being studied.

univariate analysis. The analysis of one character or variable at a time. Cf. *multivariate analysis*.

universal time. Accurate measurement of time calibrated on precise international atomic time.

universe. *n.* In statistics, the entire statistical population.

univoltine. *adj.* Having one brood or generation per year or season. Cf. *multivoltine*.

unken reflex. A warning display that is characteristic of certain species of anurans (esp. *Bombina* spp.) in which the back is arched, the hind legs raised over the body with the soles upward, and the head is pulled back, all aiding the display of colorful ventral surfaces. (Syn. *warning attitude*)

unordered transformation series. A set of homologous characters for which there are two or more possible evolutionary pathways of modification. Cf. *ordered transformation series*.

unpolarized transformation series. A transformation series in which the ancestral and derived conditions of the character have not been established. Cf. *polarized transformation series*.

unsaturated. *adj.* Reference to fatty acid molecules having some carbon-carbon double bonds.

unspecialized pseudautotomy. See *pseudautotomy*.

unstable equilibrium. In population biology, reference to a population equilibrium that is not restored if the population is disturbed, usually due to temporary environmental changes.

unstable mutation. A mutation having a high frequency of reversion.

unstirred layer. See *boundary layer*.

unweighted pair group method. In taxonomy, a method of clustering *OTUs* in which the smallest branches of the dendrogram are joined first, and the two largest branches are joined last. Each step groups together that pair of OTUs or pair of OTU groups showing the greatest similarity to each other.

upper labial. See *labial*.

upper marginal. See Fig. 4.

up-regulation. *n.* Control of physiological activity by an increase in receptor density in a target cell membrane. Cf. *down-regulation*.

upwelling. *n.* An upward movement of cold nutrient-rich water from depths of the ocean or a lake.

ura. *n.* The postanal part of the body. See *postano-coccyx, tail*.

uracil. *n.* A pyrimidine base and component of DNA.

Uraeotyphlidae. *n.* A clade (family or subfamily) of relatively primitive caecilians represented by a single genus and five species in southern India.

urate. *n.* A salt of *uric acid*.

urceolate. *adj.* Shaped like an urn.

urea. *n.* A nitrogenous waste product of protein metabolism in animals. In many reptiles and a few amphibians inhabiting arid, semi-arid, or marine environments, urea is converted to *uric acid* prior to excretion. The latter compound can be excreted with minimal expenditure of water, whereas urea is excreted in solution (urine) and requires greater water loss to rid the body of nitrogenous waste.

ureotelic. *adj.* (*n.* **-ism**) Excreting nitrogen in the form of urea.

ureter. *n.* The metanephric collecting duct that transports urine from the kidney to the urinary bladder or cloaca. (Syn. *metanephric duct*) See Fig. 38.

uric acid. A crystalline waste product of nitrogen metabolism and associated with the urine or feces of birds, many reptiles, and a few amphibians. Uric acid is poorly soluble in water and can be excreted in a precipitated form with minimal water loss. See *urate* and *urea*.

uricotelic. *adj.* (*n.* **-ism**) Excreting nitrogen in the form of uric acid or urate salts.

urinary bladder. A thin, membranous structure that provides storage of urine formed in the kidney before it is voided from the body via the cloaca. Bladders are present in amphibians, chelonians, and many lizards, but are absent in crocodilians and snakes. While most lizards have a bladder, it is rudimentary in some and absent in some others. In amphibians and tortoises the bladder is a major site of fluid and ion storage during periods of drought. See Fig. 38.

urine. *n.* Filtered blood plasma that is modified in composition by the kidney and eventually discharged from the cloaca as liquid or semi-solid excretory waste.

uriniferous tubule. The functional unit of the kidney, consisting of the nephron and collecting tubule. See *nephron*.

urobranchial shaft. A median rod of cartilage connecting the *thyroid bone* to the rest of the *hyoid apparatus* in some urodele larvae. This structure disappears during metamorphosis.

urodael diverticulum. Either of two muscular, notched (weakly bilobed) outgrowths of the dorsal wall of the *urodeum* in certain species of lizards, located above the *coprodeum* between the kidneys and the cloaca.

urodeal gland. Tubular glands producing secretions that empty into urodeal folds via numerous small orifices in various species of female lizards. These glands become active during the breeding season and are thought to secrete pheromones via the cloaca that stimulate courtship in males.

Urodela. *n.* The clade (order) that includes typical salamanders having four limbs and a long tail. An alternative name is *Caudata*.

urodele. *n.* Reference to any newt or salamander, derived from the taxonomic name *Urodela*, which usually is applied to extant salamanders.

urodeum, urodaeum. *n.* The midregion of the cloaca into which the ducts from the kidneys and gonads drain. (Syn. *urinary chamber*)

urogenital. *adj.* Pertaining to urinary (excretory) and genital (reproductive) organs.

urogenital duct. The combined duct from kidneys and gonads. See also *ductus deferens*.

urohyal. *n.* See *entoglossal*.

Uropeltidae. *n.* A clade (family) of generally burrowing snakes called *Shieldtail Snakes*, distributed in India, Sri Lanka, southern Asia, and the IndoAustralian Archipelago. Two subfamilies are sometimes recognized: *Cylindrophiinae* with eight species in a single genus, and *Uropeltinae* with eight genera and about 47 species.

Uropeltinae. *n.* A subfamily of *Uropeltidae*.

uropeltid. *n.* A member of, or pertaining to, *Uropeltidae*.

uropygial gland. A gland located at the base of the tail in *birds*, responsible for secreting lipids and proteins that are collected by the side of the beak and spread over the feathers. Preening the feathers with this secretion makes the feathers comparatively waterproof.

urostege. *n.* See *subcaudal*.

urostyle. *n.* A solid, rodlike bone formed by fused posterior vertebrae, forming the posterior end of the spine in anurans (but see *os coccygeum*). Similarly, the term has also been used for the fused terminal vertebrae in many salamanders, some turtles, and some Rattlesnakes.

urostege. *n.* A synonym for *subcaudal* in earlier literature. See Fig. 13.

urotomy. *n.* A general term to describe all types of tail breakage, including *pseudautotomy*, *autotomy*, and incidental or accidental tail loss. Definition proposed by J.M. Savage & J.B. Slowinski (*Biol. J. Linn. Soc.* 57:129–194, 1996).

Urwirbelfortsatz. *n.* A primordial vertebral process.

US gallon. A unit of volume equal to 0.8327 *Imperial gallon*.

Ussing chamber. A fluid-filled chamber used to suspend tissue for the purpose of measuring its epithelial transport properties.

uterine chamber. 1. In amphibians, see *ovotheca*. 2. In reptiles, the lumen of a uterus containing eggs or embryos.

uterine dilation. See *ovotheca*.

uterine milk. A secretion produced by deep columnar epithelial cells within compound racemose glands in the wall of the oviduct. It consists largely of lipid and is ingested by fetal caecilians in viviparous species.

uterine tube. The part of the reptilian oviduct between the *infundibulum* and *uterus*. This structure may be involved with sperm storage in squamates and albumen secretion in crocodilians and turtles. See J.E. Girling, *J. Exp. Zool.* 293:141–170, 2002. (Syn. *tuba, glandular region, albumen-secreting portion, magnum*)

uterodome. *n.* Apical cellular protrusions that replace or partially replace microvilli on uterine epithelial cells during the postovulatory secretory phase of the reproductive cycle. Because of their sensitivity to ovarian hormones, uterodomes are useful indicators of the endocrine status of mammals, and are assumed to be so in reptiles.

uterus. *n.* A glandular part of the *oviduct* anterior to the vagina and posterior to the tube, responsible for the production of fibrous and calcium components of the eggshell in reptiles. This region also interacts with the placenta in viviparous species. Usage of this term generally is restricted to that part of the oviduct in which eggs are retained in viviparous species, but this was not previously a strictly held criterion in earlier herpetological literature. With respect to amphibians, various authors have equated the term with *ovotheca*.

utricle. *n.* See *utriculus*.

utricule. *n.* See *utriculus*.

utriculus. *n.* The larger of two sac-like divisions of the membranous labyrinth of the inner ear, associated with equilibrium sense in vertebrates. (Syn. *utricle, utricule*)

UV. See *ultraviolet radiation*.

UVB. Ultraviolet radiation in the range 280–315 nm, important to synthesis of vitamin D.

uvea. *n.* The middle layer of the eye, comprised of the *choroid, ciliary body,* and *iris*.

vaccine. *n.* A substance that stimulates development of active immunity against a specific infectious agent or against a harmful product of an organism when injected into an animal's body.

vacuole. *n.* A fluid-filled, membrane-limited cavity or vacuole in the cytoplasm of a cell.

vagility. *n.* (*adj.* **vagile**) The ability to move from place to place.

vagina. *n.* The most caudal segment of the oviduct, which functions as both a copulatory and birth canal in viviparous vertebrates. (Syn. *vaginal pouch*)

vagina dentalis or ***dentis.*** A term for the sheath of tissue that covers the fang in many venomous snakes.

vaginal pouch. See *vagina*.

vagus nerve. The tenth cranial nerve that carries sensory and motor nerve fibers related to the buccal and pharyngeal regions, in addition to parasympathetic and afferent fibers to and from the viscera of the thorax and abdominal regions.

valid name. The name applied under the *International Code of Zoological Nomenclature* for a given taxon. All other available names based on that taxon are junior synonyms or rejected names. Cf. *invalid name*.

vallicepobufagin. *n.* A bufogenin isolated from *Bufo valliceps*.

value. *n.* See *lightness*.

valve. *n.* A structure that restricts movement of a fluid (usually blood or air) in one direction, but permits movement in the opposite direction.

van der Waals forces. Localized, relatively weak attractions between atoms and molecules with hydrophobic properties.

van't Hoff equation. An equation used to calculate the Q_{10} of a biological function:

$$Q_{10} = [k_2 - k_1]^{10/(t2-t1)}$$

where k_1 and k_2 are rate or reaction constants at temperatures t_1 and t_2, respectively.

vapor density. See *humidity*.

vapor pressure. The pressure exerted by a vapor. See *water vapor pressure*.

var. Abbreviation of *variety*.

varanid. *n.* A member of, or pertaining to, *Varanidae*.

Varanidae. *n.* A clade (family) of small to very large lizards known colloquially as "Monitor" or "Goanna" lizards, including the Komodo Dragon (*Varanus komodoensis*). There is one genus and about 40 species distributed in Africa and across southern Asia to China, and through the IndoAustralian Archipelago to Australia.

Varanoidea. *n.* A clade of lizards that includes the families *Varanidae, Lanthanotidae,* and *Helodermatidae.*

variability, coefficient of. In statistics, the standard deviation s, expressed as a percentage of the mean, $\bar{x}$, (CV = [s • 100]/$\bar{x}$) and used to compare the amount of variation in populations having different means.

variable. *adj., n.* 1. Any attribute of an organism that can have different values in different circumstances. 2. Not conforming to type. 3. A symbol or term to which a number of different numerical values may be assigned.

variance. *n.* A statistical measure of the spread or variation about the mean of a set of observations, denoted by s^2 and the sample estimate σ^2, calculated as

$$s^2 = \Sigma(x - \bar{x})^2 / (n - 1)$$

Whereas the *mean* measures the central tendency of observations, the variance measures the extent to which those observations differ from the central tendency. The varance is usually expresses as the *standard error* or *standard deviation* of the mean, based on the average squared deviation.

variance ratio. The ratio of two independent estimates of population variance, $F = s_1^2/s_2^2$.

variance to mean ratio. The simple ratio of *variance* to *mean*, used as an approximate test for agreement with the Poisson distribution, when it equals unity.

variance-ratio test. See *F-test*.

variant. *n.* Any individual or group showing marked deviation from type, such as in form, quality or behavior.

variate. *n.* A specific quantitative value of a *variable*.

variation, coefficient of. See *coefficient of variation*.

varicosity. *n.* A swelling along the length of a vessel or fiber.

variety. *n.* See *strain*.

vasa vasorum. The smaller arteries and veins that supply nutrients to and remove waste products from the tissues in the walls of large blood vessels.

vascular. *adj.* Of, or pertaining to, blood vessels.

vascularization. *n.* Reference to development, formation, or a system of blood vessels.

vascular remodeling. See *remodeling*.

vascular system. See *blood vascular system, cardiovascular system.*

vasculitis. *n.* Inflammation of a blood vessel.

vasculopathy. *n.* Any disorder of blood vessels.

vas deferens. See *ductus deferens.*

vas efferens. (*pl.* ***vasa efferentia***) Synonym for *ductulus efferens* and *ductulus epididymidis,* as used by several authors.

vasoactive intestinal peptide. (VIP) A peptide hormone that regulates gastric secretions during the intestinal phase of digestion. VIP found in amphibians and reptiles (alligators) seems to have varying effects on endocrine regulation. In amphibians VIP stimulates release of growth hormone, while in reptiles VIP inhibits prolactin release and stimulates TSH release.

vasoconstriction. *n.* Active narrowing of blood vessels due to contraction of smooth muscle. The action decreases vascular volume and increases vascular resistance.

vasodilation, vasodilatation. *n.* The widening of blood vessels, either actively or passively. The action increases vascular volume and decreases vascular resistance.

vasomotor. *adj.* Reference to autonomic control of smooth muscle contraction in the walls of blood vessels.

vasopressor. *n.* 1. Stimulating contraction of vascular smooth muscle, particularly at the level of arterioles. 2. A vasopressor agent.

vasotocin. *n.* See *antidiuretic hormone.*

VDS. See *venom delivery system.*

vector. *n.* 1. A carrier, usually with reference to an animal that transfers a parasite or pathogen from host to host. 2. A quantity having direction and magnitude, such as force or velocity. Also, a line that represents such a quantity by its length and orientation. 3. In statistics, a quantity whose complete description requires two or more numbers.

vegetal hemisphere. The hemisphere of an amphibian egg that is farthest from the nucleus and rich in yolk.

vegetal pole. The center of the region of the egg that normally develops into the nutritive structures. In amphibians, this region is usually without pigmentation. In cleidoic eggs, this is the region rich in yolk that provides nutrition to the developing embryo. Cf. *animal pole.*

vegetation. *n.* The total plant life or cover in an area or region.

vein. *n.* A blood vessel that carries blood from tissue capillary beds toward the heart.

velar plate. Reference to various folds or flaps that develop from the pharyngeal wall in the branchial regions of anuran larvae. These collectively act as a filter between the pharynx and the gill chamber.

veld, veldt. *n.* The open temperate grasslands of southern Africa, typically having scattered trees and shrubs.

velocity. *n.* Speed of motion, expressed in units of distance per time (e.g., km/h). The rate of change of position in a given direction. Cf. *flow.*

velum. *n.* A glandular organ between the buccal cavity and the pharynx of tadpoles, functioning to regulate water flow into the pharynx.

velum palatinum, velum palati. A transverse fold of tissue on the palate in the rear of the mouth of crocodilians, meeting a similar fold (*basihyal valve*) on the tongue and preventing water from entering the glottis when a submerged animal opens its mouth.

Velvet Geckos. Vernacular name for species of Geckos belonging to the genus *Oedura.*

venae cavae. See *precaval, postcaval veins.*

venated. *adj.* Veined, as with colors or by furrows.

venation. *n.* A network of vein-like markings.

venenation. *n.* See *envenomation.*

Venn diagram. A graphical representation of taxonomic relationships, using internested fields that correspond to the internodes of a phylogenetic tree.

venom. *n.* A toxic or potentially toxic substance normally secreted by an animal and injected by a sting or bite. Highly evolved snake venoms are a complex mixture of hydrolytic enzymes and systemic toxins. Cf. *poison, toxin.*

venom apparatus. See *venom delivery system.*

venom canal. The tubular or central lumen of the fang of a venomous snake.

venom delivery system. (VDS) With reference to snakes, the VDS includes the venom, an associated gland used for production and storage, and the muscles and specialized teeth that are used for delivery (Fig. 22). Among extant reptiles generally, two lineages have been known to have evolved VDS, the advanced snakes and helodermatid lizards. However, venom toxins have been identified in two additional lizard lineages (varanids and Iguania; B.G. Fry et al., *Nature* 439:584–588, 2006). Fry et al. demonstrate that all lineages possessing toxin-secreting oral glands form a clade and suggest a single early origin of the venom system in lizards and snakes. (Syn. *venom apparatus*)

venom duct. The canal communicating from the venom gland to the entrance lumen of the fang of a venomous snake. See Fig. 22.

venom gland. In general usage, one of several gland types, derived from one of the salivary glands of reptiles, that secretes venom in a venomous reptile. In venomous snakes, the gland is paired and located between the eye and the angle of the jaw and conveys venom to each of the fangs by means of the venom duct (Fig. 22). Because venom and Duvernoy's glands both arise from a primordium that also gives rise to the maxillary dental lamina, there has been a recent trend to identify all venom-producing glands of snakes as "venom" glands (B.G. Fry et al., *Rapid Communications in Mass Spectrometry* 17:2047–2062, 2003). In Beaded Lizards (Helodermatidae), the *mandibular gland* is specialized for venom production and discharges its secretions through ducts in the mucous membrane of the lower jaw. See also *accessory gland, gland of Gabe,* Fig. 20.

venomous. *adj.* Pertaining to an animal that possesses *venom* and envenomates other animals by use of a fang or sting.

venous shunt. A direct connection between *arterioles* and *venules*, bypassing a capillary network.

vent. *n.* The external opening of the *cloaca.* (Syn. *cloacal slit*)

vent flap. A fleshy flap of various shapes that extends from the body wall posteriorly and ventral to the *vent tube* of tadpoles. This structure sometimes encloses the hind limb buds. (Syn. *anal flap*)

venter. *n.* The abdomen or complete undersurface of an animal. (Syn. *ventrum*)

ventilation. *n.* In respiratory physiology, this term refers to the process of moving air or water between the gas exchanger (lung or gill) and ambient medium. (Syn. *breathing, irrigation* in the case of moving water)

vent position index. (VPI) A radiographic measurement of vent position based on the fractional number of caudal vertebrae between the sacrocaudal joint (taken as zero) and a metal pin marking the vent position. This measurement is used as a size-independent index of vent position for a given specimen. See R.W. Blob, *Copeia* 1998:792–801, 1998.

ventrad. *adv.* Toward the venter.

ventral. *adj., n.* 1. Pertaining to the abdomen or venter; directed toward or situated on the belly surface, in contrast to *dorsal.* Cf. *dorsal.* 2. Any of the large scales or scutes on the belly surface of a snake, between the head and anal plate or (some authors) tail tip. (Syn. *abdominal, gastrostege, scutum*) See *transverse ventral plates* and Fig. 13.

ventral arch. See *hemal arch.*

ventral collar. A distinctive row of scales running transversely across the throat anterior to the forelimbs, present in the majority of crocodilians.

ventral keel. A median keel on the ventral scales of some snakes, particularly some of the *Sea Snakes.*

ventral rib. A term variously used in reference to *pleural rib, gastralia,* or *parasternal.*

ventral root. A nerve trunk that leaves the spinal cord near its ventral surface and contains only motor axons.

ventral scale count. A count of ventral scales (scutes) in snakes, usually counted to, but not including, the anal plate. A generally accepted standard method of counting was proposed by H.G. Dowling (*British J. Herpetol.* 1:97–99, 1951) in which the starting point is defined as "the first plate bordered on both sides by the first row of dorsals."

ventral patch. See *pelvic patch.*

ventral seat patch. See *pelvic patch.*

ventral velum. A flap that covers the *branchial food trap* and acts as a valve important to irrigation of gill structures in microphagous tadpoles. See R.J. Wassersug, *Occas. Pap. Mus. Nat. Hist. Univ. Kansas* 48:1–23, 1976.

ventricle. *n.* 1. Generally, a small cavity or chamber. 2. The muscular chamber of the heart that receives venous blood and pumps it into outflow vessels in which it is distributed to the lungs and body. With exception of crocodilians, the ventricle is a single structure in amphibians and reptiles. See Figs. 36, 37. 3. A chamber of the brain.

ventriculus. *n.* The distal, muscular compartment of the crocodilian stomach. (Syn. *gizzard*)

ventrolateral. *adj.* Of or pertaining to a position intermediate between the lower surface and lateral surface of a body or structure.

ventrolateral fold. A fold of skin running along the outside edge of the ventral surface of the body in some lizards.

ventrum. *n.* The lower surface of the body or body structures. (Syn. *venter*)

vent shuffling. A behavior observed during courtship of ambystomid salamanders, when following "*waltz*" behavior the male slowly shakes the posterior parts of the body and tail from side to side with an undulating motion. The female may also shake her tail, and these behaviors precede the transfer of a spermatophore. Term is from S.J. Arnold (*Z. Tierpsychol.* 42:247–300, 1976). See also *hula dance.*

vent tube. A tube for voiding feces, having many variations of morphology in tadpoles.

venule. *n.* A small blood vessel that connects a capillary bed with a vein.

Verhulst logistic equation. See *logistic equation.*

vermicide. *n.* Any medication used to eliminate intestinal worms.

vermicular, vermiculate. *adj.* Marked with very delicate grooves, wavy lines, or streaks resembling the tracks of worms. Often used with reference to a pattern of *epidermal microsculpturing.* See R.M. Price *J. Herpetol.* 16:294–306, 1982.

vermiculite. *n.* A porous, loosely structured form of mica that is produced for insulation and other purposes, and is used as a cage substrate or egg incubation medium by herpetoculturists.

vermiform. *adj.* Wormlike; extremely elongate.

vernal. *adj.* Of or pertaining to the season of spring. Cf. *aestival, autumnal, hibernal.*

vernal pool. A temporary pool that is formed during spring from meltwater or flood water.

verruca. (*pl. -ae*) *n.* A wart or elevation on the skin. H. Elias and J. Shapiro (*Amer. Mus. Novit.* 1819:1–27, 1957) developed a nomenclature for the surface features of frog skins using this term in conjunction with appropriate modifiers.

verruca epithelialis. Epidermal thickenings that project above the skin of amphibians (H. Elias and J. Shapiro, *Amer. Mus. Novitates,* No. 1819:1–27, 1957).

verrucae hydrophilicae. A subcategorical term (see *verruca*) proposed by R.C. Drewes et al. (*J. Comp. Physiology* B,

116:257–67, 1977) to be used in reference to the warts or protuberances on ventral skin of anurans. Each *verruca* is provided with glands and a specific vascular plexus.

verrucate. *adj.* Having prominent wart-like protuberances. Used with reference to a pattern of *epidermal microsculpturing*. See R.M. Price *J. Herpetol.* 16:294–306, 1982.

verrucous, verrucose. *adj.* Having rough surfaces.

versant. *n.* The slope of a mountain or mountain chain; or general declination of a region.

vertebra. (*pl.* –ae) *n.* One of many bones firmly joined into a segmental backbone that defines the major body axis of vertebrates (*vertebral column*).

vertebral. *adj., n.* 1. Generally, reference to the vertebral column or middorsal line of the body. 2. The term is commonly used to denote any of the scale rows located on the middorsal line of the body in snakes and lizards, and is sometimes used in reference to the *centrals* and *neurals* of the turtle carapace. (Syn. *central, neural*) See Figs. 4, 5.

vertebral column. The spinal skeleton consisting of a series of vertebrae extending from the skull to the tip of the tail. (Syn. *backbone*)

vertebral crest. A single row of elevated, enlarged scales running along the vertebral region in some lizards.

vertebral process. The elevated ridge extending along the center of the back of some snakes.

vertebral stripe. A longitudinal middorsal stripe.

vertebral *venous plexus*. See *spinal vein*.

Vertebrata. *n.* The subphylum including all vertebrates in the phylum *Chordata*.

vertebrate. *n.* Any member of the subphylum *Vertebrata*, comprising all animals that have a vertebral column, including fishes, amphibians, reptiles, birds, and mammals.

vertical. *adj., n.* 1. Perpendicular to the plane of the horizon and used commonly in herpetology with reference to a pupil that is, in light, a narrow, elliptical opening with orientation from top to bottom of the eye, in contrast to horizontal. 2. Antiquated reference by early authors to the *frontal* in cephalic plates of snakes and to the *frontal* scale on the head of turtles.

vertical evolution. A process whereby an ancestral species changes through time without splitting and becomes distinguished as a new species. (Syn. *phyletic evolution, anagenesis*)

vertical life table. A life table in which the survival probabilities are estimated indirectly by comparing the relative sizes of consecutive age classes. This analysis assumes the population has achieved a *stationary age distribution*.

vertical replacement. A mode of tooth replacement in which a new tooth lies at the base, or beneath, a functional tooth, the pair forming a vertical row. The erupting tooth moves directly into a vacated socket of a shedding tooth.

vertical septum. An incomplete septum running in a dorsoventral plane in the ventricle of noncrocodilian reptiles, partially separating the *cavum arteriosum* from the *cavum venosum*. Cf. *horizontal septum*.

vertical transmission. The transmission of a trait, disease, etc., from parent to offspring across a placenta.

verticil. *n.* See *whorl*.

verticillate. *adj.* Arranged in whorls.

VES. See *visual encounter survey.*

vesicle. *n.* A small sac or space.

vesicula seminalis. See *ampulla ductus deferens.*

vespertine. *adj.* Reference to the evening. (Syn. *crepuscular*)

vestibular. *adj.* 1. In anatomy, relating to, or designating, an opening at the entrance to a canal. 2. Relating to, or involving, the inner ear.

vestibular apparatus. The collective equilibrial organs in the inner ear of a vertebrate.

vestibule. *n.* A space or cavity at the entrance to a canal.

vestibulo-hyoid mechanism. See *auditory mechanism.*

vestibulo-quadrate mechanism. See *auditory mechanism.*

vestibulo-scapular mechanism. See *auditory mechanism.*

vestibulo-squamosal mechanism. See *auditory mechanism.*

vestibulo-tympanic mechanism. See *auditory mechanism.*

vestibulum *n.* (*pl. -a*) Latin for *vestibule*.

vestige. *n.* (*adj.* **-ial**) The remnant of a structure that functioned in a previous evolutionary stage of species or individual development.

viable. *adj.* Living, or capable of maintaining an independent existence.

vicariance. *n., adj.* Reference to the separation or division of a group of organisms by a geographic barrier, with the result that the original group differentiates into new varieties or species.

vicariance biogeography. A methodological approach to biogeography that emanated from phylogenetic systematics and historical biogeography. The method is aimed at discovering and testing shared, ancestor-descendant relationships of organisms occupying areas of endemism, where divergence involves replicated events related to a single isolating event separating ancestral species that subsequently diverge into descendant species. Cf. *dispersal biogeography.*

vicariance distribution. A biogeographical distribution having gaps and therefore discontinuous, relative to a previously continuous distribution.

vicariants. *n.* Closely related taxa that are isolated geographically from one another by formation of a natural barrier.

vicarious species. See *vicars.*

vicars. *n.* Closely related and ecologically equivalent forms having disjunct distributions and tending to be mutually exclusive. (Syn. *vicarious species*)

Victoranura. *n.* A new monophyletic taxon of anurans, sister to *Ptychadenidae* within the new taxon *Natatanura*, and composed of *Ceratobatrachidae* and the new taxon

Telmatobatrachia (D.R. Frost et al., *Bull. Amer. Mus. Nat. Hist.* 297:1–370, 2006).

videlicet. (*viz.*) Latin, meaning "namely."

vidian canal. A pathway for the Vidian nerve and artery in the *basisphenoid bone*.

Vietnamese Wood Turtle. Vernacular name for the Asian turtle species *Geomyda spengleri*.

vigor. *n.* The intensity of growth and metabolic activity of an organism or population.

Villier's Blind Snake. Vernacular name for the West African leptotyphlopid species *Rhinoleptus koniagui*.

villiform. *adj.* Shaped like a *villus*.

villose. *adj.* Covered with *villi*.

villosity. *n.* Earlier reference to *villus*.

villus. (*pl.* -**i**) *n.* Generally, a small, vascular, fingerlike protrusion of tissue, as from the free surface of a membrane, and typically present as multitudinous threadlike projections. This term applies to the numerous absorptive structures of the intestine. Cf. *microvillus*.

Vinesnakes, Vine Snakes. *n.* Collective vernacular name for various arboreal species of snakes belonging to the New World genera *Oxybelis, Uromacer, Xenoxybelis,* and *Imantodes,* and the Old World genera *Ahaetulla, Langaha, Imantodes*, and *Lycodryas*. See also *Twigsnakes*.

violet cell. A basophilic cell of the anuran pituitary, present in the cephalic part of the *pars distalis*. The name was coined in reference to the color seen after azan staining. (Syn. *lavender cell*)

Viper Boa. Vernacular name for the boid species *Candoia aspera*, which inhabits humid lowland areas of New Guinea and the neighboring Bismarck Archipelago.

Viperidae. *n.* A clade (family) of familiar, venomous snakes comprising three subfamilies with 17–20 genera and approximately 228 species. *Viperinae* (pitless vipers) are distributed throughout much of Africa, Europe, and Asia, *Crotalinae* (pit vipers) range throughout southern North America, Central America, South America, and much of southern Asia, and *Azemiopinae* is a separate clade of Asian pitless viper that seems more closely related to crotalines than to viperines.

viperid. *n.* A member of, or pertaining to, *Viperidae*.

Viperinae. *n.* A subfamily of *Viperidae,* known as "*true vipers*."

vipers. *n.* Collective vernacular name for numerous species of snakes belonging to the Viperidae, especially Viperinae. The name applies especially to species belonging to the genus *Vipera*.

vipoxin. *n.* One of the rare known adrenergic toxins extracted from snake venoms.

viridobufagin. *n.* A *bufogenin* isolated from *Pseudepidalea* (formerly *Bufo*) *viridis*.

viridobufotenine. *n.* A *bufotenine* derived from the skin of *Pseudepidalea* (formerly *Bufo*) *viridis*.

viridobufotoxin. *n.* A *bufotoxin* isolated from *Pseudepidalea* (formerly *Bufo*) *viridis*.

virology. *n.* The study of viral diseases.

virulence. *n.* (*adj.* **–ent**) The pathogenecity of a microorganism. This term was originally used for viruses and then was extended to other pathogens.

virus. *n.* Any member of a large group of small infectious agents that rely totally on living cells for reproduction and metabolism.

viscera. (*pl.*) *n.* Collective reference to the internal soft organs of the major body cavities, especially the abdominal cavity. See *viscus*.

visceral arch. Cartilaginous or bony bars of the visceral skeleton, including the mandibular arch, hyoid arch, and branchial arches.

visceral peritoneum. See *peritoneum*.

visceral pouch. Pharngeal endodermal evaginations between the visceral arches from which the Eustachian tube is derived and gill slits open.

viscoelasticity. *n.* Water-damped elasticity. A property of a material wherein water decreases elastic recoil and retards both deformation and recoil processes. Viscoelasticity is largely responsible for nonlinear shapes of stress-strain curves of biomaterials.

viscometer. *n.* An instrument used for measuring the viscosity of a fluid.

viscosity. *n.* (*adj.* **viscous**) A physical property of fluids that impedes the tendency for adjacent layers of a fluid to move past each other ("stickyness"). Internal friction in a fluid causing resistance to flow.

viscus. (*pl. viscera*) *n.* (*adj.* **–al**) The large interior organs of the major body cavities, especially the abdomen.

Visible Implant Fluorescent Elastomer. (VIE) An injectable material that hardens beneath the skin and has been used for marking amphibians (S. Nauwelaerts et al., *Herpetol. Rev.* 31:154–155, 2000; M.A. Ralston Marold, *Herpetol. Rev.* 32:91–92, 2001). The elastomer can be viewed beneath the skin, sometimes requiring blue light for darkly-pigmented individuals.

visible light. Light of wavelengths generally between 380 and 780 nm.

vision. *n.* Sight. The sensory perception of light and light-mediated images.

visual encounter survey. (VES) A widely used method for sampling amphibians and reptiles, involving one or more observers who search a defined area for animals for a specified amount of time. Usually the number of individuals of a species counted is standardized by time or area searched to determine the relative abundance of species. See M.L. Crump and N.J. Scott, pp. 84–92 in W.R. Heyer et al. (eds.), *Measuring and Monitoring Biological Diversity: Standard Methods for Amphibians*, 1994; T.M. Doan, *J. Herpetol.* 37:72-–81, 2003.

visual purple. See *rhodopsin*.

Vitali organ. See *paratympanic organ*.

vital stain or **staining.** A stain introduced into a living organism and selectively taken up by specific tissues or cellular structures.

vitamin. *n.* An organic compound that is required in relatively very small amounts by an organism for its normal growth, development, and maintenance.

vitamin D. Fat-soluble compounds required in trace amounts for normal bone metabolism. In many terrestrial vertebrates, some exposure to sunlight and ultraviolet radiation is required for production of adequate amounts of vitamin D. Vitamin D production is one likely function of basking in amphibians and reptiles.

vitelline. *adj.* Of or pertaining to the yolk of an egg or ovum.

vitelline membrane. *n.* The cytoplamic membrane covering the egg cell. In amphibian eggs, this membrane separates from the egg cytoplasm at fertilization, and the space is filled with fluid. Subsequent to this event, some authors call the membrane the *fertilization membrane* while others retain *vitelline membrane*.

vitellogenesis. *n.* The production of yolk or "yolking" of an egg in the ovary. Deposition of fat reserves in the ovarian follicles.

vitellogenin. *n.* A protein that is synthesized in, and secreted from, the liver and incorporated into the yolk spheres of developing oocytes. Vitellogenin synthesis is largely under control of the steroid hormone *estrogen*, although other hormones may modulate the vitellogenic response. See *lipovitellins* and *phosvitins*.

vitellus. *n.* The yolk of the egg.

vitta. *n.* (*adj.* **-al**) A stripe or band of color.

vivarium. (*pl.* **-ia**) *n.* A tank or cage in which live animals are kept, especially amphibians or reptiles and usually with semi-natural surroundings provided within the enclosure.

viviparity. *n.* (*adj.* **viviparous**) The retention of embryos within the oviduct until development is complete, thus giving birth to developed young as opposed to laying eggs. In the past, viviparity has been defined with consideration of whether a placenta is involved (*placental* or *"true" viviparity*) or not (*pseudoviviparity*). Most workers now prefer a broad definition of viviparity that includes numerous modes of live-bearing. Cf. *oviparity*, *ovoviviparity*.

viz. See *videlicet*.

VNO. See *vomeronasal organ*.

$\dot{V}_{O_2}$ Symbol which represents the rate of oxygen consumption (usually measured as volume per unit time) in physiological literature.

vocal cords. Paired thickenings on the inner wall of the larynx of anurans that produce vibrations when air from the lungs passes over them.

vocalization. *n.* The making of sounds associated with the mouth, sometimes used synonymously with *call* in reference to anurans.

vocal pouch. See *vocal sac*.

vocal sac or **sack.** An elastic, membranous, and inflatable pouch associated with the mouth or throat of male anurans, used as a resonating chamber to amplify sounds during calling. The structure may be single or paired, visibly affecting the chin or throat or sides of the neck, respectively. (Syn. *gular sac, resonating organ, resonating sac, vocal pouch, vocal vesicle*)

vocal slit. A valvular opening in the floor of the mouth in male anurans, leading to the vocal sac.

vocal vesicle. An antiquated term used by many early authors in reference to *vocal sack*.

vol. Abbreviation for *volume*.

volant. *adj.* Adapted for, or capable of, sustained flying or gliding.

volatile. *adj.* Tendency of a chemical to vaporize readily at normal temperatures and pressure.

volar. *adj.* Reference to the palm of the hand or sole of the feet.

volcanism. *n.* Volcanic activity.

volplaning surface. A structure providing an increased surface area that allows an animal to glide from a higher to lower site by reducing the angle of descent.

volt. (V) *n.* A unit of electromotive force, or electric potential, equal to the force required to induce 1 ampere (A) of current to flow through a resistance of 1 ohm (Ω).

voltage. (E or V) *n.* The electromotive force, or electric potential, expressed in volts.

voltine. *adj.* Reference to the number of broods or generations per year or per season. See *univoltine, multivoltine*.

volume. *n.* The capacity of a structure, related to the cube of a linear dimension and usually expressed in units of μl, ml or L.

voluntary maximum tolerance. The maximum temperature an *ectotherm* will tolerate before it seeks retreat to conditions that result in lowering of body temperature. Term coined by R.B Cowles and C.M. Bogert (*Bull. Amer. Mus. Nat. Hist.* 83:261–296, 1944).

voluntary minimum. The minimum low temperature that will be tolerated by an ectotherm before it seeks conditions that result in elevation of body temperature or that minimize the rate of cooling. Term coined by R.B. Cowles and C.M. Bogert (*Bull. Amer. Mus. Nat. Hist.* 83:261–296, 1944).

voluntary nervous system. See *somatic nervous system*.

vomer. *n.* A dermal bone lying immediately behind the *premaxillary* in the roof of the mouth, forming the first in a paired series of bones that comprise the palate in reptiles and amphibians. Some authors have preferred the term *prevomer* to indicate nonhomology with the vomer of mammals, but the homology issue is not clear. See also *vomero-palatine*.

vomerine teeth. Short, conical projections of the vomer, lo-

vomeronasal fenestra. cated near the internal nasal openings in the palate of amphibians.

vomeronasal fenestra. The ventral opening from the vomeronasal organ into the mouth of snakes and lizards.

vomeronasal organ. (VNO) A specialized chemosensory chamber that forms a major sense organ in many reptiles. In squamates it consists of a convoluted blind sac opening into the mouth anterior to the choanae. It is lined with sensory and ciliated epithelium which detect odorant particles brought into the mouth on the tongue, and the afferent sensory information is conveyed to the olfactory bulb of the telencephalon via branches of the olfactory nerve. The term also applies to an accessory olfactory system in amphibians, but the organ is absent in adult crocodilians. There is not a clear concensus at this time whether a homologue of the VNO is present in chelonians. (Syn. *Jacobson's organ, naso-vomeral organ*)

vomero-palatine. *n.* A bone in the roof of the mouth of many amphibians, formed by fusion of the *vomer* and *palatine* bones.

vomeropalatine teeth. Equivalent to *vomerine teeth*.

vomodor. *n.* A sensation related to detection of chemicals by sensory cells of the *vomeronasal organ*. Cf. *odor*.

von Baer's law. The generalization that developoment proceeds from a generalized condition to a specialized condition, such that earlier embryonic stages of related organisms are identical and distinguishing features develop later during ontogeny.

-vorous. Suffix meaning "feeding upon."

vorzugstemperatur. *n.* **(VT)** An older term used for *preferred substratum temperature*.

voucher specimen. Any specimen that is identified by a recognized authority for the purposes of forming a reference collection. A voucher specimen may also be deposited in a museum to verify the species identification of animals used in a particular study, especially when the species is unusual or exotic.

VPI. See *vent position index*.

vulcanism. *n.* Volcanic activity.

vulgarobufotoxin. *n.* A *bufotoxin* isolated from *Bufo vulgaris*.

vulnerable. (V or **VU)** *adj.* An IUCN criterion for threatened species where a taxon is known to be at high risk of extinction in the wild in the medium-term future. Cf. *endangered, critically endangered, threatened*.

W

Wagler's Palm Viper. See *Temple Pitviper*.

Wagler's Snake. Vernacular name for the South American colubrid species *Waglerophis merremii*. See also *False Pitvipers*.

Wagner tree. A dendrogram based on the so-called "Wagner algorithm." See J.S. Farris, *Syst. Biol.* 19:83–92, 1970; E.O. Wiley, *Phylogenetics: The Theory and Practice of Phylogenetic Systematics,* 1981.

walking trot. Locomotion wherein the forelimbs and hindlimbs on opposite sides of the body strike the ground at the same time, while the other two feet are off the ground.

Wallacea. *n.* The transition zone between the Oriental and Australian zoogeographical regions, bounded to the east by *Weber's line* and to the west by *Wallace's line*—comprising Sulawesi, Lombok, Flores, and Timor.

Wallace effect. See *reproductive isolating mechanism*. (Syn. *Wallace's hypothesis*)

Wallace's Line. A biogeographic boundary separating Asiatic and Australasian faunas at a hypothetical line passing between Bali and Lombok, Borneo and Sulawesi, and east of the Philippines. Named after the English naturalist Alfred Russel Wallace, who in 1859 drew attention to a zone of contact between two entirely distinct terrestrial biotas (called the *Oriental* and *Australian* biogeographic realms).

Wallace's realms. The major zoogeographical divisions, namely Australian, Ethiopian, Nearctic, Neotropical, Oriental, and Palearctic.

Wall Geckos. Vernacular name for species of Geckos belonging to the genus *Tarentola*.

Wall Lizards. Collective vernacular name for various Old World lizard species belonging to the *Lacertidae* (sometimes called "true lizards"), especially the genus *Podarcis*.

waltz. *n.* A circling behavior engaged by a courting pair of *Ambystoma* spp. salamanders, as each sex attempts to bring its snout in contact with the other's cloaca. Cf. *tail walk*.

Wandering Garter Snake. Vernacular name for the subspecies of the Western Terrestrial Garter Snake, *Thamnophis elegans vagrans*.

Wandering Salamander. Vernacular name for the American species *Aneides vagrans*.

warm-blooded. *adj.* An obsolete term sometimes used inappropriately to describe the homeostatic condition of body temperatures in birds and mammals. This term should be replaced by *endothermic* or *homeothermic*. See *endotherm* and *endothermy*.

warning attitude. See *unken reflex*.

warning call, chirp, croak, or **vibration.** See *male-release call*.

warning coloration. Conspicuous colors or markings on an animal that is distasteful, poisonous, or otherwise defended against predators.

warning display. Generally, any stereotyped display that functions to deter potential predators. The term has been used most often with reference to neck spreading behaviors of cobras, related elapid species of snakes, and various other nonvenomous snakes.

wart. *n.* A *verruca* or epidermal proliferation on the body surface. The term also may be used in reference to viral epidermal tumors and glandular regions where the skin is elevated. (Syn. *pustule*)

Wart Frogs. Vernacular name for species of south Asian frogs belonging to the ranid genus *Limnonectes*.

Wart Snake. Vernacular name for the acrochordid snake *Acrochordus javanicus*. (Syn. *Filesnake, Elephant Trunk Snake*)

washout shunting. See *intracardiac shunts*.

water absorption response. Name given to a behavioral response of partially dehydrated frogs which, when placed on a moist substrate, splay the hind legs and press the ventral surfaces of the hind legs and abdomen against the substrate (W.T. Stille, *Copeia* 1958:217–218, 1958). See also *seat patch down*.

water balance. The difference between the rate of water intake and rate of water loss of an organism. See *balance*.

water capacity. The amount of water that a soil can retain against gravity, usually expressed as a percentage per volume of soil.

Water Cobra. Vernacular name for cobras of the genus *Boulengerina* inhabiting shorelines of lakes and feeding primarily on fish in forested regions of central Africa.

water column. The vertical column of water in a body of water, ocean or lake.

water conserving posture. A position or posture adopted by anuran amphibians in which the ventral surfaces are pressed firmly against the substratum and the limbs are tucked or folded tightly against the animal's sides (W.T. Stille, *Copeia* 1958:217–218, 1958). The posture minimizes exposed skin surfaces and thereby minimizes evaporative water losses. It is especially common and pronounced in arboreal frog species.

water cycle. See *hydrologic cycle*.

Waterdogs. *n.* Vernacular name for several American proteid salamanders belonging to the genus *Necturus*. See also *Mudpuppies*.

Water Dragon. Vernacular name for Australasian species of agamid lizards belonging to the genera *Lophognathus* and *Physignathus*.

Water Frog. Vernacular name for species of South American frogs belonging to the leptodactylid genus *Telmatobius*.

Water-holding Frog. Vernacular name for species of Australian frogs belonging to the hylid genus *Cyclorana*, especially the species *C. platycephala*.

Water Moccasin. An alternate, commonly used name for the Cottonmouth (*Agkistrodon piscivorus*). Usage of *Cottonmouth* is preferred. See *Moccasin*.

Water Monitor. Vernacular name for the Asian varanid species *Varanus salvator*.

water potential. The thermodynamic state of water within a compartment, equal to the difference in free energy per unit volume between metrically bound, pressurized, or osmotically constrained water and that of pure water. Water potential is the potential energy of water per unit mass and predicts the direction in which water will flow, governed by solute concentration, structure within a soil matrix (if any), and applied pressure. This concept and expression for the "driving force" moving water is most commonly used in plant literature, although the concepts can be applied to animal physiology just as well.

water saturation deficit. (WSD) The relative water content of a tissue calculated as the difference between the water content under conditions of saturation and the actual water content, expressed as a percentage of the saturation water content. Cf. *saturation deficit*.

watershed. *n.* In American usage, this refers to a unit of landscape where all water drains toward some common exit point (i.e., a drainage area). In European usage, such units are referred to as "*catchments.*"

Watersnakes, Water Snakes. *n.* Collective vernacular name for American natricine colubrid snakes of the genus *Nerodia*, various natricine *Keelbacks*, homalopsine species of the genera *Enhydris, Homalopsis,* xenodontines of the genus *Helicops* and *Hydrops*, and Central American *Tropical Water Snakes* of the genus *Hydromorphus*. See also *Boodontini*.

water table. The upper boundary of ground water.

water vapor conductance. See *conductance, gaseous*.

water vapor pressure. The pressure exerted by water vapor, expressed as the partial pressure of water vapor in atmospheric air. See *saturation vapor pressure*.

water vapor resistance. See *resistance*.

watt. (W) A derived SI unit of electric power, equivalent to the work performed at 1 joule (J) per second.

wattle. *n.* A distensible or erectile flap extending from the neck or throat in certain lizards, used by the male in courtship display.

wave. *n.* An undulation or ridge on the surface of a fluid, driven by wind in natural water systems and usually moving over the water surface.

wave drag. Drag on a swimming animal caused by formation of surface waves.

wavelength. *n.* The horizontal distance between two wave crests (or two troughs).

wavering. *n.* A category of *headbob* display in *Anolis* lizards, consisting of vertical oscillations of the head when a lizard is stalking prey.

wax. *n.* Any ester of fatty acids or long-chain monohydroxyalcohols.

Wax Frogs. Vernacular name for species of African frogs belonging to the hyperoliid genus *Cryptohylax*.

WCH. See *World Congress of Herpetology*.

WCMC. See *World Conservation Monitoring Centre*.

WCRP. *World Climate Research Programme*.

weather. *n.* Local, short-term atmospheric conditions. Cf. *climate*.

weathering. *n.* The collective group of processes, especially wind and water action, that change or decompose rock at or near Earth's surface. See *erosion*.

web, webbing. *n.* A thin, sheetlike membrane, usually extending between, and connecting, adjacent digits (per common usage).

webbing formula. An expression that quantifies the extent of webbing between fingers and toes of anuran amphibians. For a current summary of webbing notations see R.D. MacCulloch, *Herpetol. Rev.* 35:140–142, 2004.

Weber's line. An imaginary line separating the Oriental and Australian zoogeographical regions, drawn to the east of Wallace's line and passing between Maluku (Moluccas) and Sulawesi to the north, and between Timor and the Kei Islands to the south.

Web-toed salamander. Collective vernacular name for plethodontid species belonging to the genus *Hydromantes*.

wedge-bone. *n.* Reference to *intercentrum*, used by H. Gadow (*Amphibia and Reptiles*, p. 283, 1901).

wedged preocular. A condition in which the lower *preocular* projects downward between adjoining labial scales in snakes of the genus *Coluber*.

weighting. *n.* 1. In systematics, a method of attaching different importance values to different characters. 2. Altering the value of a function relative to others.

welt. *n.* A flattened mass of dense tissue found on the distal end of the hemipenis in many lizards. Term was used by G.K. Noble.

Western Alligator Lizard. Vernacular name for American anguid species belonging to the genus *Elgaria*. Cf. *Eastern Alligator Lizard*.

Western Aquatic Garter Snake. Vernacular name for the American colubrid species *Thamnophis couchii*.

western blotting. A technique similar to *Southern blotting* in which a radioactively labeled antibody is used to identify a specific protein. (Syn. *immunoblotting*) See *northern blotting, Southern blotting*.

Western Brown Snake. Vernacular name for the Australian elapid species *Pseudonaja nuchalis*.

Western Coral Snake. See *Sonoran Coral Snake*.

Western Diamondback Rattlesnake. Vernacular name for the American pitviper *Crotalus atrox*.

Western Fence Lizard. Vernacular name for the American iguanid species *Sceloporus occidentalis*.

Western Pond Turtle. Vernacular name for the western American chelonian species *Actinemys marmorata*.

Western Rattlesnake. Vernacular name for the American pitviper *Crotalus oreganus*. Numerous subspecific races of this species have other names, mostly associated with geographic regions where they occur (Grand Canyon Rattlesnake, Northern Pacific Rattlesnake, Great Basin Rattlesnake, etc.).

Western Ribbon Snake. Vernacular name for the American colubrid species *Thamnophis proximus*.

Western Skink. Vernacular name for the American scincid lizard *Eumeces skiltonianus*. Note that H.M. Smith (*J. Kansas Herpetol.* 14:15, 2005) recently proposed that *Plestiodon* be a replacement name for the genus *Eumeces* in North America.

Western Spadefoot. Collective vernacular name for American scaphiopodid species belonging to the genus *Spea*. Cf. *Southern Spadefoot*. See also *Spadefoot Toad*.

Western Terrestrial Garter Snake. Vernacular name for the American colubrid species *Thamnophis elegans*.

Western Toad. Vernacular name for the North American bufonid species *Anaxyus* (formerly *Bufo*) *boreas*. For taxonomic revision see D.R. Frost et al., *Bull. Amer. Mus. Nat. Hist.* 297:1–370, 2006.

Western Whiptail. Vernacular name for the American teiid species *Cnemidophorus* (= *Aspidoscelis*) *tigris*.

wetland. *n.* A comprehensive term for lands that are transitional between terrestrial and aquatic systems where the water table is usually at or near the surface, or the land is covered by shallow water at least part of the year. As a result, these areas have characteristic soils and vegetation. Examples are salt marshes, bogs, swamps, prairie potholes, and vernal pools. Wetlands provide important habitats for many species of amphibians and reptiles.

Whipping Frog. Vernacular name for species of Asian tree frogs belonging to the genus *Polypedates*. These frogs construct foam nests similar to those of *Chiromantis*.

Whipsnake, Whip Snake. *n.* A collective vernacular name for colubrid snakes of the North American genera *Masticophis* and *Coluber*, South American genus *Chironius*, Asian genus *Ahaetulla*, certain Old World species of *Elaphe*, and members of the Australian elapid genera *Demansia* and *Rhinoplocephalus*.

Whiptails or **Whip-tailed Lizards.** *n.* A collective common name for various species of teiid lizards, closely related to *Racerunners*. These are common to temperate and tropical habitats of the Americas and closely resemble Wall and Sand Lizards of the Old World. The name is used particularly for species belonging to the genus *Cnemidophorus* and also to Neotropical teiid lizards of the genus *Ameiva*. These genera have been shown to be paraphyletic, and species of *Cnemidophorus* north of Panama have been placed in the resurrected genus *Aspidoscelis*.

White-bellied Mangrove Snake. Vernacular name for the Asian homalopsine snake *Fordonia leucobalia*, which specializes in eating crabs.

white blood cell. *n.* See *leukocyte*.

White-lipped Frog. Collective vernacular name for leptodactylid species of anurans belonging to the genus *Leptodactylus*, distributed in Middle and South America.

White-lipped Tree Viper. Vernacular name for the Asian pitviper *Trimeresurus albolabris*.

white matter. Tissue of the brain and spinal cord that is composed predominantly of nerve fibers, many of which are *myelinated*. Cf. *gray matter*.

white muscle. A classification of muscle based on color, which does not adequately consider the full range of complexity and function. White muscle tends to be less vascularized and contain less myoglobin than *red muscle*, but it contracts very rapidly. White muscle tends to contain relatively fewer aerobic fibers than does red muscle, and therefore is less resistant to fatigue. Cf. *red muscle*.

White's Tree Frog. Vernacular name for the Australasian hylid species *Pelodryas caerulea* (formerly *Litoria caerulea*).

whorl. *n.* A symmetrical series of scales that encircle a single caudal segment in some lizards. Usually, such scales are very distinct from others. (Syn. *verticil*)

Whorltail Iguanas. Collective vernacular name for species of iguanid lizards belonging to the South American genus *Stenocercus*.

Wilcoxon two-sample test. A nonparametric method in which sample counts are ranked in a single sequence, then used to test whether two independent random samples are drawn from populations having the same distribution.

wildlife corridor. See *corridor*.

wildlife species. See *evolutionary significant unit*.

wild type. The natural or typical form of an organism or gene, arbitrarily designated as standard or normal for comparison with aberrant or mutant individuals or alleles.

window. *n.* 1. The transparent portion of the lower eyelid of certain lizards, especially common among skinks. 2. This term is used in connection with membranous structures that transmit vibrational stimuli between air or tissue and liquid of the inner ear (*oval window, round window*).

windward. *adj.* The direction or side from which the wind blows. Cf. *leeward*.

wing or **wing membrane.** *n.* 1. The lateral projection of individual elements of the plastron of turtles that contribute to forming the *bridge*. 2. The broad sheet of integument, with supporting elongate ribs, that extends from the forelimb to hindlimb and is used in gliding by the lizard *Draco*.

wing loading. The weight of a flyer divided by the area of its airfoils.

wing process. See *alary process*.

Wolffian duct. The primitive kidney duct, which, in amniotes, becomes the embryonic duct of the mesonephric kidney that differentiates into the *ductus deferens* in adult male amphibians and reptiles, and degenerates in females. (Syn. *archinephric duct; mesonephric duct*)

Wolffian regeneration. Reference to regrowth of an eye lens following extirpation in some urodeles, described by R. Rugh in 1962. Budding is presumably from the dorsal rim of the iris.

Wolf Snake. Collective vernacular name for Old World species of colubrid snakes belonging to the African genera *Cryptolycus, Lycophidion,* and the Asian genus *Lycodon*. The name relates to the possession of enlarged front teeth, which are used to catch Skinks or other smooth-scaled lizards.

Woma. *n.* Vernacular name for the Australian boid species *Aspidites ramsayi*.

Wood Frog. Vernacular name for the North American species *Lithobates sylvaticus* (formerly *Rana sylvatica*) and for South American leptodactylid species of *Batrachyla*. For recent taxonomic revision see D.R. Frost et al., *Bull. Amer. Mus. Nat. Hist.* 297:1–370, 2006.

Woodhouse's Toad. Vernacular name for the bufonid species *Anaxyrus* (formerly *Bufo*) *woodhousii*. For taxonomic revision see D.R. Frost et al., *Bull. Amer. Mus. Nat. Hist.* 297:1–370, 2006.

woodland. *n.* An area of vegetation dominated by a more or less closed stand of short trees.

Woodland Racer. Vernacular name for South American species of colubrid snakes belonging to the genus *Drymoluber*.

Woodland Salamander. Collective vernacular name for North American species of plethodontid salamanders belonging to the genus *Plethodon*.

Wood Snake. Vernacular name for several species of Boas belonging to the family *Tropidophiidae*. Also called *Dwarf Boas*.

Wood Turtle. Vernacular name for various species of emydid turtles, especially those belonging to the neotropical genus *Rhinoclemmys* and the American species *Glyptemys* (formerly *Clemmys*) *insculpta*. See also *Vietnamese Wood Turtle*.

work. *n.* 1. Force exerted upon an object over a distance: force × distance. 2. Reference to a scientific publication.

working hypothesis. A hypothesis that provides a basis for future experimentation.

World Congress of Herpetology. (WCH) A global organization founded in 1982 to promote international interest, collaboration, and cooperation in herpetology. The WCH holds international meetings every 3-5 years. Current URL: http://www.worldcongressofherpetology.org/

World Conservation Monitoring Centre. (WCMC) The *World Conservation Monitoring Centre*, a collaborative venture between IUCN, the United Nations Environment Programme (UNEP), and the Worldwide Fund for Nature (WWF). The WCMC maintains a global overview database of threatened species, habitats and protected areas, and provides information, reports, and analyses of global biodiversity for a wide range of users.

World Conservation Union. See *International Union for the Conservation of Nature and Natural Resources*.

World Wide Fund for Nature. (WWF) An organization organized to promote global environmental conservation of biological diversity, sustainable use of natural resources, and effective environmental education. Formerly the World Wildlife Fund, which is still used by WWF-US and WWF-Canada. Current URL: http://www.panda.org/

World Wildlife Fund. See *World Wide Fund for Nature*.

worm or **tongue worm.** *n.* A colloquial term for the bifurcate fleshy projection of the tongue used by the Alligator Snapping Turtle (*Macroclemys temminckii*) to lure fish into its open mouth.

Worm Lizard. Collective vernacular name for various members of the suborder *Amphisbaenia* (e.g., *Rhineura*). The name also is sometimes applied to Australian pygopodid lizards of the genus *Aprasia*, and to South American anguid lizards of the genus *Ophiodes*. See also *Two-Legged Worm Lizards*.

Worm Snakes. Collective vernacular name for species of American colubrid snakes belonging to the genus *Carphophis*. The name also is used in reference to *Blind Snakes*, family Leptotyphlopidae, including genera *Typhlops, Ramphotyphlops, Grypotyphlops*.

woven bone. *Primary bone*.

Wright's inbreeding coefficient. (F) 1. The probability that two allelic genes united in a zygote are both descended from a gene found in an ancestor common to both parents. 2. The proportion of loci at which an individual is homozygous.

Writhing Skink. Vernacular name for species of scincid lizards belonging to the genus *Lygosoma*.

WWF. *World Wide Fund for Nature*, formerly *World Wildlife Fund*.

X

xanth(o)-. Word element (Gr.) meaning "yellow."
xanthic. *adj.* Yellow.
xanthism, xanthochroism. An aberrant condition characterized by excessive presence of yellow pigment.
xantholeucophore. *n.* See *xanthophore*.
xanthophore. *n.* A lipophore in which oil droplets have a yellow, orange, or yellowish hue. (Syn. *golden cell, golden pigment cell, xantholeucophore*)
Xantusiidae. *n.* A clade (family) of lizards that have relatively flat bodies and heads, lack movable eyelids, and are relatively sedentary and secretive. Three genera and 18 species are known from southwestern United States to Panama.
xantusiid. *n, adj.* A member of, or pertaining to, *Xantusiidae*.
x-axis. *n.* The horizontal axis of a graph. (Syn. *abscissa*)
xbar. ($\bar{x}$) The *mean* (of x).
-xene. Suffix meaning "guest," "visitor."
Xenoanura. *n.* 1. A newly designated taxon, sister with *Sokolanura* within the *Lalagobatrachia* (D.R. Frost et al., *Bull. Amer. Mus. Nat. Hist.* 297:1–370, 2006). This is a monophyletic crown taxon containing *Pipidae* and *Rhinophrynidae* (and presumably a number of fossil taxa including palaeobatrachids). 2. Synonym for Orton's (1953) Type 1 tadpole. See P.H. Starrett, pp. 251–271 in J.L. Vial, ed., *Evolutionary Biology of Anurans. Contemporary Research on Major Problems,* 1973.
xenobiotic. *n.* An organic or inorganic chemical that is foreign (*allochthonous*) to biological systems.
Xenodermatinae. *n.* A clade of poorly known colubrid snakes, distributed in southeast Asia, the Indonesian Archipelago, and represented by about 20 species in four genera.
Xenodontinae. *n.* A subfamily of Colubridae, the name formerly used for a majority of endemic Neotropical colubrid species of snakes. A number of genera of snakes are not resolved with respect to placement either in Xenodontinae or Dipsadinae. This group is given familial status in alternate classification schemes (e.g., J.T. Collins, *J. Kans. Herpetol.* 19:18–20, 2006; N. Vidal et al., *Comptes Rendus Biologies* 330:182–187, 2007).
xenograft. *n.* A transplantation of tissue between individuals belonging to different species or genera, or which are greatly different. See also *heterograft*.
Xenopeltidae. *n.* A clade (family) that includes two species of burrowing, nocturnal snakes that inhabit rain forest in India and southern China to Borneo and Celebes. Known as *Sunbeam Snakes*.
xenopeltid. *n.* A member of, or pertaining to, *Xenopeltidae*.
Xenophidiidae. *n.* A clade (family) of tropical, nocturnal, and secretive snakes; two species of a single genus are distributed in Sabah (Borneo) and part of the central Malaysian peninsula.
xenoplastic. *adj.* Reference to *xenograft*. See also *heteroplastic*.
xenopsin. *n.* Peptides isolated from the skin of *Xenopus laevis*, having broad-spectrum antimicrobial activity but different from *magainins*.
Xenosalamandroidei. *n.* A newly designated monophyletic taxon of salamanders containing the *Amphiumidae* and *Plethodontidae* (D.R. Frost et al., *Bull. Amer. Mus. Nat. Hist.* 297:1–370, 2006).
Xenosauridae. *n.* A clade (family) of rock-dwelling lizards belonging to a single genus with six species inhabiting southern Mexico and Guatemala.
Xenosyneunitanura. *n.* A new monophyletic taxon of anurans, sister to the new taxon *Laurentobatrachia* within the new taxon *Afrobatrachia* and composed of *Hemisotidae* and *Brevicipitidae*. See D.R. Frost et al., *Bull. Amer. Mus. Nat. Hist.* 297:1–370, 2006.
xeric. *adj.* Reference to dry or arid conditions, lacking in moisture.
xeric pattern. One of three breeding patterns described by A.N. Bragg (*Researches on the Amphibia of Oklahoma*, p. 59–93, Univ. Oklahoma Press, 1950). Basically a noncyclic pattern in which individuals respond strongly to rainfall.
xero-. Prefix meaning "dry."
xerocolous. *adj.* Living under dry conditions.
xerophilic. (-ous) *adj.* Of or pertaining to plants and animals that require little or no environmental moisture to survive.
xiphi-. Prefix meaning "shaped like a sword."
xiphiplastron. *n.* (*pl.* **-plastra**). The posterior pair of bones in the plastron of turtles. See Fig. 7.
xiphisternal rod. Either of two elongate cartilaginous elements making up the *xiphisternum* in Geckos.
xiphisternum. *n.* The most posterior element of the sternum. (Syn. *xiphoid cartilage, ziphoid process*) See Fig. 3.
xiphoid cartilage. See *xiphisternum*.
xiphoid process. See *xiphisternum*.

X-ray diffraction. A method of examining the crystalline structure of a substance using the pattern of scattered X-rays.

xyphiplastron. *n.* Alternate spelling for *xiphiplastron.*

Ya. Abbreviation meaning "years ago." Cf. *Ma.*

Yacare. *n.* See *Jacare.*

yard. *n.* A derived *fps* unit of length equal to 0.9144 m.

Yarrow's Spiny Lizard. Vernacular name for the phrynosomatine iguanid species of American lizard *Sceloporus jarrovii.*

yaw. *n.* Rotation of a swimmer or flyer around its vertical (dorsoventral) axis.

y-axis. The vertical axis of a graph. (Syn. *ordinate*)

ybp. Years before present.

year class. Reference to all the animals in a population that were born or hatched during a particular reproductive season.

Yellow-bellied Sea Snake, Yellowbelly Sea Snake. Vernacular name for the largely pelagic elapid Sea Snake *Pelamis platurus.* Also called *Black and Yellow Sea Snake.*

Yellow-jaw Tommygoff. Vernacular name for the *Terciopelo* or *Fer-de-Lance*, used in Belize.

Yellow-legged Frog. Vernacular name for the North American anuran species *Rana boylii* (Foothill Yellow-legged Frog) and *Rana muscosa* (Mountain Yellow-legged Frog).

Yellow-lipped Sea Krait. See *Sea Kraits.*

yolk. *n.* A dense, yellowish mass of nutritive fluid that serves as a stored nutrient for the ovum or embryo.

yolk platelet. One of the rather large oval granules of yolk found in amphibian eggs, densely distributed at the vegetal pole.

yolk plug. A mass of yolk that plugs the *blastopore* during *gastrulation* in amphibians. (Syn. *Rusconian plug*)

yolk sac. An extraembryonic membrane formed of *endoderm* and *mesoderm* that covers the mass of yolk in an embryo.

yolk sac placenta. A placenta in which the wall of the yolk sac forms the embryonic contribution to the placenta, characteristically in most viviparous lizards. For terminology discussion see J.R. Stewart & D.G. Blackburn, *Copeia* 1988:839–852, 1988. See also *choriovitelline placenta.*

Yosemite Toad. Vernacular name for the California species of bufonid anuran *Anaxyrus* (formerly *Bufo*) *canorus.* For recent taxonomic revision see D.R. Frost et al., *Bull. Amer. Mus. Nat. Hist.* 297:1–370, 2006.

young generation. See *epidermal generation.*

ypsiloid apparatus. See *ypsiloid cartilage.*

ypsiloid bone or **cartilage.** A Y-shaped cartilage or bone located immediately anterior to the pelvis, to which it can be articulated in various salamanders. It is conjectured to have a hydrostatic function and to be an independent structure. (Syn. *abdominal sternum, epipubic cartilage, marsupial cartilage, ypsiloid apparatus*)

Z

zahnreihe. (*pl.* **-hen**) *n.* A single series or sequence of replacement teeth, progressing from anterior to posterior part of the jaw. Several *zahnreihen* may occur together as "waves" of teeth development.

Zebra-tailed Lizard, Zebratail. A fast running iguanid lizard species (*Callisaurus draconoides*) common to desert areas of the southwestern United States.

zeitgeber. *n.* An environmental factor that entrains a biological rhythm.

zenith. *n.* The point on the celestial sphere directly above the observer, or above a given point on the surface.

zero-order kinetics. Reference to enzymatic reactions in which the formation of product proceeds at a linear rate with time, and the rate will increase if additional substrate is added. Cf. *first-order kinetics.*

zero sum assumption. An aspect of the *Red Queen hypothesis*, based in the assumption that beneficial effects enjoyed by a species during evolutionary change is precisely matched by the sum of negative effects experienced by other species in the community. The zero sum assumption is based partly in the idea that the total resources available in the system is constant, and the rate of evolution also will be constant.

zeugopodium. *n.* The middle region of a limb. Cf. *autopodium, stylopodium.*

zip fastener model (of skin shedding). A concept involving interdigitation between cell membranes of the clear layer of the outer generation and the *oberhautchen* of the (new) inner generation of epidermis. There is increased loss of intracellular fluids at the time of shedding that softens the outer generation and allows the shedding complex to "unzip" at ecdysis. See P.F.A. Maderson, *J. Zool.* (Lond) 119:39–50, 1966; P.F.A. Maderson et al., *J. Morphol.* 236:1–24, 1998.

ziphodont. *adj.* A type of tooth having an inwardly curved, serrated shape. Found in some dinosaurs, lizards, and crocodilians.

ziphoid process. See *xiphisternum.*

-zoic. Suffix meaning "animal."

-zoite. Suffix meaning "animal."

zonal bone. Bone material resulting from a pattern of development that involves slow to moderate growth at intermittent periods. Thus, *zonal growth.*

zone. *n.* An area or subdivision of a biogeographical region that has a characteristic biota.

zonula. *n.* Zone.

zonula adherens. A form of desmosome in epithelial cells that forms a belt of cell-to-cell adhesion associated with tight junctions.

zonula occludens. A series of tight junctions between epithelial cells, usually having a ring-shaped configuration. These physically occlude the extracellular pathways across the epithelium.

zoo blotting. A technique used to evaluate the degree to which a DNA sequence (usually a coding sequence) is conserved during evolution. The method involves testing for hybridization of a probe complementary to the sequence in one species against the DNA isolated from cells of a variety of other species.

zoogeographic realms or **regions.** Divisions of land masses according to their distinctive faunas.

zoogeography. *n.* Study of the geographical distribution of animals, their causes, and evolutionary histories.

zoology. *n.* The study of animals.

zoonose. (*pl.* **–es**) *n.* A condition when infectious agents cross over from animal(s) to human(s).

zooplankton. *n.* See *plankton.*

zootoxin. *n.* Any toxic secretion of an animal; but this term has been used solely in reference to snake venoms in older herpetological literature.

zootype. *n.* A term used by J.M. Slack et al. (*Nature* 361:490–492, 1993) to define a particular spatial pattern of gene expression shared by all animals during embryonic development. The concept is old but provides a gene-based definition of an animal. The concept has been extended as recognition that a common set of *Hox-genes* have been found in all animal phyla investigated, and their expression in the same anteroposterior order is considered a defining character, or *synapomorphy.*

Zopilota. *n.* See *Mussurana.*

zwitterion. *n.* A dipolar molecule that carries both negatively and positively ionized or ionizable sites.

zygantrum. (*pl.* **–a**) *n.* The cavity above and lateral to the neural canal on the posterior face of a snake vertebra, which articulates with the *zygosphene* of the next posterior vertebra. This articulation increases the resistance to torsion in the vertebral column of snakes.

zygapophysis. *n.* Each of four processes on the border of the neural arch which articulates with its opposite member on adjacent vertebrae, providing additional strength to the articulation of the centrum and preventing undue torsion. A pair of anterior zygapophyses (called *prezygapophyses*) are always directed upwards and slightly inward, whereas the posterior zygapophyses (called *postzygapophyses*) are always directed downward and slightly outward. The anterior and posterior elements articulate with elements on adjacent neural arches and limit flexion and torsion of the vertebral column. (Syn. *processus articular*)

zygodactylous. *adj.* Generally, reference to having toes in pairs, two in front and two behind. This term has been used by herpetologists to describe the foot of the chameleon, in which the digits are fused in bundles of two or three.

zygomatic arch. *n.* A bar, usually narrow, comprised of the *jugal* and *squamosal* bones in extinct therapsid reptiles and formed below the temporal opening of the skull. These are remnants of the former skull sidewall and comprise the lower bony margin of the single temporal fenestra found in mammals and their therapsid ancestors. The bar becomes bowed outward to accommodate a large masseter muscle in mammals.

zygosphene. *n.* Either of a pair of anterior accessory articulating surfaces on the vertebrae of some lizards and all snakes, which articulates with the *zygantra* of the next anterior vertebra.

zygote. *n.* A fertilized female gamete; the diploid cell that results from the fusion of a male and female gamete.

zymogen, zymogen granules. *n.* See *proenzyme*.

Table 1.

Eon	Era	Period	Epoch		million years to present
Phanerozoic	Cenozoic	Quaternary		Holocene (recent)	0.01
				Pleistocene	1.8
		Tertiary	Neogene	Pliocene	5.2
				Miocene	23.8
			Paleogene	Oligocene	33.5
				Eocene	55.6
				Paleocene	65.0
	Mesozoic	Cretaceous			144
		Jurassic			206
		Triassic			251
	Paleozoic	Permian			290
		Carboniferous			354
		Devonian			409
		Silurian			439
		Ordovician			500
		Cambrian			543
Proterozoic					2500
Archean (Archeozoic)					3800

Table 2. International system (SI) of units, conversions, constant, and definitions.

Physical Quantity	Name of unit	Symbol for unit	Definition
Basic SI units			
Length	meter	m	
Area	square meter	m^2	
Volume	cubic meter	m^3	
Mass	kilogram	kg	
Density	kilogram per cubic meter	$kg\ m^{-3}$	
Time	second	s	
Speed	meter per second	$m\ s^{-1}$	
Electric current	ampere	A	
Temperature	Kelvin	K	
Luminous intensity	candela	cd	
Derived SI units			
Acceleration	meter per second squared	$m\ s^{-2}$	
Activity	1 per second	s^{-1}	
Biomass Density	kilogram per square meter	$kg\ m^{-2}$	
Electric capacitance	farad	F	$A\ s\ V^{-1}$
Electric charge	coulomb	C	$A\ s$
Electric field strength	volt per meter	$V\ m^{-1}$	
Electrical resistance	ohm	Ω	$V\ A^{-1}$
Energy Flux, Productivity	watt per square meter	$W\ m^{-2}$	
Entropy	joule per Kelvin	$J\ K^{-1}$	
Force	Newton	N	$kg\ m\ s^{-2}$
Frequency	Hertz	Hz	s^{-1}
Illumination	lux	lx	$lm\ m^{-2}$
Luminance	candela per square meter	$cd\ m^2$	
Membrane Tension	Newton per meter	$N\ m^{-1}$	
Moment of Inertia	kilogram per meter-squared	$kg\ m^2$	
Power, Energy consumption	watt	W	$J\ s^{-1}$
Pressure, Stress	pascal	Pa	$N\ m^{-2}$
Voltage, potential difference	volt	V	$W\ A^{-1}$
Work, energy, heat	joule	J	$N\ m$

Table 3. Metric prefixes and multipliers.

Prefix	Abbreviation	Multiplier
tera-	T	10^{12}
giga-	G	10^{9}
mega-	M	10^{6}
kilo-	k	10^{3}
hecto-	h	10^{2}
deca-	da	10
deci-	d	10^{-1}
centi-	c	10^{-2}
milli-	m	10^{-3}
micro-	μ	10^{-6}
nano-	n	10^{-9}
pico-	p	10^{-12}
femto-	f	10^{-15}
atto-	a	10^{-18}

Figures

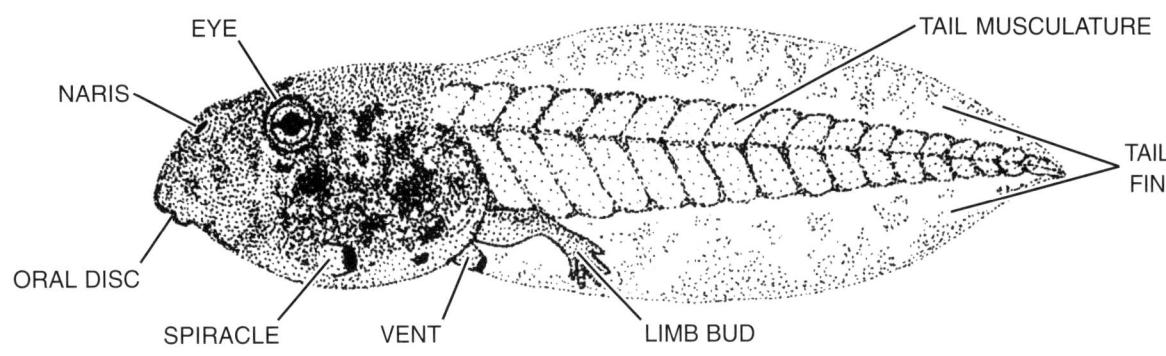

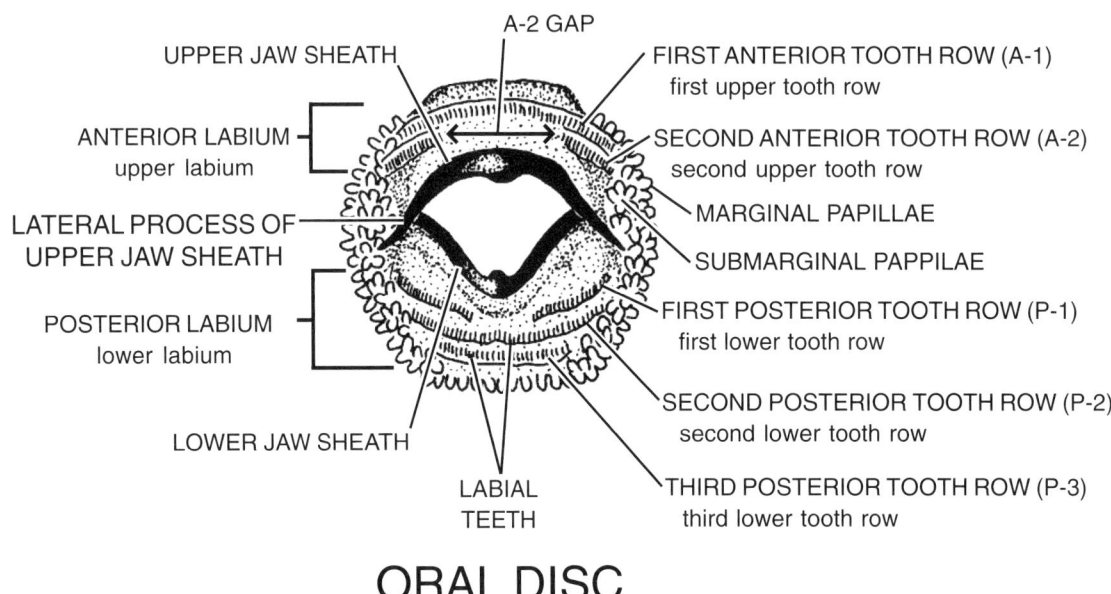

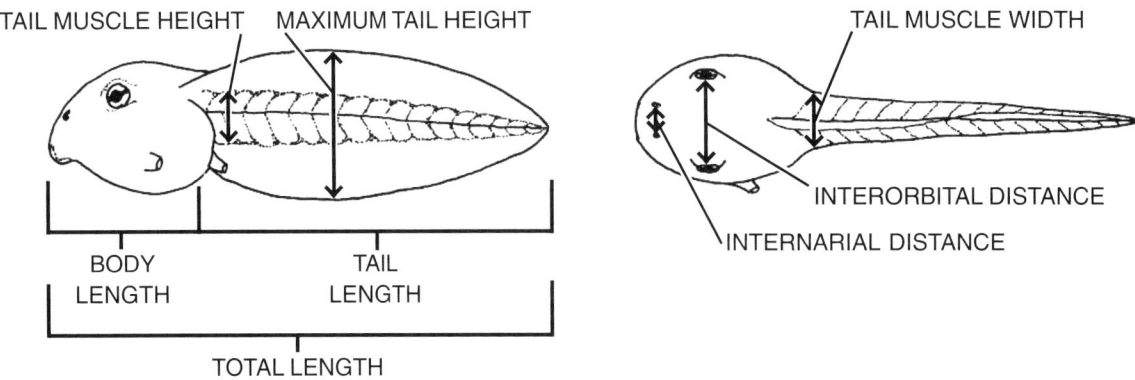

Figure 1. Schematic anuran tadpole showing features of mouth, head, and body with terminology related to diagnostic features and measurements.

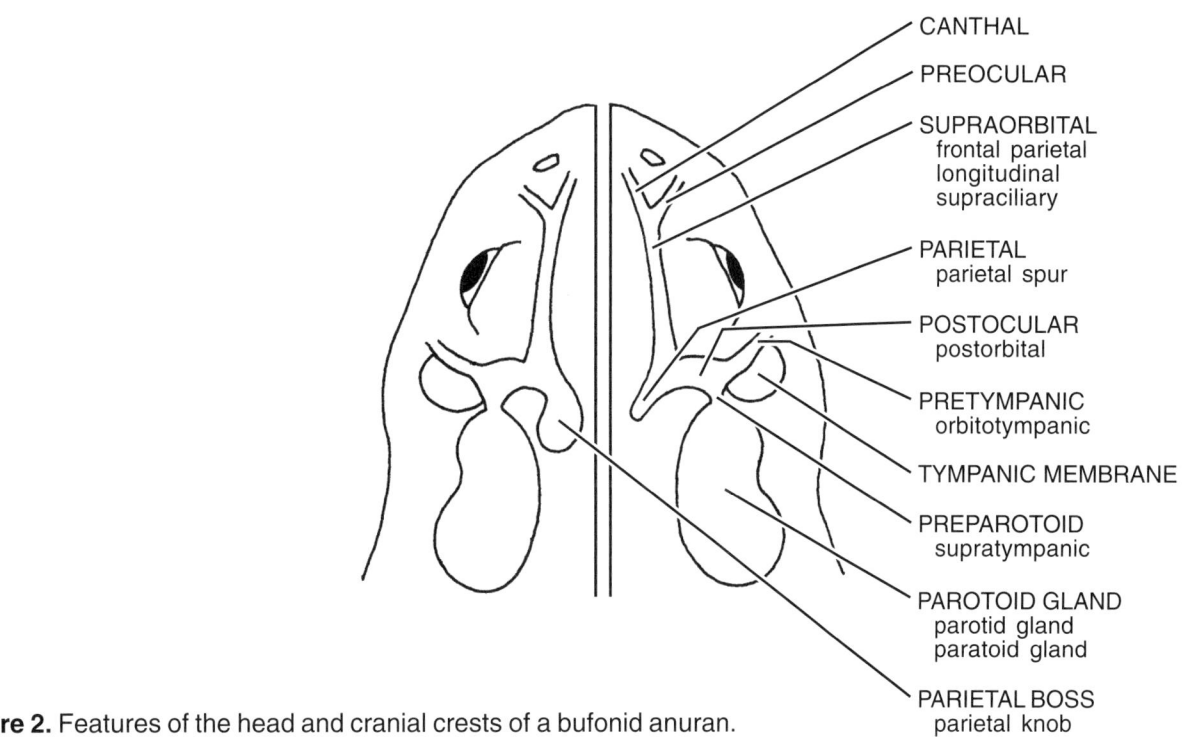

Figure 2. Features of the head and cranial crests of a bufonid anuran.

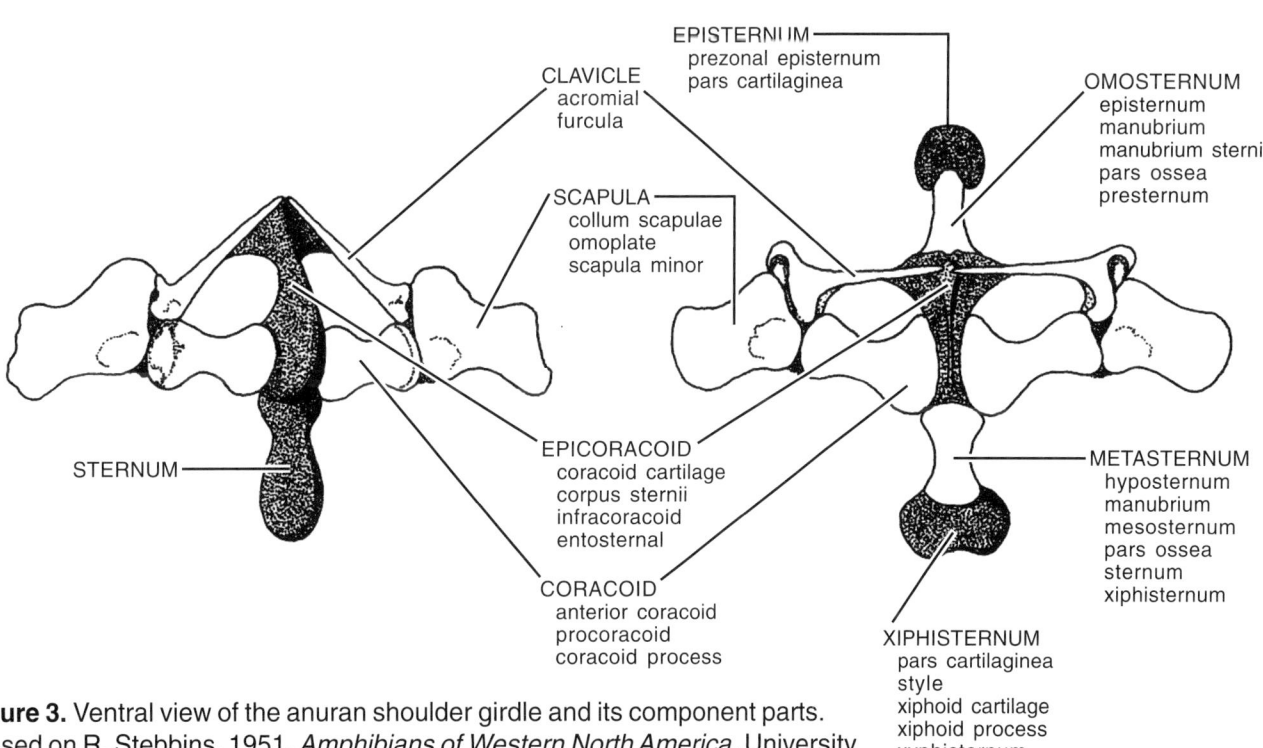

Figure 3. Ventral view of the anuran shoulder girdle and its component parts. (Based on R. Stebbins, 1951, *Amphibians of Western North America*, University of California Press, p. 183, Fig. 2.)

Figures

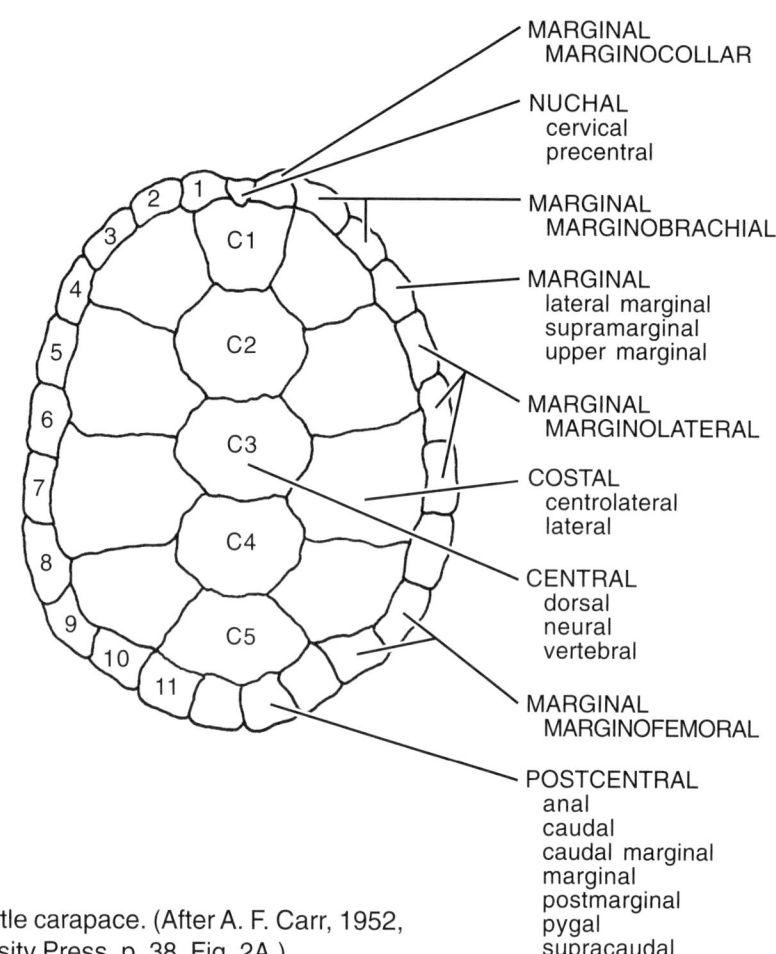

Figure 4. Epidermal laminae of a turtle carapace. (After A. F. Carr, 1952, *Handbook of Turtles*, Cornell University Press, p. 38, Fig. 2A.)

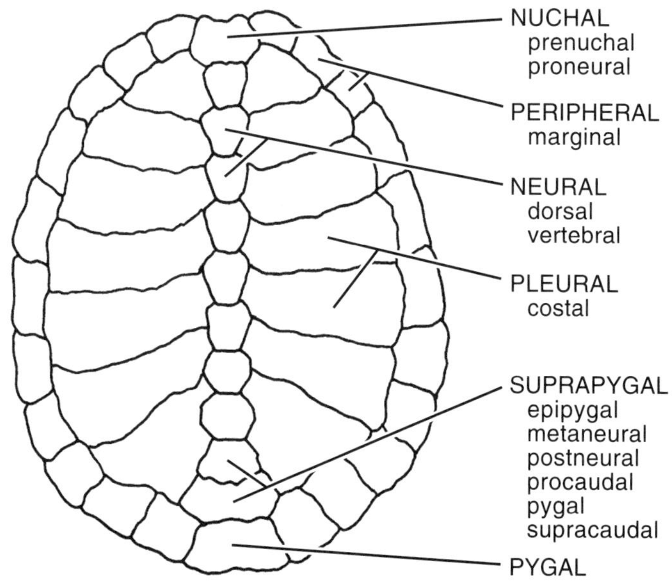

Figure 5. Bony elements of the turtle carapace. (After A. F. Carr, 1952, *Handbook of Turtles*, Cornell University Press, p. 38, Fig. 2C.)

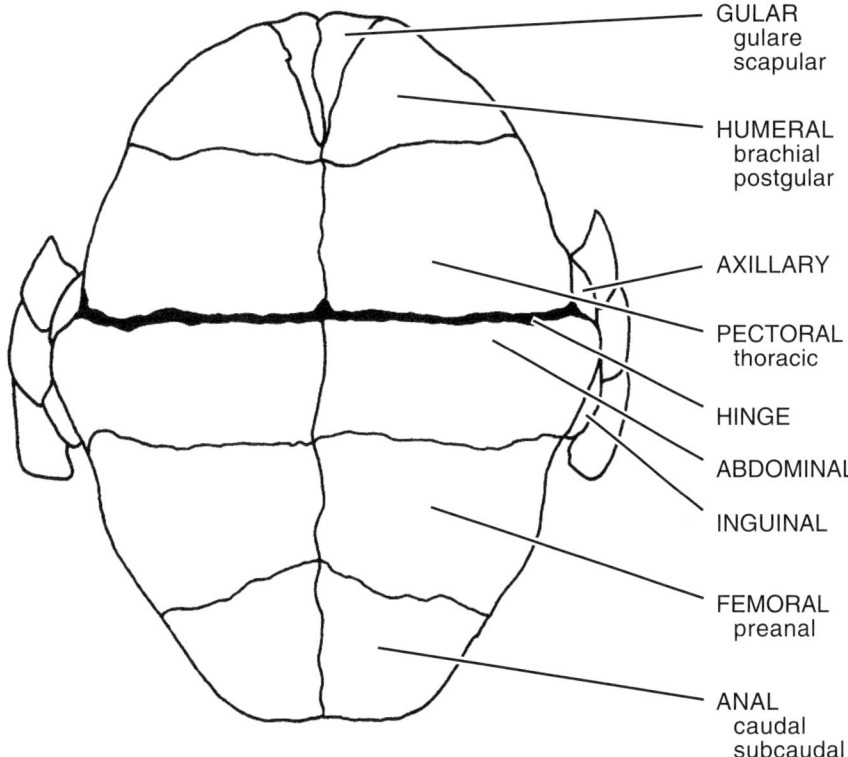

Figure 6. Epidermal laminae of the turtle plastron. (After A. F. Carr, 1952, *Handbook of Turtles*, Cornell University Press, p. 38, Fig. 2D.)

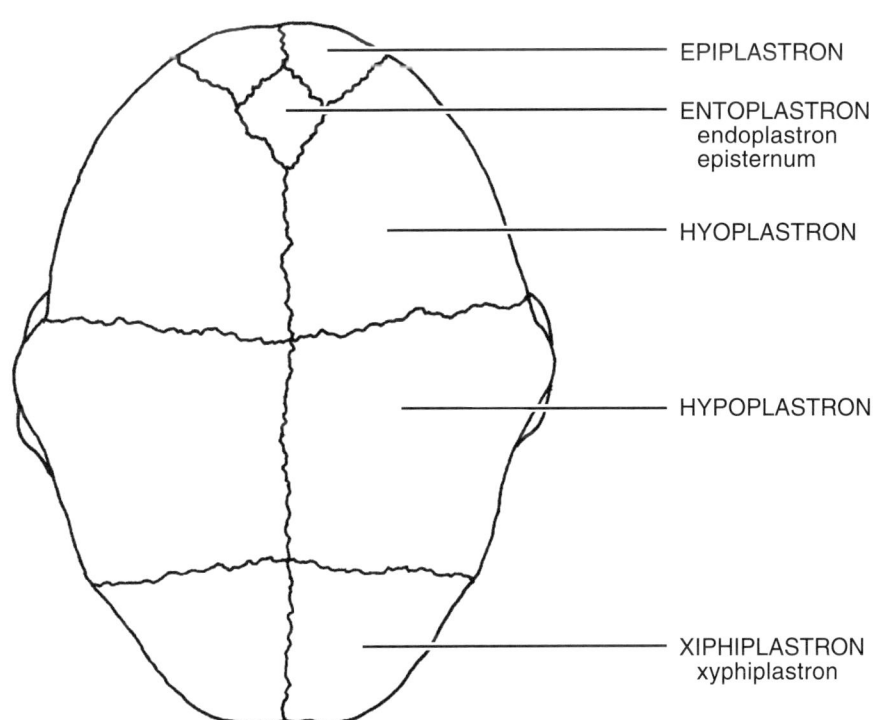

Figure 7. Bony elements of the turtle plastron. (After A. F. Carr, 1952, *Handbook of Turtles*, Cornell University Press, p. 38, Fig. 2B.)

Figures

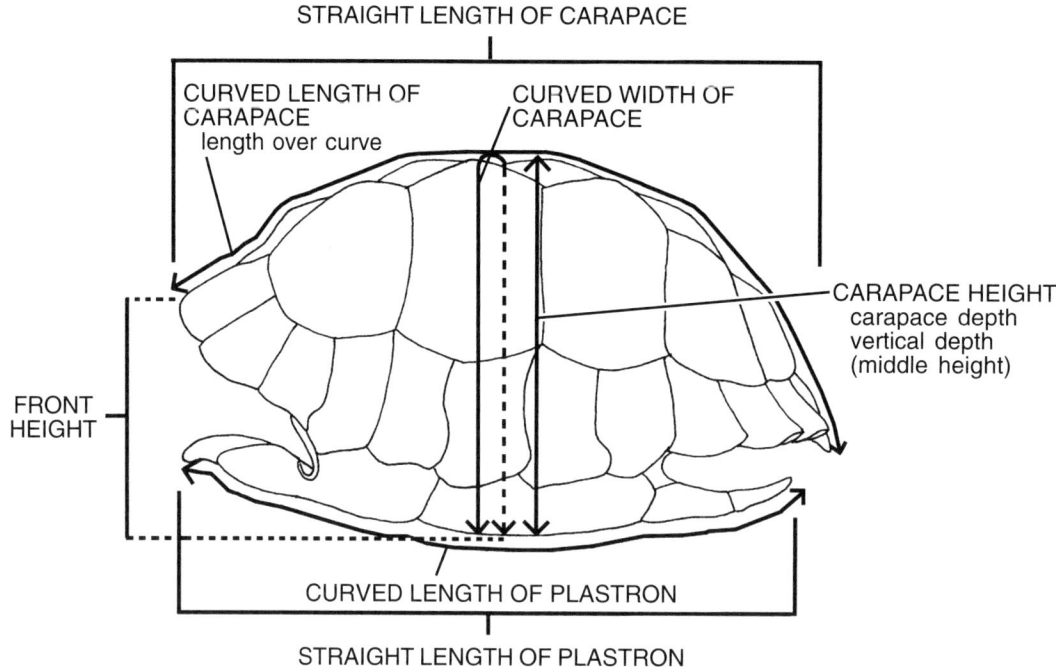

Figure 8. Standard measurements pertaining to the turtle shell, lateral view.

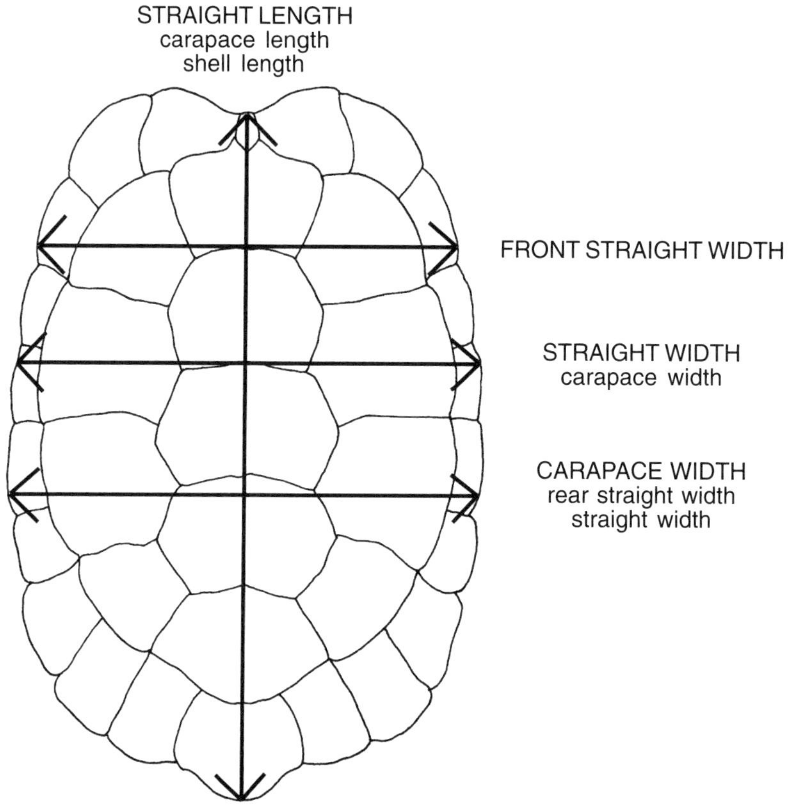

Figure 9. Standard measurements pertaining to the turtle shell, dorsal view.

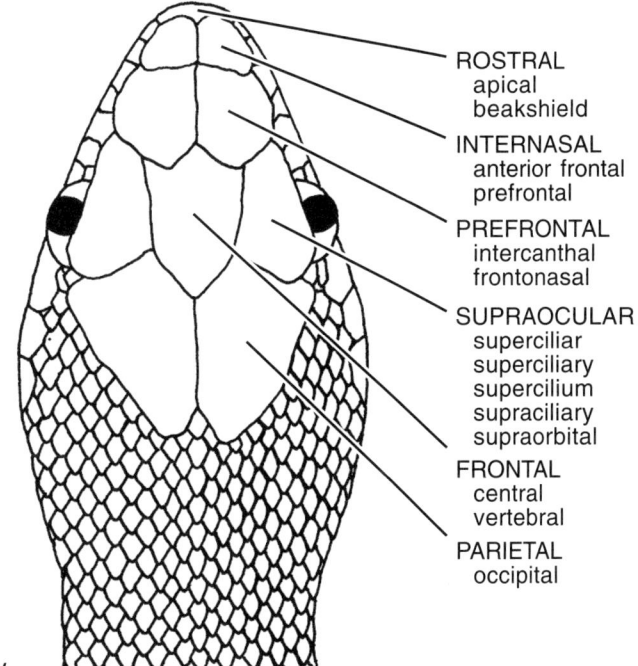

Figure 10. Dorsal view of the head of a snake, with associated scale terminology.

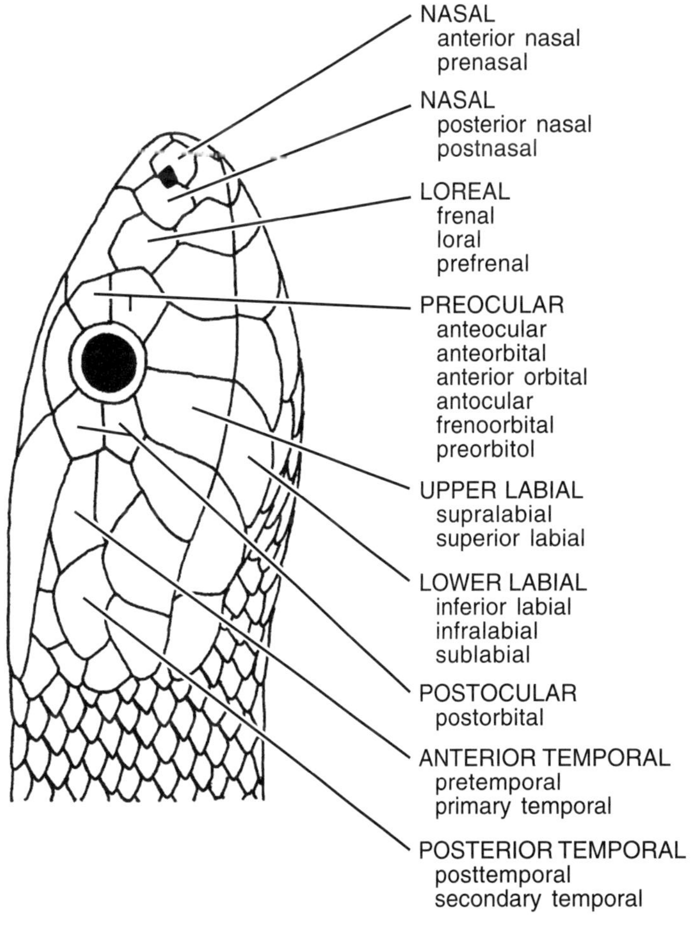

Figure 11. Lateral view of the head of a snake, with associated scale terminology.

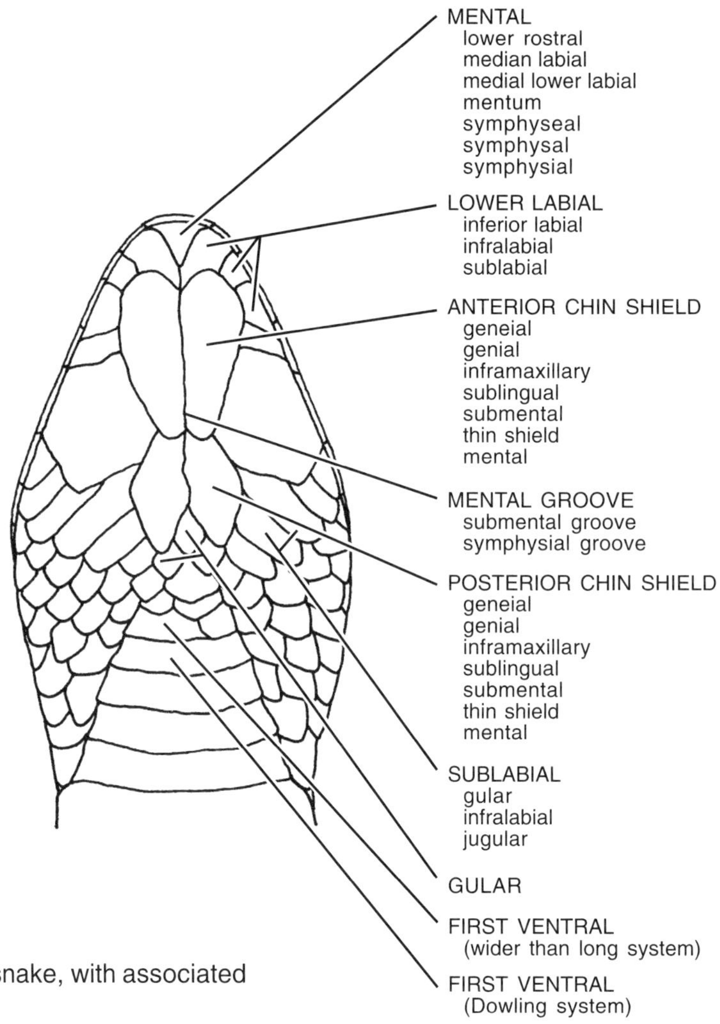

Figures

Figure 12. Ventral view of the head of a snake, with associated scale terminology.

- MENTAL
  - lower rostral
  - median labial
  - medial lower labial
  - mentum
  - symphyseal
  - symphysal
  - symphysial
- LOWER LABIAL
  - inferior labial
  - infralabial
  - sublabial
- ANTERIOR CHIN SHIELD
  - geneial
  - genial
  - inframaxillary
  - sublingual
  - submental
  - thin shield
  - mental
- MENTAL GROOVE
  - submental groove
  - symphysial groove
- POSTERIOR CHIN SHIELD
  - geneial
  - genial
  - inframaxillary
  - sublingual
  - submental
  - thin shield
  - mental
- SUBLABIAL
  - gular
  - infralabial
  - jugular
- GULAR
- FIRST VENTRAL (wider than long system)
- FIRST VENTRAL (Dowling system)

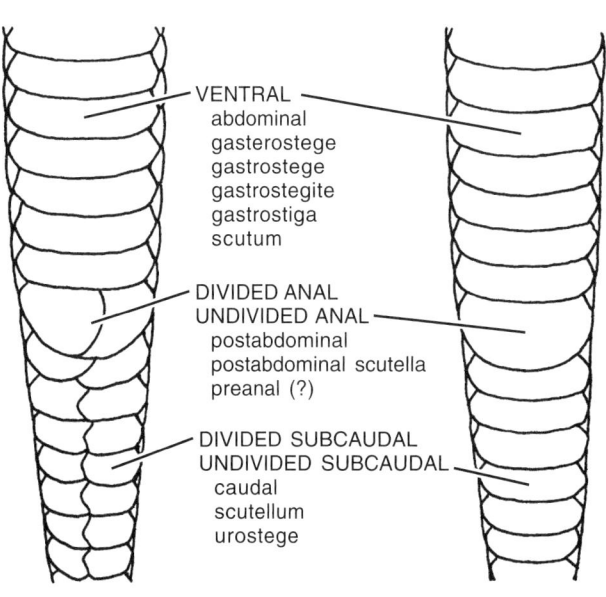

Figure 13. Ventral view of the tail of snakes, with associated terminology for scutes.

- VENTRAL
  - abdominal
  - gasterostege
  - gastrostege
  - gastrostegite
  - gastrostiga
  - scutum
- DIVIDED ANAL
- UNDIVIDED ANAL
  - postabdominal
  - postabdominal scutella
  - preanal (?)
- DIVIDED SUBCAUDAL
- UNDIVIDED SUBCAUDAL
  - caudal
  - scutellum
  - urostege

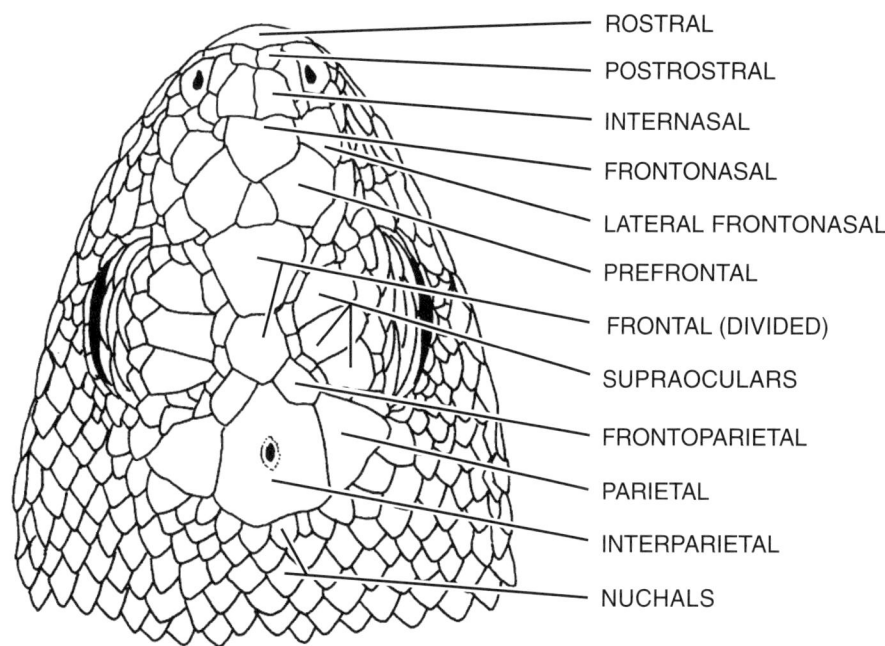

Figure 14. Terminology associated with the dorsal head scales of lizards. (After L. L. Grismer, 2002, *Amphibians and Reptiles of Baja California*, University of California Press, Berkeley, p. 47.)

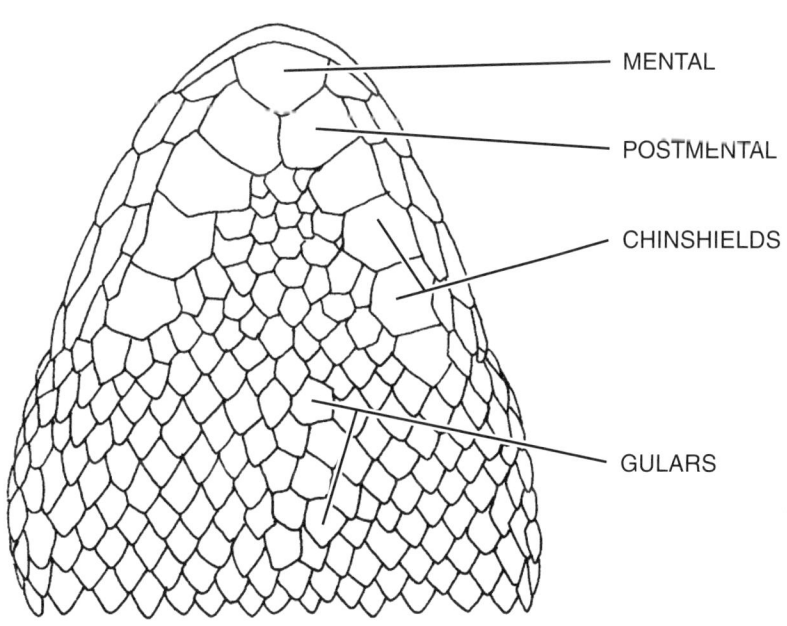

Figure 15. Terminology associated with the ventral head scales of lizards. (After L. L. Grismer, 2002, *Amphibians and Reptiles of Baja California*, University of California Press, Berkeley, p. 47.)

Figures

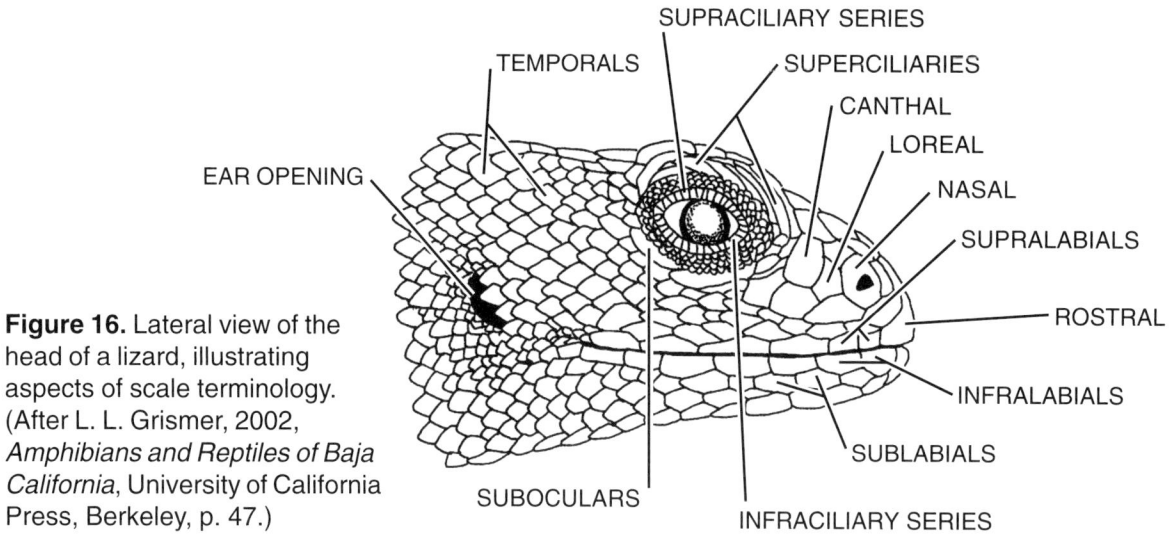

Figure 16. Lateral view of the head of a lizard, illustrating aspects of scale terminology. (After L. L. Grismer, 2002, *Amphibians and Reptiles of Baja California*, University of California Press, Berkeley, p. 47.)

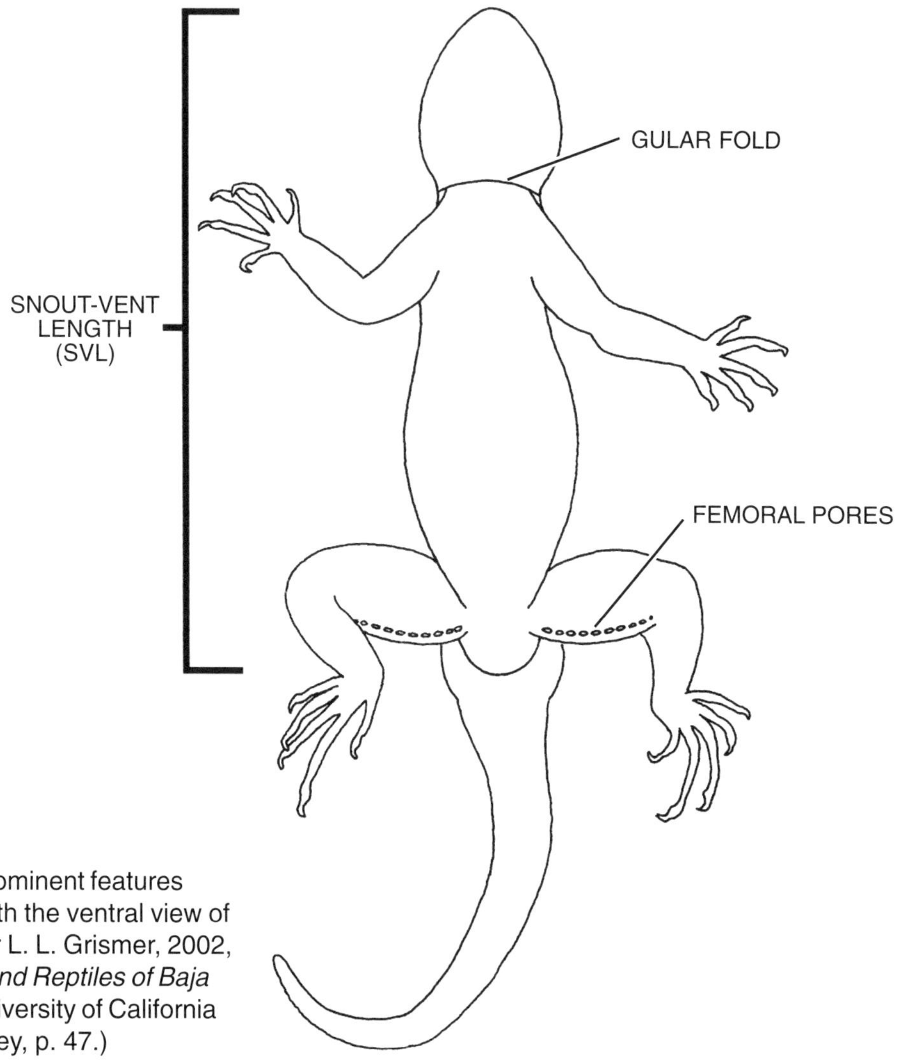

Figure 17. Prominent features associated with the ventral view of a lizard. (After L. L. Grismer, 2002, *Amphibians and Reptiles of Baja California*, University of California Press, Berkeley, p. 47.)

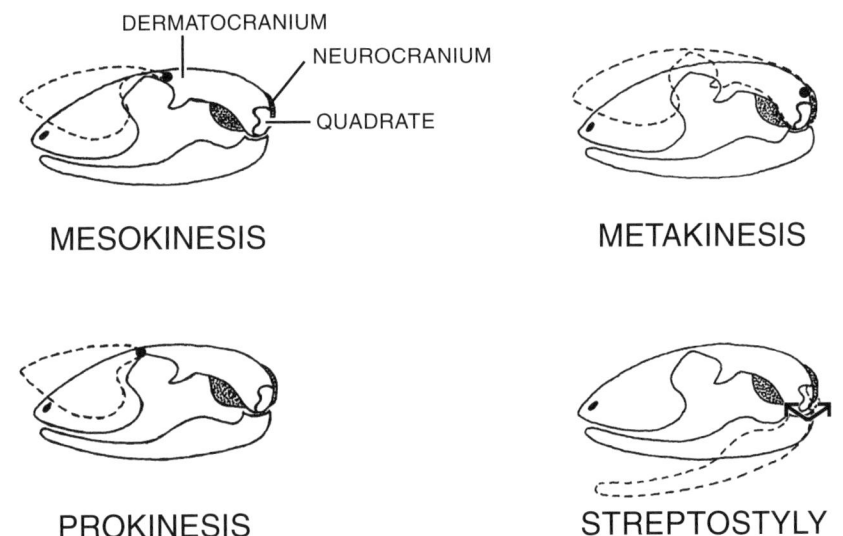

Figure 18. Schematic illustration of the patterns of cranial kinesis in squamate reptiles. (After K. V. Kardong, 2006, *Vertebrates: Comparative Anatomy, Function, Evolution.* 4th Ed. McGraw-Hill, Fig. 7.36)

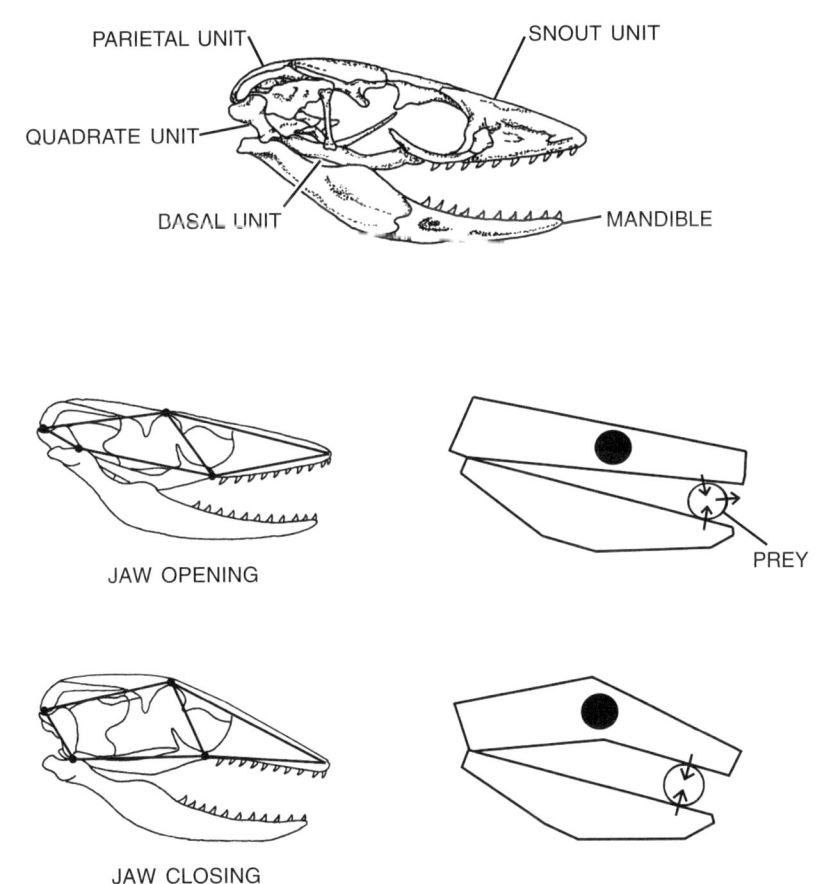

Figure 19. Kinesis and prey handling of lizard skull. Note the change in angle of tooth rows in the lower figure, which better directs the force vectors to secure the prey. (After K. V. Kardong, 2006, *Vertebrates: Comparative Anatomy, Function, Evolution.* 4th Ed. McGraw-Hill, Fig. 7.40)

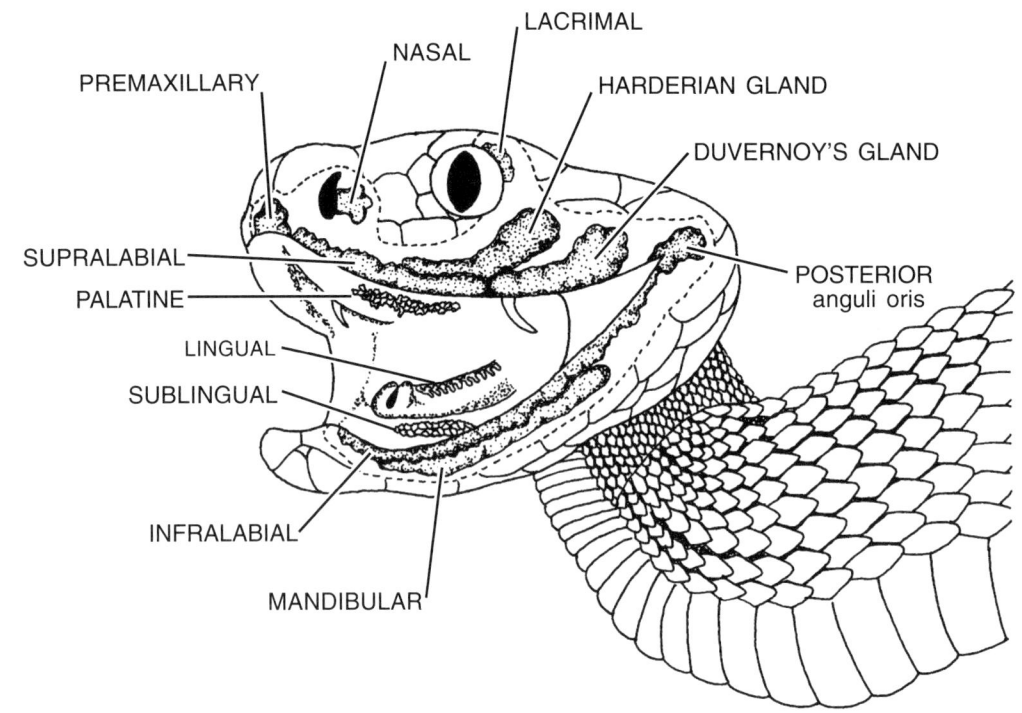

Figure 20. Oral glands of a generalized reptile. Not all glands are present in all species. (After K. V. Kardong, 2006, *Vertebrates: Comparative Anatomy, Function, Evolution.* 4th Ed. McGraw-Hill, Fig. 13.35)

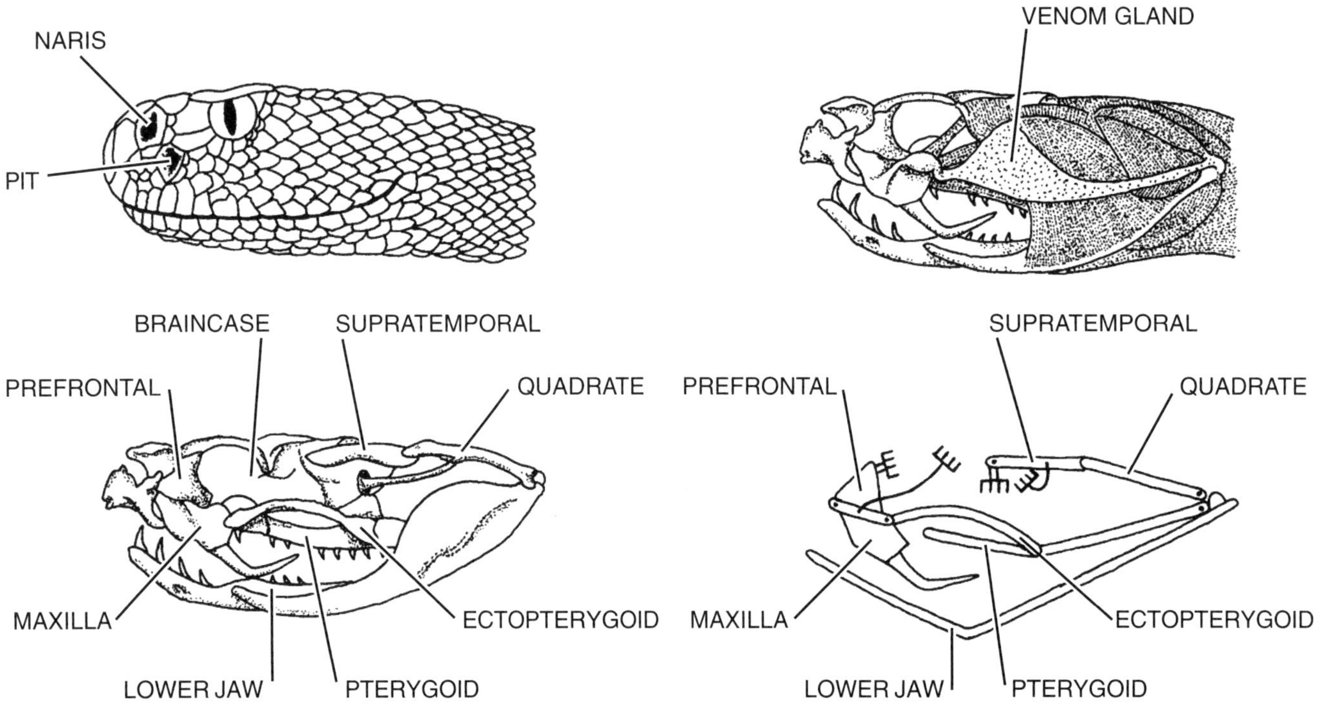

Figure 21. Muscles, bones, and biomechanical model of the movable skull elements of a snake. (After K. V. Kardong, 2006, *Vertebrates: Comparative Anatomy, Function, Evolution* 4th Ed. McGraw-Hill, Fig. 7.43)

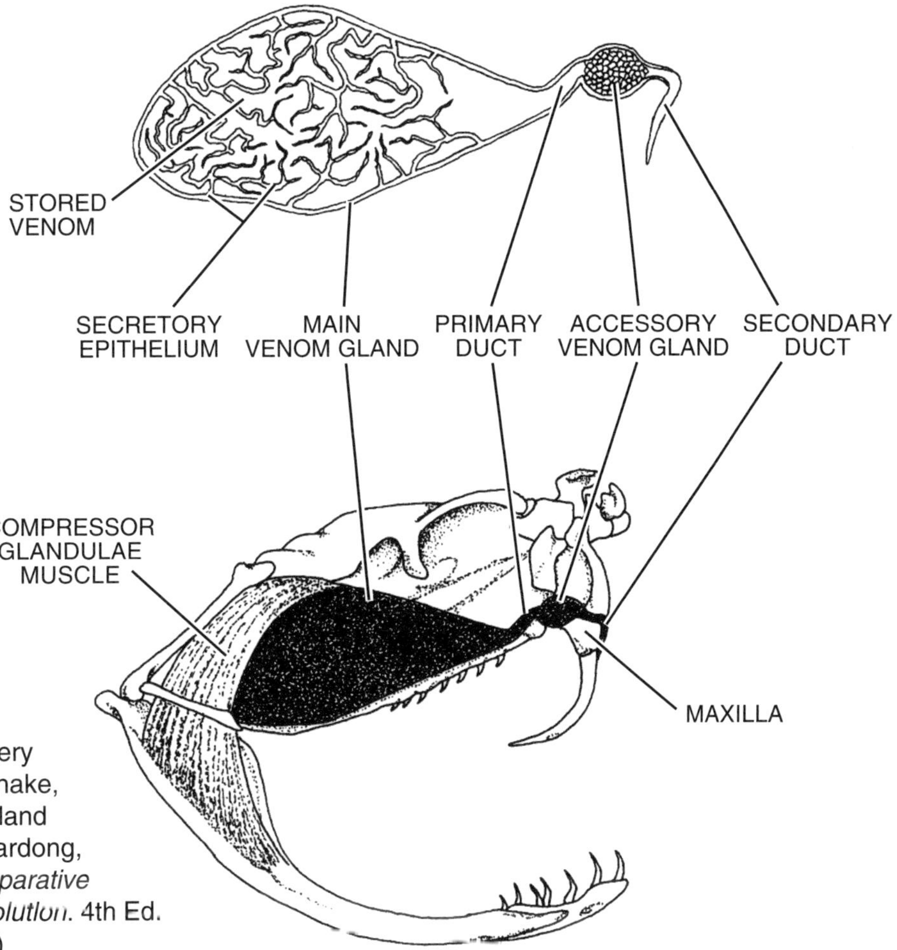

Figure 22. Venom delivery apparatus of a viperid snake, emphasizing fang and gland structure. (After K. V. Kardong, 2006, *Vertebrates: Comparative Anatomy, Function, Evolution.* 4th Ed. McGraw-Hill, Fig. 13.36)

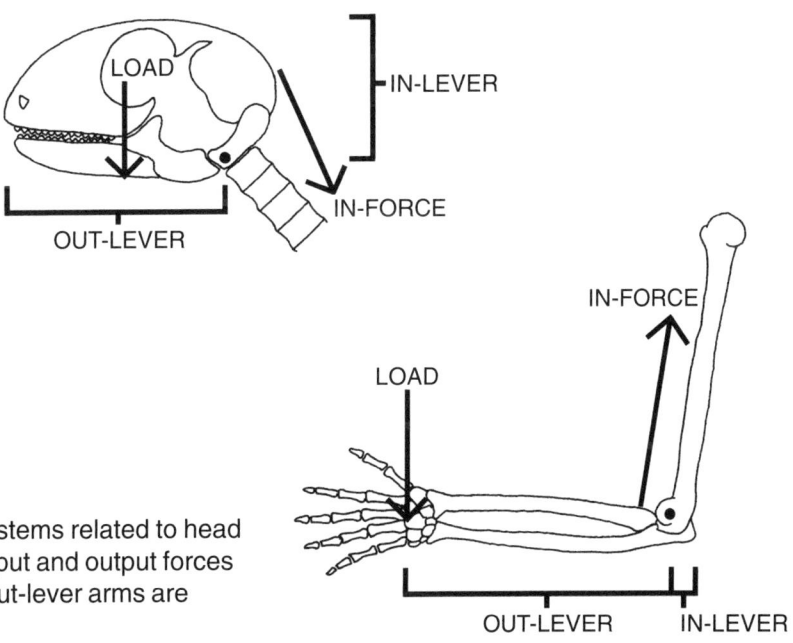

Figure 23. Two lever systems related to head and arm movements. Input and output forces as well as in-lever and out-lever arms are shown.

Figures

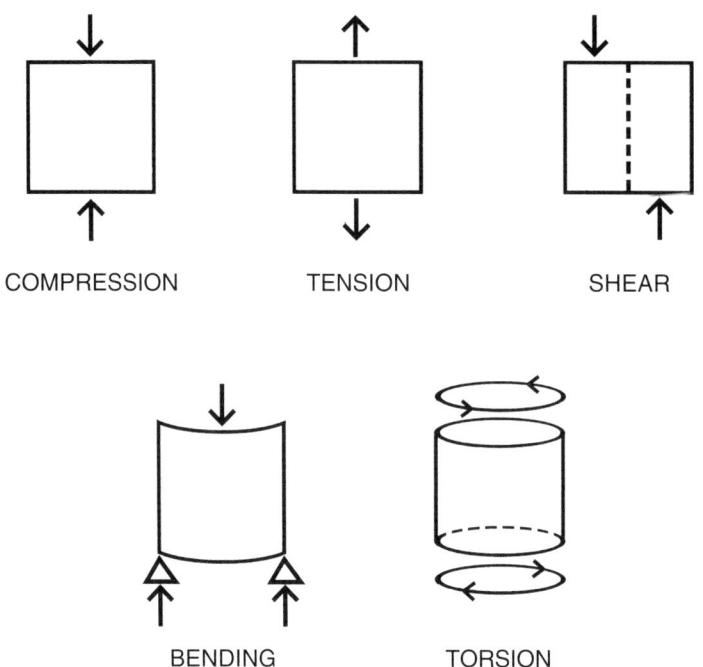

Figure 24. Forces and deformations. Arrows indicate the directions of force application. (After C. Gans, 1974, *Biomechanics. An Approach to Vertebrate Biology.* J.B. Lippincott Co., Philadelphia, Box 2-2)

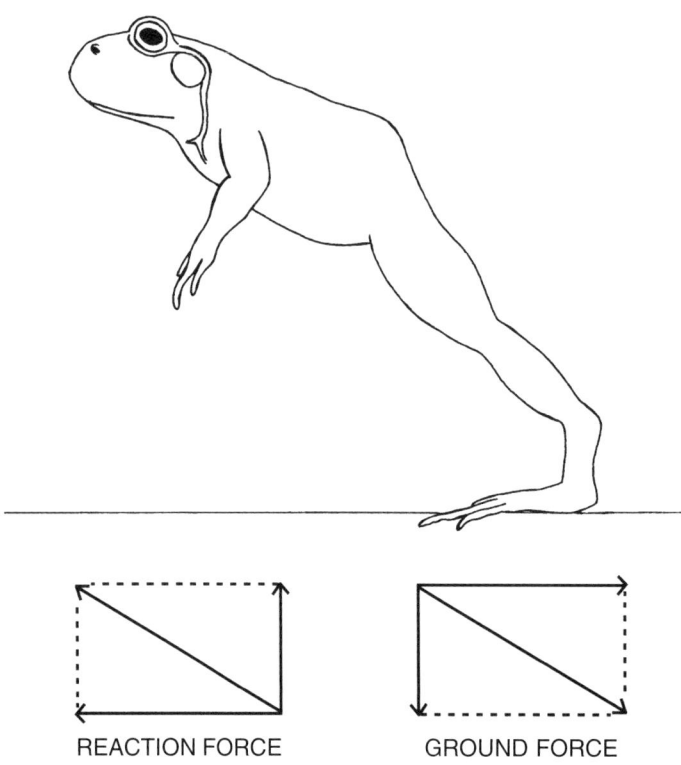

Figure 25. Forces of motion exhibited by a jumping frog. (After K. V. Kardong, 2006, *Vertebrates: Comparative Anatomy, Function, Evolution.* 4th Ed. McGraw-Hill, Fig. 4.22)

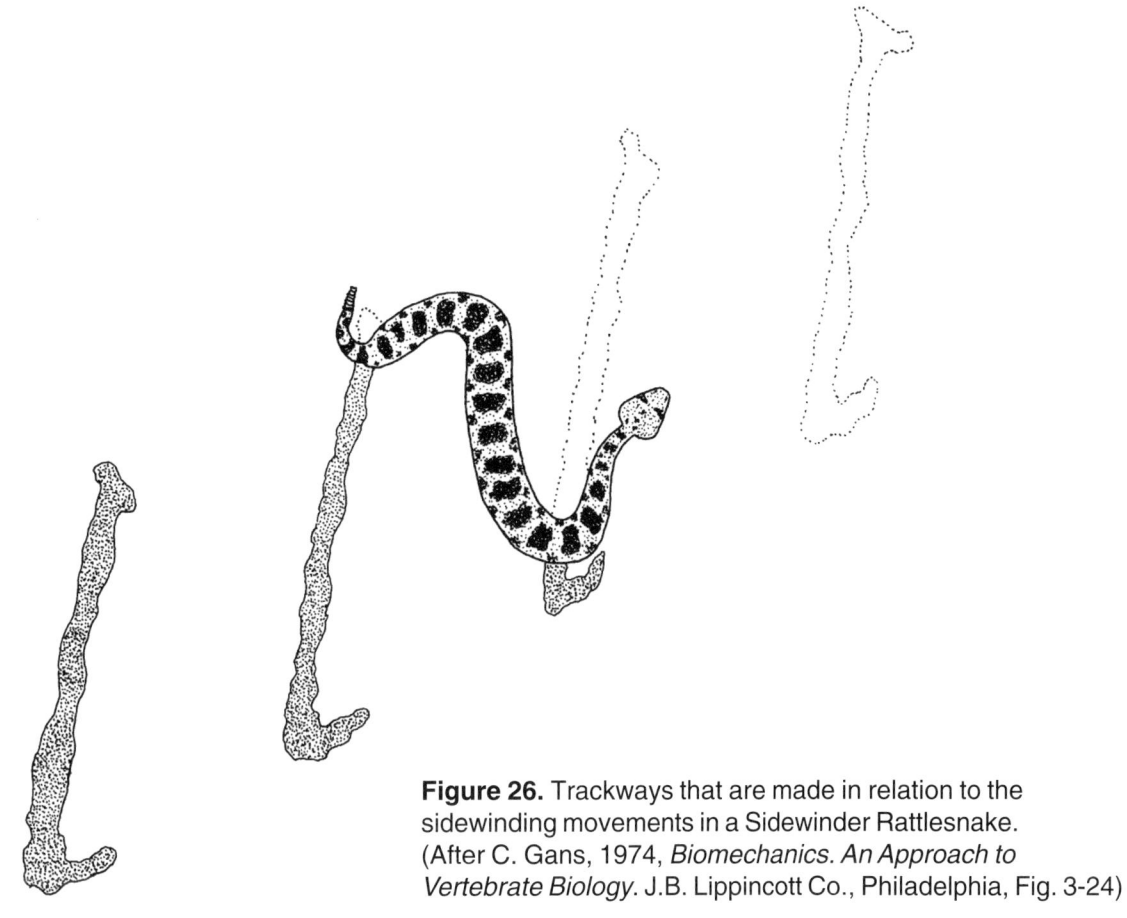

Figure 26. Trackways that are made in relation to the sidewinding movements in a Sidewinder Rattlesnake. (After C. Gans, 1974, *Biomechanics. An Approach to Vertebrate Biology.* J.B. Lippincott Co., Philadelphia, Fig. 3-24)

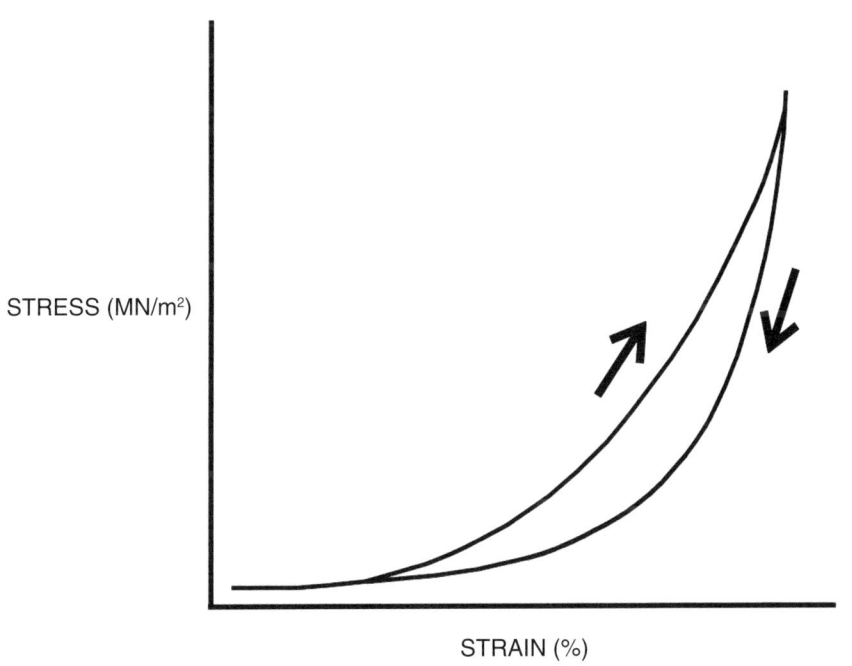

Figure 27. Stress vs. strain curves illustrating *hysteresis* of a hypothetical material response.

Figures

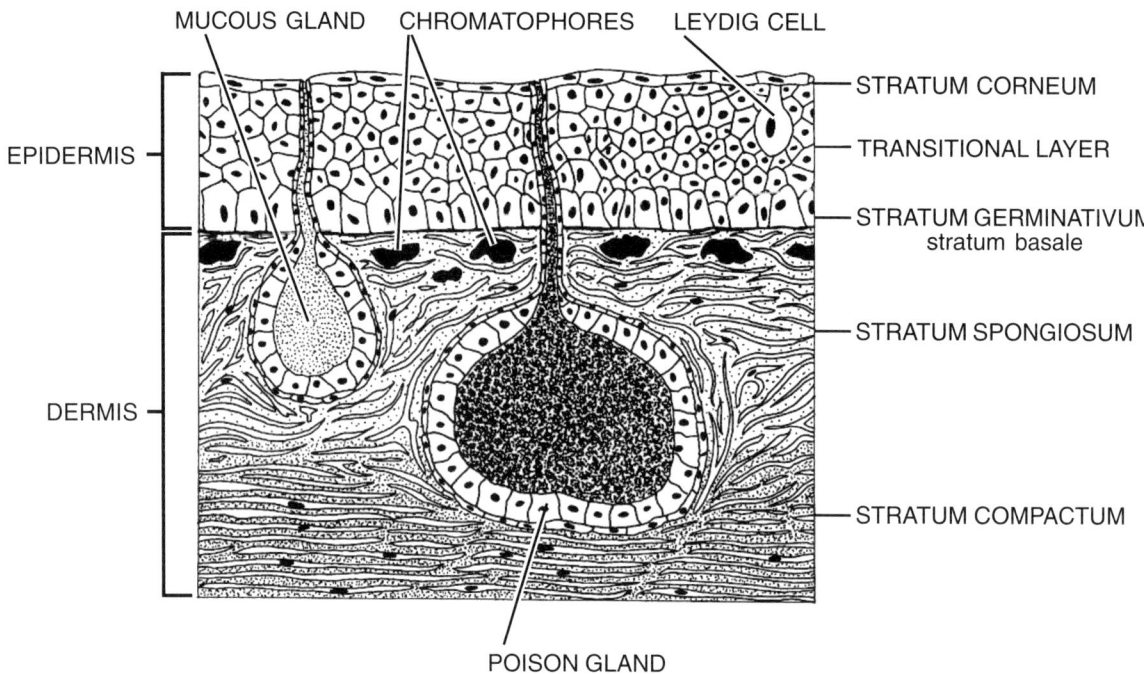

Figure 28. Schematic histological section through amphibian skin, illustrating epidermis, dermis, cutaneous glands, and other structures. (After K. V. Kardong, 2006, *Vertebrates: Comparative Anatomy, Function, Evolution.* 4th Ed. McGraw-Hill, Fig. 6.12)

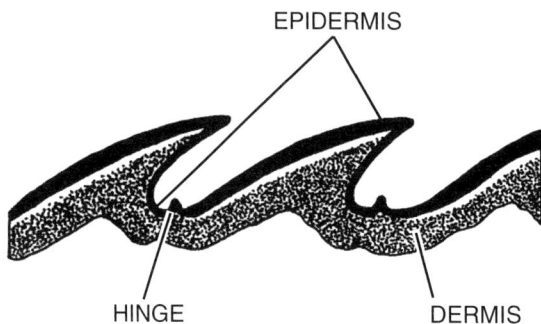

Figure 29. Generalized epidermal scales characteristic of reptiles. (After K. V. Kardong, 2006, *Vertebrates: Comparative Anatomy, Function, Evolution.* 4th Ed. McGraw-Hill, Fig. 6.13)

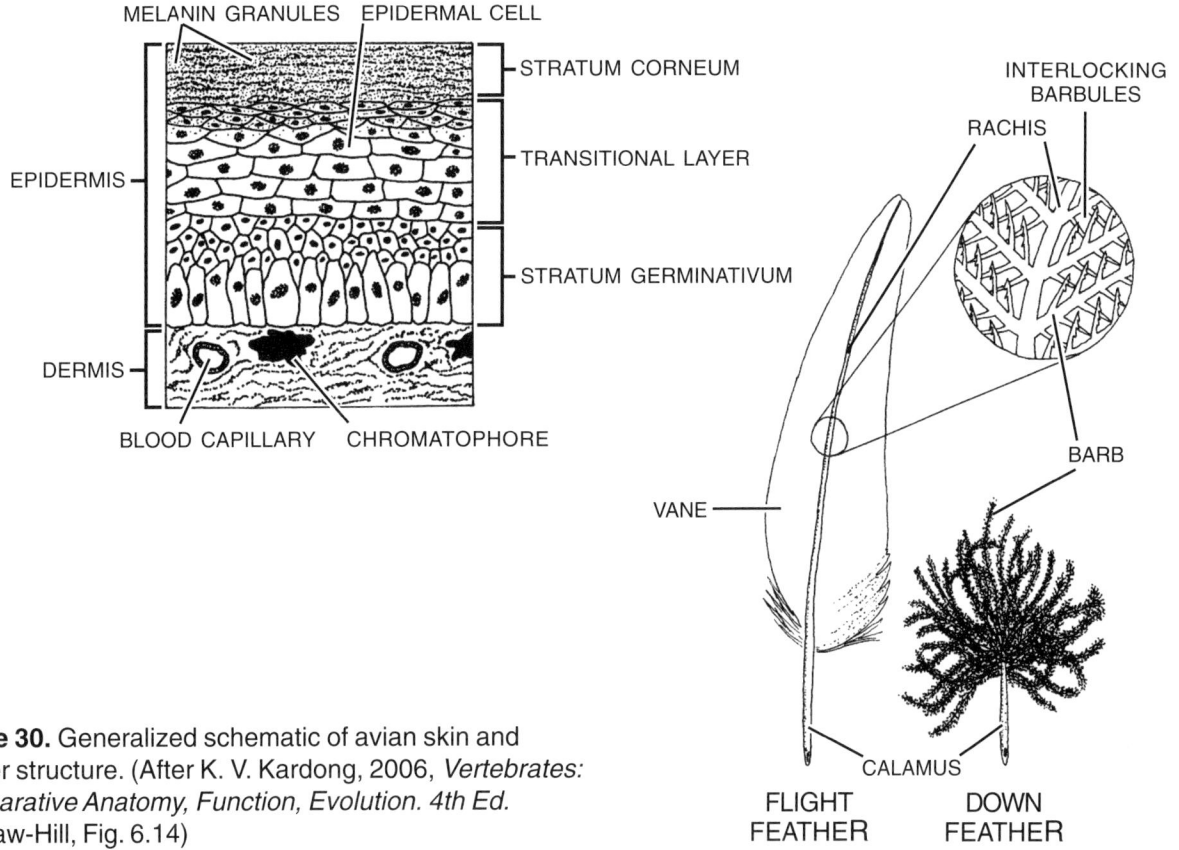

Figure 30. Generalized schematic of avian skin and feather structure. (After K. V. Kardong, 2006, *Vertebrates: Comparative Anatomy, Function, Evolution.* 4th Ed. McGraw-Hill, Fig. 6.14)

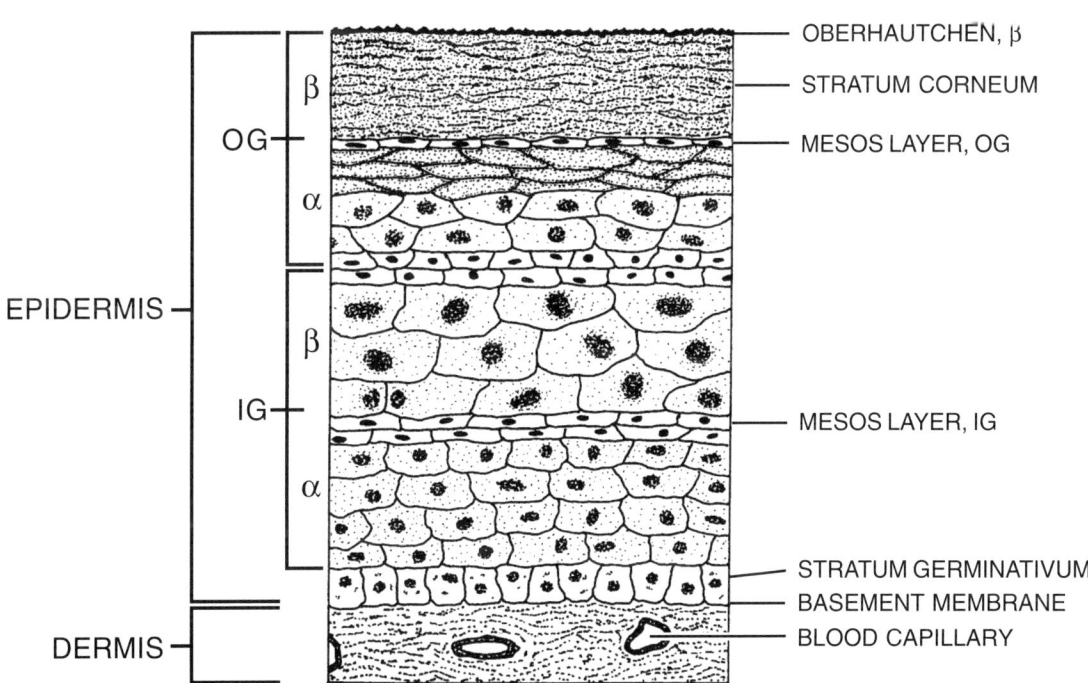

Figure 31. Generalized schematic of squamate skin illustrating both inner (IG) and outer (OG) generations of epidermis, which separate at ecdysis. Alpha and beta keratins are also illustrated.

Figures

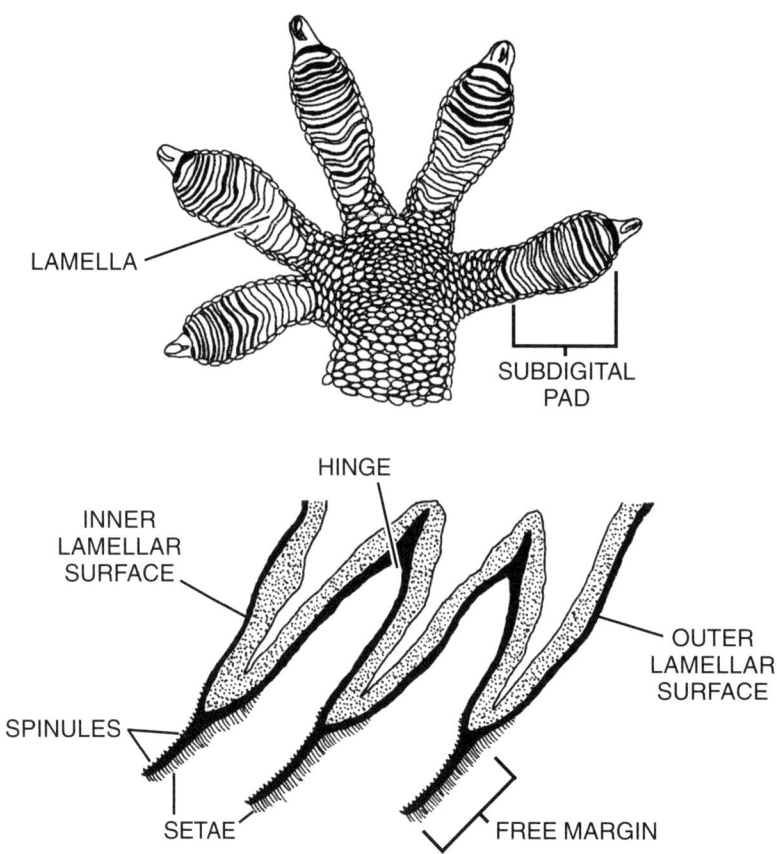

Figure 32. Generalized gekkonid or anoline toe pads, showing details of the lamellar structure.

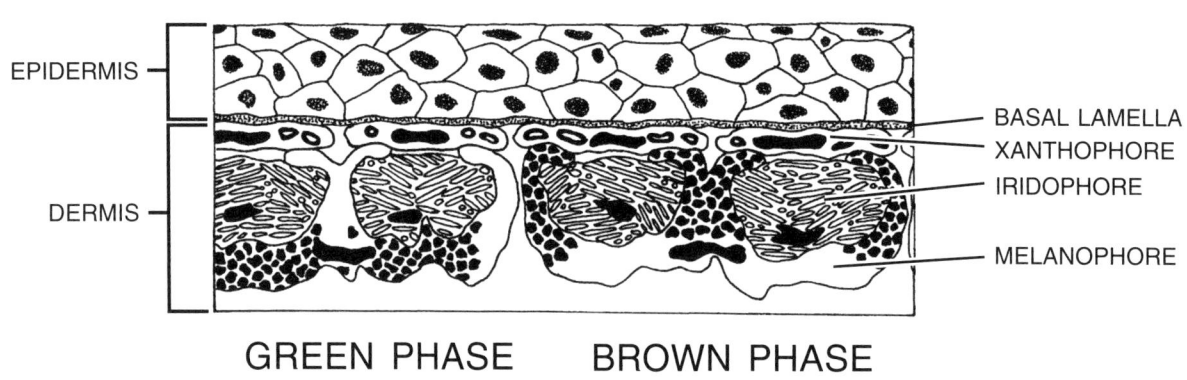

Figure 33. Dermal chromatophore unit, illustrating changes that occur during color change in *Anolis carolinensis*.

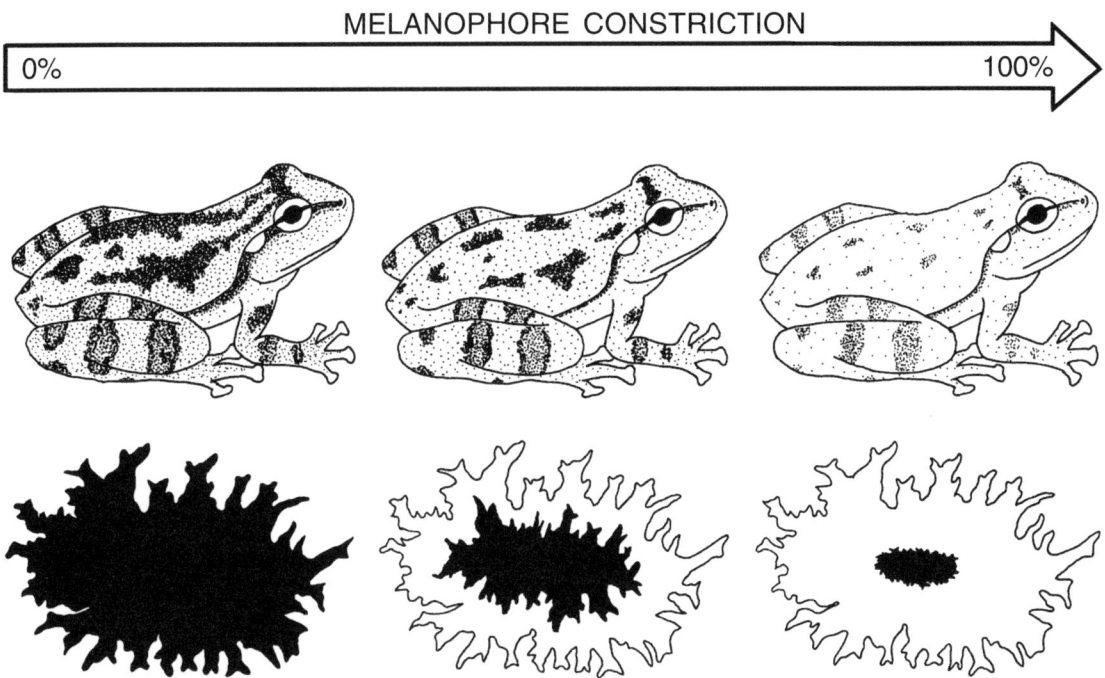

Figure 34. Melanophores of frog skin, illustrating relative states of dispersion or concentration of melanin, resulting in darkening or lightening of skin respectively.

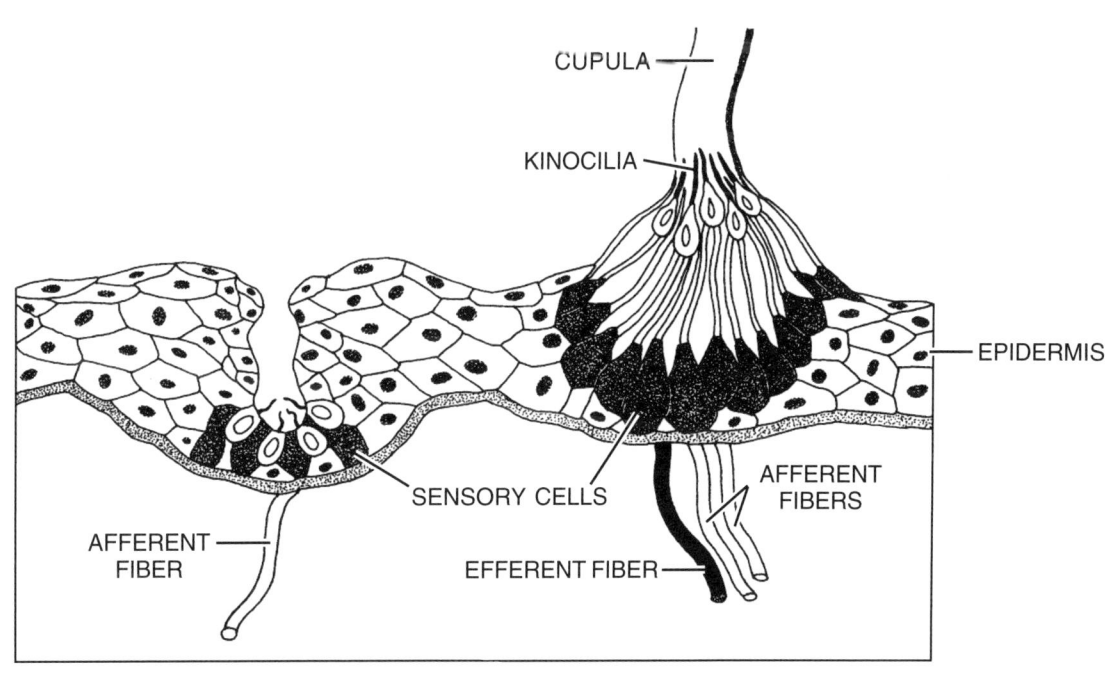

Figure 35. Morphology of ampullary organs and neuromasts in a newt. (Redrawn from B. Fritzsch and U. Wahnschaffe, 1983, *Cell Tissue Res.* 229:483-503)

Figures

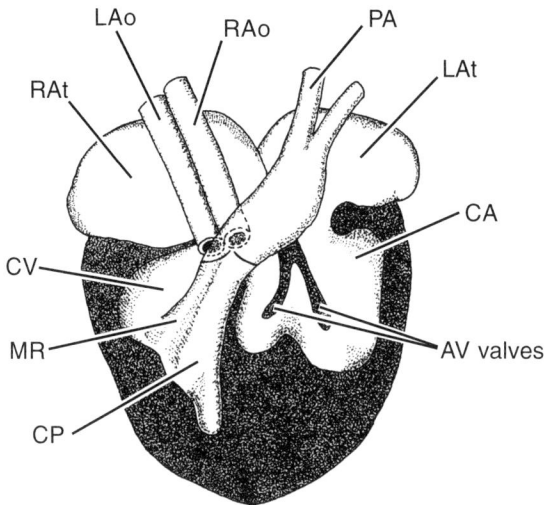

Figure 36. Heart of a non-crocodilian reptile. AV valves, atrioventricular valves; CA, cavum arteriosum; CP, cavum pulmonale; CV, cavum venosum; IS, interventricular septum; LAo, left aortic arch; Lat, left atrium; MR, muscular ridge; PA, pulmonary artery; RAo, right aortic arch; RAt, right atrium. (After J.W. Hicks, 2002, *News Physiol. Sci.* 17-241-245)

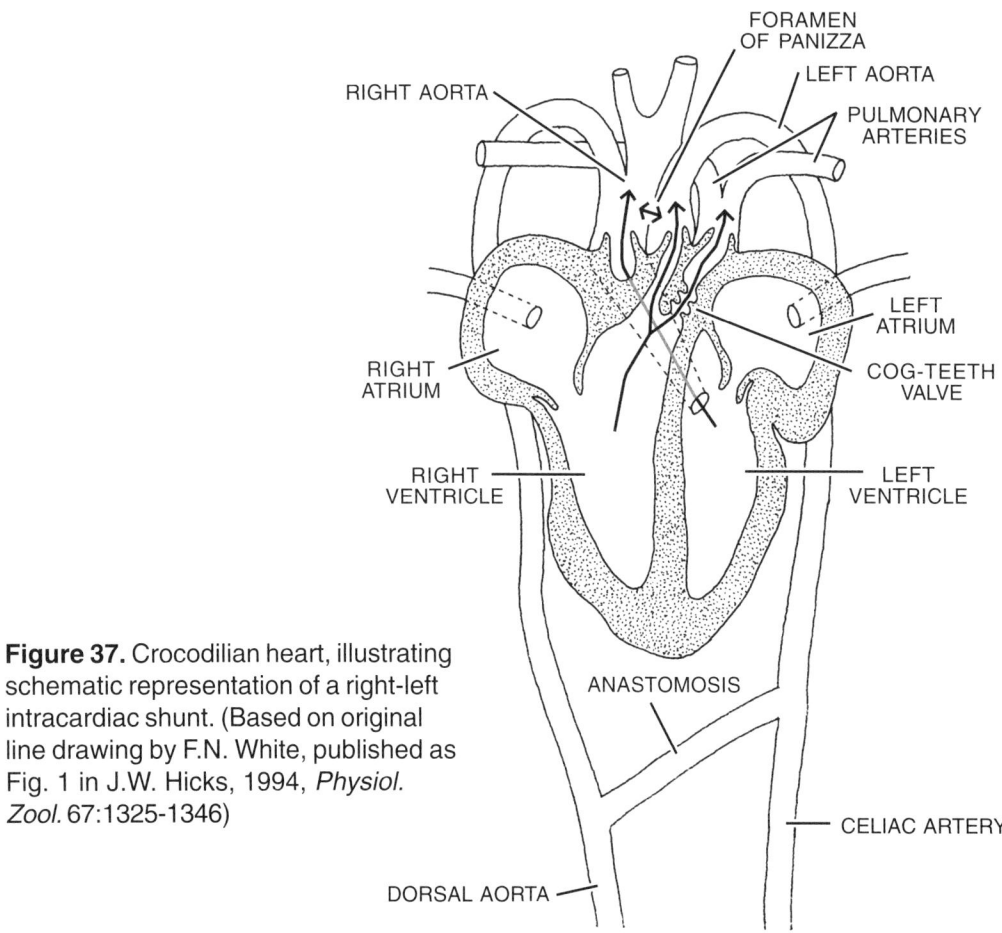

Figure 37. Crocodilian heart, illustrating schematic representation of a right-left intracardiac shunt. (Based on original line drawing by F.N. White, published as Fig. 1 in J.W. Hicks, 1994, *Physiol. Zool.* 67:1325-1346)

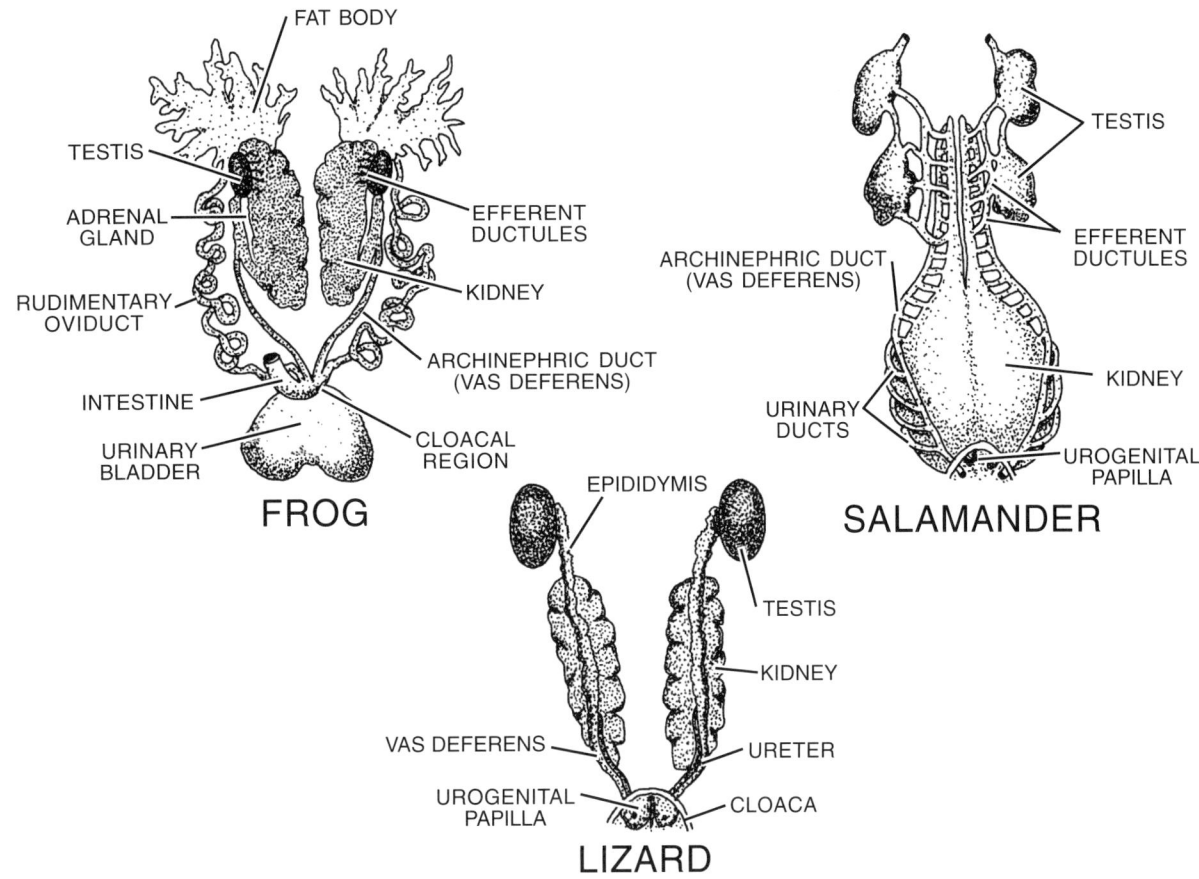

Figure 38. Ventral view of urogenital systems of male amphibians and a male lizard. (After K. V. Kardong, 2006, *Vertebrates: Comparative Anatomy, Function, Evolution.* 4th Ed. McGraw-Hill, Fig.14.36)

Figure 39. Characteristic postures during male combat between two male rattlesnakes.

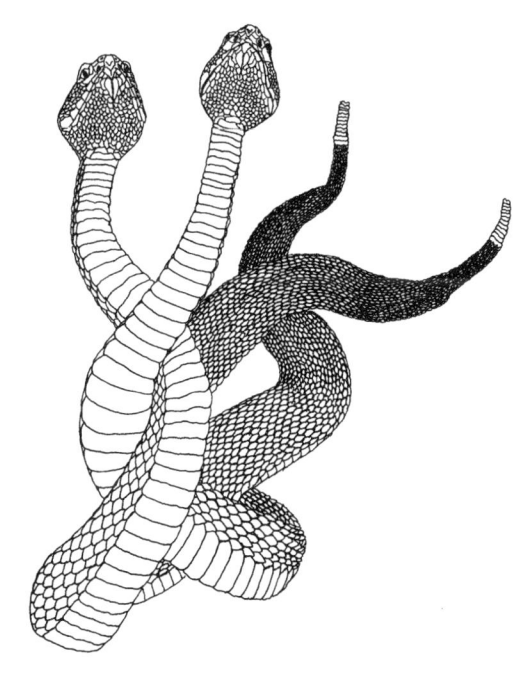

Figures

Figure 40. Schematic dewlap of a male anoline lizard.

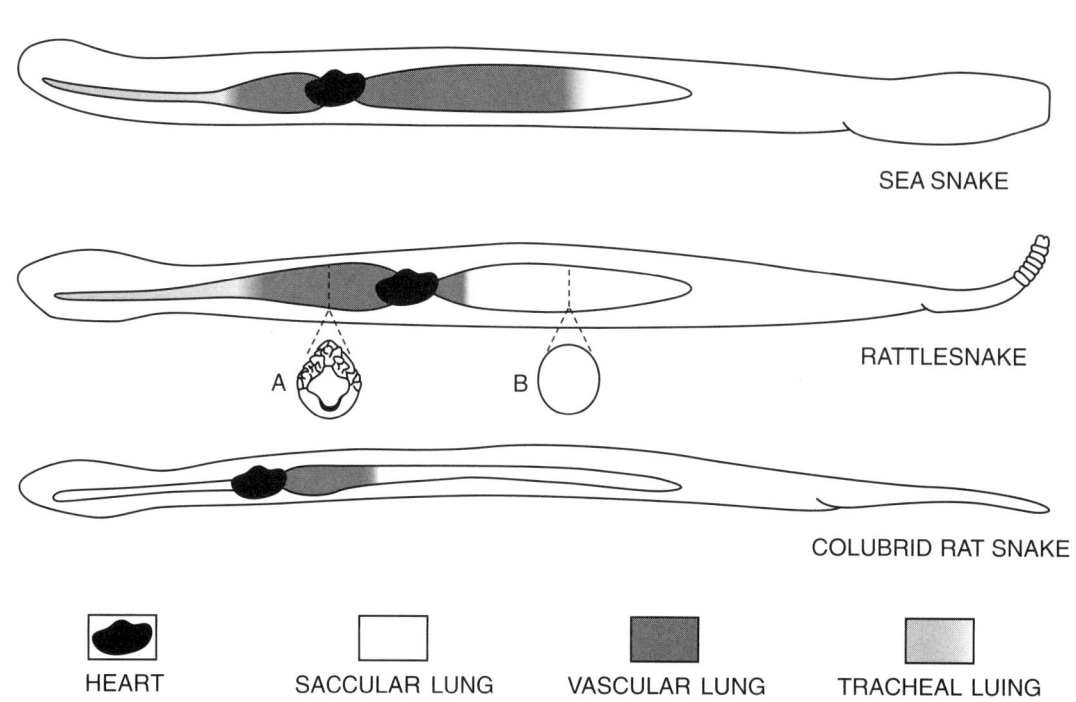

Figure 41. Schematic representation of the variation of lung structure in snakes. The insert cross-sections illustrate the secondary and tertiary subdivisions of the *faveoli* and the cartilaginous *trachea* (top and bottom of A) relative to the simple membrane that lines the saccular lung (B). After H.B. Lillywhite, *Amer.Zool.* 27:81–95, 1987, Fig. 6.